Lecture Notes in Control and Information Sciences

Edited by M. Thoma and A. Wyner

IIASA 81

Stochastic Optimization

Proceedings of the International Conference, Kiev, 1984

Edited by

V. I. Arkin, A. Shiraev, R. Wets

Springer-Verlag
Berlin Heidelberg New York Tokyo

Editors
Dr. Vadim I. Arkin
Central Economic Mathematical Institute
USSR Academy of Sciences
Prospect 60 Let Octyabria, 9
117312 Moscow, USSR

A. Shiraev
Steklov Mathematical Institute
Ul. Vavilova 41
Moscow V-333, USSR

Prof. R. Wets
Dept. of Mathematics
University of California
Davis, California 95616, USA

ISBN 3-540-16659-9 Springer-Verlag Berlin Heidelberg New York Tokyo
ISBN 0-387-16659-9 Springer-Verlag New York Heidelberg Berlin Tokyo

Printed in Germany

Offsetprinting: Mercedes-Druck, Berlin
Binding: B. Helm, Berlin
2161/3020-543210

FOREWORD

This volume includes the Proceedings of the International Conference on *Stochastic Optimization* held at Kiev, USSR in September 1984. The conference was organised by the Committee for Systems Analysis of the USSR Academy of Sciences, International Institute for Applied Systems Analysis, the Academy of Sciences of the Ukrainian SSR and V. Glushkov Institute of Cybernetics.

The purpose of the conference was to survey the latest developments in the field of controlled stochastic processes, stochastic programming, control under incomplete information and applications of stochastic optimization techniques to problems in economics, engineering, modeling of energy systems, etc.

Up to now, all these approaches to handle uncertainty followed an independent development, but recently it became apparent that they are interconnected in a number of ways. This process was stimulated by the development of new powerful mathematical tools. For instance, martingale techniques originally developed in stochastic analysis are extensively used in the theory of controlled stochastic processes and for proving convergence of stochastic programming methods. The theory of measurable multifunctions (set-valued maps) primarily used in mathematical economics, is now one of the main tools for the analysis of the dynamics controlled systems differential games, etc. Convex analysis is now widely used in stochastic optimization, but it was first applied to deterministic extremal problems.

On the other hand, new aplications appeared in which it is necessary to consider the problems of identification, filtering, control and large scale optimization simultaneously. This also leads to the integration of different approaches of stochastic optimization. Therefore, it was decided to bring together scientists from these fields and an international programme committee was formed. This committee included representatives from differnt fields:

V.S. Michalevich (USSR, Chairman)
A. Wierzbicki (Poland, Deputy Chairman)
V.I. Arkin (USSR)
K. Aström (Sweden)
A. Bensoussan (France)
D. Blackwell (USA)
A. Veinot (USA)
R. Wets (USA)
Yu. M. Ermoliev (USSR)
A.B. Kurzhanskii (USSR)
A. Prekopa (Hungary)
A.V. Skorokhod (USSR)
A.N. Shiriaev (USSR)

More than 240 scientists from 20 countries participated in the conference. The "Systems and Decision Sciences Programme" of the International Institute for Applied Systems Analysis greatly contributed to the organisation of the conference and to the preparation of the conference materials for publication. In recent years this Institute has been involved in a collaborative research project on stochastic optimization involving scientists from different countries. This collaboration was very important in achieving the high level of the conference.

The conference reflected a number of recent important developments in stochastic optimization, notably new results in control theory with incomplete information, stochastic maximum principle, new numerical techniques for stochastic programming and related software, application of probabilistic methods to the modeling of the economy.

The contributions to this volume are divided into three categories:

1. Controlled stochastic processes
2. Stochastic extremal problems
3. Stochatic optimization problems with incomplete information.

Laxenburg, July 1985

V. Arkin
A.N. Shiriaev
R. Wets

TABLE OF CONTENTS

Section I: Controlled Stochastic Processes

A martingale approach to partially observable controlled stochastic systems — 3
R.J. Chitashvili (USSR)

On the limiting distribution of extremum points for certain stochastic optimization models — 17
A.Ya. Dorogovtsev and *A.G. Kukush (USSR)*

The structure of persistently nearly-optimal strategies in stochastic dynamic programming problems — 22
E.A. Fainberg (USSR)

On the derivation of a filtering equation for a non-observable semimartingale — 32
L.I. Galčuk (USSR)

On the representation of functionals of a Wiener sheet by stochastic integrals — 37
J.I. Gihman (USSR)

The maximum principle for optimal control of diffusions with partial information — 50
U.G. Haussmann (Canada)

Explicit solution of a consumption/investment problem — 59
I. Karatzas, J. Lehoczky, S. Sethi and *S. Shreve (USA)*

On the asymptotic behavior of some optimal estimatès of parameters of nonlinear regression functions — 70
P.S. Knopov (USSR)

On the ε-optimal control of a stochastic integral equation with an unknown parameter — 79
A.M. Kolodiy (USSR)

Some properties of value functions for controlled diffusion processes — 88
N.V. Krylov (USSR)

Stochastic control with state constraints and non-linear elliptic equations with infinite boundary conditions — 96
J.-M. Lasry (France)

On the weak convergence of controlled semi-martingales 107
N.L. Lazrieva (USSR)

Estimation of parameters and control of systems with unknown parameters 118
S.Ya. Mahno (USSR)

On recursive approximations with error bounds in nonlinear filtering 127
G.B. Di Masi, W.J. Runggaldier and *B. Armellin (Italy)*

On approximations to discrete-time stochastic control problems 136
G.B. Di Masi, W.J. Runggaldier and *F. Chiariello (Italy)*

On lexicographical optimality criteria in controlled Markov chains 148
G.I. Mirzashvili (USSR)

Canonical correlations, Hankel operators and Markovian representations of multivariate stationary Gaussian processes 157
Michele Pavon (Italy)

The maximum principle in stochastic problems with non-fixed random control time 169
M.T. Saksonov (USSR)

Optimal control of stochastic integral equations 178
L.E. Shaikhet (USSR)

Some direct methods for computing optimal estimators for forecasting and filtering problems involving stochastic processes 188
A.D. Shatashvili (USSR)

On functional equations of discrete dynamic programming 201
Karel Sladký (CSSR)

Risk-sensitive and Hamiltonian formulations in optimal control 213
P. Whittle (UK)

Martingales in survival analysis 220
A.I. Yashin (USSR)

Markov decision processes with both continuous and impulsive control 234
A.A. Yushkevich (USSR)

Section II: Stochastic Extremal Problems

Stochastic programming methods: convergence and non-asymptotic estimation of the convergence rate 249
Ya.I. A'lber and *S.V. Shilman (USSR)*

Solution of a stochastic programming problem concerning the distribution of water resources 258
I.A. Aleksandrov, V.P. Bulatov, S.B. Ognivtsev and *F.I. Yereshko (USSR)*

Limit theorems for processes generated by stochastic optimization algorithms 265
V.V. Anisimov (USSR)

On the structure of optimality criteria in stochastic optimization models 275
V.I. Arkin and *S.A. Smolyak (USSR)*

Strong laws for a class of path-dependent stochastic processes with applications 287
B. Arthur (UK), Y. Ermoliev and *Y. Kaniovski (USSR)*

The generalized extremum in the class of discontinuous functions and finitely additive integration 301
V.D. Batuhtin and *A.G. Chentsov (USSR)*

Convex multivalued mappings and stochastic models of the dynamics of economic systems 309
N.N. Bordunov (USSR)

Stability in stochastic programming – probabilistic constraints 314
Jitka Dupačova (CSSR)

Duality in improper mathematical programming problems under uncertainty 326
I.I. Eremin and *A.A. Vatolin (USSR)*

Equilibrium states of monotonic operators and equilibrium trajectories in stochastic economic models 334
I.V. Evstigneev (USSR)

Finite horizon approximates of infinite horizon stochastic programs 339
Sjur D. Flam (Norway) and *Roger J.-B. Wets (USA)*

Stochastic optimization techniques for finding optimal submeasures 351
A. Gaivoronski (USSR)

Strong consistency theorems related to stochastic quasi-Newton methods 364
L. Gerencsér (Hungary)

Stochastic gradient methods for optimizing electrical transportation networks 373
M. Goursat, J.P. Quadrat and *M. Viot (France)*

On the functional dependence between the available information and the chosen optimality principle 388
V.I. Ivanenko and *V.A. Labkovskiy (USSR)*

Uncertainty in stochastic programming 393
Vlasta Kancova (CSSR)

Stochastic programming models for safety stock allocation 402
Péter Kelle (Hungary)

Direct averaging and perturbed test function methods for weak convergence 412
Harold J. Kushner (USA)

On the approximation of stochastic convex programming problems 427
R. Lepp (USSR)

Extremal problems with probability measures, functionally closed preorders and strong stochastic dominance 435
V.L. Levin (USSR)

Expected value versus probability of ruin strategies 448
L.C. Maclean and *W.T. Ziemba (Canada)*

Controlled random search procedures for global optimization 457
K. Marti (FRG)

On Bayesian methods in nondifferential and stochastic programming 475
J.B. Mockus (USSR)

On stochastic programming in Hilbert space 487
N.M. Novikova (USSR)

Reduction of risk using a differentiated approach 496
I. Petersen (USSR)

A stochastic lake eutrophication management model 501
J. Pintér and *L. Somlyódy (Hungary)*

A dynamic model of market behavior 513
I.G. Pospelov (USSR)

Recursive stochastic gradient procedures in the presence of dependent noise 522
A.S. Poznyak (USSR)

Random search as a method for optimization and adaption 534
L.A. Rastrigin (USSR)

Linear-quadratic programming problems with stochastic penalties: the finite generation algorithm 545
R.T. Rockafellar and *R.J.-B. Wets (USA)*

Convergence of stochastic infima: equi-semicontinuity 561
G. Salinetti (Italy)

Growth rates and optimal paths in stochastic models of expanding economies 576
A.D. Slastnikov and *E.L. Presman (USSR)*

Extremum problems depending on a random parameter 585
E. Tamm (USSR)

Adaptive control of parameters in gradient algorithms for stochastic optimization 591
S.P. Urjas'ev (USSR)

Stochastic models and methods of optimal planning 602
A.I. Yastremskii (USSR)

Section III: Problems with Incomplete Information

Differential inclusions and controlled systems: properties of solutions 611
A.V. Bogatyrjov (USSR)

Guaranteed estimation of reachable sets for controlled systems 619
F.L. Chernousko (USSR)

Methods of group pursuit 632
A.A. Chikrij (USSR)

An averaging principle for optimal control problems with singular perturbations 641
V.G. Gaitsgory (USSR)

On a certain class of inverse problems in control system dynamics 650
M.I. Gusev (USSR)

Simultaneous estimation of states and parameters in control systems with incomplete data 657
N.F. Kirichenko and *A.S. Slabospitsky (USSR)*

Approximate solutions of differential games using mixed strategies 669
A.F. Kleimenov, V.S. Patsko and *V.N. Ushakov (USSR)*

On the solution sets for uncertain systems with phase constraints 675
A.B. Kurzhanskii (USSR)

Existence of a value for a general zero-sum mixed game 688
J.P. Lepeltier (France)

Positional modeling of stochastic control in dynamical systems 696
Yu.S. Osipov and *A.V. Krjazhimskii (USSR)*

Use of the H-convex set method in differential games 705
V.V. Ostapenko (USSR)

A linear differential pursuit game 712
L.S. Pontryagin (USSR)

Methods of constructing guaranteed estimates of parameters of linear systems and their statistical properties 719
B.N. Pshenichnyj and *V.G. Pokotilo (USSR)*

Stochastic and deterministic control. Differential inequalities 728
N.N. Subbotina, A.I. Subbotin and *V.E. Tret'jakov (USSR)*

The search for singular extremals 738
M.I. Zelikin (USSR)

On the smoothness of the Bellmann function in optimal control problems with incomplete data 747
L.F. Zelikina (USSR)

Section I
Controlled Stochastic Processes

A MARTINGALE APPROACH TO PARTIALLY OBSERVABLE CONTROLLED STOCHASTIC SYSTEMS

R.J. Chitashvili
Mathematical Institute
Byelorussian Academy of Sciences
USSR

1. INTRODUCTION

A number of stochastic optimization problems can be related to the problem of optimal absolutely continuous change of measure in a space with filtering. The choice of a control in such problems is equivalent to the choice of some absolutely continuous transformation of the original basic measure. This covers, for instance, the control of the transition coefficient in non-generated diffusion-type processes.

In [1] we consider the problem of constructing probability measures corresponding to admissible controls, and the Hamiltonian for general control problems with complete information.

In the case of control problems with incomplete information, substantial difficulties arise in the construction of the Hamiltonian, which is basically needed in order to test the optimality of a given control by making separate tests of each of its values for every fixed moment and the corresponding history of the controlled process.

While the adjoint process in the Hamiltonian expression for the necessary optimality condition is simple enough, the structure of the adjoint process for the sufficient optimality condition is non-trivial; the formal assertion of the existence of such a process will be illustrated here by an example based on the discrete time case.

2. DEFINITION OF THE CONTROL PROCESS. SOME ASSUMPTIONS AND NOTATION

Let (Ω, F, G, P) be a probability space with the filters $F = (F_t)$, $G = (G_t)$, $0 \leq t \leq T$, $G_t \subset F_t$, satisfying the usual conditions, and $P^A = (P^a, a \in A)$ be a family of probabilities on F_T which are equivalent to P, $P^a \sim P$.

The elements $a \in A$ are interpreted as actions, P^a is a distribution corresponding to the action a; F_t is the σ-algebra of the events which describe the state of the controlled system up to time t; G_t is the σ-algebra of the events which are observed up to time t. It is assumed that the space of actions A is finite.

The possibility of choosing a sequence of actions according to the accumulated information is expressed by the introduction of a class U of controls $u \in U$, where $u = (u_t)$, $0 \le t \le T$, $u_t(\omega)$ is a G-adapted (measurable for every t with respect to G_t) predictable process taking values in A.

Following the construction scheme given in [1], the class P^u of probabilities corresponding to the controls is defined in terms of densities.

Let $\rho^a = (\rho_t^a)$, $0 \le t \le T$, denote the local density of the measure P^a with respect to P, i.e.,

$$\rho_t^a = \mathrm{d}P_t^a / \mathrm{d}P_t \quad ,$$

where $P_t^a = P^a \mid F_t$, $P_t = P \mid F_t$ are contractions of measures on the σ-algebra F_t, $0 \le t \le T$.

Suppose for simplicity that $P_0^a = P_0$ and, hence, $\rho_0^a = 1$, $a \in A$ (this is a common initial condition).

It is well-known (see [2,3]) that ρ^a can be represented in the form of an exponential martingale

$$\rho^a = \varepsilon(M^a) \quad ,$$

where M^a is a local martingale with respect to the flow F and the measure P, i.e., $M^a \in \mathbf{M}_{\mathrm{loc}}(F, P)$ and ρ^a is a solution of the linear equation

$$\mathrm{d}\rho_t^a = \rho_{t-}^a \, \mathrm{d}M_t^a \, , \; \rho_0^a = 1 \quad .$$

Set $M^A = (M^a, a \in A)$, $R^A = (\rho^a, a \in A)$; for convenience we shall sometimes use the more detailed expression $M_t^a = M(t, a)$. We can define $M^u \in \mathbf{M}_{\mathrm{loc}}(F, P)$ for $u \in U$ as a sum of stochastic integrals

$$M_t^u = \sum_a \int_0^t I_{[u_s = a]} \mathrm{d}M_s^a \quad , \tag{1}$$

Now the class P^u is defined by the elements

$$P^u = \rho^u \cdot P \quad (P^u(B) = \int_B \rho^u \mathrm{d}p \; , \; B \in F_T) \quad .$$

Hence the construction scheme for the class of measures P^u is represented by the following chain of transitions:

$$P^A \to R^A \to M^A \to M^u \to R^u \to P^u \tag{2}$$

and the determining element of this chain is (1) for M^u.

In the general case, for non-finite (uncountable) A, (1) is replaced by a line integral with respect to the family M^A along the curve u (see [1]).

The mathematical expectation with respect to the measure P^u is denoted by E^u and that with respect to P by E.

We assume that all M^a are square-integrable, $M^a \in \mathbf{M}^2(F, P)$, and that

$$\sum_a <M^a>_T \leq c < \infty \quad ,$$

where $<\cdot>$ is a square characteristic which implies (by virtue of the finiteness of A and the condition $\Delta M_t^a > -1$, $0 \leq t \leq T$, resulting from $P^a \sim P$, see [3]) that $\rho^u > 0$, $\mathrm{E}\rho_T^u = 1$, the ρ^u are square-integrable and $P^u \sim P$, $u \in U$.

For the semi-martingale x with respect to the filter F and the measure P represented by $x = V + m$, where V is a predictable process with bounded variation and $m \in \mathbf{M}^2_{loc}(F, P)$, the expression $<x, M^u)$ is assumed to be equivalent to $<m, M^u>$.

For an F-adapted process V with integrable variation, $V^{P,G}$ denotes a dual G-predictable projection, i.e., a G-predictable process with integrable variation such that

$$\mathrm{E}(V_t \mid G_t) = V_t^{P,G} + m_t \; , \; 0 \leq t \leq T \quad ,$$

and $m \in \mathbf{M}(G, P)$. The dual predictable projection with respect to the measure P^u is denoted by $V^{P^u,G}$.

The relation between the projections with respect to the measures P^u and P is given by

$$V^{P^u,G} = (\hat{\rho}^u_-)^{-1} \cdot (\rho^u_- \cdot V)^{P,G} \; , \; v_0 = 0 \tag{3}$$

or, to be more precise,

$$V_t^{P^u,G} = \int_0^t (\hat{\rho}^u_{s-})^{-1} \mathrm{d}(\int_0^s \rho^u_{\tau-} \mathrm{d}V_\tau)^{P,G} \quad ,$$

where $\hat{\rho}^u$ is a contraction of the density ρ^u on the filter G

$$\hat{\rho}_t^u = \mathrm{E}(\rho_t^u \mid G_t) \quad .$$

Of course, $\hat{\rho}^u$ may also be represented as an exponential martingale $\hat{\rho}^u = \varepsilon(\hat{M}^u)$ using some martingale $\hat{M}^u$. However, it should be noted that in this case the representation $\hat{M}^u = \sum_a I_{[u=a]} \cdot \hat{M}^a$ does not hold.

Finally, we assume that $\sum_a <M^a>$ is dominated by some G-adapted increasing process K and that

$$\max_a \mathrm{d}<m, M^a>_t^{P,G}$$

is equivalent to

$$\max_a (\mathrm{d}<m, M^a>_t^{P,G} / \mathrm{d}K_t) \cdot \mathrm{d}K_t \quad .$$

3. ESTIMATORS OF CONTROLS. LINEAR EQUATIONS

Let η be some F_T-measurable bounded random variable, and consider the problem of maximizing the mathematical expectation $\mathrm{E}^u \eta$. We shall introduce processes which estimate the quality of the control on segments of the time interval $[t, T]$:

$$S(t, u) = \mathrm{E}^u(\eta \mid F_t), \; u \in U$$

$$S^v(t, u) = \mathrm{E}^v(s(t, u) \mid G_t) = \mathrm{E}^v(\mathrm{E}^u(\eta \mid F_t) \mid G_t), \; u, v \in U \quad .$$

Here $S(t, u)$ is the conditional expected reward at time t with complete observations if the past is fixed, and $S^v(t, u)$ is the conditional expected reward with partial observation if the control v was exerted before time t: with a fixed σ-algebra of observable events, the future average expected reward depends not only on the history of the system states, but also on the history of the control.

Estimators $S(t, u)$ and $S^v(t, u)$ can be defined as the solutions of certain stochastic linear equations.

LEMMA 1. *The estimator* $S(t, u)$ *is the unique solution of*

$$\mathrm{d}S(t, u) = -\mathrm{d}<s(\cdot, u), M^u>_t + \mathrm{d}m_t^u \quad , \tag{4}$$

with a boundary condition at the end of the interval $s(T, u) = \eta$, *given that* $m^u \in \mathbf{M}(F, P)$.

In this case the martingale m^u is uniquely defined by the equation and the boundary condition.

It is also possible to write some relations which on the one hand are equivalent to (4), and on the other, are decompositions of Doob or Riecz-type estimators $S(t, u)$ with respect to measures P or P^v for some $v \in U$:

$$\mathrm{d}S(t, u) = -\mathrm{d}<s(\cdot, u), M^u - M^v>_t + \mathrm{d}m_t^{u,v}, \; m^{u,v} \in \mathbf{M}(F, p^v) \quad , \tag{5}$$

$$S(t, u) = \mathrm{E}(\eta + \int_t^T \mathrm{d}<s(\cdot, u), M^u>_s \mid F_t) \quad , \tag{6}$$

$$S(t,u) = s(t,v) + \mathrm{E}^{v}\left(\int_t^T \mathrm{d}\langle s(\cdot,u), M^{u} - M^{v}\rangle_s \mid F_t\right) \quad . \tag{7}$$

All of these relations may be derived using the generalized form of Girsanov's theorem.

The following analogues of these relations hold for the estimator $S^{v}(t,u)$, $u \in U$, $v \in U$:

LEMMA 2. *The estimator* $S^{v}(t,u)$, $u \in U$, $v \in U$, *is uniquely defined by*

$$\mathrm{d}S^{v}(t,u) = -(\hat{\rho}^{v}_{t-})^{-1}\mathrm{d}(\rho^{v}_{-}\cdot \langle s(\cdot,u), M^{u} - M^{v}\rangle)^{P,G}_{t} +$$

$$+ \mathrm{d}\langle s^{v}(\cdot,u), \hat{M}^{v}\rangle_t + \mathrm{d}m^{u,v}_t \, , \; S^{v}(T,u) = \mathrm{E}^{v}(\eta \mid G_T) \quad , \tag{8}$$

$$m^{u,v} \in \mathbf{M}(G,P) \quad ,$$

or by relations equivalent to (8):

$$\mathrm{d}S^{v}(t,u) = -\mathrm{d}\langle s^{v}(\cdot,u), M^{u} - M^{v}\rangle^{p^{v},G}_{t} + \mathrm{d}m^{v}_t \, , \; m^{v} \in \mathbf{M}(G,p^{v}) \tag{9}$$

$$S^{v}(t,u) = S^{v}(t,v) + (\hat{\rho}^{v}_{t})^{-1}\mathrm{E}\left(\int_t^T \mathrm{d}(\rho^{v}_{-}\cdot \langle s(\cdot,u), M^{u} - M^{v}\rangle^{P,G}_{s} \mid G_t\right) \tag{10}$$

$$S^{v}(t,u) = S^{v}(t,v) + \mathrm{E}^{v}\left(\int_t^T \mathrm{d}\langle s(\cdot,u), M^{u} - M^{v}\rangle^{p^{v},G}_{s} \mid G_t\right) \quad . \tag{11}$$

We shall call $S(t,u)$ and $S^{v}(t,u)$ the complete and partial estimators of the control, respectively.

4. THE HAMILTONIAN. NECESSARY OPTIMALITY CONDITIONS

The need for an optimality condition arises in the following way. Suppose that while controlling the system we reach time t. The problem is to make the best choice of the control value u_t, taking into account both the information contained in the observed events from G_t and the previous control values (v_s, $s < t$).

It seems natural to maximize the rate of growth of the regular component in the decomposition of the estimator $S^{v}(t,u)$ when choosing u_t and, hence, by virtue of (8) or (9), to choose for u_t an action a for which the Hamiltonian

$$h_t(\rho^{v}, s(\cdot,u), a) = \frac{\mathrm{d}}{\mathrm{d}K_t}\langle s(\cdot,u), M^{a}\rangle^{p^{v},G}_{t} = (\hat{\rho}^{v}_{t-})^{-1}\frac{\mathrm{d}}{\mathrm{d}K_t} \tag{12}$$

$$(\rho^{v}_{-}\cdot \langle s(\cdot,u), M^{a}\rangle)^{P,G}_{t}$$

achieves its maximum value. But h also contains the complete estimators $s(\cdot,u)$ which depends on future values of the control. Since the value of u_t is to be chosen with regard to future optimal control, (12) should contain a complete estimator which corresponds to the optimal control.

Such heuristic reasoning expressing the idea of dynamic programming actually leads to the necessary optimality condition in the case of partial observations if the past control $(v_s, s < t)$ in (12) is also assumed to be optimal.

THEOREM 1 (Necessary optimality condition). *Let* u^* *be optimal, i.e.,* $E^{u^*}\eta = \sup_u E^u \eta$. *Then*

$$\max_a h_t(\rho^*, \psi, a) = h_t(\rho^*, \psi, u_t^*), \ \mu\text{–a.e.} \quad , \tag{13}$$

where μ *is a measure on the G-predictable subset B of the space* $\Omega \times [0,T]$ *and is defined by the process K (see* [3]*):*

$$\mu(B) = E \int_0^T I_B(\omega, t) dK_t \quad ,$$

and ρ^*, ψ *are solutions of the following system of equations:*

$$d\rho_t^* = \rho_{t-}^* - dM_t^*, \ \rho_0^* = 1$$

$$d\psi_t = -d\langle \psi, M^{u^*} \rangle_t + dm_t^*, \ \psi_T = \eta \quad , \tag{14}$$

$$m^* \in \mathbf{M}(F, P) \quad .$$

In addition $\rho^* = \rho^{u^*}$, $\psi = s(\cdot, u^*)$.

Maximization condition (13) and the system of equations (14) constitute the maximum principle, where the density ρ^{u^*} and the complete estimator $s(\cdot, u^*)$ represent the optimal trajectory of the controlled process and the adjoint process, respectively.

The main steps in the proof of the assertion are given briefly below. First, we establish an inequality between the partial estimators for all $u \in U$ and t:

$$s^{u^*}(t, u^*) \geq s^{u^*}(t, u) \text{ a.s.} \quad .$$

Consequently, for any $u \in U$, t, $s \leq t$, we have

$$E^{u^*}(E^u(\psi_t - \psi_s | F_s) | G_s) \leq 0 \text{ a.s.} \quad .$$

Thus the step-wise process

$$x_t^\Delta = -\sum_{s \leq t} E^{u^*}(E^u(\psi_s - \psi_{s-\Delta} | F_{s-\Delta}) | G_{s-\Delta})$$

increases with respect to t and the process

$$y_t^\Delta = E^{u^*}(\sum_{s \leq t} E^u(\psi_s - \psi_{s-\Delta} | F_{s-\Delta}) | G_t)$$

is a supermartingale with respect to G and the measure P^{u^*}.

On the other hand, from Lemma 1 we have that the process

$$\sum_{s \le t} \mathrm{E}^{u}(\psi_s - \psi_{s-\Delta} \mid F_{s-\Delta})$$

converges to $- \langle \psi, M^{u^*} - M^{u} \rangle$.

The result is that the process

$$y_t = \lim_{\Delta \to 0} y_t^{\Delta} = \mathrm{E}^{u^*}\Big(\int_t^T \mathrm{d} \langle \psi, M^{u^*} - M^{u} \rangle_s \mid G_t\Big)$$

turns out to be a supermartingale and so the predictable projection

$$\langle \psi, M^{u^*} - M^{u} \rangle^{p^{u^*}, G}$$

is an increasing process for all $u \in U$. Application of (3) then leads to (13).

The necessary optimality condition of form (13) for diffusion-type processes was established in [4] and [5]. In this case the martingales M^{a} which define the measure densities corresponding to the actions $a \in A$ are expressed by the stochastic integrals

$$M_t^{a} = \int_0^t f(s, w^1, a) \mathrm{d}w_s^1 + \int_0^t g(s, w^1, a) \mathrm{d}w_s^2$$

with respect to the Wiener processes w^1, w^2. Here the non-anticipative functionals f and g represent drift coefficients of the observable and non-observable components, respectively, with σ-algebra flows induced by (w^1, w^2). w^2 can be used for filters F and G with the basic measure P which represents the distribution (w^1, w^2). The control problem thus formulated for a process with observable and non-observable components (y, x) given by a system of equations

$$\mathrm{d}x_t = f(t, x, a)\mathrm{d}t + \mathrm{d}w_t^1$$

$$\mathrm{d}y_t = g(t, x, a)\mathrm{d}t + \mathrm{d}w_t^2$$

is then covered by the general scheme.

5. BELLMAN'S EQUATION. CONTROL DEFECT FORMULA. SUFFICIENT OPTIMALITY CONDITION FOR COMPLETE OBSERVATIONS ([1])

If $F_t = G_t$, $0 \le t \le T$, condition (13) is reduced to

$$\max_a h_t(S, a) = \max_a \mathrm{d}\langle S, M^{a} \rangle_t / \mathrm{d}K_t = h_t(S, u_t^*) \quad , \tag{15}$$

and for

$$S(t) = s(t, u^*) = \sup_u S(t, u) = \psi_t \quad ,$$

equation (4) takes the form of a non-linear equation:

$$dS(t) = -\max_a d\langle S, M^a\rangle_t + dm_t \ , \ m \in \mathbf{M}(F, P) \ , \ S(T) = \eta \quad . \tag{16}$$

From (7), with $u = u^*$, $v = u$, we also have

$$S(t) - S(t, u) = E^u\left(\int_t^T \max_a d\langle S, M^a - M^u\rangle_s \mid F_t\right) \quad . \tag{17}$$

Equation (16) is Bellman's equation for the value S and (17) is the control defect formula. This shows that the cause of the non-optimality of the control is simply the accumulated effect (integral) of the differences

$$\max_a h_t(S, a) - h_t(S, u_t)$$

for every t. One of the consequences of the defect formula is that condition (15) is sufficient for the optimality of u^*.

6. SUFFICIENT OPTIMALITY CONDITION IN DISCRETE TIME. ADJOINT PROCESS

The reason why the necessary condition (13), (15) is also sufficient in the case of complete observations is that the complete estimator $S(t, u)$ is independent of the previous controls. In the case when $F_t = G_t$, $0 \le t \le T$, the optimal control u^* on the whole interval $[0,T]$ also turns out to be optimal on the interval $[t, T]$ regardless of the values of previous controls, i.e., it follows from $E^{u^*}\eta = \sup_u E^u \eta$ that

$$s(t) = \sup_u E^u(\eta \mid F_t) = \sup_u S^v(t, u) = \sup_u S(t, u) = S(t, u^*) \quad .$$

In the partially observable case the situation is different. Returning to the arguments for the derivation of the Hamiltonian expression, we replace ψ in (12) by the adjoint process, which is estimated taking into account the fact that the optimal control in the future ($u_s^*, s > t$) will depend on past fixed values ($v_s, s < t$).

Consideration of the discrete time case helps us to find the exact expression for this process.

Let F_t and G_t be piecewise constant on the intervals $\Delta > 0$, ($F_t = F_{n\Delta}$, $n\Delta \le t < (n+1)\Delta$, $G_t = G_{n\Delta}$, $n\Delta \le t < (n+1)\Delta$, $n = 1,2, \ldots, N$) and let us consider only discrete times $t = 0, \Delta, 2\Delta, \ldots, T = N\Delta$. The martingales $M_t^a = M(t, a)$, $a \in A$, which define the densities ρ^a may be expressed in terms of transition probability densities

$$q(t, a) = \rho_t^a \mid \rho_{t-\Delta}^a \ , \ 0 < t \le T \ , \ a \in A$$

which represent conditional densities, i.e., for $B \in F_t$

$$P^a(B \mid F_{t-\Delta}) = \int_B q(t, a)P(d\omega \mid F_{t-\Delta}), \; 0 \le t \le T \quad .$$

Here

$$M_t^a = \sum_{s \le t} (q(s, a) - 1), \; M_t^u = \sum_{s \le t} (q(s, u_s) - 1)$$

and the densities ρ^a, ρ^u are defined by the linear equations

$$\Delta\rho_t^a = \rho_t^a - \rho_{t-\Delta}^a = \rho_{t-\Delta}^a \Delta M(t, a), \; \Delta\rho_t^u = \rho_{t-\Delta}^u \Delta M(t \cdot u_t) \quad .$$

Linear equation (4) for the complete estimator $S(t, u)$ is equivalent to the common recursive relation

$$\Delta S(t, u) = -\Delta\langle s(\cdot, u), M^u\rangle_t + \Delta m_t^u = -\mathrm{E}(\Delta S(t, u)(q(t, u_1) - 1) \mid F_{t-\Delta}) + \Delta m_t^u \quad ,$$

where $\mathrm{E}(\Delta S(t, u)\Delta M(t, u_t) \mid F_{t-\Delta})$ is an increment in the mutual characteristic $\langle s(\cdot, u), M^u\rangle$ and $\Delta m_t^u = \Delta S(t, u) - \mathrm{E}(\Delta S(t, u) \mid F_{t-\Delta})$ is an increment in the martingale component of $s(\cdot, u)$. Rearranging the summands we have

$$S(t - \Delta, u) = \mathrm{E}(S(t, u)q(t, u_t) \mid F_{t-\Delta}), \; S(T, u) = \eta \quad . \tag{18}$$

Similarly, the non-linear equation for the value (16) takes the form of a common recursive relation in dynamic programming:

$$S(t - \Delta) = \max_a \mathrm{E}(s(t)q(t, a) \mid F_{t-\Delta}), \; S(T) = \eta \tag{19}$$

and the Hamiltonian $h_t(s, a)$ (with respect to the counting process $K_t = [\frac{t}{\Delta}]$) is expressed as

$$h_t(s, a) = \Delta\langle S, M^a\rangle_t = \mathrm{E}(s(t)(q(t, a) - 1) \mid F_{t-\Delta}) \quad .$$

The discrete version of relations (8) or (9) for a partial estimator of $S^v(t, u)$ can immediately be obtained from (18):

$$\begin{aligned} S^v(t - \Delta, u) &= \mathrm{E}^v(S(t - \Delta, u) \mid G_{t-\Delta}) = \\ &= \mathrm{E}^v(\mathrm{E}(S(t, u)q(t, u_t) \mid F_{t-\Delta}) \mid G_{t-\Delta}) = \\ &= (\hat{\rho}_{t-\Delta}^v)^{-1}\mathrm{E}(\rho_{t-\Delta}^v S(t, u)q(t, u_t) \mid G_{t-\Delta}) \quad . \end{aligned} \tag{20}$$

This relation is fundamental for reasons explained below. Note that, on the right-hand side of (20), the control values $(v_s, s < t)$, u_t, $(u_s, s > t)$ are contained in the expressions ρ^v, $q(t, u_t)$ and $S(t, u)$, respectively.

Let $\varphi^*(t, v)$ denote the (G-predictable) admissible control on a segment of the time interval $[t, T]$

$$\varphi^*(t, v) = (\varphi_s^*(t, v), \; t \le s \le T)$$

which maximizes the partial estimator $S^v(t-\Delta, u)$ for given $v \in U$:

$$\sup_u S^v(t-\Delta, u) = S^v(t-\Delta, \varphi^*(t, v)) \quad .$$

(There is a certain inconsistency in the indices due to the fact that in order to incorporate the discrete time case into the general continuous scheme we have to shift the control index to the right: u_t is $G_{t-\Delta}$-measurable and actually corresponds to the time $t-\Delta$.)

The value $u_t^*(v) = \varphi_t^*(t, v)$ clearly represents the best action at the t-th step with regard to the history of the control and optimal behavior in the future.

Since

$$\sup_u S^v(t, u) = (\hat{\rho}_{t-\Delta}^v)^{-1} \mathrm{E}(\rho_{t-\Delta}^v S(t, \varphi^*(t, v)) q(t, u_t^*(v)) \mid G_{t-\Delta}) \quad ,$$

and $\max_{x,y} f(x, y) = f(x^*, y^*)$ implies $\max_x f(x, y^*) = f(x^*, y^*)$, it follows that the expression which should be maximized with respect to $a \in A$ in order to obtain the value $u_t^*(v)$ is

$$(\hat{\rho}_{t-\Delta}^v)^{-1} \mathrm{E}(\rho_{t-\Delta}^v S(t, \varphi^*(t, v)) q(t, a) \mid G_{t-\Delta}) \quad . \tag{21}$$

The most unsuitable component in (21) is the random variable $S(t, \varphi^*(t, v))$, which by its construction is equal to the complete estimator $S(t, u)$ from the time t when the control $\varphi^*(t, v)$ (actually, that part of $\varphi^*(t, v)$ from $t+\Delta$ to T) is exerted, and optimal with respect to the partial estimator $S^v(t-\Delta, u)$ from the moment $t-\Delta$.

We shall now transform expression (21) into a form which can be generalized to the continuous time case.

The maximization condition (21) does not change if instead of $S(t, \varphi^*(t, v))$ and $q(t, a)$ we insert

$$\Delta\psi_t^v = S(t, \varphi^*(t, v)) - S(t-\Delta, \varphi^*(t, v)) \text{ and } \Delta M(t, a) = q(t, a) - 1 \quad .$$

Then (21) is reduced to maximization of the expression

$$h_t(\rho^v, \psi^v, a) = (\hat{\rho}_{t-\Delta}^v)^{-1} \mathrm{E}(\rho_{t-\Delta}^v \Delta\psi_t^v \Delta M(t, a) \mid G_{t-\Delta}) \quad ,$$

which is a discrete analogue of Hamiltonian (12) in which the process

$$\psi_t^v = \sum_{s \le t} (S(s, \varphi^*(s, v)) - S(s-\Delta, \varphi^*(s, v))) \tag{22}$$

represents the adjoint process.

Making the substitution $u = \varphi^*(t, v)$ in (18) leads to a recursive equation for the process ψ_t^v. This can be given, for convenience, in a form similar to (9):

$$\Delta\psi_t^v = -\mathrm{E}(\Delta\psi_t^v(\Delta M(t, u_t^*(v)) - \Delta M(t, v_t)) \mid F_{t-\Delta}) + \Delta m_t^{u,v} \quad , \tag{23}$$

where $u_t^*(v)$ is defined by

$$\max_a h_t(\rho^v, \psi^v, a) = h_t(\rho^v, \psi^v, u_t^*(v)) \quad . \tag{24}$$

In order to derive an equation for the partial value $S^v(t) = \sup_u S^v(t, u)$, consider the proces

$$Z_t^v = S(0, \varphi^*(0, v)) + \sum_{s \le t} (S(s, \varphi^*(s + \Delta, v)) - S(s, \varphi^*(s, v))) \quad .$$

It can easily be seen that

$$c_t^v = \mathrm{E}^v(\Delta Z_t^v \mid G_{t-\Delta}) = \mathrm{E}^v(S(t, \varphi^*(t + \Delta, v)) - S(t, \varphi^*(t, v)) \mid G_{t-\Delta}) =$$

$$= \mathrm{E}^v(S^v(t) - S^v(t, \varphi^*(t, v)) \mid G_{t-\Delta}) \ge 0 \quad ,$$

and hence the process $\hat{Z}_t^v = \mathrm{E}^v(Z_t^v \mid G_t)$ is a submartingale with respect to the filter G and the measure P^v.

Now (23), (24) and the obvious relation

$$S^v(t) = \mathrm{E}^v(S^v(t, \varphi^*(t + \Delta, v)) \mid G_t) = \mathrm{E}^v(\psi_t^v + Z_t^v \mid G_t)$$

leads to an equation for $S^v(t)$:

$$\Delta S^v(t) = \max_a (\hat{\rho}_{t-\Delta}^v)^{-1}\mathrm{E}(\rho_{t-\Delta}^v \Delta\psi_t^v \Delta(M(t, a)$$

$$- M(t, v_t)) \mid G_{t-\Delta}) + C_t^v + \Delta M_t^v \tag{25}$$

and the defect formula

$$S^v(t) - S^v(t, v_t) = \mathrm{E}^v(\sum_{s=t}^{T} (\max_a h_s(\rho^v, \psi^v, a) - h_s(\rho^v, \psi^v, v_s)) - C_s^v) \mid G_t) \tag{26}$$

which is equivalent to (25).

In order to obtain the optimality condition in a necessary and sufficient form we shall make an additional transformation. It is evident from (26) that $C_t^v \le \bar{h}_t^v = \max_a h_t(\rho^v, \psi^v, a) - h_t(\rho^v, \psi^v, v_t)$. Let us consider a new proces

$$\bar{\psi}_t^v = \sum_{s \le t} (1 - C_s^v / \bar{h}_s^v) \Delta \psi_s^v \quad .$$

Since $C_t^v / \bar{h}_t^v$ is G-measurable, $\bar{\psi}^v$ also satisfies (23) and (24). Now the defect formula (26) takes the form

$$S^v(t) - S^v(t, v) = \mathrm{E}^v\Big(\sum_{s=t}^{T} (\max_a h_s(\rho^v, \bar{\psi}^v, a) - h_s(\rho^v, \psi^v, v_s)) \mid G_t \Big) \quad . \tag{27}$$

This second formula implies that the condition

$$\max_a h_t(\rho^v, \bar{\psi}^v, a) = h_t(\rho^v, \bar{\psi}^v, v_t)$$

is necessary and sufficient for the optimality of v.

7. SUFFICIENT OPTIMALITY CONDITION FOR THE GENERAL CASE

THEOREM 2. *For every $v \in U$ there exists a semi-martingale $\bar{\psi}^v$ such that the defect formula*

$$S^v(t) - S^v(t, v) = \mathrm{E}^v\Big(\int_t^T (\max_a h_s(\rho^v, \bar{\psi}^v, a) - h_s(\rho^v, \bar{\psi}^v, v_s)) \mathrm{d}K_s \mid G_t \Big) \tag{28}$$

holds, where

$$h_t(\rho^v, \bar{\psi}^v, a) = (\hat{\rho}_{t-}^v)^{-1} \frac{\mathrm{d}}{\mathrm{d}K_t} (\rho_-^v \cdot \langle \bar{\psi}^v, M^a \rangle)_t^{P,G} \quad .$$

The necessary and sufficient control optimality condition is

$$\max_a h_t(\rho^{u^*}, \bar{\psi}^{u^*}, a) = h_t(\rho^{u^*}, \bar{\psi}^{u^*}, u_t^*) \quad . \tag{29}$$

The main steps of the proof are as follows:

(a) Identify the controls $u \in U$ with densities ρ^u, $u \in U$, where the set U is a subset of the Hilbert space.

(b) For every $v \in U$, a measurable mapping φ^* exists such that

$$\varphi^*(\cdot, v) \colon \Omega \times [0,T] \to U \, , \ \sup_u S^v(t-, u) = S^v(t-, \varphi^*(t, v)) \quad .$$

(c) A stochastic line integral with respect to the family $(s(\cdot, u),\ u \in U)$ is defined on the curve $\varphi^*(\cdot, u)$:

$$\psi_t^v = \int_0^t S(\mathrm{d}s, \varphi^*(s, v)) \quad . \tag{30}$$

(d) The process $\hat{Z}_t^v = S^v(t) - \mathrm{E}^v(\psi_t^v \mid G_t)$ represents a submartingale with respect to the filter G and the measure P^v; the increasing process C^v in the decomposition $\hat{Z}^v = C^v + m$, $m \in \mathbf{M}(G, P^v)$ is absolutely continuous with respect to the process K and

$$c_t^v = \mathrm{d}C_t^v / \mathrm{d}K_t \leq \bar{h}_t^v = \max_a h_t(\rho^v, \psi^v, a) - h_t(\rho^v, \psi^v, v_t) \quad .$$

(e) The semimartingale with respect to the measure P and the filter F

$$\bar{\psi}_t^v = \int_0^t (1 - c_s^v / \bar{h}_s^v) \mathrm{d}\psi_s^v$$

satisfies the equation

$$\mathrm{d}\bar{\psi}_t^v = -\mathrm{d}<\bar{\psi}^v, \mathbf{M}(\cdot, u^*(v))>_t + \mathrm{d}\bar{m}_t^v, \ \bar{m}^v \in \mathbf{M}(F, P) \quad , \tag{31}$$

where $u_t^*(v)$ maximizes $h_t(\rho^v, \bar{\psi}^v, a)$.

(f) Either (28) or the following equivalent differential equation for a partial value $S^v(t)$:

$$\mathrm{d}S^v(t) = -(\max_a h_t(\rho^v, \bar{\psi}^v, a) - h_t(\rho^v, \bar{\psi}^v, v_+))\mathrm{d}K_t + \mathrm{d}<S^v, \hat{M}^v>_t + \mathrm{d}m_t^v \tag{32}$$

holds for $m^v \in \mathbf{M}(G, P)$.

The main point in the construction of the adjoint process $\bar{\psi}^v$ is the definition of the line integral (30) which generalizes the discrete time expression (22). (The general approach to the line integral was proposed in [1].)

Assuming that A is finite or $\sum_a <M^a>_T \leq c < \infty$ is bounded simplifies the proof of (b).

For the optimal control u^*, we have $\bar{\psi}_t^{u^*} = \psi_t^{u^*} = \psi_t = S(t, u^*)$ and (29) is reduced to the necessary condition (13). Equation (31) is transformed into (16) and in the case of complete observations we have $\bar{\psi}_t^v = s(t)$.

In contrast to the optimality conditions given in [6] for diffusion-type processes, only the t-th (last) control value is checked at the t-th step in condition (25). In [6] optimality is tested using the expression

$$-(\max_a h_t(\rho^v, \bar{\psi}^v, a) - h_t(\rho^v, \bar{\psi}^v, v_t)) + \frac{\mathrm{d}}{\mathrm{d}K_t} <s^v, \hat{M}^v>_t \quad .$$

Finally, it seems interesting to find a relation between the Hamiltonian construction considered in this paper, and the construction based on the Bellman–Mortense equations for the value function in the Markov case which is considered elsewhere.

REFERENCES

1. R.J. Chitashvili. Martingale ideology in the theory of controlled stochastic processes. *Lecture Notes in Mathematics*, Vol. 1021. Springer-Verlag, Berlin, 1982.

2. Yu.M. Kabanov, R.Sh. Liptser and A.N. Shiryaev. Absolute continuity and singularity of absolutely continuous probability distributions. Matematicheski Sbornik, 10(149)(1978).

3. J. Jacod. Calcul stochastique et problemes de martingales, *Lecture Notes in Mathematics*, Vol. 714. Springer-Verlag, Berlin, 1979.

4. R.I. Elliot. The optimal control of a stochastic system. *SIAM Journal of Control and Optimization*, 15 (1977).

5. U.G. Haussmann. On the stochastic maximum principle. *SIAM Journal on Control and Optimization*, 16 (1978).

6. M.H.A. Davis and P. Varaiya. Dynamic programming conditions for observable stochastic systems. *SIAM Journal on Control and Optimization*, 11 (1973).

ON THE LIMITING DISTRIBUTION OF EXTREMUM POINTS FOR CERTAIN STOCHASTIC OPTIMIZATION MODELS

A.Ya. Dorogovtsev and A.G. Kukush
Kiev State University, Kiev, USSR

1. INTRODUCTION

The asymptotic behavior of infinite-dimensional parameter estimates has been studied in many papers (see [1–3] and the references therein). Particular attention has been paid to the *consistency* or *risk* of the estimates. However, the problem of obtaining weak convergence conditions for suitably normalized estimates turns out to be rather complicated and has not been studied in any detail (although some specific cases are treated in [1,3]). This paper gives weak convergence conditions for the normalized estimate of the drift coefficient for the simplest stochastic differential equation. We assume that the unknown drift coefficient belongs to a certain compact subset of the space of continuous functions with uniform norm. The drift coefficient is estimated by maximizing the likelihood ratio over finite-dimensional projections of this compact set. A similar method for constructing non-parametric regression estimates was proposed in [2,3]. In [1], the properties of the drift coefficient estimate obtained by maximizing the likelihood ratio over the whole compact set were explored; here we calculate the estimate more simply, although the result is somewhat less strict than in [1].

2. PROBLEM STATEMENT

For fixed numbers $\alpha > 2$ and $L > 0$ let

$$K = K(L) = \{f : \mathbf{R} \to \mathbf{R} \mid \forall t \in \mathbb{R}: f(t+2\pi) = f(t) \quad ;$$

$$|a_0(f)| \le L \, , \, k^\alpha \, |a_k(f)| \le L \, , \, k^\alpha \, |b_k(f)| \le L \, , \, k \in \mathbb{N}\} \quad ,$$

where $a_i(f)$, $i \ge 0$, and $b_j(f)$, $j \ge 1$, are Fourier coefficients of the function f with respect to the sequence

$$\frac{1}{2} \, , \, \cos t \, , \, \sin t \, , \ldots , \, \cos(nt) \, , \, \sin(nt) \, , \ldots ; \; t \in [0, 2\pi] \quad .$$

In addition, let K_0 be the set $K(L_0)$ for a certain fixed and possibly unknown value $L_0 < L$. Note that $K \subset C^1([0,2\pi])$ and that K contains all compact subsets of the space $C([0,2\pi])$ with uniform norm.

Let $(\Omega, \mathbf{F}, P)$ be a probability space and $\{w(t), t \geq 0\}$ be a standard Wiener process. We consider the problem of estimating an unknown but fixed function $s_0 \in K_0$ on the basis of observations on the segment $[0,T]$, $T > 0$, of the values of a process $\{x(t), t \geq 0\}$ with a stochastic differential of the form

$$dx(t) = s_0(t)dt + dw(t), \ t > 0 \quad .$$

Note that estimates of s_0 which are consistent as $T \to +\infty$ were first considered in [4].

Let $\pi_k : K \to K$ be the map defined as follows:

$$\pi_k(f)(t) = \frac{1}{2} a_0(f) + \sum_{i=1}^{k} (a_i(f) \cos(it) + b_i(f) \sin(it)) \quad ,$$

$$f \in K, \ k \in \mathbb{N}, \ t \in \mathbb{R} \quad .$$

Let $\{n(m), m \geq 0\}$, $n(0) = 1$, and $\{k(m), m \geq 1\}$ be fixed increasing sequences of natural numbers. For $T \geq 2\pi$ we define $m(T) \in \mathbb{N}$ such that

$$2\pi n(m(T)) \leq T < 2\pi n(m(T)+1) \quad .$$

For each $T \geq 2\pi$ we define the estimate s_T of the function s_0 by means of the log-likelihood function

$$Q_T(s) = \frac{1}{T} \int_0^T s(t)dx(t) - \frac{1}{2T} \int_0^T s^2(t)dt, \ s \in K \quad ,$$

as some value from $\pi_{k(m(T))}K$ which satisfies the condition

$$Q_T(s_T) = \max \{Q_T(s) \mid s \in \pi_{k(m(T))}K\} \quad . \tag{1}$$

Simple extension of the results given in [5] proves that for each $T \geq 2\pi$ the value s_T may be chosen to be a random element with values in $C([0,2\pi])$ (see also [1]). The main results of the present paper are summarized in the following theorem:

THEOREM 1. *Assume that for some $\beta < 1/4$ we have*

$$\lim_{m \to \infty} (n^{\beta}(m)/k^{\alpha}(m)) > 0 \quad . \tag{2}$$

Then the following statements hold a.s. *as $T \to +\infty$:*

1. $T^{2\gamma} \int_0^T (s_T(t) - \pi_{k(m(T))} s_0(t))^2 dt \to 0 \quad \forall \gamma < 1/4 \quad ;$

2. $\max_{t\in R} |s_T(t) - s_0(t)| \to 0$;

3. $T^\delta \max_{t\in R} |s_T(t) - \pi_{k(m(T))} s_0(t)| \to 0 \quad \forall \delta < (\alpha-1)/(2(2\alpha-1))$.

Moreover, the net of measures corresponding to the family of random processes

$$\{w_T(t) := (\frac{T}{2\pi})^{1/2} \int_0^t (s_T(u) - \pi_{k(m(T))} s_0(u))du \ , \ t \in [0,2\pi]\} \ , \ T \geq 2\pi$$

converges weakly to the measure corresponding to the standard Wiener process on the segment $[0,2\pi]$ *as* $T \to +\infty$.

3. PROOF OF THE THEOREM

Statements 1–3 of the theorem are concerned with strict consistency and also with the rate of convergence of the estimate to an unknown value. The proofs of these statements are similar to those given in [1, § 2, Chap. IV] and are omitted here. To prove the last part of the theorem we consider the sequence of values T: $T(n) = 2\pi n$, $n \in \mathbb{N}$, $m(T(n)) = m_n$. It follows from the inclusion $s_0 \in K_0$, condition (2) and statement 1 of the theorem that the element $\pi_{k(m_n)} s_{T(n)}$ is not an extreme point of the convex set $\pi_{k(m_n)} K$ for all $n \geq N(\omega)$. Thus

$$(Q_{T(n)} \pi_{k(m_n)})'(s_{T(n)}) = 0 \ , \ n \geq N(\omega) \quad ,$$

and the derivative is the Frechet derivative of a real function defined on $L_2([0,2\pi])$. Using Taylor's formula, for $n \geq N(\omega)$ and $h \in L_2([0,2\pi])$ we have

$$(Q_{T(n)} \pi_{k(m_n)})'(s_0)h + \langle (Q_{T(n)} \pi_{k(m_n)})''(s_0)(s_{T(n)} - s_0) \ , \ h \rangle = 0 \quad , \tag{3}$$

where

$$(Q_{T(n)} \pi_{k(m_n)})'(s_0)h = -\frac{1}{2\pi n} \int_0^{2\pi n} (\pi_{k(m_n)} h)(t) dw(t) =$$

$$= -\frac{1}{2\pi n} \int_0^{2\pi} (\pi_{k(m_n)} h)(t) \sum_{i=0}^{n-1} dw(t + 2i\pi) \quad ; \tag{4}$$

$$\langle (Q_{T(n)} \pi_{k(m_n)})''(s_0)(s_{T(n)} - s_0) \ , \ h \rangle = \frac{1}{2\pi n} \cdot \tag{5}$$

$$\cdot \int_0^{2\pi n} (\pi_{k(m_n)} s_{T(n)}(t) - \pi_{k(m_n)} s_0(t)) \pi_{k(m_n)} h(t) dt = \frac{1}{2\pi} \int_0^{2\pi} (s_{T(n)} - \pi_{k(m_n)} s_0) h \, dt \quad .$$

The function h in (3)–(5) is the function from $L_2([0,2\pi])$ periodically continued on

R.

Now consider the operator

$$A: L_2([0,2\pi]) \to C([0,2\pi]) \quad ,$$

$$A(h)(\tau) = \int_0^\tau h(t)dt \ , \ \tau \in [0,2\pi] \ , \ h \in L_2([0,2\pi]) \quad .$$

The conjugate operator A^* acts from a space $M([0,2\pi])$ of signed finite measure on $[0,2\pi]$ to $L_2([0,2\pi])$:

$$A^*(\mu)(s) = \int_s^{2\pi} \mu(dt) \ , \ s \in [0,2\pi] \ , \ \mu \in M([0,2\pi]) \quad .$$

Let $\{e_p\}$ be a sequence of functions of the trigonometric sequence from Section 2 which are orthonormal on $[0,2\pi]$. For $h = A^*\mu$ we have

$$\int_0^{2\pi} \pi_{k(m_n)} h \, dw(t) = \sum_{p=1}^{2k(m_n)+1} \int_0^{2\pi} e_p(t)dw(t) \int_0^{2\pi} A^*\mu e_p(s)ds =$$

$$= \sum_{p=1}^{2k(m_n)+1} \int_0^{2\pi} e_p(t)dw(t) \int_0^{2\pi} \left(\int_0^v e_p(s)ds\right)\mu(dv) = \langle \mu \, , \sum_{p=1}^{2k(m_n)+1} \int_0^{2\pi} e_p(t)dw(t) \int_0^{\cdot} e_p(s)ds \rangle \quad .$$

The right-hand side of (4) may now be transformed to

$$-\frac{1}{2\pi n} \langle \mu \, , \sum_{i=0}^{n-1} \sum_{p=1}^{2k(m_n)+1} \int_0^{2\pi} e_p(t)dw(t+2i\pi) \int_0^{\cdot} e_p(s)ds \rangle \quad , \tag{6}$$

and the right-hand side of (5) for $h = A^*\mu$ may be written

$$\frac{1}{2\pi} \langle \mu \, , A(s_{T(n)} - \pi_{k(m_n)} s_0) \rangle \quad . \tag{7}$$

It follows from (6), (7) and (3) that for $n \geq N(\omega)$ we have

$$\int_0^v (s_{T(n)}(t) - \pi_{k(m_n)}(s_0)(t))dt =$$

$$= \frac{1}{n} \sum_{i=0}^{n-1} \sum_{p=1}^{2k(m_n)+1} \int_0^{2\pi} e_p(t)dw(t+2i\pi) \int_0^v e_p(s)ds \ , \ v \in [0,2\pi] \quad ,$$

or

$$w_{T(n)}(v) = \frac{1}{\sqrt{n}} \sum_{i=0}^{n-1} \sum_{p=1}^{2k(m_n)+1} \int_0^{2\pi} e_p(t)dw(t+2i\pi) \int_0^v e_p(s)ds \ , \ v \in [0,2\pi] \quad .$$

Let $\rho_n(v)$, $v \in [0,2\pi]$, be the right-hand side of (8). Then for all $n \geq 1$, the process $\{\rho_n(v), 0 \leq v \leq 2\pi\}$ is a zero mean Gaussian process such that

$$E(\rho_n(v_1) - \rho_n(v_2))^2 = \sum_{p=1}^{2k(m_n)+1} \left(\int_{v_1}^{v_2} e_p(s)ds\right)^2, \ 0 \leq v_1 \leq v_2 \leq 2\pi \quad .$$

Hence,

$$E\rho_n^2(v) \to v, \ n \to \infty; \ v \in [0,2\pi]$$

$$E(\rho_n(v_1) - \rho_n(v_2))^2 \leq |v_1 - v_2|, \ \{v_1, v_2\} \subset [0,2\pi] \quad .$$

The convergence of the finite-dimensional distributions of this process to those of the Wiener process is obvious; the compactness of the distributions in the space $C([0,2\pi])$ follows from the inequality for the fourth moments and the Kolmogorov compactness condition. In addition, the difference between the left and right-hand sides of (8) tends to zero a.s. uniformly on v because the number $N(\omega)$ does not depend upon $V \in [0,2\pi]$. Then $w_{T(n)} \to w$ in distribution on $C([0,2\pi])$, $n \to \infty$, thus proving the theorem.

Remark. The theorem also holds for the estimates obtained by maximizing Q_T over the whole compact set K, replacing $\pi_{k(m_n)}s_0$ by s_0. However, it is necessary to have $\alpha > \varepsilon$ for this (see [1]).

REFERENCES

1. A.Ya. Dorogovtsev. *The Theory of Estimation of Parameters of Random Processes.* Visha shkola, Kiev, 1982 (in Russian).

2. A.S. Nemirovskii, B.T. Polak and A.B. Tsibakov. Estimates of the maximum likelihood type for non-parametric regression (in Russian). *Doklady Akademii Nauk SSSR*, 273(6)(1983) 1310–1314.

3. A.Ya. Dorogovtsev. Estimating the mean value parameter of the measure in Hilbert space (in Russian). *Teoria Sluchainih Processov*, 11(1983)28–31.

4. I.Sh. Ibramhalilov and A.V. Skorohod. The derivation of the mean value of a Wiener process observed over an infinite interval (in Russian). *Teoria Verojatnosti i ee Primeneniya*, 18(4)(1973) 21–28.

5. R.I. Jennrich. Asymptotic properties of nonlinear squares estimators. *Annals of Mathematical Statistics*, 2(1969) 633–643.

THE STRUCTURE OF PERSISTENTLY NEARLY-OPTIMAL STRATEGIES IN STOCHASTIC DYNAMIC PROGRAMMING PROBLEMS

E.A. Fainberg
Department of Applied Mathematics,
Moscow Institute of Transport Engineers (MIIT)
Moscow, USSR

1. INTRODUCTION

This paper deals with the structure of persistently nearly-optimal strategies in discrete-time countable-state stochastic dynamic programming models. For models with finite state and action sets the problem has been completely solved by Blackwell [1] and Krylov [2]. Using different methods, they both proved the existence of stationary optimal strategies. However, stationary optimal (or even nearly-optimal) strategies may not exist for models with infinite action sets [3,4]. In this connection the existence of stationary nearly-optimal strategies has been proved for certain classes of models (for example, positive models [5–7], models with finite or compact action sets [8,9], strongly convergent models [9–11], and contracting models [9,12]).

In the general case Fainberg and Sonin [13,14] have proved that if the value function of the model is replaced by the value function of the class of stationary strategies then uniformly nearly-optimal strategies exist. In Section 2 of this paper we consider this result and its application to various special classes of models (strongly convergent, contracting, etc.).

The class of Markov strategies is a natural extension of the class of stationary strategies. Van der Wal [15], Sonin [16], Fainberg and Sonin [13,14], and Van Dawen [17] have proved the existence of persistently nearly-optimal Markov strategies.

Another approach to the extension of the class of stationary strategies for finite state sets has been offered by Everett [18] and developed by Chitashvily [19,20]. These authors considered strategies which were stationary on a subset of the state space. For countable state models the existence of good strategies of this type has been established by Fainberg and Sonin [21] and Fainberg [22]. This approach allows us to prove results (see Section 3 of this paper) which are more general than those concerning the existence of good Markov strategies.

Consider a Markov decision model $\mu = \{X, A, A(\cdot), p, r\}$, where (i) X is a countable state space; (ii) A is a set of actions which is assumed to be endowed with a σ-field $\mathbf{A}$ containing all single to n sets; (iii) $A(x)$, $x \in X$, is the set of admissible actions if the model is in state x, $A(\cdot) \in \mathbf{A}$; (iv) $p(z \mid x, a)$ is a transition probability, $p(z \mid x, a) \geq 0$ and $\sum_{z \in X} p(z \mid x, a) \leq 1$; (v) $r(x, a)$ is a reward function, $-\infty \leq r(x, a) < +\infty$, $x, z \in X$, $a \in A$. Functions p and r are assumed to be measurable in a. We shall write $H_n = (X \times A)^n \times X$, $n = 0,1, \ldots, \infty$, and $H = \bigcup_{0 \leq n < \infty} H_n$. Products of σ-fields 2^X and $\mathbf{A}$ generate σ-fields $\mathbf{F}_n$ and $\mathbf{F}$ on H_n and H, respectively.

Consider three sets of strategies: the set of all (possibly randomized and history-dependent) strategies Π, the set of all (non-randomized) Markov strategies M, and the set of (non-randomized) stationary strategies S, $S \subseteq M \subseteq \Pi$.

As usual the pair $x \in X$ and $\pi \in \Pi$ defines the measure P_x^π on $(H_\infty, \mathbf{F}_\infty)$. Expectations with respect to P_x^π are denoted by E_x^π. We shall consider the total expected reward criterion

$$w^\pi(x) = \mathrm{E}_x^\pi \sum_{i=0}^{\infty} r(x_i, a_i) \quad . \tag{1}$$

The standard general convergence condition is assumed throughout: for each $x \in X$, $\pi \in \Pi$

$$\mathrm{E}_x^\pi \sum_{i=0}^{\infty} r^+(x_i, a_i) < \infty \quad ,$$

where $g^+ = \max(g, 0)$, $g^- = \min(g, 0)$ for any number g.

For $\pi \in \Pi$ and $h = (x_0 a_0 \cdots x_n) \in H$, $n = 0,1, \ldots,$ we define the strategy $\pi[h]$ as follows:

$$\pi[h]_m(\cdot \mid h') = \pi_{m+n}(x_0 a_0 \cdots x_{n-1} a_{n-1} x_0' a_0' \cdots x_m')$$

for any $h' = (x_0' a_0' \cdots x_m') \in H$, $m = 0,1, \ldots,$ Note that $\varphi[h] = \varphi$ for $\varphi \in S$. Let $w^\pi(h_n) = w^{\pi[h_n]}(x_n)$.

For $\Delta \subseteq \Pi$, $h \in H$ we write $v_\Delta(h) = \sup\{w^\pi(h) : \pi \in \Delta\}$. Let $V = V_\Pi$, $s = V_S$. Note that $v(h_n) = v(x_n)$, $s(h_n) = s(x_n)$ for any $h_n = (x_0 a_0 \cdots x_n) \in H$. If the function r is replaced by r^+ (or r^-) in (1), we will write v_+ (or v_-) instead of v.

Let $g : X \to [0, \infty)$. A strategy π is said to be *g-optimal* if $w^\pi(x) \geq v(x) - g(x)$ for any $x \in X$. A strategy π is said to be *persistently g-optimal* if $w^\pi(h_n) \geq v(x_n) - g(x_n)$ for any $h_n = (x_0 a_0 \cdots x_n) \in H$. Note that every persistently g-optimal strategy is g-optimal, and that every stationary g-optimal strategy is persistently g-optimal.

Let $Q[0,n]$, $n=0,1,\ldots,\infty$, denote the set of all Markov times with respect to the flow $\{\mathbf{F}_m\}_{m=0}^{\infty}$ such that $\tau=\tau(h_\infty)\le n$, $h_\infty\in H_\infty$.

Define

$$d(x)=\sup_{\varphi\in S}\lim_{n\to\infty}\inf_{\tau\in Q[0,n]}\mathrm{E}_x^{\pi}S(x_\tau),\ x\in X\quad,$$

$$X_\Delta=\{x\in X:\ w^{\pi}(x)=v_\Delta(x)\text{ for some }\pi\in\Delta\},\ \Delta\subseteq\Pi\quad.$$

For functions $g:X\to[-\infty,+\infty)$ such that $g\le V_+$ and for $a\in A$, consider the following operators (where $g\le f$ implies $g(x)\le f(x)$ for any $x\in X$):

$$P^a g(x)=\sum_{z\in X}p(z\mid x,a)g(z),\ Pg(x)=\sup_{a\in A(x)}P^a g(x)\quad,$$

$$T^a g(x)=r(x,a)+P^a g(x),\ Tg(x)=\sup_{a\in A(x)}T^a g(x)\quad.$$

Let $L_0(g,X')$ be the set of all functions $l:X\to[0,+\infty)$ such that $l(x)=0$ for $x\in X'$ and $l(x)>0$, $l(x)\ge\max\{g(x),Pl(x)\}$ for $x\in X\setminus X'$, where $X'\subseteq X$ and $g:X\to[-\infty,+\infty)$. Since $(v_++1)1_{X\setminus X'}\in L_0(v,X')$, we have that $L_0(v,X')\ne\phi$. Note that $d\le s\le v$ and $L_0(d,X')\supseteq L_0(s,X')\supseteq L_0(v,X')$.

1.1. STATIONARY NEARLY-OPTIMAL STRATEGIES

THEOREM 1 (Fainberg and Sonin [14]). *For any $\varepsilon>0$ and any $l\in L_0(d,X_S)$ there exists a stationary strategy φ such that $w^{\varphi}\ge s-\varepsilon l$.*

Theorem 1 has been proved for $l\in L_0(s,X_S)\subseteq L_0(d,X_S)$ in [13]. This theorem implies the result obtained by Van Dawen and Schäl [23] and Van Dawen [17], in which the equality $s=v$ is assumed. Theorem 1 also implies $s=Ts$ [13, Lemma 2.2]. The following corollary gives a general method for proving the existence of stationary (uniformly) nearly-optimal strategies.

COROLLARY 1. *If $s=v$ and $l\in L_0(d,X_S)$, then for any $\varepsilon>0$ there exists a stationary εl-optimal strategy.*

Note that if $s=v$ then $X_S=X_\Pi$ [22, Lemma 3.9]. The following result has been used in the proof of Theorem 1:

THEOREM 2 (Van der Wal [9], Theorem 2.22). *If $|A(x)|<\infty$ for any $x\in X$, then $s=v$.*

Various generalizations of Theorem 2 are given in [8,21,22,24,25].

It has been shown in [13] using Corollary 1 that Theorem 1 implies the result obtained by Ornstein [7] and Frid [6] and that Theorem 1 allows us to extend the result obtained by Van der Wal [25]. We shall now consider applications of Theorem 1 to strongly convergent and contracting models.

THEOREM 3. *If we have*

$$\limsup_{n\to\infty} E_x^\pi s(x_n) \ge 0 \tag{2}$$

for some $x \in X$ and for any $\pi \in M$, then $s(x) = v(x)$.

Proof. If $v(x) = -\infty$ then $s(x) = -\infty$. let $v(x) > -\infty$. Fix an arbitrary $\varepsilon > 0$. Using the equality $v_M = v$ [26–28] we can choose $\pi \in M$ such that $w^\pi(x) \ge v(x) - \varepsilon/4$. Consider an integer n such that

$$E_x^\pi \sum_{i=n}^{\infty} r^+(x_i, a_i) \le \varepsilon/4 \,,\; E_x^\pi s(x_n) \ge -\varepsilon/4 \quad .$$

Let $l \in L_0(d, \phi)$. Choose $\delta > 0$ such that $\delta l(x) \le \varepsilon/4$. Using Theorem 1 we consider $\psi \in S$ such that $w^\psi \ge s - \delta l$. Let σ be a non-randomized strategy defined as follows:

$$\sigma_i(x_0 a_0 \cdots x_i) = \begin{cases} \pi_i(x_i) & , i < n \quad , \\ \psi(x_i) & , i \ge n \quad . \end{cases}$$

Then

$$w^\sigma(x) = E_x^\pi\{\sum_{i=1}^{n-1} r(x_i, a_i) + w^\psi(x_n)\} \ge E_x^\pi\{\sum_{i=1}^{n-1} r(x_i, a_i) + s(x_n) - \delta l(x_n)\}$$

$$\ge E_x^\pi \sum_{i=1}^{n-1} r(x_i, a_i) + E_x^\pi s(x_n) - \delta l(x) \ge E_x^\pi \sum_{i=1}^{n-1} r(x_i a_i) - \varepsilon/2$$

(note that $l(x) \ge E_x^\pi l(x_n)$).

On the other hand,

$$w^\pi(x) = E_x^\pi\{\sum_{i=0}^{n-1} r(x_i, a_i) + \sum_{i=1}^{\infty} r(x_i, a_i)\} \le$$

$$\le E_x^\pi \sum_{i=0}^{n-1} r(x_i, a_i) + E_x^\pi \sum_{i=n}^{\infty} r^+(x_i, a_i) \le E_x^\pi \sum_{i=0}^{n-1} r(x_i, a_i) + \varepsilon/4 \quad .$$

Consequently

$$w^\sigma(x) \ge w^\pi(x) - 3\varepsilon/4 \ge v(x) - \varepsilon \quad .$$

Theorem 2 implies $s(x) \ge w^\sigma(x)$. Since $\varepsilon > 0$ is arbitrary, $s(x) = v(x)$.

COROLLARY 2. *If* (2) *holds for any $x \in \{x \in X: v(x) > -\infty\}$ and any $\pi \in M$, then there exists a stationary εl-optimal strategy for any $\varepsilon > 0$ and any $l \in L_0(d, X_\Pi)$.*

Corollary 2 generalizes existing results for strongly convergent models [4,9–11], i.e., models satisfying the conditions

$$z^*(x) = \sup_{\pi\in\Gamma} \mathrm{E}_x^{\pi} \sum_{n=0}^{\infty} |r(x_n, a_n)| < \infty \ , \ x \in X \ , \tag{3}$$

$$\lim_{n\to\infty} \sup_{\pi\in M} \sum_{t=n}^{\infty} \mathrm{E}_x^{\pi} |r(x_t, a_t)| = 0 \ , \ x \in X \ . \tag{4}$$

It is obvious that (2) is weaker than (4). If (3) holds then $z^* \cdot 1_{X\setminus X_s} \in L_0(d, X_S)$.

The assumptions of Corollary 2 and condition (3) thus imply that for each $\varepsilon > 0$ there exists a stationary εz^*-optimal strategy (this result is a generalization of Theorem 4.3 in [4], see also [9,10]). The assumptions of Corollary 2 and the condition $d \le K < \infty$ (note that $d \le s \le v$) imply that for each $\varepsilon > 0$ there exists a stationary ε-optimal strategy.

Theorem 3 implies the following result:

THEOREM 4. *If* $s \ge 0$ *then* $s = v$.

Note that we also have $s = v$ in the following two cases:

THEOREM 5 (Van der Wal [25]). *If* $v > 0$ *then* $s = v$.

THEOREM 6 ([22]). *If* $v < 0$ *then* $s = v$.

Corollary 1 allows us to prove some new results for contracting models. For example, the equality $s = v$ has been proved for model I from [9, chapter 5]. However, the existence of stationary ε-optimal strategies is not proved in [9], only the existence of stationary $\varepsilon\mu$-optimal strategies, where μ may be unbounded. But the inequality $d \le 0$ is proved for this model in [9, (5.10)]. Since $1_{X\setminus X_\Pi} \in L_0(0, X_\Pi)$, Corollary 1 implies the existence of stationary ε-optimal strategies for model I from [9, Chapter 5].

Note that for deterministic models Theorem 1 is valid for functions $l \in L_0(0, X_S)$ [13]. This fact allows us to prove the existence of good persistently εl-optimal strategies for $l \in L_0(0, X_\Pi)$ when the model is deterministic.

However, it may happen that $d(x) > 0$ for deterministic models. It would be interesting to identify a function d^* such that (i) it is possible to replace d by d^* in Theorem 1; (ii) $d^* \le d$; (iii) $d^* \le 0$ for deterministic models. The question of whether such a function exists is still open. For instance, it is not clear if Theorem 1 is valid for

$$d^*(x) = \lim_{N\to\infty} \sup_{\varphi\in S} \lim_{n\to\infty} \sup_{\tau\in Q[0,n]} \mathrm{E}_x^{\varphi} 1\{s(x_\tau) \ge N\} s(x_\tau) \ .$$

2. PERSISTENTLY NEARLY-OPTIMAL STRATEGIES

Let B be a countable (or finite) set, $f: H \to B$.

Definition 1 ([24]). A non-randomized strategy φ is said to be (f, B)-generated if $\varphi_n(h_n) = \varphi(f(h_n), x_n)$ for any $h_n = x_0 a_0 \cdots x_n \in H$, $n = 0,1,\ldots$ Let B^f denote the set of all (f, B)-generated strategies.

Condition 1 ([14]). f and B generate a transitive statistic, see Shiryaev [29]). If $x_n = x'_m$ and $f(h_n) = f(h'_m)$ for some h_n, $h'_m \in H$, $n, m = 0,1,\ldots,$ then $f(h_n a z) = f(h'_m a z)$ for any $a \in A$, $z \in X$.

Examples of (f, B)-generated strategies satisfying Condition 1 (stationary, Markov and others) are given in [14]. We shall consider one such example here.

Example 1 (Strategies of renewal on Y). Let $Y \subseteq X$. For $h_n \in H$, $n = 0,1,\ldots,$ we define

$$\vartheta(h_n) = \begin{cases} \max\{i \geq 0: x_i \in Y, i \leq n\}, & \text{if } x_i \in Y \text{ for some } i \leq n, \\ +\infty & \text{, if } x_i \notin Y \text{ for any } i \leq n. \end{cases}$$

A non-randomized strategy φ is called a strategy of renewal on Y if there are maps $\varphi: Y \times X \times \{0,1,\ldots\} \to A$ and $\varphi': (X \setminus Y) \times \{0,1,\ldots\} \to A$ such that for any $h_n \in H$, $n = 0,1,\ldots,$ we have

$$\varphi_n(h_n) = \begin{cases} \varphi(x_{\vartheta(h_n)}, x_n, n - \vartheta(h_n)), & \vartheta(h_n) \leq n, \\ \varphi'(x_n, n), & \vartheta(h_n) = +\infty. \end{cases}$$

Let R^Y denote the set of all strategies of renewal on Y. Let $B = (Y \cup \{y\}) \times \{0,1,\ldots\}$, where $y \notin Y$ is an arbitrary point. Set

$$f(h_n) = \begin{cases} (x_{\vartheta(h_n)}, n - \vartheta(h_n)), & \vartheta(h_n) \leq n, \\ (y, n), & \vartheta(h_n) = +\infty. \end{cases}$$

Then if $f(h_n) = (x, m)$, where $x \in Y \cup \{y\}$, $m = 0,1,\ldots,$ we have $f(h_n a z) = (x, m+1)$ for $z \in X \setminus Y$ and $f(h_n a z) = (z, 0)$ for $z \in Y$.

A result ([14, Theorem 4.1]) similar to Theorem 1 has been proved for (f, B)-generated strategies satisfying Condition 1. We shall not derive this result here, but a corollary of this theorem (Lemma 1) for the class R^Y is given below.

For $\Delta \subseteq \Pi$ we shall write

$$d'_\Delta(x) = \sup_{\pi \in \Delta} \lim_{n \to \infty} \inf_{\tau \in Q[0,n]} E_x^\pi V(x_\tau), \; x \in X,$$

$$X^+ = \{x \in X: v(x) > 0\}\ , X^- = \{x \in X: v(x) < 0\}\ , X^0 = \{x \in X: v(x) = 0\} \quad ,$$

$$A^c(x) = \{a \in A(x): T^a v(x) = v(x)\}\ , x \in X \quad ,$$

$$X^c = \{x \in X: A^c(x) \neq \phi\}\ , X^* = X^+ \cup X^- \cup (X^0 \cap X^c) \quad .$$

LEMMA 1 ([22, Corollary 3.3]). *Let $\Delta = R^Y$, where $X^+ \subset Y \subset X$, and $v_\Delta(h_n) = v(x_n)$ for any $h_n \in H$. Then for any $\varepsilon > 0$ and any $l \in L_0(d'_\Delta, X_\Pi)$ there exists a persistently εl-optimal strategy $\varphi \in \Delta$.*

Let $Z \subset X$ and suppose that there exist functions $r^i(x, a)$, $i = 1,2$, measurable in a such that (i) $r = r^1 + r^2$, $r^2 \geq 0$, $v^2 < +\infty$, where v^i is the value of the model $\mu^i = \{X, A, A(\cdot), p, r^i\}$, $i = 1,2$; (ii) $Z = \{x \in X: v^1(x) < 0\}$. For example, for $r^1 = r^-$, $r^2 = r^+$ we have $Z = \{x \in X: v_-(x) < 0\}$ and for $r^1 = r$, $r^2 = 0$ we have $Z = X^-$.

LEMMA 2 ([22], Theorem 5.2, Corollary 3.2). *Let $X^+ \subset Y \subset X^+ \cup X^- \cup Z$ and $\Delta = R^Y$. Then for any $\varepsilon > 0$ and any $l \in L_0(d'_\Delta, X_\Pi)$ there exists a persistently εl-optimal strategy $\varphi \in \Delta$.*

The proof of Lemma 2 is based on Lemma 1, Theorems 5 and 6, and the construction of the embedded model described in [24].

A non-randomized strategy φ is said to be stationary on Y, $Y \subset X$, if $\varphi_n(h_n) = \varphi(x_n)$ for $x_n \in Y$ (the set of all such strategies is denoted by S^Y). Note that $R^Y \subset S^Y$, $R^X = S^X = S$ and write $\Delta^* = R^{X+} \cup X^- \cup Z \cap S^{X^*}$.

THEOREM 7 ([22, Theorem 2.1]). *For any $l \in L_0(d'_{\Delta^*}, X_\Pi)$ and any $\varepsilon > 0$ there exists a persistently εl-optimal strategy $\varphi \in \Delta^*$.*

COROLLARY 3. *Let $\Delta = S^{X^* \cup Z}$. Then $v_\Delta = v$.*

Example 2.1 in [22] shows that $v_\Delta \neq v$ for $\Delta = R^{X^* \cup Z}$.

Condition 2. For any two histories h_n, $h_m \in$, H, $n, m = 0,1, \ldots$, such that $m > n$, $h_m = h_n a_n x_{n+1} \cdots x_m$ and $x_m = x_n$, we have $f(h_n) \neq f(h_m)$.

Examples of classes B^f satisfying Condition 2 (Markov, tracking and others) can be found in [24]. The following theorem generalizes various results on the existence of nearly-optimal strategies which are stationary, Markov or stationary on subsets (see [21], where the weaker result is given).

THEOREM 8 ([22, Theorem 2.2], see also [21, Theorem 2.1]). *Let f and B satisfy Condition 2. Then for any $\varepsilon > 0$ and any $l \in L_0(d'_{\Delta^*}, X_\Pi)$ there exists a persistently εl-optimal strategy $\varphi \in S^{X^* \cup Z} \cap B^f$.*

We shall consider a method which allows us to extend sets for which it is possible to assert the existence of good stationary actions. Let the model

$\bar{\mu} = \{X, A, \bar{A}(\cdot), p, r\}$, where $\bar{A}(\cdot) \subseteq A(\cdot)$, be given and the value of the model $\bar{\mu}$ coincide with v. Let $\bar{X}^*$ and $\bar{Z}$ denote the analogs of the sets X^* and Z for model $\bar{\mu}$. Then we can define $\Delta^* = R^{X^+ \cup X^- \bar{Z}} \cap S^{\bar{X}^*}$ in Theorems 7 and 8, and $\varphi \in S^{\bar{X}^* \cup \bar{Z}} \cap B^f$ in Theorem 8. The following example shows that this really implies the extension of classes of sets for which the existence of good stationary actions can be proved (see also the example in [19]).

Example 2. Let $X = \{-1,0,1\}$, $A(\cdot) = A = [-1,0) \cup (0,1]$, $p(\cdot \mid 0, a) = 0$, $r(0, a) = -1$ for any $a \in A$; $p(0 \mid i, a) = a$, $p(i \mid i, a) = 1 - a$, $r(i, a) = 0$ for $i = \pm 1$, $a \in (0,1]$ and $p(-i \mid i, a) = 1$, $r(i, a) = a$ for $i = \pm 1$, $a \in [-1,0)$. Then it is possible to verify that $Z = \phi$ and Theorems 1 and 2 state that good stationary actions exist only for $X^* = \{0\}$. However, if we consider $\bar{A}(-1) = (0,1]$, $\bar{A}(1) = [-1,0)$ then it is possible to choose $\bar{Z} = \{1\}$ so that persistently nearly-optimal strategies exist which are stationary on $\{0,1\}$. Setting $\bar{A}(-1) = [-1,0)$, $\bar{A}(1) = (0,1]$, we find that persistently nearly-optimal strategies exist which are stationary on $\{-1,0\}$. (Since $s \neq v$, there are no good stationary strategies in this example.)

We shall now return to the question of the existence of stationary nearly-optimal strategies.

THEOREM 9. *If* $s(x) = 0$ *for* $x \in X^0 \setminus (X^c \cup Z)$ *then* $s = v$.

Proof. Fix $x \in X$, $\varepsilon > 0$, $l \in L_0(d, X_S)$. Let $\varphi \in S^{X^* \cup Z}$ and $w^\varphi(x) \geq v(x) - \varepsilon$ (see Corollary 3), and let $\psi \in S$ and $w^\psi \geq s - \varepsilon l$ (see Theorem 1). Consider a strategy σ which coincides with φ before the first encounter with $X^0 \setminus (X^c \cup Z)$ and with ψ thereafter. Then $w^\sigma(x) \geq v(x) - \varepsilon(1 + l(x))$. Since $\varepsilon > 0$ is arbitrary, Theorem 2 implies $s(x) \geq v(x)$.

ACKNOWLEDGEMENT

I wish to thank Dr. I.M. Sonin for helpful discussions and many valuable suggestions.

REFERENCES

1. D. Blackwell. Discrete dynamic programming. *Annals of Mathematical Statistics*, 33(1962) 719–726.

2. N.V. Krylov. Construction of an optimal strategy for a finite controlled chain. *Theory of Probability and its Application*, 10(1965) 45–54.

3. E.B. Dynkin and A.A. Yushkevich. *Controlled Markov Processes*. Springer-Verlag, Berlin 1979.

4. J. van der Wal and J. Wessels. *On the Use of Information in Markov Decision Processes*. Eindhoven University of Technology, Memorandum-COSOR 81-20, 1981.

5. D. Blackwell. Positive dynamic programming. In *Proceedings of the 5th Berkeley Symposium on Mathematical Statistics and Probability*, 1967, pp. 415–418.

6. E.B. Frid. On a problem of D. Blackwell from the theory of dynamic programming. *Theory of Probability and its Applications*, 15(1970) 719–722.

7. D. Ornstein. On the existence of stationary optimal strategies. *Proceedings of the American Mathematical Society*, 20(1969) 563–569.

8. M. Schäl. Stationary policies in dynamic programming models under compactness assumptions. *Mathematics of Operational Research*, 8(1983) 366–372.

9. J. van der Wal. *Stochastic Dynamic Programming*. Mathematical Centre Tracts 139, Mathematical Centre, Amsterdam, 1981.

10. K. van Hee, A. Hordijk and J. van der Wal. Successive approximations for convergent dynamic programming. In *Markov Decision Theory*, Mathemaical Centre Tracts 93, Mathematical Centre, Amsterdam, 1977, p. 183–211.

11. N.L. Lazrieva. On the existence of an ε-optimal stationary strategy for controlled Markov processes with a general additive critrion (in Russian). In *Studies in Probability Theory and Mathematical Statistics*, Tbilisi, 1982.

12. J. van Nunen. *Contracting Markov Decision Processes*. Mathematical Centre Tracts 71, Mathematical Centre, Amsterdam, 1976.

13. E.A. Fainberg and I.M. Sonin. Stationary and Markov policies in countable state dynamic programming. *Lecture Notes in Mathematics*, 1021(1983) 111–129.

14. E.A. Fainberg and I.M. Sonin. Persistently nearly-optimal strategies in stochastic dynamic programming. In R.S. Liptser, N.V. Krylov and A.A. Novikov (Eds.), Proceedings of the Steklov Seminar *Statistics and Control of Random Processes*, 1984. To appear in Lecture Notes in Control and Information Sciences.

15. J. van der Wal. On uniformly nearly-optimal Markov strategies. In *Operational Research Proceedings*, Springer-Verlag, Berlin, 1983, pp. 461–467.

16. I.M. Sonin. The existence of a uniformly nearly-optimal Markov strategy for a controlled Markov chain with a countable state space (in Russian). In *Models and Methods of Stochastic Optimization*, Central Economic–Mathematical Institute, Moscow, USSR, 1984, pp. 213–232.

17. R. van Dawen. *Stationäre Politiken in Stochastischen Entscheidungsmodellen*. Dissertation, Universität Bonn, 1984.

18. H. Everett. Recursive games. *Annals of Mathematical Statistics*, 39(1957) 47–78.

19. R.Ja. Chitashvily. On the existence of ε-optimal stationary policies for a controlled Markov chain (in Russian). *Communications of the Academy of Sciences of the Georgian SSR*, 83(1976) 549–552.

20. A.A. Yushkevich and R.Ja. Chitashvili. Controlled random sequences and Markov chains (in Russian). *Uspekhi Matematicheskie Nauk*, 37(1982) 213–242.

21. I.M. Sonin and E.A. Fainberg. Sufficient classes of policies in countable state Markov decision chains with a total reward criterion (in Russian). *Doklady Academii Nauk USSR*, 275(1984) 806–809.

22. E.A. Fainberg. On sufficient classes of strategies in dynamic programming. Submitted to *Theory of Probability and its Application*.

23. R. van Dawen and M. Schäl. On the existence of stationary optimal policies in Markov decision models. *Zeitschrift für Angewandte Mathematik und Mechanik*, 63(1983) 404–405.

24. E.A. Fainberg. On some classes of policies in dynamic programming. Submitted to *Theory of Probability and its Applications*.

25. J. van der Wal. On uniformly nearly-optimal stationary strategies. *Mathematics of Operational Research*, 9(1984) 290–300.

26. E.A. Fainberg. Nonrandomized Markov and semi-Markov policies in dynamic programming. *Theory of Probabillity and its Application*, 27(1)(1982).

27. E.A. Fainberg. Markov decision processes with arbitrary real-valued criteria. *Theory of Probability and its Application*, 27(3)(1982).

28. K. van Hee. Markov strategies in dynamic programming. *Mathematics of Operational Research*, 3(1978) 37–41.

29. A.N. Shiryaev. *Optimal Stopping Rules*. Springer-Verlag, New York, 1978.

ON THE DERIVATION OF A FILTERING EQUATION FOR A NON-OBSERVABLE SEMIMARTINGALE

L.I. Galčuk
Kiev State University

Consider a probability space $(\Omega, \mathbf{F}, \mathrm{P})$ with a non-decreasing family of σ-algebras $F = (\mathbf{F}_t)$, $G = (G_t)$, $t \in R_+$, $G_t \leq \mathbf{F}_t$ satisfying the usual conditions. Let the σ-algebra G_t be generated by observations up to the time t. suppose that the process $\vartheta = (\vartheta_t)$, $t \in R_+$, describes the system but cannot be observed. We have to derive a recursive equation (a filtering equation) for the process $\pi(\vartheta) = (\pi_t(\vartheta))$, $t \in R_+$, where $\pi_t(\vartheta) = \mathrm{E}[\vartheta_t \mid G_t]$.

We shall assume that the trajectories of the processes under consideration are right-continuous and limited on the left.

Let $\mathbf{A}(F)$, $\mathbf{M}(F)$ denote the spaces of integrable variation processes and martingales with respect to the family F. The statement $X \in \mathbf{A}_{\mathrm{loc}}(F)$ (and analogously, $X \in \mathbf{M}_{\mathrm{loc}}(F)$) implies the existence of a sequence of F-stopping times (T_n), $n \in |\mathbf{N}|$, $T_n \uparrow \infty$ a.s. and that the process terminates $X^{T_n} \in \mathbf{A}(F)$, $\forall n \in |\mathbf{N}|$. We shall use $\mathbf{O}(G)$ and $\mathbf{P}(G)$ to denote the optional and predictable σ-algebras on $\Omega \times R_+$ corresponding to the family G. We also use $\mathrm{T}(G)$ to denote the family of all G-stopping times. (See [1] for further details of the concepts and notation used here.)

Assumption 1. The process ϑ is an F-semimartingale with the decomposition

$$\vartheta = \vartheta_0 + A + M \quad ,$$

where $\vartheta_0 \in \mathbf{F}_0$, $\mathrm{E}|\vartheta_0| < \infty$, $A \in \mathbf{A}_{\mathrm{loc}}(F)$, $M \in \mathbf{M}_{\mathrm{loc}}(F)$, and there is a sequence (T_n), $n \in |\mathbf{N}|$, of G-stopping times such that $A^{T_n} \in \mathbf{A}(F)$, $M^{T_n} \in \mathbf{M}(F)$, $\forall n \in |\mathbf{N}|$.

Definition. Let Y be a non-negative measurable process. The $\mathbf{O}(G)$-measurable process ${}^0Y$ is called the *G-optional projection of the process Y* if

$$\mathrm{E}Y_T I_{T<\infty} = \mathrm{E}\,{}^0Y_T I_T < \infty, \ \forall T \in \mathrm{T}(G) \quad .$$

The process ${}^0Y$ exists and is unique, and in addition

$${}^0Y_T I_{T<\infty} = \mathrm{E}[Y_T \mid G_T] I_{T<\infty} \text{ a.s. } \forall T \in \mathrm{T}(G) \quad .$$

We see that at finite times $T \in \mathrm{T}(G)$, the process $\pi(\vartheta)$ coincides with ${}^0\vartheta$. A more precise statement is given by Lemma 1.

LEMMA 1. *Let $Z = (Z_t)$, $t \in R_+$, be a measurable process with right-continuous trajectories which are limited on the left. Let the family of random variables $\{Z_T I_{T<\infty}, T \in \mathrm{T}(G)\}$, be uniformly integrable and $\pi(Z)$ denote a right-continuous modification of the process $(\mathrm{E}\{Z_t \mid G_t])$, $t \in R_+$. Then $\pi(Z) = {}^0Z$ up to indistinguishability.*

Proof. It suffices to show that for any $T \in \mathrm{T}(G)$ we have

$$\pi_T(Z) I_{T<\infty} = \mathrm{E}[Z_T I_{T<\infty} \mid G_T] \quad \text{a.s.}$$

We put $T^{(n)} = k2^{-n}$ if $(k-1)2^{-n} \le T < k2^{-n}$, $\infty^{(n)} = \infty$, and have $T^{(n)} \in \mathrm{T}(G)$ $T^{(n)} > T$, $T^{(n)} \downarrow T$. Also, for $A \in G_T$ we have

$$\mathrm{E} I_A \pi_{T^{(n)}}(Z) I_{T^{(n)}<\infty} = \sum_k \mathrm{E} I_A I_{T^{(n)}=k2^{-n}} \pi_{k2^{-n}}(Z) =$$

$$= \sum_k \mathrm{E}\mathrm{E}[I_A I_{T^{(n)}=k2^{-n}} Z_{k2^{-n}} \mid G_{k2^{-n}}] \quad ,$$

since $A \cap \{T^{(n)} = k2^{-n}\} \in G_{k2^{-n}}$. Hence

$$\mathrm{E} I_A I_{T^{(n)}<\infty} \pi_{T^{(n)}}(Z) = \mathrm{E} I_A Z_{T^{(n)}} I_{T^{(n)}<\infty}$$

or

$$\pi_{T^{(n)}}(Z) I_{T^{(n)}<\infty} = \mathrm{E}[Z_{T^{(n)}} I_{T^{(n)}<\infty} \mid G_{T^{(n)}}] \quad \text{a.s.}$$

Let us take the limit as $n \to \infty$. Since the process is right-continuous we have the following expression on the left-hand side:

$$\lim_{n\to\infty} \pi_{T^{(n)}}(Z) I_{T^{(n)}<\infty} = \pi_T(Z) I_{T<\infty} \quad .$$

On the right-hand side we have

$$\lim_{n\to\infty} \mathrm{E}(Z)_{T^{(n)}} I_{T^{(n)}<\infty} \mid G_{T^{(n)}}] = \mathrm{E}[Z_T I_{T<\infty} \mid G_T] \quad \text{a.s.}$$

because of the uniform integrability of the family $\{Z_T I_{T<\infty}, T \in \mathrm{T}(G)\}$ and since $G_{T^{(n)}} \downarrow G_T$. □

Remark. Suppose that the process Z does not satisfy the uniform integrability condition assumed in the lemma. Suppose also that there is a sequence $(T_n) \in \mathrm{T}(G)$, $T_n \uparrow \infty$, a.s. such that for any n there exists a uniformly integrable family $\{Z_{T\wedge T_n} I_{T<\infty}$, $T \in \mathrm{T}(G)\}$. Then $\pi(Z) = {}^0Z$ up to indistinguishability.

We can now establish the nature of the process ${}^0\vartheta$.

LEMMA 2. *Let $\vartheta = (\vartheta_t)$, $t \in R_+$, be an F-semimartingale satisfying Assumption 1. Then the process ${}^0\vartheta$ is a G-semimartingale with decomposition*

$$ {}^0\vartheta = ({}^0\vartheta)_0 + \alpha + Y \,, \ \alpha \in \mathbf{P}(G) \cap \mathbf{A}_{\mathrm{loc}}(G) \,, \ Y \in \mathbf{M}_{\mathrm{loc}}(G) \quad . $$

Proof. First we shall suppose that the process ϑ can be decomposed as follows:

$$ \vartheta = \vartheta_0 + A + M \quad , $$

where $\mathrm{E}|\vartheta_0| < \infty$, $A \in \mathbf{A}(F)$, $M \in \mathbf{M}(F)$, $A_0 = M_0 = 0$.

The optional projection is linear, so

$$ {}^0\vartheta = {}^0(\vartheta_0) + {}^0A + {}^0M \quad . $$

The definition of optional projection implies that ${}^0(\vartheta_0) \in \mathbf{M}(G)$, ${}^0M \in \mathbf{M}(G)$. Since $A \in \mathbf{A}(F)$, we have $A = A^{(1)} - A^{(2)}$, where $A^{(i)}$, $i = 1,2$, is an increasing integrable process. We know that

$$ {}^0A = {}^0A^{(1)} - {}^0A^{(2)} $$

The process ${}^0A^{(i)}$, $i = 1,2$, is a G-submartingale: for $s \le t$ we have

$$ \mathrm{E}[{}^0A_t^{(i)} \,|\, G_s] = \mathrm{E}[{}^0A_s^{(i)} - {}^0A_s^{(i)} \,|\, G_s] + {}^0A_s^{(i)} = $$

$$ = \mathrm{E}[A_t^{(i)} - A_s^{(i)} \,|\, G_s] + {}^0A_s^{(i)} \ge {}^0A_s^{(i)} \quad . $$

Since $A^{(i)}$ is an integrable process, ${}^0A^{(i)}$, $i = 1,2$, is a submartingale of the class (D). Hence the process ${}^0A$ is the difference of two G-submartingales of the class (D). From the Doob–Meyer decomposition, there is a $\mathbf{P}(G)$-measurable process $\alpha = (\alpha_t)$ with integrable variation such that ${}^0A - \alpha \in \mathbf{M}(G)$. Thus

$$ {}^0\vartheta = ({}^0\vartheta)_0 + \alpha + Y \quad , $$

where $\alpha \in \mathbf{P}(G) \cap \mathbf{A}(G)$, $Y \in \mathbf{M}(G)$, $\alpha_0 = Y_0 = 0$, and

$$ Y = {}^0(\vartheta_0) - ({}^0A - \alpha) + {}^0M \quad . $$

This leads to a general case by combination.

Now we consider a special case. Suppose there is a collection of continuous martingales $(X^{(1)}, \ldots, X^{(n)}) \in \mathbf{M}(G)$ and a random integer-valued measure μ on the product of spaces $R_+ \times R_1$, $\mathbf{B}(R_+) \otimes \mathbf{B}(R_1)$, with a $\mathbf{P}(G)$-compensator ν such that any martingale $Z \in \mathbf{M}(G)$ can be represented in the form

$$Z = \sum_{i=1}^{n} f^{(i)} \cdot X^{(i)} + g * (\mu - \nu) \quad . \tag{1}$$

Here functions $f^{(1)}, \ldots, f^{(n)}$ are $\mathbf{P}(G)$-measurable and function g is $\tilde{\mathbf{P}}(G) = \mathbf{P}(G) \otimes \mathbf{B}(R_1)$-measurable. The functions $f^{(1)}, \ldots, f^{(n)}$ are then defined by the system of equations

$$Cf = \begin{bmatrix} \dfrac{d<Z,X^{(1)}>}{d<X,X>} \\ \cdot \\ \cdot \\ \cdot \\ \dfrac{d<Z,X^{(n)}>}{d<X,X>} \end{bmatrix}, \quad f = \begin{bmatrix} f^{(1)} \\ \cdot \\ \cdot \\ \cdot \\ f^{(n)} \end{bmatrix}, \quad C = \left\| \frac{d<X^{(i)},X^{(j)}>}{d<X,X>} \right\|_{i,j=1,\ldots,n} , \tag{2}$$

$$<X , X> = \sum_{i=1}^{n} <X^{(i)} , X^{(i)}> \quad .$$

The function g is defined by

$$g(s , x) = V(s , x) + \frac{\hat{V}(s)}{1-a_s} I_{a_s<1} \quad , \tag{3}$$

where $a_s = \nu(\{s\} , R_1)$, $V = M_\mu[\Delta Z \mid \tilde{\mathbf{P}}(G)]$, and $\hat{V}(s) = \int_{R_1} v(s , x)\nu(\{s\} , dx)$ (see [1]). $M_\mu[\Delta Z \mid \tilde{\mathbf{P}}(G)]$ is a $\tilde{\mathbf{F}}(G)$-measurable function such that for any $\tilde{\mathbf{P}}(G)$-measurable non-negative function φ the following equality holds:

$$\mathrm{E}(M_\mu[\Delta Z \mid \tilde{\mathbf{P}}(G)]\varphi) * \mu_\infty = \mathrm{E}((\Delta Z)\varphi) * \mu_\infty \quad .$$

THEOREM 1. *Let the F-semimartingale ϑ have a decomposition $\vartheta = \vartheta_0 + A + M$, $A \in \mathbf{A}_{loc}(F)$, $M \in \mathbf{M}_{loc}(F)$ and satisfy Assumption 1. Let the representation* (1) *hold for any martingale $Z \in \mathbf{M}(G)$. Then*

$$\pi(\vartheta) = \pi_0(\vartheta) + \alpha + \sum_{i=1}^{n} f^{(i)} \cdot X^{(i)} + W * (\mu - \nu) \quad ,$$

where α is the process from Lemma 2, and the functions $f^{(1)}, \ldots, f^{(n)}$ are defined by system (2) *with the process Z replaced by $Y = {}^0(\vartheta_0) - ({}^0\vartheta)_0 + ({}^0A - \alpha) + {}^0M$,*

$$W = U + \frac{\hat{u}}{1-a} I_{a<1} , \quad U = M_\mu[\vartheta \mid \tilde{\mathbf{P}}(G)] - \pi_-(\vartheta) - \Delta\alpha \quad ,$$

$$\hat{u}(t) = \int_{R_1} u(t , x)\nu(\{t\} , dx) \quad .$$

The proof follows directly from Lemma 2 and the representation of the martingale Y in the form (1)–(3).

Remark. The representation of martingales (1) was first assumed by Grigelionis. A filtering equation for semimartingales that can be represented as the sum of a martingale and a continuous process of bounded variation is presented by him. In the general case the martingale Y from Lemma 2 can be decomposed into (i) a component belonging to a subspace generated by given processes $X^{(1)}, \dots, X^{(n)}$ and measure μ, and (ii) a component orthogonal to this subspace:

THEOREM 2. *Let the F-semimartingale ϑ have a decomposition $\vartheta = \vartheta_0 + A + M$, $A \in \mathbf{A}_{loc}(F)$, $M \in \mathbf{M}_{loc}(F)$, and satisfy Assumption 1. Let there exist continuous martingales $X^{(1)}, \dots, X^{(n)} \in \mathbf{M}(G)$ and an integer-valued random measure μ on $\mathbf{B}(R_+) \otimes \mathbf{B}(R_1)$ with $\mathbf{P}(G)$-compensator ν. Then*

$$\pi(\vartheta) = \pi_0(\vartheta) + \alpha + \sum_{i=1}^{n} f^{(i)} \cdot X^{(i)} + W * (\mu - \nu) + h * \mu + Z \quad ,$$

where $h = \Delta Y - M_\mu[\Delta Y \mid \tilde{\mathbf{P}}(G)]$, α, $f^{(1)}, \dots, f^{(n)}$, W, Y are defined in Theorem 1, $Z \in \mathbf{M}_{loc}(G)$, Z, is orthogonal to martingales $X^{(1)}, \dots, X^{(n)}$ and without jumps on the measure support μ.

Remark. Yor [2] suggests another way of deriving the filtering equation. When the martingale Y from Lemma 2 belongs to the Hilbert space $\mathbf{M}^2(G)$ of square-integrable martingales and the space $\mathbf{M}^2(G)$ is separable, the martingale Y can be decomposed with respect to the basis of the space $\mathbf{M}^2(G)$.

REFERENCES

1. J. Jacod. Calcul stochastique et problemes des martingales. *Lecture Notes in Mathematics*, Vol. 714, 1979.

2. M. Yor. Sur la theorie du filtrage. *Lecture Notes in Mathematics*, Vol. 876, 1981, pp. 239-280.

ON THE REPRESENTATION OF FUNCTIONALS OF A WIENER SHEET BY STOCHASTIC INTEGRALS

J.I. Gihman
Institute of Applied Mathematics and Mechanics,
Academy of Sciences of the Ukrainian SSR

Any measurable functional $F(w)$ of a Wiener process can be represented by a stochastic integral of the form

$$F(w) = \int_0^1 g(s)dw(s) + EF(w) \quad . \tag{1}$$

Here $g(s)$ is an $(\mathbf{F}_s^w)$-adapted function such that

$$\int_0^1 g^2(s)ds < \infty \quad \text{a.s.}$$

and $\mathbf{F}_s^w$ denotes the σ-algebra generated by variables $w(t)$, $t \in [0,s]$. When applying this theorem, in particular for the optimal control of stochastic systems, it is important to have a "good" expression for the function $g(s)$. The following result has been obtained (see [1]).

Let $F(x)$ be a Fréchet-differentiable functional defined on a space C of continuous functions $x = x(t)$, $t \in [0,1]$. A derivative $\nabla F(x)$ of this functional at any point is a continuous linear functional; we shall denote its values on the function $y = y(t)$ by $(\nabla F(x), y)$.

From Riesz's theorem regarding the σ-algebra $\mathbf{B}[0,1]$ of Borelian sets of interval $[0,1]$, there exists a measure $\lambda(ds) = \lambda(x;ds)$ such that

$$(\nabla F(x), y) = \int_0^1 y(s)\lambda(x;ds) \quad .$$

Suppose that the following condition holds: for some positive constants K, δ, α

$$|F(x+y) - F(x) - (\nabla F(x), y)| \le K \|y\|^{1+\delta}(1+\|x\|^{\delta})(1+\|y\|^{\alpha}) \,,\ \forall x, y \in C(z) \quad . \tag{2}$$

Here $\|x\|$ is a norm in the space C, i.e.,

$$\|x\| = \max_{0 \le t \le 1} |x(t)| \quad .$$

THEOREM 1. *If the functional F satisfies condition* (2), *then*

$$F(w) = \int_0^1 \mathrm{E}[\lambda(w;]t,1])/\mathbf{F}_t^w]w(dt) \quad \text{a.s.} \quad . \tag{3}$$

Formula (3) represents a particular linearization of the functional F. In this paper we shall consider the generalization of formula (3) to functionals of two variables.

We must first introduce some notation. Let R^2 be a set of pairs $u = (s, t)$ of real numbers s and t ordered in some natural manner. We shall consider a Wiener sheet $w = w(u) = w(s, t)$, where $u \in [0, S] \times [0, T]$.

Let

$$D_u = D_{st} = \{u' : u \le u' \le \bar{u}_0\}, \ \bar{u}_0 = (S, T) \quad ;$$

$$D_u^{--} = D_{st}^{--} = \{u' : 0 \le u' \le u\} \quad ;$$

$$D_u^{+-} = D_{st}^{+-} = \{u' : s \le s' \le S, 0 < t' \le t\} \quad ;$$

$$D_u^{-+} = D_{st}^{-+} = \{u' : 0 \le s' \le s, t < t' \le T\} \quad ;$$

$$D = D_{00} \quad .$$

For each function $f(u)$ let

$$f\,]u', u''] = f(s'', t'') - f(s', t'') - f(s'', t') + f(s', t') \quad \text{for } u' < u'' \quad .$$

Let $C(D)$ be a space of continuous functions on a rectangle $D = [0, u_0]$, $\Omega = C(D) \times C(D)$.

Introduce into Ω a measure P induced by a 2-dimensional Wiener sheet $(w(u), w'(u))$, where w' is a Wiener sheet independent of w, and a filter for the σ-algebra $(\mathbf{F}_u)_{u \in D}$, where $\mathbf{F}_u$ is a σ-algebra generated by random variables $\{w(u'), u' \in D_u^{--}\}$.

Let $x \in C(D)$, i.e., $x = x(u)$, $u \in D$, be a continuous function and $F(x)$ be a continuous functional on $C(D)$. We shall also consider this functional as a functional on Ω. Moreover, if $\omega = (x, y)$, $(x, y) \in C(D) \times C(D)$, then $F(\omega) = F(x)$. The following theorem is due to Wong and Zakai [2].

THEOREM 2. *Any square-integrable* $\mathbf{F}_{u_0}$*-measurable functional* $F(w)$ *can be represented in the following form:*

$$F(w) = \mathrm{E}F(w) + \int_D g(u)w(du) + \overset{WZ}{\int\int_{D\times D}} G(u',u'')w(du')w(du'') \quad . \tag{4}$$

The first integral in (4) is an Ito integral with respect to a two-parameter Wiener sheet, while the second is known as a stochastic integral of the second type, which we shall call a *W–Z integral* (Wong–Zakai integral). For a description of this integral, see [2] or [3]. We wish to obtain formulae for the functions g and G which are analogous to those derived from (3). It turns out that we shall be using the second differential of the functional F.

Suppose that $F(x)$, $x \in C(D)$, is twice Fréchet-differentiable. This means that for all $x, h \in C(D)$ we have

$$F(x+h) - F(x) = (\nabla F(x), h) + \frac{1}{2}(\nabla^2 F(x)h, h) + R(x,h) \quad , \tag{5}$$

where $|R(x,h)| = o(\|h\|^2)$, $\|h\| = \max_{u\in D} |h(u)|$, $\nabla F(x)$ is a linear functional in $C(D)$ for a fixed x, and $\nabla^2 F(x)$ is a continuous bilinear functional. Moreover, $(\nabla F(x), h)$ is a value of the functional on the function h, and $(\nabla^2 F(x)h_1, h_2)$ is a value of the functional $\nabla^2 F(x)$ on the pair h_1, h_2, $h_i \in C(D)$.

According to Riesz's theorem,

$$(\nabla F(x), h) = \int_D h(u)\lambda(x;du) \quad , \tag{6}$$

$$(\nabla^2 F(x)h, h) = \int_D\int_D h(u')h(u'')\mu(x;du',du'') \quad , \tag{7}$$

where $\lambda(x;\cdot)$ is a measure on D, and $\mu(x;\cdot,\cdot)$ is a measure on D^2.

Suppose that there exist positive numbers K, β, δ such that

$$\|\nabla^2 F(x) - \nabla^2 F(y)\| \le K\|x-y\|^{\delta}(1+\|x\|^{\beta}+\|y\|^{\beta}) \quad . \tag{8}$$

Without loss of generality we can suppose that $\beta \ge 1$, $\delta \in]0,1]$. It is easily shown that the inequalities

$$\|\nabla^2 F(x)\| \le K(1+\|x\|^{\beta+\delta}) \;,\; \|\nabla F(x)\| \le K(1+\|x\|^{1+\beta+\delta}) \quad ,$$
$$|R(x,h)| \le K(1+\|x\|^{\beta}+\|h\|^{\beta})\|h\|^{2+\delta} \tag{9}$$

follow from (4). Note that $\|x\|$ is the norm of the element x in the space C, $\|\nabla^2 F(x)\|$ is the norm of a bilinear functional, and $\|\nabla F(x)\|$ is the norm of a linear functional on C. The constant K here and elsewhere denotes a constant independent of the elements in C, and its value may change during the course of this study. Instead of $F(x)$ we shall write $F(w)$, where w is the first component of the pair $\omega = (w, w')$. From the Doob–Cairoli inequality $E\|w\|^p \le c_p < \infty$ $\forall p > 1$ so that the values $\|\nabla^2 F(w)\|$, $\|\nabla F(w)\|$, $|R(w, w')|$ possess finite moments of any rank.

THEOREM 3. *If $F(x)$ is a twice-differentiable functional satisfying condition* (4), *then*

$$F(w) = EF(w) + \int_D \bar{\lambda}(u) w(du) + \frac{1}{2} \overset{WZ}{\int\int_{D\times D}} \nu(u', u'') w(du') w(du'') \quad , \qquad (10)$$

where

$$\bar{\lambda}(u) = E[\lambda(w; D_u) / \mathbf{F}_u] \quad ,$$

$$\nu(u', u'') = \begin{cases} E[\mu(w, D_{u'} \times D_{u''}) / \mathbf{F}_{s't''}] & \text{for } s'' \le s' \text{ and } t' \le t'' \\ E[\mu(w, D_{u'} \times D_{u''}) / \mathbf{F}_{s''t'}] & \text{for } s' \le s'' \text{ and } t'' \le t' \end{cases}$$

and $\nu(u', u'') = 0$ *in other cases.*

Proof. We utilize Clark's technique (see [1]), which uses the second differential of the functional F. Subdivide a rectangle D into partial rectangles $\square_{ij}$ at the points $(s_i, t_j) = u_{ij}$, $0 = s_0 < s_1 < \cdots < s_n = S$, $0 = t_0 < t_1 < \cdots < t_n = T$, taking the lengths of the intervals $\Delta s_k =]s_k, s_{k+1}]$ ($\Delta t_k =]t_k, t_{k+1}]$) to be equal.

Write $D_{ij}^{\pm\pm} = D_{u_{ij}}^{\pm\pm}$ and

$$W_{ij}(s, t) = \begin{cases} w(s, t) & \text{when } (s, t) \in D_{ij}^{--} \\ w(s_i, t) + w'(s, t) - w'(s_i, t) & \text{when } (s, t) \in D_{ij}^{+-} \\ w(s, t_j) + w'(s, t) - w'(s, t_j) & \text{when } (s, t) \in D_{ij}^{-+} \\ w(s_i, t_j) + w'(s, t) - w'(s_i, t_j) & \text{when } (s, t) \in D_{ij} \end{cases} \quad .$$

A stochastic measure corresponding to the function W_{ij} coincides with a measure w in domain D_{ij}^{--} and with a measure w' outside domain D_{ij}^{--}. The sheet W_{ij} is also stochastically equivalent to w. Further,

$$E[F(w) / \mathbf{F}_{ij}] = E[F(W_{ij}) / \mathbf{F}_{ij}] = E[F(W_{ij}) / \mathbf{F}_{nn}] \quad .$$

Here $\mathbf{F}_{ij} \overset{\text{Def}}{=} \mathbf{F}_{s_i t_j}$. Let

$$\gamma_{ij} = E[F(w) / \mathbf{F}_{ij}] \quad ,$$

$$\square \gamma_{ij} = \gamma_{i+1,j+1} - \gamma_{i,j+1} - \gamma_{i+1,j} + \gamma_{ij} \quad .$$

Obviously, $\gamma_{ij} = \mathrm{E}[F(W_{ij})/\mathbf{F}_{nn}]$ so that

$$\square\gamma_{ij} = \mathrm{E}[F(W_{i+1,j+1}) - F(W_{i,j+1}) - F(W_{i+1,j}) + F(W_{ij})/\mathbf{F}_{nn}] \quad .$$

In addition,

$$\sum_{(i,j)=(0,0)}^{(n-1,n-1)} \square\gamma_{ij} = \gamma_{nn} - \gamma_{0n} - \gamma_{n0} + \gamma_{00}$$

and

$$\gamma_{nn} = \mathrm{E}[F(w)/\mathbf{F}_{nn}] = F(w)\,,\; \gamma_{n0} = \gamma_{0n} = \gamma_{00} = \mathrm{E}[F(w)/\mathbf{F}_{00}] = \mathrm{E}F(w) \quad .$$

Thus,

$$F(w) = \sum_{i,j=(0,0)}^{(n-1,n-1)} \square\gamma_{ij} + \mathrm{E}F(w) \quad .$$

It can easily be seen that

$$\mathrm{E}[\square\gamma_{ij}/\mathbf{F}_{in}] = \mathrm{E}[\square\gamma_{ij}/\mathbf{F}_{nj}] = \mathrm{E}[\square\gamma_{ij}/\mathbf{F}_{ij}] = 0 \quad .$$

Setting

$$\tilde{\delta}_{ij} = \mathrm{E}[\delta_{ij}/\mathbf{F}_{nn}] - \mathrm{E}[\delta_{ij}/\mathbf{F}_{in}] - \mathrm{E}[\delta_{ij}/\mathbf{F}_{nj}] + \mathrm{E}[\delta_{ij}/\mathbf{F}_{ij}] \quad , \tag{11}$$

we introduce a transposition $\{\tilde{\delta}_{ij}\,;\, i,j = 1,2,\ldots,n\}$ for an arbitrary series of random variables $\{\delta_{ij}\,;\, i,j = 0,1,2,\ldots,n\}$. It follows from the previous equalities that

$$F(w) = \sum_{i,j=0}^{n-1} \square\tilde{\gamma}_{ij} + \mathrm{E}F(w) \quad . \tag{12}$$

We shall consider the limit as $n \to \infty$ in this formula.

In line with the decomposition of the functional F into the sum of three terms in (5) we let

$$\square\tilde{\gamma}_{ij} = \square\tilde{\gamma}'_{ij} + \square\tilde{\gamma}''_{ij} + \square\tilde{\gamma}'''_{ij} \quad ,$$

where

$$\square\gamma'_{ij} = \mathrm{E}[(\nabla F(W_{ij})\,,\square W_{ij})/\mathbf{F}_{nn}] \quad , \tag{13}$$

$$\square W_{ij} = W_{i+1,j+1} - W_{i+1,j} - W_{i,j+1} + W_{i,j} \quad ,$$

$$\square\gamma''_{ij} = \frac{1}{2}\mathrm{E}[\alpha_{ij} - \delta'_{ij} - \delta''_{ij}/\mathbf{F}_{nn}] \quad , \tag{14}$$

$$\alpha_{ij} = (\nabla^2 F(W_{ij})\,,\, W_{i+1,j+1} - W_{ij}\,,\, W_{i+1,j+1} - W_{ij}) \quad , \tag{15}$$

$$\delta'_{ij} = (\nabla^2 F(W_{ij}) , W_{i+1,j} - W_{ij} , W_{i+1,j} - W_{ij}) \quad , \tag{16}$$

$$\delta''_{ij} = (\nabla^2 F(W_{ij}) , W_{i,j+1} - W_{ij} , W_{i,j+1} - W_{ij}) \quad , \tag{17}$$

$$\square\delta''_{ij} = \mathrm{E}[R(W_{ij} , W_{i+1,j+1} - W_{ij}) - R(W_{ij} , W_{i+1,j} - W_{ij}) - R(W_{ij} , W_{i,j+1} - W_{ij}) / \mathbf{F}_{nn}] \tag{18}$$

We shall consider each of the terms $\square\tilde{\gamma}'_{ij}, \square\tilde{\gamma}''_{ij}, \square\tilde{\gamma}'''_{ij}$ separately.

1. First we shall obtain an expression for the function $\square W_{ij}$. Subdivide a rectangle D into nine rectangles $D^{kl} = D^{kl}_{ij}$, for each (i , j), $k, l = 0,1,2$:

$D^{00} = \{u: s \in [0, s_i] , t \in [0, t_j]\}$;

$D^{01} = \{u: s \in [0, s_i] , t \in]t_j , t_{j+1}]\}$;

$D^{02} = \{u: s \in [0, s_i] , t \in]t_{j+1}, T]\}$;

$D^{10} = \{u: s \in]s_i , s_{i+1}] , t \in [0, t_j]\}$;

$D^{11} = \{u: s \in]s_i , s_{i+1}] , t \in]t_j , t_{j+1}]\}$;

$D^{12} = \{u: s \in]s_i , s_{i+1}] , t \in]t_{j+1}, T]\}$;

$D^{20} = \{u: (s , t) \in]s_{i+1}, S] \times [0, t_j]\}$;

$D^{21} = \{u: (s , t) \in]s_{i+1}, S] \times]t_j , t_{j+1}]\}$;

$D^{22} = \{u: (s , t) \in]s_{i+1}, S] \times]t_{j+1}, T]\}$.

Simple calculations show that

$$\square W_{ij}(u) = I(D_{ij} , u)[w_{ij}(u) - w'_{ij}(u)] = I(D_{ij} , u)(w(\square_{ij}) - w^{*}_{ij}(u)) \quad , \tag{19}$$

where $I(A , u)$ is an indicator of set A, and

$$w_{ij}(u) = w]u \wedge u_{ij} , u \wedge u_{i+1,j+1}] , \; w'_{ij}(u) = w']u \wedge u_{ij} , u \wedge u_{i+1,j+1} \quad ,$$

$$(u \wedge u') = (\min (s , s') , \min (t , t')) , \; w^{*}_{ij}(u) = w''_{ij}(u) + w'_{ij}(u)$$

and $w''_{ij} = w(\square_{ij}) - w_{ij}(u)$.

Applying the expression for $(\nabla F , h)$ we obtain

$$\square\gamma'_{ij} = w(\square_{ij})\mathrm{E}[\lambda(w , D_{ij}) / \mathbf{F}_{ij}] - \mathrm{E}[\int_{D_{ij}} \lambda(W_{ij} ; du) w^{*}_{ij}(u) / \mathbf{F}_{nn}] \quad .$$

Since $\mathrm{E}[\int_{D_{ij}} \lambda(W_{ij} ; du) w''(u) / \mathbf{F}_{in}] = 0$, we have

$$\mathrm{E}[\square\gamma'_{ij} / \mathbf{F}_{in}] = -\mathrm{E}[\int_{D_{ij}} \lambda(W_{ij} , du) w'_{ij}(u) / \mathbf{F}_{nn}] \quad .$$

It follows from symmetry that $E[\Box\gamma'_{ij}/\mathbf{F}_{in}] = E[\Box\gamma'_{ij}/\mathbf{F}_{jn}] = E[\Box\gamma'_{ij}/\mathbf{F}_{ij}]$. Introducing the function $\lambda_u(A) = E[\lambda(w;A)/\mathbf{F}_u]$ we may write an expression for $\tilde{\Box\gamma}'_{ij}$ in the form

$$\tilde{\Box\gamma}'_{ij} = w(\Box_{ij})\lambda_{ij}(w, D_{ij}) - \int_{D_{ij}} \lambda_{ij}(du)w''_{ij}(u) \quad ,$$

where $\lambda_{ij}(A) = \lambda_{u_{ij}}(A)$. Integrating by parts we obtain

$$\tilde{\Box\gamma}'_{ij} = w(\Box_{ij})\lambda_{ij}(D_{ij}) - \varphi'_{ij} - \varphi''_{ij} - \varphi'''_{ij} \quad , \tag{20}$$

where

$$\varphi'_{ij} = \int_{\Box_{ij}} \lambda_{ij}(\Box_{ij}\setminus]u \, , \, u_{i+1,j+1}]w(ds\, ,dt) \quad ,$$

$$\varphi''_{ij} = \int_{t_j}^{t_{j+1}} \lambda_{ij}(]s_{i+1}, S] \times]t_j\, , t])w(]s_i\, , s_{i+1}], dt) \quad , \tag{21}$$

$$\varphi'''_{ij} = \int_{s_i}^{s_{i+1}} \lambda_{ij}(]s_i\, , s] \times]t_{j+1}, T])w(ds\, ,]t_j\, , t_{j+1}]) \quad . \tag{22}$$

2. Consider the variables $\tilde{\Box\gamma}''_{ij}$. It can easily be seen that

$$\Box\gamma''_{ij} = \frac{1}{2}E[\alpha'_{ij} + \alpha''_{ij} + \alpha'''_{ij}/\mathbf{F}_{nn}] \quad , \tag{23}$$

where

$$\alpha'_{ij} = \iint_{D\times D} d\mu(W_{ij})\,\Box W_{ij}(u')\,\Box W_{ij}(u'') \quad , \tag{24}$$

$$\alpha''_{ij} = \beta'_{ij} + \beta''_{ij}\, ; \; \beta'_{ij} = \iint_{D\times D} d\mu(W_{ij})\,\Box W_{ij}(u')(\Delta^1 + \Delta^2)W_{ij}(u'') \quad , \tag{25}$$

$$\beta''_{ij} = \iint_{D\times D} d\mu(W_{ij})(\Delta^1 + \Delta^2)W_{ij}(u')W_{ij}(u'') \quad ,$$

$$\alpha'''_{ij} = \bar{\beta}'_{ij} + \bar{\beta}''_{ij}\, ; \; \bar{\beta}'_{ij} = \iint_{D\times D} d\mu(W_{ij})\Delta^1 W_{ij}(u')\Delta^2 W_{ij}(u'') \quad , \tag{26}$$

$$\bar{\beta}''_{ij} = \iint_{D\times D} d\mu(W_{ij})\Delta^2 W_{ij}(u')\Delta^1 W_{ij}(u'') \quad .$$

Here $d\mu(W_{ij})$ denotes $\mu(W_{ij}\, , ds'dt', ds''dt'') = \mu(W_{ij}\, , du', du'')$. After some computation we obtain

$$\tilde{\alpha}'_{ij} = \iint_{D_{ij}\times D_{ij}} [w_{ij}(u')w_{ij}(u'')]^0 d\mu_{ij}(w) - $$

(27)

$$-\mathrm{E}[\iint_{D_{ij}\times D_{ij}} (w_{ij}(u')w'_{ij}(u'') + w_{ij}(u'')w'_{ij}(u'))d\mu(W_{ij})/\mathbf{F}_{nn}] \ .$$

We shall introduce the following notations here: if ξ is a random variable, then $[\xi]^0$ is its central value, i.e., $[\xi]^0 = \xi - E\xi$ and $d\mu_{ij}(w) = \mathrm{E}[d\mu(w)/\mathbf{F}_{ij}] = \mathrm{E}[d\mu(W_{ij})/\mathbf{F}_{nn}]$. We want to find appropriate expressions for variables $\tilde{\alpha}''_{ij}$, and therefore need values for $\Delta^1 W_{ij}$ and $\Delta^2 W_{ij}$. We have the following formulas:

$$\Delta^1 W_{ij} = I_1(\bar{w}_{ij} - \bar{w}'_{ij}) \ , \ I_1 = I(D_{ij}^{+-} \cup D_{ij}) \ ,$$

$$\bar{w}_{ij} = w(]s_i, s \wedge s_{i+1}], t \wedge t_j) \ , \ \bar{w}'_{ij} = w'(]s_i, s \wedge s_{i+1}], t \wedge t_j) \ ,$$

$$\Delta^2 W_{ij} = I_2(\check{w}_{ij} - \check{w}'_{ij}) \ , \ I_2 = I(D_{ij}^{-+} \cup D_{ij}) \ ,$$

$$\check{w}_{ij} = w(s \wedge s_i,]t_j, t \wedge t_{j+1}]) \ , \ \check{w}'_{ij} = w'(s \wedge s_i,]t_j, t \wedge t_{j+1}]) \ .$$

The functions w_{ij}, $\bar{w}_{ij}$, $\check{w}_{ij}$ are w-measures of various domains which lie outside D_{ij}^{--} and thus have no common points. Therefore, for all $u \in D$ the variables $w_{ij}(u)$, $\bar{w}_{ij}(u)$, $\check{w}_{ij}(u)$ are mutually independent and also independent of the σ-algebra $\mathbf{F}_{ij}$. An analogous assertion also holds for w'_{ij}, $\bar{w}'_{ij}$ and $\check{w}'_{ij}$. On the basis of the above considerations we obtain

$$\tilde{\alpha}''_{ij} = \tilde{\beta}'_{ij} + \tilde{\beta}''_{ij} \ ,$$

where

$$\tilde{\beta}'_{ij} = \iint_{D_{ij}\times D} w_{ij}(u')(I_1\bar{w}_{ij} + I_2\check{w}_{ij})(u'')d\mu_{ij} -$$

$$- \iint_{D_{ij}\times D} w_{ij}(u')\mathrm{E}(I_1\bar{w}_{ij} + I_2\check{w}_{ij})(u'')d\mu(w)/\mathbf{F}_{ij}] \ , \qquad (28)$$

and if we rearrange the variables of integration u', u'' then $\tilde{\beta}''_{ij}$ can be expressed in an analogous form.

Now consider the variables $\tilde{\alpha}'''_{ij} = \tilde{\tilde{\beta}}'_{ij} + \tilde{\tilde{\beta}}''_{ij}$ defined by the formulae (26) and (11). For variable $\tilde{\tilde{\beta}}'_{ij}$ we obtain

$$\tilde{\tilde{\beta}}'_{ij} = \mathrm{E}[\iint_{D\times D} I_1(u')I_2(u'')\bar{w}_{ij}(u')\check{w}_{ij}(u'')d\mu(w)/\mathbf{F}_{ij}] \ ,$$

or

$$\tilde{\tilde{\beta}}'_{ij} = \iint_{D\times D} I_1(u')I_2(u'')\bar{w}_{ij}(u')\check{w}_{ij}(u'')d\mu_{ij} \ .$$

Integration by parts and some elementary transpositions lead to

$$\tilde{\beta}'_{ij} = \int_0^{t_j} \int_{s_i}^{s_{i+1}} \left(\int_0^{s_i} \int_{t_j}^{t_{j+1}} \mu_{ij}(D_{u'} \times D_{u''}) w(du'') \right) w(du') \quad . \tag{29}$$

It follows from symmetry that

$$\tilde{\beta}''_{ij} = \int_{t_j}^{t_{j+1}} \int_0^{s_i} \left(\int_0^{t_j} \int_{s_i}^{s_{i+1}} \mu_{ij}(D_{u'} \times D_{u''}) w(du'') \right) w(du') \quad . \tag{30}$$

We can now take the limit as $n \to \infty$ in equality (12).

(a) Let

$$S_i = \sum_{i,j} \square\gamma_{ij} = S_1^0 - S_1' - S_1'' - S_1''' \quad ,$$

where $S_1^0 = \sum w(\square_{ij}) \bar{\lambda}_{ij}(D_{ij})$, $S_1' = \sum_{i,j} \varphi'_{ij} \cdots, S_1''' = \sum_{i,j} \varphi'''_{ij}$. We shall prove that each of the sums S_1', S_1'', S_1''' converges to zero in the mean square. Note that the function $\lambda(x, A)$ is a countably additive measure for fixed x, and if $|\lambda|(x, A)$ represents the complete variation of this measure on a set A, then $|\lambda|(x, A) = \|\nabla F(x)\|$. Thus we obtain the estimates

$$|\lambda|(w, A) \le K(1 + \|w\|^{1+\beta+\delta}) \, , \; |\bar{\lambda}_u|(A) \le K(1 + \|w\|^{1+\beta+\delta}) \quad ,$$

so that the stochastic measure $|\bar{\lambda}_u|(A)$ is uniformly bounded as a function of (u, A) and admits final moments of any rank. Analogous estimates also exist for the complete variation $|\mu|(w, A)$ of a measure $\mu(w, A)$:

$$|\mu|(w, A) \le K(1 + \|w\|^{\beta+\delta}) \quad .$$

Now consider the sum S''_{ij}. It can easily be seen that the elements of this sum are orthogonal. Hence

$$\mathrm{E}|S_1'|^2 = \sum_{i,j} \mathrm{E} \int\int_{\square_{ij}} \lambda(\square_{ij} /]u \, , \, u_{i+1,j+1}]) w(ds, dt)|^2 =$$

$$= \sum_{i,j} \mathrm{E} \int\int_{\square_{ij}} \lambda^2(\square_{ij} /]u \, , u_{i+1,j+1}]) ds\,dt \le |\square_{ij}| \mathrm{E}|\lambda|^2(D)$$

(it was established above that the area $|\square_{ij}|$ of the rectangle $\square_{ij}$ is independent of (i, j)).

$\mathrm{E}(S_1'')^2$ and $\mathrm{E}(S_1''')^2$ can be estimated in a similar way:

$$\mathrm{E}(S_1'')^2 \le \mathrm{E} \sum_{i,j} \int_{t_j}^{t_{j+1}} \lambda^2(]s_i \, ; S] \times]t_j \, , t]) \Delta s_i \, dt$$

$$\leq \mathrm{E}\sum_{i,j} \Delta s_i \, \Delta t_j \, \lambda^2(]s_{i+1}, S] \times]t_j, t_{j+1}]) \leq \Delta t_j S \mathrm{E}\lambda^2(D) \to 0 \quad ,$$

$$\mathrm{E}(S_1''')^2 \leq \Delta s_i \, T \mathrm{E}\lambda^2(D) \to 0 \quad .$$

Now we shall consider the sum S_1^0 and show that it converges to the integral

$$J = \iint_D \lambda_u(D_u) w(ds, dt) \quad .$$

To do this we shall write the difference $J - S_1^0$ in the form

$$J - S_1^0 = \iint_D (\lambda_u(D_u) - \lambda_u^{(n)}(D_u)) w(ds, dt) \quad ,$$

where $\lambda_u^{(n)}(D_u) = \lambda_{ij}(D_{ij})$ for $u \in \Box_{ij}$. Then

$$\mathrm{E}(J - S_1^0)^2 = \mathrm{E}\iint_D (\lambda_u^{(n)}(D) - \lambda_u(D_u))^2 ds\, dt \leq$$

$$\leq 2\mathrm{E}\iint_D (\lambda_u'^{(n)}(D_u) - \lambda_u(D_u))^2 ds\, dt + 2\mathrm{E}\iint_D \mu_n^2(u) ds\, dt \quad ,$$

where

$$\lambda_u'^{(n)}(D_u) = \mathrm{E}[\lambda(D_u)/\mathbf{F}_{ij}] \, , \; \mu_n(u) = \mathrm{E}[\lambda(D_{ij}/D_u)/\mathbf{F}_{ij}] \text{ for } u \in \Box_{ij} \quad .$$

Thus $\lambda_u'^{(n)}(D_u) - \lambda_u(D_u) = \mathrm{E}[\lambda(D_u)/\mathbf{F}_{ij}] - \mathrm{E}[\lambda(D_u)/\mathbf{F}_u]$. If the σ-algebra $\mathbf{F}_u = \mathbf{F}_u^w$ is left-continuous, then for each u we have $\mathrm{E}[\lambda(D_u)/\mathbf{F}_{ij}] \to \mathrm{E}[\lambda(D_u)/\mathbf{F}_u]$ a.s. as $n \to \infty$. Moreover, the variables $(\lambda_u'^{(n)}(D_u) - \lambda_u(D_u))^2$ admit an integrable majorant. Hence

$$\mathrm{E}\iint_D (\lambda_u'^{(n)}(D_u) - \lambda_u(D_u))^2 ds\, dt \to 0 \quad .$$

Further,

$$\mathrm{E}\iint_D \mu_n^2(u) ds\, dt \leq \mathrm{E}\iint_D \lambda^2(D_{i(u)j(u)} \backslash D_u) ds\, dt \quad ,$$

where $(i(u), j(u))$ are the indexes of the rectangle $\Box_{ij}$ for the point u. A set of those u for which $\lambda(D_{i(u)j(u)})$ does not tend to $\lambda(D_u)$ is contained in the sum of the countable number of lines parallel to the coordinate axes in R_+^2; this set has Lebesgue measure 0. Therefore $\iint_D \lambda^2(D_{i(u)j(u)} \backslash D_u) ds\, dt \to 0$ and if the integrands admit an integrable majorant, then

$$\mathrm{E}\iint_D \lambda^2(D_{i(u)j(u)} \backslash D_u) ds\, dt \to 0 \quad .$$

Thus, it is proved that $\lim \sum \square\gamma'_{ij} = J$.

(b) We shall now find the limit of the sums

$$\sum_{i,j} \square\gamma''_{ij} = \frac{1}{2}(\sum_{i,j} \tilde{\alpha}'_{ij} + \sum_{i,j} \tilde{\alpha}''_{ij} + \sum_{i,j} \tilde{\alpha}'''_{ij}) \quad ,$$

as $n \to \infty$.

From expression (27) for the variables $\tilde{\alpha}'_{ij}$ we have $\sum \tilde{\alpha}'_{ij} = S'_2 - S'_3 - S'_4$, where

$$S'_2 = \sum_{i,j} \iint_{D_{ij} \times D_{ij}} [w_{ij}(u') - w_{ij}(u'')]^0 \mathrm{d}\mu_{ij} \quad ,$$

$$S'_3 = \sum_{i,j} \mathrm{E}[\iint_{D_{ij} \times D_{ij}} w_{ij}(u')w'_{ij}(u'')\mathrm{d}\mu(W_{ij})/\mathbf{F}_{nn}] \quad ,$$

$$S'_4 = \sum_{i,j} \mathrm{E}[\iint_{D_{ij} \times D_{ij}} w_{ij}(u'')w'_{ij}(u')\mathrm{d}\mu(W_{ij})/\mathbf{F}_{nn}] \quad .$$

The terms in these sums admit finite moments of any rank, where $w_{ij}(u)$ is a Wiener measure of a certain rectangle in $\square_{ij}$, and μ_{ij} is an $\mathbf{F}_{ij}$-measurable function. Therefore, when $(i\,,j) \neq (i'\,,j')$ and either $i < i'$ or $j < j'$ we have $(S_{ij} = \mathbf{F}_{in} \vee \mathbf{F}_{nj})$

$$\mathrm{E} \iint_{D_{ij} \times D_{ij}} [w_{ij}(u')w_{ij}(u'')]^0 \mathrm{d}\mu_{ij} \cdot \iint_{D_{i'j'} \times D_{i'j'}} [w_{i'j'}(u')w_{i'j'}(u'')]^0 \mathrm{d}\mu_{i'j'} =$$

$$= \mathrm{E}\{\iint_{D_{ij} \times D_{ij}} \iint_{D_{i'j'} \times D_{i'j'}} \mathrm{E}[w_{i'j'}(u')w_{i'j'}(u'')]^0 \cdot [w_{ij}(u')w_{ij}(u'')]^0 \mathrm{d}\mu_{i'j'}\mathrm{d}\mu_{ij} = 0 \quad ,$$

since the random variable $w_{ij}(u')w_{i'j'}(u'')$ is independent of the σ-algebra S_{ij}, and $(\tilde{w}_{ij}(u)\tilde{w}_{ij}(u'')\mathrm{d}\mu_{i'j'}\mathrm{d}\mu_{ij}$ is S_{ij}-measurable. Thus, the terms in the sum S'_2 are orthogonal. Hence

$$\mathrm{E}(S'_2)^2 = \sum_{i,j} \mathrm{E}(\iint_{D_{ij} \times D_{ij}} [\tilde{w}_{ij}(u')\tilde{w}_{ij}(u'')]^0 \mathrm{d}\mu_{ij})^2 \leq |\square_{ij}|\mathrm{E}\,|\mu|^2(w\,;D \times D)ST \quad ,$$

i.e., $\mathrm{E}(S'_2)^2 \to 0$.

The sums S'_3 and S'_4 may be treated in the same manner. We obtain

$$\lim S'_k = 0\,,\; k = 2,3,4 \quad .$$

(c) Analogously it can be shown that $\lim \sum_{i,j} \tilde{\alpha}''_{ij} = 0$.

(d) We shall now compute a limit for the sum $S_5 = \sum_{i,j} \tilde{\alpha}'''_{ij}$, $\tilde{\alpha}'''_{ij} = \tilde{\tilde{\beta}}'_{ij} + \tilde{\tilde{\beta}}''_{ij}$, where $\tilde{\tilde{\beta}}'_{ij}$ and $\tilde{\tilde{\beta}}''_{ij}$ are given by formulas (29) and (30). Let

$$S'_5 = \sum_{i,j} \tilde{\tilde{\beta}}'_{ij}\,,\; S''_5 = \sum_{i,j} \tilde{\tilde{\beta}}'_{ij} \quad .$$

Let

$$\nu(u',u'') = \nu(s',t',s'',t'') \underset{Def}{=} \mathrm{E}[\mu(w\,;D_{s't'}\times D_{s''t''})/\mathbf{F}_{s't''}] \quad ,$$

for $s'' \le s'$ and $t' \le t''$;

$$\nu(u',u'') = \nu(s',t',s'',t'') = \mathrm{E}[\mu(w\,;D_{s't'}\times D_{s''t''})/\mathbf{F}_{s''t'}] \quad ,$$

for $s' \le s''$; and $t'' \le t'$ and $\nu(s',t',s'',t'') = 0$ if neither of these relations is valid, i.e., either $u' \le u''$ or $u'' < u'$.

Let

$$\nu_n(u',u'') = \mathrm{E}[\mu(w\,;D_{s't'}\times D_{s''t''})/\mathbf{F}_{ij}]$$

for $s'' \le s'$, $t' \le t''$ and $s' \in]s_i\,,\,s_{i+1}]$, $t'' \in]t_j\,,\,t_{j+1}]$;

$$\nu_n(u',u'') = \mathrm{E}[\mu(w\,;D_{s't'}\times D_{s''t''})/\mathbf{F}_{ij}]$$

for $s' \le s''$, $t'' \le t'$ and $s'' \in]s_i\,,s_{i+1}]$, $t' \in]t_j\,,t_{j+1}]$, $i,j = 1,2,\dots,n$, $\nu_n(u',u'') = 0$, when $u' < u''$ or $u'' < u'$. since the current $\mathbf{F}_{(s,t)}$ is left-continuous, we have $\nu_n(u',u'') \to \nu(u',u'')$ a.s. for any u',u'' from D. We then have

$$S_5 = S_5' + S_5'' = \overset{\mathrm{WZ}}{\underset{D\times D}{\int\int}} \nu_n(u',u'')w(du')w(du'') \quad ,$$

where $\overset{\mathrm{WZ}}{\underset{D\times D}{\int\int}}$ is Wong–Zakai integral of the second type.

Since $\mathrm{E}|\nu(u',u'') - \nu_n(u',u'')|^k \le c_k$, $k \ge 2$, we obtain

$$\mathrm{E}\,|\nu(u',u'') - \nu_n(u',u'')|^2 \to 0 \quad \text{for all } u',u'' \in D$$

and

$$\mathrm{E}\,|S_5 - \underset{D\times D}{\int\int} \nu(u',u'')w(du')w(du'')|^2 =$$

$$= \underset{D\times D}{\int\int} \mathrm{E}\,|\nu(u',u'') - \nu_n(u',u'')|^2 ds'dt'ds''dt'' \to 0 \quad ,$$

as $n \to \infty$. Thus

$$\lim S_5 = \underset{D\times D}{\int\int} \nu(u',u'')w(du')w(du'') \quad ,$$

and

$$\lim S_5 = \lim \sum_{i,j} \square\gamma_{ij} = \frac{1}{2} \underset{D\times D}{\int\int} \nu(u',u'')w(du')w(du'') \quad .$$

(e) Finally, we shall estimate the remainder terms $\square\tilde{\gamma}_{ij}'''$.

Let

$$\square\tilde{\gamma}_{ij}''' = \tilde{\rho}_{ij}^1 + \tilde{\rho}_{ij}^2 + \tilde{\rho}_{ij}^3 \quad ,$$

where

$$\rho_{ij}^1 = R(W_{ij}, \Delta^1 W_{ij}) , \ \rho_{ih}^2 = R(W_{ij}, \Delta^2 W_{ij}) , \ \rho_{i,j}^3 = R(W_{i,j}, W_{i+1j+1} - W_{ij}) \quad .$$

LEMMA 1. $\lim \sum_{i,j=0}^{n-1} \square\tilde{\gamma}_{ij}''' = 0.$

Proof. We have

$$\mathrm{E}\,(\sum_{i,j} \rho_{ij}^1)^2 = \sum_{i,j} \mathrm{E}(\rho_{ij}^1)^2 \quad ,$$

$$\mathrm{E}(\mathrm{E}[\rho_{ij}^1 / \mathrm{F}_{nn}])^2 \leq \mathrm{E}(\rho_{ij}^1)^2 \leq K\{\mathrm{E}[1 + \|W_{ij}\|^{8\beta} +$$

$$+ \|\Delta^1 W_{ij}\|^{8\beta}] \times \mathrm{E}\|\Delta^1 W_{ij}\|^{8+4\delta}\}^{1/2} \leq K \ |\Delta s_i|^{2+\delta} \quad .$$

Thus,

$$\lim \sum_{i,j} \rho_{ij}^1 = 0 , \ \lim \sum_{i,j} \rho_{ij}^k = 0 , \ k = 2,3 \quad .$$

This proves the lemma and also completes the proof of Theorem 3.

REFERENCES

1. J.M.C. Clark. The representation of functionals of Brownian motion by stochastic integrals. *Annals of Mathematical Statistics*, 41(1970)1282–1295.

2. E. Wong and M. Zakai. Martingales and stochastic integrals for processes with multidimensional parameters. *Zeithschrift für Wahrscheinlichkeitstheorie und Verwandte Gebiete*, 29(1974)109–122.

3. R. Cairoli and J.B. Walsh. Stochastic integrals in the plane. *Acta Mathematica*, 134(1973)111-123.

THE MAXIMUM PRINCIPLE FOR OPTIMAL CONTROL OF DIFFUSIONS WITH PARTIAL INFORMATION[1]

U. G. Haussmann
Mathematics Department
University of British Columbia
Vancouver, Canada V6T 1Y4

1. INTRODUCTION

Necessary conditions for the problem

$$\min\{J(u): u \in \mathcal{U}\}$$

$$J(u) = E\, c(x_T) \tag{1.1}$$

$$dx_t = f(t,x_t,u(t,y))dt + \sigma(t,x_t)dw_t,\; x_o = \bar{x}_o, \tag{1.2}$$

$$dy_t = h(t,x_t)dt + d\bar{w}_t,\; y_o = 0, \tag{1.3}$$

have recently been given by Bensoussan (1983), assuming much differentiability and boundedness of the data as well as uniform non-degeneracy of $a(t,x) = \sigma(t,x)\sigma(t,x)'$ ('denotes transpose), and convexity and compactness of the set of control points U. In this work we relax most of these hypotheses. We can add without difficulty a cost term $\int_o^T \ell(t,x_t,u(t,y))dt$, but more difficult problem where constraints are present will be treated elsewhere. Note that our method is based on taking strong variations of u and thus precludes allowing σ to depend on u, since a variation of u active for a period of time ε must give rise to a perturbation in x_t of the order ε. We only sketch proofs in this article, details will be given elsewhere.

Bensoussan (1983) uses the stochastic Zakai equation to define the state of the separated problem, and then the method of weak variations (Gateau differentiability) to obtain necessary conditions. Much effort is devoted to obtaining an (abstract) representation of the adjoint process, somewhat similar to the work of Bismut (1978) for the case with complete observation. Our approach consists of using weak solutions of the robust (non-stochastic) form of the Zakai equation (Haussmann 1985) as well as using strong variations. This allows us to find an explicit representation (in terms of

---

[1]This research was supported by NSERC under grant A8051.

the data of the problem) of the adjoint process much as in the deterministic case and the case with complete observation (Haussmann 1981). Although this procedure requires regularity of the data including non-degeneracy of a, the final representation does not, so that regularization can be used to obtain the result under our weaker hypotheses.

In section two we give the setting of the problem and the assumptions as well as some notation. In section three we apply the results from filtering theory to give a representation of the adjoint process and to compute the perturbation of J due to a strong variation in u. The maximum principle is given in section four and in addition some remarks concerning the adjoint process are made.

2. THE PROBLEM

The following hypotheses are made. U, the set of control points, is a Borel set in some euclidean space.

(A_1) $f: [o,T] \times R^n \times U \to R^n$ is Borel measurable, $u \to f(t,x,u)$ is continuous $\forall(t,x)$, $x \to f(t,x,u)$ is C^1 $\forall(t,u)$,

$$(1 + |x| + |u|)^{-1} |f(t,x,u)| + f_x(t,x,u)| \leq K_1;$$

(A_2) $\sigma: [o,T] \times R^n \to R^n \otimes R^m$ is Borel measurable, $x \to \sigma(t,x)$ is C^1 $\forall t$,

$$(1 + |x|)^{-1}|\sigma(t,x)| + |\sigma_x(t,x)| \leq K_2;$$

(A_3) $h: [o,T] \times R^n \to R^d$ is Borel measurable, $x \to h(t,x)$ is C^1 $\forall t$, with modulus of continuity which is uniform in (t,x) in a compact set,

$$(1 + |x|)^{-1} |h(t,x)| + |h_x(t,x)| \leq K_3;$$

(A_4) $c: R^n \to R$ is C^1,

$$|c(x)| + |c_x(x)| \leq K_4(1 + |x|^q), \text{ some } q \in [o,\infty);$$

(A_5) $E|\overline{x}_o|^q < \infty$.

Note that c_x denotes the vector $\partial c/\partial x_i = (c_{x_i})$ and h_x denotes the matrix $(h^i_{x_j})$ where h^i is the ith component of h. Finally σ_x denotes the tensor with entries $\sigma^{ij}_{x_k}$ if $\sigma = (\sigma^{ij})$. We write $C(o,T; R^d)$ for the continuous functions

$C[o,T] \to R^d$ and we denote its canonical Borel filtration by $\{G_t^d\}$. For $\eta \in C(o,T;\ R^d)$ let

$$\|\eta\|_t = \sup\{\eta(s):\ o \leq s \leq t\}.$$

The set of admissible controls is

$$\mathcal{U} = \{u:[o,t] \times C(o,T;\ R^d) \to U, \text{ Borel, adapted, } |u(t,\eta)| \leq K_u(1 + \|\eta\|_t),$$

some $K_u < \infty\}$.

Let P_o denote the distribution of $\bar{x}_o$. Let (Ω, F, P) be the canonical space of $(\bar{x}_o, w, y)$ where (w,y) is an $m + d$ dimensional standard Brownian motion independent of $\bar{x}_o$. Then

$$\Omega = R^n \times C(o,T;\ R^m) \times C(o,T;\ R^d)$$

and $P = P_o \times P_w^m \times P_w^d$ where P_w^d is Wiener measure on $C(o,T;\ R^d)$. For $u \in \mathcal{U}$ let $\tilde{P}^u$ be the law of (x^u, y) where x^u is the unique (strong) solution on (Ω, F, P) of

$$dx_t = f(t,x_t,u(t,y))dt + \sigma(t,x_t)dw_t, \quad x_o = \bar{x}_o, \tag{2.1}$$

and let $\tilde{P}_{sx}^u$ be the law of (x^{usx}, y) where x^{usx} is the unique (strong) solution of

$$dx_t = f(t,x_t,u(t,y))dt + \sigma(t,x_t)dw_t, \quad t \geq s,$$

$$x_t = x \quad , \quad t \leq s, \tag{2.1'}$$

and let $\tilde{P}_{sx}^{u\eta}$ be the law of $x^{u\eta sx}$, solution of

$$dx_t = f(t,x_t,u(t,\eta))dt + \sigma(t,x_t)dw_t, \quad t \geq s$$

$$x_t = x \quad , \quad t \leq s. \tag{2.2}$$

Note that $\tilde{P}_{sx}^{u\eta}$ is a regular conditional probability distribution of $\tilde{P}_{sx}^{u\eta}|F^Y$, where F^Y is the subalgebra of F generated by y.

Let $(\tilde{\Omega}, \tilde{F}, \tilde{P}^u)$ be the canonical space of (x^u, y), so that $\tilde{\Omega} = C(o,T;\ R^n) \times C(o,T;\ R^d)$. Define

$$\tilde{Z}_t^s = \exp\{\int_s^t h(\theta,x_\theta)\cdot dy_\theta - \frac{1}{2}\int_s^t |h(\theta,x_\theta)|^2 d\theta\} \tag{2.3}$$

on $(\tilde{\Omega}, \tilde{F}, \tilde{P}_{sx}^u)$. $\tilde{Z}_t^o$ can also be defined, in a consistent manner, on $(\tilde{\Omega}, \tilde{F}, \tilde{P}^u)$, not just $(\tilde{\Omega}, \tilde{F}, \tilde{P}_{ox}^u)$. If $dP^u \equiv \tilde{Z}_T^o\, d\tilde{P}^u$, then (x_t, y_t) is a solution of (1.2), (1.3) on $(\tilde{\Omega}, \tilde{F}, P^u)$, or similarly (and ambiguously, but this causes no confusion) if $dP^u \equiv Z_T^{ou} dP$ (where Z_t^{su} is defined by (2.3) with x_θ replaced by

x_θ^{usx}), then (x_t^u, y_t) is a solution of (1.2), (1.3) on (Ω, F, P^u). Since we have weak uniqueness of the solutions of (1.2), (1.3) we are able to switch from $\tilde{\Omega}$ to Ω and vice versa.

In what follows we shall use some concepts from differential equations for which we now introduce some notation. Let $H = L^2(R^n)$ with inner product $\langle v,g\rangle$. Define

$$H^1 = \{v \in H: \ \partial_i v \in H, \ i = 1,\ldots,n\}$$

where $\partial_i v$ is the x_i component of the distributional derivative of $v(x)$. Let H^{-1} be the dual of H^1 with pairing (v,g) for $v \in H^{-1}$, $g \in H^1$. We use the convention that repeated indices are summed. We use the following notation

$$a_t^{ij}: x \to a^{ij}(t,x)$$

$$b_t^{u\eta i}: x \to f^i(t,x,u(t,\eta)) - \frac{1}{2} a^{ij}(t,x)[\eta(t)\cdot h(t,x)]_{x_j} - \frac{1}{2} a^{ij}(t,x)_{x_j}$$

$$\eta_t = \eta(t) \quad , \ h_t: x \to h(t,x) \ \partial_o h_t \ : x \to \partial_o h(t,x),$$

where ∂_o $h(t,x)$ is the distributional derivative of h (as a function on R^{n+1}) with respect to t. Finally define

$$W(o,T) = \{v \in L^2(o,T;H^1): \frac{dw}{dt} \in L^2(o,T;H^{-1})\}.$$

Here $\frac{dv}{dt}$ denotes the distributional derivative of $t \to v(t) \in H^1$.

3. <u>THE ADJOINT PROCESS</u>

We shall compute here the perturbation in $J(u)$ generated by a strong variation of $\hat{u}$, an optimal control whose existence we assume (since we are looking for necessary conditions).

Throughout this section we assume (A_1) - (A_5), and <u>until further notice</u> also

$$\begin{aligned} &|f(t,x,u)| + |\sigma(t,x)| + |h(t,x)| \leq K \\ &a(t,x) \geq \alpha I, \quad \alpha > 0, \\ &\|\partial_o h(t,x)\|_{L^\infty((o,T)\times R^n)} \leq K \\ &\|c\|_{L^2(R^n)} \leq K. \end{aligned} \tag{3.1}$$

On H^1 we can define a bilinear form.

$$\begin{aligned} A_t^{u\eta}(\mu,v) = &-\frac{1}{2}\langle a_t^{ij}\partial_i\mu, \partial_j v\rangle + \langle b_t^{u\eta i}\partial_i\mu, v\rangle + \frac{1}{2}\langle \partial_i(\eta_t \cdot h_t)a_t^{ij}\mu, \partial_j v\rangle \\ &- \langle[\partial_i(\eta_t\cdot h_t)b_t^{u\eta i} + \eta_t\cdot\partial_o h_t + \frac{1}{2}|h_t|^2]\mu, v\rangle \end{aligned} \tag{3.2}$$

Lemma 3.1 <u>There exists a unique solution</u> $\mu_t^{u\eta}$ <u>in</u> W(o,T) <u>of</u>

$$\left(\frac{d\mu_t}{dt}, v\right) + A_t^{u\eta}(\mu_t, v) = 0 \qquad v \in H^1, \tag{3.3}$$

$$\mu_T = c\,\exp(\eta_T \cdot h_T). \tag{3.4}$$

<u>Moreover if</u>

$$P_o \text{ has a density } p_o \text{ such that } p_o \in L^2(R^d), \tag{3.5}$$

<u>then</u>

$$J(u) = \tilde{E}_w < \mu_o^{u\eta}, p_o >. \tag{3.6}$$

Here $\tilde{E}_w$ is expectation with respect to P_w^d. The lemma follows from Haussmann (1985) corollaries 2.1 and 5.1. Note that $\mu_t^{u\eta}(x)\exp[-\eta_t \cdot h_t(x)]$ is the expected cost to go from t, given the observation is $\{\eta_s : o \le s \le T\}$ and $x_t^u = x$.

On $(C(o,T;\,R^n), G_T^n, \tilde{P}_{sx}^{u\eta})$ let $\tilde{\Phi}_{sx}^{u\eta}(t)$ be the fundamental matrix solution of

$$d\xi_t = f_x(t,x_t,u(t,\eta))\xi_t dt + \sigma_x^{(i)}(t,x_t)\xi_t dw_t^i \tag{3.7}$$

where $\sigma^{(i)}$ is the ith column of σ, so that each column of $\tilde{\Phi}_{sx}^{u\eta}$ is a solution of (3.7) with $\tilde{\Phi}_{sx}^{u\eta}(s) = I$. Moreover this can be done consistently so that $\tilde{\Phi}_{sx}^{uy}$, considered on $(\tilde{\Omega},\tilde{F},\tilde{P}_{sx})$, is the fundamental matrix of

$$d\xi_t = f_x(t,x_t,u(t,y))\xi_t dt + \sigma_x^{(i)}(t,x_t)\xi_t^i dw_t^i. \tag{3.7)'$$

Let $\nu_t^{u\eta} = \mu_t^{u\eta}\exp(-\eta_t \cdot h_t)$. Observe that μ_t, hence ν_t, depends on the past of η through $u(t,\eta)$, and on the future of η through the dynamics (3.2), (3.3). We wish to compute $\tilde{E}_w\{\nabla\nu_s^{u\eta}(x)\,|G_s^d\}$

Lemma 3.2. <u>For each</u> s, a.e. x, w.p.1

$$\tilde{E}_w\{\nabla\nu_s^{u\eta}(x)|G_s^d\}(y) = \tilde{E}_{sx}^u\{[c_x(x_T)\tilde{\Phi}_{sx}^{uy}(T) + c(x_T)\int_s^T h_x^i(t,x_t)\tilde{\Phi}_{sx}^{uy}(t)(dy_t^i - h^i(t,x_t)dt)]\,\tilde{Z}_T^s|\tilde{F}_s^y\} \tag{3.8}$$

We observe that the right side of (3.7) is w.p.1 equal to

$$\tilde{q}_s^u(x)' \equiv E_{sx}^u\{c_x(x_T)\tilde{\Phi}_{sx}^{uy}(T) + c(x_T)\int_s^T h_x^i(t,x_t)\tilde{\Phi}_{sx}^{uy}(t)d\overline{w}_t^i|\tilde{F}_s^y\} \tag{3.9}$$

where E_{sx}^u is expectation with respect to P_{sx}^u, $dP_{sx}^u = \tilde{Z}_T^s\, d\tilde{P}_{sx}^u$.

"Proof" : From Haussmann (1985), corollary 3.2, we have

$$\nu_s^{u\eta}(x) = \tilde{E}_{sx}^{u}\{c(x_T)\tilde{Z}_T^{s}|\tilde{F}^{y}\}(\eta)$$

or switching back to (Ω, F, P)

$$\nu_s^{u\eta}(x) = E\{c(x_T^{usx})Z_T^{su}|F^{y}\}(\eta)$$

so that

$$\tilde{E}_w\{\nu_s^{u\eta}(x)|G_s^d\}(y) = E\{c(x_T^{usx})Z_T^{su}|F_s^{y}\}$$

If $\Phi_{sx}^{u\eta}(t)$ is defined as is $\tilde{\Phi}_{sx}^{u\eta}(t)$ but on (Ω,F,P), and if $\xi_{sx}^{u\eta}(t)$ is the solution of (3.7) on (Ω,F,P) with x_t replaced by $x_t^{u\eta sx}$ and with $\xi_{sx}^{u\eta}(s) = \xi$ (fixed), then

$$\tilde{E}_w\{\nu_s^{u\eta}(x+\xi)|G_s^d\}(y) = \tilde{E}_w\{\nu_s^{u\eta}(x)|G_s^d\}(y) + E\{c_x(x_T^{usx})\Phi_{sx}^{uy}(T)\xi Z_T^{su}$$
$$+ c(x_T^{usx})Z_T^{su}\int_s^T h_x^i(t,x_t^{usx})\Phi_{sx}^{uy}(t)\xi[dy_t^i - h^i(t,x_t^{usx})dt]\ |F_s^{y}\}$$
$$+ o(|\xi|).$$

The result follows.

Fix s, ε. Then a <u>strong variation</u> (corresponding to (s,ε)) is an element $u \in U$ such that

$$u(t,\eta) = \begin{cases} \bar{u}(t,\eta) & s \leq t < s+\varepsilon \\ & , \\ \hat{u}(t,\eta) & \text{otherwise} \end{cases} \tag{3.10}$$

Let u be a strong variation of u corresponding to (s,ε). Let $\hat{\tilde{q}}_s(x) = \hat{\tilde{q}}_s^u(x)$ (similarly without ~ if on (Ω,F,P)) and let

$G_t^u(x) = \hat{\tilde{q}}_t(x)\cdot f(t,x,u(t,y))$ and $\hat{G}_t(x) = G_t^{\hat{u}}(x)$.

Theorem 3.1 <u>Assume</u> $(A_1) - (A_5)$, (3.1) <u>and</u> (3.5). <u>Then</u>

$$J(u) - J(\hat{u}) = \hat{E}\int_s^{s+\varepsilon}[G_t^u(x_t) - \hat{G}_t(x_t)]dt + o(\varepsilon) \tag{3.11}$$

<u>where</u> E <u>denotes expectation</u> w.r.t. $P = P^{\hat{u}}$.

"Proof": If $\mu_t^{\varepsilon\eta} \equiv \mu_t^{u\eta} - \hat{\mu}_t^{\eta}$, then $\mu_t^{\varepsilon\eta} = 0$ if $t \geq s+\varepsilon$, and

$$\left(\frac{d\mu^{\varepsilon\eta}}{dt}, v\right) + \hat{A}_t^{\eta}(\mu_t^{\varepsilon\eta}, v) = \begin{cases} 0 & \text{if } t < s \\ -\langle\nabla\nu_t^{u\eta}\Delta f\, e^{\eta_t\cdot h_t}, v\rangle & \text{if } s \leq t \leq s+\varepsilon \end{cases}$$

where $\Delta f = f(t,\cdot,u(t,\eta)) - f(t,\cdot,\hat{u}(t,\eta))$. Then

$$J(u) - J(\hat{u}) = \tilde{E}_w < \mu_o^{\varepsilon\eta}, p_o>$$

$$= \tilde{E}_w < \mu_s^{\varepsilon\eta}, \hat{\rho}_s^{\eta} e^{-y_s \cdot h_s}>$$

$$= \tilde{E}_w < \hat{\tilde{E}}_{sx}\{ \int_s^{s+\varepsilon} \nabla v_t^{uy}(x_t)\Delta f \tilde{Z}_t^s \, dt | \tilde{F}_s^y\}(\eta), \hat{\rho}_s^{\eta} >$$

$$= \hat{\tilde{E}}\{ \int_s^{s+\varepsilon} \tilde{q}_t^u(x_t)\Delta f \, dt \tilde{Z}_T^o\}. \quad (3.12)$$

$\hat{\rho}_s^{\eta}$ is the unnormalized conditional density of x_t given F_s^y on $(\tilde{\Omega},\tilde{F},\tilde{P})$. Regularity of $\tilde{q}_t^u(x)$ allows us to replace it by $\hat{\tilde{q}}_t(x)$ in (3.12) to obtain the result.

Corollary 3.1 Assume $(A_1) - (A_5)$. Then

$$J(u) - J(\hat{u}) = \hat{E} \int_s^{s+\varepsilon} \hat{q}_t(\hat{x}_t)[f(t,\hat{x}_t,u(t,y) - f(t,\hat{x}_t,\hat{u}(t,y))]dt + o(\varepsilon). \quad (3.13)$$

Here we have shifted back to (Ω,F,P) and we use $\hat{x} = x^{\hat{u}}$, $\hat{P} = P^{\hat{u}}$, c.f. (2.1) and §2, and $\hat{q}$, the equivalent on (Ω,F,P) of $\hat{\tilde{q}}$. The proof is by regularization since both extremities of (3.12) are well defined without (3.1) and (3.5).

4. THE MAXIMUM PRINCIPLE

The required necessary condition follows readily from (3.13) except that it holds only a.e. (s), so we need to reduce the set of strong variations to a countably generated set. Write $\tilde{U}_s$ for $L^1(C(o,T; R^d),G_s^d,\tilde{P}_w;U)$ so that $\tilde{U}_s$ is separable, and let V be a countable dense subset of $\tilde{U}_T$ such that each element of V is bounded. Define the measurable map

$$i_s: (C(o,T; R^d), G_s^d) \to (C(o,T; R^d), G_T^d)$$

by $(i_s\eta)(t) = \eta(t \wedge s)$ where $t \wedge s = \min\{t,s\}$. If $V_s = V \circ i_s$ then V_s is a countable dense subset of $\tilde{U}_s$ and

$$\overline{U} = \{u: u(t,\eta) = v \circ i_t(\eta), \quad v \in V\}$$

is a countable subset of U. Now let $U_{s\varepsilon}$ be the set of strong variations, c.f. (3.10), corresponding to (s,ε) with $\overline{u} \in U$.

We define

$$H(t,x,u,p) = p\cdot f(t,x,u)$$

$$\hat{p}_t(x)' = -\hat{E}_{tx}\{c_x(\hat{x}_T)\hat{\Phi}_t(T) + c(\hat{x}_T) \int_t^T h_x^i(s,\hat{x}_s)\hat{\Phi}_t(s)d\overline{w}_s^i | F_t^y\}$$

$$d\hat{\Phi}_t(s) = f_x(s,\hat{x}_s,\hat{u}(s,y))\hat{\Phi}_t(s)ds + \sigma_x^{(i)}(s,\hat{x}_s)\hat{\Phi}_t(s)dw_s^i, \ s \geq t$$

$$d\overline{w}_t = dy_t - h(t,\hat{x}_t)dt.$$

Theorem 4.1 Assume $(A_1) - (A_5)$. If $\hat{u}$ is optimal then there exists a null set N such that for $t \notin N$, $u \in U$,

$$\hat{E}\{H(t,\hat{x}_t,u,\hat{p}_t(\hat{x}_t))|F_t^y\} \\ \leq \hat{E}\{H(t,\hat{x}_t,\hat{u}(t,y),\hat{p}_t(\hat{x}_t)|F_t^y\} \text{ w.p.1.} \tag{4.1}$$

Proof: From (3.13) it follows that for t not in a null set $N(\bar{u})$

$$0 \geq \hat{E}\{H(t,\hat{x}_t,v_t(y),\hat{p}_t(\hat{x}_t)) - H(t,\hat{x}_t,\hat{u}(t,y),\hat{p}_t(\hat{x}_t))\},$$

for all $v_t \in V_t$. N.b. $u(t,y) = v_t(y) = (v \circ i_t)(y)$ if $u \in U_{t\varepsilon}$.

By denseness this holds for all $v_t \in \tilde{U}_t$, hence (4.1) follows with

$N = \bigcup_{v \in U} N(\bar{u})$.

We conclude with some remarks about the adjoint process. Observe that

$$\tilde{E}_w\{\hat{v}_s^\eta(x)|G_s^d\}(y) = E\{c(x_T)\,\tilde{Z}_T^s|x_s = x,\ y_t: o \leq t \leq s\},$$

$$= v_s^y(x)$$

the value function, i.e. the expected optimal cost to go, for the problem with partial observation, given the past observations and the present state. From (3.9) and lemma 3.2 $\hat{p}_t(\hat{x}_t)' = -\nabla v_s^y(\hat{x}_t)$, w.p.1, so as usual the adjoint process is the negative of the gradient of the value function.

If we consider instead the equivalent separated problem, (Haussmann 1985, section 5), the optimal cost to go given the past observations and the present "state" ρ is

$$\tilde{V}_s^y(\rho) = \tilde{E}_w\{<c, \hat{\rho}_T> |\tilde{F}_s^y\}$$

with

$$d\hat{\rho}_t = \hat{L}_t^* \hat{\rho}_t\, dt + \hat{\rho}_t h_t \cdot dy_t, \quad t \geq s,$$

$$\hat{\rho}_s = \rho$$

$$\hat{L}_t^* v = \frac{1}{2}\partial_i\partial_j(a_t^{ij}v) - \partial_i(f^i(t,\cdot,u(t,y))v).$$

Then it can be shown that

$$\tilde{V}_s^y(\rho) = <V_s^y, \rho>,$$

so that $\nabla\tilde{V}_s^y = <V_s^y,\cdot>$, i.e. the gradient of $\tilde{V}_s^y$ is (represented by) the value function of the original problem.

For the separated problem we define the Hamiltonian by $(p \in L^2(R^n)^*)$

$$\tilde{H}(t,\rho,u,p) = p(L_t^{u*}\rho),$$

where for $u \in U$

$$L_t^{u*}v = \frac{1}{2}\partial_i\partial_j(a_t^{ij}v) - \partial_i(f^i(t,\cdot,u)v)$$

so that formally

$$\begin{aligned}\tilde{H}(t,\rho,u,-\nabla \tilde{V}_t^y) &= -\langle V_t^y, L_t^{u*}\rho\rangle \\ &= -\langle L_t^u V_t^y, \rho\rangle \\ &= -\frac{1}{2}\langle a_t^{ij}\partial_i\partial_j V_t^y, \rho\rangle - \langle f^i(t,\cdot,u)\partial_i V_t^y, \rho\rangle \\ &= \phi_t^y + \langle H(t,\cdot,u,\hat{p}_t(\cdot)),\rho\rangle.\end{aligned}$$

Since ϕ is independent of u, then again

$$\begin{aligned}\tilde{H}(t,\hat{\rho}_t,u,-\nabla \tilde{V}_t^y) &= \phi_t^y + \hat{E}\{H(t,\hat{x}_t,u,\hat{p}_t(\hat{x}_t)|F_t^y\} \\ &\leq \phi_t^y + \hat{E}\{H(t,\hat{x}_t,\hat{u}(t,y),\hat{p}_t(\hat{x}_t))|F_t^y\} \\ &= \tilde{H}(t,\hat{\rho}_t,\hat{u}(t,y),-\nabla \tilde{V}_t^y) \text{ w.p.1},\end{aligned}$$

i.e. $\hat{u}$ maximizes the Hamiltonian, and the adjoint process is the negative of the gradient of the value function. The state space is, however, a functions space.

5. REFERENCES

Bensoussan, A. (1983). Maximum principle and dynamic programming approaches of the optimal control of partially observed diffusions, Stochastics, 9: 169-222.

Bismut, J.M. (1978). An introductory approach to duality in optimal stochastic control, SIAM Review, 20: 62-78.

Haussmann, U.G. (1981). On the adjoint process for optimal control of diffusion processes, SICOPT, 19: 221-243.

Haussmann, U.G. (1985). L'équation de Zakai et le problème séparé du contrôle optimal stochastique, to appear in Séminaire de Probabilité, Lecture Notes in Mathematics.

EXPLICIT SOLUTION OF A CONSUMPTION/INVESTMENT PROBLEM

I. Karatzas J. Lehoczky, S. Sethi, S. Shreve

1. INTRODUCTION

This paper solves an optimal stochastic control problem which arises in finance. Specifically, we characterize the optimal consumption and investment policies of an individual who allocates his wealth into two investments, one which is deterministic with rate of increase r, while the other is given by a log Brownian motion process with rate of increase $\alpha \neq r$ and variance σ^2. The individual seeks to maximize $E_x(\int_0^\infty e^{-\beta t} U(c_t)dt)$, where $c_t \geq 0$ represents the consumption rate, β is a discount factor, and U is a utility function. We assume π_t represents an investment control and denotes the fraction of wealth allocated to the log Brownian motion investment. The controls $\{c_t, t \geq 0\}$ and $\{\pi_t, t \geq 0\}$ give rise to a wealth process $\{x(t), t \geq 0\}$ which satisfies the Itô stochastic differential equation

$$dx(t) = [(\alpha - r)\pi_t x(t) + (rx(t) - c_t)]dt + x(t)\pi_t \sigma dw_t, \quad x(0) = x \geq 0, \tag{1.1}$$

where $\{w_t, t \geq 0\}$ in a standard Brownian motion.

The model requires some assumption concerning what investment and consumption options are available *if* wealth reaches zero, since further consumption would result in negative wealth. We refer to the state of having zero wealth as *bankruptcy*. Many bankruptcy models are possible, and we incorporate all these models into one by stopping the problem and assigning a value P when bankruptcy occurs. Specifically, we define

$$T_0 = \inf\{t \geq 0 \mid x(t) = 0\}. \tag{1.2}$$

The individual thus seeks to maximize

$$E_x\left(\int_0^{T_0} e^{-\beta t} U(c_t)\,dt + Pe^{-\beta T_0}\right). \tag{1.3}$$

The parameter P is arbitrary; however, the value $P = U(0)/\beta$, the "natural payment", plays a distinguished role. We will find that the optimal policies for all $P \le U(0)/\beta$ are the same. The natural payment $U(0)/\beta$ corresponds to the individual consuming zero forever after bankruptcy occurs.

This paper is an abreviated version of [2], in which a more general treatment is given. In [2], multiple and more general risky investments are considered. The case $\alpha = r$ was considered in [3]. The finance background for this kind of problem can be found in [3, 4, and 5].

2. ASSUMPTIONS

We assume that U is a real-valued function, strictly increasing, strictly concave and C^3 on $(0,\infty)$. We set $U(0) = \lim_{c\downarrow 0} U(c)$ and $U'(0) = \lim_{c\downarrow 0} U'(c)$. Note that $U(0)$ may be $-\infty$ and $U'(0)$ may be $+\infty$. Furthermore, we impose the condition

$$\lim_{c\to\infty} \frac{U(c)}{c} = 0. \tag{2.1}$$

Define $\gamma = \frac{1}{2}\left(\frac{\alpha-r}{\sigma}\right)^2$, and recall $\gamma > 0$. Consider the equation

$$\gamma\lambda^2 - (r-\beta-\gamma)\lambda - r = 0, \tag{2.2}$$

and let $\lambda_+ > 0$ and $\lambda_- < 0$ be its two solutions. We assume

$$\int_c^\infty (U'/\theta))^{-\lambda_-}\,d\theta < \infty, \quad \forall c > 0, \tag{2.3}$$

a condition stronger than (2.1). We note that (2.3) is a sufficient condition for the value function to be finite. Condition (2.3) is also a necessary condition when U is of the HARA class (see [2]).

Let $\{W_t, \mathcal{F}_t, t \ge 0\}$ be a standard Brownian motion on a probability space $(\Omega,\mathcal{F},P)$, where $\{\mathcal{F}_t, t \ge 0\}$ is a nondecreasing, right-continuous family of σ-fields. An admissible consumption process $\{c_t, t \ge 0\}$ is a non-negative process adapted to $\{\mathcal{F}_t, t \geq 0)$ which satisfies almost surely

$$\int_0^t c_s ds < \infty, \qquad t \geq 0. \tag{2.4}$$

The investment process $\{\pi_t, t \geq 0\}$ is an $\{\mathcal{F}_t\}$ adapted process. We define the $\{\mathcal{F}_t\}$ stopping time

$$T(\pi) = \sup \{t \geq 0 \mid \int_0^t \pi_s^2 ds < \infty\},$$

and call an investment policy <u>admissible at x</u> if $T(\pi) = \infty$ or $T_0 < T(\pi)$ or $\lim_{t \uparrow T(\pi)} x(t)$ exists and is equal to zero. In [2], it is shown that (1.1) has a solution for finite $t \leq T_0$. Furthermore, the integral and expectation in (1.3) are well-defined.

For given admissible c and π, we define

$$V_{c(\cdot),\pi(\cdot)}(x) = E_x(\int_0^{T_0} e^{-Bt} U(c_t)dt + Pe^{-BT_0}), \text{ and}$$
$$V^*(x) = \sup_{c(\cdot),\pi(\cdot)} V_{c(\cdot),\pi(\cdot)}(x), \quad x \geq 0 \tag{2.5}$$

We note $V^*(0) = P$.

3. SUMMARY OF RESULTS

In this paper, we present a closed form solution of the problem posed in sections 1 and 2. We summarize the qualitative results here.

(i) If $P \geq \frac{1}{\beta} \lim_{c\to\infty} U(c)$, one should consume to bankruptcy quickly, and $V^*(x) = P$. There is no optimal policy, since instantaneous bankruptcy cannot be achieved.

(ii) If $P \leq \frac{1}{\beta} U(0)$ and $U'(0) = \infty$, the optimal consumption is never zero but is not bounded away from zero, and the optimal wealth process does not lead to bankruptcy.

(iii) If $P > \frac{1}{\beta} U(0)$ and $U'(0) = \infty$, the optimal consumption is bounded below by a positive constant a, and the optimal wealth process leads to bankruptcy with positive probability. The probability of bankruptcy is equal to one if and only if $\beta \geq r + \gamma$.

(iv) If $U'(0)$ is finite and

$$P < P^* \triangleq \frac{1}{\beta} U(0) - \frac{(U'(0))^{\lambda_-+1}}{\beta\lambda_-} \int_0^\infty \frac{d\theta}{(U'(\theta))^{\lambda_-}}, \tag{3.1}$$

then for low levels of wealth the optimal consumption is identically zero. If $P \le \frac{1}{\beta} U(0)$, bankruptcy occurs with positive probability which is equal to one if and only if $\beta \ge r + \gamma$. If $U'(0)$ is finite and $P \ge P^*$, the optimal consumption is never zero and is bounded away from zero if and only if $P > P^*$. There is a positive probability of bankruptcy, and this probability is equal to one if and only if $\beta \ge r + \gamma$.

4. THE BELLMAN EQUATION

The Bellman equation for our stochastic optimal control problem is given by

$$\beta V(x) = \sup_{c \ge 0, \pi} [((\alpha - r)\pi x + (rx - c))V'(x) + \frac{1}{2}\sigma^2\pi^2 x^2 V''(x) + U(c)], \quad x > 0, \tag{4.1}$$

$$V(0) = P.$$

We prove in [2] the following standard theorem

Theorem 4.1

Let P be finite and $V : (0,\infty) \to (P,\infty)$ be a C^2 function satisfying (4.1). Then $V(x) \ge V^*(x)$, $x > 0$, provided either

(i) $|U(0)| < \infty$, or

(ii) $U(0) = -\infty$, but under any consumption/investment policy,

$$E_x \int_0^{T_0} e^{-\beta t} U^+(c_t)\,dt < \infty, \text{ where } U^+ = \max(0,U).$$

The assumption of unconstrained π (unlimited borrowing) allows the individual to create a risky investment with arbitrarily large variance. For initial wealth x_2 with $0 < x_1 < x_2 < x_3$, he can thus exit the interval $[x_1,x_3]$ in an arbitrarily short time with nearly linear exit probabilities. Consequently, V^* satisfies

$$V^*(x_2) \ge \frac{x_3 - x_2}{x_3 - x_1} V^*(x_1) + \frac{x_2 - x_1}{x_3 - x_1} V^*(x_3)$$

which establishes the concavity of V^* when it is finite. We thus solve (4.1) under the assumption $V'' < 0$, and we verify after the fact that this assumption is justified. The maximizing π is

$$\pi = \frac{-(\alpha-r)V'(x)}{\sigma^2 x V''(x)}, \tag{4.2}$$

and when the constraint $c \geq 0$ is slack, the maximizing $c = C(x)$ satisfies

$$V'(x) = U'(C(x)). \tag{4.3}$$

We assume $C(\cdot)$ has an inverse function $X(\cdot)$. Differentiation of (4.1) with respect to c and (4.3) with respect to x and substitution of (4.2) and (4.3) yield a linear, second-order ordinary differential equation

$$\gamma X''(c) = [(r-\beta-2\gamma)\frac{U''(c)}{U'(c)} + \frac{\gamma U'''(c)}{U''(c)}]X'(c) + [\frac{U''(c)}{U'(c)}]^2(rX(c)-c). \tag{4.4}$$

The general solution of (4.4) is given by

$$B(U'(c))^{\lambda_+} + \hat{B}(U'(c))^{\lambda_-} + X(c;a) \tag{4.5}$$

where λ_+ and λ_- are defined by (2.2) and

$$X(c;a) = \frac{c}{r} - \frac{1}{\gamma(\lambda_+-\lambda_-)}\{\frac{(U'(c))^{\lambda_+}}{\lambda_+}\int_a^c (U'(\theta))^{-\lambda_+}d\theta + \frac{(U'(c))^{\lambda_-}}{\lambda_-}\int_c^\infty (U'(\theta))^{-\lambda_-}d\theta\}. \tag{4.6}$$

We define $X(c;a,B)$, $c > a$ by (4.5) with $\hat{B} = 0$. This eliminates the term $(U'(c))^{\lambda_-}$ (which grows too rapidly). We set $X(a;a,B) = \lim_{c\downarrow a} X(c;a,B)$. We can show for $a \geq 0$ and $B \leq 0$ that $X'(.;a,B) > 0$ on (a,∞), and so $X(.;a,B)$ maps $[a,\infty)$ onto $[X(a;a,B),\infty)$. Moreover, its inverse function $C(\cdot;a,B)$ is C^2, increasing and maps $[X(a;a,B),\infty)$ onto $[a,\infty)$.

5. CANDIDATE OPTIMAL POLICIES AND THEIR PERFORMANCE EVALUATION

We fix $a \geq 0$, $B \leq 0$ and use the functions $X(c) = X(c;a,B)$, $C(x) = C(x;a,B)$ of §4 to create a policy $\{c_t, \pi_t, t \geq 0\}$.

We will ultimately select $a \geq 0$, $B \leq 0$ so that $X(a) \geq 0$. Let $\xi = X(a)$ and assume $\xi \geq 0$. Given $x_0 > \xi$, we define a wealth process $X(\cdot)$ by (1.1), where

$$c_t = C(x_t), \quad \pi_t = -\frac{(\alpha-r)U'(c_t)}{\sigma^2 x_t U''(c_t)C'(x_t)}, \quad 0 \leq t \leq T_\xi. \tag{5.1}$$

We derive a stochastic differential equation for $y_t = U'(c_t)$ using Itô's rule and (5.1) and (4.4) applied to (1.1). We find

$$dy_t = -(r-\beta)y_t dt - \frac{\alpha-r}{\sigma} y_t dw_t, \quad 0 \leq t \leq T_\xi. \tag{5.2}$$

We let U' have inverse I and solve (5.2) to obtain the candidate optimal policies

$$c_t = I(U'(c_0)\exp[-(r-\beta+\gamma)t - \frac{(\alpha-r)}{\sigma} W_t), \quad 0 \leq t \leq T_\xi, \tag{5.3}$$

$$\pi_t = \frac{-(\alpha-r)}{\sigma^2} \frac{U'(c_t)}{U''(c_t)} \frac{X'(c_t)}{X(c_t)}, \quad 0 \leq t \leq T_\xi. \tag{5.4}$$

Note that

$$T_\xi \triangleq \inf\{t \geq 0 : x_t = \xi\} = \inf\{t \geq 0 : c_t = a\} = \inf\{t \geq 0 : y_t = U'(a)\}.$$

If $a = 0$ and $U'(0) = \infty$, then $T_\xi = \infty$ a.s., because y_t does not explode. Under such conditions, bankruptcy cannot occur. However, if $\xi = 0$ and $U'(a)$ is finite, bankruptcy will occur when y_t rises to $U'(a)$. This happens almost surely if $\beta \geq r + \gamma$, and with positive probability less than one if $\beta < r + \gamma$,

To evaluate the expected return associated with (5.3), (5.4), we let $\bar{v} = V_{c(\cdot),\pi(\cdot)}(\xi)$, and we assume $\bar{v}$ is finite. For $x_0 > \xi$, let

$$V_{c(\cdot),\pi(\cdot)}(x_0) = H(c_0) \triangleq E_{x_0}[\int_0^{T_\xi} e^{-\beta t}U(c_t)dt + \bar{v}e^{-\beta T_\xi}]. \tag{5.5}$$

Using Theorem 13.16 of Dynkin [1] (sometimes called the Feynman-Kac formula) we can show that if H is well defined and finite, then it is C^2 on (a,∞), satisfies

$$\beta H(c) = \frac{-U'(c)}{U''(c)}\left[r - \beta + \gamma \frac{U'(c)U'''(c)}{(U''(c))^2}\right]H'(c) + \gamma\left(\frac{U'(c)}{U''(c)}\right)^2 H''(c) \tag{5.6}$$

$$+ U(c), \quad c > a$$

and, if $U'(a) < \infty$,

$$\lim_{c \downarrow a} H(c) = \bar{v}. \tag{5.7}$$

The general solution to (5.6) is

$$J(c;a,A,\hat{A}) \triangleq A(U'(c))^{\rho_+} + \hat{A}(U'(c))^{\rho_-} + J_0(c;a) \tag{5.8}$$

where $\rho_{\pm} = 1 + \lambda_{\pm}$ are the roots of the equation

$$\gamma\rho^2 - (r-\beta+\gamma)\rho - \beta = 0, \tag{5.9}$$

and a particular solution to (5.6) is

$$J_0(c;a) = \frac{U(c)}{\beta} - \frac{1}{\gamma(\rho_+-\rho_-)}\left\{\frac{(U'(c))^{\rho_+}}{\rho_+}\int_a^c \frac{d\theta}{(U'(\theta))^{\lambda_+}}\right. \tag{5.10}$$

$$\left. + \frac{(U'(c))^{\rho_-}}{\rho_-}\int_c^\infty \frac{d\theta}{(U'(\theta))^{\lambda_-}}\right\}.$$

In [2], we prove the following theorems:

<u>Theorem 5.1</u> If $U'(a) < \infty$ and $H(c)$ given by (5.5) is well-defined and finite for all $c > a$, then $H(c) = J(c;a,A,0)$, $c > a$, where A satisfies

$$A(U'(a))^{\rho_+} + \frac{U(a)}{\beta} - \frac{(U'(a))^{\rho_-}}{\gamma\rho_-(\rho_+-\rho_-)}\int_a^\infty (U'(\theta))^{-\lambda_-}d\theta = \bar{v}. \tag{5.11}$$

<u>Theorem 5.2</u> If $a = 0$ and $U'(0) = \infty$ and if $H(c)$ is well-defined and finite for every $c > 0$, then $H(c) = J(c;0,0,0)$, $c > 0$.

6. SOLUTION WHEN THE CONSUMPTION CONSTRAINT IS INACTIVE

In this section, we summarize the results for all cases when the consumption constraint can be ignored. Given $a \geq 0$ and $B \leq 0$, we obtained a class of feedback policies c and π given by (4.5) with $\hat{B} = 0$ and (5.1). These policies yield expected return $J(C(x;a,B);a,A,0)$, and so $J(\cdot) \leq V^*(\cdot)$. We seek particular choices of a and B for which J satisfies (4.1), and therefore, by Theorem 4.1, $J(\cdot) \geq V^*(\cdot)$. As a result we will have then explicitly obtained V^* and the optimal policies. In [2] we establish the following theorem, which shows that the appropriate choice of A is $\lambda_+ B/\rho_+$.

Theorem 6.1 For $a \geq 0$, $B \leq 0$, the function

$$V(x;a,B) \triangleq J(C(x;a,B);a,\lambda_+ B/\rho_+,0), \quad x > X(a;a,B) \tag{6.1}$$

satisfies the Bellman equation (4.1).

We can now solve the case studied by Merton [4].

Theorem 6.2 Assume $U'(0) = \infty$ and $P \leq U(0)/\beta$. Then V^* is obtained by setting $a = 0$ and $B = 0$ in (6.1).

To obtain an explicit solution in other cases, we introduce the strictly decreasing function

$$F(c) = \frac{-(U'(c))^{\rho_-}}{\gamma\lambda_-\rho_-} \int_c^\infty (U'(\theta))^{-\lambda_-} d\theta - \frac{\rho_+}{\beta} U(c) + \frac{\lambda_+}{r} cU'(c), \quad c > 0, \tag{6.2}$$

and seek positive solutions of the equation

$$F(c) = -\rho_+ P. \tag{6.3}$$

We prove in [2] that (6.3) has at most one positive solution, and this occurs if and only if $F(0) + \rho_+ P > 0$ where $F(0) \triangleq \lim_{c \downarrow 0} F(c)$. Given a positive solution, c^*, of (6.3), we take $a = c^*$ and B given by

$$B(U'(a))^{\lambda_+} + \frac{a}{r} - \frac{(U'(a))^{\lambda_-}}{\gamma\lambda_-(\lambda_+ - \lambda_-)} \int_a^\infty (U'(\theta))^{-\lambda_-} d\theta = 0. \tag{6.4}$$

Moreover, when $F(0) + \rho_+ P \leq 0$ and both $U(0)$ and $U'(0)$ are finite, we define

$$P^* = -F(0)/\rho_+ . \tag{6.5}$$

We can now state the following

Theorem 6.3

If any (i)-(iv) hold, then V^* given by (6.1) with $a = c^*$ and B is given by (6.4)

(i) $U(0) = -\infty$, P finite,

(ii) $U(0)$ finite, $U'(0) = \infty$, $P > \frac{1}{\beta}U(0)$,

(iii) $U(0)$ and $U'(0)$ finite and $P > P^*$;

(iv) $U(0)$ and $U'(0)$ finite $P = P^*$. Here we set $c^* = 0$.

7. SOLUTION WHEN THE CONSUMPTION CONSTRAINT IS ACTIVE

The only remaining case is that of finite $U'(0)$ and $P < P^*$. Under these conditions, we establish the existence of $\bar{x} > 0$ and $B \leq 0$ such that the optimal consumption is given by

$$c = \begin{cases} 0 & 0 \leq x \leq \bar{x} \\ C(x;0,B), & x \geq \bar{x} \end{cases} \tag{7.1}$$

where $C(\bar{x};0,B) = 0$ and $C(x;0,B) > 0$ for $x > \bar{x}$. When $P \leq \frac{1}{\beta} U(0)$, we will have $B = 0$; when $\frac{1}{\beta} U(0) < P < P^*$, we will have

$$B = -\frac{\beta}{\gamma(\lambda_+ - \lambda_-)\bar{y}^{\rho_+}} [P - \frac{1}{B} U(0)], \tag{7.2}$$

where $\bar{y}$ is defined by

$$\bar{y}^{\rho_-} = -\beta\lambda_- [P - \frac{1}{\beta} U(0)] [\int_0^\infty \frac{d\theta}{(U'(\theta))^{\lambda_-}}]^{-1}. \tag{7.3}$$

Also,

$$\bar{x} = B(U'(0))^{\lambda_+} - \frac{(U'(0))^{\lambda_-}}{\gamma\lambda_-(\lambda_+ - \lambda_-)} \int_0^\infty \frac{d\theta}{(U'(\theta))^{\lambda_-}} \tag{7.4}$$

In the preceding sections, we chose consumption as an intermediate variable. This is no longer appropriate as the mapping described in (7.1) is not invertible. We will therefore allow $y = \frac{d}{dx}V^*(x)$ to play the role of intermediate variable. We will discover that V^* is strictly concave, so the mapping from x to $\frac{d}{dx}V^*(x)$ is invertible. Moreover, when $x \geq \bar{x}$, we will have wealth and optimal consumption related by $\frac{d}{dx}V^*(x) = U'(C)$, so $y = U'(c)$ for $x \geq \bar{x}$.

Let us recall the function $I : (0,U'(0)] \to [0,\infty)$ which is the inverse of U'. We extend I by setting $I \equiv 0$ on $[U'(0),\infty)$. If V is C^2 and strictly concave. The Bellman equation (4.1) can be written as

$$\beta V(x) = \frac{-\gamma(V'(x))^2}{V''(x)} + [rx - I(V'(x))]V'(x) + U(I(V'(x))), \quad x > 0 \qquad (7.5)$$

By analogy with (5.8) with $a = 0$ and $c = I(y)$, we define for $A \leq 0$, $B \leq 0$,

$$\mathcal{X}(y;B) = By^{\lambda_+} + \frac{1}{r} I(y)$$
$$- \frac{1}{\gamma(\lambda_+ - \lambda_-)}\{\frac{y^{\lambda_+}}{\lambda_+}\int_0^{I(y)} (U'(\theta))^{-\lambda_+}d\theta + \frac{y^{\lambda_-}}{\lambda_-}\int_{I(y)}^{\infty} (U'(\theta))^{-\lambda_-}d\theta\}, \quad y > 0 \qquad (7.6)$$

$$\mathcal{J}(y;A) = Ay^{\rho_+} + \frac{1}{\beta} U(I(y))$$
$$- \frac{1}{\gamma(\lambda_+ - \lambda_-)}\{\frac{y^{\rho_+}}{\rho_+}\int_0^{I(y)} (U'(\theta))^{-\lambda_+}d\theta + \frac{y^{\rho_-}}{\rho_-}\int_{I(y)}^{\infty} (U'(\theta))^{-\lambda_-}d\theta\}, \quad y > 0 \qquad (7.7)$$

In [2] it is shown that $\mathcal{X}$ has an inverse

$$\mathcal{Y}(\cdot;B) : (0,\infty) \xrightarrow[\text{onto}]{} (0,\infty) \text{ if } B = 0, \text{ and}$$
$$\mathcal{Y}(\cdot;B) : [0,\infty) \xrightarrow[\text{onto}]{} (0,\bar{y}) \text{ if } B < 0. \qquad (7.8)$$

By analogy with Theorem 6.1, we have

<u>Theorem 7.1</u> Assume $P < P^*$, (7.1)-(7.4), and define

$$V(x) = \mathcal{J}(\mathcal{Y}(x;B);\frac{\lambda_+}{\rho_+}B), \quad x > 0. \qquad (7.9)$$

The function V is strictly increasing, strictly concave, satisfies the Bellman equation (7.5) and $\lim_{x \downarrow 0} V(x) = P$.

We can now state the explicit solution when the consumption constraint is active.

Theorem 7.2 Assume $U(0)$ and $U'(0)$ are finite.

(i) If $P = U(0)/\beta$, then $V^*(x) = \mathcal{J}(\mathcal{Y}(x;0);0)$, $x > 0$

(ii) If $U(0)/\beta < P < P^*$, then $V^*(x) = V(x)$ of (7.9) with B given by (7.2) and (7.3).

Theorems 6.2, 6.3, and 7.2 provide a complete explicit solution to the consumption/investment problem. These results have been specialized to the HARA utility function case in [2].

REFERENCES

[1] E. B. Dynkin, Markov Processes, Vol. II, Academic Press, New York, 1965.

[2] I. Karatzas, J. Lehoczky, S. Sethi, and S. Shreve, Explicit solution of a general consumption/portfolio problem, To appear: Math. Operations Research.

[3] J. Lehoczky, S. Sethi, and S. Shreve, Optimal consumption and investment policies allowing consumption constraints and bankruptcy, Math. Operations Research 8 (1983), 613-636.

[4] R. C. Merton, Optimum consumption and portfolio rules in a continuous time model, J. Economic Theory 3 (1971), 373-413.

[5] P. A. Samuelson, Lifetime portfolio selection by dynamic stochastic programming, Rev. Econ. Statist. 51 (1969), 239-246.

ACKNOWLEDGEMENT

Research was supported in part by the following grants: National Science Foundation MCS-8202210 to Carnegie-Mellon University (Lehoczky and Shreve), MCS-8103435 to Columbia University (Karatzas), and NSERC-A4619 and SSHRC 410-83-0888 to University of Toronto (Sethi).

ON THE ASYMPTOTIC BEHAVIOR OF SOME OPTIMAL ESTIMATES OF PARAMETERS OF NONLINEAR REGRESSION FUNCTIONS

P.S. Knopov
V.M. Glushkov Institute of Cybernetics
Ukrainian Academy of Sciences
252207 Kiev 207, USSR

This paper is concerned with periodic estimates of the unknown parameters of a given deterministic signal of known structure, observed in the presence of random noise.

We shall make the following assumptions:

1. Let $n(t)$, $t \in R^1$ be a stationary random process with trajectories $\mathrm{E}n(t) = 0$, $\mathrm{E}n(s+t)n(s) = r(t)$ which are continuous with probability 1 and satisfying the following conditions:

$$\sup_{A \in F^t_{-\infty}, B \in F^{\infty}_{t+\tau}} |\mathrm{P}(AB) - \mathrm{P}(A)\mathrm{P}(B)| \le \frac{c}{\tau^{1+\varepsilon}}, \ \tau > 0, \ \varepsilon > 0, \ c > 0 \quad ,$$

where $F^b_a = \sigma\{n(t), t \in [a, b]\}$ is the least σ-algebra generated by the random process $n(t)$, $t \in [a, b]$.

2. For some $\delta > \frac{4}{\varepsilon}$ we have $\mathrm{E}|n(t)|^{4+\delta} < \infty$.

3. Let $\varphi(t)$ be an almost periodic function of the form

$$\varphi(t) = \sum_{k=-\infty}^{\infty} C_k e^{i\lambda_k t} \quad ,$$

where C_k and λ_k satisfy the conditions

$$\sum_{k=-\infty}^{\infty} |C_k| < \infty, \ \lambda_k \ge 0 \text{ for } k \ge 0, \ \lambda_l \ge \lambda_k \text{ for } l \ge k > 0 \quad ,$$

$$C_k = \bar{C}_{-k}, \ \lambda_k = -\lambda_{-k}, \ |\lambda_k - \lambda_l| \ge \Delta > 0 \text{ for } l \ne k \quad .$$

We study the problem of estimating unknown parameters A_0 and ω_0 of the following observed random process

$$x(t) = A_0\varphi(\omega_0 t) + n(t)\,,\ t \in [0,T] \quad .$$

Consider the functional

$$Q_T(\omega) = \left|\frac{2}{T}\int_0^T x(t)e^{i\omega t}dt\right|^2$$

Let ω_T be the value of ω for which $Q_T(\omega)$ attains its maximum value. Then the following statement holds.

THEOREM 1. *Let assumptions 1–3 be satisfied and*

$$|C_{i_0}| > |C_i|\,,\ i \neq \pm i_0\,,\ i_0 > 0 \quad . \tag{1}$$

Then

$$\bar{\omega}_T = \frac{\omega_T}{\lambda_{i_0}} \longrightarrow 0 \text{ as } T \longrightarrow \infty$$

with probability 1.

Let us briefly consider the main steps in the proof. Having fixed $\omega \neq 0$, we consider the behavior of the value $Q_T(\omega)$ as $T \longrightarrow \infty$:

$$Q_T(\omega) = \left|\frac{2}{T}\int_0^T x(t)e^{i\omega t}dt\right|^2 = \left|\frac{2}{T}\int_0^T [A_0\varphi(\omega_0 t) + n(t)]e^{i\omega t}dt\right|^2 =$$

$$= \left|\frac{2}{T}\int_0^T A_0\varphi(\omega_0 t)e^{i\omega t}dt\right|^2 + I_T(\omega) \quad ,$$

$$I_T(\omega) = \frac{4}{T^2}\left|\int_0^T n(t)e^{i\omega t}dt\right|^2 + 2\mathrm{Re}\int_0^T \varphi(\omega_0 t)e^{i\omega t}\int_0^T n(t)e^{-i\omega t}dt \quad .$$

From assumption 3 we have

$$\frac{1}{T}\left|\int_0^T \varphi(\omega_0 t)e^{i\omega t}dt\right| \leq C \quad .$$

From [1] it follows that

$$\sup_\omega |I_T(\omega)| \longrightarrow 0 \text{ as } T \longrightarrow \infty \text{ with probability 1} \quad . \tag{2}$$

Set

$$\Phi_T(\omega_0, \omega) = \frac{2A_0}{T}\int_0^T \varphi(\omega_0 t)e^{i\omega t}dt \quad .$$

From (2) we have

$$\overline{\lim_{T\to\infty}} \sup_{|\lambda_{i_0}\omega_0-\omega|\geq\delta} Q_T(\omega) \leq \lim_{T\to\infty} \sup_{|\lambda_{i_0}\omega_0-\omega|\geq\delta} |\Phi_T(\omega_0,\omega)|^2 .$$

By means of simple formulae it is possible to show that for $0 < \delta < \Delta\omega_0/2$, the following inequality holds with probability 1:

$$\overline{\lim_{T\to\infty}} \sup_{|\lambda_{i_0}\omega_0-\omega|\geq\delta} |\Phi_T(\omega_0,\omega)|^2 < 4A_0^2 \, |C_{i_0}|^2 . \tag{3}$$

At the same time it is not difficult to see that with probability 1

$$\lim_{T\to\infty} Q_T(\lambda_{i_0}\omega_0) = 4A_0^2 \, |C_{i_0}|^2 .$$

Let $E = \{e\}$ be the space of elementary events and

$$\Psi = \{e\colon \overline{\lim_{T\to\infty}} \sup_{\omega\in\Phi_\delta} Q_T(\omega) < \lim_{T\to\infty} Q_T(\lambda_{i_0}\omega_0) = L \, , \, L < \infty\} ,$$

where

$$\Phi_\delta = \{\omega\colon |\lambda_{i_0}\omega_0 - \omega| \geq \delta\} \, , \, 0 < \delta < \Delta\omega_0/2 .$$

From the above considerations we have $P\{\Psi\} = 1$. Now suppose that $\bar{\omega}_T \not\mapsto \omega_0$ with probability 1. Let $\Psi_1 = \{e\colon \bar{\omega}_T \not\mapsto \omega_0\}$ and the elementary event $e \in \Psi_1 \cap \Psi$.

For this event there exists a subsequence $T_k \longrightarrow \infty$ for $k \longrightarrow \infty$ such that

$$\bar{\omega}_{T_k(e)}(e) \longrightarrow \bar{\omega}'(e) \neq \omega_0 \, , \, 0 \leq \bar{\omega}'(e) \leq \infty .$$

Take

$$0 < \delta(e) < \min\left(|\omega'(e) - \lambda_{i_0}\omega_0| \, , \, \frac{\Delta\omega_0}{2}\right) .$$

Then, from

$$\overline{\lim_{k\to\infty}} \sup_{\omega\in\Phi_{\delta(e)}} Q^{(e)}_{T_k(e)}(\omega) < \lim_{k\to\infty} Q^{(e)}_{T_k(e)}(\lambda_{i_0}\omega_0) ,$$

we have

$$\lim_{k\to\infty} Q^{(e)}_{T_k(e)}(\omega_{T_k(e)}(e)) < \lim_{k\to\infty} Q^{(e)}_{T_k(e)}(\lambda_{i_0}\omega_0) .$$

On the other hand, by definition

$$Q^{(e)}_{T_k(e)}(\omega_{T_k(e)}(e)) \geq Q^{(e)}_{T_k(e)}(\lambda_{i_0}\omega_0) ,$$

so that

$$\lim_{k\to\infty} Q_{T_k(e)}^{(e)}(\lambda_{i_0}\omega_0) = L \ , \ L > \overline{\lim_{k\to\infty}}\ Q_{T_k(e)}^{(e)}(\omega_{T_k(e)}(e)) \quad .$$

This leads to a contradiction. Therefore, $P\{\Psi_1\} = 0$ and $\overline{\omega}_T \to \omega_0$ as $T \to \infty$ with probability 1. This proves the theorem. The stronger statement is also true.

THEOREM 2. *Under the conditions of Theorem 1,*

$$T\left(\frac{\omega_T}{\lambda_{i_0}} - \omega_0\right) \to 0 \text{ as } T \to \infty \text{ with probability 1} \quad .$$

The proof is based on the relation

$$\lim_{T\to\infty} Q_T(\omega_T) = \lim_{T\to\infty} Q_T(\lambda_{i_0}\omega_0) = 4A_0\,|C_{i_0}|^2 \quad , \tag{4}$$

which holds with probability 1.

From (4) it follows that, when the conditions of Theorem 1 are satisfied, the value

$$A_T = \frac{1}{2}\,|C_{i_0}|^{-1}\,Q_T^{1/2}(\omega_T)$$

represents a strongly consistent estimate of the parameter A_0.

Now we shall turn our attention to the asymptotic distribution of the values ω_T and A_T.

THEOREM 3. *Let assumptions 1–3, condition (1) and* $f(\lambda_{i_0}\omega_0) > 0$ *be satisfied, where* $f(\lambda)$ *is the spectral density of the random process* $n(t)$*. Then the value*

$$T^{3/2}(\omega_T - \lambda_{i_0}\omega_0)$$

is asymptotically normal with a mean value of zero and dispersion

$$\sigma^2 \doteq 3\cdot 2^6 \pi A_0^{-2}\,|C_{i_0}|^{-2} f(\lambda_{i_0}\omega_0) \quad .$$

The proof is based on two lemmas which we state without proof.

LEMMA 1. *Let the conditions of Theorem 3 be satisfied. Then we can write*

$$T^{-1/2}\,\frac{\partial Q_T(\lambda_{i_0}\omega_0)}{\partial\omega} = \zeta_{T1} + \zeta_{T2} \quad ,$$

where ζ_{T1} *is an asymptotically normal random value with zero mean value and dispersion*

$$\sigma^2 = \frac{16}{3}\pi_1 A_0^2\,|C_{i_0}|^2 f(\lambda_{i_0}\omega_0) \quad ,$$

and $\zeta_{T2} \to 0$ *as* $T \to \infty$ *by probability.*

LEMMA 2. *Let assumptions 1–3 and condition (1) be satisfied. Then for any random value* $\check{\omega}_T$ *satisfying the inequality*

$$|\check{\omega}_T - \lambda_{i_0}\omega_0| < |\omega_T - \lambda_{i_0}\omega_0|$$

with probability 1, we have

$$\frac{1}{T^2}\frac{\partial^2 q_T(\check{\omega}_T)}{\partial\omega^2} \to -\frac{1}{6}|C_{i_0}|^2 \,, \; T \to \infty$$

by probability.

Now let us prove the statement given in Theorem 3.

Since $\omega_T/\lambda_{i_0} \to \omega_0$ as $T \to \infty$ with probability 1, ω_T will be an inner point of the semi-axis $(0, \infty)$ with probability approaching 1 for $T \to \infty$. With the same probability $Q_T'(\omega_T) = 0$, and the equality

$$Q_T'(\lambda_{i_0}\omega_0) + Q_T''(\check{\omega}_T)(\omega_T - \lambda_{i_0}\omega_0) = 0 \tag{5}$$

holds, where some random value ω_T satisfies the inequality

$$|\check{\omega}_T - \lambda_{i_0}\omega_0| \le |\omega_T - \lambda_{i_0}\omega_0| \,, \; T > 0 \quad ,$$

with probability 1.

From (5) it follows that

$$\omega_T - \lambda_{i_0}\omega_0 = -\frac{Q_T'(\lambda_{i_0}\omega_0)}{Q_T''(\check{\omega}_T)} \quad . \tag{6}$$

Equality (6) is equivalent to

$$T^{3/2}(\omega_T - \lambda_{i_0}\omega_0) = \frac{T^{-1/2}Q_T'(\lambda_{i_0}\omega_0)}{T^{-2}Q_T''(\check{\omega}_T)} \quad . \tag{7}$$

By virtue of Lemma 2, the denominator of the right-hand side of (7) tends by probability to the value

$$A = -\frac{|C_{i_0}|^2}{6} A_0^2 \quad .$$

Making use of Lemma 1, we obtain confirmation of the theorem.

Using Lemma 2 and Theorems 2 and 3 it is possible to prove a similar result for the estimate A_T. Let us formulate this result.

THEOREM 4. *Let assumptions 1–3, condition (1) and $f(\lambda_{i_0}\omega_0) > 0$ be satisfied. Then the value*

$$\xi_T = \sqrt{T}(A_T - A_0)$$

is asymptotically normal with parameters $(0\,,\ \pi\,|C_{i_0}|^{-2} f(\lambda_{i_0}\omega_0))$.

Now consider periodic estimates of the second type obtained by maximizing the functional

$$\tilde{Q}_T(\omega) = \left|\frac{2}{T}\int_0^T x(t)\,\varphi(\omega t)dt\right|^2 .$$

We shall assume that the unknown parameter ω satisfies $\omega \in (\underline{\omega}\,,\overline{\omega})$, $\underline{\omega} > 0$, $\overline{\omega} < \infty$.

We choose the value $\tilde{\omega}_T \in [\underline{\omega}\,,\overline{\omega}]$ as an estimate of ω_0, where $\tilde{Q}_T(\omega)$ represents the maximum value. We now introduce some statements regarding the asymptotic behaviour of the estimates $\tilde{\omega}_T$ as $T \to \infty$. The proofs of these statements follow the same pattern as those for the values of ω_T.

THEOREM 5. *Let assumptions 1–3 be satisfied. Then, with probability 1, $T(\tilde{\omega}_T - \omega_0) \to 0$ as $T \to \infty$.*

THEOREM 6. Let assumptions 1–3 be satisfied, and

$$\sum_{j=-\infty}^{\infty} |\lambda_j C_j| < \infty\,,\quad \sum_{j=-\infty}^{\infty} |C_j|^2 f(\lambda_j\omega_0) > 0 .$$

Then the value

$$\xi_T = T^{3/2}(\tilde{\omega}_T - \omega_0)$$

is asymptotically normal with zero mean and dispersion

$$\sigma^2 = 6\pi A_0^{-2}\Big[\sum_{\nu=-\infty}^{\infty} \lambda_\nu^2 |C_\nu|^2\Big]^{-1} \sum_{j=-\infty}^{\infty} \lambda_j^2 |C_j|^2 f(\lambda_j\omega_0) .$$

Let

$$\tilde{A}_T = \frac{1}{2}\Big(\sum_{\nu=-\infty}^{\infty} |C_\nu|^2\Big)^{-1}\tilde{Q}_T^{1/2}(\tilde{\omega}_T) .$$

The above statements about strong consistency and asymptotic normality also hold for $\tilde{A}_T$.

THEOREM 7. *Let assmptions 1–3 be satisfied. Then the value $\tilde{A}_T$ is a strongly consistent estimate of the parameter A_0.*

THEOREM 8. *Let the conditions of Theorem 6 be fulfilled. Then the value*

$$\alpha_T = \sqrt{T}(\tilde{A}_T - A_0)$$

is asymptotically normal with zero mean and dispersion

$$\sigma^2 = 2\pi \sum_{k=-\infty}^{\infty} |C_k|^2 f(\lambda_k \omega_0)\left(\sum_{k=-\infty}^{\infty} |C_k|^2\right)^{-2} .$$

Now consider periodic estimates of the parametrs in the space R^2. We shall make the following assumptions:

4. Let $n(s,t)$, $(s,t) \in R^2$ be a homogeneous random field such that $En(s,t) = 0$, and which satisfies the following conditions [2]:

$$\sup_{\substack{A \in \Theta(s) \\ B \in \Theta(F)}} |P(AB) - P(A)P(B)| \le \Psi(d(s,F)) \le \frac{C}{d^{2+\varepsilon}} ,$$

where $\Theta(s)$ is a σ-algebra generated by a random field

$$n(s,t),\ (s,t) \in S,\ d(s,F) = \inf\{||x-y||,\ x \in S,\ y \in F\},\ \Psi(d) \downarrow 0,\ d \to \infty .$$

Here $||x - y||$ is the euclidean distance between the elements x and y.

5. For some $\delta > \frac{8}{\varepsilon}$ we have $E|n(s,t)|^{4+\delta} < \infty$.

6. $\varphi(s,t)$ is a real function which is 2π-periodic with respect to both variables and takes the form:

$$\varphi(s,t) = \sum_{k,l=-\infty}^{\infty} C_{kl} e^{i(ks+lt)} .$$

where

$$\sum_{k,l=-\infty}^{\infty} |C_{kl}| < \infty .$$

Consider a random field

$$x(s,t) = A_0 \varphi(\omega_{10}s, \omega_{20}t) + n(s,t)$$

observed in the domain $D_T = [0,T] \times [0,T]$.

It is necessary to estimate $\omega_0 = (\omega_{10}, \omega_{20})$ from the observation of $x(s,t)$ in D_T. Consider the functionals

$$Q_T(\omega_1, \omega_2) = \left|\frac{4}{T_2} \int_0^T \int_0^T x(s,t) e^{i(\omega_1 s + \omega_2 t)} ds\,dt\right|^2 .$$

Let $\omega_T = (\omega_{1T}, \omega_{2T})$ be the value of $\omega = (\omega_1, \omega_2)$ in D_T at which the functional $Q_T(\omega_1, \omega_2)$ attains its maximum value. Then the following statement holds:

THEOREM 9. *Let assumptions 4–6 and*

$$|C_{i_0 k_0}| > |C_{ik}| \, , \, (i_0, k_0) \neq \pm (i, k) \, , \, i_0 > 0 \, , \, k_0 > 0 \tag{8}$$

be satisfied.

Then

$$T\left(\frac{\omega_{1T}}{i_0} - \omega_{10}\right) \to 0 \, , \, T\left(\frac{\omega_{2T}}{k_0} - \omega_{20}\right) \to 0 \text{ as } T \to \infty$$

with probability 1.

Let

$$A_T = \frac{1}{4} |C_{i_0 k_0}|^{-1} Q_T^{1/2}(\omega_{1T}, \omega_{2T}) \quad .$$

THEOREM 10. *Let assumptions 4–6 and condition (8) be satisfied. Then $A_T \to A_0$ as $T \to \infty$ with probability 1.*

We shall now consider periodic estimtes of the second type.

Let

$$\tilde{Q}_T(\omega_1, \omega_2) = \left| \frac{4}{T^2} \int_0^T \int_0^T x(s, t) \varphi(\omega_1 s, \omega_2 t) ds\, dt \right|^2 \quad .$$

Assume that the unknown two-dimensional parameter $\omega_0 = (\omega_{10}, \omega_{20})$ belongs to the domain $\Omega = \{\omega: (\omega_1, \omega_2), \ 0 < \underline{\omega}_1 < \omega_1 < \bar{\omega}_1 < \infty, \ 0 < \underline{\omega}_2 < \omega_2 < \bar{\omega}_2 < \infty\}$. As an estimate of $\omega_0 \in \Omega$ we take the value of $\tilde{\omega}_T \in \bar{\Omega} = \{\omega = (\omega_1, \omega_2), \ 0 < \underline{\omega}_1 \leq \omega_1 \leq \bar{\omega}_1 < \infty, \ 0 < \underline{\omega}_2 \leq \omega_2 \leq \bar{\omega}_2 < \infty\}$ for which the functional $\tilde{Q}_T(\omega_1, \omega_2)$ attains its maximum value.

THEOREM 11. *Let assumptions 4–6 be satisfied. Then $\tilde{\omega}_T \to \omega_0$ as $T \to \infty$ with probability 1.*

Let

$$\tilde{A}_T = \frac{1}{4} \left[\sum_{k,j=-\infty}^{\infty} |C_{k_j}|^2 \right]^{-1} Q_T^{1/2}(\tilde{\omega}_{1T}, \tilde{\omega}_{2T}) \quad .$$

THEOREM 12. *Let assumptions 4–6 be satisfied. Then $\tilde{A}_T \to A_0$ as $T \to \infty$ with probability 1.*

The proofs of the statements made in Theorems 9–12 are analogous to those in the one-dimensional case. Theorems on the asymptotic normality of the above estimates also exists, but their formulations are much more unwieldy and we shall not consider them here.

REFERENCES

1. A.V. Ivanov. On the solution of a problem concerning the detection of latent periodicities (in Russian). *Theory of Probabilities and Mathematical Statistics*, 20(1979) 44–60.

2. N.N. Leonenko and M.I. Jadrenko. Central limit theorem for homogeneous and isotropic random fields (in Russian). *DAN of the Ukr. SSR, Ser.A*, 4(1975) 314–316.

ON THE ε-OPTIMAL CONTROL OF A STOCHASTIC INTEGRAL EQUATION WITH AN UNKNOWN PARAMETER

A.M. Kolodiy
Institute of Applied Mathematics and Mechanics,
Ukrainian Academy of Sciences, Donezk, USSR

1. INTRODUCTION

In this paper we shall consider an ε-optimal control problem leading to the solution of Ito–Volterra stochastic integral equations with coefficients which do not depend on a random parameter. The solution of this control problem will be obtained by means of an auxiliary optimal control problem which can be constructed for any controlled process in discrete time. This method is similar to that used in [1] to construct an ε-optimal control function for solution of stochastic differential equations with known coefficients.

2. NOTATION

We shall use the following notation:

$(\Omega, \mathbf{F}, P)$ is a complete probability space.

$(\mathbf{F}_t)_{t\geq 0}$ is an increasing right-continuous sequence of complete σ-algebras of $\mathbf{F}$.

$(w(t), t \geq 0)$ is an m-dimensional $(\mathbf{F}_t)$-Wiener process.

$(\nu(t, A), t \geq 0, A \in B(R^d \setminus \{0\}))$ is an $(\mathbf{F}_t)$-Poisson measure with $\mathrm{E}\nu(t, A) = tq(A)$ [1]. Assume that w and ν are mutually independent. Let $\tilde{\nu}(t, A) = \nu(t, A) - tq(A)$.

D is the space of all cadlag functions $g\colon [0,T] \to R^d$ (g is a cadlag function if it is continuous on the right and has finite limits on the left; $\|g\|_t = \sup_{r \leq t} |g(r)|$;

$$|||g|||_t = \{\int_0^t |g(s)|^2 \mathrm{d}K(s)\}^{1/2} + \sum_{i=1}^{\infty} c_i \, |g((t-s_i)^+)| \quad ,$$

where $K(\cdot)$ is a monotonically non-decreasing right-continuous function, $c_i \geq 0$, $\sum c_i = 1$, $s_i \in [0,T]$, $K = K(T) - K(0) < \infty$.

α_t is the minimal σ-algebra generated by cylinder sets in D with bases over $[0,T]$.

$\varphi(t)$ is a strongly increasing continuous function with $\varphi(0)=0$ and $\sum_1^\infty 2^{n/p}\varphi(2^{-n})<\infty$, where $p>1$.

H_p, $p>1$, is the space of $(\mathbf{F}_t)$-adapted processes ξ with trajectories in D and for which $\mathrm{E}\|\xi\|_t^p<\infty$.

If $f(\alpha,y)$ is any function with $y\in R^d$, $\alpha\in\mathbf{A}$, where $\mathbf{A}$ is an arbitrary set, then

$$[f(\alpha)]_p=\{\int|f(\alpha,y)|^2q(dy)\}^{1/2}+\{\int|f(\alpha,y)^pq(dy)\}^{1/p} \quad .$$

3. THE EXISTENCE AND UNIQUENESS OF A SOLUTION

Let ξ be any process. Then we define $\omega_{t,p}(\Delta,\xi)$ for arbitrary $\Delta>0$, $p>1$, $t>0$ as follows:

$$\omega_{t,p}(\Delta,\xi)=\sup\{(\mathrm{E}\,|\xi(t_1)-\xi(t_2)|^p)^{1/p}:|t_1-t_2|\le\Delta\,;\,t_1vt_2\le t\} \quad .$$

Utilizing methods from [2] we obtain the following assertion:

LEMMA 1. *Let ξ be a measurable process with*

$$\int_0^t(\mathrm{E}\,|\xi(s)|^p)^{1/p}ds<\infty\,;\,\sum_1^\infty 2^{n/p}\omega_{t,p}(2^{-n},\xi)<\infty$$

for any $p>1$ and all $t>0$. Then process ξ has a continuous modification $\tilde\xi$ and

$$(\mathrm{E}\|\tilde\xi\|_t^p)^{1/p}\le c\int_0^t(\mathrm{E}\,|\xi(s)|^p)^{1/p}ds+L_p(t)\sum_1^\infty 2^{n/p}\omega_{t,p}(2^{-n},\xi) \quad ,$$

where c is any positive constant and $L_p(t)$ is a monotonically non-decreasing (for fixed p) positive function.

The proof of Lemma 1 may easily be obtained from the proof of Theorem 2 in [2] with some simple transformations.

LEMMA 2. *Let $\beta(t,s)$ and $\gamma(t,s,y)$ be random, measurable, $\mathbf{F}_s$-adapted functions with values in $R^d\otimes R^m$ and R^d, respectively. Assume that*

$$|\beta(t,s)|+[\gamma(t,s)]_p\le F(t,s)$$

$$|\beta((t-\Delta)^+,s)-\beta(t,s)|+[\gamma((t-\Delta)^+,s)-\gamma(t,s)]_p\le F(t,s)\varphi(\Delta) \quad ,$$

where $F(t,s)$ is a random, measurable function which is monotonically non-decreasing in t and such that $\mathrm{E}\int_0^t F^p(t,s)ds<\infty$. Then the process

$$\zeta(t) = \int_0^t \beta(t,s)dw(s) + \int_0^t \int \gamma(t,s,y)\tilde{\nu}(ds,dy)$$

has a modification with cadlag trajectories and

$$E\|\zeta\|_t^p \leq c(t)E\int_0^t F^p(t,s)ds \quad ,$$

where $c(t) = c(t,p,\varphi(\cdot))$ *is a locally bounded function.*

Proof. It is well known [3,4] that the trajectories of the process

$$\kappa(t) = \int_0^t \beta(s,s)dw(s) + \int_0^t \int \gamma(s,s,y)\tilde{\nu}(ds,dy)$$

are càdlàg and that

$$E\|\kappa\|_t^p \leq c'(t)E\int_0^t (|\beta(s,s)|^p + [\gamma(s,s)]_p^p)ds \quad ,$$

where $c'(t) = c'(t,p)$ is a locally bounded function. It may easily be proved that the process $\kappa_1 = \zeta - \kappa$ satisfies the conditions of Lemma 1. Thus κ_1 is continuous and

$$E\|\kappa_1\|_t^p \leq c''(t)E\int_0^t F^p(t,s)ds \quad ,$$

where $c''(t) = c''(t,p,\varphi(\cdot))$ is a locally bounded function.

Consider the stochastic integral equation

$$\xi(t) = \xi(0) + \int_0^t a(t,s,\xi)ds + \int_0^t b(t,s,\xi)dw(s) + \int_0^t \int c(t,s,\xi,y)\tilde{\nu}(ds,dy)\,,\ t \in [0,T] \quad , \tag{1}$$

where $\xi(0)$ is an $\mathbf{F}_0$-measurable d-dimensional vector; random functions $a(t,s,g)$, $b(t,s,g)$ and $c(t,s,g,y)$ are measurable on the combination of the variables, and for all t, s as functions of $(g,\omega) \in D \times \Omega$ are measurable with respect to the σ-algebra $\alpha_s \times \mathbf{F}_s$; $a(\cdot)$ and $c(\cdot)$ take values in R^d; $b(\cdot)$ takes values in $R^d \otimes R^m$.

Using Lemma 2 and the usual methods employed in theorems on the existence and uniqueness of solutions of stochastic integral equations [3,5], it is possible to prove the following theorem:

THEOREM 1. *Assume that there exists a function* φ *and a number* $p \geq 2$ *which satisfy the conditions given in Section 2 and in addition:*

(a) $\mathrm{E}\,|\xi(0)|^p < \infty$;

(b) $|a(t,s,g)| + |b(t,s,g)| + [c(t,s,g)]_p \le L(1+\|g\|_s)$;

$|a(t+\Delta,s,g) - a(t,s,g)| + |b(t+\Delta,s,g) - b(t,s,g)| +$

$+ [c(t+\Delta,s,g) - c(t,s,g)]_p \le L(1+\|g\|_s)\varphi(\Delta)$;

(c) $|a(t,s,g) - a(t,s,g')| + |b(t,s,g) - b(t,s,g')| +$

$+ [c(t,s,g) - c(t,s,g')]_p \le L\|g-g'\|_s$;

$|b(t+\Delta,s,g) - b(t+\Delta,s,g') - b(t,s,g) + b(t,s,g')| + [c(t+\Delta,s,g) -$

$- c(t+\Delta,s,g') - c(t,s,g) + c(t,s,g')]_p \le L\|g-g'\|_s\varphi(\Delta)$.

Then there exists one and only one process $\xi \in H_p$ which satisfies the stochastic integral equation (1).

4. THE ε-OPTIMAL CONTROL

Consider a controlled process ξ with trajectories in D, which satisfies the stochastic integral equation

$$\xi(t) = J(t,\xi,\eta) \overset{\Delta}{=} \xi(0) + \int_0^t \lambda(t,s,\xi,\eta(s),\zeta,ds)\,,\ t \in [0,T]\quad, \tag{2}$$

where

$$\lambda(t,s,g,u,z,ds) = a(t,s,g,u,z)ds + b(t,s,g,u,z)dw(s) +$$

$$+ \int c(t,s,g,u,z,y)\tilde{\nu}(ds,dy)\quad;$$

η is a control process with values in U (U is a compact subset of R^m); ζ is an $\mathbf{F}_0$-measurable variable with values in a metric, complete, separable space Z and with a known distribution $p(\cdot)$; functions $a(t,s,g,u,z)$, $b(t,s,g,u,z)$ and $c(t,s,g,u,z,y)$ are non-random, measurable on the combination of the variables, and for all (t,s,y) as functions of (g,u,z) are measurable with respect to the σ-algebra $\alpha_s \times B(U) \times B(Z)$. Assume either that $\xi(0)$ and ζ are independent or that the conditional distribution of $\xi(0)$ with respect to ζ has a continuous and strongly positive density.

Suppose that $a(\cdot)$, $b(\cdot)$ and $c(\cdot)$ satisfy the conditions (a) and (b) of Theorem 1 uniformly in (u,z) and that

$$|a(t,s,g,u,z) - a(t,s,g',u',z)| + |b(t,s,g,u,z) - b(t,s,g',u',z)| +$$

$$+ [c(t,s,g,u,z) - c(t,s,g',u',z)]_p \le L(\|g-g'\|_s + |u'-u|)$$

$$|b(t+\Delta,s,g,u,z) - b(t+\Delta,s,g',u',z) - b(t,s,g,u,z) + b(t,s,g',u',z)| +$$

$$+[c(t+\Delta,s,g,u,z)-c(t+\Delta,s,g',u',z)-c(t,s,g,u,z)+c(t,s,g',u',z)]_p \le$$

$$\le L(|||g-g'|||_s + |u-u'|)\varphi(\Delta) \quad .$$

Using Lemma 2 it can easily be proved that if (i) ξ and ξ' belong to H_p and (ii) η and η' are measurable, $(\mathbf{F}_t)$-adapted processes with $\mathrm{E}\int_0^T (|\eta(t)|+|\eta'(t)|)^p dt < \infty$, then

$$\mathrm{E}|||J(\cdot,\xi,\eta)|||_t^p \le c_1 + c_2 \int_0^t \mathrm{E}|||\xi|||_s^p ds \tag{3}$$

$$\mathrm{E}|||J(\cdot,\xi,\eta) - J(\cdot,\xi',\eta')|||_t^p \le c_3 \int_0^t \mathrm{E}(|||\xi-\xi'|||_s^p + |\eta(s)-\eta'(s)|^p) ds \quad , \tag{4}$$

where $c_i = c_i(L, T, p, \varphi(\cdot)) \ge 0$, $i = 1,2,3$.

Let F denote the class of all functions $f(t,g)$: $[0,T] \times \mathbf{D} \to U$ which are left-continuous in t and α_t-measurable in g (for t fixed) and which satisfy the Lipschitz condition with respect to the seminorm $|||\cdot|||$.

Let δ denote a subdivision of the interval $[0,T]$ with dividing points $t_0 = 0 < t_1 < \cdots < t_n = T$, and let $[\delta] = \max\,(t_{i+1}-t_i)$. Let F_δ denote the class of functionals

$$f_\delta(t,g) = f_0(g) I_{\{0\}}^{(t)} + \sum_{i=1}^{n-1} f_i(g) I_{]t_i,t_{i+1}]}^{(t)} \quad ,$$

where the $f_i(g)$ are α_{t_i}-measurable with values in U.

For an arbitrary δ and $f_\delta \in F_\delta$ we define

$$\hat{\xi}(t) = \xi_i \,,\ t \in [t_i, t_{i+1}[\,,\ i = \overline{0,n-1} \,;\ \hat{\xi}(T) = \xi_n$$

$$\xi_0 = \xi(0) + |\delta|^3 \mu_0 \chi \,;\ \xi_i = J(t_i, f_\delta(\cdot,\hat{\xi})) + |\delta|^3 (\chi\mu_0 + \sum_{r=0}^{i-1} \mu_{ir}) \,,\ i = \overline{1,n} \tag{5}$$

$$\bar{\xi}(t) = J(t,\hat{\xi},f_\delta(\cdot,\hat{\xi})) + |\delta|^3 (\chi\mu_0 + \sum_{i=1}^{n-1}\sum_{r=0}^{i-1} \mu_{ir} I_{[t_i,t_{i+1}[}^{(t)} + \sum_{r=0}^{n-1} \mu_{n,r} I_{\{T\}}^{(t)}) \quad , \tag{6}$$

where μ_0 and $\mu_{i,r}$ are mutually independent, normally-distributed random vectors with zero means and single covariance matrixes; $\chi = 1$ if $\xi(0)$ and ζ are independent and $\chi = 0$ otherwise.

Let $\tilde{\mathbf{U}}$ denote the space of processes $\eta(t) = f(t,\xi)$, where $f \in F$ and ξ is a solution of the equation $\xi(t) = J(t,\xi,f(\cdot,\xi))$. Denote by $\mathbf{U}_\delta$ the space of processes

$\eta(t) = f_\delta(t, \hat{\xi})$, where $f_\delta \in F_\delta$ and $\hat{\xi}$ is defined by (5). Let $\mathbf{U}_0 = \bigcup_\delta \mathbf{U}_\delta$. The class of admissible controls $\mathbf{U}$ is given by $\mathbf{U} = \tilde{\mathbf{U}} \cup \mathbf{U}_0$.

For each $\varepsilon > 0$ we wish to find a control $\eta_\varepsilon \in \mathbf{U}$ such that $T(\eta_\varepsilon) < \inf \{T(\eta), \eta \in \mathbf{U}\} + \varepsilon$ where $T(\eta) = E\,\Psi(\xi, \eta)$ is the cost functional and $\Psi(\cdot,\cdot)$ is any measurable functional. Assume that $\Psi(\cdot,\cdot)$ is bounded and continuous with respect to the metric

$$\rho[(g, f), (g', f')] = \|g - g'\|_T + \int_0^T |f(t) - f'(t)|\,dt \quad .$$

LEMMA 3. *Let $\hat{\xi}$ and $\bar{\xi}$ be defined by* (5) *and* (6), *and ξ^δ be a solution of the equation $\xi^\delta(t) = J(t, \xi^\delta, f_\delta(\cdot, \hat{\xi}))$. Then*

$$E\,\|\xi^\delta - \bar{\xi}\|_T^p \le c(|\delta| \vee \varphi^p(|\delta|)) \quad ,$$

where c is independent of δ and f_δ.

Proof. From (3) we have $\|\hat{\xi}\|_t^p \le \|\bar{\xi}\|_t^p \le c'$. Looking at the inequalities for the moments of stochastic integrals in [3], we have $\sup_t E\,|\hat{\xi}(t) - \bar{\xi}(t)|^p \le c''(|\delta| \vee \varphi^p(|\delta|))$. Moreover, c' and c'' are independent of δ and f_δ. It follows from inequality (4) that

$$E\,\|\xi^\delta - \bar{\xi}\|_t^p \le c_3(K^{1/2}+1)^p \int_0^t E\,\|\xi^\delta - \bar{\xi}\|_s^p ds \; +$$

$$+\, 3^{p-1} c_3 T(K^{p/2}+1) \sup_t E\,|\hat{\xi}(t) - \bar{\xi}(t)|^p \quad .$$

The application of Gronwall's lemma completes the proof.

THEOREM 2. $\inf \{T(\eta), \eta \in \mathbf{U}\} = \inf \{T(\eta), \eta \in \mathbf{U}_0\}$.

Proof. Let $\eta(t) = f(t, \xi)$ be any control from $\tilde{\mathbf{U}}$. We define f_δ as $f_\delta(t, g) = f(t_i, g)$ for $t \in\,]t_i, t_{i+1}]$, $i = \overline{0, n-1}$, $f_\delta(0, g) = f(0, g)$. Then $f_\delta \in F_\delta$ and $f_\delta(t, g) \to f(t, g)$ as $|\delta| \to 0$. For δ and f_δ we define $\hat{\xi}$ and ξ^δ using (5) and the equation $\xi^\delta(t) = J(t, \xi^\delta, f_\delta(\cdot, \hat{\xi}))$. Let $\eta^\delta(t) = f_\delta(t, \hat{\xi})$. From inequality (4) we obtain

$$E\,\|\xi - \xi^\delta\|_t^p \le \hat{c}\Big(\int_0^t E\,|f(s, \xi) - f_\delta(s, \xi)|^p ds + \int_0^t E\,\|\xi - \xi^\delta\|_s^p ds \; +$$

$$+ \sup_t E\,|\xi^\delta(t) - \hat{\xi}(t)|^p\Big) \quad ,$$

where $\hat{c}$ is independent of η and δ. Therefore

$$\lim_{|\delta| \to 0} E\,\|\xi - \xi^\delta\|_T^p = 0 \, , \; p - \lim_{|\delta| \to 0} |\eta(t) - \eta^\delta(t)| = 0 \quad .$$

Hence $\lim_{|\delta|\to 0} \mathrm{E}\,|\Psi(\xi^\delta,\eta^\delta)-\Psi(\xi,\eta)| = 0$. □

For fixed δ we consider the controlled process $(\xi_i,\eta_i)_{i=\overline{0,n}}$ with performance index $\mathrm{E}\Psi_\delta(\xi_0,\dots,\xi_n,\eta_0,\dots,\eta_n)$, where (ξ_i) is defined by (5), $\eta_i = f_i(\hat{\xi})$; $\Psi_\delta(x_0,\dots,x_n,u_0,\dots,u_n) = \Psi(\hat{x},\hat{u})$; $\hat{x}(t) = x_i$ for $t \in [t_i, t_{i+1}[$, $\hat{x}(T) = x_n$; $\hat{u}(t) = u_i$ for $t \in]t_i, t_{i+1}]$, $\hat{u}(0) = u_0$.

Let (ξ_i^u) be the sequence defined by (5) for $f_\delta(t,\cdot) = \hat{u}(t)$. Define the random sequences $\tilde{\xi}_{j,i}$ and $\xi_{j,i}$ as follows:

$$\tilde{\xi}_{ji} = x_{ji} + \int_{t_i}^{t_{i+1}} \lambda_j(s, x_0,\dots,x_i,u_i,z,ds) + |\delta|^3\mu_{ji}\ ,\ j=\overline{i+1,n}\ ,\ i=\overline{0,n-1}$$

$$\tilde{\xi}_{i+1,i} = \xi_{i+1,i}\ ,\ i=\overline{0,n-1}$$

$$\xi_{ji} = x_{ji} + \int_{t_i}^{t_j} \lambda_j(s, x_0,\dots,x_i,\xi_{i+1,i},\dots,\xi_{j-1,i};\hat{u}(s),z,ds) +$$

$$+\ |\delta|^3 \sum_{r=i}^{j-1} \mu_{jr}\ ,\ j=\overline{i+2,n}\ ,\ i=\overline{0,n-2}\ .$$

Here and elsewhere $\lambda_j(s,x_0,\dots,x_{j-1},\hat{u}(s),z,ds) = \lambda(t_j,s,\hat{x},\hat{u}(s),z,ds)$, $x_{i,0} = x_0$; $x_{i,i} = x_i$. Using the independence of $\xi_{j,0}$ and (ξ_0,ζ), and the independence of $\xi_{j,i+1}$ and $\tilde{\xi}_{j,i}$, we obtain

$$\mathrm{P}\{\xi_j^u \in A_j\ ;\ j=\overline{0,r}\} = \int_Z \int_{A_0} \mathrm{P}\{\xi_{j,0}\in A_j\ ;\ j=\overline{1,r}\}\mathrm{P}\{\xi_0\in dx_0 \mid \zeta = z\} = p(dz)$$

for $r=\overline{1,n}$ and

$$\mathrm{P}\{\xi_{j,i} \in A_j\ ;\ j=\overline{i+1,r}\} =$$

$$= \int_{R^d}\cdots\int_{R^d}\int_{A_{i+1}} \mathrm{P}\{\xi_{j,i+1}\in A_j\ ;\ j=\overline{i+2,r}\}\mathrm{P}\{\tilde{\xi}_{j,i}\in dx_{j,i+1}\ ;\ j=\overline{i+1,r}\}$$

for $r=\overline{2,n}$; $i=\overline{0,r-2}$. Therefore

$$\mathrm{P}\{\xi_j^u\in A_j\ ;\ j=\overline{0,r}\} = \int_Z\int_{A_0}\int_{R^{(r-1)d}}\int_{A_1}\int_{R^{(r-2)d}}\cdots$$

$$\cdots\int_{A_{r-1}}\int_{A_r}\Big(\prod_{i=r-1}^{0}\mathrm{P}\{\tilde{\xi}_{j,i}\in dx_{j,i+1}\ ;\ j=\overline{i+1,r}\}\Big)\mathrm{P}\{\xi_0\in dx_0 \mid \zeta=z\}p(dz)\ .$$

Let $\rho_0(x_0\mid z)$ denote the density of the conditional distribution ξ_0 with respect to ζ. This density is obviously continuous in x_0 and strongly positive.

It can easily be verified that the joint distribution

$$p_{i,r}(z, u_i, x_0, \dots, x_i; x_{i+1,i}; \dots; x_{r,i}; A_{i+1}, \dots, A_r) = \mathrm{P}\{\tilde{\xi}_{j,i} \in A_j; j = \overline{i+1,r}\}$$

has a continuous (for fixed z) and strongly positive density

$$\rho_{i,r}(z, u_i, x_0, \dots, x_i; x_{i+1,i}; \dots; x_{r,i}; x_{i+1,i+1}; \dots; x_{r,i+1}) \quad .$$

The conditional distributions

$$p_r(A \mid x_0, \dots, x_{r-1}, u_0, \dots, u_{r-1}) = \mathrm{P}\{\xi_r^u \in A \mid \xi_0 = x_0, \xi_1^u = x_1, \dots, \xi_{r-1}^u = x_{r-1}\}$$

can be found as the Radon–Nikodym density of the measure. $\mathrm{P}\{\xi_r^u \in A; \xi_i^u \in dx_i; i = \overline{0,r-1}\}$ with respect to the measure $\mathrm{P}\{\xi_i^u \in dx_i; i = \overline{0,r-1}\}$. Therefore the functions

$$p_1(A \mid x_0, u_0) = \int_Z \int_A \rho_{01}(z, u_0, x_0, x_1)\rho_0(x_0 \mid z)dx_1 p(dz) \cdot \{\int_Z \rho_0(x_0 \mid z)p(dz)\}^{-1}$$

$$p_2(A \mid x_0, u_0, x_1, u_1) = \int_Z \int_{R^d} \int_A \rho_{02}(z, u_0, x_0, x_1; x_{2,1})\rho_{1,2}(z, u_1, x_0, x_1; x_{2,1}; x_2) \times$$

$$\times \rho_0(x_0 \mid z)dx_2 dx_{2,1} p(dz) \cdot \{\int_Z \rho_{0,1}(z, u_0, x_0, x_1)\rho_0(x_0 \mid z)p(dz)\}^{-1}$$

$$p_r(A \mid x_0, \dots, x_{r-1}, u_0, \dots, u_{r-1}) = \int_Z \int_{R^{dr(r-1)/2}} \int_A (\prod_{i=0}^{r-1} \rho_{i,r}(z, u_i, x_0, \dots, x_i;$$

$$x_{i+1,i}; \dots; x_{r,i}; x_{i+1,i+1}; \dots; x_{r,i+1}))\rho_0(x_0 \mid z)dx_r (\prod_{i=0}^{r-2} \prod_{j=i+2}^{r} dx_{j,i+1})p(dz) \times$$

$$\times \{\int_Z \int_{R^{d(r-1)(r-2)/2}} (\prod_{i=0}^{r-2} \rho_{i,r-1}(z, u_i, x_0, \dots, x_i; x_{i+1,i}; \dots; x_{r-1,i};$$

$$x_{i+1,i+1}; \dots; x_{r-1,i+1}))\rho_0(x_0 \mid z)(\prod_{i=0}^{r-3} \prod_{j=i+2}^{r-1} dx_{j,i+1})p(dz)\}^{-1}, \; r = \overline{3,n}$$

satisfy the following condition: $\int h(x)p_r(dx \mid x_0, \dots, x_{r-1}, u_0, \dots, u_{r-1})$ is a continuous function for an arbitrary bounded and continuous function $h(x)$.

Now, if (ξ_i) corresponds to the control $\eta_i = f_i(\hat{\xi})$, then by analogy with previous arguments we can find $\mathrm{P}\{\xi_r \in A \mid \xi_0 = x_0, \dots, x_{r-1} = x_{r-1}\}$. At this point we can note that if $u_i = f_i(\hat{x}) = \bar{f}_i(x_0, \dots, x_i)$, then

$$\mathrm{P}\{\xi_r \in A \mid \xi_0 = x_0, \dots, \xi_{r-1} = x_{r-1}\} =$$

$$p_r(A \mid x_0, \dots, x_{r-1}, \bar{f}_0(x_0), \dots, \bar{f}_{r-1}(x_0, \dots, x_{r-1})) \quad .$$

Hence, when studying the optimal control problem for process (ξ_i, η_i), one can consider this controlled process to be defined by the conditional distributions $p_r(\cdot \mid \cdot)$ [1]. Function $\Psi_\delta(\cdot)$ and distributions $p_r(\cdot \mid \cdot)$ satisfy the conditions of Theorem 1.5 in [1]. Therefore, there exists a control $\eta_i^* = f_i^*(\hat{\xi})$ which minimizes the criterion $E\Psi_\delta(\xi_0, \ldots, \xi_n, \eta_0, \ldots, \eta_n)$. Functions f_i^* are defined by (1.18) and (1.20) from [1].

THEOREM 3. *Let*

$$|\Psi(g, f) - \Psi(g', f)| \le \psi(r(g, g')) \quad ,$$

where $r(\cdot \mid \cdot)$ *is the Skorohod metric in D and* $\psi(t)$ *is a bounded, positive function for which* $\psi(t) \to 0$ *as* $t \downarrow 0$. *then*

$$\inf T(\eta) = \lim_{|\delta| \to 0} E\Psi_\delta(\xi_0, \ldots, \xi_n, \eta_0^*, \ldots, \eta_n^*) \quad ,$$

i.e., *the process*

$$\eta^*(t) = \eta_0^* I_{\{0\}}^{(t)} + \sum_{i=0}^{n-1} \eta_i^* I_{]t_i, t_{i+1}]}^{(t)}$$

is the ε-optimal control for the solution of equation (2), *provided that* $|\delta|$ *is sufficiently small.*

The proof of Theorem 3 is based on Lemma 3 and Theorem 2, and is analogous to that of Lemma 3.13 and Theorem 3.18 in [1].

REFERENCES

1. I.I. Gihman and A.V. Skorohod. *Controlled Random Processes.* Naukova Dumka, Kiev, 1977 (in Russian).

2. I.A. Ibragimov. Conditions for smoothing the trajectories of random functions (in Russian). *Teoria Veroyatnosti i ee Primenenia,* 28(2) (1983)229–250.

3. I.I. Gihman and A.V. Skorohod. *Stochastic Differential Equations and Their Applications.* Naukova Dumka, Kiev, 1982 (in Russian).

4. A.A. Novikov. Martingale equations and inequalities and their applications in nonlinear bounded problems for random processes (in Russian). *Matemeticheskie Zametki,* 35(3) (1984)455–471.

5. A.M. Kolodiy. Existence of solutions to the Ito–Volterra integral equations with continuous and locally-integral trajectories (in Russian). *Teoria Sluchainih Protsessov,* (12) (1984) 32–40.

SOME PROPERTIES OF VALUE FUNCTIONS FOR CONTROLLED DIFFUSION PROCESSES

N.V. Krylov
Moscow State University
Moscow, USSR

1. INTRODUCTION

This paper is concerned with the general properties of value functions for controlled diffusion processes halted on the boundary ∂D of a given domain $D \subset E_d$. Our definitions are similar to those used in the theory of Markov processes and differ from the definitions given in [1,2], where the process is stopped when it first encounters $E_d \setminus \bar{D}$.

2. NON-HOMOGENEOUS CASE

We shall begin by considering the non-homogeneous case. Let E_d be a Euclidean space of dimension d, $(\Omega, \mathbf{F}, P)$ be a complete probability space, $(w_t, \mathbf{F}_t)$ be a d_1-dimensional Wiener process on this space, and A be a separable metric space. Suppose that for all $\alpha \in A$, $t \in (-\infty, \infty)$, $x \in E_d$ we are given a $d \times d_1$ matrix $\sigma(\alpha, t, x)$, a d-vector $b(\alpha, t, x)$ and real-valued $c^\alpha(t, x) \geq 0$, $f^\alpha(t, x)$, $g(t, x)$. Let σ, b, c, f, g be Borel, bounded, continuous in (α, x), and uniformly continuous in x with respect to α for every t; g be equicontinuous in (t, x); and σ, b be Lipschitzian in x with constants independent of α, t. Recall (see [3, §3.1]) that $\mathbf{A}$ is the set of all progressively measurable A-valued processes, that the process $\alpha \in \mathbf{A}$ is called a *strategy*, and that, for $\alpha \in \mathbf{A}$, $s \in (-\infty, \infty)$, $x \in E_d$, the solution of $dx_t = \sigma(\alpha_t, s+t, x_t) \times dw_t + b(\alpha_t, s+t, x_t)dt$, $x_0 = x$ is denoted by $x_t^{\alpha,s,x}$. Some additional notation from [3, §3.1] is also used.

Fix a bounded domain $Q \subset E_{d+1} = \{(t, x): t \in (-\infty, \infty), x \in E_d\}$ and define

$$\tau_Q = \tau_Q^{\alpha,s,x} = \inf \{t \geq 0 ; (s+t, x_t^{\alpha,s,x}) \notin Q\} \quad ,$$

$$\varphi_t = \varphi_t^{\alpha,s,x} = \int_0^t c^{\alpha_\tau}(x+\tau, x_\tau^{\alpha,s,x}) d\tau \quad ,$$

$$v_Q^\alpha(s,x) = E_{s,x}^\alpha[\int_0^{\tau_Q} e^{-\varphi_t} f^{\alpha_t}(s+t)dt +$$

$$+ e^{-\varphi_{\tau_Q}} g(s+\tau_Q, x_{\tau_Q})],\ v_Q(x,x) = \sup_{\alpha \in \mathbf{A}} v_Q^\alpha(s,x) \quad .$$

The properties of the value function v_Q will be studied using methods developed in [4]. In fact our results are completely analogous to those of [4]. We shall therefore suppose that there exists an $\underline{\alpha} \in A$ such that for the strategy $\underline{\alpha} \in \mathbf{A}$ given by $\underline{\alpha}_t(\omega) \equiv \underline{\alpha}$ and for all $(s,x) \in Q$ the process

$$e^{-\varphi_{t\wedge\tau_Q}} g(s+t\wedge\tau_Q, x_{t\wedge\tau_Q}) + \int_0^{t\wedge\tau_Q} e^{-\varphi_t} f^{\underline{\alpha}}(s+t, x_t)dt \qquad (1)$$

is an $\mathbf{F}_t$-submartingale, where $(\tau_Q, x_t, \varphi_t) = (\tau_Q^{\underline{\alpha},s,x}, x_t^{\underline{\alpha},s,x}, \varphi_t^{\underline{\alpha},s,x})$.

As in the proof of Theorem 1.1 in [4], it can be shown that if Q^1, Q^2 are domains, $Q^1 \subset Q^2 \subset Q$, then $g \le v_{Q^1} \le v_{Q^2}$. Hence $v_{Q(n)}$ increases for domains $Q(n) \subset Q(n+1) \subset Q$, $n = 1,2,\ldots$ and $v \le v_Q$. On the other hand it is obvious that $v_{Q(n)}^\alpha \to v_Q^\alpha$ for every $\alpha \in \mathbf{A}$ if we also have $Q = \bigcup_n Q(n)$, and this implies that $v_{Q(n)} \uparrow v_Q$.

It is useful to note that (1) is a submartingale if, for example, g has derivatives g_t, g_x, g_{xx} which are continuous in Q and $L^{\underline{\alpha}}g + f^{\underline{\alpha}} \ge 0$ in Q. This follows immediately from the Ito formula.

In what follows we suppose that $Q \subset H_T := (0,T) \times E_d$, $T < \infty$. The value functions corresponding to H_T are well-known (see [3]) so it is natural to use them to approximate v_Q. One way to do this is based on the growth of the stopping intensity near ∂Q. This method was developed in [4].

LEMMA 1. *Let the domains* $Q(n)$, $n = 1,2\ldots$, *satisfy* $\bar{Q}(n) \subset Q(n+1)$, $Q = \bigcup_n Q(n)$. *Construct smooth functions* C_n *on* E_d *such that* $c_n = 0$ *on* $Q(n)$, $c_n = 1$ *on* $E_d \setminus Q(n+1)$, $0 \le c_n \le 1$. *For* $m \ge 0$, $s \in [0,T]$ *define*

$$\varphi_t^{\alpha,s,x}(n,m) = m \int_0^t c_n(s+\tau, x_\tau^{\alpha,s,x})d\tau + \varphi_t^{\alpha,s,x}$$

$$u^{n,m}(s,x) = \sup_{\alpha \in \mathbf{A}} E_{s,x}^\alpha \{g(T, x_{T-s})e^{-\varphi_{T-s}(n,m)} +$$

$$+ \int_0^{T-s} [f^{\alpha_t}(s+t, x_t) + mc_n g(s+t, x_t)]e^{-\varphi_t(n,m)}dt\} \quad .$$

Then $v_{Q(n)} - \varepsilon(n,m) \le v^{n,m} \le v_{Q(n+1)} + \varepsilon(n,m)$ *in* H_T, *where*, $\varepsilon(n,m)$ *are independent of* (s, x) *and* $\varepsilon(n, m) \to 0$ *for* $m \to \infty$. *Consequently, in* H_T:

$$v_Q = \lim_{n\to\infty} \underline{\lim}_{m\to\infty} v^{n,m} = \lim_{n\to\infty} \overline{\lim}_{m\to\infty} v^{n,m} \quad .$$

Proof. We shall write

$$\varepsilon_1(n, m) = \varepsilon_1 = \sup\{|v^{n,m}(s, x) - g(s, x)| \; ; \; \bar{\bar{H}}_T \setminus Q(n+1)\} \quad .$$

The function g is equicontinuous and σ, b, c, f are bounded, so it is easy to see that $\varepsilon_1(n, m) \to 0$ if $m \to \infty$. From the lemma in Appendix 2 of [3], formula (1) will again yield a submartingale if φ, f^{α} are replaced by $\varphi(n, m)$, $f^{\alpha} + mc_n g$. Hence the Bellman principle (see [3, §3.1]) implies

$$\begin{aligned} v^{n,m}(s, x) &\geq \mathrm{E}^{\alpha}_{s,x}\{v^{n,m}(s + \tau(n+1), x_{\tau(n+1)})e^{-\varphi_{\tau(n+1)}(n,m)} + \\ &+ \int_0^{\tau(n+1)} [f^{\alpha} + mc_n g](s + t, x_t)e^{-\varphi_t(n,m)}dt\} \geq \\ &\geq \mathrm{E}^{\alpha}_{s,x}\{g(s + \tau(n+1), x_{\tau(n+1)})e^{-\varphi_{\tau(n+1)}(n,m)} + \\ &+ \int_0^{\tau(n+1)} [f^{\alpha} + mc_n g](s + t, x_t)e^{-\varphi_t(n,m)}dt\} - \varepsilon_1 \geq g(s, x) - \varepsilon_1 \quad , \end{aligned} \tag{2}$$

where $\tau(n+1) = \tau_{Q(n+1)}$. Again by the Bellman principle, we have

$$\begin{aligned} v^{n,m}(s, x) &= \sup_{\alpha\in\mathbf{A}} \mathrm{E}^{\alpha}_{s,x}\{v^{n,m}(s + \tau(n), x_{\tau(n)})e^{-\varphi_{\tau(n)}} + \\ &\int_0^{\tau(n)} f^{\alpha_t}(s + t, x_t)e^{-\varphi_t}dt\} \geq v_{Q(n)}(s, x) - \varepsilon_1 \quad . \end{aligned}$$

We shall now prove that $v^{n,m} \leq v_{Q(n+1)} + \varepsilon(n,m)$. From Lemmas 3.3.5 and 3.3.7 of [3], the process

$$\begin{aligned} \kappa_t = \kappa_t^{\alpha,s,x} :&= v^{n,m}(s + t, x_t)e^{-\varphi_t(n,m)} + \\ &+ \int_0^t [f^{\alpha_\tau} + mc_n g](s + \tau, x_\tau)e^{-\varphi_\tau(n,m)}d\tau \end{aligned} \tag{3}$$

is a continuous supermartingale on $[0, T-s]$ for every $\alpha \in \mathbf{A}$, $(s, x) \in \bar{H}_T$, where $(x_t, \varphi_t(n, m)) = (x_t^{\alpha,s,x}, \varphi_t^{\alpha,s,x}(n, m))$. Hence the lemma from Appendix 2 of [3] implies that

$$\rho_t = \rho_t^{\alpha,s,x} := v^{n,m}(s + t, x_t)e^{-\varphi_t} + \int_0^t [f^{\alpha_\tau} + mc_n(g - v^{n,m})](s + \tau, x_\tau)e^{-\varphi_\tau}d\tau$$

is also a supermartingale and

$$v^{n,m}(s,x) \geq E^{\alpha}_{s,x}\rho_{\tau(n+1)} \geq e^{mT}[E^{\alpha}_{s,x}\kappa_{T-s} - v^{n,m}(s,x)] + v^{n,m}(s,x) \quad .$$

The upper bound of the last expression over $\alpha \in \mathbf{A}$ is equal to $v^{n,m}(s,x)$ by definition. Therefore

$$v^{n,m}(s,x) = \sup_{\alpha\in\mathbf{A}} E^{\alpha}_{s,x}\rho_{\tau(n+1)} \leq v_{Q(n+1)}(s,x) + \tag{4}$$
$$+ \varepsilon_1 + T\sup\{m(g - v^{n,m})_+ ; Q(n+1)\} \quad .$$

To complete the proof it is sufficient to show that the last term tends to zero if $m \to \infty$. Using the submartingale property of (1) (see also (2)) and the Bellman principle we find

$$g(s,x) \leq \sup_{\alpha\in\mathbf{A}} E^{\alpha}_{s,x}\{g(s+\tau_Q, x_{\tau_Q})e^{-\varphi_{\tau_Q}(n,m)} +$$
$$+ \int_0^{\tau_Q}[f^{\alpha_t} + mc_n g](s+t, x_t)e^{-\varphi_t(n,m)}dt\} \quad ,$$

$$v^{n,m}(s,x) = \sup_{\alpha\in\mathbf{A}} E^{\alpha}_{s,x}\{v^{n,m}(s+\tau_Q, x_{\tau_Q})e^{-\varphi_{\tau_Q}(n,m)} +$$
$$+ \int_0^{\tau_Q}[f^{\alpha_t} + mc_n g](s+t, x_t)e^{-\varphi_t(n,m)}dt\} \quad ,$$

$$g(s,x) - v^{n,m}(s,x) \leq \sup_{\alpha\in\mathbf{A}} E^{\alpha}_{s,x}[g - v^{n,m}](s+\tau_Q, x_{\tau_Q}) \times$$
$$\times e^{-\varphi_{\tau_Q}(n,m)} \leq N \sup_{\alpha\in\mathbf{A}} E^{\alpha}_{s,x} e^{-\varphi_{\tau_Q}(n,m)} \leq N \sup_{\alpha\in\mathbf{A}} E^{\alpha}_{s,x} e^{-m(\tau_Q - \tau_{Q(n+1)})} \quad ,$$

where N is independent of s, x, n, m. We now use estimates of the moments of the stochastic integrals. Then for $(s,x) \in Q(n+1)$, $\rho_n := \operatorname{dist}(Q(n+1), \partial Q)$, $m \geq 4\rho_n^{-2}$ it follows that

$$E^{\alpha}_{s,x} e^{-m(\tau_Q - \tau_{Q(n+1)})} \leq e^{-\sqrt{m}} + P^{\alpha}_{s,x}\{\tau_Q - \tau_{Q(n+1)} \leq \frac{1}{\sqrt{m}}\} \leq$$
$$\leq e^{-\sqrt{m}} + P^{\alpha}_{s,x}\{\sup_{\tau_{Q(n+1)} \leq t \leq \tau_{Q(n+1)} + \frac{1}{\sqrt{m}}} |x_t - x_{\tau_{Q(n+1)}}| \geq \frac{1}{2}\rho_n\} \leq$$
$$\leq e^{-\sqrt{m}} + 2^8 \rho_n^{-8} N_m^{-2} \quad ,$$

where N is independent of s, x, n, m. Thus the last term in (4) tends to zero and the lemma is proved.

Analogously to [4], the following deductions can be made from this lemma:

COROLLARY 1. *The function v_Q is uniquely defined by σ, b, c, f, g, Q, A and will not alter if the probability space or the Wiener process is changed.*

COROLLARY 2. *v_Q is lower-semicontinuous in (s, x).*

The following fact is also useful:

COROLLARY 3. *Let $\alpha \in \mathbf{A}$, $(s, x) \in H_T$, and $\tau_t = \tau_t(w)$ be real, bounded, and progressively measurable. Then the process*

$$\gamma_t := v_Q(s + t \wedge \tau, x_{t\wedge\tau}) e^{-\varphi_{t\wedge\tau} - \psi_{t\wedge\tau}} + \tag{5}$$
$$+ \int_0^{t\wedge\tau} [f^{\alpha_u} + \tau_u v_Q](s + u, x_u) e^{-\varphi_u - \psi_u} du$$

is a supermartingale for $t \in [0, T - s]$ with (a.e.) *right-continuous trajectories, where $(x_u, \varphi_u) = (x_u^{\alpha,s,x}, \varphi_u^{\alpha,s,x})$, $\psi_u = \int_0^u \tau_p \, dp$, $\tau = \tau_Q$. Furthermore, for every stopping time χ we have*

$$v_Q(s, x) \ge \mathrm{E}\gamma_\chi \ge e^{N(T-s)}[v_Q^\alpha(s, x) - v_Q(s, x)] + v_Q(s, x) \quad , \tag{6}$$

where $N = \sup\{\tau_t^-(w) \, ; \, t \ge 0, \, w \in \Omega\}$.

The proof of Corollary 3 for general τ is reduced to the case $\tau \equiv 0$ as in the lemma from Appendix 2 of [3]. In fact this lemma has actually been proved for continuous supermartingales but the arguments also apply, at least in our case, to supermartingales which are only right-continuous. Therefore it suffices to prove only the first assertion of the corollary for $\tau \equiv 0$.

As noted above, the process κ_t from (3) is a continuous supermartingale with respect to $\{\mathbf{F}_t\}$. It is evidently also a supermartingale with respect to $\{\mathbf{F}_{t+}\}$. The equality $c_n = 0$ in $Q(n)$ and the Doob theorem imply that for $k \ge n$

$$v^{k,m}(s + t \wedge \tau(n), x_{t\wedge\tau(n)}) e^{-\varphi_{t\wedge\tau(n)}} + \int_0^{t\wedge\tau(n)} f^{\alpha_\tau}(s + \tau, x_\tau) e^{-\varphi_\tau} d\tau$$

is an $\mathbf{F}_{t+}$-supermartingale. Letting $m \to \infty$, $k \to \infty$, the Fatou theorem implies that

$$v_Q(s + t \wedge \tau(n), x_{t\wedge\tau(n)}) e^{-\varphi_{t\wedge\tau(n)}} + \int_0^{t\wedge\tau(n)} f^{\alpha_\tau}(s + \tau, x_\tau) e^{-\varphi_\tau} d\tau \tag{7}$$

is an $\mathbf{F}_{t+}$-supermartingale; from Corollary 2, its trajectory is lower-semicontinuous with respect to t. Now the arguments of Ray and Meyer (see [5, Chap. VI, §2, Theorem 16]) may be applied after obvious modifications to prove the right-continuity of (7) (a.e.). The right-continuity of γ_t follows immediately from the fact that $\{t < \tau\} = \bigcup_n \{t \le \tau(n)\}$, $v_Q(s+\tau\wedge T, x_{\tau\wedge T}) = g(s+\tau\wedge T, x_{\tau\wedge T})$ if $\tau \ge T$. Application of the Fatou theorem to (7) proves that γ_t is a supermartingale (if $r \equiv 0$).

The right-continuity of γ_t for $t = 0$ and Corollary 2 lead to:

COROLLARY 4. *v_Q is lower-continuous on Q:*

$$v_Q(s,x) = \varliminf_{(t,y)\to(s,x)} v_Q(t,y) \quad .$$

Taking the upper bounds in (6) over $\alpha \in \mathbf{A}$, we obtain:

COROLLARY 5 (The Bellman principle). *Let $(s,x) \in H_T$, and suppose that for every $\alpha \in \mathbf{A}$ we are given an $\mathbf{F}_t$-stopping time $\chi^\alpha \le \tau_Q^{\alpha,s,x}$ and a progressively measurable bounded process $r_t^\alpha(w)$. Suppose that $r_t^\alpha(w)$ is bounded from below as a function of (α, t, w). Then*

$$v_q(s,x) = \sup_{\alpha\in\mathbf{A}} \mathrm{E}^\alpha_{s,x} \{ v_Q(s+\chi, x_\chi) e^{-\varphi_\chi - \psi_\chi} +$$

$$+ \int_0^\chi [f^{\alpha_u}(s+u, x_u) + r_u v_Q(s+u, x_u)] e^{-\varphi_u - \psi_u} du \} \quad ,$$

where $\psi_u^\alpha = \int_0^u r_p^\alpha dp$.

3. HOMOGENEOUS CASE

We shall now consider the homogeneous case. Suppose that σ, b, c, f, g are independent of t and satisfy the conditions of Section 2; that $c \ge \varepsilon$ with a constant $\varepsilon > 0$, D is a bounded domain in E_d and the process (1) is a submartingale, with the first exit time of $x_t^{\alpha,s,x}$ from D substituted for τ_Q. It is clear that $x_t^{\alpha,s,x}$ is independent of s and coincides (a.e.) with $x_t^{\alpha,x} := x_t^{\alpha,0,x}$. In the same way, $\varphi_t^{\alpha,s,x} = \varphi_t^{\alpha,0,x} =: \varphi_t^{\alpha,x}$ and so on.

Define

$$\tau_D = \tau_D^{\alpha,x} = \inf\{t \ge 0 \,;\, x_t^{\alpha,x} \notin D\} \quad ,$$

$$v_D^\alpha(x) = \mathrm{E}_x^\alpha \left[\int_0^{\tau_D} e^{-\varphi_t} f^{\alpha_t}(x_t) dt + e^{-\varphi_{\tau_D}} g(x_{\tau_D}) \right] \quad ,$$

$$v_D(x) = \sup_{\alpha \in \mathbf{A}} v_D^\alpha(x) \quad .$$

From the condition $c \geq \varepsilon$ it follows that the v_D^α are well-defined and uniformly bounded, and that v_D is bounded. Moreover, for every $s > 0$ we have

$$\sup_{\alpha \in \mathbf{A}} |v_D^\alpha(x) - v^\alpha_{(0,n)\times D}(s,x)| \to 0 , \; v_{(0,n)\times D}(s,x) \uparrow v_D(x) \tag{8}$$

if $n \to \infty$ (the increase of $v_{(0,n)\times D}$ in n is proved in Section 2 after the introduction of process (1)).

THEOREM

(a) *The function v_D is uniquely defined by σ, b, c, f, g, D, A. It does not change if the probability space, filtration $\{\mathbf{F}_t\}$ or Wiener process are changed.*

(b) *The function $v_D(x)$ is lower-semicontinuous in x.*

(c) *Let $\alpha \in \mathbf{A}$, $x \in D$,* and $\tau_t = \tau_t(w)$ *be a real, bounded, progressively measurable process. Then*

$$\gamma_t := v_D(x_{t\wedge\tau})e^{-\varphi_{t\wedge\tau}-\psi_{t\wedge\tau}} + \int_0^{t\wedge\tau} [f^{\alpha_u} + \tau_u v_D](x_u)e^{-\varphi_u-\psi_u}du \tag{9}$$

is a càdlàg supermartingale for $t \geq 0$ (a.e.), *where* $\psi_u = \int_0^u \tau_p dp$, $(x_u, \varphi_u, \tau) = (x_u^{\alpha,x}, \varphi_u^{\alpha,x}\tau_D^\alpha, x)$. *Moreover, for every stopping time χ we have*

$$v_D(x) \geq E\gamma_\chi \geq e^{NT}(v_D^\alpha(x) - v_D(x)) + v_D(x)$$

if $NT < \infty$, *where* $N = \sup\{\tau_t^-(w) ; \; t \geq 0, \; w \in \Omega\}$, $T = \sup\{\chi(w) ; \; w \in \Omega\}$ $(0 \cdot \infty = 0)$.

(d) *v_D is lower-continuous on D.*

(e) *Let $x \in D$ and suppose that for every $\alpha \in \mathbf{A}$ we are given a stopping time $\chi^\alpha \leq \tau_D^{\alpha,x}$ and a progressively measurable bounded process τ_t^α. Suppose that $N^\alpha T^\alpha$ is bounded in α, where* $N^\alpha = \sup_{t,w}(\tau_t^\alpha)^-$, $T^\alpha = \sup_w \chi^\alpha$ $(0 \cdot \infty = 0)$.

Then for $\psi_t^\alpha := \int_0^t \tau_u^\alpha du$ *we have*

$$v_D(x) = \sup_{\alpha \in D} E_x^\alpha \{ v_D(x_\chi)e^{-\varphi_\chi - \psi_\chi} +$$

$$+ \int_0^\chi [f^{\alpha_u} + \tau_u v_D](x_u)e^{-\varphi_u-\psi_u}du \} \quad .$$

The proof of this theorem is obviously based on the results and methods of Section 2 and on the second formula in (8). The only point that should be explained is the right continuity of γ_t. It is sufficient to consider the case $\tau_t \equiv 0$; in this case γ_t from (9) is the upper bound over n of the increasing sequence of right-continuous supermartingales γ_t^n constructed by formula (5), starting with $v_Q = v_{(0,n)\times D}$. From the Ray–Meyer theorem, γ_t is then right-continuous (a.e.).

REFERENCES

1. P.L. Lions. On the Hamilton–Jacobi–Bellman equations. *Acta Applicandae Mathematicae*, 1(1983) 17–41.

2. P.L. Lions. Optimal control of diffusion processes and Hamilton–Jacobi–Bellman equations. Part I: The dynamic programming principle and applications. *Communications in Partial Differential Equations*, 8(10)(1983) 1101–1174.

3. N.V. Krylov. *Controlled Diffusion Processes*. Springer-Verlag, Berlin, 1980.

4. N.V. Krylov. On controlled diffusion processes with unbounded coefficients. *Izvestiya Academii Nauk SSSR, Seriya Matematicheskaya*, 45(1981) 734–759. (English translation in *Mathematics of the USSR – Izvestiya*, 19(1982) 41–64.)

5. P.-A. Meyer. *Probabilités et potential*. Hermann, Paris, 1966.

STOCHASTIC CONTROL WITH STATE CONSTRAINTS AND NON-LINEAR ELLIPTIC EQUATIONS WITH INFINITE BOUNDARY CONDITIONS

Jean-Michel Lasry
CEREMADE, Université Paris-Dauphine

1. A STATE CONSTRAINT PROBLEM IN STOCHASTIC CONTROL

A large part of stochastic control theory is devoted to problems where the state $X(t)$ is a diffusion process driven by a stochastic differential equation :

$$dX = a(X)dt + dB_t$$

where B_t is a Brownian motion and where the drift $a(X)$ is the feedback control of this random dynamic.

For some problems the state $X(t)$ is allowed to take any values in $\mathbb{R}^N$ while in other problems the state should remain in some bounded domain $\overline{\Omega}$. In this later case there is a non-zero probability that the state $X(t)$ reach the boundary if the control $a(\cdot)$ is bounded. One must then precise what will be the dynamic at the boundary - and the cost involved. There are mainly two usual cases : the first one is to introduce reflected motion at the boundary, the second one is to introduce the exit time T - i.e. : the first time the state escape from the open set Ω and to decide to stop everything then and there at time T . The total cost take account of the inside motion and of the boundary phenomena : a cost can be attached to the "number" of reflection or to the "sale off" in the case of the stopped dynamic at exit time T .

Contrasting with this two previous usual treatments of the boundary to the state domain, we will be concerned here by the mathematical modeling of a stringent constraint on the state $X(t)$: we will ask that the drift control $a(\cdot)$ should prevent the state $X(t)$ to reach the boundary $\partial\Omega$ of Ω . Such situations arise in practical cases when there is no available reflection mechanism at any cost, nor any possible stop and "sale off" at the boundary. Then the feedback control $a(\cdot)$ should be designed in order to keep the state $X(t)$ in the open domain Ω, off the boundary $\partial\Omega$ - i.e. : admissible controls are those for which the exit time T is almost surely $T \equiv +\infty$.

This state constrained problem is a natural modelisation of many situations, as for example problems involving security threshold. Suprinsingly this question has not yet been studied for diffusion process with drift control. One of our result - all the results herein come from a joint work with Pierre-Louis Lions - might explain this lack : we will see that low cost on the drift control $a(\cdot)$ combined with a (classical) high final cost (sale off cost) at the boundary leads to the same optimal strategy as our problem with state constraints. On the other side, when the drift cost is high there is an optimal strategy for our problem which differs from any solution of any problem with a classical treatment of the boundary. In this case the state constraints problem could be viewed as a stopping time problem with infinite final cost - which implies that any reasonnable control should insure $T \equiv +\infty$ almost surely.

2. THE CORRESPONDING HAMILTON-JACOBI-BELLMAN EQUATION

On the side of partial differential equations (P.D.E.) our problem turn out to be a quasi-linear elliptic problem in a smooth bounded open domain with singular boundary conditions. The connection between drift control of diffusion process and quasi-linear elliptic problems is now classic : H. Flemming started the theory on the grounds of Hamilton and Jacobi, old deterministic theory and of R. Bellman dynamic programming. There is a large, diverse, lively litterature on this topic (see bibliography in [1], [2], [3], [8]) and of course a lot of conferences on it in this book.

To be more precise it is time to introduce notations. As yet said is a smooth open set of $\mathbb{R}^N$. Denote B_t a standard Brownian motion in $\mathbb{R}^N$. Let x in Ω be the initial value of the state X. Let $a \in C^1(\Omega)$ be the drift feedback. The motion of the state X is governed by the stochastic differential equation

$$dX = a(X)dt + \sqrt{2}\, dB_t \tag{2.1}$$

with <u>initial condition</u>

$$X(0) = x \tag{2.2}$$

The choice of the square root $\sqrt{2}$ in (1) is a normalisation only made to avoid later supplementary constants (recall "$E|dB_t|^2 = dt$" where E stands for the expected value).

Denote by T the <u>exit time</u> from Ω .

Let $g \in C(\overline{\Omega} \times \mathbb{R}^N)$ be the <u>Lagrangian function</u> and let $\varphi \in C(\partial\Omega)$ be the sale off cost or <u>final cost</u> at the boundary. Both functions g and φ are supposed bounded below.

Let $\lambda > 0$ be some given constant later refered as the <u>discount factor</u>.

Finally let $J(x,a)$ be the <u>cost functionnal</u> defined by

$$J(x,a) = E\left\{\int_0^T e^{-\lambda t} g(x(t),a(x(t)))dt + e^{-\lambda T}\varphi(x(T))\right\}$$

This cost J is the sum of two terms : an integral term which represent a <u>distributed cost</u> taking account of the motion inside Ω , and a final cost.

Suppose for simplicity that the functions f and φ are smooth and let us make the natural assumption that $g(y,\cdot)$ is strictly convex and coercive, more precisely suppose that :

$$g''_{yy}(x,y) \quad \text{is positive defined for all } x \text{ in } \overline{\Omega}, \text{ and for all } y \text{ in } \mathbb{R}^N \tag{2.4}$$

$$g(x,y)|y|^{-1} \to +\infty \quad \text{when} \quad |y| \to \infty \quad (\text{uniformly in } x) \tag{2.5}$$

Introduce, like in the deterministic Hamilton-Jacobi theory, the Hamiltonian h defined by

$$h(x,p) = -\inf \{p\alpha + g(x,\alpha) \mid \alpha \in \mathbb{R}^N\} \tag{2.6}$$

Note that from the smoothness of g , and assumptions (4) and (5) one can deduce that h belongs to $C^2(\overline{\Omega} \times \mathbb{R}^N)$.

Then a now classical result that the Bellman function defined by :

$$u(x) \quad = \quad \operatorname{Inf} \{J(x,a) \mid a \in C^1(\bar{\Omega})\} \tag{2.7}$$

is the unique solution of the quasilinear equation

$$-\Delta u(x) + h(x,\nabla u(x)) + \lambda u(x) \quad = \quad 0 \qquad \text{for all} \quad x \quad \text{in} \quad \Omega \tag{2.8}$$

with boundary condition

$$u = \varphi \qquad \text{on} \qquad \partial\Omega \tag{2.9}$$

(Recall that smoothness and above assumptions (2.4), (2.5) could be weakened a lot, see for example [8] and the bibliography therein).

With some similarity to the reverse sided style of the existence proofs of the direct method of calculus of variations which start by existence of weak solutions continuing by a posteriori regularity to end up by existence of a unique classic solution, the proof of the above stated theorem goes backward : one first prove existence of solution of (2.8), (2.9), then their regularity to end up by the deduction that any solution of (2.8), (2.9) should be equal to the Bellman function defined by (2.7) - and hence unique.

Let us recall another essential feature of this theory. The optimal feedback control is unique and can computed from u by the formula

$$a(x) \quad = \quad \operatorname{Arg} \cdot \inf \{\nabla u(x) \cdot \alpha + g(x,\alpha) \mid \alpha \in \mathbb{R}^N\} \tag{2.10}$$

or by the equivalent formula :

$$\nabla u(x)\alpha + g'_\alpha(x,\alpha) \quad = \quad 0 \qquad \text{for} \quad \alpha = a(x) \tag{2.11}$$

By Fenchel equivalence $(p\alpha+g'(x,\alpha) = 0 \Leftrightarrow \alpha = -h'_p(x,p))$ this gives another definition of the optimal feedback a :

$$a(x) \quad = \quad -h'_p(x,\nabla u(x)) \qquad \forall\, x \in \Omega \tag{2.12}$$

Now let us come to our specific problem which is to enter our state constraint. <u>Admissible feedback</u> will function $a \in C^1(\Omega)$ such that the exit time verifies $\mathcal{T} \equiv +\infty$ almost surely. We will denote $\mathcal{C}$ the set of such admissible controls a . Now to define the Bellman function for the constraint replace the above definition (2.7) of u by the following one :

$$u(x) = \operatorname{Inf} \{J(x,a) \mid a \in C\} \tag{2.13}$$

The question now arising is : what boundary condition will replace the previous one, namely : $u = \varphi$ on $\partial\Omega$? It turns out that there are several possible candidates - at least the three following ones :

$$u = +\infty \quad \text{on} \quad \partial\Omega \quad (\text{i.e. : } u(x) \to +\infty \text{ when } x \to \partial\Omega) \tag{2.14}$$

$$\frac{\partial u}{\partial n} = +\infty \quad \text{on} \quad \partial\Omega \quad \left(\frac{\partial}{\partial n} \text{ is exterior normal derivation}\right) \tag{2.15}$$

u is the lowest solution of (2.8) which is greater than any bounded solution of (2.8) (2.16)

Let us explain why each of these conditions sounds, at least at first sight, to be reasonable modelisation of the state constraint.

For example the first one (2.14) introduce an infinite value of the Bellman function u at the points where the constraint is not satisfied, which regarding the minimisation problem looks economicaly reasonable. But a closer economical look leads to the fact that the value should blow up to $+\infty$ near the state constraint only if it is costly to drive the state off $\partial\Omega$. This reasonning will be just confirmed by the comparison of theorem 1 and 2 of section 2 : in the first one large drifts are costly (q is large) and (2.14) holds, while in the second theorem - with less expansive large drifts - (2.14) no more hold.

So on one hand (2.16) might seem the best because it always hold - at least in the case that we have studied. On the other hand the other conditions - (2.14) and (2.15) - will provide more specific information when they will characterize the Bellman function.

All this heuristic considerations will find there mathematical counter-part in the theorems of the next section.

3. RESULTS

Let us recall that the results given in this paper are part of a joint work with P.L. Lions [6], [7].

We will now describe the results in the special case where the

Lagrange function which enter in the cost (2.3) is of the following type :

$$g(x,y) = f(x) + m|y|^q \quad , \quad \forall\, x \in \bar{\Omega} \; , \; \forall\, y \in \mathbb{R}^N \quad (\text{with} \quad q > 1\,) \qquad (3.1)$$

(Same type of results also hold for more general Lagrange functions, see [7]).

The function f in (3.1) will be supposed to belongs to $C^1(\Omega)$ for simplicity (and $(mq)^p m(q-1) = 1$ to avoid later supplementary constraints in (3.2)).

The value of q will play a crucial role in the choice of the "good" singular boundary condition, between (2.14), (2.15), (2.16).

Due to (3.1), the Hamiltonian function h defined by (2.6) is now :

$$h(x,z) \;=\; -f(x) + |z|^p \quad , \quad \forall\, x \in \Omega \; , \; \forall\, y \in \mathbb{R}^N \qquad (3.2)$$

with

$$1/p + 1/q \;=\; 1 \qquad (3.3)$$

So the Hamilton-Jacobi-Bellman (HJB) equation (2.8) is now :

$$-\,\Delta u + |\nabla u|^p + \lambda u \;=\; f(x) \qquad \forall\, x \in \Omega \qquad (3.4)$$

Note that we do not intend to detail here the best possible regularity results, as our main interest is in the boundary behaviour of u : hence f is C^1 in Ω and solutions of (2.4) will mean classical ones, i.e. : $u \in C^2(\Omega)$, at least.

The results on existence and unicity for this equation (3.4) under the singular boundary conditions (2.14), (2.15) or (2.16) will differ very much according to the behaviour of the function f near the boundary and to the value of the constant $q > 1$. Let us consider emphasize that these results are related to strictly non linear phenomena and do not hold in the homogeneous case $p = 1$ (which cen be considered as "almost linear").

We will first give the P.D.E.'s theorems (they have their own interest ; they belong to the growing flow of works on non linear P.D.E. with singular boundary conditions). Then we will give the results concerning the related stochastic problems.

The hypothesis in the next theorems can be expressed briefly (and roughly speaking) as follow :

- theorem 1 deals with large q (i.e.: $q \geq 2$) and bounded f.
- theorem 2 deals with small q (i.e.: $1 < q < 2$) and bounded f.
- theorem 3 deals with rapidly blowing up f.
- theorem 4 deals with non discounted case.
- theorem 5 and 6 gives stochastic control interpretation of the previous theorem.

In the first three theorems one can see that unicity of the solution u of the HJB equation (3.4) is obtained under weak boundary conditions : this is obvious for theorem 3 where there are almost no boundary conditions, except the condition that u is bounded below in Ω, note also that condition (2.5) is weak due to the fact that the speed of convergence of $u(x)$ to $+\infty$ when $d(x)$ tends to 0 is specified only a posteriori.

The first two following theorems illustrate the possible appearance of various singular conditions at the boundary as was announced before.

Finally the theorems 5 and 6 show that the Bellman function of the stochastic control problem with state constraint is the unique solution of the HJB with the suitable singular boundary condition.

Theorem 1. Let $q \geq 2$ (so that $1 < p \leq 2$ (see (3.3)). Let $f \in C^1(\overline{\Omega})$. Then there exists a unique solution u to the Hamilton-Jacobi-Bellman equation (3.4) such that

$$u(x) \to +\infty \qquad \text{when} \quad d(x) \to 0 \tag{3.5}$$

where d denote the distance to the boundary

$$d(x) = \text{dist}\,(x, \partial\Omega) \qquad \text{for all} \quad x \in \Omega \tag{3.6}$$

Theorem 2. Let $q < 2$ so that $p > 2$. Let $f \in C^1(\overline{\Omega})$. Then all the solutions $u \in C^2(\Omega)$ of HJB equation (3.4) are bounded in Ω, and have a continuous extension to $\overline{\Omega}$.

There is a unique maximum solution u of (3.4), i.e. which satisfies :

$$u \geq v \qquad \text{for any solution } v \text{ of (3.4)} \tag{3.7}$$

Theorem 3. Let $f \in C^1(\Omega)$, $p > 1$ and

$$f(x)\,[\,d(x)\,]^{\beta} \to c_1 > 0 \qquad \text{when} \quad d(x) \to 0 \tag{3.8}$$

where β is a constant such that $\beta > p$ and $\beta > q$. Then there exists a unique bounded below solution of H.J.B. equation (3.4).

Actually we prove that this unique bounded below solution satisfies the boundary condition (3.5).

Let us now turn to a result about the "non-discounted" case, i.e. : to the study of the limit $\lambda \downarrow 0$ $(\lambda > 0)$. Such a study is relevant (according to the stochastic interpretation) when the state remains in a bounded domain due to the geometry of the domain (see [5]), to reflection on the boundary (see [4]) or to induced "no escape" cases ([10] in $\mathbb{R}^N$) , and which the case here (our state will live in Ω for all time $t \geqslant 0$).

Theorem 4. Let $q \geqslant 2$ (hence $1 < p \leqslant 2$). Let $f \in C^1(\bar{\Omega})$. Let u_λ be the unique solution of H.J.B. equation (2.4) which verifies (2.4) (see theorem 1). Given some point $x_o \in \Omega$ there exists $u_o \in C^2(\Omega)$ and $\delta \in \mathbb{R}$ such that

$$\lambda u_\lambda(x) \to \delta \qquad \forall\, x \in \Omega \tag{3.9}$$

$$[u_\lambda(x) - u_\lambda(x_o)] \to v(x) \qquad \forall\, x \in \Omega \tag{3.10}$$

and (v_o, δ) is the unique solution of the following system :

$$- \Delta v_o + |\nabla v|^p + \delta = f \qquad \text{in } \Omega \tag{3.11}$$

$$v_o(x) \to +\infty \qquad \text{when} \quad d(x) \to 0 \quad \text{(see (2.6))} \tag{3.12}$$

$$v_o(x_o) = 0 \tag{3.13}$$

(choosing another x_o would just lead to replace v_o by v_o + const. and leave δ unchanged).

Recall that due to the special form of function g the formula (1.8) reduces to

$$a(x) = - p|\nabla u|^{p-2} \nabla u(x) \qquad \forall\, x \in \Omega \tag{3.14}$$

Theorem 5. Let u be the Bellman function of the stochastic control problem with state constraint (as defined by (2.7)). Then

1) under the hypothesis of theorem 1, u is the unique solution of H.J.B. equation (3.4) which satisfies (3.5) and the feedback given by (3.14) is the unique optimal feedback.

2) under the hypothesis of theorem 2, u is the unique solution of H.J.B. equation (3.4) which satisfies (3.7).

3) under the hypothesis of theorem 3, u is the unique solution of H.J.B. equation (3.4) which is bounded below, and the feedback a given by (3.14) is the unique optimal feedback.

<u>Theorem 6.</u> Let $\delta_o \in \mathbb{R}$ be the value of the following problem of stochastic control of the main value cost (3.16) - under state constraints (compare to (2.7)) :

$$\delta_o = \inf \{\mu(a) \mid a \in C\} \tag{3.15}$$

the cost function μ is defined by :

$$\mu(a) = \limsup_{T \to +\infty} E \frac{1}{T} \int_0^T (f(X_t) + m|a(X_t)|^q)\, dt \tag{3.16}$$

Then under the hypothesis of theorem 4, δ_o is equal to the constant δ of theorem 4 and the feedback a given by (3.14) is the unique optimal feedback.

The proof of these theorems (see [7]) relies on various comparison arguments with numerous ad-hoc sub- or super-solutions together with a repeated use of the following estimate due to P.L. Lions [7].

<u>Theorem 7.</u> If a function $u \in C^2(\omega)$ verifies for some $p > 1$:

$$-\Delta u + |\nabla u|^p = f \qquad \text{in } \omega$$

with $f \in L^\infty(\omega)$, then it satisfies in ω , $\bar{\omega}' \subset \omega$

$$|\nabla u|_{L^\infty(\omega')} \leqslant c$$

where the constant c depends only on ω' and $|f|_{L^\infty(\omega)}$.

Note that $p > 1$ is an important feature of this estimate which stop to be true for $p = 1$ - i.e. : this estimate is related to non-linearity. A first insight on this estimate comes out from the proof of the one dimensional case.

An explicit example :

Let us finish by explicit example. In the special case $p = q = 2$, the "ergodic" problem $\lambda = 0$ can be reduced to a linear problem through the transformation $v = -\log u$. This enable both explicit solution (see below) and new proofs of some log-convexity results (see [7]).

As an example of explicit solution let us consider the problem

$$\text{Minimize } \{\mu(a) \mid a \in C\} \tag{3.17}$$

where μ is defined by

$$\mu(a) = \lim_{T \to +\infty} \frac{1}{T} E \int_0^T \alpha(t)^2 \, dt \tag{3.18}$$

with $\alpha(t) = a(X_t)$ where X is the solution of :

$$dX = a(X_t)dt + \sqrt{2}\, dB_t \quad , \quad \forall\, t \geq 0 \tag{3.19}$$

$$X(0) = x_0 \tag{3.20}$$

and where C is the set of feedback such that the state X_t satisfies (almost surely)

$$-1 < X_t < +1 \qquad \text{for all } t \geq 0 \tag{3.21}$$

Then the optimal feedback for this problem is given by :

$$\underline{a}(x) = -2\pi \operatorname{tg} \pi x \qquad \forall\, x \in \,]-1,+1[\tag{3.22}$$

and the optimal value is

$$\mu(\underline{a}) = \inf_{a \in C} \mu(a) = 4\pi^2$$

This is a consequence of theorems 4 and 6 and of the possibility in this special case $(p = q = 2\ ,\ f = 0\ ,\ \Omega =]-1,+1[\,)$ to compute the solution of equation (3.11), (3.12), (3.13), (3.14) which reduce here to

$$-v'' + v'^2 + \delta = 0 \qquad \text{on }]-1,+1[\tag{3.23}$$

$$v(0) = 0 \quad , \quad v(x) \to +\infty \qquad \text{when } x \to \pm 1 \tag{3.24}$$

$$\underline{a}(x) = -2v'(x) \tag{3.25}$$

(note that the choice of m in (3.16) does not change the problem hence the value of $\underline{a}$ here, due to $f = 0$).

REFERENCES

[1] A. Bensoussan and J.L. Lions : Applications des inéquations variationnelles en contrôle stochastique. Dunod, Paris, 1978.

[2] A. Bensoussan and J.L. Lions : Contrôle impulsionnel et inéquations quasi-variationnelles. Dunod, Paris, 1982.

[3] W.H. Fleming and R. Rishel : Deterministic and stochastic optimal control. Springer Berlin, 1975.

[4] F. Gimbert : Problème ergodique pour les équations quasi-linéaires avec conditions de Neumann. Thèse de 3e cycle, Université Paris-Dauphine, 1984, and to appear in J. Funct. Anal..

[5] J.M. Lasry : Contributions à la théorie de contrôle, Thèse, Paris, 1975.

[6] J.M. Lasry and P.L. Lions : Equations elliptiques non linéaires avec conditions aux limites infinies et contrôle stochastique avec contraintes d'état, C.R. Acad. Sc., Paris, t. 299, série I, n° 7, 1984, p. 213-216.

[7] J.M. Lasry and P.L. Lions : Non linear elliptic equations with singular boundary conditions and stochastic control with state constraint, preprint, "Cahiers du CEREMADE", n° , Université Paris-Dauphine, 75775 Paris Cedex 16.

[8] P.L. Lions : Optimal control of diffusion processes and Hamilton-Jacobi-Bellman equations, Parts 1,2, Comm. in P.D.E., 8, 1983, p. 1101-1174, p. 1229-1276.

[9] P.L. Lions and B. Perthame : Quasi-variational inequalities and ergodic impulse control, to appear.

[10] R. Tarres : Coercive properties of the optimal feedback for some stochastic control in $\mathbb{R}^N$, to appear in SIAM, june 1985.

ON THE WEAK CONVERGENCE OF CONTROLLED SEMI-MARTINGALES

N.L. Lazrieva
Moscow State University

1. BASIC NOTIONS, PROBLEM FORMULATION, AND STATEMENT OF THE MAIN RESULTS

In this paper we shall use the definition of a controlled process introduced in [1].

Let $(\Omega, \mathbf{F}, P)$ be a probability space with the filter $F = (\mathbf{F}_t, \mathbf{F}_s < \mathbf{F}_t \subset \mathbf{F}_T = \mathbf{F}, s \leq t \leq T)$ satisfying the usual conditions. The events in σ-algebra $\mathbf{F}_t$ are assumed to occur before time t.

Let $(A, \mathbf{A})$ be a measurable space, where the set A is interpreted as a set of solutions $a \in A$. Each element is associated with a probability P^a on $\mathbf{F}_T$, such that $P^a \sim P$. Let $\mathbf{P}^A = \{P^a, a \in A\}$. The opportunity to choose a certain action based on the accumulated information is expressed by the introduction of a class of controls $\bar{\mathbf{u}}$ consisting of elements $u = \{u(t) = u(t, \omega), t \in [0,T]\}$ which represent mappings of $\Omega \times [0,T]$ into A adapted to the filter F (i.e., $u_t \in \mathbf{F}_t$).

To formulate the controlled process we have to construct the measures P^u corresponding to the strategies $u \in \bar{\mathbf{u}}$, with the natural requirement $P^u = P^a$ for $u \equiv a$. The measures are constructed in the following manner:

Let $\rho_t^a = dP_t^a / dP_t$ be a local density. It is well-known from general martingale theory (see, e.g., [2] and [3]) that the density ρ^a can be represented as the exponential martingale $\varepsilon(M^a)$ of some martingale $M^a = \{M(t, a), t \in [0,T]\}$, i.e., as a solution of the Dolean–Dade equation

$$d\rho_t^a = \rho_{t-}^a dM_t^a \quad ,$$

The measures P^u, $u \in \mathbf{U}$, are given by means of local densities $P^u = \rho^u P$ which, in turn, are the solutions of the equation

$$d\rho_t^u = \rho_{t-}^u dM_t^u \quad ,$$

where $M^u = \{M_t^u, t \in [0,T]\}$ is a stochastic line integral with respect to the class

$\{M^a, a \in A\}$ along the curve $u \in \mathbf{U}$, and $\mathbf{U}$ is the class of admissible strategies, i.e., $\mathbf{U} = \{u : a$ stochastic line integral M^u exists and $\mathrm{E}\varepsilon_T(M^u) = 1\}$.

The stochastic line integral was first considered by Gikhman and Skorohod (see [4]) in the case where the square characteristic of the martingale M^a satisfies the Lipschitz and linear growth conditions with respect to a. The definition of the line integral for the more general situation is given in [1]. The integral $M_t^u = \int_0^t M(\mathrm{ds}, u_s)$ is defined in the following way. The continuous part $M^{u,c}$ is constructed using the Kunita–Watanabe technique, i.e., a continuous martingale $M^{u,c}$ is found such that for any $m \in M_{\mathrm{loc}}^c$ the mutual square characteristic $\langle M^{u,c}, m\rangle$ is given by the equality $\langle M^{u,c}, m\rangle_t = \int_0^t K(\mathrm{ds}, u_s)$, where $K(s, a) = \langle M^{a,c}, m\rangle$. For the discontinuous part $M^{u,d}$ we look for a pure discontinuous martingale whose jump at time t is equal to $M(t, u_t) - M(t_-, u_t)$.

We shall now give two examples which illustrate the notion of the stochastic line integral.

Example 1. $A = \{a_1, a_2, \dots\}$. Then

$$M_t^u = \sum_i \int_0^t I[u_s = a_i] M(\mathrm{ds}, a_i) \quad .$$

Example 2. Let the measure P have the property of integral representation. Then there exist (i) a vector martingale $m = (m^i)$ with continuous components, and (ii) an integer-valued measure μ on a vector space of jumps E, ε, such that for any M^a, $a \in A$, we have the representation

$$M_t^a = \int_0^t (\varphi(s, a), \mathrm{d}m_s) + \int_0^t \int_E \gamma(s, x, a)(\mu - \nu)(\mathrm{ds}, \mathrm{d}x) \quad , \tag{1}$$

where ν is the compensator of the measure μ. Then

$$M_t^u = \int_0^t (\varphi(s, u_s), \mathrm{d}m_s) + \int_0^t \int_E \gamma(s, x, u_s)(\mu - \nu)(\mathrm{ds}, \mathrm{d}x) \quad .$$

We can now formulate the optimization problem (or the problem of optimal absolutely continuous change of measure) as the problem of maximizing the functional $S^u = \mathrm{E}^u \eta$ with respect to the class $\mathbf{U}$, i.e., $s^u = \mathrm{E}^u \eta \to \max_u$, where η is some $\mathbf{F}_t$-measurable random variable with $\mathrm{E}|\eta| < \infty$ and E^u is the expectation with respect to the measure P^u.

Define

$$S = \sup_{u \in \mathbf{U}} \mathrm{E}^u \eta \quad .$$

We shall denote the optimal strategy by u^*, i.e.,

$$S = S^{u^*} = \sup_{u \in \mathbf{U}} \mathrm{E}^u \eta \quad ,$$

and the so-called value process by S_t:

$$S_t = \sup_{u \in \mathbf{U}} \mathrm{E}^u (\eta \mid \mathbf{F}_t) \quad .$$

We shall now consider the problem of the convergence of controlled processes. Let a sequence of sets $\mathbf{P}_n^A = \{P_n^a, a \in A\}$, $n \geq 0$, $P_n \sim P$, be given, together with corresponding families of martingales $M_n^a \in \mathbf{M}_{\mathrm{loc}}^2$, $n \geq 0$. Let $\mathbf{P}_n^u$, $n \geq 0$, be a sequence of sets of measures P_n^u, $n \geq 0$, corresponding to admissible strategies, i.e., $\mathbf{P}_n^u = \{P_n^u = \rho_n^u \cdot P$, $u \in \mathbf{U}\}$, where $\rho_n^u = \varepsilon(M_n^u)$, $M_n^u = \int_0^t M_n(\mathrm{d}s, u_s)$. We study the conditions under which the closeness of the classes $\mathbf{P}_n^a$, $n \geq s$, to class $\mathbf{P}_0^A$ leads to the convergence of the values

$$S_n = \sup_{u \in \mathbf{U}} S_n^u \rightarrow S_0 = \sup_{u \in \mathbf{U}} S_0^u \, , \, n \rightarrow \infty \quad ,$$

and also to the convergence

$$S_0^{u_n^*} \rightarrow S_0 , \, n \rightarrow \infty \quad ,$$

where U_n^* is the optimal strategy in the n-th problem. The second of the above convergence shows that the optimal strategy obtained in the approximated problem is close to the optimal strategy for the initial problem.

We shall illustrate the above statements by looking at a diffusion-type process with a controlled drift coefficient. In this case the measures P_n^a, $a \in A$, are distributions of the weak solutions of the stochastic differential equation

$$\mathrm{d}x_t = f_n(t, x_t, a)\mathrm{d}t + \mathrm{d}w_t \, , \, n = 0,1,2, \ldots \quad ,$$

and the value convergence conditions may be expressed in terms of the convergence of the drift coefficients of the n-th problem to the coefficient of the initial problem. In particular, one of the results stated below for the general case has the following form: if

$$\sup_{n, u \in \mathbf{U}} \left(\int_0^T f_n^2(s, x_s, u_s)\mathrm{d}s \right) \leq c \, , \, p^w \, - \text{a.s.} \quad ,$$

where p^w is a Wiener measure on $(C_{[0,T]}, \mathbf{B}_T)$ and $x = \{x_s, s \in [0,T]\}$ is a coordinate process, then $(A) \Longrightarrow \sup_{u \in \mathbf{U}} \mathrm{Var}\,(P_n^u, P_0^u)_T \to \infty$, $n \to \infty$, where

$$(A) = \sup_{u \in \mathbf{U}} P^w\{\int_0^T (f_n(s, x_s, u_s) - f_0(s, x_s, u_s))^2 ds \geq \varepsilon\} \to 0, \; n \to \infty \quad . \tag{2}$$

In the above case the measures P_n^a become dominated by the Wiener measure P^w with local density

$$\rho_{n,t}^a = dP_{n,t}^a / dP_t^w = \varepsilon_t(M^a) \quad ,$$

where

$$M_n(t, a) = \int_0^t f_n(s, x_s, a) dx_s \quad .$$

The measures P_n^u are also dominated by the Wiener measure, the local densities being exponential martingales $\varepsilon(M_n^u)$, where

$$M_{n,t}^u = \int_0^t f_n(s, x_s, u_s) dx_s \quad .$$

It can be seen from equation (2) that sufficient conditions for the convergence of the controlled processes may be expressed in terms of the convergence of square characteristics of the martingales M_n^a, $n \geq 1$.

It turns out that the conditions for the convergence of controlled processes retain the same form for the more general problem of optimal absolutely continuous change of measure.

THEOREM 1. *Let the condition*

$$\sup_{n,u} \langle M_n^u \rangle_T \leq C \tag{3}$$

hold p-a.s., *where* C *is some constant. Then*

$$(A) \Longrightarrow \sup_{u \in \mathbf{U}} \mathrm{Var}\,(P_n^u, P_0^u)_T \to 0, \; n \to \infty \quad ,$$

where

$$(A): \sup_{u \in \mathbf{U}} P\{\langle M_0^u - M_n^u \rangle_T \geq \varepsilon\} \to 0, \; n \to \infty \quad .$$

In the case of integral representation, i.e., when the martingales M_n^a are given by (1) with functions $\varphi_n(s, a)$ and $\gamma_n(s, x, a)$, condition (A) has the form

$$\sup_{u \in U} P(|\int_0^T \sum_{i,j} (\varphi_n^i - \varphi_0^i)(\varphi_n^j - \varphi_0^j) d\langle m^i, m^j \rangle_s +$$

$$+\int_0^T \int_E (\gamma_n(s,x,u_s) - \gamma_0(s,x,u_s))^2 \nu(ds,dx) - \sum_{s\le t} (\gamma_n - \gamma_0)^2 | \ge \varepsilon\} \to 0 \quad , \quad (4)$$

where

$$\hat{\gamma} = \int_E \gamma(s,x,u_s)\nu(\{s\},dx) \quad .$$

Note that the "integral representation" scheme is arrived at by considering the control of processes satisfying stochastic equations with Wiener and Poisson parts (control of the drift and the jump parts), as well as the control problem in discrete time.

Condition (A) ensures a uniform (with respect to controls) convergence of the measures P_n^u to P_0^u with respect to variations, but is too strong for convergence of values.

THEOREM 2. *Let condition* (3) *of Theorem 1 hold. In addition, let* $|\eta| \le c$, *P-a.s. Then*

$$(B) \Longrightarrow S_n \to S_0 \, , \, n \to \infty \quad ,$$

where

$$(B)\colon \sup_{u \in \mathbf{U}} P\{| <m, M_n^u - M_0^u>_T | \ge \varepsilon\} \to 0 \, , \, n \to \infty$$

for any $m \in \mathbf{M}_{\text{loc}}^2$ *and* $<m, M>$ *is a mutual characteristic of the martingales* m *and* M.

Condition (B) is a condition for weak (uniform with respect to u) convergence of the martingales M_n^u to M_0^u. In the case of a diffusion-type process with controlled drift, condition (B) has the following form:

$$\sup_{u \in \mathbf{U}} P^w(|\int_0^T \varphi(s,x)(f_n(s,x_s,u_s) - f_0(s,x_s,u_s))ds| \ge \varepsilon) \to 0 \, , \, n \to \infty \quad ,$$

for all square-integrable functions φ.

In the case of integral representation the convergence condition takes an analogous form.

To obtain a convergence condition for approximately optimal controls, i.e., the convergence $S_0^{u_n^*} \to S_0$, it is necessary to strengthen condition (B).

THEOREM 3. *If the conditions of Theorem 2 are satisfied, then*

$$(C) \Longrightarrow S_0^{u_n^*} \to S_0 \, , \, n \to \infty \quad ,$$

where

$$(C): \sup_{u \in \mathbf{U}} P(|\langle m^u, M_n^u - M_0^u \rangle_T| \geq \varepsilon\} \to 0, \; n \to \infty$$

for any set of square-integrable martingales $\{m^u, u \in \mathbf{U}\}$.

2. AUXILIARY RESULTS

As in the case of controlled Markov diffusion-type processes, the proof of all statements in this paper is based on Bellman's equations for values.

A formal expression of the optimality principle is given in [1] for controlled processes with the structure outlined in Section 1, i.e., it is proved that a process of value S_t, $t \in [0,T]$, is a solution of the following non-linear stochastic equation with a boundary condition at the end of the interval $[0,T]$:

$$dS_t = dm_t^* - \sup_{a \in A} d\langle s, M^a \rangle_t, \; S_T = \eta \quad , \tag{5}$$

Here m^* is a P-martingale which is a solution of the integral equation

$$m_t^* = \mathrm{E}(\eta + \sup_{u \in \mathbf{U}} \langle m^*, M^u \rangle_T \mid \mathbf{F}_t) \quad , \tag{6}$$

and the expression $\sup_{a \in A} d\langle s, M^a \rangle_t$ is explained below. The process S_t, $t \in [0,T]$, is a special semi-martingale with a martingale part m^*. Then $\langle s, M^a \rangle_t = \langle m^*, M^a \rangle_t$ and

$$\sup_{a \in A} d\langle s, M^a \rangle_t = \sup_{a \in A} k(m^*, t, a) d\langle m^* \rangle_t \quad , \tag{7}$$

where

$$k(m^*, t, a) = d\langle m^*, M^a \rangle_t \mid d\langle m^* \rangle_t \quad .$$

For instance, in the case of a diffusion-type process with a controlled drift coefficient, we have

$$d\xi_t^u = f(t, u_t(\xi^u), \xi^u)dt + dw_t, \; \xi_0 = 0$$

$$m_t^* = S(0,0) + \int_0^t \psi(s, dw_s), \; \psi_t = \frac{\partial}{\partial x} S(t,x)|_{x=w_t} \quad ,$$

$$S_t = \sup_u \mathrm{E}^u(\vartheta(x_T) \mid \mathbf{F}_t) = S(t, w_t) \quad ,$$

and $S(t,x)$ is a solution of Bellman's equation

$$\frac{\partial}{\partial t} S(t,x) + \sup_a (\frac{\partial}{\partial x} S(t,x) f(t,x,a)) + \frac{1}{2} \frac{\partial^2}{\partial x^2} S(t,x) = 0, \; S(T,x) = \vartheta(x) \quad .$$

Equation (5) is an analogue of the familiar recursive relations arising in discrete time dynamic programming:

$$\Delta S_{t+1} = \sup_{a \in A} \mathrm{E}^{a} (S_{t+1} \mid \mathbf{F}_t) \, , \, S_T = \eta$$

which can be written in the following equivalent form:

$$\Delta S_{t+1} = \Delta m_{t+1} - \sup_{a \in A} \Delta \langle S \, , M^{a} \rangle_{t+1} \, , \, S_T = \eta \quad , \tag{8}$$

where

$$\Delta M_t^a = q_t^a \, , \, q_t^a = \rho_t^a / \rho_{t-1}^a \, , \, \rho_t^a = \mathrm{d}P_t^a / \mathrm{d}P_t \quad .$$

Equations (5) and (8) provide the basis for estimating the value difference between the approximated and initial problems, i.e., the modulus of the difference $S_{n,t} - S_{0,t}$. Indeed, the relation

$$dS_{n,t} = dm^{*}_{n,t} - \sup_{a \in A} d \langle S_n \, , M_n^a \rangle_t \, , \, S_{n,T} = \eta \, , n \geq 0 \quad ,$$

leads easily to the inequality

$$d(S_{0,t} - S_{n,t}) \leq d(\text{mar } t) - \inf_{a \in A} (d \langle S_0 - S_n \, , M_n^a \rangle_t + d \langle S_0 , M_0^a - M_n^a \rangle_t \quad ,$$

$$S_{0,T} - S_{n,T} = 0 \quad .$$

It can readily be shown that the solution of this inequality has the form

$$S_{0,T} - S_{n,t} \geq -\sup_{u \in \mathbf{U}} \mathrm{E}_n^u (\mid \langle S_0 , M_0^u - M_n^u \rangle_t^T \mid \mid \mathbf{F}_t) \quad .$$

It is also easy to estimate

$$S_{0,t} - S_{n,t} \leq \sup_{u \in \mathbf{U}} \mathrm{E}_n^u (\mid \langle S_0 , M_0^u - M_n^u \rangle_t^T \mid \mid \mathbf{F}_t) \quad .$$

Then, finally,

$$|S_{0,t} - S_{n,t}| \leq \sup_{u \in \mathbf{U}} \mathrm{E}_n (\mid \langle S_0 , M_0^u - M_n^u \rangle_t^T \mid \mid \mathbf{F}_t) \quad . \tag{9}$$

We shall also need an expression for the difference in benefit resulting from the fixed strategy $u \in \mathbf{U}$ in the approximated and initial control problems, i.e.,

$$S_{0,t}^u - s_{n,T}^u = \mathrm{E}_0^u(\eta \mid \mathbf{F}_t) - \mathrm{E}_n^u(\eta \mid \mathbf{F}_t) \quad .$$

Using Girsanov's theorem it is not difficult to check the validity of the relation

$$S_{0,t}^u - S_{n,t}^u = \mathrm{E}_n^u (\langle S_0^u \, , M_0^u - M_n^u \rangle_t^T \mid \mathbf{F}_t) \quad , \tag{10}$$

where $<S, M>_t^T = <S, M>_T - <S, M>_t$.

We shall use the following facts:

$$E <m^*>_T < \infty \tag{11}$$

$$\sup_{u \in \mathbf{U}} E<m^u>_T < \infty \quad , \tag{12}$$

where m^* is a martingale satisfying equation (6) and m^u is a martingale which appears in the decomposition of the special semi-martingale $S_t^u = E^u(\eta \mid \mathbf{F}_t)$:

$$S_t^u = m_t^u + A_t^u \quad .$$

We shall now demonstrate the validity of (11). It is shown in [1] that the martingale m^* can be constructed using the sequential approximation technique in the following way. Let

$$\sup_{u \in \mathbf{U}} <M^u>_T \le c < \infty \quad .$$

Define the Markov moments $0 = \tau_0 < \tau_1 < \cdots < \tau_n$ using the relations

$$\sup_{u \in \mathbf{U}} (<M^u>_{\tau_{i+1}} - <M^u>_{\tau_i}) \le c/n < 1 \quad ,$$

and construct a sequence of martingales m^i, $i = 0,1, \ldots, n$, which are solutions of the equations

$$m_t^i = E(m_{\tau_{i+1}}^{i+1} - \sup_{u \in \mathbf{U}} <M^u, m^i>_{\tau_i}^{\tau_{i+1}} \mid \mathbf{F}_t) = V_t(m^i) \tag{13}$$

as the limits of sequential approximations $m^{i,l}$, $l \to \infty$,

$$m_t^{i,l+1} = V_t(m^{i,l}) \, , \; l = 0,1, \ldots \quad . \tag{14}$$

The convergence of this procedure and the uniqueness of the solution of equation (13) follow from the estimate

$$\| m^{i,l+1} - m^{i,l} \|^2 = E(m_{\tau_{i+1}}^{i,l+1} - m_{\tau_{i+1}}^{i,l})^2 \le \frac{c}{n} \| m^{i,l} - m^{i,l-1} \|^2 \quad , \tag{15}$$

which can easily be checked for every $0 < i \le n$. It may readily be seen that the expression defined by the relation

$$m_t^* = \begin{cases} m_t^0 & , \tau_0 \le t < \tau_i \\ m_t^i + m_{\tau_i}^* & , \tau_i \le t < \tau_{i+1} \quad , 1 \le i \le n \end{cases}$$

is a solution of equation (6).

From (15) we can derive the estimate

$$E\,\|m^{i,l+1}\|^2 \le (c/n)^l E(m^{i+1}_{\tau_{i+1}})^2$$

which, in turn, leads to (11).

Relation (12) may be proved in a similar manner.

3. PROOFS OF THE THEOREMS

Proof of Theorem 1. We have

$$\sup_{u\in U} \mathrm{Var}\,(P^u_{0,T}, P^u_{n,T}) = \sup_{u\in U}\sup_{|\eta|\le c} |E^u_0\eta - E^u_n\eta| \le$$

$$\le \sup_{|\eta|\le c}\sup_{u\in U} |E^u_0\eta - E^u_n\eta| = \sup_{|\eta|\le c}\sup_{u\in U} |S^u_0 - S^u_n| \quad . \tag{16}$$

From formula (10) we have

$$\sup_{u\in U} \mathrm{Var}\,(P^u_{0,T}, P^u_{n,T}) \le \sup_{|\eta|\le c}\sup_{u\in U} E^u_n(<S^u, M^u_0 - M^u_n>_T) \le$$

$$\sup_{|\eta|\le c, u} E(\rho^u_{n,T}<m^u>^{1/2}_T<M^u_0 - M_n>^{\frac{1}{2}}_T \quad , \tag{17}$$

since $<S^u, m> = <m^u, M>$ for any P-martingale.

Note that by virtue of condition (3), $E(\rho^u_{n,T})^2 \le e^c$ ([2]). The sequence $\rho^u_{n,T}$, $n \ge 1$, is therefore uniformly integrable. By virtue of (12) the sequence $\rho^u_{n,t}<m^u>^{1/2}_T$ is also uniformly integrable. The assertion of Theorem 1 now follows from (17), condition (A) and the fact that the sequence $\rho^u_{n,T}<m^u>^{1/2}_T<M^u_0 - M^u_n>_T$, $n \ge 1$, converges to zero with respect to the measure P and is bounded by a uniformly integrable sequence.

Proof of Theorem 2. We have the estimate

$$|S_{0,T} - S_{n,t}| \le \sup_{u\in U} E^u_n(|<S_0, M^u_0 - M^u_n>^T_t| \mid F_t) \quad . \tag{18}$$

Consequently,

$$|S_0 - S_n| \le \sup_{u\in U} E^u_n(|<S, M^u_0 - M^u_n>_T|) = \sup_{u\in U} E(\rho^u_{n,T}<m^*, M^u_0 - M^u_0 - M^u_n>_T)$$

and the assertion of the theorem immediately follows from condition (B), relation (11) and the fact that the sequence $\rho^u_{n,T}<m^*, M^u_0 - M^u_n>_T$, $n \ge 1$, is bounded by the uniformly integrable sequence $k(c)\rho^u_{n,t}<m^*>^{1/2}_T$.

Remark 1. Using arguments similar to the above, it can easily be shown that under the conditions of Theorem 2 we have $S_{n,T} \to S_{0,t}$ with respect to the measure P for every $t \in [0,T]$.

Proof of Theorem 3. We know that

$$|S_0 - S_0^{u_n^*}| \le |S_0 - S_n| + |S_n - S_0^{u_n^*}| \quad .$$

But since $S_n = S_n^{u_n^*}$ we have

$$|S_0 - S_0^{u_n^*}| \le |S_0 - S_n| + |S_n^{u_n^*} - S_0^{u_n^*}| \quad .$$

By virtue of Theorem 2, the first term in the last inequality converges to zero. With regard to the second term, we have

$$|S_n^{u_n^*} - S_0^{u_n^*}| \le \sup_{u \in \mathbf{U}} |S_n^u - S_0^u| \le \sup_{u \in \mathbf{U}} \mathrm{E}_n^u(|<S_0^u, M_0^u - M_n^u>_T|)$$

and by arguments similar to those used above, it can be proved that under the conditons of the theorem

$$(B) \Longrightarrow |S_n^{u_n^*} - S_0^{u_n^*}| \to 0 \, , \, n \to \infty \quad .$$

This completes the proof of Theorem 3.

Example 1. We shall illustrate the use of the theorem for the convergence of controlled processes by applying it to the problem of discrete approximations. The Eulerian approximation of diffusion-type equations (smoothing) is a particular case of the approximation scheme in which the martingales M_0^a have representation (1) and the martingales M_n^a are given by (1) with the functions φ_0 and γ_0 replaced by φ_n and γ_n defined as follows:

$$\varphi_n(t, a) = \varphi_0(t_{ni}, a) \, , \; t_{ni} \le t < t_{n(i+1)} \quad ,$$

$$\gamma_n(t, a) = \gamma_0(t_n, x, a) \, , \; t_{ni} \le t < t_{n(i+1)} \quad ,$$

where $\{t_{ni}, i = \overline{1,n}\}$ is a subdivision of the interval $[0,T]$ such that $\Delta_n = \max_{0 \le i \le n} |t_{n(i+1)} - t_{ni}| \to 0 \, , \, n \to \infty$.

Applying the results of the present paper, we can obtain sufficient conditions for the convergence of the controlled processes $\{P_n^u = \varepsilon(M_n^u) \cdot P \, , \, u \in \mathbf{U}\}$ to the process $\{P_0^u = \varepsilon(M_0^u) \cdot P \, , \, u \in \mathbf{U}\}$ in terms of the continuity of the functions φ_0 and γ_0. Thus, for instance, if the martingales w^i in representation (1) are orthogonal and $\nu(\{t\}, \mathrm{E}) = 0$, then in order to have

$$\sup_{u \in \mathbf{U}} \text{Var}\,(P_n^u, P_0^u)_T \to 0\,,\; n \to \infty \quad ,$$

it is sufficient that

$$\omega^{\varphi_i}(\Delta) = \sup_{|t-s|} \sup_{a \in A} |\varphi_0^i(t\,,a) - \varphi_0^i(s\,,a)|^2 \overset{P}{\to} 0 \quad ,$$

$$\int_E \omega^\gamma(\Delta, x)\nu([0,T]\,,dx) \overset{P}{\to} 0 \quad ,$$

where $\omega^\gamma(\Delta, x) = \sup_{|t-s| \le \Delta} \sup_{a \in A} |\gamma_0(s\,,x\,,a) - \gamma_0(t\,,x\,,a)|^2$.

The problems of constructing approximations for controlled processes have been considered by a number of authors (Kushner, Gikhman, Skorohod, Praguruskas, Cristopheit, etc.).

The convergence of controlled Markov chains to diffusion-type Markov processes with controlled drift coefficients is studied in [5], where, in addition to value convergence, convergence of the approximated optimal strategy was also proved.

REFERENCES

1. R.Ya. Chitashvili. Martingale ideology in the theory of controlled stochastic processes. In: K. Ito and Y.V. Prokhorov (Eds.), Probability theory and mathematical statistics. Proceedings, 4th USSR-Japan Symposium, Tbilisi, August 1982. *Lecture Notes in Mathematics*, Vol. 1021. Springer-Verlag, Berlin, Heidelberg, New York, Tokyo, 1983, pp. 73-92.

2. Yu.M. Kabanov, R.S. Liptser and A.N. Shiryaev. Absolute continuity and singularity of locally absolutely continuous probability distributions, I (in Russian). *Matematicheskie Sbornik*, 107(3)(1973)364–415.

3. J. Jacod. Calcul stochastique et problemés de martingales. *Lecture Notes in Mathematics*, Vol. 714. Springer-Verlag, Berlin, Heidelberg, New York, 1979.

4. I.I. Gikhman and A.V. Skorohod. *Theory of Random Processes*, Vol. 3. Nauka, Moscow, 1975 (in Russian).

5. N.L. Lazrieva. Construction of ε-optimal controls for controlled Markov processes (in Russian). *Trudy IPM TGU.*

ESTIMATION OF PARAMETERS AND CONTROL OF SYSTEMS WITH UNKNOWN PARAMETERS

S.Ya. Mahno
Institute of Applied Mathematics and Mechanics,
Academy of Sciences of the Ukrainian SSR, Kiev, USSR

At present we have a well-developed theory for estimating the unknown parameters of completely observed stochastic systems [1]. However, real objects do not always permit direct observation, and in connection with this arises the additional problem of parameter estimation under indirect observations. In this paper we shall consider parameter estimations in a partially observed system and will study the properties of maximum likelihood estimators. In particular, we shall obtain formulas for computing their shifts and mean-square observations, and give a condition for asymptotic normality. We shall also solve (for a linear stochastic system) the control problem which arises when the equation of motion of an object contains non-random unknown parameter and obtain direct formulas for optimal control.

The stochastic processes and variables considered here are defined on the main probability space. We shall use $E\{\cdot\}$ and $E\{\cdot/\cdot\}$ to denote mathematical expectation and conditional mathematical expectation, respectively.

Let (η_t, ξ_t) be a partially observable random process satisfying the following system of recursive equations:

$$\eta_{t+1} = a_0(t,\xi) + a_1(t,\xi)\vartheta + a_2(t,\xi)\eta_t + b_1(t,\xi)\varepsilon_1(t+1) + b_2(t,\xi)\varepsilon_2(t+1) \quad ,$$

$$\xi_{t+1} = A_0(t,\xi) + A_1(t,\xi)\eta_t + B_1(t,\xi)\varepsilon_1(t+1) + B_2(t,\xi)\varepsilon_2(t+1) \quad , \tag{1}$$

$$\xi_0 = 0 \, , \quad t = 0,1,\ldots,T-1 \quad .$$

Here $\varepsilon_1(t) = (\varepsilon_{11}(t),\ldots, \varepsilon_{1e}(t))$ and $\varepsilon_2(t) = (\varepsilon_{21}(t),\ldots, \varepsilon_{2q}(t))$ are independent components. Each of these components is Gaussian $N(0,1)$. The stochastic process η_t is a vector of dimension n, the process ξ_t is a vector of dimension m and ϑ is an unknown parameter of dimension k. The coefficients of the equations are non-anticipative functionals.

The problem is to estimate the parameter ϑ using the results of observations ξ_t. We will suppose that the following conditions are satisfied:

1. If $g(t,\xi)$ is one of the functionals $a_0^i(t,\xi)$, $a_1^{ij}(t,\xi)$, $A_0^i(t,\xi)$, $b_r^{ij}(t,\xi)$, $B_r^{ij}(t,\xi)$, $r=1,2$, then

$$\mathrm{E}\,|g(t,\xi)|^2 < \infty,\ t=0,1,\dots,T \quad .$$

2. With probability 1, we have

$$|a_2^{ij}(t,\xi)| + |A_1^{ij}(t,-\xi)| \leq C \quad .$$

3. The distribution η_0 is Gaussian $N(m_0,\gamma_0)$.

Set $m_t = \mathrm{E}\{\eta_t\,|F_t^\xi\}$, $\gamma_t = \mathrm{E}\{(\eta_t - m_t)(\eta_t - m_t)^* / F_t^\xi\}$, $F_t^\xi = \sigma\{\xi_0,\dots,\xi_t\}$. It is known (see [2]) that if conditions 1–3 are satisfied, then the random processes m_t and γ_t satisfy the equations:

$$m_{t+1} = a_0(t,\xi) + (a_2(t,\xi) - [b\cdot B + a_2\gamma_t A_1^*][B\cdot B + A_1\gamma_t A_1^*]^{-1} \times$$

$$\times A_1(t,\xi))m_t + a_1(t,\xi)\vartheta + [b\cdot B + a_2\gamma_t A_1^*][B\cdot B + A_1\gamma_t A_1^*]^{-1}(t,\xi)(\xi_{t+1} - A_0(t,\xi)) \quad ,$$

(2)

$$\gamma_{t+1} = a_2(t,\xi)\gamma_t a_2^*(t,\xi) + b\cdot b + [b\cdot B + a_2\gamma_t A_t^*][B\cdot B + a_2\gamma_t A_1^*] \times$$

$$\times [b\cdot B + a_2\gamma_t A_1^*]^*(t,\xi) \quad .$$

We used here the notation from [2]. In particular, $b\cdot b = b_1 b_1^* + b_2 b_2^*$, $b\cdot B = b_1 B_1^* + b_2 B_2^*$, $B\cdot B = B_1 B_1^* + B_2 B_2^*$, and A^* is conjugate matrix to A. Note that γ_t is a non-anticipative measurable functional independent of ϑ. From equation (2) we have

$$m_{t+1} = a_{01}(t,\xi) + a_{11}(t,\xi)\vartheta + g(t)m_0 + \delta(t+1,\xi) \quad , \tag{3}$$

where

$$a_{01}(t,\xi) = \sum_{s=0}^{t} g_{s+1}(t)[a_0(s,\xi) - c(s,\xi)A_0(s,\xi)] \quad ,$$

$$a_{11}(t,\xi) = \sum_{s=0}^{t} g_{s+1}(t)a_1(s,\xi) \quad ,$$

$$\delta(t+1,\xi) = \sum_{s=0}^{t} g_{s+1}(t)c(s,\xi)\xi_{s+1} \quad ,$$

$$g_u(t) = \prod_{s=u}^{t} [a_2(s,\xi) - c(s,\xi)A_1(s,\xi)],\ g_{t+1}(t) = 1 \quad ,$$

$$g(t) = g_0(t),\ g_0(-1) = 1 \quad ,$$

$$c(t,\xi) = [b \cdot B + a_2 \gamma_t A_1^*][B \cdot B + A_1 \gamma_t A_1^*]^{-1}(t,\xi) \quad .$$

We shall assume that sums of the type $\sum_{s=0}^{-1}$ are equal to zero. From the relation between ξ_t and m_t (see [2]), (13.78)):

$$\xi_{t+1} = A_0(t,\xi) + A_1(t,\xi)m_t + [B \cdot B + A_1 \gamma_t A_1^*]^{1/2}(t,\xi)\varepsilon_3(t+1)$$

and the equality (3), we have

$$\xi_{t+1} = A_{01}(t,\xi) + A_{11}(t,\xi)\vartheta + B_{11}(t,\xi)\varepsilon_3(t+1) \quad , \tag{4}$$

for

$$A_{01}(t,\xi) = A_0(t,\xi) + A_1(t,\xi)a_{01}(t-1,\xi) + A_1(t,\xi)g(t-1)m_0 + A_1(t,\xi)\delta(t,\xi) \quad ,$$

$$A_{11}(t,\xi) = A_1(t,\xi)a_{11}(t-1,\xi) \quad ,$$

$$B_{11}(t,\xi) = [B \cdot B + A_1 \gamma_t A_1^*]^{1/2}(t,\xi) \quad .$$

From (4), the estimate of ϑ can be represented as follows:

$$\vartheta_t = \left(\sum_{s=1}^{t-1} A_{11}^*(s,\xi)(B_{11}B_{11}^*)^{-1}(s,\xi)A_{11}(s,\xi)\right)^{-1} \times \\ \times \sum_{s=1}^{t-1} A_{11}(s,\xi)(B_{11}B_{11}^*)^{-1}(s,\xi)(\xi_{s+1} - A_{01}(s,\xi)) \quad . \tag{5}$$

Notice that the sumation in (5) starts at one ($A_{11}(0,\xi) = 0$). This may be explained by the fact that the dependence of ξ_t on ϑ only comes into effect with moment $t = 2$.

Define

$$\sigma(t,\xi) = \sum_{s=1}^{t-1} A_{11}^*(s,\xi)(B_{11}B_{11}^*)^{-1}(s,\xi)A_{11}(s,\xi) \quad ,$$

$$\Gamma(t,\xi) = \sum_{s=1}^{t-1} A_{11}^*(s,\xi)(B_{11}B_{11}^*)^{-1}(s,\xi)B_{11}(s,\xi) \quad ,$$

$$M(t,\xi) = \sum_{s=1}^{t-1} A_{11}^*(s,\xi)(B_{11}B_{11}^*)^{-1}(s,\xi)B_{11}(s,\xi)\varepsilon_3(s+1) \quad ,$$

and let σ_{ij}, Γ_{ij}, M_i be the elements of the matrices σ, Γ and vector M, respectively.

LEMMA 1. *Let $\delta_{ij}(t,\xi)$, $i,j = \overline{1,k}$, be a non-anticipative functional and the following conditions hold for every $\vartheta_1^i < \vartheta_2^i$, $i = \overline{1,k}$ (written $\vartheta_1 < \vartheta_2$ below for simplicity):*

$$\sup_{\vartheta_1 \le \vartheta \le \vartheta_2} \mathrm{E}_\vartheta \delta_{ij}^{\alpha}(t\,,\xi) < \infty \quad ,$$

$$\sup_{\vartheta_1 \le \vartheta \le \vartheta_2} \mathrm{E}_\vartheta \sigma_{ij}^{\beta}(t\,,\xi) < \infty \quad , \tag{6}$$

$$\sup_{\vartheta_1 \le \vartheta \le \vartheta_2} \mathrm{E}_\vartheta \Gamma_{ij}^{\beta}(t\,,\xi) < \infty$$

for $\alpha = \beta = 4$. *Then the function* $\mathrm{E}_\vartheta \delta_{ij}(t\,,\xi)$ *is differentiable on* ϑ *and*

$$\frac{\partial}{\partial \vartheta_j} \mathrm{E}_\vartheta \delta_{ij}(t\,,\xi) = \mathrm{E}_\vartheta \delta_{ij}(t\,,\xi) M_j(t\,,\xi) \quad .$$

If the inequalities (6) also hold for $\alpha = 2\beta = 16$, then $\mathrm{E}_\vartheta \delta_{ij}(t\,,\xi)$ will be twice differentiable in ϑ and

$$\sum_{i,j=1}^{k} \frac{\partial^2}{\partial \vartheta_i \, \partial \vartheta_j} \mathrm{E}_\vartheta \delta_{ij}(t\,,\xi) = \mathrm{E}_\vartheta M^*(t\,,\xi)\delta(t\,,\xi)M(t\,,\xi) - \mathrm{E}_\vartheta \mathrm{Sp}\,\delta(t\,,\xi)\sigma(t\,,\xi) \quad ,$$

where SpA is the trace of matrix A.

Consider the properties of estimator (5).

THEOREM 1. *If the function* $\varphi(t)$ *is such that* $\lim_{t\to\infty} \varphi(t) = \infty$ *and* $\lim_{t\to\infty} \frac{1}{\varphi(t)} \sigma(t\,,\xi) = \sigma$, *where matrix* σ *is nondegenerate, then the estimator* ϑ_T *is consistent and the vector* $\sqrt{\varphi(T)}(\vartheta_T - \vartheta)$ *is asymptotically normal* $N(0\,,\sigma^{-1})$. *Moreover, if the inequalities* (6) *hold for* $\alpha = \beta = 4$ *and* $\delta(t\,,\xi) = \sigma^{-1}(t\,,\xi)$, *then the bias of the estimator may be defined by the formula*

$$\mathrm{E}[\vartheta_T^i - \vartheta^i] = \sum_{j=1}^{k} \frac{\partial}{\partial \vartheta_j} \mathrm{E}_\vartheta \delta_{ij}(T\,,\xi) \quad .$$

If inequalities (6) *hold for* $\alpha = 2\beta = 16$ *and* $\delta(t\,,\xi) = \sigma^{-2}(t\,,\xi)$, *then*

$$\mathrm{E}(\vartheta_T - \vartheta)(\vartheta_T - \vartheta)^* = \sum_{i,j=1}^{k} \frac{\partial^2}{\partial \vartheta_i \, \partial \vartheta_j} \mathrm{E}_\vartheta \delta_{ij}(T\,,\xi) + \mathrm{E}_\vartheta \mathrm{Sp}\,\sigma^{-1}(T\,,\xi) \quad .$$

These assertions are proved in [3].

COROLLARY. *Let a partially observable random process be described by equations*

$$\eta_{t+1} = a_0(t\,,\xi) + a_1(t)\vartheta + a_2(t)\eta_t + b_1(t)\varepsilon_1(t+1) + b_2(t)\varepsilon_2(t+1) \quad ,$$

$$\xi_{t+1} = A_0(t\,,\xi) + A_1(t)\eta_t + B_1(t)\varepsilon_1(t+1) + B_2(t)\varepsilon_2(t+1) \quad , \tag{7}$$

$$\xi_0 = 0\,,\ t = 0,1\,,\ldots,\ T-1 \quad .$$

Then the estimator (5) *for partially observable process* (7) *is unbiased and consistent.*

Now consider the following control problem. Let a partially observable random process (η_t, ξ_t) be described by the equations:

$$\eta_{t+1} = a_1(t)\vartheta + a_2(t)\eta_t + \alpha_t u_t + b(t)\varepsilon_1(t+1) \quad ,$$

$$\xi_{t+1} = A(t)\eta_t + B(t)\varepsilon_2(t+1) \quad , \tag{8}$$

with initial conditions $\xi_0 = 0$, η_0, where η_0 is a Gaussian vector with parameter (m_0, γ_0). The problem is to choose the control u_t which minimizes the cost functional

$$I[u] = \mathrm{E}[\sum_{t=0}^{T-1} (\eta_t^* L_t \eta_t + u_t^* N_t u_t) + \eta_T^* L_T \eta_T] \quad . \tag{9}$$

The dimensions of the vectors in (8) are assumed to be the same as in (1) (the system (8) is a particular case of (1) with $a_0(t, \xi) = \alpha_t u_t$), and the dimensions of $u_t = (u_t^1, \dots, u_t^r)$ is equal to r.

The class of admissible controls u consists of controls u_t which are $\sigma\{\xi_0, \xi_1, \dots, \xi_t\}$-measurable functionals for any $t = 0,1, \dots, T-1$, explicitly independent of ϑ, and for which

$$\sum_{t=1}^{T} \mathrm{E}|u_t^t|^2 < \infty \, , \; t = 0,1, \dots, T-1 \quad .$$

Assume $\alpha_0 = \alpha_1 = 0$ (i.e., an observer starts to control the system only after receiving some information about the unknown parameter). The cost functional $I[u]$ may be represented in the following way (see [2], (14.71)):

$$I[u] = I_1[u] + \sum_{t=0}^{T} \mathrm{Sp} L_t^{1/2} \gamma_t L_t^{1/2} \quad ,$$

$$I_1[u] = \mathrm{E}[\sum_{t=0}^{T-1} (m_t^* L_t m_t + u_t^* N_t u_t) + m_T^* L_T m_T] \quad ,$$

where the processes m_t and γ_t are defined by (2). In this case γ_t is a non-random matrix. Using Lemma 13.5 from [2], we may write an equation for m_t:

$$m_{t+1} = a_1(t)\vartheta + a_2(t)m_t + \alpha_t u_t + D_t \varepsilon_3(t+1) \quad ,$$

where $D_t = a_2(t)\gamma_t A^*(t) \, [B(t)B^*(t) + A(t)\gamma_t A^*(t)]^{1/2}$.

Introduce a function $V_t^\vartheta(x) = x^* P_t x + \vartheta^* Q_t^* x + x^* Q_t \vartheta + \vartheta^* R_t \vartheta + K_t$, where P_t, Q_t, R_t are matrix functions and K_t is a scalar function. These functions may be defined as solutions of the following recursive equations:

$$P_t = L_t + a_2^*(t)P_{t+1}a_2(t) - a_2^*(t)P_{t+1}\alpha_t[N_t + \alpha_t^*P_{t+1}\alpha_t]^+ \times$$

$$\times \alpha_t^* P_{t+1}a_2(t)\,,\ P_T = L_T \quad ,$$

$$Q_t = a_2^*(t)P_{t+1}a_1(t) + a_2^*(t)Q_{t+1} - a_2^*(t)P_{t+1}\alpha_t[N_t + \alpha_t^*P_{t+1}\alpha_t] \times$$

$$\times \alpha_t^*(Q_{t+1} + P_{t+1}a_1(t))\,,\ Q_T = 0 \quad ,$$

$$R_t = a_1^*(t)P_{t+1}a_1(t) + a_1^*(t)Q_{t+1} + Q_{t+1}^*a_1(t) - (Q_{t+1} + P_{t+1}a_1(t))^* \times$$

$$\times \alpha_t(N_t + \alpha_t^*P_{t+1}\alpha_t)^+\alpha_t^*(Q_{t+1} + P_{t+1}a_1(t))\,,\ R_T = 0 \quad ,$$

$$K_t = K_{t+1} + \mathrm{Sp}D_t^*P_{t+1}D_t\,,\ K_T = 0 \quad ,$$

where A^+ is the pseudo-inverse of matrix A.

Notice that we cannot use a random process m_t in the control problem since it is dependent on an unknown parameter ϑ. We shall therefore act in the following way. Define the stochastic process x_t as the solution of the recursive equations:

$$x_0 = m_0 \quad ,$$

$$x_{t+1} = [a_2(t) - a_2(t)\gamma_t A^*(t)](B(t)B^*(t) + A(t)\gamma_t A^*(t))^{-1} \times \tag{10}$$

$$\times A(t)]x_t + \alpha_t u_t + a_2(t)\gamma_t A^*(t)(B(t)B^*(t) + A(t)\gamma_t A^*(t))^{-1}\xi_{t+1} \quad .$$

Then

$$m_{t+1} = a_{11}(t)\vartheta + x_{t+1} \quad . \tag{11}$$

Making use of (11) we can easily obtain

$$\mathrm{E}[V_{t+1}^{\vartheta}(m_{t+1}) - V_t^{\vartheta}(m_t)] = -\mathrm{E}[m_t^*L_t m_t + u_t^*N_t u_t] +$$

$$+ \mathrm{E}[u_t + (\alpha_t^*P_{t+1}\alpha_t + N_t)^+\alpha_t^*P_{t+1}a_2(t)x_t + (\alpha_t^*P_{t+1}\alpha_t + N_t)^+ \times$$

$$\times \alpha_t^*(P_{t+1}a_2(t)a_{11}(t-1) + P_{t+1}a_1(t) + Q_{t+1})\vartheta]^*(\alpha_t^*P_{t+1}\alpha_t + N_t) \times$$

$$\times [u_t + (\alpha_t^*P_{t+1}\alpha_t + N_t)^+\alpha_t^*P_{t+1}a_2(t)x_t + (\alpha_t^*P_{t+1}\alpha_t + N_t)^+ \times$$

$$\times \alpha_t^*(P_{t+1}a_2(t)a_{11}(t-1) + P_{t+1}a_1(t) + Q_{t+1})\vartheta] \quad .$$

From this relation we then have

$$I_1[u] = V_0^{\vartheta}(m_0) + \sum_{t=0}^{T-1} \mathrm{E}[u_t + (\alpha_t^* P_{t+1}\alpha_t + N_t)^+ \alpha_t^* P_{t+1} a_2(t) x_t +$$

$$+ (\alpha_t^* P_{t+1}\alpha_t + N_t)^+ \alpha_t^* (P_{t+1} a_2(t) a_{11}(t-1) + P_{t+1} a_1(t) + Q_{t+1})\vartheta]^* \times \tag{12}$$

$$\times (\alpha_t^* P_{t+1}\alpha_t + N_t)[u_t + (\alpha_t^* P_{t+1}\alpha_t + N_t)^+ \alpha_t^* P_{t+1} a_2(t) x_t +$$

$$+ (\alpha_t^* P_{t+1}\alpha_t + N_t)^+ \alpha_t (P_{t+1} a_2(t) a_{11}(t-1) + P_{t+1} a_1(t) + Q_{t+1})\vartheta] \quad .$$

We shall now show that the optimal control has the following form:

$$u_t^0 = -(\alpha_t^* P_{t+1}\alpha_t + N_t)^+ \alpha_t^* P_{t+1} a_2(t) x_t - (\alpha_t^* P_{t+1}\alpha_t + N_t)^+ \times$$

$$\times \alpha_t^* (P_{t+1} a_2(t) a_{11}(t-1) + P_{t+1} a_1(t) + Q_{t+1})\vartheta_t \ , \ t \geq 2 \ , \ u_0^0 = u_1^0 = 0 \quad , \tag{13}$$

where

$$\vartheta_t = \Big(\sum_{s=1}^{t-1} a_{11}^*(s-1) A^*(s)(B_{11}(s) B_{11}^*(s))^{-1} A(s) a_{11}(s-1)\Big)^{-1} \times$$

$$\times \sum_{s=1}^{t-1} a_{11}^*(s-1) A^*(s)(B_{11}(s) B_{11}^*(s))^{-1}(\xi_{s+1} - A(s) x_s) \quad . \tag{14}$$

The estimator (14) is none other than the estimator (5) expressed in terms of an auxiliary process x_t.

Let u be an arbitrary admissible control. Assume that

$$\bar{u}_t = -(\alpha_t^* P_{t+1}\alpha_t + N_t)^{1/2}[u_t + (\alpha_t^* P_{t+1}\alpha_t + N_t)^+ \alpha_t^* P_{t+1} a_2(t) x_t] \quad ,$$

$$T_t = (\alpha_t^* P_{t+1}\alpha_t + N_t)^{1/2}(\alpha_t^* P_{t+1}\alpha_t + N_t)^+ \alpha_t^* (P_{t+1} a_2(t) a_{11}(t-1) + P_{t+1} a_1(t) + Q_{t+1}) \quad .$$

We then write the relation (12) in the form

$$I_1[u] = V_0^{\vartheta}(m_0) + \sum_{t=0}^{T-1} \mathrm{E} u_t^* N_t u_t + \sum_{t=2}^{T-1} \mathrm{E}[T_t\vartheta - T_t T_t^+ \bar{u}_t + T_t T_t^+ \bar{u}_t - \bar{u}_t]^* \times$$

$$\times [T_t\vartheta - T_t T_t^+ \bar{u}_t + T_t T_t^+ \bar{u}_t - \bar{u}_t] \quad .$$

Utilizing the properties of pseudo-inverse matrices $(AA^+)^* = AA^+$, $A^*AA^+ = A^*$, $AAA^+ = A$ [2], we transform the last expression to give

$$I_1[u] = V_0^{\vartheta}(m_0) + \sum_{t=0}^{1} \mathrm{E} u_t^* N_t u_t + \sum_{t=2}^{T-1} \mathrm{E}\bar{u}_t^* (T_t T_t^+ - \mathbb{I})^* (T_t T_t^+ - \mathbb{I})\bar{u}_t$$

$$+\sum_{t=2}^{T-1} \mathrm{E}(\vartheta - T_t^{+}\bar{u}_t)^{*} T_t^{*}(\vartheta - T_t^{+}\bar{u}_t) \quad ,$$

where $\mathbb{I}$ is the unit matrix.

Hence, for any control u we have

$$I_1[u] \geq V_0^{\vartheta}(m_0) + \sum_{t=2}^{T-1} \mathrm{E}(\vartheta - T_t^{+}\bar{u}_t)^{*} T_t^{*} T_t (\vartheta - T_t^{+}\bar{u}_t) \quad .$$

In view of the effectiveness of the estimator ϑ_t (see the Corollary) the sum on the right-hand side is no less than $\sum_{t=2}^{T-1} \mathrm{E}(\vartheta - \vartheta_t)^{*} T_t^{*} T_t (\vartheta - \vartheta_t)$. Thus, for any control u we have

$$I_1[u] \geq V_0^{\vartheta}(m_0) + \sum_{t=2}^{T-1} \mathrm{E}(\vartheta - \vartheta_t)^{*} T_t^{*} T_t (\vartheta - \vartheta_t) \quad .$$

Apply Theorem 1. Then

$$I_1[u] \geq V_0^{\vartheta}(m_0) + \sum_{t=2}^{T-1} \mathrm{Sp}\sigma^{-1}(t)(P_{t+1}a_2(t)a_{11}(t-1) +$$

$$+ P_{t+1}a_1(t) + Q_{t+1})\alpha_t(\alpha_t^{*}P_{t+1}\alpha_t + N_t)^{+}\alpha_t^{*}(P_{t+1}a_2(t)a_{11}(t-1) + P_{t+1}a_1(t) + Q_{t+1}) \quad .$$

It is clear from (12) that the equality is satisfied for the control (13). This proves the following theorem:

THEOREM 2. *An optimal control from the class of admissible controls for the system* (8), (9) *is defined by equality* (13), *where x_t can be found from* (10) *and ϑ_t from* (14). *The optimal cost functional is then*

$$I^0 = V_0^{\vartheta}(m_0) + \sum_{t=0}^{T} \mathrm{Sp}L_t^{1/2}\gamma_t L_t^{1/2} + \sum_{t=2}^{T-1} \mathrm{Sp}\sigma^{-1}(t)(P_{t+1}a_2(t)a_{11}(t-1) +$$

$$+ P_{t+1}a_1(t) + Q_{t+1})^{*}\alpha_t(\alpha_t^{*}P_{t+1}\alpha_t + N_t)^{+}\alpha_t^{*}(P_{t+1}a_2(t)a_{11}(t-1) +$$

$$+ P_{t+1}a_1(t) + Q_{t+1}) \quad .$$

REFERENCES

1. I.A. Ibragimov and R.Z. Hasminski. *Asymptotic Theory of Estimation*. Nauka, Moscow, 1979 (in Russian).
2. R.S. Liptzer and A.N. Shiryaev. *Statistics of Random Processes*. Nauka, Moscow, 1974 (in Russian).

3. S.Ya. Mahno. The estimation of the drift parameter of the non-observed component of partially observed random processes (in Russian). *Problemi Peredatchi Informatsii*, 16(4) (1980) 36–40.

ON RECURSIVE APPROXIMATIONS WITH ERROR BOUNDS IN NONLINEAR FILTERING

G.B. Di Masi, W.J. Runggaldier*, B. Armellin*

Istituto di Elettrotecnica e di Elettronica, Università di Padova and CNR-LADSEB, 35100 Padova, Italy

* Seminario matematico, Università di Padova, 35100 Padova, Italy

1. INTRODUCTION

We consider the following discrete-time nonlinear filtering problem: a partially observable process (x_t, y_t), $x_t, y_t \in \mathbb{R}$, with x_t the unobservable and y_t the observable components, is given for $t=0,1,\dots,T$ on some probability space $(\Omega, \mathcal{F}, P)$ by

$$x_{t+1} = a(x_t) + v_{t+1} \quad ; \quad x_o = v_o \tag{1.1.a}$$

$$y_t = c(x_t) + w_t \quad ; \quad y_o = w_o \tag{1.1.b}$$

where $\{v_t\}$ and $\{w_t\}$ are independent standard white Gaussian noises.

Given a measurable function f, the filtering problem consists in computing for each $t=1,\dots,T$, assuming it exists, the least squares estimate of $f(x_t)$, given the observations up to time t, namely

$$E\{f(x_t) | \mathcal{F}_t^y\} \tag{1.2}$$

where $\mathcal{F}_t^y := \sigma\{y_s | s \leq t\}$.

More generally, the filtering problem can be formulated as follows: given a Markov process x_t with known transition densities $p(x_t|x_{t-1})$ and an observable process y_t, characterized by a known conditional density $p(y_t|x_t)$ it is desired to compute for each $t=1,\dots,T$ the filtering density $p(x_t|y^t)$ where $y^t := \{y_o, y_1, \dots, y_t\}$.

A solution to this problem can be obtained by means of the recursive Bayes formula

$$p(x_t|y^t) = \frac{p(y_t|x_t)\; p(x_t|y^{t-1})}{\int p(y_t|x_t)p(x_t|y^{t-1})dx_t} =$$
$$= \frac{p(y_t|x_t)\int p(x_t|x_{t-1})p(x_{t-1}|y^{t-1})dx_{t-1}}{\int p(y_t|x_t)\int p(x_t|x_{t-1})p(x_{t-1}|y^{t-1})dx_{t-1}dx_t} \tag{1.3}$$

However, there is an inherent computational difficulty with this formula due to the fact that the integral

$$\int p(x_t|x_{t-1})p(x_{t-1}|y^{t-1})dx_{t-1}$$

is parametrized by $x_t \in \mathbb{R}$.

This difficulty disappears in all those situations when

$$p(x_t|x_{t-1}) = \sum_{i=o}^{n} \phi_i(x_t)\psi_i(x_{t-1}) \tag{1.4}$$

In fact, letting $\propto$ denote proportionality, it is easily seen that, with (1.4), $p(x_t|y^t)$ can actually be computed by means of (1.3) resulting in

$$p(x_t|y^t) \propto \sum_{i=o}^{n} d_i(y^{t-1})p(y_t|x_t)\phi_i(x_t); \qquad t=1,\dots,T \tag{1.5}$$

where the vector $d(y^{t-1})$ of the coefficients in the combination can be recursively obtained as

$$d_i(y^o) = \int \psi_i(x_o)p(x_o)dx_o; \qquad i=0,\dots,n \tag{1.6.a}$$

$$d(y^t) = d(y^{t-1})B(y_t); \qquad t \geqq 1 \tag{1.6.b}$$

with $B(y_t) = \{b_{ij}(y_t)\}_{i,j=0,\dots,n}$ where

$$b_{ij}(y_t) = \int \psi_j(x_t)p(y_t|x_t)\phi_i(x_t)dx_t \tag{1.6.c}$$

Notice that if in (1.4) we have n=0, the process x_t reduces to a sequence of i.i.d. random variables and the filtering problem reduces to a sequence of standard Bayesian estimation problems for x_t, where for all t the prior distribution of x_t is given by $p(x_t)=\phi(x_t)$.

The purpose of the present study is to exploit the computational advantage resulting from (1.4) in order to approximate $p(x_t|y^t)$ by means of approximating densities $p_n(x_t|y^t)$, $n\geq 1$, that can be explicitly computed in a recursive way. Such $p_n(x_t|y^t)$ will be obtained by means of the recursive

Bayes formula (1.3) using approximations to $p(x_t|x_{t-1})$ given by suitable non negative functions $p_n(x_t|x_{t-1})$ of the form (1.4). Furthermore the approximation will be such that an explicitly computable bound can be obtained for an appropriate approximation error; if in addition $f(\cdot)$ does not grow more than exponentially, then also $E\{f(x_t)|\mathscr{F}_t^y\}$ can be approximated by $\int f(x_t)p_n(x_t|y^t)dx_t$ with a corresponding error bound.

Approximations to the nonlinear filtering problem have been considered by various authors (for approaches concerning continuous-time nonlinear filtering problems see e.g. (Kushner 1977), (Clark 1978), (Davis 1981), (Di Masi and Runggaldier 1981), (Le Gland 1981), (Picard 1984), (Talay 1984)). These approximations do not completely solve the practically important problem of obtaining explicit error bounds. In (Di Masi and Runggaldier 1982) an approximation to a discrete-time nonlinear filtering problem with explicit error bounds has been derived and in (Di Masi and Runggaldier 1985) this approximation has been extended also to continuous-time problems. While in (Di Masi and Runggaldier 1982) the approximation is obtained by approximating the model (1.1)-(1.2), here we follow the alternative approach of directly approximating the solution to the recursive Bayes formula (1.3).

In the next Section 2 we shall show that under suitable assumptions an approximation $p_n(x_t|x_{t-1})\geq 0$ of the type (1.4) leads to corresponding approximations to $p(x_t|y^t)$ as well as $E\{f(x_t)|\mathscr{F}_t^y\}$ with explicit error bounds that go to zero as $n\to\infty$. In the final Section 3 we shall give two examples of suitable approximations $p_n(x_t|x_{t-1})$ of the type (1.4).

2. CONVERGENCE OF THE APPROXIMATION AND ERROR BOUNDS

As mentioned in the introduction, our purpose here is to provide approximations to the filtering density $p(x_t|y^t)$ as well as the corresponding filter $E\{f(x_t)|\mathscr{F}_t^y\}$.

To this end it will be convenient to provide approximations to $p(x_t|y^t)$ in a suitable weighted norm of the type.

$$\| g \|_\alpha := \int \alpha(x)\,|g(x)|\,dx \tag{2.1}$$

In what follows we shall choose $\alpha(x) = \exp[\underline{\alpha}|x|]$, $(\alpha>0)$, as this will enable us to approximate $E\{f(x_t)|\mathcal{F}_t^y\}$ for all those $f(\cdot)$ for which $|\exp[-\alpha|x|]f(x)|<+\infty$; in particular, it will allow the approximation of all the conditional moments, as long as they exist. Using explicit upper bounds, we shall show that the convergence of $p_n(x_t|x_{t-1})\geq 0$ to $p(x_t|x_{t-1})$ in the norm $\|\cdot\|_\alpha$ implies the convergence in the same norm of $p_n(x_t|y^t)$ to $p(x_t|y^t)$ (Proposition 2.1 below) as well as the convergence of the conditional moments (Corollary 2.1).

We shall need the following assumtions: there exist $V(y_t)$, U, W, Z, Z_n such that

A.1 : $0<V(y_t)\leq p(y_t|x_t)\leq U$ for all x_t

A.2 : $\int \inf_{x_{t-1}} p_n(x_t|x_{t-1})dx_t \geq W>0$

A.3 : $\|p_n(x_t|x_{t-1})\|_\alpha \leq Z$ for all x_{t-1}

A.4 : $\|p(x_t|x_{t-1}) - p_n(x_t|x_{t-1})\|_\alpha \leq Z_n$ for all x_{t-1}, with $\lim_{n\to\infty} Z_n=0$

From A.1 and A.3, using (1.3) and the fact that $p_n(x_t|x_{t-1})$ is nonnegative, we immediatley have by induction

Lemma 2.1. For all t, $p_n(x_t|y^t)$ is a density function, i.e.

$$p_n(x_t|y^t)\geq 0 \quad \text{and} \quad \int p_n(x_t|y^t)dx_t = 1$$

We then immediately have from the assumptions

Lemma 2.2. For all $t\geq 1$

$$\|p_n(x_t|y^{t-1})\|_\alpha \leq Z$$

and, letting for $t\geq 1$

$$K(y^t) := \int p(y_t|x_t)p(x_t|y^{t-1})dx_t \tag{2.2}$$

$$K_n(y^t) := \int p(y_t|x_t)p_n(x_t|y^{t-1})dx_t \tag{2.3}$$

Lemma 2.3. For all $t\geq 1$

$$K(y^t)\geq V(y_t); \quad K_n(y^t)\geq WV(y_t)$$

We now prove the main result of this section

Proposition 2.1. Under A.1-A.4, for all $t\geq 1$

$$\| p(x_t|y^t)-p_n(x_t|y^t)\|_\alpha \leq Z_n \sum_{s=1}^{t} (2U^2W^{-1}Z^2)^s \prod_{u=t-s+1}^{t} V^{-2}(y_u)$$

Proof: Using (1.3), the definitions (2.2) and (2.3), the assumptions A.1-A.4, as well as Lemmas 2.2 and 2.3, and assuming without loss of generality Z>1, we have

$$\| p(x_t|y^t)-p_n(x_t|y^t)\|_\alpha \leq$$

$$\leq \int \alpha(x_t)p(y_t|x_t) \left| \frac{p(x_t|y^{t-1})}{K(y^t)} - \frac{p_n(x_t|y^{t-1})}{K_n(y^t)} \right| dx_t \leq$$

$$\leq \frac{U}{K(y^t)K_n(y^t)} \int \alpha(x_t) \Big[K_n(y^t)\,|p(x_t|y^{t-1})-p_n(x_t|y^{t-1})| +$$

$$+ |K(y^t)-K_n(y^t)|\ |p_n(x_t|y^{t-1})| \Big] dx_t \leq$$

$$\leq \frac{U^2}{K(y^t)K_n(y^t)} \Big[\Big|\int p_n(x_t|y^{t-1})dx_t\Big| \int \alpha(x_t)|p(x_t|y^{t-1})-p_n(x_t|y^{t-1})|dx_t +$$

$$+ \int |p(x_t|y^{t-1})-p_n(x_t|y^{t-1})|dx_t \cdot \int \alpha(x_t)|p_n(x_t|y^{t-1})|dx_t \Big] \leq$$

$$\leq \frac{2U^2}{K(y^t)K_n(y^t)} \| p_n(x_t|y^{t-1})\|_\alpha \ \| p(x_t|y^{t-1})-p_n(x_t|y^{t-1})\|_\alpha \leq$$

$$\leq \frac{2U^2}{K(y^t)K_n(y^t)} \| p_n(x_t|y^{t-1})\|_\alpha \ \cdot$$

$$\cdot \Big[\int \| p(x_t|x_{t-1})-p_n(x_t|x_{t-1})\|_\alpha \ p(x_{t-1}|y^{t-1})dx_{t-1} +$$

$$+ \| p_n(x_t|x_{t-1})\|_\alpha \ \| p(x_{t-1}|y^{t-1})-p_n(x_{t-1}|y^{t-1})\|_\alpha \Big] \leq$$

$$\leq 2U^2V^{-2}(y_t)W^{-1}Z\Big[Z_n + Z\| p(x_{t-1}|y^{t-1})-p_n(x_{t-1}|y^{t-1})\|_\alpha\Big] \leq$$

$$\leq 2U^2V^{-2}(y_t)W^{-1}Z^2\Big[Z_n + \| p(x_{t-1}|y^{t-1})-p_n(x_{t-1}|y^{t-1})\|_\alpha\Big]$$

from which the conclusion follows.

Corollary 2.1. Under A.1-A.4, letting M>0 be such that $|\exp[-\alpha|x|]f(x)| \leq M$, we have for all $t \geq 1$

$$|E\{f(x_t)|\mathcal{F}_t^y\} - \int f(x_t)p_n(x_t|y^t)dx_t| \leq MZ_n \sum_{s=1}^{t} (2U^2W^{-1}Z^2)^s \prod_{u=t-s+1}^{t} V^{-2}(y_u)$$

3. EXAMPLES OF APPROXIMATIONS

In this section we present two examples of how to obtain nonnegative approximations $p_n(x_t|x_{t-1})$ to $p(x_t|x_{t-1})$ of type (1.4) and satisfying A.2-A.4, when x_t is given by model (1.1) on which we make the following assumptions:

H.1 : $|a(x)| \leq A < +\infty$

H.2 : $|c(x)| \leq C < +\infty$

Notice that, due to the normalization in (1.3), we can take $p(y_t|x_t) = \exp\left[-\frac{1}{2}(y_t - c(x_t))^2\right]$ so that H.2 immediately implies A.1 with $V(y_t) = \exp\left[-\frac{1}{2}(|y_t|+C)^2\right]$ and U=1.

Example 1. In this example we assume the following strenghtening of. H.1, namely that a(•) in (1.1.a) can be uniformly approximated by step functions. Then, denoting by $I_{\Pi_i}(x)$ the indicator function of the interval Π_i let

$$a_n(x) = \sum_{i=o}^{n} a_i I_{\Pi_i}(x) \tag{3.1}$$

be a sequence of step functions such that

$$\| a(x) - a_n(x) \| \leq A_n \xrightarrow{n\to\infty} 0 \tag{3.2.a}$$

$$\| a_n(x) \| \leq A \tag{3.2.b}$$

where $\|\cdot\|$ denotes the sup-norm.

We now let for $n \in \mathbb{N}$

$$p_n(x_t|x_{t-1}) = \frac{1}{\sqrt{2\pi}} \exp\left[-\frac{1}{2}(x_t - \sum_{i=o}^{n} a_i I_{\Pi_i}(x_{t-1}))^2\right] =$$

$$= \sum_{i=o}^{n} \frac{1}{\sqrt{2\pi}} \exp\left[-\frac{1}{2}(x_t - a_i)^2\right] I_{\Pi_i}(x_{t-1}) \tag{3.3}$$

which is nonnegative and of type (1.4).

It remains to show that for $p_n(x_t|x_{t-1})$ as defined by (3.3) the assumptions A.2-A.4 are fulfilled. Notice that A.2 is only needed to prove the second inequality in Lemma 2.3. Since in this case $p_n(x_t|x_{t-1})$ is actually a density, this second inequality can be proved as the first one so that A.2, although true, is not needed here. Assumption A.3 can be immediately verified with

$$Z=2\exp\left[\frac{1}{2}(A+\alpha)^2\right] \tag{3.4}$$

For assumption A.4 we have the following

<u>Proposition 3.1</u>. For all $t\geq 1$ and all possible values of x_{t-1}, we have

$$\| p(x_t|x_{t-1})-p_n(x_t|x_{t-1})\|_\alpha \leq Z_n$$

with

$$Z_n=4(A+1)\exp\left[\frac{1}{2}(A+\alpha+1)^2\right]A_n \tag{3.5}$$

Proof: Using the inequalities $|e^x-e^y|\leq|x-y|(e^x+e^y)$, $|x|\leq e^{|x|}$, $1\leq e^{|x|}$ and $\| a^2(x)-a_n^2(x)\|\leq 2A\| a(x)-a_n(x)\|$, we have

$$\| p(x_t|x_{t-1})-p_n(x_t|x_{t-1})\|_\alpha \leq$$

$$\leq \frac{1}{\sqrt{2\pi}}\cdot\int \exp\left[\alpha|x_t|\right]\exp\left[-\frac{1}{2}(x_t^2-2x_t a(x_{t-1})+a^2(x_{t-1})\right]\cdot$$

$$\cdot\exp\left[|x_t|\right]\cdot\left[|a(x_{t-1})-a_n(x_{t-1})|+\frac{1}{2}|a^2(x_{t-1})-a_n^2(x_{t-1})|\right]dx_t +$$

$$+\frac{1}{\sqrt{2\pi}}\int \exp\left[\alpha|x_t|\right]\exp\left[-\frac{1}{2}(x_t^2+2x_t a_n(x_{t-1})+a_n^2(x_{t-1}))\right]\cdot$$

$$\cdot\exp\left[|x_t|\right]\left[|a(x_{t-1})-a_n(x_{t-1})|+\frac{1}{2}|a^2(x_{t-1})-a_n^2(x_{t-1})|\right]dx_t \leq$$

$$\leq \frac{2}{\sqrt{2\pi}}\int \exp\left[\alpha|x_t|-\frac{1}{2}|x_t|^2+A|x_t|+|x_t|\right]\cdot$$

$$\cdot (A+1)|a(x_{t-1})-a_n(x_{t-1})|dx_t \leq$$

$$\leq 2\exp\left[\frac{1}{2}(A+\alpha+1)^2\right](A+1)A_n\cdot$$

$$\cdot\frac{1}{\sqrt{2\pi}}\int \exp\left[-\frac{1}{2}(|x_t|-(A+\alpha+1))^2\right]dx_t \leq$$

$$\leq 4(A+1)\exp\left[\frac{1}{2}(A+\alpha+1)^2\right]A_n$$

<u>Remark 3.1</u>: Notice that the approximation of $p(x_t|x_{t-1})$ by $p_n(x_t|x_{t-1})$ as given in (3.3) corresponds to an approximation of the entire model (1.1) by a model of the same type where $a(\cdot)$ is replaced by $a_n(\cdot)$ (see (Di Masi and Runggaldier 1985), where such approach is used for the approximation of continuous-time nonlinear filtering problems).

<u>Example 2</u>. In this example we shall consider an approximation that is based

on the following Taylor-series approximation of the exponential function ($n \in \mathbb{N}$)

$$e^x \cong \tilde{a}_n(x) = \sum_{i=o}^{2n} \frac{x^i}{i!} \tag{3.6}$$

As it can be easily verified, we have

$$\begin{aligned} 1 \leqq \tilde{a}_n(x) \leqq e^x \qquad & \text{for} \quad x \geqq 0 \\ e^x < \tilde{a}_n(x) < e^{|x|} \qquad & \text{for} \quad x < 0 \end{aligned} \tag{3.7}$$

We now consider the following approximation to the transition density

$$p_n(x_t|x_{t-1}) = \frac{1}{\sqrt{2\pi}} \exp\left[-\frac{1}{2}(x^2+a^2(x_{t-1})\right] \cdot \tilde{a}_n(x_t a(x_{t-1})) \tag{3.8}$$

which is nonnegative and of type (1.4).

Due to (3.7) we have

$$\exp\left[-A|x_t|\right] \leqq \tilde{a}_n(x_t a(x_{t-1})) \leqq \exp\left[A|x_t|\right] \tag{3.9}$$

so that, as can be easily seen, A.2 and A.3 are satisfied with

$$W = 2\left[1-\Phi(A)\right] \tag{3.10}$$

$$Z = 2\exp\left[\frac{1}{2}(A+\alpha)^2\right] \tag{3.11}$$

where Φ is the standard normal cumulative. Finally, for A.4 we have

Proposition 3.2. For all $t \geqq 1$ and all possible values of x_{t-1}, we have

$$\| p(x_t|x_{t-1}) - p_n(x_t|x_{t-1}) \|_\alpha \leqq Z_n$$

with

$$Z_n = \frac{2A^{2n+1}}{(2n+1)!} \exp\left[\frac{1}{2}(A+\alpha+1)^2\right]$$

Proof: Using the expression for the error in a Taylor-series approximation and the inequality $x^n \leqq \exp|x|$, we have

$$\| p(x_t|x_{t-1}) - p_n(x_t|x_{t-1}) \|_\alpha \leqq \frac{1}{\sqrt{2\pi}} \int \exp\left[\alpha|x_t|\right] \exp\left[-\frac{1}{2}x_t^2\right] \cdot$$

$$\cdot \frac{|x_t^{2n+1} a^{2n+1}(x_{t-1})|}{(2n+1)!} \exp\left[|x_t a(x_{t-1})|\right] dx_t \leqq$$

$$\leqq \frac{A^{2n+1}}{(2n+1)!} \cdot \frac{1}{\sqrt{2\pi}} \int \exp\left[-\frac{1}{2}x_t^2 + (A+\alpha+1)|x_t|\right] dx_t \leqq$$

$$\leq \frac{A^{2n+1}}{(2n+1)!} \exp\left[\frac{1}{2}(A+\alpha+1)^2\right] \cdot \frac{1}{\sqrt{2\pi}} \int \exp\left[-\frac{1}{2}(|x_t| - (A+\alpha+1))^2\right] dx_t \leq$$

$$\leq \frac{2A^{2n+1}}{(2n+1)!} \exp\left[\frac{1}{2}(A+\alpha+1)^2\right]$$

REFERENCES

Clark, J.M.C. (1978). The design of robust approximations to the stochastic differential equations of nonlinear filtering. In J.K.Skwirzynski (ed.), Communication systems and random process theory. Sijthoff and Noordhoff.

Davis, M.H.A. (1981). New approach to filtering for nonlinear systems, I.E.E. Proc., Part D 128: 166-172.

Di Masi, G.B., and Runggaldier, W.J. (1981). Continuous time approximations for the nonlinear filtering problem. Appl. Math. Optim. 7: 233-245.

Di Masi, G.B., and Runggaldier, W.J. (1982). Approximations and bounds for discrete-time nonlinear filtering. In A.Bensoussan, J.L. Lions (eds.) Analysis and Optimization of Systems: L.N. in Control and Info.Sci. 44. Springer-Verlag.

Di Masi, G.B., Pratelli, M., and Runggaldier, W.J. (1985). An approximation for the nonlinear filtering problem, with error bound, Stochastics (to appear).

Kushner, H.J. (1977). Probability methods for approximations in stochastic control and for elliptic equations. Academic Press, New York.

Le Gland, F. (1981). Estimation de paramètres dans les processus stochastiques en observation incomplète. Thèse, Université Paris IX.

Picard, J. (1984). Approximation of nonlinear filtering problems and order of convergence. In H.Korezlioglu, G.Mazziotto, J.Szpirglas (eds.) Filtering and control of random processes. L.N. in Control and Info.Sci. 61. Springer-Verlag.

Talay, D. (1984). Efficient numerical schemes for the approximation of expectations of functionals of the solution of S.D.E. and applications. In H.Korezlioglu, G.Mazziotto, J.Szpirglas (eds.). Filtering and control of random processes. L.N. in Control and Info.Sci. 61. Springer-Verlag.

ON APPROXIMATIONS TO DISCRETE-TIME STOCHASTIC CONTROL PROBLEMS

G.B. Di Masi, W.J. Runggaldier*, F. Chiariello*

Istituto di Elettrotecnica e di Elettronica, Università di Padova and CNR-LADSEB, 35100 Padova, Italy.

* Seminario Matematico, Università di Padova, 35100 Padova, Italy.

1. INTRODUCTION

Consider the following discrete-time stochastic control problem: a partially observable process $\{x_t, y_t\}$, $x_t, y_t \in R$ with x_t the unobservable an y_t the observable components, is given for $t = 0,1,\dots,T$ on some probability space $\{\Omega, \mathcal{F}, P\}$ by

$$x_{t+1} = a(x_t, u_t) + \sigma(x_t)v_{t+1} \quad ; \quad x_o = v_o \tag{1.1a}$$

$$y_t = c(x_t) + w_t \quad ; \quad y_o = w_o \tag{1.1b}$$

where $\{v_t\}$ and $\{w_t\}$ are independent standard white Gaussian noises and $\{u_t\}$ is a sequence of admissible controls, namely such that u_t takes values in a given set $U \subset R$ and depends only on past and present observations $y^t := \{y_o,\dots,y_t\}$ and past controls $u^{t-1} := \{u_o,\dots,u_{t-1}\}$. Defining the value function

$$v(u^{T-1}) := E\{\sum_{t=0}^{T-1} r(x_t, u_t\} + b(x_T) \tag{1.2}$$

where $r(x,u)$ and $b(x)$ are given functions, it is desired to find, for any given $\varepsilon > 0$, and ε-optimal control $\{u_t^\varepsilon\}$, i.e. an admissible control such that

$$v(u^{\varepsilon T-1}) \leq \inf_{u^{T-1}} v(u^{T-1}) + \varepsilon$$

where the inf is over all admissible controls u^{T-1}.

The usual approach to this problem consists in transforming it into an equivalent complete-observation problem by taking as new state at time t the conditional distribution of x_t given y^t and u^{t-1} (see e.g. (Dynkin and Yushkevich 1979). The major difficulty that arises with this approach is that the new state takes values in an infinite-dimensional space, namely the space of all probability distributions over the real line.

There are however particular classes of stochastic control problem for which the state of the equivalent complete-observation problem can be taken as a finite set of conditional probabilities. One possible class of this type is given by

$$x_{t+1} = \sum_{i,k=1}^{n} a_i(k) I_{B_i}(x_t) I_{U_k}(u_t) + \sum_{i=1}^{n} \sigma_i I_{B_i}(x_t) v_{t+1} \tag{1.3a}$$

$$y_t = \sum_{i=1}^{n} c_i I_{B_i}(x_t) + w_t \tag{1.3b}$$

$$v(u^{T-1}) = E\{ \sum_{t=0}^{T-1} \left[\sum_{i,k=1}^{n} r_i(k) I_{B_i}(x_t) I_{U_k}(u_t) \right] + \sum_{i=1}^{n} b_i I_{B_i}(x_T)\} \tag{1.4}$$

where $a_i(k)$, $r_i(k)$, σ_i, b_i, c_i $(i,k = 1,\dots,n)$ are given real numbers, $\{B_i\}$ is a finite partition of the real line into intervals, $\{U_k\}$ is a class of disjoint intervals on R and the admissible control set is given by $U = \bigcup_k U_k$. In other words, this class consists of problems (1.1) - (1.2) with a,r,σ,b,c step functions, and U a finite union of intervals.

It is clear from the particular structure of (1.3) - (1.4) that the conditional probabilities $\pi_t^i := P\{x_t \in B_i | y^t, u^{t-1}\}$, $i = 1,\dots,n$ contain all the information on the past history (y^t, u^{t-1}) which is relevant for control purposes so that the vector $\pi_t = [\pi_t^1,\dots,\pi_t^n]$ can be taken as state variable of the equivalent complete-observation control problem. It is also clear from (1.3) - (1.4) that the choice of a particular value for the control u_t

reduces to the choice of a $k = 1,\dots,n$ so that in what follows we shall consider $U = \{1,\dots,n\}$.

Furthermore, exploiting the particular structure of (1.3) (see (Di Masi and Runggaldier 1983), it is possible to determine the transition law for π_t; in fact, using the recursive Bayes formula, it is easily seen that

$$\pi^j_{t+1} = \Gamma^j(\pi_t, y_{t+1}, u_t) := \frac{\sum_{i=1}^n \pi^i_t p_{ij}(u_t) f_j(y_{t+1})}{\sum_{i=1}^n \sum_{h=1}^n \pi^i_t p_{ih}(u_t) f_h(y_{t+1})} \tag{1.5}$$

where

$$p_{ij}(u_t) = 1/\sqrt{2\pi} \int_{B_j} \exp\left[-\frac{(x-a_i(u_t))^2}{2\sigma_i^2}\right] dx \tag{1.6}$$

$$f_j(y_{t+1}) = 1/\sqrt{2\pi} \exp\left[-\frac{1}{2}(y_{t+1}-c_j)^2\right] \tag{1.7}$$

and the initial condition is given by

$$\pi^j_o = P\{x_o \in B_j\}, \quad j = 1,\dots,n . \tag{1.8}$$

Notice that the denominator in (1.5) is the conditional density $g(y_{t+1}|y^t, u^{t-1})$.

The equivalent complete-observation problem, which in the sequel we shall refer to as problem (P), is characterized by the state space $\Pi = \{\pi | \pi^i \in [0,1], \ i = 1,\dots,n; \sum_i \pi^i = 1\}$ by the state-transition law

$$\pi_{t+1} = \Gamma(\pi_t, y_{t+1}, u_t) \tag{1.9}$$

as given in (1.5), by admissible control sequences $\{u_t\}$ such that $u_t \in U$ depends only on π_t, and cost functions given, with abuse of notation, by

$$r(\pi_t, u_t) = E\{r(x_t, u_t)|y^t, u^{t-1}\} = \sum_{i=1}^n \pi^i_t r_i(u_t) \tag{1.10a}$$

$$b(\pi_T) = E\{b(x_T)|y^T, u^{T-1}\} = \sum_{i=1}^n \pi^i_T b_i \tag{1.10b}$$

It is worth remarking that the particular class (1.3) - (1.4) is not only interesting in itself, but proved also useful for the approximation of rather general problems of the form (1.1) - (1.2). In fact it is shown in (Di Masi and Ruggaldier 1983) that ε-optimal controls for problems (1.3) - (1.4) lead, under suitable assumptions, to ε-optimal controls for problems (1.1) - (1.2).

However the derivation of ε-optimal controls for (1.3) - (1.4) is by no means trivial since the equivalent complete-observation problem (P) has still an infinite state space.

It is the aim of the present paper to provide a method for obtaining ε-optimal controls for problem (P) and consequently for (1.3) - (1.4). Notice first that, although (P) has the state taking values in a finite-dimensional space, its possible values are still infinite. In analogy to some recent work concerning Markovian decision problems by Bertsekas (1975), Hinderer (1979) and Whitt (1978), our approach consists in approximating problem (P) by a problem $(\bar{P})$ whose state space is finite and for which, since the control set is also finite, an optimal control can be actually computed. This control, suitably extended to the entire state space of (P), is shown to be ε-optimal for (P). A direct application to problem (P) of the methods in (Bertsekas 1975), (Hinderer 1979) and (Whitt 1978) leads to various difficulties, in particular that of determining the transition law of the approximating finite-state problem. Therefore it appears more convenient to exploit the partially observable nature of the original problem (1.3) - (1.4) and to first discretize the observation process y_t.

In the next Section 2 we shall describe our approximation approach, while in Section 3 we show its convergence properties and derive an algorithm for the computation of the value of the approximating control.

2. THE APPROXIMATION TECHNIQUE

For $m \in N$ and $\bar{m} := 2m^2 + 2$, let $\{Y_i,\ i = 1,\dots,\bar{m}\}$ be the partition on R given by

$$Y_1 = (-\infty, -m)$$

$$Y_i = \left[-m + \frac{i-2}{m}, -m + \frac{i-1}{m}\right), \quad i = 2, \ldots, \bar{m}-1 \tag{2.1}$$

$$Y_{\bar{m}} = [m, +\infty)$$

and let $\{\eta_i : \eta_i \in Y_i,\ i = 1, \ldots, \bar{m}\}$ be a set of representative elements of the partition. Then, defining the projection

$$\bar{y}(y) = \sum_{i=1}^{\bar{m}} \eta_i I_{Y_i}(y) \tag{2.2}$$

we consider the process $\bar{y}_t$ given by

$$\bar{y}_t = \bar{y}(y_t) \tag{2.3}$$

with y_t given by (1.3b), and we set, with abuse of notation, $\bar{y}(y^t) := \bar{y}^t$. Furthermore, let

$$\bar{f}_j(\eta_i) := P\{\bar{y}_t = \eta_i | x_t \in B_j\} = 1/\sqrt{2\pi} \int_{Y_i} \exp\left[-\frac{1}{2}(y-c_j)^2\right] dy \tag{2.4}$$

Analogously to what has been done in the Introduction for the definition of problem (P), we now consider the vector $\{\pi_t^i,\ i = 1, \ldots, n\}$ of conditional probabilities $\bar{\pi}_t^i = P\{x_t \in B_i | \bar{y}^t, u^{t-1}\}$ and take as approximating problem for (P) the problem $(\bar{P})$ characterized by the finite state space $\bar{\Pi}$ given by the vectors $\bar{\pi}_t$, by state transition law

$$\bar{\pi}_{t+1} = \bar{\Gamma}(\bar{\pi}_t, \bar{y}_{t+1}, u_t) \tag{2.5}$$

with $\bar{\Gamma}$ as the Γ in (1.5) with $f_j(y)$ replaced by $\bar{f}_j(\bar{y})$ as defined in (2.4), by admissible control sequences $\{u_t\}$ such that $u_t \in U$ depends only on $\bar{\pi}_t$ and cost functions given, with abuse of notation, by

$$r(\bar{\pi}_t, u_t) = E\{r(x_t, u_t) | \bar{y}^t, u^{t-1}\} = \sum_{i=1}^{n} \bar{\pi}_t^i r_i(u_t) \tag{2.6a}$$

$$b(\bar{\pi}_T) = E\{b(x_T) | \bar{y}^T, u^{T-1}\} = \sum_{i=1}^{n} \bar{\pi}_T^i b_i \tag{2.6b}$$

Problem $(\bar{P})$ admits an optimal control $\bar{u}_t(\bar{\pi}_t) = \bar{u}_t(\bar{y}^t, u^{t-1})$ which can be derived via dynamic programming or, using the fact that the observation and state spaces are finite, via the algorithm proposed by Smallwood and Sondik (1973). This control can be extended to every history (y^t, u^{t-1}) of the original problem (1.3) - (1.4) letting

$$\bar{u}_t(y^t, u^{t-1}) = \bar{u}_t(\bar{y}(y^t), u^{t-1}) \tag{2.7}$$

In the next section we shall show that $\bar{u}_t(\bar{y}(y^t), \bar{u}^{t-1})$ is an approximate solution to problem (P) and consequently to the original problem (1.3)-(1.4).

3. PROPERTIES OF THE APPROXIMATION

In what follows we shall denote by $v(u_t^{T-1}; y^t, u^{t-1})$ and $\bar{v}(u_t^{T-1}; \bar{y}^t, u^{t-1})$ the cost-to-go functions at time t for (P) and $(\bar{P})$ respectively, $v(u^{T-1})$ is the value function (1.2) for problem (P) and $\bar{v}(u^{T-1})$ is the value function for $(\bar{P})$. Moreover, let $v(y^t, u^{t-1})$ and $\bar{v}(\bar{y}^t, u^{t-1})$ be the optimal cost-to-go at time t for (P) and $(\bar{P})$ respectively. In particular, for $t = 0$, the optimal cost-to-go is the optimal value, which for problems (P) and $(\bar{P})$ will be denoted by v and $\bar{v}$ respectively.

3.1 Convergence results

The main purpose of this subsection is to show that as the partition on the observation space becomes finer and finer, i.e. as m in (2.1) goes to infinity, the value $v(\bar{u}^{T-1})$, namely the value for (P) of the optimal control for $(\bar{P})$, converges to the optimal value v. This result is obtained in the corollary to Theorems 3.1 and 3.2 below. The proofs of these theorems are based on some preliminary results, the most important of which is Proposition 3.1 and whose proofs can be found in (Di Masi and Runggaldier 1983).

Lemma 3.1. With Γ^j as defined in (1.5), we have for all $\pi_1, \pi_2 \in \Pi$, $u \in U$ and $y \in R$

$$\max_j |\Gamma^j(\pi_1, y, u) - \Gamma^j(\pi_2, y, u)| \leq L \max_i |\pi_1^i - \pi_2^i| , \tag{3.1}$$

where

$$L = \frac{n}{(\min_{hiu} p_{ih}(u))^2} \tag{3.2}$$

□

Lemma 3.2. With Γ^j and $\bar{\Gamma}^j$ as defined in (1.5) and (2.5) respectively, we have for any j=1,...,n and all $\pi \in \Pi$, $u \in U$

$$\lim_{m\to\infty} \sup_y |\Gamma^j(\pi,y,u) - \bar{\Gamma}^j(\pi,\bar{y}(y),u| = 0$$

□

In what follows, when convenient, we shall make explicit the dependence of π_t and $\bar{\pi}_t$ on (y^t,u^{t-1}) and $(\bar{y}^t,u^{t-1})$ respectively. Then letting

$$V_t^m := \sup_{y^t} \max_{u^{t-1}} \max_i |\pi^i(y^t,u^{t-1}) - \bar{\pi}^i(\bar{y}(y^t),u^{t-1})| \tag{3.3}$$

we have following

Proposition 3.1: For any $t = 0,\dots,T$

$$\lim_{m\to\infty} V_t^m = 0$$

□

Letting $B := \max_{ik}\{b_i, r_i(k)\}$, we now have

Theorem 3.1. For any $t = 0,\dots,T$ and all y^t,u^{t-1}

$$|v(y^t,u^{t-1}) - \bar{v}(\bar{y}^t,u^{t-1})| \leq U_t^m \tag{3.4}$$

where U_t^m is defined recursively by

$$U_T^m = n\,B\,V_T^m \tag{3.5a}$$

$$U_t^m = n\,B(T-t+1)V_t^m + U_{t+1}^m \tag{3.5b}$$

with V_t^m given by (3.3).

Proof: The proof proceeds by backward induction. For t=T we have

$$|v(y^T,u^{T-1})-\bar{v}(\bar{y}^T,u^{T-1})| = |\sum_i b_i(\pi_t^i(y^T,u^{T-1})-\bar{\pi}_t^i(y^T,u^{T-1}))| \leq n\,B\,V_T^m \quad (3.6)$$

so that (3.4) holds for t=T with U_T^m given by (3.5a).

Assume now that (3.4) holds for t+1. Then, using the optimality equations of dynamic programming and the properties of the control $\{\bar{u}_t\}$ optimal for $(\bar{P})$

$$\begin{aligned}
& v(y^t,u^{t-1}) - \bar{v}(\bar{y}^t,u^{t-1}) \leq \\
& \leq \sum_i \pi_t^i(y^t,u^{t-1})r_i(\bar{u}_t) + \sum_{i,h,\ell} \pi_t^i(y^t,u^{t-1})p_{ih}(\bar{u}_t)\,\cdot \\
& \cdot \int_{Y_\ell} v((y^t,y_{t+1}),(u^{t-1},\bar{u}_t))f_h(y_{t+1})dy_{t+1} - \\
& - \sum_i \bar{\pi}_t^i(\bar{y}^t,u^{t-1})r_i(\bar{u}_t) + \sum_{i,h,\ell} \bar{\pi}_t^i(\bar{y}^t,u^{t-1})p_{ih}(\bar{u}_t)\,\cdot \\
& \cdot\, \bar{v}((\bar{y}^t,\eta_\ell),(u^{t-1},\bar{u}_t))\bar{f}_h(\eta_\ell) \leq \\
& \leq \sum_i |(\pi_t^i(y^t,u^{t-1}) - \bar{\pi}_t^i(\bar{y}^t,u^{t-1}))r_i(\bar{u}_t)| + \\
& + \sum_{i,h,\ell} |\pi_t^i(y^t,u^{t-1})-\bar{\pi}_t^i(\bar{y}^t,u^{t-1})|p_{ih}(\bar{u}_t)\int_{Y_\ell} |v((y^t,y_{t+1}),(u^{t-1},\bar{u}_t))|f_h(y_{t+1})dy_{t+1} + \\
& + \sum_{i,h,\ell} \bar{\pi}_t^i(\bar{y}^t,u^{t-1})p_{ih}(\bar{u}_t)\,\cdot \\
& \cdot \int_{Y_\ell} |v((y^t,y_{t+1}),(u^{t-1},\bar{u}_t)) - \bar{v}((\bar{y}^t,\bar{y}_{t+1}),(u^{t-1},\bar{u}_t))| \cdot f_h(y_{t+1})dy_{t+1} \leq \\
& \leq n\,B\,V_t^m + n\,B(T-t)V_t^m + U_{t+1}^m \qquad (3.7)
\end{aligned}$$

where in the last inequality we have used the fact that $|v(y^{t+1},u^t)| \leq B(T-t)$ for all y^{t+1},u^t. Furthermore, denoting by $\tilde{u}_t$ a control such that $r(\pi_t,\tilde{u}_t) + E\{v(y^{t+1},(u^{t-1},\tilde{u}_t))|y^t,u^{t-1}\} \leq v(y^t,u^{t-1}) + \delta$ for $\delta > 0$, we have

$$\begin{aligned}
& \bar{v}(\bar{y}^t,u^{t-1}) - v(y^t,u^{t-1}) \leq \\
& \leq \sum_i \bar{\pi}_t^i(\bar{y}^t,u^{t-1})r_i(\tilde{u}_t) + \sum_{i,h,\ell} \bar{\pi}_t^i(\bar{y}^t,u^{t-1})p_{ih}(\tilde{u}_t)\,\cdot \\
& \cdot\, \bar{v}((\bar{y}^t,\eta_\ell),(u^{t-1},\tilde{u}_t))\bar{f}_h(\eta_\ell) -
\end{aligned}$$

$$- \sum_i \pi_t^i(y^t,u^{t-1}) r_i(\tilde{u}_t) - \sum_{i,h,\ell} \pi_t^i(y^t,u^{t-1}) p_{ih}(\tilde{u}_t) \cdot$$

$$\cdot \int_{Y_\ell} v((y^t,y_{t+1}),(u^{t-1},\tilde{u}_t)) f_h(y_{t+1}) dy_{t+1} + \delta \leq$$

$$\leq \sum_i |(\bar{\pi}_t^i(\bar{y}^t,u^{t-1}) - \pi_t^i(y^t,u^{t-1})) r_i(\tilde{u}_t)| +$$

$$+ \sum_{i,h,\ell} |\bar{\pi}_t^i(\bar{y}^t,u^{t-1}) - \pi_t^i(y^t,u^{t-1})| p_{ih}(\tilde{u}_t) |\bar{v}((\bar{y}^t,\eta_\ell),(u^{t-1},\tilde{u}_t)) \bar{f}_h(\eta_\ell) +$$

$$+ \sum_{i,h,\ell} \pi_t^i(y^t,u^{t-1}) p_{ih}(\tilde{u}_t) \cdot$$

$$\cdot \int_{Y_\ell} |\bar{v}((\bar{y}^t,\bar{y}_{t+1}),(u^{t-1},\tilde{u}_t)) - v((y^t,y_{t+1}),(u^{t-1},\tilde{u}_t))| f_h(y_{t+1}) dy_{t+1} +$$

$$+ \delta \leq n B V_t^m + n B(T-t) V_t^m + U_{t+1}^m + \delta \qquad (3.8)$$

By the arbitrariness of δ, (3.7) and (3.8) show that (3.4) holds for t with U_t^m given by (3.5b). □

Using arguments analogous to those used in the first part of the proof of Theorem 3.1, it is possible to prove the following

<u>Theorem 3.2.</u> For any t=0,...,T and all y^t,u^{t-1}

$$|v(\bar{u}_t^{T-1}; y^t,u^{t-1}) - \bar{v}(\bar{y}^t,u^{t-1})| \leq U_t^m$$

where U_t^m is given by (3.5). □

From Theorems 3.1 and 3.2, (3.5) and Proposition 3.1, we immediately have the following

<u>Corollary 3.1</u>

$$|v(\bar{u}^{T-1}) - v| \leq 2U_o^m$$

where U_o^m can be obtained using (3.5). Furthermore $U_o^m \to 0$ as $m \to \infty$.

□

<u>Remark</u>: Corollary 3.1 shows convergence of $v(\bar{u}^{T-1})$ to v. By suitably modifying Lemma 3.2 and Proposition 3.1, we could obtain an upper bound for $|v(\bar{u}^{T-1}) - v|$. However, this bound would depend on L in (3.2) which, by

(1.6), is in general very large so that the bound would not be sharp.

3.2 Computation of the approximating value

As pointed out in the final remark of the proceeding subsection, our approximation is not particularly effective for the actual determination of ε-optimal controls. However, the next proposition shows that our procedure allows the explicit computation of the value $v(\bar{u}^{T-1})$, which provides further information on the quality of the suboptimal control $\{\bar{u}_t\}$.

Proposition 3.2. For any $t = 0,\ldots,T$ we have

$$v(\bar{u}_t^{T-1};y^t,\bar{u}^{t-1}) = \sum_i \pi_t^i(y^t,\bar{u}^{t-1})\rho^i(\bar{y}^t) \tag{3.9}$$

$$E\{v(\bar{u}_t^{T-1};y^t,\bar{u}^{t-1})|y^{t-1}\} = \sum_i \pi_{t-1}^i(y^{t-1},\bar{u}^{t-2})\beta^i(\bar{y}^{t-1}) \tag{3.10}$$

where $\rho^i(\bar{y}^t)$ and $\beta^i(\bar{y}^{t-1})$ $(i = 1,\ldots,n)$ are defined recursively by

$$\rho^i(\bar{y}^T) = b_i \tag{3.11a}$$

$$\beta^i(\bar{y}^{t-1}) = \sum_{h=1}^{\bar{m}}\sum_{j=1}^{n} p_{ij}(\bar{u}_{t-1}(\bar{y}^{t-1},\bar{u}^{t-2}))\cdot\bar{f}_j(\eta_h)\rho^j(\bar{y}^{t-1},\eta_h) \tag{3.11b}$$

$$\rho^i(\bar{y}^{t-1}) = r_i(\bar{u}_{t-1}(\bar{y}^{t-1},\bar{u}^{t-2}) + \beta^i(\bar{y}^{t-1}) \tag{3.11c}$$

Proof: The proof proceeds by backward induction. We have

$$v(y^T,\bar{u}^{T-1}) = \sum_i \pi_T^i(y^T,\bar{u}^{T-1})b_i \tag{3.12}$$

$$E\{v(y^T,\bar{u}^{T-1})|y^{T-1}\} = \sum_i b_i E\{\pi_T^i(y^T,\bar{u}^{T-1})|y^{T-1}\} =$$

$$= \sum_i b_i \int \Gamma^j(\pi_{T-1}(y^{T-1},\bar{u}^{T-2}),y_T,\bar{u}_{T-1})\cdot$$

$$\cdot\ 1/\sqrt{2\pi}\sum_{j,h}\pi_{T-1}^j(y^{T-1},\bar{u}^{T-2})p_{jh}(\bar{u}_{T-1})f_h(y_T)dy_T =$$

$$= \sum_i b_i \sum_j \pi_{T-1}^j(y^{T-1},\bar{u}^{T-2})p_{ji}(\bar{u}_{T-1}) \tag{3.13}$$

Equations (3.12) and (3.13) show that (3.9) and (3.10) hold t=T with $\rho^i(\bar{y}^T)$ and $\beta^i(\bar{y}^{T-1})$ given by (3.11a) and (3.11b).

Assume now that (3.9) and (3.10) hold for t+1. Then we have

$$v(\bar{u}_t^{T-1};y^t,\bar{u}^{t-1}) = \sum_i \pi_t^i(y^t,\bar{u}^{t-1})r_i(\bar{u}_t(\bar{y}^t,\bar{u}^{t-1})) +$$

$$+ E\{v(\bar{u}_{t+1}^{T-1};y^{t+1},\bar{u}^t|y^t\} =$$

$$= \sum_i \pi_t^i(y^t,\bar{u}^{t-1})(r_i(\bar{u}_t(\bar{y}^t,\bar{u}^{t-1})) + \beta^i(\bar{y}^t)) \tag{3.14}$$

Furthermore

$$E\{v(\bar{u}_{t+1}^{T-1};y^{t+1},\bar{u}^t)|y^t\} =$$

$$= E\{\sum_i \pi_{t+1}^i(y^{t+1},\bar{u}^t)\rho^i(\bar{y}^{t+1})|y^t\} =$$

$$= \sum_{i,\ell}\int_{Y_\ell} \Gamma^i(\pi_t^i(y^t,\bar{u}^{t-1}),y_{t+1},\bar{u}_t)\cdot \rho_i(\bar{y}^t,\eta_\ell)\cdot$$

$$\cdot \sum_{jh} \pi_t^j(y^t,\bar{u}^{t-1})p_{jh}(\bar{u}_t)f_h(y_{t+1})dy_{t+1} =$$

$$= \sum_j \pi_t^j(y^t,\bar{u}^{t-1})\sum_{i,\ell} p_{ji}(\bar{u}_t)\bar{f}_i(\eta_\ell)\rho_i(\bar{y}^t,\eta_\ell) \tag{3.15}$$

Equations (3.14) and (3.15) show that (3.9) and (3.10) hold for t with $\rho^i(\bar{y}^t)$ and $\beta^i(\bar{y}^{t-1})$ given by (3.11b) and (3.11c).

□

Notice that, since $p_{ij}(u)$ and $\bar{f}_j(\eta_h)$ are already available from the computations leading to the optimal control $\{\bar{u}_t\}$ for the approximating problem, Proposition 3.2 immediately provides a recursive algorithm to determine $v(\bar{u}^{T-1})$.

REFERENCES

Bertsekas D.P., (1975), Convergence of discretization procedures in dynamic programming. IEEE Trans. AC 20: 415-419.

Di Masi G.B. and Runggaldier W.J., (1983), An approach to discrete-time control problems under partial observation. LADSEB-CNR Int. Rep. 83/02.

Dynkin E.B. and Yushkevich A.A., (1979), Controlled Markov processes. Springer-Verlag, New York.

Hinderer K., (1979), On approximate solutions to finite-stage dynamic programs. In: Dynamic programming and its applications. (M. Puterman ed.), Academic Press, New York.

Smallwood R.D. and Sondik E.J., (1973), The optimal control of partially observable Markov processes over a finite horizon. Operations Res. 11: 1071-1088.

Whitt W., (1978), Approximations of dynamic programs - I. Math. Operations Res. 3: 231-243.

ON LEXICOGRAPHICAL OPTIMALITY CRITERIA IN CONTROLLED MARKOV CHAINS

G.I. Mirzashvili
Mathematical Institute
Tbilisi, USSR

We shall consider a homogeneous Markov chain with a finite set of states S, an arbitrary set of controls A and a family of transitional probabilities

$$\mathrm{P} = \{P^a_{ss'} : P^a_{ss'} \geq 0 \, , \sum_{s'} P^a_{ss'} = 1 \, , s \, , s' \in S \, , a \in A\} \quad .$$

The set of all admissible strategies will be denoted by π, and the set of stationary (Markov, non-randomized, homogeneous) strategies by F.

We shall first consider one possible approach, which leads to the consideration of a control problem with an infinite horizon and different optimality criteria.

The simplest problem is a control problem with a discounted optimality criterion:

$$R^\pi_\delta(s) = \mathrm{E}^\pi_s \sum_{n=0}^{\infty} (1-\delta)^n \tau(s_n \, , a_n) \, , \ s \in S \quad ,$$

which can be written in vector form as

$$R^\pi_\delta = \mathrm{E}^\pi \sum_{n=0}^{\infty} (1-\delta)^n \tau(s_n \, , a_n) \quad ,$$

where $\tau(s \, , a)$ is a bounded numerical function, $s \in S$, $a \in A$; the E^π_s, $s \in S$, are mathematical expectations with respect to measures P^π_s, $s \in S$, $\pi \in \Pi$, on the space of histories of the controlled chain $(s_0, a_0, s_1, a_1 \, , \ldots , s_n \, , a_n \, , \ldots)$ starting from $s_0 = s$, $s \in S$, and generated by the strategies $\pi \in \Pi$ and the family P. The factor $(1-\delta)$, $0 < \delta \leq 1$, represents the discounting coefficient, where δ can be interpreted as the probability of a break at the k-th step unless there has been a break earlier. If τ denotes the time of such a break, distributed geometrically with parameter δ, then

$$R^\pi_\delta = \mathrm{E}^\pi \sum_{n=0}^{\tau} \tau(s_n \, , a_n) \quad .$$

This expression shows that a control problem with a discounted optimality criterion is not in fact a problem with an infinite control horizon. The control horizon becomes infinite if $\delta = 0$, but in this case the criterion R_δ^π in general becomes indefinite. This was the reason for the development of two approaches to the study of problems with infinite control horizons: one involving an additive criterion $\mathrm{E}^\pi \sum_{n=0}^{\infty} \tau(s_n, a_n)$ assuming not only the existence of cost but also other conditions, and the other using a function of average expected cost per unit time

$$\overline{\lim_{N\to\infty}} \frac{1}{N} \mathrm{E}^\pi \sum_{n=0}^{N-1} \tau(s_n, a_n)$$

(a lower limit is also possible) assuming different conditions for the chain and the sets S and A.

An approach generalizing these two would imply a generalized understanding of summation in the additive optimality criterion, and the replacement of the usual sum by a generalized Cesáro or Abelian sum of a different order. The average expected cost per unit time is a Cesáro sum of order -1 (the Cesáro limit) and the additive criterion, when it is well-defined, coincides with the Cesáro or Abelian sum of order zero. Thus two classes of (k, ε)-optimality criterion ($k = -1,0,1, \ldots, \varepsilon \ge 0$) can be suggested: the Abelian criterion and the Cesáro one.

Definition 1. The strategy $\pi \in \Pi$ is (k, ε)-optimal with respect to $\overline{R}_k$, $k = -1,0,1, \ldots, \varepsilon \ge 0$, if for every $\pi' \in \Pi$ we have

$$\overline{\lim_{\delta\to\infty}} (R_{\delta,k}^{\pi'}(s) - R_{\delta,k}^{\pi}(s)) \le \varepsilon, \ s \in S \quad ,$$

where

$$R_{\delta,k}^{\pi}(s) = \mathrm{E}_s^\pi \sum_{n=0}^{\infty} \delta^{-k}(1-\delta)^n \tau(s_n, a_n), \ 0 < \delta \le 1 \quad .$$

Definition 2. The strategy $\pi \in \Pi$ is (k, ε)-optimal with respect to $\overline{V}_k$, $k = 1,0,1, \ldots, \varepsilon \ge 0$, if for every $\pi' \in \Pi$ we have

$$\overline{\lim_{N\to\infty}} (V_{N,k}^{\pi'}(s) - V_{N,k}^{\pi}(s)) \le \varepsilon, \ s \in S \quad ,$$

where

$$V_{N,k}^{\pi}(s) = \frac{1}{N} \mathrm{E}_s^\pi \sum_{n_{-1}=0}^{N-1} \cdots \sum_{n_k=0}^{n_{k-1}} \tau(s_{n_k}, a_{n_k} \quad .$$

We shall call $(k, 0)$-optimal strategies *k-optimal*, $k = 1,0,1, \ldots$. Replacing upper limits by lower limits gives the criteria $\underline{R}_k$ and $\underline{V}_k$, $k = 1,0,1, \ldots$. The criteria $\overline{R}_k$, $k = -1,0,1$

, . . . , are considered, for instance, in [1,2] for finite S and A, and in [3] for finite S and arbitrary A. The criterion $\bar{V}_0$ was proposed by Veinott in [2].

In addition to providing a generalized understanding of summation, the new criteria also have another interesting property: they enable us to define the optimal strategy without introducing the concept of cost. This involves some inconvenience – in particular, control problems involving the above criteria cannot be solved using traditional techniques, i.e., by means of Bellman's equation.

Such difficulties can be partly overcome by the following expansion of R_δ^f, $f \in F$, in a Laurent series in the neighborhood of $\delta = 0$:

$$R_\delta^f = \sum_{k=-1}^{\infty} \delta^k \rho_k^f \tag{1}$$

(see [3]). The coefficients $(\rho_k^f)_{k \geq -1}$ come from the Markov chain transition matrix P^f obtained using the strategy $f \in F$ and the function $\tau(s, a)$; $\rho_{-1}^f = \bar{P}^f \tau^f$ is the average expected cost per unit time, $\rho_0^f = H^f \tau^f$, $\rho_m^f = (-1)^m (P^f H^f)^m H^f \tau^f$, $m = 1,2, \ldots$. Here τ^f is the column vector $(\tau(s, f(s)))_{s \in S}$, and $\bar{P}^f$ and H^f are, respectively, the marix of stationary probabilities and the basis matrix corresponding to P^f and related to it in the following manner:

$$\bar{P}^f = \lim_{N \to \infty} \frac{1}{N} \sum_{n=0}^{N-1} (P^f)^n = \lim_{\delta \to 0} \delta \sum_{n=0}^{\infty} (1-\delta)^n (P^f)^n ,$$

$$H^f = \lim_{N \to \infty} \frac{1}{N} \sum_{n=0}^{N-1} \sum_{k=0}^{n} ((P^f)^k - \bar{P}^f) = (I - P^f + \bar{P}^f)^{-1} - \bar{P}^f = \lim_{\delta \to 0} \sum_{k=0}^{\infty} (1-\delta)^k ((P^f)^n - \bar{P}^f) .$$

Expansion (1) shows that, at least in a class of stationary strategies, the optimal strategy with respect to $\bar{R}_k$ or $\underline{R}_k$, $k = 1,0,1, \ldots$ ($\bar{R}_k$ and $\underline{R}_k$ coincide in F) is to sequentially (lexicographically) optimize the components in the sequence $(\rho_m^f)_{-1 \leq m \leq k}$. To make our statements more precise we shall need the following definitions.

Definition 3. We say that a sequence of S-dimensional vectors $(X_m)_{m \geq -1}$ is lexicographically larger than a sequence of S-dimensional vectors $(Y_m)_{m \geq -1}$, i.e., $(X_m)_{m \geq -1}$, $(Y_m)_{m \geq -1}$, if for every $s \in S$ the first non-zero element of the sequence $(X_m(s) - Y_m(s))_{m \geq -1}$ is positive.

Definition 4. Let $\{(X_m^i)_{m \geq -1}, i \in I\}$ be a set of sequences of S-dimensional vectors. Let $\overline{\sup_{i \in I}} (X_m^i)_{m \geq -1}$ be the supremum with respect to lexicographical ordering – this is a sequence of S-dimensional vectors $(X_m^*)_{m \geq -1}$, the s-th components of which are recursively defined by the following relation:

$$X_m^*(s) = \inf_{\varepsilon \geq 0} \{\sup X_m^i(s) : X_k^i(s) \geq X_k^*(s) - \varepsilon , -1 \leq k \leq m \leq -1\} .$$

Definition 5. The strategy $\pi^* \in \Pi$ which corresponds to a set of sequences of S-dimensional vectors $(X_m^\pi)_{-1 \le m \le k}$ depending on strategies $\pi \in \Pi$, is (R, ε)-optimal with respect to the lexicographical optimality criterion if for every $\pi' \in \Pi$

$$(X_m^{\pi^*})_{-1 \le m \le k-1} \quad (X_m^{\pi'})_{-1 \le m \le k-1} \quad ,$$

and if for some $s \in S$ and some $\pi' \in \Pi$

$$(X_m^{\pi^*}(s))_{-1 \le m \le k-1} = (X_m^{\pi'}(s))_{-1 \le m \le k-1} \quad ,$$

so that

$$X_k^{\pi^*}(s) \ge X_k^{\pi'}(s) - \varepsilon \quad .$$

The above statement can now be reformulated as a theorem, making use of the above definitions.

THEOREM 1. *The optimality criteria* $\bar{R}_k$, $\underline{R}_k$, $\bar{V}_k$, $\underline{V}_k$ *and lexicographical optimality criterion* ρ_k, $k = -1,0,1, \ldots,$ *are strongly equivalent in* F, *which implies that for an arbitrary* $\varepsilon \ge 0$, $f_1, f_2 \in F$ *and* $k = -1,0,1, \ldots,$ *the following assertions are equivalent*:

(a) $(\rho_m^{f_1})_{-1 \le m \le k-1} \quad (\rho_m^{f_2})_{-1 \le m \le k-1}$,

and if $(\rho_m^{f_1}(s))_{-1 \le m \le k-1} = (\rho_m^{f_2}(s))_{-1 \le m \le k-1}$, then $\rho_k^{f_1}(s) \ge \rho_k^{f_2}(s) - \varepsilon$;

(b) $\lim_{\delta \to 0} (R_{\delta,k}^{f_1}(s) - R_{\delta,k}^{f_2}(s)) \ge -\varepsilon$, $s \in S$;

(c) $\lim_{N \to \infty} (V_{N,k}^{f_1}(s) - V_{N,k}^{f_2}(s)) \ge -\varepsilon$, $s \in S$.

The equivalence of (a) and (b) follows from expansion (1) and (a) and (c) are equivalent by virtue of the following lemma, which gives an analogue of expansion (1) for $V_{N,k}^f$, $f \in F$.

LEMMA 1. *For any integer-valued* $k \ge 1$ *and* $f \in F$ *we have*

$$V_{N,k}^f = \frac{1}{N} \sum_{m=-1}^{k} \begin{pmatrix} N + k - m \\ k - m + 1 \end{pmatrix} \rho_m^f + o(1) \, , \; N \to \infty \quad . \tag{2}$$

Proof. For $K = -1$, equation (2) implies that

$$\lim_{N \to \infty} V_{N,-1}^f = \rho_{-1}^f \quad .$$

For $k = 0$ the proof of (2) follows from the expression $\rho_0^f = H^f \tau^f$ and the definition of H^f. Indeed,

$$\rho_0^f(s) = \lim_{N\to\infty} \mathrm{E}_s^f \frac{1}{N} \sum_{n=0}^{N-1} \sum_{k=0}^{n} [\tau(s_k, a_k) - \rho_{-1}^f(s)] = \lim_{N\to\infty} [V_{N,0}^f(s) - \frac{1}{N}\frac{N+1}{2}\rho_{-1}^f(s)] \quad .$$

Hence $V_{N,0}^f(s) = \frac{1}{N}[\begin{pmatrix} N \\ 1 \end{pmatrix}\rho_0^f(s) + \begin{pmatrix} N+1 \\ 2 \end{pmatrix}\rho_{-1}^f(s)] + o(1)$, $N \to \infty$. An analogous proof can be constructed for $K \geq 1$.

When investigating control problems with the lexicographical optimality criterion ρ_k, $k = -1,0,1, \ldots,$ it is possible to use the lexicographical version of Bellman's equation

$$(Z_m)_{-1 \le m \le k} = \overline{\sup_{f \in F}}\, T_k^f (Z_m)_{-1\le m\le k}, \; k = -1,0,1, \ldots,$$

where the operator T_k^f, $k = -1,0,1,\ldots,$ transforms the sequence of vectors $(Z_m)_{-1\le m\le k}$ into a sequence of vectors of the following form:

$$(P^f Z_{-1}, P^f Z_0 + \tau^f - P^f Z_{-1}, P^f Z_1 - P^f Z_0, P^f Z_2 - P^f Z_1, \ldots, P^f Z_k - P^f Z_{k-1}) \quad .$$

The solution of this equation is the cost $\overline{\sup_{f\in F}}\,(\rho_m^f)_{-1\le m\le k}$ and, in general, the problem can be solved in the usual way (see [3,4]). Thus we have solved the problem of finding the optimal strategy in class F for the criteria $\underline{R}_k$, $\overline{R}_k$, $\underline{V}_k$, $\overline{V}_k$, $k = -1,0,1, \ldots$.

It would be expected that the problem of finding a stationary strategy which is (k, ε)-optimal in the class of all strategies Π for the criteria $\underline{R}_k$, $\overline{R}_k$, $\underline{V}_k$, $\overline{V}_k$, $k = -1,0,1 \cdots$ would be facilitated by considering control problems in which the lexicographical optimality criterion ρ_k, $k = -1,0,1, \ldots$ is generalized to the class of general strategies Π, and by establishing the relation between them.

In order to do this we shall give four different expressions for the sequence $(\rho_m^f)_{m\ge -1}$, $f \in F$, which coincide with each other in the class F (for convenience, however, they are denoted by different symbols).

$$A_{-1}^f = \lim_{\delta\to 0} \delta R_\delta^f, \; A_k^f = \lim_{\delta \to 0} \delta^{-k}(R_\delta^f - \sum_{n=-1}^{k-1} \delta^n A_n^f), \; k \ge 0$$

$$\alpha_{-1}^f = \lim_{N\to\infty} V_{N,-1}^f, \; \alpha_k^f = \lim_{N\to\infty} (V_{N,k}^f - \frac{1}{N} \sum_{m=-1}^{k-1} \begin{pmatrix} N+k-m \\ k-m+1 \end{pmatrix} \alpha_m^f), \; k \ge 0$$

$$B_{-1}^f = \lim_{\delta\to\infty} \mathrm{E}^f \delta \sum_{n=0}^{\infty} (1-\delta)^n \tau(s_n, a_n)$$

$$\beta_{-1}^f = \lim_{N\to\infty} \frac{1}{N} \mathrm{E}^f \sum_{n=0}^{N-1} \tau(s_n, a_n)$$

$$B_0^f(s) = \lim_{\delta\to 0} \mathrm{E}_s^f \sum_{n=0}^{\infty} (1-\delta)^n [\tau(s_n, a_n) - B_{-1}^f(s, s_{n+1})]$$

$$\beta_0^f(s) = \lim_{N\to\infty} \frac{1}{N} \mathrm{E}_s^f \sum_{n=0}^{N-1} \sum_{k=0}^{n} [\tau(s_k, a_k) - \underline{\beta}_{-1}^f(s, s_{k+1})]$$

$$B_k^f(s) = \lim_{\delta\to 0} \mathrm{E}_s^f \sum_{n=0}^{\infty} -(1-\delta)^n B_{k-1}^f(s, s_{n+1}), \; k \geq 1$$

$$\beta_k^f(s) = \lim_{N\to\infty} \frac{1}{N} \mathrm{E}_s^f \sum_{n=0}^{N-1} \sum_{m=0}^{n} -\beta_{k-1}^f(s, s_{m+1}), \; k \geq 1 \quad ,$$

where

$$\underline{B}_{-1}^f(s, s_{n+1}) = \lim_{\delta\to 0} \mathrm{E}_s^f(\delta \sum_{k=0}^{\infty} (1-\delta)^k \tau(s_{n+k+1}, a_{n+k+1}) / (s_{n+1}), \; n \geq 0$$

$$\underline{\beta}_{-1}^f(s, s_{n+1}) = \lim_{N\to\infty} \frac{1}{N} \mathrm{E}_s^f(\sum_{k=0}^{N-1} \tau(s_{n+k+1}, a_{n+k+1}) / (s_{n+1}), \; n \geq 0 \quad ,$$

and $B_k^f(s, s_{n+1})$, $\beta_k^f(s, s_{n+1})$, $k \geq 0$, $n \geq 0$, are defined in an analogous way. In the class of general strategies Π each of the above expressions has two forms – one with upper and the other with lower limits. The resulting eight sequences of functions of the criteria $(\underline{A}_m^\pi)_{m\geq -1}$, $(\bar{A}_m^\pi)_{m\geq -1}$, $(\bar{\alpha}_m^\pi)_{m\geq -1}$, $(\underline{\alpha}_m^\pi)_{m\geq -1}$, $(\bar{B}_m^\pi)_{m\geq -1}$, $(\underline{B}_m^\pi)_{m\geq -1}$, $(\bar{\beta}_m^\pi)_{m\geq -1}$, $(\underline{\beta}_m^\tau)_{m\geq -1}$ will generally be different. They define (see Definition 5) the desired lexicographical optimality criterion and together with the criteria $\underline{R}_k$, $\bar{R}_k$, $\underline{V}_k$, $\bar{V}_k$, $k = 1,0,1, \ldots$ produce the combination of optimality criteria that we are interested in:

$$\underline{A}_k, \bar{A}_k, \underline{\alpha}_k, \bar{\alpha}_k, \underline{B}_k, \bar{B}_k, \underline{\beta}_k, \bar{\beta}_k, \underline{R}_k, \bar{R}_k, \underline{V}_k, \bar{V}_k \quad . \tag{3}$$

We shall now partially order the criteria (3).

Definition 6. We say that the optimality criterion X is stronger than the optimality criterion Y, i.e., $X > Y$, if for an arbitrary $\varepsilon > 0$ a stationary strategy which is ε-optimal with respect to X is also ε-optimal with respect to Y.

Definition 7. The optimality criteria X and Y are equivalent, i.e., $X \sim Y$, if $X > Y$ and $Y > X$ hold simultaneously.

THEOREM 2. *The optimality criteria* (3) *have the following partial order (as specified in Definition 6)*:

$$\underline{A}_k < \underline{R}_k < \underline{B}_k < \bar{B}_k < \bar{R}_k < \bar{A}_k \tag{4}$$

$$\underline{\alpha}_k < \underline{V}_k < \underline{\beta}_k < \underline{B}_k < \bar{B}_k < \bar{R}_k < \bar{A}_k \tag{5}$$

$$\underline{\alpha}_k < \underline{V}_k < \underline{\beta}_k < \bar{\beta}_k < \bar{V}_k < \bar{\alpha}_k \tag{6}$$

$$\underline{A}_k < \underline{R}_k < \underline{B}_k < \bar{B}_k < \bar{\beta}_k < \bar{V}_k < \bar{\alpha}_k \quad . \tag{7}$$

The proof of Theorem 2 is rather long, so we shall only give an outline of the proof and point out the main steps. The assertions of the theorem emerge from the following sequence of relations:

(a) $\underline{A}_k < \underline{R}_k < \bar{R}_k < \bar{A}_k$;

(b) $\underline{B}_k < \underline{B}_k < \bar{B}_k < \bar{R}_k$;

(c) $\underline{\alpha}_k < \underline{V}_k < \bar{V}_k < \bar{\alpha}_k$;

(d) $\underline{V}_k < \underline{\beta}_k < \bar{\beta}_k < \bar{V}_k$;

(e) $\underline{\beta}_k < \underline{B}_k < \bar{B}_k < \bar{\beta}_k$.

The "Abelian" criteria are compared in (a) and (b), and the "Cesáro" criteria in (c) and (d). Relation (e) allows us to compare these different sets of criteria with each other. The main aim with regard to (a), (b), (c), (d) is to obtain analogues of the series expansions (1) and (2) for R_δ^π, $V_{N,k}^\pi$. Note that these analogues are interesting in themselves, since for non-stationary strategies the expansion in series of R_δ^π is not a consequence of matrix theory. One such analogue has the following form:

$$\sum_{m=-1}^{k} \delta^m \underline{A}_m^\pi(s) + \delta^k \lambda(\delta) \leq R_\delta^\pi(s) \leq \sum_{m=-1}^{k-1} \delta^m \underline{B}_m^\pi(s) + \delta^k \underline{B}_k^{\pi,\delta}(s)\ ,\ \delta \to 0\quad ,$$

where $\lambda(\delta) \to 0$ as $\delta \to 0$, $\underline{\lim_{\delta \to 0} B_k^{\pi,\delta}} = \underline{B_k^\pi}$, $k = -1,0,1, \ldots$. The left-hand side of this inequality can easily be proved using the definition of the sequence $(\underline{A}_m^\pi)_{m \geq -1}$; the right-hand side is obtained if the inequality

$$\mathrm{E}_s^\pi \underline{B_{k-1}^\pi}(s, s_{n+1}) \leq \sum_{m=0}^{k} \binom{m+n}{m} \underline{B_{k-m-1}^\pi}(s) - \mathrm{E}_s^\pi \sum_{n_{-1}=0}^{n} \cdots \sum_{n_{k-2}=0}^{n_{k-3}} \tau(s_{n_{k-2}}, a_{n_{k-2}})\quad ,$$

is used in the definition of $\underline{B}_k^\pi$. This inequality may be proved by induction. Finally, (e) may be proved by means of well-known inequalities relating Cesáro and Abelian sums (see [5]).

Theorems 1 and 2 lead to the following conditional theorem:

THEOREM 3. *If a stationary strategy exists which is* ε*-optimal with respect to some criterion from* (3), *for any* $\varepsilon > 0$, *then this criterion is equivalent to all criteria weaker than itself.*

Theorem 3 permits us to achieve our aim. In what follows we assume that A is a compact subset of a Polish space and that the standard continuity conditions hold. The following lemma is proved in [3] by applying lexicographical Bellman equation techniques:

LEMMA 2. *Let the following hold:*

$$(\rho_m^*)_{-1\le m\le k} = \overline{\sup_{f\in F}}\,(\rho_m^f)_{-1\le m\le k}\ ,\ k = -1,0,1\ \cdots\ \textit{is finite}$$

(*C*)

$$F_{k-1} = \{f \in F\colon (\rho_m^f)_{-1\le m\le k-1} = (\rho_m^*)_{-1\le m\le k-1}\}\ ,\ k = -1,0\ ,\ \ldots\ ,\ \textit{in non-empty}$$

$$\rho_m^* = \lim_{\delta\to 0} \delta^{-m}[\sup_{\pi\in\Pi_{m-1}} R_\delta^\pi - \sum_{i=-1}^{m-1} \delta^i \rho_i^*]\ ,\ -1 \le m \le k \quad .$$

Then Π_{m-1} *is a set of* $(m-1)$*-optimal strategies with respect to* $\bar{A}_{m-1}$, $0 \le m \le k$.

It follows directly from Lemma 2 that under the above conditions

$$(\rho_m^*)_{-1\le m\le k} = \overline{\sup_{\pi\in\Pi}}\,(\bar{\bar{A}}_m^\pi)_{-1\le m\le k}\ ,\ k = -1,0,1\ ,\ldots \quad . \tag{8}$$

The existence of $(k\,,\varepsilon)$-optimal stationary strategies with respect to $\bar{A}_k$, $k = -1,0,1\ ,\ \ldots\ ,\ \varepsilon > 0$, follows in turn from (8) under conditions (*C*). And, finally, Theorem 3 makes it possible to state the following result:

THEOREM 4. *Under conditions (C), all the criteria* (4), (5) *are equivalent, and there exists a* $(k\,,\varepsilon)$*-optimal stationary strategy* $k = -1,0,1\ ,\ \ldots\ ,\ \varepsilon > 0$, *with respect to each of them.*

The following theorem also holds.

THEOREM 5. *Under the condition*

$$\sup_{f\in F} \|H^f\| < \infty \tag{9}$$

all the criteria (4), (5) *are equivalent, and there exists a* $(k\,,\varepsilon)$*-optimal stationary strategy with respect to each of them for all integer-valued* $k \ge -1$.

Condition (9), which is stronger than (*C*), is at the same time weaker than the uniform Markov condition (positiveness of the matrix $(P^f)^n$ for sufficiently large n and all $f \in F$) often imposed on the controlled chain.

Fainberg's results [6] show that there exists an ε-optimal stationary strategy with respect to criterion $\bar{\alpha}_{-1}$ for any $\varepsilon > 0$, so that, by Theorem 3, all (-1)-criteria from (3) are equivalent. Using the martingale approach to controlled Markov chains, it can be shown that under condition (C) equation (8) holds for the criterion $\bar{\alpha}_0$. This implies that all 0-criteria from (3) are equivalent and that for each of them (including $\bar{V}_0$) an ε-optimal stationary strategy exists.

We are interested in the existence of ε-optimal stationary strategies ($\varepsilon > 0$) for criteria (3). It turns out that under conditions (C) such strategies exist for the criteria (4), (5) and also for the criteria (6), (7) with $k = -1,0$. It is hoped that this is also the case for the criteria (6), (7) with $k > 0$. However the conditions cannot be improved because there are well-known examples showing that if these conditions are not satisfied then ε-optimal stationary strategies may not exist, e.g., for the criterion $\bar{R}_{-1}$. The equivalence of the criteria (in terms of Definition 6) has been actively used in the course of the proof.

It is possible to give a more general definition of the equivalence of the criteria, in which stationary strategies are replaced by more general strategies, but this is not necessary for our purposes.

When there are no ε-optimal stationary strategies (e.g., if F is non-compact) a more general equivalence definition should be used and one should consider ε-optimal non-stationary (e.g., Markov) strategies.

Among various papers dealing with the equivalence of criteria we should mention [7], where the equivalence of $\bar{R}_k$ and $\bar{V}_k$ $k = -1,0,1$,... is proved, in a broad sense, for finite S and A. A similar result for $\bar{V}_0$ and $\bar{R}_0$ (predicted by Veinott) was obtained earlier in [8]. And, finally, the equivalence of the same criteria in a class of Markov strategies is proved in [9] for countable S with restrictions on the chain, a result which leads, among other things, to our condition (9).

REFERENCES

1. D. Blackwell. Discrete dynamic programming. *Annals of Mathematical Statistics*, 33(1962)719–726.
2. A.F. Veinott, Jr. On finding optimal policies in discrete dynamic programming with no discounting. *Annals of Mathematical Statistics*, 37(1966)1284–1294.
3. R.Ya. Chitashvili. Controlled finite Markov chains with an arbitrary set of controls (in Russian). *Teoriya Veroyatnosti i ee Primeneniya*, 20(4)(1975)855–864.
4. A.A. Yushkevich and R.Ya. Chitashvili. Controlled random sequences and Markov chains. *Russian Mathematical Surveys* (translated from *Uspekhi Matematicheskie Nauk*), 37(6)(1982)239–274.
5. B.V. Widder. *The Laplace Transform*. Princeton, 1946.
6. E.A. Fainberg. On the existence of stationary ε-optimal strategies for finite controlled Markov chains (in Russian). *Teoriya Veroyatnosti i ee Primeneniya*, 23(2)(1978)313–330.
7. K. Sladky. On the set of optimal controls for Markov chains with rewards. *Kybernetica*, 10(4)(1974)350–364.
8. E. Denardo and B. Miller. An optimality condition for discrete dynamic programming with no discounting. *Annals of Mathematical Statistics*, 39(4)(1968)1220–1227.
9. A. Hordijk and K. Sladky. Sensitive optimallity crieria in countable state dynamic programming. *Mathematics of Operations Research*, 2(1)(1977)1–14.

CANONICAL CORRELATIONS, HANKEL OPERATORS AND MARKOVIAN REPRESENTATIONS OF MULTIVARIATE STATIONARY GAUSSIAN PROCESSES

Michele Pavon[1]
LADSEB-CNR, Corso Stati Uniti 4, 35020 Padova, Italy

1. INTRODUCTION

Let $Y=\{y(k);k \in Z\}$ be a centered, real or complex, m-dimensional Gaussian process. We assume that y is stationary and regular /Roz/. Then y admits the spectral representation

$$y(k) = \int_{-\pi}^{\pi} e^{ikt} d\hat{y}(t)$$

where $d\hat{y}$ is a vector orthogonal stochastic measure satisfying

$$E\{d\hat{y}d\hat{y}^*\} = (2\pi)^{-1} f(e^{it})dt,$$

f(.) being the <u>spectral density</u> matrix-function and star denoting transposition plus conjugation. For the sake of simplicity we suppose that y is regular of full rank, namely that f has rank m a.e. on the unit circle T. In the case when f is a matrix of rational functions y admits finite dimensional <u>Markovian representations</u> such as

(1.a) $$x(k+1) = Ax(k)+Bu(k),$$

(1.b) $$y(k) = Cx(k)+Du(k),$$

where u is normalized Gaussian white noise sequence of dimension $p \geq m$ and the eigenvalues of A are in the unit disk D. The stationary Markov process x, which is required to be of smallest possible dimension, is called the <u>state</u> process of the model (1). The <u>Markovian representation problem</u> consists in characterizing all models (1) given y and, possibly, an exogenous process (see /LPP/-/Ruc-2/ and references therein).

The purpose of this paper is to study the dependence between the future of y and the past of a noise u driving one of its minimal Markovian representations. We do not restrict ourselves to the rational case, however, in view of the abstract theory covering infinite-dimensional Markovian representation developed by Lindquist-Picci /LiP/ and Ruckebusch /Ruc-1/.

---

[1] This research was conducted at the Institut für Mathematische Stochastik, Universität Hamburg, Hamburg, West Germany with support provided by an Alexander von Humboldt-Stiftung fellowship.

As is well known, the problem of characterizing the dependence of the future at time $k \geq o$ $\{y(k),y(k+1),\ldots\}$ of a stationary process y on its past at time zero $\{y(-1),y(-2),\ldots\}$ in terms of its spectral measure has received considerable attention in the past because of its importance for the prediction and ergodic theories of Gaussian processes, se /HeS-1/, /HeS-2/, /Yag/, /Roz/, /DyM/, /PeK/; /JeB / /JeB/ . Our problem turns out to be a twofold generalization of the above since we consider multivariable processes and we recover the classical problem in the case when $u=u_-$ the <u>innovations</u> of y. This because for Gaussian processes the most interesting types of dependence can all be characterized in terms of the corresponding <u>Gaussian spaces</u> /Nev/ and the spaces induced by the past of u_- and the past of y coincide.

Our motivation for considering such a generalization of the classical problem is manyfold. On the one hand we like to extract information on a particular representation of y which may prove useful when we need to approximate such a representation by low-order models /JoH/-/Pav/ as well as when we try to identify the particular representation at hand from input-output data as in /Aka, Section 1/. Indeed, if u are the normalized innovations of a process z and we can estimate the spectrum of z, the spectrum of y and the <u>canonical correlation coefficients</u> induced by past of z - future of y, our results provide information of the u driven Markovian representation of y. On the other hand, if the past of z represents the information available to us in order to predict the future of y, our analysis permits to characterize the best predictable functionals of the future of y /Yag/ and provides a guideline for the approximate prediction problem, see Section 4. Finally we would like to point out that, when specialized to the past of y - future of y dependence, our results appear, to the best of our knowledge, as the first function-theoretic characterizations of the classes of multivariable stationary Gaussian processes satisfying certain regularity conditions among which Rosenblatt's celebrated strong mixing condition /Ros/ (see Corollary 1). The derivation relies on some results from vectorial Hankel operator theory /AAk/,/Pag/.

The outline of the paper goes as follows. In the next Section we recall some definitions and basic results from Markovian representation theory. In Section 3 we describe the past of u - future of y dependence in terms of certain (generalized) canonical correlation coefficients. The various conditions are then rephrased in Section 4 in terms of Hankel operators and their symbols.

The rational spectral density case has been studied in /Pav/.

2. BACKGROUND MATERIAL

We denote by L^2_{mxp} the Hilbert space of C^{mxp} valued, measurable functions F defined on T such that $||F(e^{it})||$ is square integrable on T, where $||F(e^{it})||^2$ is the largest eigenvalue of $F(e^{it})^* F(e^{it})$. We define the Hardy class H^2_{mxp} $(\bar{H}^2_{mxp})$ as the subspace of L^2_{mxp} of all functions whose Fourier coefficients of negative (positive) index vanish. Every function in $H^2_{mxp}(\bar{H}^2_{mxp})$ possesses an analytic extension into D (into $|z| > 1$) from which it may be recovered by strong non-tangential limits /Fuh/. When p=1 we simply write L^2_m, H^2_m, etc. A function $F \in H_{mxm}$ $(\bar{H}_{mxm})$ is called outer (conjugate outer or minimum phase) if $\{Fh; h \in H^2_m\} = H^2_m$ $(\{Fh; h \in \bar{H}^2_m\} = \bar{H}^2_m)$.

Under the present assumptions the spectral density f admits factorizations of the form

$$f(e^{it}) = W(e^{it})W(e^{it})^*, \tag{2}$$

where W is mxp, $p \geq m$, and stable, i.e. $W \in \bar{H}^2_{mxp}$. Among such factors there exists a conjugate outer mxm factor W_- which is unique up to multiplication on the right by a constant unitary matrix /Hel, p.122/.

Then the m-dimensional process u_- defined by

$$u_-(k) = \int_{-\pi}^{\pi} e^{ikt} W_-(e^{it})^{-1} d\hat{y}(t) \tag{3}$$

is the innovations process of y satisfying $H_k^-(u_-) = Y_k^-$ for all k, where $H_k^-(u)$ and Y_k^- are the Gaussian spaces induced by the components of $\{u(j); j<k\}$ and $\{y(j); j<k\}$, respectively. Let us also introduce the spaces $H_k^+(u), Y_k^+$, $H(u)$ and Y induced by the components of $\{u(j); j\geq k\}$, $\{y(j); j\geq k\}$, $\{u(k); k\in Z\}$ and $\{y(k); k\in Z\}$, respectively. When k = o we delete the subscript. We assume that y is regular also in the reverse time direction which amounts to $\bigcap_{k\geq o} Y_k^+ = \{o\}$. Then f also admits factorizations (2) where $W\in H^2_{mxp}$. Among these there exists an essentially unique mxm outer factor $\bar{W}_+$ (the notation comes from /LiP/). Then the normalized backward innovations $\bar{u}_+$ of y are given by

$$\bar{u}_+(k) = \int_{-\pi}^{\pi} e^{ikt} \bar{W}_+(e^{it})^{-1} d\hat{y}(t) \tag{4}$$

and satisfy

$$H_k^+(\bar{u}_+) = Y_k^+ \quad , \quad k\in Z. \tag{5}$$

Finally we assume that y is a strictly noncyclic process. This is equivalent to the fact that $\bar{W}_+$ has a meromorphic pseudocontinuation of bounded type to the outside of the unit disk, see /Ruc-2/ where further information on the significance of this assumption may be found.

Let u be a p-dimensional normalized white noise and W a stable mxp spectral factor such that

$$y(k) = \sum_{j=o}^{\infty} W_j u(k-j), \tag{6}$$

where $W(e^{it}) = \sum_{j=o}^{\infty} W_j e^{-ijt}$ is the Fourier representation of W. Let X be the Gaussian space induced by the orthogonal projections of elements of Y^+ onto $H^-(u)$. Then X induces a Markovian representation of y in the following sense. Let U be the unitary operator on H(u) which shifts the coordinates, i.e. $U(u_r(k)) = u_r(k+1)$ and let $X_k = U^k(X)$. Then we have the following conditional orthogonality properties:

(i) $$(\bigvee_{k \leq o} X_k) \perp (\bigvee_{k \geq o} X_k) \mid X ,$$

(ii) $$Y^- \perp Y^+ \mid X ,$$

/Ruc-1/ where $\vee$ stands for closed sum.

Conditions (i) and (ii) may be shown to be the natural infinite dimensional counterparts of (1) where X plays the role of x(o) /Ruc-1/. Under the present assumptions X is regular, coregular and observable /Ruc-2/. We shall also assume that it is minimal, i.e. that there exists no proper closed subspace of X which satisfies (i) and (ii). We then have a true infinite-dimensional counterpart of (1). We shall study the problem of characterizing the dependence between $H^-(u)$ and Y_k^+, $k \geq o$ in terms of W. It is immediate that the dependence of Y_k^+ on $H^-(u)$ only occurs through the state X so that equivalently we study the dependence of Y_k^+ on X.

3. CANONICAL CORRELATIONS.

Consider two separable Gaussian spaces H_1 and H_2 of random variables defined on a common probability space and let $K := H_1 \vee H_2$. Let B be the operator $P_2P_1P_2$ mapping K into K where $P_1(P_2)$ is the orthogonal projection in K onto $H_1(H_2)$. Clearly B is a self-adjoint, positive contraction. The square roots σ_i of its (at most countably many) nonzero eigenvalues of finite multiplicity σ_i^2, where $1 \geq \sigma_o \geq \sigma_1 \geq \ldots$, are called the canonical correlation coefficients of the pair (H_1, H_2). The square root σ_∞ of the supremum of the essential spectrum of B /GoK/ is called the essential correlation coefficient. It is readily seen that our definition agrees with the classical one /Kul/ in the case when H_1 and H_2 are finite dimensional. The first canonical coefficient $\sigma_o = ||B||^{\frac{1}{2}}$ is called the <u>maximal correlation coefficient</u> of H_1 and H_2.

Let us consider the family of pairs of Gaussian spaces $(H_1, H_2(k))$, $k \geq o$. Such a family is called <u>completely regular</u> if the corresponding sequence $\{\sigma_o(k)\}$ tends to zero as k tends to infinity. Because of the Gaussian assumption a completely regular family satisfies Rosenblatt's strong mixing condition /Roz, p.186/ which has been successfully applied to limit theorems for weakly dependent random variables /Ros/, /Roz, p.191/.

In the case when B is compact we define the mutual <u>information</u> of H_1 and H_2 by

(7) $$I(H_1,H_2) = -\tfrac{1}{2} \operatorname{logdet} (I-B) = -\tfrac{1}{2} \sum_{i=o}^{\infty} \log(1-\sigma_i^2).$$

In view of the results of Gelfand and Yaglom /GeY/ our definition is seen to be consistent with the usual one. Moreover we have that $I(H_1,H_2)$ is finite if and only if B is a strict contraction with finite trace. We say that $(H_1,H_2(k)), k \geq o$, is <u>informationally regular</u> if $I(H_1,H_2(k)) \to o$ as $k \to \infty$. This type of regularity is stronger than complete regularity and equivalent to <u>absolute regularity</u>, see /IbR, Chapter IV/.

<u>Proposition 1</u>. The family $(H_1,H_2(k))$ is completely regular if and only if $\bigcap_{k \geq o} H_2(k) = \{o\}$ and B corresponding to $(H_1,H_2(o))$ is compact. The family $(H_1,H_2(k))$ is informationally regular if and only if it is completely regular and B has finite trace.

<u>Proof</u>. A proof may easily be constructed along the same lines as in Theorems 3 and 6 in /IbR, Chapter IV/. //

Let u and W be as at the end of Section 2. We shall denote by B_k the B operators corresponding to $(H^-(u), Y_k^+)$.

4. GENERALIZED PAST-FUTURE DEPENDENCE VIA HANKEL OPERATORS.

Let $F \in L^{\infty}_{p\times m}$ and denote by P_+ the orthogonal projection in L^2_p onto H^2_p (Riesz projection). Then the Hankel operator with symbol F denoted by H_F is defined as a bounded operator from H^2_m into $L^2_p \ominus H^2_p$ by

$$H_F(h) := (I-P_+)(Fh), h \in H^2_m,$$

where I is the identity on L^2_p, see /AAK/. Let $F(e^{it}) = \sum_{k=-\infty}^{\infty} F_k e^{ikt}$ be the Fourier representation of F. Then it is readily seen that H_F is unitarily equivalent to the operator induced on $l^2_+(C^m)$ by the infinite block-Hankel matrix $H_F = (F_{-r-s+1})^{\infty}_{r,s=1}$. According to a generalization of Nehari's theorem the norm of H_F satisfies

$$(8) \qquad ||H_F|| = \text{dist}_{L^{\infty}_{p\times m}} (F, H^{\infty}_{p\times m}),$$

/AAK/.

Let H_1, H_2 be Hilbert spaces and $A: H_1 \to H_2$ be a linear bounded operator. Then the singular numbers $s_n(A)$, $n \geq o$, are defined by

$$s_n(A) = \inf\{||A-C||; C: H_1 \to H_2, \text{ rank } C \leq n\}.$$

When A is compact $\{s_n(A), n \geq o\}$ coincide with the eigenvalues of $(A^*A)^{\frac{1}{2}}$ /PeK/ (here star denotes adjoint).

Let $x \in H(u)$. Then x admits a representation

$$(9) \qquad x = \int_{-\pi}^{\pi} \hat{x}(e^{it})' d\hat{u}(t),$$

where $d\hat{u}$ is the vector stochastic measure of u and prime denotes transposition. We define the unitary operator T_u from H(u) to L^2_p by $T_u(x) = \hat{x}$. Similarly we define $T_{\bar{u}_+}$ from Y to L^2_m. We are now ready to establish a strict connection between the operators B_k and certain Hankel operators which allow us to describe the dependence between $H^-(u)$ and Y^+_k in an effective fashion.

<u>Lemma 1</u> Let P^- and P^+_k be the orthogonal projections in $K = H^-(u) \vee Y^+_k$ onto $H^-(u)$ and Y^+_k, respectively. Then

(10) $$P^- P_k^+ = (T_u)^{-1} H_{z^k(W')(\bar{W}'_+)^{-1}} P_+ z^{-k} T_{\bar{u}_+} .$$

Proof. First observe that $P^- = (T_u)^{-1}(I-P_+)T_u$ and $P_k^+ = (T_{\bar{u}_+})^{-1} z^k P_+ z^{-k} T_{\bar{u}_+}$. Next let $x \in Y$ and $\hat{x} = T_{\bar{u}_+}(x)$. Since $d\hat{u}_+ = \bar{W}_+^{-1} W d\hat{u}$ we see that $T_u(T_{\bar{u}_+})^{-1}$ acts on the column function $\hat{x}$ as multiplication on the left by $(W')(\bar{W}'_+)^{-1}$ and the conclusion follows. //

Let us denote by G the function $(W')(\bar{W}'_+)^{-1}$ which has norm one a.e. on T. As observed by S. Mitter /Mit/, the function G may be viewed as a scattering matrix according to the abstract Lax-Phillips theory /LaP/, see also /AdA/.

Theorem 1 The spaces $H^-(u)$ and Y_k^+ are at a positive angle, i.e. $\sigma_o(k) < 1$, if and only if one of the following equivalent conditions is verified:

(i) $$\| H_{z^k G} \| < 1 ;$$

(ii) $$\operatorname{dist}_{L^\infty_{pxm}} (G, z^{-k} H^\infty_{pxm}) < 1.$$

Proof. The result follows at once from (10) and (8). //

This result may be viewed, in the case $W = W_-$, as a generalization of some results of Helson and Szegö, and Helson and Sarason, /HeS-1/, /HeS-2/, which are, however, phrased in terms of the spectral density.

Theorem 2 Let $\{\sigma_i\}$, $i \geq o$ be the canonical correlation coefficients of $(H^-(u), Y_k^+)$. Then

(11a) $$\sigma_i = s_i(H_{z^k G}) \quad , \quad i=0,1,\ldots$$

(11b) $$\sigma_\infty = \| H_{z^k G} \|_e ,$$

where $\|\cdot\|_e$ denotes the essential norm of an operator; i.e. its distance from the closed subspace of compact operators.

Proof. Since T_u and $z^{-k}T_{\bar{u}_+}$ are unitary operators we get from (10) that B_k and $(H_{z^kG})^*H_{z^kG}$ are unitarily equivalent. //

Let us assume that $\sigma_o(k)^2$ is an eigenvalue of B_k of finite multiplicity. Then it is possible to characterize the best predictable functionals /Yag/, /JeB-1/ on Y_k^+ given $H^-(u)$ introducing the <u>canonical components</u> corresponding to $\sigma_o(k)$. We describe the procedure for $k = o$. Let $p \in H_m^2$ be an eigenvector of norm one for $(H_G)^*H_G$ corresponding to the eigenvalue σ_o^2 and let $q := (\sigma_o)^{-1}H_G(p)$. Also let $\eta = T_{\bar{u}_+}^{-1}(p)$ and $\xi = T_u^{-1}(q)$. Then $\eta \in Y^+, \xi \in H^-(u)$, $||\eta|| = ||\xi|| = 1$ and

$$\text{(12)} \qquad E\{\eta\xi\} = (\sigma_o)^{-1} \langle p, (H_G)^*H_G\, p\rangle_{H_m^2} = \sigma_o,$$

the latter following immediately from the fact that $T_u(\eta) = W'(\bar{W}'_+)^{-1}\bar{p}$.
The elements η and ξ are called <u>canonical variables</u> corresponding to the canonical correlation coefficient σ_o. It is immediate that η is a best predictable functional on the basis of $H^-(u)$ which is unique if and only if σ_o^2 has multiplicity one. Otherwise starting with another unit eigenvector p_1 corresponding to σ_o^2 and orthogonal to p one gets in the same way another pair of canonical variables corresponding to σ_o. The same procedure may then be applied to σ_1, σ_2 and so on until, after obtaining a finite or infinite number of pairs of canonical variables one reaches σ_∞ where the procedure stops. Further details may be found in /JeB/. Notice that the canonical variables ξ so obtained belong to X and that they allow to solve the approximate prediction problem where one tries to optimally predict the elements of Y^+ on the basis of a preassigned number of elements in $H^-(u)$.

In order to take advantage of a result from operator theory we now assume that W is mxm. We remark that this is equivalent to assuming that $H(u) \subset Y$, i.e. that the Markovian representation is <u>internal</u> or <u>output induced</u>, see e.g. /Ruc-2/.

Theorem 3 The family $(H^-(u), Y_k^+)$ is completely regular if and only if

(13) $$G \varepsilon (H^\infty_{mxm} + C_{mxm}(T)),$$

where $C_{mxm}(T)$ are the mxm continuous functions on T. It is also informationally regular if and only if

(14) $$\sum_{k=1}^{\infty} k \text{ trace } (C^*_{-k} C_{-k}) < \infty ,$$

where $G(e^{it}) = \sum_{k=-\infty}^{\infty} C_k e^{ikt}$.

Proof. It follows from Proposition 1 and Theorem 2 that $(H^-(u), Y^+_k)$, $k \geq 0$, is completely regular if and only if H_G is compact; we now get condition (13) from a generalization of Hartman's theorem due to Page /Pag/. It also follows from Proposition 1 and Theorem 2 that the family is informationally regular if and only if H_G is Hilbert-Schmidt. This property is quickly seen to be equivalent to (14).//

We now record an important particular case of the above result.

Corollary 1 The process y is completely regular (equivalently it satisfies the strong mixing condition) if and only if

(15) $$\bar{W}_+^{-1} W_- \varepsilon (H^\infty_{mxm} + C_{mxm}(T)).$$

In the case when H_G is Hilbert-Schmidt with norm less than one the information $I(H^-(u), Y^+_k)$ is finite and given by

$$I(H^-(u), Y^+_k) = -\tfrac{1}{2} \sum_{i=0}^{\infty} \log(1 - s_i(H_z k_G)^2).$$

The function G also yields some information on the singular values of $H_{W'}$ which, as argued in /Pav/, may be useful in some situations for the Hankel-norm approximation of Markovian representations.

Theorem 4 Suppose that $f \varepsilon L^\infty_{mxm}$. Then

$$s_i(H_{W'}) \leq ||f||^{\frac{1}{2}}_{L^\infty_{mxm}} s_i(H_G) , \quad i = 0, 1, \ldots$$

The proof is a trivial extension of that of Theorem 5 in /Pav/ and is therefore omitted.

REFERENCES

/AdA/ V.M. Adamjan and D.Z. Arov, On unitary couplings of semi-unitary operators, Amer. Math. Soc. Transl. (2) vol.95, 1970, p.75-129.

/AAK/ V.M. Adamjan, D.Z. Arov and M.G. Krein, Infinite Hankel block matrices and related extension problems, Amer. Math. Soc. Transl. (2) vol.111, 1978, p.133-156.

/Aka/ H. Akaike, Stochastic theory of minimal realizations, I.E.E.E. Trans. Automatic Control, Ac-19, 1974, p.667-674.

/DyM/ H. Dym and H.P. McKean, Gaussian Processes, Function Theory, and the Inverse Spectral Problem, Academic Press, New York, 1976.

/Fuh/ P.A. Fuhrmann, Linear Operators and Systems in Hilbert Space, McGraw-Hill, New York, 1981.

/GeY/ I.M. Gelfand and A.M. Yaglom, Calculation of the amount of information about a random function contained in another such function,Amer. Math. Soc. Transl. (2) vol.12, 1959, p.199-246.

/GoK/ I.C. Gohberg and M.G. Krein, Introduction to the theory of linear non selfadjoint operators, Translation of Math. Monographs, vol.18, Amer. Math. Soc., Providence, R.I., 1969.

/Hel/ H. Helson, Lectures on invariant subspaces, Academic Press, New York, 1964.

/HeS-1/ H. Helson and G. Szegö, A problem in prediction theory, Ann. Mat. Pura App. 51, 1960, p.107-138.

/HeS-2/ H. Helson and D.E. Sarason, Past and future, Math. Scand. 21, 1967, p.5-16. Addendum by D.E. Sarason, Math. Scand. 30, 1972, p.62-64.

/IbR/ I.A. Ibragimov and Y.A. Rozanov, Gaussian Random Processes, Springer, New York, 1978.

/JeB/ N.P. Jewell and P. Bloomfield, Canonical correlations of past and future for time series: definitions and theory, Ann. Statist. 11, 1983, p.837-847.

/JBB/ N.P. Jewell, P. Bloomfield and F.C. Bartmann, Canonical correlations of past and future for time series: bounds and computation, Ann. Statist. 11, 1983, p.848-855.

/JoH/ E.A. Jonckheere and J.W. Helton, Power spectrum reduction by optimal Hankel norm approximation of the phase of the outer spectral factor, Proc. Amer. Control Conf., San Diego, CA, June 1984.

/Kul/ S. Kullback, Information Theory and Statistics, Dover, New York,1968.

/LaP/ P.D. Lax and R.S. Phillips, Scattering Theory, Academic Press, New York, 1967.

/Lip/ A. Lindquist and G. Picci, State space models for Gaussian stochastic processes; in Stochastic Systems: The Mathematics of Filtering and Identification and Applications, M. Hazewinkel and J.C. Willems Eds., Reidel Publ. Co. Dordrecht, 1981, p.169-204.

/LPP/ A. Lindquist, M. Pavon and G. Picci, Recent trends in stochastic realization theory, in Prediction Theory and Harmonic Analysis - The Pesi Masani Volume, V. Mandrekar and H. Salehi Eds., North Holland, Amsterdam, 1983, p.201-224.

/Mit/ S.K. Mitter, Personal communication, 1980 and talk delivered at this conference.

/Nev/ J. Neveu, Processus Aléatoires Gaussiens, Presses de l'Université de Montréal, 1968.

/Pag/ L.B. Page, Bounded and compact vectorial Hankel operators, Trans. Amer. Math. Soc. 150, 1970, p.529-539.

/Pav/ M. Pavon, Canonical correlations of past inputs and future outputs for linear stochastic systems, Systems & Control Letters 4, 1984, p.209-215.

/PeK/ V.V. Peller and S.V. Khrushchev, Hankel operators, best approximations, and stationary Gaussian processes, Russian Math. Surveys 37, 1982, p.61-144.

/Ros/ M. Rosenblatt, A central limit theory and a strong mixing condition, Proc. Nat. Acad. Sci. U.S.A. 42, 1956, p.43-47.

/Roz/ Y.A. Rozanov, Stationary Random Processes, Holden-Day, San Francisco, 1967.

/Ruc-1/ G. Ruckebusch, Théorie géométrique de la Représentation Markovienne, Ann. Inst. Henri Poincaré, Section B, XVI, 1980, p.225-297.

/Ruc-2/ G. Ruckebusch, Markovian representations and spectral factorizations of stationary Gaussian processes, in Prediction Theory and Harmonic Analysis - The Pesi Masani Volume, V. Mandrekar and H. Salehi Eds., North Holland, Amsterdam, 1983, p.275-307.

/Yag/ A.M. Yaglom, Stationary Gaussian processes satisfying the strong mixing and best predictable functionals, Proc. Int. Research Seminar of the Statistical Laboratory, Univ. of Calif. Berkeley, 1963, p.241-252, Springer-Verlag, New York.

THE MAXIMUM PRINCIPLE IN STOCHASTIC PROBLEMS WITH NON-FIXED RANDOM CONTROL TIME

M.T. Saksonov
Tadjik Pedgogical Institute
Dushanbe, USSR

In this paper we shall consider the problem of controlling strong solutions of a system of stochastic differential equations under finite-dimensional constraints. There are two control parameters in the problem: the control itself, which is an adapted measurable function with values in the given set, and a random stopping time.

Our final result is that one additional optimality condition appears in the system of necessary conditions [1]: a Hamiltonian function is equal to zero at the end of the optimal control time.

Let a one-dimensional Wiener process $(w_t(\mathbf{F}_t))$ be defined on a complete probability space $(\Omega, \mathbf{F}, P)$. We shall assume that $\mathbf{F}_t = \mathbf{F}_t^w \cup N$, $\mathbf{F}_t^w = \sigma\{w_s, s \le t\}$, and that N is a family of sets of P-zero measure in $\mathbf{F}$. Consider the following control problem:

$$Q_0(\mathrm{E}x_T) \longrightarrow \min \tag{1}$$

$$\mathrm{d}x_t = f(t, x_t, u_t)\mathrm{d}t + \sigma(t, x_t)\mathrm{d}w_t, \; x_0 = 0 \tag{2}$$

$$Q_1(\mathrm{E}x_T) \in C \tag{3}$$

$$u_t \in U$$

$$Q_0\colon R^n \longrightarrow R^1, \; Q_1\colon R^n \longrightarrow R^l, \; f\colon [0,T_c] \times \Omega \times R^n \times R^k \longrightarrow R^n \tag{4}$$

$$\sigma\colon [0,T_c] \times \Omega \times R^n \longrightarrow R^n, \; U \subset R^k, \; C \subset R^l,$$

where T_c is a positive number. The control u_t is a progressively measurable function taking values in U. The problem is to minimize a functional (1) on the set of values (x_t, u_t, T) satisfying conditions (2)–(4), where u_t is a control and T is a stopping time with respect to $\mathbf{F}_t$.

Remark 1. In our case all stopping times are predictable [2].

We shall assume that the functions f, σ are continuous with respect to the aggregate variable (t, x, u), uniformly continuous in u, adapted to $\mathbf{F}_t$ and continuously differentiable in x. Then there exists a constant $K > 0$ such that

$$\|f_x'\| + \|\sigma_x'\| \le K, \ \|f\| + \|\sigma\| \le K(1 + \|x\|) \quad .$$

Moreover, the function $Q = (Q_0, Q_1)$ is continuously differentiable. C is a closed convex set and U is a compact set. Let $H(t, p, x, u) = \langle p, f(t, x, u)\rangle$, where $p \in R^n$ and $\langle , \rangle$ denotes a scalar product.

THEOREM 1. *Let* (x_t^*, u_t^*, T^*), $0 < T^* < T_c$, *be the solution of* (1)–(4) *and our assumptions be satisfied. Then there exist functions* p_t, h_t *and a vector* $\lambda = (\lambda_0, \lambda_1)$ *such that*

(a) $\lambda_0 \ge 0$, λ_1 *is normal to* C at the point $Q_1(\mathrm{E}x_{T^*}^*)$, $\lambda_0^2 + \|\lambda\|^2 = 1$, $\mathrm{E}\sup_{0<t<T_c} \|p_t\|^s < \infty$, $\forall$ $s > 0$, $\mathrm{E}\int_0^{T_c} h_t^2 \mathrm{d}t < \infty$. *These functions are measurable and adapted with respect to* $\mathbf{F}_t$ *and are the solutions of the inverse stochastic equation*

$$\mathrm{d}p_t = (-f_x'^*(t, x_t^*, u_t^*)p_t - \sigma_x'^*(t, x_t^*)h_t)\mathrm{d}t + h_t \mathrm{d}w_t, \ p_{T^*} = -Q'^*(\mathrm{E}x_{T^*}^*) \cdot \lambda \quad .$$

The formula for solutions of this equation is given in [1].

(b) $\sup_{u \in U} H(t, p_t, x_t^*, u) = H(t, p_t, x_t^*, u_t^*)$, $l \times P$ a.s. if $t < T^*$.

(c) *The function* $H(t, p_t, x_t^*, u_t^*)$ *is continuous in* t *on* $[0, T^*]$ *if the appropriate modification of* u_t *has been chosen, and moreover* $H(T^*, p_{T^*}, u_{T^*}) = 0$.

Some of our assumptions are not essential to prove the theorem. It is possible to derive analogous results for the m-dimensional Wiener process, and moreover to show that there exist some vector coordinates for which the result does not require growth conditions or uniformly Lipschitz conditions.

The theorem can be proved in three stages. The first stage involves the construction and investigation of sequences of auxiliary problems. The second is a proof of the second assertion of the theorem, while the third is a proof of the final assertion of the theorem.

It is obvious that an optimal control cannot be unique because it is arbitrary if $t > T^*(\omega)$. Thus, for the sake of convenience, we shall assume that $u^*(t, \omega) = u^*(2T - t, \omega)$ if $t > T^*(\omega)$.

In accordance with the above, let $\tau_1^*, \ldots, \tau_n^* \cdots$ be a sequence of stopping times predicting T^*. Let τ_m^* be fixed. Consider problem (1)–(4) with (4) replaced by (4m):

$$u_t \in U_t^m \quad , \tag{4m}$$

where $U_t^m = U$ if $t < \tau_m^*$; $U_t^m = u_t^*$ if $t > \tau_m^*$. Obviously (x_t^*, u_t^*, T^*) is also a solution to this problem. Let $d[(u^1(\cdot), T_1), (u^2(\cdot), T_2)] = l \times P\{(t, \omega): u_t^1(\omega) \neq u_t^2(\omega)\} + \mathrm{E}\,|T_1 - T_2|$ be a metric on the set of control parameters of the problem (1)–(4m). Without significantly complicating the proof in [3], it is easy to see that this leads to a Polish space, which we shall denote by X. This will be needed in the next theorem [3].

THEOREM 2 (Ekeland). *Let V be a Polish space with metric d, and $F: V \to R^1$ be a continuous function. For $u^0 \in V$ and $\varepsilon_0 > 0$, let $\inf_V F \le F(u^0) \le \inf_V F + \varepsilon_0$. Then for all $\kappa > 0$ there exists a $v^0 \in V$ such that $F(v^0) < F(u^0)$; $d(u^0, v^0) < \kappa$, $F(v) \ge F(v^0) - \frac{\varepsilon}{\kappa} d(v^0, v)$; $\forall v \in V$.*

Take $\tilde{c}_j = \{(x_0, x) \in R^{l+1}, \quad x_0 \le Q_0(\mathrm{E}x_{T^*}^*) - 1/j, \quad x \in c\}$ and let $G_j(x_0 x) = \rho((x_0, x), \tilde{c}_j)$ be the distance between $(x_0 x)$ and $\tilde{c}_j$. Let x_t^u denote the phase variable corresponding to control u_t. From our assumptions the function $F(u(\cdot), T) = G_j(Q(\mathrm{E}x_T^u))$ is contintuous on X. Moreover, $0 \le G_j(Q(\mathrm{E}x_{T^*}^{u^*})) \le 1/j$.

Using Ekeland's theorem with $\varepsilon_j = 1/j$, $\kappa_j = 1/\sqrt{j}$ and setting

$$\eta(u, v) = \begin{cases} 0, & u = v \\ 1, & u \neq v \end{cases},$$

we deduce that there must exist a control ${}^{(m)}u_t^j$, a stopping time ${}^{(m)}T_j$, and a phase variable ${}^{(m)}x_t^j = x_t^{{}^{(m)}u^j}$ that are solutions of the problem

$$G_j(Q(\mathrm{E}x_T)) + \kappa_j \cdot \mathrm{E}y_T + \kappa_j \mathrm{E}\,|T - T_j| \to \min \tag{5}$$

$$dy_t = \eta(u_t, u_t^j)dt, \; y_0 = 0 \tag{6}$$

$$dx_t = f(t, x_t, u_t)dt + \sigma(t, x_t)dw_t, \; x_0 = 0 \tag{7}$$

$$u_t \in U_t^m . \tag{8}$$

The function G_j is continuously differentiable in the neighborhood of the point $Q(\mathrm{E}x_{T_j})$ because it lies outside the set $\tilde{C}_j$. (For the properties of this function see [4, II, Theorem 3.16].) We shall make use of an additional index (m) to indicate values coming from problem (1)–(4m). The following lemma is the simplest necessary condition for stopping time T_j to be optimal for problem (5)–(8).

LEMMA 1.

(a) *We have*

$$\overline{\lim_{k \to \infty}} \frac{{}^{(m)}\mu^j}{\mathrm{E}\,|I_A(T_j - \tau_k^j)|} - \mathrm{E}\int_{\tau_k^j}^{T_j} I_A \cdot f(t, {}^{(m)}x_t^j, u_t^j)dt \le {}^{(m)}\kappa_j \tag{9}$$

for all sequences of stopping times $\bar{t}^j_k$ *predicting* T_j *on the set* $A \in \mathbf{F}_{T_j}$ *such that* $A \subset \{\omega\colon \tau^j_k > \tau_m\}$.

(b) *We have*

$$\lim_{k\to\infty} \frac{{}^{(m)}\mu^j}{\mathrm{E}|T_j - \tau^j_k|} \cdot \mathrm{E}\int_{T_j}^{\tau^j_k} f(t, {}^{(m)}x^j_t, u^*_t)dt \geq -{}^{(m)}\kappa^j \tag{10}$$

for all decreasing sequences of stopping times τ^j_k *tending to* T_j. *Here* I_A *is an indicator function of the set A and*

$${}^{(m)}\mu^j = G_j'(Q(\mathrm{E}\,{}^{(m)}x^j_{T_j})) \cdot Q_x'(\mathrm{E}\,{}^{(m)}x^j_{T_j}) \quad . \tag{11}$$

If we consider problem (5)–(8) with fixed ${}^{(m)}T_j$, then we obtain the problem that was considered earlier. The fact that T_j is a random variable is not important. For all controls u_t from the maximum principle in integral form [5, Lemma 7], it follows that

$$\mathrm{E}\int_0^{{}^{(m)}T_j} (<{}^{(m)}p^j_t, f(t, {}^{(m)}x^j_t, u_t) - f(t, {}^{(m)}x^j_t, {}^{(m)}u^j_t)> - \kappa_j\eta(u_t, {}^{(m)}u^j_t)]dt \leq 0 \tag{12}$$

where

$${}^{(m)}p^j_t = (-{}^{(m)}\Phi^j_t)^{*-1}\,\mathrm{E}[{}^{(m)}\Phi^{j*}_t \cdot {}^{(m)}\mu^j / \mathbf{F}_t]\,,\ {}^{(m)}p^j_t = \text{const}\,,\ t \geq T_j \tag{13}$$

and ${}^{(m)}\Phi^j_t$ is a matrix solution of the system

$$d\,{}^{(m)}\Phi^j_t = f_x'(t, {}^{(m)}x^j_t, {}^{(m)}u^j_t)\,{}^{(m)}\Phi^j_t dt + \sigma_x'(t, {}^{(m)}x^j_t)\,{}^{(m)}\Phi^j_t dw_t\,,\ {}^{(m)}\Phi^j_0 = I \quad .$$

Without loss of generality we can assume that

$$G_j'(Q(\mathrm{E}\,{}^{(m)}x^j_{T_j})) \longrightarrow {}^{(m)}\lambda^*$$

as $j \longrightarrow \infty$, where ${}^{(m)}\lambda \in R^{l+1}$. From the properties of functions G_j we have that $\|{}^{(m)}\lambda\| = 1$ and, moreover, ${}^{(m)}\lambda$ is normal to the set $\hat{C}_\infty = \{(x_0, x)\colon x_0 \leq 0\,,\ x \in C\}$ at the point $Q(\mathrm{E}x_{T^*})$. This means that if ${}^{(m)}\lambda$ is written as ${}^{(m)}\lambda = ({}^{(m)}\lambda_0, {}^{(m)}\lambda_1)$, where ${}^{(m)}\lambda_1 \in R^l$, then ${}^{(m)}\lambda_0 \geq 0$, ${}^{(m)}\lambda_1$ is normal to the set C at the point $Q_1(\mathrm{E}x^*_{T^*})$. We shall use the notation ${}^{(m)}\mu = {}^{(m)}\lambda^* \cdot Q_x'(\mathrm{E}x^*_{T^*})$,

$${}^{(m)}p_t = -(\Phi^*_t)^{-1} \cdot \mathrm{E}[\Phi^*_{T^*} \cdot {}^{(m)}\mu^* / \mathbf{F}_t]\,,\ {}^{(m)}p_t = {}^{(m)}p_{T^*}\,,\ t > T^* \quad . \tag{14}$$

Let Φ_t be a matrix solution of the system

$$d\Phi_t = f_x'(t, x^*_t, u^*_t)\Phi_t dt + \sigma_x'(t, x^*_t)dw_t\,,\ \Phi_0 = I \quad .$$

It is easy to see that under our assumptions we have

$$\mathrm{E}\sup_{0\le t\le T_c}\{\|{}^{(m)}x_t^j - x_t^*\|^s + \|{}^{(m)}\Phi_t^j - \Phi\|^s + \|{}^{(m)}p_t^j - {}^{(m)}p_t\|^s\} + \|{}^{(m)}\mu^j - {}^{(m)}\mu\| \xrightarrow{j\to\infty} 0 \qquad (15)$$

for all $s > 1$. Taking the limit in (12) we find that the integral maximum principle

$$\mathrm{E}\int_0^{T^*} \langle {}^{(m)}p_t , f(t, x_t^*, u_t) - f(t, x_t^*, u_t^*)\rangle \, dt \le 0 \qquad (16)$$

holds for all u_t from problem (1)–(4).

Up to now we have considered problems with a restricted set of controls. Without loss of generality we can suppose that ${}^{(m)}\lambda \to \lambda$, $\|\lambda\| = 1$ and that λ has the same property of normality to C as each of the ${}^{(m)}\lambda$. Let $\mu = \lim_{m\to\infty} {}^{(m)}\mu$. Obviously $\mu = \lambda^* Q'(\mathrm{E}x_{T^*}^*)$ and

$$p_t = -\Phi_t^{*\,-1}\cdot \mathrm{E}[\Phi_{T^*}^*\cdot \mu^* / \mathbf{F}_t],\ p_t = p_{T^*} \ \text{ if } t > T^* .$$

Taking the limit with respect to m in (16) for every control u_t of problem (1)–(4), we obtain

$$\mathrm{E}\int_0^{T^*} \langle p_t , f(t, x_t^*, u_t) - f(t, x_t^*, u_t^*)\rangle dt \le 0 . \qquad (17)$$

From the integral maximum principle, using standard methods based on measurable choice theorems [5], we obtain a pointwise maximum principle

$$\sup_{u\in U} H(t, \omega, x_t^*, u) = H(t, \omega, x_t^*, u_t^*) \qquad (18)$$

$(l \times p)$-a.s. on the set $t \le T^*(\omega)$.

Now from [4, II, Lemma 3.5] it follows that if the corresponding modification of the optimal control has been chosen, the Hamiltonian $H(t, \omega, x_t^*, u_t^*)$ is continuous in t on $[0, T^*]$ p-a.s. We shall assume that the optimal control has been chosen in this way.

To complete the proof of the theorem we must show that the Hamiltonian is equal to zero at time T^*. We shall suppose that there exists a set Ω_0, $p(\Omega_0) = \delta > 0$, and $\varepsilon_0 > 0$ such that for every $\omega \in \Omega_0$ we have $\langle -p_{T^*}, f(T^*, \omega, x_{T^*}^*, u_{T^*}^*)\rangle > \varepsilon_0$ and argue by contradiction. Without loss of generality we can suppose that there exists a constant K_1 such that $\sup_{\omega\in\Omega_0}\sup_{0<t\le T^*} \|x_t^*\| \le K_1$ and that $\Omega_0 \in \mathbf{F}_{T^*}$. Let m be sufficiently large that

$$\|{}^{(m)}\mu - \mu\| \le \frac{\varepsilon_0}{20K(K_1+1)} . \qquad (19)$$

If j is sufficiently large then

$$^{(m)}\kappa_j < \frac{\varepsilon_0}{10}, \ \mathrm{E} \sup_{0 \le t \le T_c} \|x_t^* - {}^{(m)}x_t^j\| \le \frac{\varepsilon_0 \cdot \delta}{80\|\mu\| \cdot K} = \varepsilon_1 \quad . \tag{20}$$

$^{(m)}\kappa_j$ appears in the conditions of Lemma 1. Moreover, there exists a set Ω_δ^j, $p(\Omega_\delta^j) > \frac{\delta}{4}$, on which either (a) $^{(m)}T_j < T^*$, or (b) $^{(m)}T_j = T^*$, or (c) $^{(m)}T_j > T^*$ and the following inequalities hold:

$$^{(m)}T_j > \tau_m + c_0 \quad \text{(where } c_0 \text{ is a constant)}$$

$$< -P_{^{(m)}T_j}, f({}^{(m)}T_j, x^*_{^{(m)}T_j}, u^*_{T_j}) > \ > \frac{19}{20}\varepsilon_0 \tag{21}$$

$$\| -P_{^{(m)}T_j} - \mu^* \| < \frac{\varepsilon_0}{20K(K_1+1)}, \ \sup_{t \le {}^{(m)}T_j} \|x_t^*\| \le 2K_1 \tag{22}$$

$$\| {}^{(m)}\mu^j - {}^{(m)}\mu \| \le \frac{\varepsilon_0}{20K(K_1+1)}, \ \| {}^{(m)}\mu^j \| \le 2\|\mu\| \quad . \tag{23}$$

It is clear that Ω_δ^j can be chosen such that $\Omega_\delta^j \in \mathbf{F}_{^{(m)}T_j}$. Let us consider three cases (a), (b), (c) separately and argue by contradiction. We shall begin by assuming that (a) holds on the set Ω_δ^j for the stopping time $^{(m)}T_j$. It is easy to see, from (20), that there exists a set B_j of positive measure $B_j \subset \Omega_\delta^j$ on which

$$\sup_{t < T_c} \|x_t^* - {}^{(m)}x_t^j\| \le \frac{4\varepsilon_1}{\delta} \quad .$$

Let s_k be an increasing sequence of stopping times predicting $^{(m)}T_j$. Consider an increasing sequence of sets

$$\Omega_\varepsilon^k = \{\omega, \sup_{s_k < t < {}^{(m)}T_j} \| -p_t - \mu^* \| < \frac{\varepsilon_0}{20K(K_1+1)} ; \sup_{s_k < t < {}^{(m)}T_j} \|x_t^* - {}^{(m)}x_t^j\| < \frac{4\varepsilon_1}{\delta}$$

$$\inf_{s_k < t < {}^{(m)}T_j} < -p_t, f(t, x_t^*, u_t^*) > \ > \frac{9}{10}\varepsilon_0, \ \omega \in \Omega_\delta^j\} \quad .$$

It is clear that $\Omega_3^k \in \mathbf{F}_{^{(m)}T_j}$. If k_1 is sufficiently large then $p(\Omega_3^{k_1}) > 0$. If k is sufficiently large, then $\Omega_3^{k_1} \in \mathbf{F}_{s_k}$ because the family $\mathbf{F}_t$ is quasicontinuous.

Let $\tilde{s}_k$ be a sequence of stopping times

$$\tilde{s}_k = \begin{cases} s_k, & \omega \in \Omega_3^{k_1} \\ {}^{(m)}T_j, & \omega \notin \Omega_3^{k_1} \end{cases} \quad .$$

We can obviously assume that $\tilde{s}_k > \tau_m$ if $\omega \in \Omega_3^{k_1}$. Our aim is to demonstrate a contradiction with Lemma 1. Indeed,

$$\frac{1}{\mathrm{E}|\tilde{s}_k - {}^{(m)}T_j|}\,\mathrm{E}\int_{\tilde{s}_k}^{{}^{(m)}T_j} \langle {}^{(m)}\mu^j, f(t, {}^{(m)}x_t^j, u_t^*)\rangle\, dt =$$

$$\frac{1}{\mathrm{E}|\tilde{s}_k - {}^{(m)}T_j|}\,\mathrm{E}\int_{\tilde{s}_k}^{{}^{(m)}T_j} \langle {}^{(m)}\mu^j, f(t, {}^{(m)}x_t^j, u_t^*)\rangle - \langle p_t, f(t, x_t^*, u_t^*)\rangle\, dt +$$

$$+\frac{1}{\mathrm{E}|\tilde{s}_k - {}^{(m)}T_j|}\,\mathrm{E}\int_{\tilde{s}_k}^{{}^{(m)}T_j} \langle -p_t, f(t, x_t^*, u_t^*)\rangle\, dt \quad .$$

From the choice of Ω_3, the second term in the sum is greater than $\frac{9}{10}\,\varepsilon_0$. Let us estimate the integrand in the first term:

$$|\langle {}^{(m)}\mu^j \cdot f(t, {}^{(m)}x_t^j, u_t^*)\rangle + \langle p_t, f(t, x_t^*, u_t^*)\rangle| \le$$

$$\le \|{}^{(m)}\mu^j\| \cdot \|f(t, {}^{(m)}x_t^j, u_t^*) - f(t, x_t^*, u_t^*)\| + |\langle {}^{(m)}\mu^{j*} + p_t, f(t, x_t^*, u_t^*)\rangle| \le$$

$$2\|\mu\| \cdot K \cdot \frac{4\varepsilon_1}{\delta} + (\|{}^{(m)}\mu^j - {}^{(m)}\mu\| + \|{}^{(m)}\mu - \mu\| + \|\mu + p_t\|) \cdot 2K(K_1+1) \le \frac{\varepsilon_0}{10} + \frac{3\varepsilon_0}{10} = \frac{2}{5}\,\varepsilon_0 \quad .$$

Thus for every $\tilde{s}_k$ we have

$$\frac{1}{\mathrm{E}|\tilde{s}_k - {}^{(m)}T_j|}\,\mathrm{E}\int_{\tilde{s}_k}^{{}^{(m)}T_j} \langle {}^{(m)}\mu^j, f(t, {}^{(m)}x_t^j, u_t^*)\rangle\, dt \ge \frac{9}{10}\,\varepsilon_0 - \frac{2}{5}\,\varepsilon_0 = \frac{\varepsilon_0}{2} \quad ,$$

and at the same time ${}^{(m)}\kappa_j < \frac{\varepsilon_0}{10}$. This contradicts Lemma 1.

Let us now suppose that there is no set with a measure greater than $\delta/4$ which satisfies (a) starting at some j_0. If (b) holds for an infinite subsequence j_k then all the previous reasoning is still valid and a contradiction is obtained in an analogous way. Let us consider this case. Suppose that for all j starting from some number j_0 there is a set Ω_0^j of measure greater than $\delta/4$ on which ${}^{(m)}T_j > T^*$. We shall now use the symmetry of the optimal control with respect to the moment T^*. Introduce the new stochastic functions

$$\begin{aligned} {}^{(s)}x_t^* &= \begin{cases} x_t^*, & t \notin]]T^*,\ T^*+1/s[[\\ x_{2T^*-t}^*, & t \in]]T^*,\ T^*+1/s[[\end{cases} \\ {}^{(s)}p_t^* &= \begin{cases} p_t, & t \notin]]T^*,\ T^*+1/s[[\\ p_{2T^*-t}, & t \in]]T^*,\ T^*+1/s[[\end{cases} \end{aligned} \qquad (24)$$

Choosing s and j sufficiently large we have

$$E \sup_{0<t\le T_c} \|{}^{(s)}x_t^* - {}^{(m)}x_t^j\| \le \varepsilon_1$$

and there exists a set Ω_0^j, $p(\Omega_0^j) > \delta/4$, on which $T^* < {}^{(m)}T_j < T^* + 1/s$ and (21)–(23) hold with x_t^* and p_t^* replaced by ${}^{(s)}x_t^*$ and ${}^{(s)}p_t^*$. In this case, and if $\omega \in \Omega_0^j$, $t < {}^{(m)}T_j$, the functions ${}^{(s)}x_t$, ${}^{(s)}p_t^*$ and $<{}^{(s)}p_t^*, f(t, {}^{(s)}x_t^*, u_t^*)>$ are continuous in t. A contradiction with Lemma 1 can then be obtained by reasoning as in case (a) above.

We can thus show that the Hamiltonian cannot be negative at the end of the control time. Let us also show that it cannot be positive. Let $<-p_{T^*}, f(t, x_t^*, u_{T^*}^*)> < -\varepsilon_0$ on some set Ω_0, $p(\Omega_0) = \delta > 0$. Once again, we shall argue by contradiction. Without loss of generality we can assume that there exists a K_1 such that $\sup_{0\le t\le T^*} \|x_t^*\| < K_1 \ \forall \omega \in \Omega_0$. Choose m such that (19) holds. We shall now choose ${}^{(m)}T_j$. For the sake of simplicity we shall concentrate on case (a): for the infinite sequence j_k there exists a set $\Omega_0^{j_k}$, $p(\Omega_0^{j_k}) > \delta/4$, on which $\tau_m + c_0 < T_j^m < T^*$. Case (c) leads to a contradiction in the same way as (a) by substituting function (24), while case (b) is similar to case (c).

Thus from our assumptions there exists a number j such that (20) holds, and a set Ω_0^j, $p(\Omega_0^j) > \delta/4$, on which $\tau_m + c_0 < {}^{(m)}T_j < T^*$,

$$<-P_{{}^{(m)}T_j}, f({}^{(m)}T_j, x^*_{{}^{(m)}T_j}, u^*_{{}^{(m)}T_j})> < -\frac{19}{20}\varepsilon_0 \quad ,$$

and (22), (23) hold. Introduce the stopping time

$$\tau(\omega) = \begin{cases} {}^{(m)}T_j, & \omega \notin \Omega_0^j \\ \inf\limits_{{}^{(m)}T_j \le t \le T^*} & \{t : <-p_t, f(t, x_t^*, u_t^*)> > -\frac{9}{10}\varepsilon_0 ; \ |p_t + \mu^*| > \frac{\varepsilon_0}{20K(K_1+1)} ; \\ & \|x_t^*\| > 2K_1, \|x_t^* - {}^{(m)}x_t^j\| > \frac{4\varepsilon_1}{\delta}\}, \quad \omega \in \Omega_0^j \end{cases}$$

Using the same transformations as before it is easy to see that the second claim of Lemma 1 does not hold for the sequence of stopping times $\tau^i = \tau \wedge ({}^{(m)}T_j + 1/i)$. This concludes the proof of the theorem.

REFERENCES

1. V.I. Arkin and M.T. Saksonov. Necessary optimality conditions in control problems for stochastic differential equations. *Soviet Mathematics Doklady*, 20(1)(1979).

2. C. Dellacherie. *Capacities et processus stochastiques*. Springer-Verlag, Berlin, 1972.

3. J. Ekeland. On the variational principle. *Journal of Mathematical Analysis and Applications*, 47(1974)324–353.

4. B.N. Pshenichnii. *Convex Analysis and Extremum Problems*. Nauka, Moscow, 1980.

5. V.I. Arkin and M.T. Saksonov. On the stochastic maximum principle, and stochastic differential systems. *Proceedings of IFIP-WG*, 7/1, Springer-Verlag, Vilnius, 1978.

OPTIMAL CONTROL OF STOCHASTIC INTEGRAL EQUATIONS

L.E. Shaikhet
Institute of Mathematics and Mechanics
Donetsu, USSR

1. INTRODUCTION

In this paper we study optimal control problems involving stochastic integral equations with an integral cost functional, and obtain necessary conditions for optimality of controls in this type of problem. The optimal control for a linear equation with a quadratic cost functional is also given.

2. PROBLEM STATEMENT

Consider an optimal control problem $\{\xi_u(t), I(u), U\}$ with a trajectory $\xi_u(t)$, a cost functional $I(u)$ and a set of admissible controls U. Let u_0 be the optimal control of this problem, i.e., $I(u_0) = \inf_{u \in U} I(u)$, and u_ε be an admissible control which is "close" to u_0 for sufficiently small $\varepsilon > 0$ and identical to it for $\varepsilon = 0$. Since $I(u_\varepsilon) \geq I(u_0)$, the limit

$$I'(u_0) = \lim_{\varepsilon \to 0} \frac{1}{\varepsilon} [(u_\varepsilon) - I(u_0)] \tag{1}$$

must be nonnegative, if it exists. Thus the condition $I'(u_0) \geq 0$ is necessary for the optimality control of u_0.

Our aim is to find limit (1) for control problems with a trajectory that is described by the stochastic integral equation

$$\xi(t) = \eta(t) + \Phi(t, \vartheta_t, \xi) + \int_0^t A(t, s, \vartheta_s \xi, u(s), ds) \quad ,$$

$$\vartheta_0 \xi = \varphi_0 , \; t \in [0,T] \quad , \tag{2}$$

$$A(t, s, \varphi, u, h) = a(t, s, \varphi, u)h + b(t, s, \varphi)(w(t+h) - w(t))$$

and by the cost functional

$$I(u) = \mathrm{E}[F(\vartheta_T \xi) + \int_0^T G(s, \vartheta_s \xi, u(s))ds] \quad . \tag{3}$$

This has already been done for ordinary stochastic differential equations [1,2], stochastic differential equations of the hyperbolic type [2,3] and stochastic Volterra equations [4]. Proofs which are analogous to those given in [4] will be omitted here.

We shall first introduce some notation, assumptions and conditions. Let $\{\Omega, \sigma, P\}$ be a fixed probability space, and $\{f_t, t \in [0,T]\}$ be a family of σ-fields, $f_t \in \sigma$. $H_0(H_1)$ denotes a space of $f_0(f_t)$-adapted functions $\varphi(t)$, $t \in [-\infty, 0]([0,T])$ for which $\|\varphi\|_0^2 = \sup_{s \geq 0} \mathrm{E}|\varphi(s)|^2 < \infty$ $(\|\varphi\|_1^2 = \sup_{0 \leq t \leq T} \mathrm{E}|\varphi(t)|^2 < \infty)$; $\vartheta_t \xi(s) = \xi(t+s)$, $t \geq 0$, $s \leq 0$.

An arbitrary f_t-adapted l-dimensional function $u(t)$ for which $\|u\|_1 < \infty$ will be called an *admissible control*.

Let $D(\alpha)$ be a space of f_t-adapted functions $\varphi(t)$ for which $\mathrm{E}|\varphi(t) - \varphi(s)|^2 \leq C|t-s|^\alpha$, and $\mathbf{M}$ be a set of nondecreasing functions $K(\tau)$, $\tau \in (-\infty, 0]$, which are right-continuous, left-limited and such that $\int_{-\infty}^0 dK(\tau) < \infty$.

Let there exist a $\delta > 0$ such that a function $K(\tau)$ from $\mathbf{M}$ has a unique jump in zero on the segment $[-\zeta, 0]$. In this case we shall say that the function $K(\tau)$ has an *isolated jump* in zero. Let $dK(0) = K(0) - K(-0)$ be the size of this jump.

Let $\mathbf{M}_1$ be the subset of functions $K(\tau)$ from $\mathbf{M}$ which have an isolated jump in zero and for which $dK(0) < 1$; $\mathbf{M}_0$ is the subset of functions $K(\tau)$ from $\mathbf{M}_1$ for which $dK(0) = 0$; $\mathbf{L}$ is a set of non-negative functions $R(t, \tau)$, $t \in [0,T]$, which are nondreasing in $\tau \in [0,T]$ and for which $\sup_{0 \leq t \leq T} \int_0^t dR(t, \tau) < \infty$; $\mathbf{M}_2$ is the subset of functions $K(\tau)$ from $\mathbf{M}$ for which the nucleus $dK(\tau - t)$ has a resolvent in $\mathbf{L}$.

If X and Y are two normed spaces and $B(x)$ is an operation from X into Y, then $\nabla B(x)$ is the Gâteaux derivative with respect to x of this operation. For fixed $x_0 \in X$, $\nabla B(x_0)$ is a linear operation from X into Y. If $Y = R^1$, then $<\nabla B(x_0), x>$ is the value of the linear functional $\nabla B(x_0)$ on element $x \in X$ [5].

We shall use C to denote arbitrary positive constants.

Scalar functions $F(\varphi)$, $G(t, \varphi, u)$, n-dimensional functions $\Phi(t, \varphi)$, $a(t, s, \varphi, u)$ and an $n \times m$ matrix function $b(t, s, \varphi)$ are defined for $0 \leq s \leq t \leq T$, $u \in R^e$, $\varphi \in H$. Let $w(t)$ be an f_t-adapted m-dimensional Wiener process, where $w(t)$ and $\eta(t)$ are independent. We shall consider the following conditions:

C1. $\varphi_0 \in H_0 \cap D(\alpha_1)$.

C2. $\eta \in H_1 \cap D(\alpha_2)$.

C3. $\varphi_0(0) = \eta(0) + \Phi(0, \varphi_0)$.

C4. $u_0 \in U \cap D(\alpha_3)$.

C5. The functions $\Phi(t, \varphi)$, $a(t, s, \varphi, u)$, $b(t, s, \varphi)$ satisfy

$$|\Phi(t, \varphi)| \leq \int_{-\infty}^{0} (1 + |\varphi(\tau)|) dK_0(\tau) \quad ,$$

$$|a(t, s, \varphi, u)|^2 + |b(t, s, \varphi)|^2 \leq \int_{-\infty}^{0} (1 + |u|^2 + |\varphi(\tau)|^2) dK_1(\tau) \quad ,$$

$$|\Phi(t_1, \varphi_1) - \Phi(t_2, \varphi_2)| \leq \int_{-\infty}^{0} [|\varphi_1(\tau) - \varphi_2(\tau)| +$$

$$+ (1 + |\varphi_1(\tau)| + |\varphi_2(\tau)|) |t_1 - t_2|^{\alpha_4}] dK_0(\tau) \quad ,$$

$$|a(t_1, s_1, \varphi_1, u_1) - a(t_2, s_2, \varphi_2, u_2)|^2 \leq \int_{-\infty}^{0} [|\varphi_1(\tau) - \varphi_2(\tau)|^2 +$$

$$+ |u_1 - u_2|^2 + (1 + |u_1|^2 + |u_2|^2 + |\varphi_1(\tau)|^2 +$$

$$+ |\varphi_2(\tau)|^2)(|t_1 - t_2|^{\alpha_5} + |s_1 - s_2|^{\alpha_6})] dK_1(\tau) \quad ,$$

$$|b(t_1, s, \varphi_1) - b(t_2, s, \varphi_2)|^2 \leq \int_{-\infty}^{0} [|\varphi_1(\tau) - \varphi_2(\tau)|^2 +$$

$$+ (1 + |\varphi_1(\tau)|^2 + |\varphi_2(\tau)|^2) |t_1 - t_2|^{\alpha_7}] dK_1(\tau) \quad .$$

C6. The functions $\Phi(t, \varphi)$, (a, s, φ, u), $b(t, s, \varphi)$ have a Gâteaux derivative with respect to φ and

$$|\nabla \Phi(t, \varphi_1)\varphi_2| \leq \int_{-\infty}^{0} |\varphi_2(\tau)| dK_0(\tau) \quad ,$$

$$|\nabla a(t, s, \varphi_1, u)\varphi_2|^2 + |\nabla b(t, s, \varphi_1)\varphi_2|^2 \leq \int_{-\infty}^{0} |\varphi_2(\tau)|^2 dK_1(\tau) \quad ,$$

$$|(\nabla \Phi(t, \varphi_1) - \nabla \Phi(t, \varphi_2))\varphi_3|^2 + |(\nabla a(t, s, \varphi_1, u) - \nabla a(t, s, \varphi_2, u))\varphi_3|^2 +$$

$$+ |(\nabla b(t, s, \varphi_1) - \nabla b(t, s, \varphi_2))\varphi_3|^2 \leq \int_{-\infty}^{0} |\varphi_1(\tau) - \varphi_2(\tau)|^2 |\varphi_3(\tau)|^2 dK_1(\tau) \quad .$$

C7. The functions $F(\varphi)$ and $G(t,\varphi,u)$ satisfy

$$|F(\varphi)| + |G(t,\varphi,u)| \le \int_{-\infty}^{0} (1+|u|^2+|\varphi(\tau)|^2)\mathrm{d}K_1(\tau) \quad ,$$

$$|G(t_1,\varphi_1,u_1) - G(t_2,\varphi_2,u_2)| \le L(\varphi_1,\varphi_2,u_1,u_2)[|u_1-u_2| +$$

$$+ \int_{-\infty}^{0} |\varphi_1(\tau) - \varphi_2(\tau)|\mathrm{d}K_1(\tau) + L(\varphi_1,\varphi_2,u_1,u_2)|t_1-t_2|^{\alpha_8}] \quad ,$$

$$L^2(\varphi_1,\varphi_2,u_1,u_2) \le \int_{-\infty}^{0} (1+|u_1|^2+|u_2|^2+|\varphi_1(\tau)|^2+|\varphi_2(\tau)|^2)\mathrm{d}K_1(\tau) \quad .$$

C8. The functions $F(\varphi)$ and $G(t,\varphi,u)$ have a Gâteaux derivative with respect to φ and

$$|<\nabla F(\varphi_1),\varphi_2>| + |<\nabla G(t,\varphi_1,u),\varphi_2>| \le$$

$$\le \int_{-\infty}^{0} (1+|u| + |\varphi_1(\tau)|)|\varphi_2(\tau)|\mathrm{d}K_1(\tau) \quad .$$

In conditions C5–C8 we assume that $K_0 \in \mathbf{M}_1 \cap \mathbf{M}_2$, $K_1 \in \mathbf{M}$, and $\alpha_1, \ldots, \alpha_8$ are definite positive constants.

THEOREM 1. *Let conditions* C1–C8 *hold, the stochastic variable* v *be* $f_{t_0-\varepsilon}$*-adapted,* $\mathrm{E}|v|^2 < \infty$, *and*

$$u_\varepsilon(t) = \begin{cases} v \quad , & t \in [t_0-\varepsilon,t_0) \, , \, 0<\varepsilon<t_0<T \quad , \\ u_0(t) \, , & t \in [0,T] \setminus [t_0-\varepsilon,t_0) \quad . \end{cases} \tag{4}$$

Then the limit (1), (4) *for control problem* (2), (3) *exists and is equal to*

$$I'(u_0) = \mathrm{E}[G(t_0,\vartheta_{t_0}\xi_0,v) - G(t_0,\vartheta_{t_0}\xi_0,u_0(t_0)) +$$

$$+ <\nabla F(\vartheta_T\xi_0) \, , \, \vartheta_T q_0> + \int_{t_0}^{T} <\nabla G(s,\vartheta_s\xi_0,u_0(s)) \, , \, \vartheta_s q_0>\mathrm{d}s] \quad .$$

Here $q_0(t)$, $t \in [t_0,T]$, is a solution of the equation

$$q_0(t) = \eta_0(t) + \Phi_0(t)\vartheta_t q_0 + \int_{t_0}^{t} A_0(t,s,\mathrm{d}s)\vartheta_s q_0 \quad ,$$

where

$$\eta_0(t) = a(t,t_0,\vartheta_{t_0}\xi_0,v) - a(t,t_0,\vartheta_{t_0}\xi_0,u_0(t_0)) \quad ,$$

$$\Phi_0(t) = \nabla \Phi(t, \vartheta_t \xi_0), \ A_0(t, s, h) = \nabla a(t, s, \vartheta_s \xi_0, u_0(s), h) \quad ,$$

and $\xi_0(t)$ is a solution of equation (2) with control u_0.

3. PROOF OF THE THEOREM

To prove the theorem we need the following assertions:

LEMMA 1. *Let $y(t)$ be a non-negative function which satisfies the inequality*

$$y(t) \le \int_{-t}^{0} y(t+\tau)\mathrm{d}K(\tau) + x(t) \quad .$$

Here $x(t)$ is a non-negative, non-reduced, continuous function which is differentiable with respect to t and $K \in \mathbf{M}_0 \cap \mathbf{M}_2$. Then $y(t) \le Cx(t)$.

Proof. Let $y_0(t)$ be a solution of the equation

$$y_0(t) = \int_{-t}^{0} y_0(t+\tau)\mathrm{d}K(\tau) + x(t) \quad ,$$

which from [6] exists and is unique. Let $\mathrm{d}R(t, \tau)$ be a resolvent of the nucleus $\mathrm{d}K(\tau - t)$. Then

$$y_0(t) = x(t) + \int_0^t \mathrm{d}R(t, \tau)x(\tau) \le x(t)[1 + \int_0^t \mathrm{d}R(t, \tau)] \le Cx(t) \quad .$$

Assume that $z_0(t) = y(t)$, $z_n(t) = x(t) + \int_{-t}^{0} z_{n-1}(t+\tau)\mathrm{d}K(\tau)$, $n = 1, 2, \ldots$. It can easily be shown that $z_n(t) \ge z_{n-1}(t)$ and $\lim_{n \to \infty} z_n(t) = y_0(t)$. Therefore $y(t) = z_0(t) \le z_n(t) \le y_0(t) \le Cx(t)$, proving Lemma 1.

LEMMA 2. *Let $u \in U$ and conditions* C1–C3, C5 *hold. Then equation* (2) *has a unique solution, $\xi \in H_1 \cap D(\alpha)$, $\alpha = \min\,[1, \alpha_1, \alpha_2, 2\alpha_4, \alpha_5, \alpha_7]$.*

Proof. Let $\vartheta_0 \xi_n = \varphi_0$, $n \ge 0$, $\xi_0(t) = \eta(t)$, and

$$\xi_{n+1}(t) = \eta(t) + \Phi(t, \vartheta_t \xi_{n+1}) + \int_0^t A(t, s, \vartheta_s \xi_n, u(s), \mathrm{d}s) \quad .$$

Then

$$|\xi_{n+1}(t)|(1 - \mathrm{d}K_0(0)) \le C + |\eta(t)| + \int_{-t}^{-0} |\xi_{n+1}(t+\tau)|\mathrm{d}K_0(\tau) +$$

$$+ \int_{-\infty}^{-t} |\varphi_0(t+\tau)|\mathrm{d}K_0(\tau) + |\int_0^t A(t, s, \vartheta_s \xi_n, u(s), \mathrm{d}s)| \quad . \tag{5}$$

By virtue of Lemma 1 it can be shown (analogously to [6,7]) that $z_n(t) = \sup_{0\le s\le t} \mathrm{E}|\xi_n(s)|^2$ is uniformly bounded and $\lim_{n\to\infty} \sup_{0\le s\le t} \mathrm{E}|\xi_n(s) - \xi_{n-1}(s)|^2 = 0$. Therefore there exists a process $\xi(t)$ from H_1 for which $\lim_{n\to\infty} \mathrm{E}|\xi_n(t) - \xi(t)|^2 = 0$ uniformly on $t \in [0,T]$. And this process is the unique solution of equation (2).

We now only have to show that $\mathrm{E}|\xi(t_1) - \xi(t_2)|^2 \le C|t_1 - t_2|^\alpha$ for arbitrary t_1 and t_2 from $[0,T]$. Let $t_2 = 0$, $t_1 = t$, $z(t) = \mathrm{E}|\xi(t) - \varphi_0(0)|^2$. From (2) and C3 for $|\xi(t) - \varphi_0(0)|$ we obtain inequalities analogous to (5). By virtue of inequalities $\mathrm{E}|\xi(t+\tau) - \varphi_0(\tau)|^2 \le Q[z(t+\tau) + |\tau|^{\alpha_1}]$ we can easily obtain

$$z(t) \le C[t^\alpha + \int_{-t}^{-0} z(t+\tau)\mathrm{d}K_0(\tau)] \quad .$$

Therefore (from Lemma 1) $z(t) \le Ct^\alpha$. Assume that $t_2 = t < t_1 = t + \Delta$, $z(t) = \mathrm{E}|\xi(t+\Delta) - \xi(t)|^2$. In the same way it can be shown that $z(t) \le C\Delta^\alpha$, thus proving Lemma 2.

Let ξ_ε be a solution of equation (2) for control (4), and

$$q_\varepsilon(t) = \frac{1}{\varepsilon}(\xi_\varepsilon(t) - \xi_0(t)) \, , \; p_\varepsilon(t) = \frac{1}{\varepsilon}(\Phi(t, \vartheta_t \xi_\varepsilon) - \Phi(t, \vartheta_t \xi_0)) \quad ,$$

$$\eta_\varepsilon(t) = \frac{1}{\varepsilon} \int_{t_0-\varepsilon}^{t_0\wedge t} [A(t, s, \vartheta_s \xi_\varepsilon, v, \mathrm{d}s) - A(t, s, \vartheta_s \xi_0, u_0(s), \mathrm{d}s)] \quad ,$$

$$t_0 \wedge t = \min[t_0, t] \, , \; t \in [t_0 - \varepsilon, T]$$

$$\rho_\varepsilon(t) = \frac{1}{\varepsilon} \int_{t_0}^{t} [A(t, s, \vartheta_s \xi_\varepsilon, u_0(s), \mathrm{d}s) - A(t, s, \vartheta_s \xi_0, u_0(s), \mathrm{d}s)] \quad ,$$

$$t \in [t_0, T] \quad ,$$

$$\eta_\varepsilon(t) = 0 \, , \; t \in [0, t_0 - \varepsilon] \, , \; \rho_\varepsilon(t) = 0 \, , \; t \in [0, t_0] \quad .$$

Then $q_\varepsilon(t) = \eta_\varepsilon(t) + p_\varepsilon(t) + \rho_\varepsilon(t)$, $t \in [0,T]$. Let

$$\lambda_\varepsilon^x(t) = \xi_0(t) + x(\xi_\varepsilon(t) - \xi_0(t)) \, , \; x \in [0,1]$$

$$\Phi_\varepsilon(t) = \int_0^1 \nabla \Phi(t, \vartheta_t \lambda_\varepsilon^x)\mathrm{d}x$$

$$A_\varepsilon(t, s, h) = \int_0^1 \nabla A(t, s, \vartheta_s \lambda_\varepsilon^x, u_0(s), h)\mathrm{d}x \quad .$$

Then

$$p_\varepsilon(t) = \Phi_\varepsilon(t)\vartheta_t q_\varepsilon \quad ,$$

$$\rho_\varepsilon(t) = \int_{t_0}^{t} A_\varepsilon(t, s, ds)\vartheta_s q_\varepsilon$$

$$q_\varepsilon(t) = \eta_\varepsilon(t) + \Phi_\varepsilon(t)\vartheta_t q_\varepsilon + \int_{t_0}^{t} A_\varepsilon(t, s, ds)\vartheta_s q_\varepsilon, \; t \in [t_0, T] \quad .$$

LEMMA 3. *Let conditions* C1–C6 *hold. Then*

$$\lim_{\varepsilon \to 0} \mathrm{E}|q_\varepsilon(t) - q_0(t)|^2 = 0$$

uniformly on $t \in [t_0, T]$.

LEMMA 4. *Let conditions* C1–C8 *hold. Assume that*

$$\mu_1(\varepsilon) = \frac{1}{\varepsilon} \mathrm{E} \int_{t_0-\varepsilon}^{t_0} [G(s, \vartheta_s \xi_\varepsilon, v) - G(s, \vartheta_s \xi_0, u_0(s))]ds$$

$$\mu_2(\varepsilon) = \frac{1}{\varepsilon} \mathrm{E}[F(\vartheta_T \xi_\varepsilon) - F(\vartheta_T \xi_0)]$$

$$\mu_3(\varepsilon) = \frac{1}{\varepsilon} \mathrm{E} \int_{t_0}^{T} [G(s, \vartheta_s \xi_\varepsilon, u_0(s)) - G(s, \vartheta_s \xi_0, u_0(s))]ds \quad .$$

Then

$$\lim_{\varepsilon \to 0} \mu_1(\varepsilon) = \mathrm{E}[G(t_0, \vartheta_{t_0}\xi_0, v) - G(t_0, \vartheta_{t_0}\xi_0, u_0(t_0))]$$

$$\lim_{\varepsilon \to 0} \mu_2(\varepsilon) = \mathrm{E}\langle \nabla F(\vartheta_T \xi_0), \vartheta_T q_0 \rangle$$

$$\lim_{\varepsilon \to 0} \mu_3(\varepsilon) = \mathrm{E} \int_{t_0}^{T} \langle \nabla G(s, \vartheta_s \xi_0, u_0(s)), \vartheta_s q_0 \rangle \, ds \quad .$$

The proofs of Lemmas 3 and 4 are analogous to those given in [4].

The proof of Theorem 1 follows from Lemmas 3 and 4, since

$$\frac{1}{\varepsilon} [I(u_\varepsilon) - I(u_0)] = \sum_{i=1}^{3} \mu_i(\varepsilon) \quad .$$

Note that the theorem also holds for equations with Poisson stochastic perturbations [2,4].

4. LINEAR SYSTEMS WITH QUADRATIC COST FUNCTIONALS

Consider the optimal control problem

$$\xi(t) = \eta(t) + \int_{-\infty}^{0} dK(t,s)\xi(t+s) + \int_{0}^{t} a(t,s)u(s)ds \tag{6}$$

$$I(u) = E[\xi^*(T)H\xi(T) + \int_{0}^{T} u^*(s)N(s)u(s)ds] \quad . \tag{7}$$

Here $\eta \in H_1 \cap D(\alpha)$, $\varphi_0(0) = \eta(0) + \int_{-\infty}^{0} dK(0,s)\varphi_0(s)$, $a(t,s)$ is a non-random, bounded $n \times l$ matrix which is Hölderian for both variables, $N(s)$ is a non-random $l \times l$ matrix which is Hölderian, bounded and uniformly positively definite on s, H is a non-random, non-negative definite $n \times n$ matrix, and $K(t,s)$ is non-random $n \times n$ matrix such that

$$|dK(t,s) - dK(\tau,s)| \leq |t-\tau|^{\alpha} dK_0(s)$$

$$\sup_{0 \leq t \leq T} |dK(t,s)| \leq dK_0(s) \, , \; K_0 \in \mathbf{M}_1 \cap \mathbf{M}_2 \quad .$$

For the optimal control problem (6), (7) we have

$$I'(u_0) = E[v^* N(t_0)v - u_0^*(t_0)N(t_0)u_0(t_0) + \mathbf{Q}q_0^*(T)H\xi_0(T)]$$

$$q_0(t) = \eta_0(t) + \int_{-\infty}^{0} dK(t,s)q_0(t+s)$$

$$\eta_0(t) = a(t,t_0)(v - u_0(t_0)) \, , \; t \in [t_0, T] \quad .$$

Let $dR(t,s)$ be a resolvent of nucleus $dK(t,s-t)$, and

$$\psi(T,t,a)(\cdot,s)) = a(T,s) + \int_{t}^{T} dR(T,\tau)a(\tau,s) \quad .$$

Then $q_0(T) = \psi(T,t_0,a(\cdot,t_0))(v - u_0(t_0))$ and we easily obtain

$$I'(u_0) = E[(v - u_0(t))^* N(t)(v - u_0(t)) +$$

$$+ \mathbf{Q}(v - u_0(t))^*(N(t)u_0(t) + \psi^*(T,t,a(\cdot,t)))HE_t\xi_0(T)] \quad ,$$

noting that $E_t \cdot = E\{\cdot / f_t\}$.

Thus for $I'(u_0)$ to be non-negative it is necessary and sufficient that the optimal control has the following

$$u_0(t) = -N^{-1}(t)\psi^*(T,t,a(\cdot,t))HE_t\xi_0(T) \quad .$$

By virtue of (6), $E_t\xi_0(T)$ can be expressed as a functional of $\vartheta_t\xi_0$ (see [4]). Then we finally obtain

$$u_0(t) = \alpha(t) + p(t)\psi(T,t,\mathbb{1})\xi_0(t) + \int_0^t dR_0(t,\tau)\xi_0(\tau) \quad .$$

Here $\mathbb{1}$ is the identity matrix,

$$\alpha(t) = p(t)\psi(T,t,b(\cdot,t)) + \int_0^t Q(t,s)p(s)\psi(T,s,b(\cdot,s))ds$$

$$p(t) = -N^{-1}(t)\psi^*(T,t,a(\cdot,t))H[\mathbb{1} +$$

$$+ \int_t^T \psi(T,s,a(\cdot,s))N^{-1}(s)\psi^*(T,s,a(\cdot,s))dsH]^{-1} \quad ,$$

$Q(t,s)$ is a resolvent of the nucleus $p(t)\psi(T,t,a_t(\cdot,s))$,

$$b(t,s) = E_s(\eta(t) - \eta(s)) \; , \; a_t(\tau,s) = a(\tau,s) - a(t,s)$$

$$dR_0(t,\tau) = \int_\tau^t Q(t,s)p(s)\psi(T,s,dK_s(\cdot,\tau))ds +$$

$$+ p(t)\psi(T,t,dK_t(\cdot,\tau)) + Q(t,\tau)p(\tau)\psi(T,\tau,\mathbb{1})d\tau$$

$$dK_t(s,\tau) = dK(s,\tau-s) - dK(t,\tau-t) \quad .$$

5. CONCLUSION

Optimal control in these types of equations is a natural development of the theory of control for stochastic differential equations [8–11].

Numerous examples of applications for equations of this type demonstrate their practical importance [6,12,13].

REFERENCES

1. V.M. Warfield. A stochastic maximum principle. *SIAM Journal of Control and Optimization*, 14(1976)803–826.

2. L.E. Shaikhet. Necessary conditions of control optimality for some stochastic systems (in Russian). In the *Proceedings of the Conference on Stochastic Differential Equation Theory*, Donetsk, 1982, pp. 104–105.

3. L.E. Shaikhet. On a necessary condition of control optimality for stochastic

differential equations of hyperbolic type (in Russian). In *Theory of Stochastic Processes*, Naukova Dumka, Kiev, 1984, pp. 96–100.

4. L.E. Shaikhet. On optimal control of Volterra equations. *Problems of Control and Information Theory*, 13(3)(1984)141–152.

5. A.N. Kolmogorov and S.V. Fomin. *Elements of Function Theory and Functional Analysis*. Nauka, Moscow, 1968 (in Russian).

6. V.B. Kolmanovsky and V.R. Nosov. *Stability and Periodic Regimes of Regulated Systems with Delay*. Nauka, Moscow, 1981 (in Russian).

7. I.I. Gikhman and A.V. Skorohod. *Stochastic Differential Equations and Their Applications*. Naukova Dumka, Kiev, 1982 (in Russian).

8. H.J. Kushner. *Stochastic Stability and Control*. Academic Press, New York, 1967.

9. R.Z. Khasminsky. *Stability of Systems of Differential Equations with Stochastic Perturbations in Their Parameters* (in Russian). Nauka, Moscow, 1969.

10. I.I. Gikhman and A.V. Skorohod. *Stochastic Control Processes*. Naukova Dumka, Kiev, 1977 (in Russian).

11. F.L. Chernousko and V.B. Kolmanovsky. *Optimal Control with Stochastic Perturbations*. Nauka, Moscow, 1978 (in Russian).

12. N.H. Arutjunjan and V.B. Kolmanovsky. *The Creep Theory for Inhomogeneous Bodies*, Nauka, Moscow, 1983 (in Russian).

13. I.S. Astapov, S.M. Belocerkovsky, B.O. Kachanov and Yu.A. Kochetkov. On systems of integro-differential equations which establish the motion of a body in a continuous medium (in Russian). *Differential Equations*, 18(9)(1982)1628–1637.

SOME DIRECT METHODS FOR COMPUTING OPTIMAL ESTIMATORS FOR FORECASTING AND FILTERING PROBLEMS INVOLVING STOCHASTIC PROCESSES

A.D. Shatashvili
State University, Donezk, USSR

1. INTRODUCTION

One of the main problems in the theory of stochastic processes is to find optimal estimators for forecasting and filtering problems. The solution of linear forecasting and filtering problems given by Wiener and Kolmogorov is optimal only for Gaussian processes; for general processes the solution is optimal only for a class of linear estimators and in any given case may be far from optimal. Investigations by Wiener, Zade, Stratonovitch, Shirjaev, Liptzer, Grigelionis and others have concentrated on efficient methods for computing nonlinear estimators for the problems outlined above. However, the results obtained by these authors are related, in general, to some classes of Markovian processes. The linear theory for the forecasting and filtering of stochastic processes can be considered to be fully developed. A detailed description of these results can be found in [1]. Generalization of the linear theory to the case of optimal estimators is complicated by the need to study all possible finite-dimensional distributions of processes. Thus nonlinear problems may be solved effectively only in the case when the information about the finite-dimensional distributions of stochastic proceses is of closed form. One way of tackling this problem is to define the density of a measure of the process under consideration with respect to a certain standard measure; this is the approach used in this paper.

We shall first consider the class of stochastic processes which are solutions of differential equations of the type

$$\frac{dx(t)}{dt} + \alpha f(t, x(t)) = \xi'(t) \quad , \tag{1}$$

where α is a parameter, $\xi(t)$ is a Gaussian process, $f(t, x)$ is a nonlinear function and the equation itself is considered to exist over some finite-dimensional Euclidian or separable Hilbert space.

Here we shall propose one direct method for computing of optimal forecasting and filter for a solution of (1) and for some functionals of these solutions. In the case where α is small, the estimators obtained will be expanded to the power of this small parameter; moreover, in all cases of expansion the linear estimators will be the main terms.

Note that if the function $f(t, x(t))$ is linear with respect to $x(t)$, then $x(t)$ is a Gaussian process and the solution of the problems mentioned above follows from the general theory of linear forecasting and filtering of stochastic processes. If $\xi'(t)$ is Gaussian "white noise", then equation (1) must be regarded as an Ito stochastic differential equation; the random process $x(t)$ is a Markovian diffusion process when corresponding conditions hold for the function $f(t, x(t))$. As mentioned above, the problems involving such processes were solved by Shirjaev and others. Therefore the class of processes studied in this paper may be considered as an extension of the class of Gaussian and Markovian processes to include some non-Gaussian and non-Markovian processes.

2. OPTIMAL EXTRAPOLATION (FORECASTING) OF THE SOLUTION OF A NON-LINEAR DIFFERENTIAL EQUATION WITH GAUSSIAN PERTURBATIONS

Let a certain random process $x(t)$ be observed on the interval $[0, T]$. It is required to forecast its value at a point $T + h$, $h > 0$, in the best possible manner. To do this we choose a certain functional which is dependent on trajectories $x(t)$ on the interval $[0, T]$ and for which the mean-square deviation at $x(T+h)$ will be a minimum. These considerations can be formulated as follows: if $\mathbf{F}_T$ is a σ-algebra generated by the behaviour of process $x(s)$, $s \leq T$, then a functional of the required type will be $\mathbf{F}_T$-measurable and, consequently, to find the optimal forecast of a variable $x(T+h)$ can be interpreted as finding an $\mathbf{F}_T$-measurable random variable η for which $\mathrm{E}(\eta - x(T+h))^2$ takes its minimum value.

LEMMA 1 (see [6]). *Let $\mathbf{F}_T$ be a given σ-algebra and consider a certain random variable ξ for which $\mathrm{E}\xi^2 < \infty$. If another random variable η, is measurable with respect to the σ-algebra $\mathbf{F}_T$ and $\mathrm{E}(\eta - \xi)^2$ takes its minimal value, then*

$$\eta = \mathrm{E}(\xi / \mathbf{F}_T) \quad . \tag{2}$$

Thus, if $\mathrm{E}x^2(t) < \infty$, then the optimal forecast $\hat{x}(T+h)$ at a point $T + h$ is of the form

$$\hat{x}(T+h) = \mathrm{E}(x(T+h) / \mathbf{F}_T) \quad . \tag{3}$$

Below we shall suggest a method for computing the conditional mathematical expectation on the right-hand side of (3). Let μ_x and μ_ξ be measures generated by random processes $x(t)$ and $\xi(t)\,(0 \le t \le a)$, respectively, H be a separable Hilbert space, and $C_a(x)$ be a space of functions defined on the interval $[0,a]$ with values in H. Assume that the random processes $x(t)$ and $\xi(t)$, $t \in [0,a]$, take values in H and that sample points belong to $C_a(x)$. Let μ_x^T and μ_ξ^T denote a contraction of measures μ_x and μ_ξ over the space $C_T(x)$, $T < a$, and a density $d\mu_x^T/d\mu_x^T$ (if it exists) be denoted by $\rho_T(\cdot)$. Let $\rho_T(\cdot)$ and $\rho_{T+h}(\cdot)$ exist. Set

$$\alpha_T = \alpha_T(T+h\,,\,\mathbf{Q}) = \frac{\mathrm{E}^{(\xi)}\{\xi(T+h)\rho_{T+h}(\cdot)/\mathbf{F}_T^*\}}{\rho_T(\mathbf{Q})}\,, \tag{4}$$

where the superscript (ξ) means integration over μ_ξ, and $\mathbf{F}_T^*$ is the σ-algebra generated by $\xi(s)$, $sS \le T$.

THEOREM 1. *Let the random processes $x(t)$ and $\xi(t)$ be observed on the interval $[0,a]$. Let $\rho_a(\cdot)$, T exist and $T+h \in [0,a]$. Then*

$$\hat{x}(T+h) = \mathrm{E}\{x(T+h)/\mathbf{F}_T\} = \tag{5}$$

$$= [\rho_T(x(\cdot))]^{-1}\cdot \mathrm{E}^{(\xi)}\{\xi(T+h)\rho_{T+h}(\cdot)/\mathbf{F}_T^*\}/\,\xi(\cdot) = x(\cdot)\quad .$$

Proof. Let $\gamma_n = f(x(t_1),\ldots,x(t_n))$ be a $\mathbf{F}_T$-measurable random variable, where $f(z_1,\ldots,z_n)$ is a measurable bounded function. Utilizing the properties of conditional mathematical expectations we have the following string of equalities:

$$\mathrm{E}\alpha_T\cdot\gamma_n = \mathrm{E}\alpha_T f(x(t_1),\ldots,x(t_n)) = \mathrm{E}\alpha_T f(\xi(t_1),\ldots,\xi(t_n))\rho_T(\xi)$$

$$= \mathrm{E}_{\alpha_T} f(\xi(t_1),\ldots,\xi(t_n))\rho_T(\xi) = \mathrm{E}f(\xi(t_1),\ldots,\xi(t_n))\times$$

$$\times\frac{\mathrm{E}\{\xi(T+h)\rho_{T+h}(\xi)/\mathbf{F}_T^*\}}{\rho_T(\xi)}\rho_T(\xi) =$$

$$\tag{6}$$

$$= \mathrm{E}\mathrm{E}\{f(\xi(t_1),\ldots,\xi(t_n))\xi(T+h)\rho_{T+h}(\xi)/\mathbf{F}^*\} =$$

$$= \mathrm{E}f(\xi(t_1),\ldots,\xi(t_n))\xi(T+h)\rho_{T+h}(\xi) = \mathrm{E}f(x(t_1),\ldots,x(t_n))x(T+h) =$$

$$= \mathrm{E}x(T+h)\gamma_n\cdot\mathrm{E}_{\alpha_T}\gamma_n = \mathrm{E}x(T+h)\gamma_n\quad .$$

Formula (6) is valid for all bounded $\mathbf{F}_T$-measurable γ_n. Since any $\mathbf{F}_T$-measurable variable γ may be approximated by a sequence γ_n, we obtain

$$\mathrm{E}_{\alpha T}\gamma = \mathrm{E}x(T+h)\gamma \tag{7}$$

taking the limit in (6). The last expression supports the validity of (5), thus proving Theorem 1.

Formula (5) allows us to carry out the integration on another measure (naturally, a standard measure) for the extrapolation problem. Unfortunately, however, it does not help us to avoid the computation of the conditional mathematical expectation. However, if $\xi(t)$ is a Gaussian process, then we can make one more simplification.

Suppose that $\xi(t)$ is a Gaussian process defined on the interval $[0, a]$, and that $\mathrm{E}\xi(t) = 0$ with correlation function $R(t, s)$. It is known from the theory of linear extrapolation that on the interval $[T, T+h]$ the Gaussian process $\xi(t)$ may be represented in the form

$$\xi(t) = l_T(t, \xi(\cdot)|_0^T) + \varepsilon_T(t), \; t \in [T, T+h] \quad , \tag{8}$$

where

$$l_T(t) = l_T(t, \xi(\cdot)) = \mathrm{E}\{\xi(t)/\mathbf{F}_T^*\} \tag{9}$$

is the linear forecast of the Gaussian process $\xi(t)$ and $\varepsilon_T(t)$ is Gaussian but independent of the σ-algebra $\mathbf{F}_t^*$. Consequently, if $\xi(t)$ is the Gaussian process from Theorem 1, then (5) may be expressed in the following form:

$$\hat{x}(T+h) = \mathrm{E}\{\xi(T+h)\rho_{T+h}(\xi(\cdot)\;\mathbf{F}_T^*\}|\;\xi(\cdot) = x(\cdot)[\rho_T(x(\cdot))]^{-1} =$$

$$= [\rho_T(x(\cdot))]^{-1}\cdot \mathrm{E}\{(u_1 + \varepsilon_T(T+h))\rho_T + h(u_2, u_3 + \varepsilon_T(\cdot))\}| \tag{10}$$

$$u_1 = l_T(T+h, x(\cdot))$$
$$u_2 = x(\cdot)|_0^T \qquad ,$$
$$u_3 = l_T(\cdot, x(\cdot))$$

where $\varepsilon_T(t) = \xi(t) - l_T(t)$.

Thus, formula (10) is qualitatively different from formula (5). It not only simplifies the computing of the unconditional mathematical expectation expressed by (5) but also gives an algorithm for calculating the unconditional mathematical expectation, i.e., it allows us to compute the optimal forecast directly if the density $p_a(\cdot)$ is known. The problems of the absolute continuity ($\mu_x \ll \mu_\xi$) or equivalence ($\mu_x \sim \mu_\xi$) of measures generated by solutions of differential equations of the type (1) with respect to the measure of the Gaussian process on the right-hand side have been studied by many authors (see [2–5]) and formulas for the densities have been obtained. These formulas are given below.

Define a differential equation in H:

$$\frac{dx(t)}{dt} + \alpha f(t, x(t)) = \xi'(t), \; 0 \le t \le a, \; x(0) = \xi(0) = 0 \quad , \tag{11}$$

where α is a parameter, $\xi(t)$ is a Gaussian process with values in H, $\mathrm{E}\xi(t) = 0$, and $R^2(t, s)$ is its correlation function. If all the necessary conditions (see [2,3]) are valid, then $\mu_x \ll \mu_\xi$ and $\rho_a(\xi) = \frac{d\mu_x}{d\mu_\xi}(\cdot)$ is defined by the formula

$$\rho_a(\xi(\cdot)) = \exp\{-\alpha \int_0^a (g(t), dw(t)) - \frac{\alpha^2}{2} \int_0^a \|g(t)\|^2 dt\} \quad , \tag{12}$$

where $(\cdot,\cdot)$ and $\|\cdot\|$ are the scalar product and norm in H, respectively, and $\int_0^a (g(t), dw(t))$ is a stochastic Ito integral. The functions $g(t)$, $R(t, w)$ and the Wiener process $w(t)$ are defined by the relations

$$f(t, \xi(t)) = \int_0^a R(t, u) g(u) du \, , \; \int_0^a \|g(t)\|^2 dt < \infty, \; \xi'(t) = \int_0^a R(t, u) dw(u) \quad ,$$

$$R^2(t, s) = \int_0^a R(t, u) R(u, s) du \quad .$$

It may easily be verified that if we replace $\xi(\cdot)$ by $x(\cdot)$ in $\rho_T^0(\xi(\cdot))$, then $\rho_T(x(\cdot))$ is $\mathbf{F}_T$-measurable. Therefore, if a solution of equation (11) is observed on the interval $[0, T]$ and it is necessary to find its optimal forecast at a point $T + h$, then from formulas (10) and (12) and the relation $\rho_{T+h}^{(\xi)} = \rho_T^{(\xi)} \cdot \rho_{T+h}^{T(\xi)}$, where

$$\rho_{T+h}^{T(\xi)} = \exp\{-\alpha \int_T^{T+h} (g(t), dw(t)) - \frac{\alpha^2}{2} \int_T^{T+h} \|g(t)\|^2 dt\} \quad , \tag{13}$$

we obtain

$$\hat{x}(T+h) = \mathrm{E}\{(u_1 + \varepsilon_T(T+h))\rho_{T+h}^T(u_2, u_3 + \varepsilon_T(\cdot))\}| \tag{14}$$

$$\begin{aligned} u_1 &= l_T(T+h, x(\cdot)) \\ u_2 &= x(\cdot)|_0^T \\ u_3 &= l_T(\cdot, x(\cdot)) \end{aligned} \quad .$$

Assume now that α is a small parameter, and that $\{\lambda_k\}$ and $\{\varphi_k(t)\}$ are the eigenvalues and eigenfunctions, respectively, of the correlation operator function $R(t, s)$. We expand the formula (14) as a power series in α:

$$\exp\{-\alpha\int_T^{T+h}(g(t),dw(t))-\frac{\alpha^2}{2}\int_T^{T+h}\|g(t)\|^2dt\}=1-\alpha\int_T^{T+h}(g(t),dw(t))-$$

$$-\frac{\alpha^2}{2}\int_T^{T+h}\|g(t)\|^2dt+\frac{\alpha^2}{2}(\int_T^{T+h}g(t),dw(t)))^2+o(\alpha^2) \tag{15}$$

We then insert (15) into (14) and taking into account the fact that $\rho_{T+h}^T(\xi)$ can be written in the form

$$\rho_{T+h}^T(\xi)=\{-\sum_{k=1}^{\infty}\frac{1}{\lambda_k}[\alpha\int_T^{T+h}\int_T^{T+h}(f(t,\xi(t)),\varphi_k(t))(\varphi_k(s))\xi'(s)dtds+$$

$$+\frac{\alpha^2}{2}(\int_T^{T+h}(f(t,\xi(t)),\varphi_k(t))dt)^2]\} \tag{16}$$

and making certain calculations we obtain

$$\hat{x}(T+h)=l_T(T+h,x(\cdot)|_0^T)-\alpha\sum_{k=1}^{\infty}\frac{1}{\lambda_k}[V_1^{(k)}(T+h)+V_2^{(k)}(T+h)]+$$

$$+\frac{\alpha^2}{2}[\sum_{i,j=1}^{\infty}\frac{1}{\lambda_i\lambda_j}V_3^{(ij)}(T+h)-\sum_{k=1}^{\infty}\frac{1}{\lambda_k}V_4^{(k)}(T+h)]+o(\alpha^2)\quad, \tag{17}$$

where $l_T(T+h,x(\cdot)|_0^T)$ is the linear forecast of $x(t)$, and the variables $V_1^{(k)}$, $V_2^{(k)}$, $V_3^{(ij)}$, $V_4^{(k)}$ are defined by $f(t,x)$, $l_T(\cdot)$ and by normal distributions of known Gaussian variables $\varepsilon_T(\cdot)$. For example,

$$V_1^{(k)}=\int_{-\infty}^{\infty}\int_{-\infty}^{\infty}\int_T^{T+h}\int_T^{T+h}(f(t,l_T(t,x(\cdot))+z_1,\varphi_k(t)(l_T'(s,x(\cdot)\varphi_k(s)z_2\times$$

$$\times p_2(z_1,z_2,t,T+h)dtdsdz_1dz_2\quad, \tag{18}$$

where $p_2(z_1,z_2,t,T+h)$ is the two-dimensional normal density of the distribution of Gaussian random variables $\varepsilon_T(t)$ and $\varepsilon_T(T+h)$ (here $\varepsilon_T(t)$ and $\varepsilon_T(T+h)$ are random processes at points t and $T+h$). The main term in expansion (18) is the linear forecast; the others adjust for the effects of linearity.

The densities of the measures are calculated for a system of differential equations in H of the following type (see [5]):

$$\frac{dx(t)}{dt}-A(t)x(t)+f(t,x(t))=\eta(t)$$

$$\frac{d\xi(t)}{dt} - A(t)\xi(t) = \eta(t) \tag{19}$$

$$0 \le t \le a \ , \ x(0) = \xi(0) = 0 \quad ,$$

where $A(t)$ is a family of linear operators, generally unbounded, but they are generated from a family of operators $V(t, s)$, α is a parameter, and $\eta(t)$ is a Gaussian process $E\eta(t) = 0$ with correlation operator function $k(t, s)$. Under certain assumptions it can be proved that $\mu_k \sim \mu_\xi$ and the density $d\mu_x d\mu_\xi = \rho_a$ may be calculated from

$$\rho_a(u) = \exp\{-\alpha \int_0^a < g(s, u(\cdot)), dw(s)) > -\frac{\alpha^2}{2} \int_0^a \|g(s, u(\cdot)\|^2 ds\} \quad , \tag{20}$$

where $g(s, u(\cdot)$ and Wiener process $w(t)$ are defined by $f(t, u(\cdot)) = \int_0^a k(t, s) g(s, u(\cdot)) ds$ and $\eta(t) = \int_0^a k(t, s)\, dw(s)$. The integral $\int_0^a <\cdot, \cdot>$ in (20) should be interpreted as an expanded stochastic integral (see [5]). Thus, in formulas (10) or (14) we choose a pair of processes $x(t)$ and $\xi(t)$ as a solution of system (19), where $\xi(t)$ is a Gaussian process. Observing the process $x(t)$ on the interval $[0, T]$, we define its optimal forecast $\hat{x}(T+h)$ at the point $T+h$ by the formula (14) where

$$\rho_{T+h}^T(\xi) = \exp\{-\alpha \int_T^{T+h} < g(s, \xi(\cdot)), dw(s) > -\frac{\alpha^2}{2} \int_T^{T+h} \|g(s, \xi(\cdot))\|^2 ds)\} \quad . \tag{21}$$

2.1. OPTIMAL (NONLINEAR) FILTERS FOR SOLUTIONS OF DIFFERENTIAL EQUATIONS

Let one of the two random process $x(t)$ and $y(t)$ defined on the interval $[0, a]$ and with values in H be observed on the interval $[0, T]$, $T < a$. Let this process be $x(t)$ and process $y(t)$ be unobserved. Let $\mathbf{F}_T^{(1)}$ denote the σ-algebra generated by the random process $x(t)$ for $t \le T$. The optimal filtering problem is to construct estimators $x(t)$ of the process $y(s)$, $s \in [0, T]$, using the observed values of process $\hat{y}(s)$, such that $E(y(s) - \hat{y}(s))^2$ is minimal. It is clear that the estimator $\hat{y}(s)$ is a $\mathbf{F}_T^{(1)}$-measurable random variable. We will assume that $Ey^2(s) < \infty$. Then

$$\hat{y}(s) = E\{y(s) \mid \mathbf{F}_T^{(1)}\} \quad . \tag{22}$$

We shall pursue the idea of the density of an initial measure with respect to a certain standard measure.

Let $\mu_{x,y}$ be a measure generated by the pair of processes $x(t)$ and $y(t)$ defined in the domain $[0, a] \times [0, a]$, and $\mu_{\xi,\eta}$ be a measure generated by another pair of processes $\xi(t)$ and $\eta(t)$ defined in the same domain. We shall suppose that $\mu_{x,y} \ll \mu_{\xi,\eta}$

and that $\lambda(\xi,\eta)$ is the density, i.e., $\lambda(\xi,\eta) = d\mu_{x,y} d\mu_{\xi,\eta}$. Define

$$\beta(s\,,\xi) = \frac{E\{\eta(s)\lambda(\xi(\cdot),\eta(\cdot))\,|\,\mathbf{F}_T^{(1)^*}\}}{\frac{d\mu_x}{d\mu_\xi}(\xi(\cdot))} \quad , \tag{23}$$

where $\mathbf{F}_T^{(1)^*}$ is the σ-algebra generated by the values of process $\xi(t)$ for $t \le T$. We will now prove the following theorem:

THEOREM 2. *Let measure $\mu_{x,y}$ be absolutely continuous with respect to measure $\mu_{\xi,\eta}(\mu_{x,y} \ll \mu_{\xi,\eta})$ and $\lambda(\xi,\eta) = d\mu_{x,y}|d\mu_{\xi,\eta}$. Then for any $s \in [0,T]$ the following relation holds:*

$$\hat{y}(s) = \beta(s\,,x(\cdot))|_0^T \quad . \tag{24}$$

Suppose that $\xi(s)$ and $\eta(s)$ are a pair of Gaussian processes, and that $\bar{\eta}^*(s) = \bar{\eta}^*(s\,,\xi(\cdot))$ is a linear filter for the process $\eta(s)$ under the values of $\xi(s)$ on the interval $[0,T]$, $s \in [0,T]$. Then $\bar{\eta}^*(s)$ is a measurable variable and in this case

$$\eta(s) = \bar{\eta}^*(s) + \varepsilon(s) \quad , \tag{25}$$

where $\varepsilon(s)$ is a Gaussian process independent of the σ-algebra $\mathbf{F}_T^{*(1)}$.

Define

$$\bar{\eta}(s) = \bar{\eta}^*(s\,,\xi(\cdot))\,|\,\xi(\cdot) = x(\cdot) \quad . \tag{26}$$

Then

$$\hat{y}(s) = [\frac{d\mu_x}{d\mu_\xi}(\xi(\cdot))]^{-1}\,,\, E\{(\bar{\eta}^*(s) + \varepsilon(s))\lambda(\xi(\cdot)\,,\,\bar{\eta}^*(\cdot) + \varepsilon(\cdot))\,|\,\mathbf{F}_T^{*(1)}|\ \xi(\cdot) = x(\cdot) =$$

$$= [\frac{d\mu_x}{d\mu_\xi}(x(\cdot))]^{-1} E\{(u(s) + \varepsilon(s))\lambda(z(\cdot)\,,\,u(\cdot) + \varepsilon(\cdot))\}\ \Big|\ \begin{matrix} z(\cdot) = x(\cdot) \\ u(\cdot) = \bar{\eta}(\cdot) \end{matrix} \quad . \tag{27}$$

Taking into account the fact that $E\{\lambda(z(\cdot)\,,\,u(\cdot) + \varepsilon(\cdot)\} = \frac{d\mu_x}{d\mu_\xi}(z(\cdot))$ and making further simplifications, we have

$$\hat{y}(s) = \eta_+(s) + \frac{E\{(s)\lambda(z(\cdot),u(\cdot)+\varepsilon(\cdot))\}}{E\{\lambda(z(\cdot),u(\cdot)+\varepsilon(\cdot))\}}\ \Big|\ \begin{matrix} z(\cdot) = x(\cdot) \\ u(\cdot) = \bar{\eta}(\cdot) \end{matrix} \quad . \tag{28}$$

Now consider the system of differential equations of the type

$$\frac{dx(t)}{dt} + \alpha f_1(t, x(t), y(t)) = \xi'(t) \quad 0 \le t \le a \tag{29}$$

$$\frac{dy}{dt} + \alpha f_2(t, x(t), y(t)) = \eta'(t) \quad x(0) = y(0) = \eta(0) = \xi(0) = 0 \quad ,$$

where $\xi(t)$ and $\eta(t)$ are Gaussian processes in H, $E\xi(t) = E\eta(t) = 0$, and α is a parameter. For the space $H \times H = H_2$ this system can be rewritten as follows:

$$(f_1(t, x(t), y(t)), f_2(t, x(t), y(t)) = F(t, y(t)), 0 \le t \le a$$

$$\frac{dx(t)}{dt} + \alpha F(t, y(t)) = X'(t) \quad y(0) = X(0) = 0 \quad . \tag{30}$$

Then the measures $\mu_{x,y}$ and $\mu_{\xi,\eta}$ will coincide with the measures μ_y and μ_x. Applying the conditions of the theorem from [2] to equation (11) in order to obtain $\mu_g \ll \mu_x$ or $\mu_{x,y} \ll \mu_{\xi,\eta}$, we can write the following formula for the density:

$$\frac{d\mu_{x,y}}{d\mu_{\xi,\eta}}(\xi, \eta) = \frac{d\mu_y}{d\mu_x}(X) = \exp\{-\alpha \int_0^a (G, (dw(t))_{H_2} -$$

$$-\frac{\alpha^2}{2} \int_0^a \|G(t)\|^2{}_{H_2} dt\} = \lambda_a(\xi, \eta) \quad , \tag{31}$$

where

$$F(t, y(t)) = \int_0^a R(t, u) G(u) du \quad \int_0^a \|G(u)\|_{H_2}^2 < \infty \quad ,$$

$$X(t) = \int_0^a R(t, s) dw(s), \; R^2(t, s) = \int_0^a R(t, u) R(u, s) du \quad .$$

Here $(\cdot, \cdot)_{H_2}$ and $\|{}^0_2 \cdot \|_{H_2}$ represent the scalar product and norm, respectively, in H_2, and $R(t, s)$ is the correlation operator function for Gaussian process $X(t)$. If $\{\varphi_k(t)$ and $\{\lambda_k\}$ are vector eigenfunctions and eigenvalues of operator $R^2(t, s)$, then the density $\lambda(\xi, \eta)$ defined by formula (31) can be written in the form

$$\lambda_a(\xi, \eta) = \exp\{-\sum_{k=1}^{\infty} [\frac{\alpha}{\lambda_k} \int_0^a \int_0^a (F(t, X(t), \varphi_k(t))(X'(s), \varphi_k(s) dt ds +$$

$$+\frac{\alpha^2}{2} (\int_0^a (F(t, X(t)), \varphi_k(t)) dt)^2\} \quad . \tag{32}$$

Suppose now that only one of the components in system (29) is observed. Let this be $x(t)$. Then, to find the optimal estimator $\hat{y}(t)$ for the other component we use formula (28), setting $\lambda(\xi,\eta) = \lambda_T(\xi,\eta)$ as defined by (31) or (32). It can easily be seen that if α is small in these formulas, then the right-hand side of formula (31) can be expanded in powers of this small parameter. Moreover, it may easily be seen from (28) that the main term in this expansion is a linear filter of the process $y(t)$.

3. GENERAL FORMULAS FOR NONLINEAR EXTRAPOLATION AND FILTERING FOR FUNCTIONALS OF STOCHASTIC PROCESSES

We shall consider a certain random proces $x(t)$ defined on the interval $[0,a]$. Let $\mathbf{F}_T$ denote the σ-algebra generated by variables $x(t)$ for $s \le T$, $T \in [0,a]$. If η is a random variable which is measurable with respect to $\mathbf{F}_a$ (such a variable is called a functional of the process $x(t)$), then the best (mean-square) estimator of the variable η with respect to the σ-algebra $\kappa \subset \mathbf{F}_a$ is the κ-measurable random variable $\hat{\eta}$ for which $\mathrm{E}(\eta - \hat{\eta})$ takes its minimal value. The best estimator $\hat{\eta}$ calculated by the formula $\hat{\eta} = \mathrm{E}(\eta \mid \kappa)$ exists when $\mathrm{E}\eta^2 < \infty$.

When studying the problems of optimal extrapolation and filtering of stochastic processes, we take the variables $\eta = h(x(t))$ instead of η and consider the σ-algebra $\mathbf{F}_T^b$ generated by $b(x(s))$ instead of the σ-algebra κ, where $h(\cdot)$ and $b(\cdot)$ are certain measurable functionals defined over the space H.

We shall also assume that the measure μ_x corresponding to the function $x(t)$ is generated in the space of functions $\mathbf{L}_2\{[0,a], H\} = \mathbf{L}_2$ defined on the interval $[0,a]$ and taking values in H and $\int_0^a \|x(t)\|^2 dt < \infty$. Consider two processes $x(t)$ and $\xi(t^0)$ in the space H with second-order finite moments and let μ_x and μ_ξ be the corresponding measures in L_2. Suppose that $\mu_x \ll \mu_\xi$ or $\mu_x \sim \mu_\xi$ and that $\rho_a(\cdot) = d\mu_x$ *romand* μ_ξ is the density of measures μ_x with respect to measure μ_ξ on the interval $[0,a]$. Define

$$\gamma(t,\xi(\cdot)) = [\rho_T(\xi(\cdot))]^{-1}\mathrm{E}\{h(\xi()\rho_T(\xi(\cdot)) \mid \mathbf{F}_T^{b^*}\} \quad , \tag{33}$$

where $\mathbf{F}_T^{b^*}$ is the σ-algebra generated by $b(\xi(s))$ for $s \le T$, $t \in [0,a]$. It can be shown that

$$\hat{h}(x(t)) = \mathrm{E}\{h(x(t)) \mid \mathbf{F}_T^b\} = \gamma(x(\cdot),t) \quad . \tag{34}$$

If ξ in (33) is a Gaussian process and $b(\cdot)$ is a linear function on H, then the σ-algebra $F_T^{b^*}$ will be generated by Gaussian variables under the values of Gaussian processes $\xi(t)$ for for $t \le T$; we shall denote it by $\mathbf{F}_T^*$. Therefore, as is already known, $\xi(t) = l_T^*(t) + \varepsilon_T(t)$, where $l_T^*(t) l_T^*(t,\xi(\cdot) = \mathrm{E}\{\xi(t) \mid \mathbf{F}_T^*\}$ is the linear estimator $\xi(t)$, which is measurable with respect to σ-algebra $\mathbf{F}_T^{\bullet}$, and $\varepsilon_T(t)$ is Gaussian and independent of σ-algebra $\mathbf{F}_T^{\bullet}$. In this case formula (34) takes the form

$$\hat{h}(x(t)) = [\rho_T(\xi(\cdot)]^{-1}\mathrm{E}\{h(\xi(\cdot))\cdot\rho_t(\xi(\cdot))\mid \mathbf{F}_T^{\bullet}\}\mid \xi(\cdot)=x(\cdot) =$$

$$= l_T(t) = l_T(t, x(\cdot)) = l_T^{\bullet}(t, \xi(\cdot))\mid \xi(\cdot)=x(\cdot)\mid \qquad (35)$$

$$\begin{array}{l} u(t) = l_T(t) \\ u(\cdot) = l_T(\cdot) \\ \xi(\cdot) = x(\cdot),\ \hat{h}(x(t)) = \mathrm{E}h \end{array}$$

where $l_T(t) = l_T(t, x(\cdot)) = l_T^{\bullet}(t, \xi(\cdot))\mid \xi(\cdot) = x(\cdot)$ and $l_T(\cdot) = l_T(\cdot, x(\cdot))$.

Formula (35) is a general formula since it gives the optimal estimator $\hat{h}(x(t))$ for the function $h(x(t))$. For example, if $t < T$ in (35), then $\hat{h}(x(t))$ is the optimal forecast for the function $h(x(t))$; if $t = T$, then $\hat{h}(x(t))$ is the optimal filter for the function $h(x(t))$; and if $t < T$, then $\hat{h}(x(t))$ is the optimal interpolation of the function $h(x(t))$.

Suppose that the function $x(t)$ and the Gaussian process $\xi(t)$ are respectively a solution of equation (11) and its right-hand side (or a solution of equation (19). Let $\mu_x \ll \mu_\xi$, and the densities be of the forms (12) and (13). Consider, for example, the system of equations (19). In view of the comments made above regarding $x(t)$ we can use formula (35), replacing $\rho_T(\cdot)$ by expression (20). We therefore have

$$\hat{h}(x(t) = \frac{\mathrm{E}\{h(u(t)+\varepsilon_T(t))\exp\{-\alpha\int_0^t <g(t), dw(t)> - \frac{\alpha^2}{2}\int_0^t \|g(t)\|^2 dt\}}{\exp\{-\alpha\int_0^T <g(t), dw(t)> - \frac{\alpha^2}{2}\int_0^T \|g(t)\|^2 dt\}} \qquad (36)$$

$$\left| \begin{array}{l} u(t) = l_T(t) \\ u(\cdot) = l_T(\cdot) \\ \xi(\cdot) = x(\cdot) \end{array}\right. ,$$

where the function $g(t)$ and the Wiener process $w(t)$ are defined by the formulas (*) and (**), respectively.

Expression (36) is the most general form. The solutions of all nonlinear extrapolation and filtering problems involving stochastic processes, and depending on the choice of $h(\cdot)$, $b(\cdot)$, and point t, may be obtaned from this formula. If $g(\cdot) = h(\cdot) = x(\cdot)$, $t > T$, then (36) yields the formula for the optimal forecast of stochastic process $x(t)$; and if the random process $x(t)$ (viewed as a vector) consists of two components b and h (which may themselves be vectors, but with a dimension smaller than that of x), and $t \leq T$, then (36) yields the optimal filtering formula for one component ($n(\cdot)$) of the vector $x(\cdot)$ from observations of the other, i.e., $b(\cdot)$.

If α is small and $\{\lambda_k\}$ and $\{\varphi_k(t)\}$ are eigenvalues and vector eigenfunctions of the correlation operator function $R^2(t,s)$ of a Gaussian process, then expanding the exponents in (36) in powers of α, one obtains the expansion of the estimator $h(x(t))$ in powers of this small parameter. We shall assume below that $t = T + \beta$ in (36) and to simplify the calculation we will consider equation (1). The density $p_t(\cdot)$ in (36) will then be the same as in (12) for $a = t$.

Following this procedure, we obtain the expansion

$$\hat{h}(x(t)) = C_1 - \alpha C_2 + \alpha^2 C_3 + o(\alpha^2) \quad , \tag{37}$$

where C_1, C_2, and C_3 are defined by the linear system of algebraic equations

$$C_1 = A_1 \, , \; C_2 = A_2 - A_1 B_2 \, , \; C_3 = A_3 - 2B_2(A_2 - A_1 B_2) - A_1 B_3 \quad , \tag{38}$$

and the values A_1, A_2, A_3, B_2, B_3 are defined by the relations

$$A_1 = \mathrm{E}h(u(T+\beta) + \varepsilon_T(T+\beta)) \mid u(T+\beta) = l_T(T+\beta) \tag{39}$$

$$A_2 = \mathrm{E}\{h(u(T+\beta) + \varepsilon_T(T+\beta))[\sum_{k=1}^{\infty} \frac{1}{\lambda_k} \int_0^{T+\beta}\int_0^{T+\beta} (f(z, u(z) + \varepsilon_T(z)), \varphi_k(z))(v(s) + \tag{40}$$

$$+ \varepsilon_T'(s), \varphi_k(s))dz\,ds\,]\} \mid \begin{array}{l} u(\cdot) = l_{T_1}(\cdot) \\ v(\cdot) = l_T(\cdot) \end{array}$$

$$A_3 = \mathrm{E}\{h(u(T+\beta) + \varepsilon_T(T+\beta))[A_2^2 - \sum_{k=1}^{\infty} \frac{1}{\lambda_k}(\int_0^{T+\beta} (f(s, u(s) +$$

$$+ \varepsilon_T(s)), \varphi_k(s))ds)^2]\} \mid_{u(\cdot)=l_T(\cdot)} \tag{41}$$

$$B_2 = \sum_{k=1}^{\infty} \frac{1}{\lambda_k} \int_0^T \int_0^T (f(t, x(t)), \varphi_k(t))(x'(s), \varphi_k(s))dt\,ds \tag{42}$$

$$B_3 = B_2^2 - \sum_{k=1}^{\infty} \frac{1}{\lambda_k} (\int_0^T (f(t, x(t)), \varphi_k(t))dt)^2 \tag{43}$$

Since $x(t)$ is observed up to time T, we have that $x(t)$ and $x'(t)$ are known variables in expressions (42) and (43). Now, choosing the values of h and β in formulas (39)–(43) and using (37), we obtain an expansion of optimal estimators for concrete problems. Thus, for example, if $\beta > 0$ and $h(x(t)) = x(t)$, formula (37) yields an expansion of the optimal forecasting formula for $x(t)$; if $x(t) = x(b, h)$ and $\beta > 0$, then (37) yields an expansion of the optimal filter $h(x(t))$, entering at the point $T + \beta$, under observations $b(x(t))$ up to time T.

Moreover, formula (37) shows that for small nonlinearities the deviation of the optimal extrapolation from the linear one is of the same order as the order of nonlinearity. Thus the use of optimal estimators instead of linear ones results in an essential improvement in linear estimators.

REFERENCES

1. U.A. Rosanov. *Stationary Random Processes*. Fizmatgiz, 1963 (in Russian).
2. A.D. Shatashvili. Optimal extrapolation and filtering for one class of random processes (in Russian). *Teoria Verojatnostey i Matematicheskaya Statistica*, (2) (1970) 211-231; (3) (1970) 235-253.
3. A.D. Shatashvili. On the densities of measures which correspond to the solution of some differential equations with random functions (in Russian). *DAN SSSR*, 194(2) (1970) 275-277.
4. A.V. Skorohod and A.D. Shatashvili. On the absolute continuity of the Gaussian measure for nonlinear transformations (in Russian). *Teoria Verojatnostey i Matematicheskaya Statistica*, (15)(1975).
5. A.V. Skorohod. On one generalization of the stochastic integral (in Russian). *Teoria Verojatnostey i jeje Prinenehia*, 20(1975) 223-238.
6. J.I. Gihman and A.V. Skorohod. *Introduction to the Theory of Random Processes*. Nauka, Moscow, 1965 (in Russian).

ON FUNCTIONAL EQUATIONS OF DISCRETE DYNAMIC PROGRAMMING

Karel Sladký, Institute of Information Theory and Automation
Czechoslovak Academy of Sciences, Pod vodárenskou věží 4,
182 08 Praha 8-Libeň, Czechoslovakia

1. INTRODUCTION

We consider at discrete time points $n = 0,1,\ldots$ a system with finite state space $I = \{1,2,\ldots,N\}$ whose utility vector at time n, denoted $x(n)$ (column N-vector), obeys the following dynamic programming recursion

$$x(n+1) = \max_{f \in F} Q(f)\, x(n) = Q(\hat{f}^{(n)})\, x(n). \tag{1.1}$$

Here $x(0) > 0$ is given, $Q(f)$ is an $(N\times N)$-nonnegative matrix depending on a decision vector f (i.e. N-vector whose i-th component $f(i) \in F(i)$ specifies the decision in state i, i.e. the i-th row of the matrix $Q(f)$) and $F = F(1) \times \ldots \times F(N)$ is a finite set of all decision vectors at each time point. Recall that the set F possesses an important "product property", i.e. if $f_1, f_2 \in F$ then there exists also $f \in F$ such that $[Q(f_1)]_{i_1} = [Q(f)]_{i_1}$, $[Q(f_2)]_{i_2} = [Q(f)]_{i_2}$ for each pair $i_1, i_2 \in I$. $[A]_i$, resp. $[A]_{ij}$, denotes the i-th row, resp. ij-th element, of the matrix A. Consequently, vectorial maximum in (1.1) always exists.

Remember that $f^{(n)} \in F$ is reserved for the decision selected at time n, $\hat{f}^{(n)}$ is the maximizer in (1.1) over all $f^{(n)}$, and a policy (i.e. sequence of decision vectors) selecting $f^{(n)} = f$ is called stationary. In what follows we shall denote by $\bar{F} \subset F$ the minimal set of decision vectors possessing the "product property" and containing all decision vectors occurring infinitely often in (1.1). Observe that $\bar{F}$ depends on the considered (fixed) initial condition x(0).

Investigating Markov decision chains the respective dynamic programming recursion for calculating maximum expected rewards turns out to be a very special case of (1.1). Besides this very specific form intensively studied in the literature, the general case of (1.1) has many other interesting and useful applications (e.g. supervised linear economic models, controlled branching processes, Markov decision chains with multiplicative utility functions - cf. Sladký (1980) for details).

It can be shown (cf. Sladký (1980), Zijm (1982), Rothblum, Whittle (1982) and Chapter 35 of Whittle (1983)) that the growth of x(n) is given by an exponential as well as a polynomial part (i.e. $[x(n)]_i \approx \sigma^n n^{\nu-1}$ for some $\sigma \geqq 0$ and integer $\nu > 1$). These facts, based on the Perron-Frobenius theorem and a "uniform block-triangular decomposition" of the set $\{Q(f), f \in F\}$, are summarized in Section 2. Having found the growth rate on x(n) by employing the "uniform block-triangular decomposition" of the set $\{Q(f), f \in F\}$ we can construct polynomial bounds on respective subvectors of x(n) and, by using similar methods as in Markov decision chains, establish the asymptotic properties of x(n). The results are discussed in Section 3; notice that, unlike in Markov decision chains, considering the general case of (1.1) all coefficients in the polynomial bounds on x(n) will depend on the initial condition x(0) and boundedness of $n^{-\nu+1}x(n)$ does not imply that, for $n \to \infty$, $n^{-\nu+1}x(n)$ converges or attains its maximum in the class of stationary policies.

2. GENERAL ANALYSIS AND PRELIMINARIES

The material of this section is mostly adapted from Sladký (1980). Considering the set $\{Q(f), f \in F\}$, $Q_{ij}(f)$ denotes the submatrix of Q(f) and $\sigma_i(f)$, resp. $u_i(f) > 0$ (i.e. nonnegative, nonzero), is reserved for the spectral radius, resp. corresponding right eigenvector, of the diagonal submatrix $Q_{ii}(f)$. Recall (cf. Gantmakher (1966)) that by the well-known Perron-Frobenius theorem $u_i(f) \gg 0$ (i.e. strictly positive) if and only if (by possibly permuting rows and corresponding columns - i.e. for suitable labelling of states in our model) $Q_{ii}(f)$ can be written in the following upper block-triangular form

$$Q_{ii}(f) = \begin{bmatrix} Q_{i(11)}(f) & Q_{i(12)}(f) & \dots & Q_{i(1r)}(f) \\ 0 & Q_{i(22)}(f) & \dots & Q_{i(2r)}(f) \\ \vdots & \vdots & \ddots & \vdots \\ 0 & 0 & \dots & Q_{i(rr)}(f) \end{bmatrix} \qquad (2.1)$$

where $Q_{i(jj)}(f)$ are irreducible classes of $Q_{ii}(f)$ such that ($\sigma_{i(j)}(f)$ denotes the spectral radius of $Q_{i(jj)}(f)$)

$$\sigma_{i(j)}(f) < \sigma_i(f) \Rightarrow Q_{i(jk)}(f) \neq 0 \quad \text{at least for one } k \neq j$$

$$\sigma_{i(j)}(f) = \sigma_i(f) \Rightarrow Q_{i(jk)}(f) = 0 \quad \text{for all } k \neq j. \qquad (2.1')$$

Recall that if $\sigma_{i(j)}(f) = \sigma_i(f)$, resp. $\sigma_{i(j)} < \sigma_i(f)$, the class $Q_{i(jj)}(f)$ is called basic, resp. non-basic; using the terminology of Markov chain theory we can say that $u_i(f) \gg 0$ if and only if each non-basic class of $Q_{ii}(f)$ is accessible at least to one basic class and each basic class is not accessible to any other irreducible class of $Q_{ii}(f)$.

On the base of the Perron-Frobenius theorem we can deduce that, for suitable labelling of states, the matrix Q(f) can be written in a block-triangular form with s(f) diagonal blocks $Q_{ii}(f)$'s, each of them being of the form (2.1) and having strictly positive eigenvector $u_i(f) \gg 0$ corresponding to the spectral radius $\sigma_i(f)$ of $Q_{ii}(f)$. The decomposition can be suggested in such a way that $\sigma_1(f) \geq \sigma_2(f) \geq \dots \geq \sigma_{s(f)}(f)$ and, furthermore, $\sigma_i(f) = \sigma_{i+1}(f) \Rightarrow \nu_i(f) = \nu_{i+1}(f) + 1$, where $\nu_i(f)$ (index of $Q_{ii}(f)$) is defined as the largest number of irreducible classes having spectral radius $\sigma_i(f)$ that can occur in a chain of irreducible classes successively accessible from the class $Q_{ii}(f)$. Such a decomposition has the property that the diagonal blocks are the largest submatrices of Q(f) having strictly positive eigenvectors corresponding to their spectral radii.

Moreover, similar results can be extended to the whole set $\{Q(f), f \in F\}$, i.e. we can suggest a "fixed" decomposition such that using this decomposition each Q(f) with $f \in F$ is upper block-triangular and the diagonal submatrices possess further additional properties. These facts are very important for the analysis of our model and are therefore summarized as

Theorem 2.1. There exists some $\hat{f} \in F$ such that for suitable permutation of rows and corresponding columns (i.e. for suitable labelling of states) any $Q(f)$ with $f \in F$ is upper block-triangular, i.e.

$$Q(f) = \begin{bmatrix} Q_{11}(f) & Q_{12}(f) & \dots & Q_{1s}(f) \\ 0 & Q_{22}(f) & \dots & Q_{2s}(f) \\ \vdots & \vdots & \ddots & \vdots \\ 0 & 0 & \dots & Q_{ss}(f) \end{bmatrix} \qquad (2.2)$$

where for $\hat{f} = f$ and strictly positive eigenvectors $u_i(\hat{f})$'s

$$\sigma_1(\hat{f}) \geqq \sigma_2(\hat{f}) \geqq \dots \geqq \sigma_s(\hat{f}) \qquad (2.3)$$

$$\sigma_{i+1}(\hat{f}) = \sigma_i(\hat{f}) \Longrightarrow \nu_i(\hat{f}) = \nu_{i+1}(\hat{f}) + 1 \qquad (2.3')$$

and moreover for any $f \in F$

$$\sigma_i(\hat{f})\, u_i(\hat{f}) = Q_{ii}(\hat{f})\, u_i(\hat{f}) \geqq Q_{ii}(f)\, u_i(\hat{f})\,. \qquad (2.4)$$

Remark 2.2. Observe that if $\sigma_i(\hat{f}) > 0$, $Q_{ii}(\hat{f}) \neq 0$ can be written in a block-triangular form (2.1) fulfilling conditions (2.1'); however, it may happen that $\sigma_i(f) < \sigma_i(\hat{f})$ and also $Q_{ii}(f)$ need not be upper block-triangular for any $f \in F$ (even it may happen that $Q_{ii}(f) = 0$ for some $f \in F$). Moreover, if $\sigma_i(\hat{f}) = 0$ then for any $f \in F$ $Q_{ii}(f) = 0$, and for $\sigma_i(\hat{f}) = 0$ with $i < s$ $Q_{i,i+1}(\hat{f}) \neq 0$ and $\nu_i(\hat{f}) = s-i+1$.

Remark 2.3. The results summarized in Theorem 2.1 were originally established only under assumption $\sigma_s(\hat{f}) > 0$, however, taking into account the facts presented in Remark 2.2 we can easily verify their validity even for $\sigma_s(\hat{f}) = 0$.

From now on for any $Q(f)$ the same decomposition as in (2.2) will be considered. Using this "uniform block-triangular decomposition" in (1.1) we get for $j = 1,2,\dots,s$

$$x_i(n+1) = Q_{ii}(\hat{f}^{(n)})\, x_i(n) + \sum_{j=i+1}^{s} Q_{ij}(\hat{f}^{(n)})\, x_j(n). \qquad (2.5)$$

On the base of (2.3), (2.4), (2.5) we can evaluate the first-order approximation of the growth of $x(n)$. The results are summarized as Theorem 2.4 (we abbreviate $\sigma_i(\hat{f}), \nu_i(\hat{f})$ by σ_i, ν_i respectively, the case $\sigma_i = 0$ follows immediately using the reasoning sketched in Remarks 2.2 and 2.3).

Theorem 2.4. There exist (nonnegative) vectors k_i (being of the same dimension as $x_i(n)$) such that if $\sigma_i > 0$

$$x_i(n) \leqq k_i \, \sigma_i^n \, n^{\nu_i - 1} \quad \text{for all } n = 0,1,\ldots \,. \tag{2.6}$$

Moreover, if $\sigma_i = 0$ then $x_i \leqq k_i$ for $n = 0,1,\ldots,\nu_i - 1$, but $x_i(n) = 0$ for all $n \geqq \nu_i$.

Taking into account the results of Theorem 2.4 we can continue our study concerning asymptotic properties of (2.5) under the following general assumption:

Assumption GA. $\sigma_i \equiv \sigma_i(\hat{f}) > 0$ for all $i = 1,\ldots,s$ (of course, by (2.3) it suffices only to assume $\sigma_s(\hat{f}) > 0$).

3. POLYNOMIAL BOUNDS AND THE ASYMPTOTIC BEHAVIOUR OF THE UTILITY VECTOR

Firstly, we rewrite the dynamic programming recursion for the maximum utility in a more suitable form. Obviously, by (1.1), (2.5) we get

$$\bar{x}_i(n+1) = \bar{Q}_{ii}(\hat{f}^{(n)}) \, \bar{x}_i(n) + \sum_{j=i+1}^{i+\nu_i-1} \bar{Q}_{ij}(\hat{f}^{(n)}) \, \bar{x}_j(n) + o_i(\hat{f}^{(n)};n) \tag{3.1}$$

where

$$\bar{x}_j(n) = \sigma_i^{-n} \, x_j(n), \qquad \bar{Q}_{ij}(f) = \sigma_i^{-1} \, Q_{ij}(f) \quad \text{for } j = i,\ j > i$$

$$o_i(f;n) = \sum_{j=i+\nu_i}^{s} \bar{Q}_{ij}(f) \, \bar{x}_j(n).$$

Moreover, by (2.6) of Theorem 2.4 $o_i(f;n)$ converges to the zero vector as $n \to \infty$ and the convergence is exponential, i.e. there exist vectors $c_i' \ll 0$, $c_i'' \gg 0$ and a number $\lambda \in (0,1)$ such that

$$c_i' \lambda^n \leqq o_i(f;n) \leqq c_i'' \lambda^n \quad \text{for any } n = 0,1,\ldots \tag{3.1'}$$

(notice that (3.1′) can be fulfilled for any $\lambda \in (\sigma_{i+\nu_i}/\sigma_i, 1)$ and c_i', c_i'' selected according to the choice of λ).

Observe that the set $\{\bar{Q}_{ii}(f), f \in F\}$ is positively similar to some set of (sub)-stochastic matrices, in particular by Theorem 2.1 there exists some $u_i \gg 0$ such that for any $f \in F$ $\bar{Q}_{ii}(f) \, u_i \leqq u_i$, hence the similarity matrix U_i can be chosen as $U_i = \mathrm{diag}\{u_i\}$.

In what follows we shall also need convex combinations of $Q_{ii}(f)$'s specified by the parameter $\tilde{f}$ with generic $\tilde{F}$, i.e.

$$Q_{ii}(\tilde{f}) = \sum_{j=1}^{p} \alpha_j(\tilde{f})\, Q_{ii}(f_j), \qquad \sum_{j=1}^{p} \alpha_j(\tilde{f}) = 1$$

where $f_j \in F$ and $\alpha_j(\tilde{f})$ are positive numbers for $j = 1,\ldots,p$. Observe (cf. the Perron-Frobenius theorem and analogy to stochastic matrices) that for the matrices $Q_{ii}(f_j)$'s fulfilling condition $Q_{ii}(f_j)\,u_i = \sigma_i\, u_i$ spectral radius of the matrix $Q_{ii}(\tilde{f})$ arising by a convex combination of $Q_{ii}(f_j)$'s is again equal to σ_i, each basic class of $Q_{ii}(\tilde{f})$ is not accessible to any other class of $Q_{ii}(\tilde{f})$ (and there exists at least one such a basic class) and the periodicity of the basic class of $Q_{ii}(\tilde{f})$ is nongreater than the periodicity of each basic class of any $Q_{ii}(f_j)$ (recall that the periodicity of $Q_{ii}(\tilde{f})$ is given by the minimum integer $\varkappa = \varkappa(\tilde{f})$ such that the matrix $(Q_{ii}(\tilde{f}))^{\varkappa}$ is aperiodic and consequently $\lim (Q_{ii}(\tilde{f}))^{n\varkappa}$ exists as $n \to \infty$).

Sometimes we shall need to construct multi-step decisions in the respective dynamic programming recursion. To this order we introduce

$$Q^{(\varkappa)}(f^{(n)}) = Q(f^{(n+\varkappa-1)})\; Q(f^{(n+\varkappa-2)}) \;\ldots\; Q(f^{(n+1)})\; Q(f^{(n)})$$

where $Q^{(1)}(f^{(n)}) = Q(f^{(n)})$ and $Q^{(0)}(f^{(n)}) = I$ is an identity matrix. In particular, $Q^{(\varkappa)}(f) = (Q(f))^{\varkappa}$. Considering multi-step decisions by (3.1) we obtain

$$\bar{x}_i(n+m) = \bar{Q}_{ii}^{(m)}(\hat{f}^{(n)})\; \bar{x}_i(n) + \sum_{j=i+1}^{i+\nu_i-1} \bar{Q}_{ij}^{(m)}(\hat{f}^{(n)})\; \bar{x}_j(n) + \\ + o_i^{(m)}(\hat{f}^{(n)};n) \qquad (3.2)$$

where $Q_{ij}^{(m)}(.)$ denotes the ij-th block of the matrix $Q^{(m)}(.)$ and $o_i^{(m)}(f^{(n)};n) \to 0$ as $n \to \infty$ and the convergence is exponential.

To establish asymptotic properties of $x_i(n)$ we shall proceed by induction on $j = i+\nu_i-1$, $j = i+\nu_i-2, \ldots,$ $j = i$. Observe that (by our definition of the index ν_i) for all above j's $\sigma_j = \sigma_i$ with $\nu_j = \nu_i + i - j$; however $\sigma_j < \sigma_i$ for $j \geq i+\nu_i$. The first step of the induction procedure, i.e. establishing asymptotic properties of $x_i(n)$ with $\nu_i = 1$ and (cf. Theorem 2.4) with pure exponential growth, is relatively simple. To study asymptotic properties of $x_i(n)$ with $\nu_i > 1$ we must construct polynomial bounds on certain subsequencies of $\bar{x}_i(n)$.

3.1. Behaviour of Utilities with Pure Exponential Growth

Supposing $\nu_i = 1$ the second term on the RHS of (3.1) vanishes and (3.1) then reads

$$\bar{x}_i(n+1) = \bar{Q}_{ii}(\hat{f}^{(n)})\, \bar{x}_i(n) + o_i(\hat{f}^{(n)};n) \tag{3.3}$$

with $o_i(\hat{f}^{(n)};n) \to 0$ exponentially fast.

As it is shown in Theorem 3.1, the behaviour of $\{x_i(n)\}$ defined by (3.3) heavily depends on the periodicity of a suitable convex combination of $Q_{ii}(f)$'s with $f \in \bar{F}$, say $Q_{ii}(\tilde{f})$ with $\tilde{f} \in \tilde{F}$. It can be easily shown that by a simple algorithmic procedure we can construct $Q_{ii}(\tilde{f})$ such that

(i) For any $f \in \bar{F}$ the states that may belong to some class of $Q_{ii}(f)$ whose spectral radius is equal to σ_i are also contained within some basic class of $Q_{ii}(\tilde{f})$.

(ii) There exists no other convex combination of $Q_{ii}(f)$'s with $f \in \bar{F}$ fulfilling condition (i) whose periodicity is less than that of the matrix $Q_{ii}(\tilde{f})$.

Asymptotic properties of $\bar{x}_i(n)$ defined by the recursive relation (3.3) are summarized in

Theorem 3.1. Let $\varkappa_i$ be the period of $Q_{ii}(\tilde{f})$. Then for $\bar{x}_i(n)$ given by (3.3) there exists

$$\lim_{n\to\infty} \bar{x}_i(n\varkappa_i+m) = x_i^{(m)} \qquad \text{for } m = 0,1,\ldots,\varkappa_i-1 \tag{3.4}$$

and the convergence in (3.4) is exponential.

The proof of Theorem 3.1 is sketched in the Appendix. Remember that the sequence of decisions occurring in (3.4) need not be stationary even if $\lim x_i(n)$ for $n \to \infty$ exists. Examples can be constructed (cf. Sladký (1976)) that for $F = \{f_1, f_2\}$ in (3.3) $\hat{f}^{(n)} = f_1$ (resp. $\hat{f}^{(n)} = f_2$) if n is odd (resp. even).

3.2. Polynomial Bounds on Utilities

Supposing $\nu_i > 1$ we show how to construct polynomial bounds on the utility vector $\bar{x}_i(n)$ calculated recursively from (3.1).

To this order we shall suppose that for all $x_j(n)$ with $j=i+1, \dots, i+\nu_i-1$ there exist vectors $w_j^{(k,m)}$ (compatible with $x_j(n)$) such that for some integer $\varkappa$ and any $m = 0,1,\dots,\varkappa-1$

$$z_j(n\varkappa+m) = \bar{x}_j(n\varkappa+m) - \sum_{k=0}^{\nu_j-1} \binom{n\varkappa+m}{k} w_j^{(k,m)} \longrightarrow 0 \qquad (3.5)$$

as $n \to \infty$ and the convergence is exponential, i.e. there exist vectors $c_j' \ll 0$, $c_j'' \gg 0$ and a number $\lambda \in (0,1)$ such that

$$c_j' \lambda^n \leqq z_j(n\varkappa+m) \leqq c_j'' \lambda^n . \qquad (3.5')$$

Our reasoning will be similar to that of Sladký (1981) (in Sladký (1981) only the aperiodic case with $\varkappa = 1$ was considered).

First we write a recursive relation for $z_i(n\varkappa+m)$. By (3.2), (3.5) we get (we set $\bar{n} = n\varkappa + m$, hence $(n+1)\varkappa + m = \bar{n} + \varkappa$)

$$z_i(\bar{n}+\varkappa) = \bar{x}_i(\bar{n}+\varkappa) - \sum_{k=0}^{\nu_i-1} \binom{\bar{n}+\varkappa}{k} w_i^{(k,m)} = \bar{Q}_{ii}^{(\varkappa)}(\hat{f}^{(\bar{n})})\,\bar{x}_i(\bar{n}) +$$

$$+ \sum_{j=i+1}^{i+\nu_i-1} \bar{Q}_{ij}^{(\varkappa)}(\hat{f}^{(\bar{n})})\,\bar{x}_j(\bar{n}) + o_i^{(\varkappa)}(\hat{f}^{(\bar{n})};\bar{n}) - \sum_{k=0}^{\nu_i-1} \sum_{l=0}^{k} \binom{\bar{n}}{l}\binom{\varkappa}{k-l} w_i^{(k,m)} =$$

$$= \bar{Q}_{ii}^{(\varkappa)}(\hat{f}^{(\bar{n})})\, z_i(\bar{n}) + \sum_{j=i+1}^{i+\nu_i-1} \bar{Q}_{ij}^{(\varkappa)}(\hat{f}^{(\bar{n})}) \sum_{k=0}^{\nu_j-1} \binom{\bar{n}}{k} w_j^{(k,m)} + \bar{o}_i^{(\varkappa)}(\hat{f}^{(\bar{n})};\bar{n}) +$$

$$+ (\bar{Q}_{ii}^{(\varkappa)}(\hat{f}^{(\bar{n})}) - I) \sum_{k=0}^{\nu_i-1} \binom{\bar{n}}{k} w_i^{(k,m)} - \sum_{k=1}^{\nu_i-1} \sum_{l=0}^{k-1} \binom{\bar{n}}{l}\binom{\varkappa}{k-l} w_i^{(k,m)} \qquad (3.6)$$

($\bar{o}_i^{(\varkappa)}(\hat{f}^{(\bar{n})};\bar{n}) = o_i^{(\varkappa)}(\hat{f}^{(\bar{n})};\bar{n}) + \sum_{j=i+1}^{i+\nu_i-1} \bar{Q}_{ij}^{(\varkappa)}(\hat{f}^{(\bar{n})})\, z_j(\bar{n})$, by (3.5), (3.5′) $\bar{o}_i^{(\varkappa)}(\hat{f}^{(\bar{n})};\bar{n}) \to 0$ exponentially fast as $\bar{n} \to \infty$).

Using some algebraic manipulations for the second and the last terms on the RHS of (3.6) we have

$$\sum_{j=i+1}^{i+\nu_i-1} \bar{Q}_{ij}^{(\varkappa)}(\hat{f}^{(\bar{n})}) \sum_{k=0}^{\nu_j-1} \binom{\bar{n}}{k} w_j^{(k,m)} = \sum_{k=0}^{\nu_i-2} \binom{\bar{n}}{k} \sum_{j=i+1}^{i+\nu_i-1-k} \bar{Q}_{ij}^{(\varkappa)}(\hat{f}^{(\bar{n})})\, w_j^{(k,m)}$$

$$\sum_{k=1}^{\nu_i-1} \sum_{l=0}^{k-1} \binom{\bar{n}}{l}\binom{\varkappa}{k-l} w_i^{(k,m)} = \sum_{l=0}^{\nu_i-2} \binom{\bar{n}}{l} \sum_{p=1}^{\nu_i-1-l} \binom{\varkappa}{p} w_i^{(p+l,m)} .$$

Introducing $b_i^{(\nu_i-1,m)}(f) = (\bar{Q}_{ii}^{(\varkappa)}(f) - I)\, w_i^{(\nu_i-1,m)}$, and

$$b_i^{(k,m)}(f) = (\bar{Q}_{ii}^{(\varkappa)}(f) - I)\, w_i^{(k,m)} - \sum_{l=1}^{\nu_i-k-1} \binom{\varkappa}{l} w_i^{(k+l,m)} + \sum_{j=i+1}^{i+\nu_i-k-1} \bar{Q}_{ij}^{(\varkappa)}(f)\, w_j^{(k,m)} \qquad \text{for } k=\nu_i-2,\dots,0$$

(observe that the argument f should be considered for a sequence of $\varkappa$ decision vectors $f\in\bar{F}$) recursion (3.6) can be written as

$$z_i(\bar{n}+\varkappa) = \bar{Q}_{ii}^{(\varkappa)}(\hat{f}^{(\bar{n})})\; z_i(\bar{n}) + s_i^{(\bar{n},m)}(\hat{f}^{(\bar{n})}) + \bar{o}_i^{(\varkappa)}(\hat{f}^{(\bar{n})};\bar{n}) \qquad (3.7)$$

where

$$s_i^{(\bar{n},m)}(f^{(\bar{n})}) = \sum_{k=0}^{\nu_i-1} \binom{\bar{n}}{k} b_i^{(k,m)}(f^{(\bar{n})}) .$$

Now, by Proposition A.1 of the Appendix, on the base of given $w_j^{(k,m)}$'s we can construct $w_i^{(k,m)}$'s (successively for $k=\nu_i-1,\dots,0$) such that

$$\left.\begin{aligned} & b_i^{(\nu_i-1,m)}(f) \leqq 0 \quad \text{for any } f\in(\underbrace{\bar{F}\times\dots\times\bar{F}}_{\varkappa}) \equiv F^{(\nu_i,m)} \\ & \text{and for } k=\nu_i-2,\dots,1,0 \\ & b_i^{(k,m)}(f) \leqq 0 \quad \text{for any } f\in F^{(k+1,m)} \end{aligned}\right\} \qquad (3.8)$$

where $\{F^{(k,m)},\ k=\nu_i-1,\dots,1,0\}$ is defined recursively by

$$F^{(k,m)} = \{f\in F^{(k+1,m)}:\ b_i^{(k,m)}(f)=0\} . \qquad (3.8')$$

Moreover, according to Proposition A.1, equality holds in (3.8) at least for one $f\in F^{(0,m)}$, say $f=f_m^*$.

<u>Theorem 3.2</u>. Let for $Q_{ii}(f)$ occurring in (3.1) $\nu_i>1$ and let (3.5), (3.5′) hold for $j=i+1,\dots,i+\nu_i-1$ and any $m=0,\dots,\varkappa-1$. Then there exist vectors $w_i^{(k,m)}$ ($k=0,\dots,\nu_i-1;\ m=0,\dots,\varkappa-1$) such that (3.8), (3.8′) are fulfilled and

$$z_i(n\varkappa+m) = \bar{x}_i(n\varkappa+m) - \sum_{k=0}^{\nu_i-1} \binom{n\varkappa+m}{k} w_i^{(k,m)} \quad \text{are bounded.}$$

<u>Proof</u>. Supposing that $w_i^{(k,m)}$'s are selected such that (3.5), (3.5′) hold, as F is finite there exists $\bar{n}_i<\infty$ such that for any $\bar{n}\geq\bar{n}_i$ (where $\bar{n}=n\varkappa+m$) $s_i^{(\bar{n},m)}(f^{(\bar{n})})\leqq 0$. So by (3.7) for any $\bar{n}\geqq\bar{n}_i$ we get

$$\bar{Q}_{ii}^{(\varkappa)}(\hat{f}^{(\bar{n})})\; z_i(\bar{n}) + \bar{o}_i^{(\varkappa)}(\hat{f}^{(\bar{n})};\bar{n}) \geqq z_i(\bar{n}+\varkappa) \geqq$$

$$\geqq \bar{Q}_{ii}^{(\varkappa)}(f_m^*)\; z_i(\bar{n}) + \bar{o}_i^{(\varkappa)}(f_m^*;\bar{n}) . \qquad (3.9)$$

Recalling that by (2.4) $\bar{Q}_{ii}^{(t)}(f^{(n)})\, u_i \leq u_i$ for any $t = 1,2,\ldots$ and by (3.5′) $\bar{o}_i^{(\varkappa)}(.;\bar{n}) \to 0$ exponentially as $\bar{n} \to \infty$, on iterating (3.9) and using the above facts we conclude that $z_i(n\varkappa + m)$ is bounded in n for all $m = 0,\ldots,\varkappa - 1$. □

In virtue of boundedness of $z_i(n\varkappa + m)$, the same arguments as in the proof of Theorem 3.2 let us conclude that $s_i^{(\bar{n},m)}(\hat{f}^{(\bar{n})})$ is bounded; hence (cf. (3.7)) $s_i^{(\bar{n},m)}(\hat{f}^{(\bar{n})}) = b_i^{(0,m)}(\hat{f}^{(\bar{n})})$ for sufficiently large $\bar{n} = n\varkappa + m$. Applying these results to (3.7) we immediately get

<u>Corollary 3.3</u>. There exists $\bar{n}_i < \infty$ such that for any $\bar{n} = n\varkappa + m \geq \bar{n}_i$

$$z_i(\bar{n} + \varkappa) = \bar{Q}_{ii}^{(\varkappa)}(\hat{f}^{(\bar{n})})\, z_i(\bar{n}) + b_i^{(0,m)}(\hat{f}^{(\bar{n})}) + \bar{o}_i^{(\varkappa)}(\hat{f}^{(\bar{n})};\bar{n}) .$$

3.3. <u>Asymptotic Behaviour of the Utility Vector</u>

Having constructed in Section 3.1 polynomial bounds on $\bar{x}_i(n\varkappa + m)$, to establish asymptotic properties of $\bar{x}_i(n)$ with $\nu_i > 1$ it only suffices to show (cf. Theorem 3.2) that for suitably selected $w_i^{(0,m)}$'s and $\varkappa_i = \tilde{\varkappa}_i \varkappa$, $z_i(n\varkappa_i + m) \to 0$ for any $m = 0,\ldots,\varkappa_i - 1$. The number $\tilde{\varkappa}_i$ will heavily depend on the periodicity of a suitable convex combination of $Q_{ii}^{(\varkappa)}(f)$'s with $f \in \bar{F}$; recall that a class of $Q_{ii}^{(\varkappa)}(f)$ will be called basic, iff its spectral radius equals $\sigma_i^{\varkappa}$. In particular, considering $Q_{ii}^{(\varkappa)}(f)$'s with $f \in F^{(0,m)}$ similarly as in Section 3.1 we can construct a convex combination of $Q_{ii}^{(\varkappa)}(f)$'s with $f \in F^{(0,m)}$, say $Q_{ii}^{(\varkappa)}(\tilde{f})$, whose periodicity is equal to $\tilde{\varkappa}_i$, such that

(i) The states that may belong to some basic class of $Q_{ii}^{(\varkappa)}(f)$ with $f \in F^{(0,m)}$ are contained within some basic class of $Q_{ii}^{(\varkappa)}(\tilde{f})$.

(ii) There exists no other convex combination of $Q_{ii}^{(\varkappa)}(f)$'s with $f \in F^{(0,m)}$ fulfilling condition (i) whose periodicity is less than that of the matrix $Q_{ii}^{(\varkappa)}(\tilde{f})$.

Asymptotic properties of $\bar{x}_i(n)$ can be summarized as

<u>Theorem 3.4</u>. Let for $Q_{ii}(f)$ occurring in (3.1) $\nu_i > 1$, (3.5), (3.5′) hold for $j = i+1,\ldots,i+\nu_i - 1$ and let $\tilde{\varkappa}_i$ be the period of $Q_{ii}^{(\varkappa)}(\tilde{f})$. Then for $\varkappa_i = \tilde{\varkappa}_i \varkappa$ there exist vectors $w_i^{(0,m)}$ $(m = 0,\ldots,\varkappa_i - 1)$ such that for $n \to \infty$

$$z_i(n\varkappa_i+m) = \bar{x}_i(n\varkappa_i+m) - \sum_{k=0}^{\nu_i-1} \binom{n\varkappa_i+m}{k} w_i^{(k,m)} \longrightarrow 0 \qquad (3.10)$$

exponentially fast.

The proof of Theorem 3.4 is sketched in the Appendix (cf. Proposition A.2, observe that we can restrict ourselves to the recursion presented in Corollary 3.3 and employ estimates (3.9)).

Combining Theorems 3.1, 3.2 and 3.4 we immediately get

Corollary 3.5. For any $i = 1,\ldots,s$ there exist naturals $\varkappa_i$ and vectors $w_i^{(k,m)}$ $(k = 0,\ldots,\nu_i-1;\ m = 0,\ldots,\varkappa_i-1)$ such that (3.10) holds.

Remark 3.6. Observe that $\varkappa_i = \tilde{\varkappa}_i \ldots \tilde{\varkappa}_{i+\nu_i-1}$, where $\tilde{\varkappa}_j$'s depend on the periodicity of appropriate convex combinations of $Q_{jj}(f)$ with $f\in\bar{F}$. In particular, if all $Q_{jj}(f)$ with $f\in\bar{F}$ are aperiodic, then $\varkappa_i = 1$ and $\bar{x}_i(n)$ converges to some polynomial of degree ν_i-1. Considering the model with s=2, $Q_{11}(f)$ stochastic and $Q_{22}(f) = 1$ (hence $Q_{12}(f)$ is a column vector), then conditions (3.5), (3.5′) are trivially fulfilled, (3.1) reduces to the functional equation for maximum expected rewards of a Markov decision chain and the results of Theorem 3.4 are well-known from the dynamic programming literature.

APPENDIX

Consider $\{Q(f), f\in F\}$ such that $Q(f)\,u \leq Q(\hat{f})\,u = u$ for some $u \gg 0$, $\hat{f}\in F$ and any $f\in F$. So (cf. Theorem 2.1) $Q(f)$'s are positively similar to (sub)-stochastic matrices; $u^{(k)}$, $c^{(k)}(f)$ will denote column vectors of compatible dimensions.

Proposition A.1. Let $c^{(k)}(f)$'s and the natural $\varkappa$ be given. Then there exist $u^{(k)}$'s and $\bar{f}\in F$ such that $b^{(k)}(\bar{f}) = 0$ (for $k = 0,\ldots,r$), and for all $f\in F$ $(b^{(r)}(f), \ldots, b^{(0)}(f)) \precsim 0$ (i.e. lexicographically non-positive) where $(c^{(r)}(f) \equiv 0)$

$$b^{(k)}(f) = (Q(f) - I)\,u^{(k)} - \sum_{p=1}^{r-k} \binom{\varkappa}{p} u^{(k+p)} + c^{(k)}(f).$$

The proof can be performed by policy iterations similarly as for $\varkappa = 1$ in Sladký (1981).

Now (cf. Corollary 3.3) consider $z(n)$ (bounded in n) calculated recursively by $z(n+1) = \max_{f\in F}\left[Q(f)\,z(n) + b(f) + o(f;n)\right]$

where $b(f) \leq 0$ (with $b(f) = 0$ iff $f \in F^{O} \subset \bar{F}$) and $o(f;n) \to 0$ exponentially fast. Let $Q(\tilde{f})$ (with period $\varkappa$) be a convex combination of $Q(f)$'s with $f \in F^{O}$ such that each state belonging to some basic class of $Q(f)$ with $f \in F^{O}$ is contained within some basic class of $Q(\tilde{f})$.

Proposition A.2. For any $m = 0,\ldots,\varkappa-1$, $z(n\varkappa+m) \to z^{(m)}$ as $n \to \infty$ and the convergence is exponentially fast.

To show that $z^{(m)}$'s exist we can proceed similarly as in Schweitzer, Federgruen (1977). Then considering the $\varkappa$-step decisions in the recursive relation for $z(n\varkappa+m) - z^{(m)}$ we can show that the convergence is exponential. The first published result in this direction seems to be that of Schweitzer, Federgruen (1981), different (and simpler) proofs can be found in Zijm (1982) and Sladký (1983).

REFERENCES

Gantmakher, F.R. (1966). Teoriya matric. Second edition, Nauka, Moscow.

Rothblum, U.G. and Whittle, P. (1982). Growth optimality for branching Markov decision chains. Mathematics of Operations Research, 7(4):582-601.

Schweitzer, P.J. and Federgruen, A. (1977). The asymptotic behavior of undiscounted value iteration in Markov decision problems. Mathematics of Operations Research, 2(4):360-381.

Schweitzer, P.J. and Federgruen, A. (1981). Nonstationary Markov decision problems with converging parameters. Journal of Optimization Theory and Applications, 34(2):207-241.

Sladký, K. (1976). On dynamic programming recursions for multiplicative Markov decision chains. Mathematical Programming Study, 6:216-226.

Sladký, K. (1980). Bounds on discrete dynamic programming recursions I, models with non-negative matrices. Kybernetika, 16(6):526-547.

Sladký, K. (1981). Bounds on discrete dynamic programming recursions II, polynomial bounds on problems with block-triangular structure. Kybernetika, 17(4):310-328.

Sladký, K. (1983). Bounds on convergence rates of undiscounted Markov decision chains. In J.Beneš, L.Bakule (Eds.), Fourth Formator Symposium on Mathematical Methods for the Analysis of Large Scale Systems, Academia, Prague.

Whittle, P. (1983). Optimization over Time - Dynamic Programming and Stochastic Control, Vol II. Wiley, Interscience.

Zijm, W.H.M. (1982). Nonnegative Matrices in Dynamic Programming. Mathematical Centre Tract, Amsterdam.

RISK-SENSITIVE AND HAMILTONIAN FORMULATIONS IN OPTIMAL CONTROL

P. Whittle
Statistical Laboratory, University of Cambridge

1. INTRODUCTION

This paper is an amalgam of two sets of ideas. One is the introduction into LQG (linear/quadratic/Gaussian) theory of the notion of *risk-sensitivity* by use of an exponential-quadratic rather than a simply quadratic criterion. In this way one can introduce a degree of optimism or pessimism on the part of the controller, and so significantly generalise the classic LQG theory.

The other element is the use of an extended Hamiltonian formulation for non-Markov models. This approach has been followed consistently by Whittle (1983, Chapters 11 and 12) in the contexts of both estimation and control. It now turns out that this theory has a natural and illuminating risk-sensitive generalisation.

2. THE RISK-SENSITIVE CERTAINTY EQUIVALENCE PRINCIPLE

The validity of an appropriate certainty equivalence principle turns out to be crucial. The classic risk-neutral principle (due originally to Theil (1957)) has a direct but unobvious risk-sensitive version.

State structure is irrelevant; we give the principle in its most general finite-horizon form. We make the assumptions

(a) The actions u_t to be taken over a finite horizon $(0 \le t \le h)$ take values in finite-dimensional vector spaces.

(b) The action u_t can be a function only of observables W_t at t : previous actions $U_{t-1} = \{u_0, u_1, \ldots, u_{t-1}\}$ and observation history $Y_t = \{y_0, y_1, \ldots, y_t\}$ $(0 \le t \le h)$.

(c) The cost function $\mathbb{C}$ is a quadratic function $Q(U_{h-1}, \xi)$ of the control sequence $U_{h-1} = \{u_0, u_1, \ldots, u_{h-1}\}$ and an exogenous noise vector ξ , positive definite in U_{h-1} for all ξ .

(d) ξ is normally distributed with known parameters, independent of policy.

(e) Each y_t is a policy-independent linear function of ξ.

The vector ξ is considered to embody all exogenous stochastics of the problem: process noise, observation noise, and the stochastics of a reference path which one may wish the controlled system to follow. The actual observations y will in general depend upon control actions, but it is enough that these can be corrected to exhibit the observations as being effectively linear functions of ξ.

The conventional (risk-neutral) criterion is that one chooses policy π to minimise $E_\pi(\mathbb{C})$, where E_π is the expectation operator induced by π. The classic certainty equivalence principle is then that the optimal value of u_t is determined by minimising $Q(U_{h-1},\xi^{(t)})$ with respect to $u_t, u_{t+1}, \ldots, u_{h-1}$, where

$$\xi^{(t)} = E(\xi|Y_t) \tag{1}$$

is the optimal estimator of ξ based upon W_t. ("Optimal" in that it has minimum mean square error; it is also the maximum likelihood estimate.)

Note that the certainty equivalence principle has two features:

(i) Conversion to free form. A minimisation with respect to functions $u_\tau(W_\tau)$ $(\tau\geq t)$, constrained in that u_τ may depend only upon W_τ, is replaced by a free minimisation with respect to constants u_τ.

(ii) Separation. Optimisation of estimation and control are separated, in that the estimate (1) is derived without reference to the determination of the u_t, and the u_t are determined as they would be in the full-information case, with the simple substitution of $\xi^{(t)}$ for ξ.

Suppose now that the criterion that $E_\pi(\mathbb{C})$ be minimised is modified to the criterion that

$$\gamma_\pi(\theta) = -\frac{2}{\theta}\log E_\pi e^{-\theta\mathbb{C}/2} \tag{2}$$

be minimised with respect to π. Here θ is a scalar parameter, the *risk-sensitivity* parameter. The case $\theta=0$ corresponds to the *risk-neutral* case

$$\gamma_\pi(0) = E_\pi(\mathbb{C}) .$$

In the case $\theta > 0$ the optimiser is *risk-seeking*; he is more concerned to reduce the frequent occurrence of moderate values of $\mathfrak{C}$ than the occasional occurrence of large values. In the case $\theta < 0$ the reverse is the case, and the optimiser is *risk-averse*.

It is remarkable that the features of the well-known risk-neutral case have a version for the risk-sensitive case, although these are transformed sufficiently that they are not immediately evident. To take the analysis beyond the point reached by Jacobson (1973, 1977) required a certainty equivalence principle, which Whittle proved, first for the state-structured case (1981, 1982, 1983), and then for the general case (1985). However, the line of proof for the general case is quite clear from those proofs already published for the state-structured case.

Suppose that the exogenous noise vector ξ is normally distributed with zero mean and covariance matrix V. Define

$$\mathfrak{D} = \xi' V^{-1} \xi \tag{3}$$

recognisable as occurring in the exponent of the ξ-density, and define the *total stress*,

$$\mathfrak{S} = \mathfrak{C} + \frac{1}{\theta}\mathfrak{D} \quad . \tag{4}$$

This is a spontaneously occurring combination when one evaluates the expectation in (2). $\mathfrak{C}$ is the component of stress due to cost (e.g. to departures of u from zero) and $\mathfrak{D}$ the component due to implausibility, (i.e. to departures of ξ from zero).

We shall use the term *extremisation* to denote an operation which is minimisation when $\theta \geq 0$ and maximisation when $\theta < 0$.

Theorem 1. The risk-sensitive certainty equivalence principle. Suppose that one wishes to choose policy π *to minimise criterion (2). Then under assumptions (a)-(e) above the optimal value of* u_t *is determined by simultaneously minimising* $\mathfrak{S}$ *with respect to* $u_t, u_{t+1}, \ldots, u_{h-1}$ *and extremising it with respect to* $y_{t+1}, y_{t+2}, \ldots, y_h$ *. In words: one minimises stress with respect to all decisions currently unmade and extremises it with respect to all quantities currently unobservable.*

This principle really does reduce to the risk-neutral version as $\theta \downarrow 0$. The stress-extremising value of unobservables tends to that minimising for a given value of Y_t, this leads exactly to the estimate $\xi^{(t)}$ of ξ.

One can ask in what sense this is a certainty equivalence principle. It certainly has the property (i) above, of conversion to free form. That is, minimisation of γ_π with respect to policies π, i.e. with respect to functions $u_\tau(W_\tau)$, has been replaced by a free minimisation/extremisation of stress with respect to relevant decisions/unobservables.

It does not have property (ii), of separation. Determination of optimal control u_t and of an effective current estimate of ξ are intertwined in the minimisation/extremisation of stress.

This fact is inevitable. If "separation" is a meaningful concept at all, it must surface in another and less evident form. This it does, as demonstrated in Whittle (1981). In the state-structured case one can evaluate extremal values of past stress and future stress at time t separately, conditional on the true (but in general unknown) value of current state x_t. This achieves separation, in that one has two decoupled calculations of the familiar recursive form which separately yield an optimal condensation of data and an optimal determination of control, both parametrised by the "pivot" x_t. Let the optimal control thus determined be $u_t(x_t)$. One now recouples these two calculations by choosing x_t to extremise total stress, already determined parametrically in terms of x_t. If the estimate thus yielded is $x_t^{(t)}$ then the optimal control is $u_t(x_t^{(t)})$. This is very obviously a certainty-equivalence statement. Moreover, separation holds in the sense that optimisation of estimation and control have been decoupled by parametrisation of these sub-problems in terms of x_t.

3. THE HAMILTONIAN FORMULATION

We suppose process variable x and control variable u and observation y with respective (vector) values x_t and u_t, time t, supposed discrete. We shall assume the conventional quadratic cost function

$$\mathbb{C} = \sum_{t=h_1}^{h_2-1} (x'Rx + u'Sx + x'S'u + uQu)_t + (x'\Pi x)_{h_2} \tag{5}$$

where the notation $(\ )_t$ implies that all quantities inside the bracket are evaluated at time t. The times h_1 and h_2 are then respectively the initial point and the horizon point, at which costing and decision respectively begin and cease.

The state-structured case was analysed by Whittle (1981) and produced analogues of the Kalman filter, the forward and backward Riccati equations etc. Our aim is now to consider a non-Markov plant equation of the form

$$A(T)x_t + B(T)u_t = \varepsilon_t \tag{6}$$

Here ε is Gaussian white noise with covariance matrix N, T is the backwards translation operator and the matrix operator coefficients have the form $A(T) = \sum_{j=0}^{\infty} A_j T^j$, $B(T) = \sum_{j=1}^{\infty} B_j T^j$.

For simplicity we assume state observable at unit lag; the more general case of imperfect state observation is considered in Whittle and Kuhn (1986).

Let us write the operator $A(T) = \Sigma A_j T^j$ simply as A, and the associated operator $A(T^{-1})' = \Sigma A_j' T^{-j}$ as $\bar{A}$. The process equation (6) can then be written

$$(Ax + Bu - \varepsilon)_t = 0 \tag{7}$$

We have

$$\mathfrak{S} = \mathfrak{C} + \frac{1}{\theta} \sum_{h_1}^{h_2} (\varepsilon' N^{-1} \varepsilon)_t \tag{8}$$

where $\mathfrak{C}$ is given by (5).

If ζ_τ is a quantity defined at time τ let us use $\zeta_\tau^{(t)}$ to denote its "minimal-stress" estimate based on observations available at time t.

Theorem 2. The optimal value of u_t *is the quantity* $u_t^{(t)}$ *determined by solution of the equations*

$$\begin{bmatrix} R & S' & \bar{A} \\ S & Q & \bar{B} \\ A & B & -\theta N \end{bmatrix} \begin{bmatrix} x \\ u \\ \lambda \end{bmatrix}_\tau^{(t)} = 0 \qquad (t \le \tau < h_2) \tag{9}$$

with terminal conditions

$$\lambda_{h_2}^{(t)} + \Pi x_{h_2}^{(t)} = 0 \qquad (10)$$

$$\lambda_{\tau}^{(t)} = 0 \qquad (\tau > h_2) \qquad (11)$$

The quantity $\lambda_{\tau}^{(t)}$ *has the interpretation*

$$\varepsilon_{\tau}^{(t)} = \theta N \lambda_{\tau}^{(t)} \qquad (12)$$

The proof follows by an application of the risk-sensitive certainty-equivalence principle with the evaluation (8) of stress.

The striking feature is the symmetric nature of the linear equation system (9). Relations (9)-(11) constitute the optimality equation of a stochastic maximum principle. Indeed, the risk-sensitive certainty-equivalence principle *is* the stochastic maximum principle for this model.

Interpretation (12) is interesting: that a quantity such as λ (usually appearing as the Lagrangian multiplier associated with the constraint of a deterministic plant equation) should be related to the minimal-stress estimate of process noise.

Let us write equation (9) as

$$\Phi(T)\zeta_{\tau}^{(t)} = 0 \qquad (t \le \tau < h_2) \qquad (13)$$

Suppose that $\Phi(z)$ has canonical factorisation

$$\Phi(z) = \Phi_{-}(z)\Phi_{+}(z) \qquad (14)$$

where Φ_{+}, Φ_{+}^{-1} are analytic in $|z| \le 1$, etc. Then, under generalised controllability hypotheses $\zeta_{\tau}^{(t)}$ goes to zero sufficiently fast with increasing τ that one can conclude from (13) that

$$\Phi_{+}(T)\zeta_{\tau}^{(t)} = 0 \qquad (\tau \ge t) \qquad (15)$$

in the infinite horizon. Equation (15) for $\tau = t$ determines $u_t^{(t)}$, the optimal control at t, explicitly.

If state observation is imperfect then one deduces a pair of coupled systems of type (9), both reducible as in (15) under generalised controllability/observability hypotheses (see Whittle and Kuhn (1986)).

REFERENCES

Jacobson, D.H. (1973). Optimal stochastic linear systems with exponential performance criteria and their relation to deterministic differential games. IEEE Trans. Automatic Control. AC-18: 124-131.

Jacobson, D.H. (1977). Extensions of linear-quadratic control, optimization and matrix theory. Academic Press.

Theil, H. (1957). A note on certainty equivalence in dynamic planning. Econometrica, 25: 346-349.

Whittle, P. (1981). Risk-sensitive linear/quadratic/Gaussian control. Adv. Appl. Prob., 13: 764-777.

Whittle, P. (1982). Optimization over time, Vol.1. Wiley Interscience.

Whittle, P. (1983). Prediction and regulation by linear least square methods. (2nd edition of 1963 volume). University of Minnesota Press.

Whittle, P. (1985). The risk-sensitive certainty equivalence principle. In Essays in Time Series and Allied Processes. Published by Applied Probability Trust.

Whittle, P. and Kuhn, J. (1986). A Hamiltonian formulation of risk-sensitive linear/quadratic/Gaussian control. Int. J. Control. (To appear).

MARTINGALES IN SURVIVAL ANALYSIS

Anatoli I. Yashin
IIASA, A-2361 Laxenburg, Austria

1. INTRODUCTION

The problems of explaining the observed trends in mortality, morbidity and other kinds of individuals' transitions generated the numerous attempts of incorporating the covariates into the survival models. First models use the deterministic constant factors as explanatory variables (Cox 1972, 1975; Bailey 1983). Gradually it became clear that the random and dynamic nature of the covariates should also be taken into account (Anderson and Gill 1981; Prentice and Self 1983). This understanding has led to the fact that the notion of random intensity became widely used in the analysis of the asymptotic properties of the maximum likelihood and Cox-regression estimators (Cox and Oakes 1984; Elandt-Johnson and Johnson 1979).

Having the clear intuitive sense the notion of random intensity can be introduced in different ways. The traditional way is to define the intensity in terms of probability distributions of the failure time (Barlow and Proschan 1975; Lawless 1982; Nelson 1982). Another way appeals to the martingale theory (Jacod 1979) and defines the intensity in terms of the predictable process, called "compensator" (Liptzer and Shiryaev 1978; Jacod 1975). For the deterministic rates and simple cases of stochastic intensities there are already results that establish a one-to-one correspondence between two definitions. The correspondence is reached by the probabilistic representation results for compensator (Liptzer and Shiryaev 1978; Jacod 1975). Martingale theory guarantees the existence of the predictable compensator in more general cases. However the results on the probabilistic representation similar to simple cases are still unknown. Meanwhile such representation is crucial, for instance, in the analysis of the relations between the duration of the life cycle of some unit and stochastically changing influential variables. This paper shows the result of such representation for some particular case. The generalization on the more general situations is straightforward. The consideration will use some basic notions of a "general theory of processes" (Jacod 1979; Dellacherie 1972).

2. PRELIMINARIES

Let T be the stopping time defined on some probability space $(\Omega, H, \mathbf{H}, P)$, where $\mathbf{H} = (H_t)_{t \geq 0}$ is the right-continuous nondecreasing family of σ-algebras in Ω, such that $H_\infty = H$ and H_0 is completed by P-zero sets from H. If distribution of T is absolutely continuous the traditional definition of the intensity $\lambda(t)$, related to the stopping time T is as follows

$$\lambda(t) = \frac{\frac{d}{dt} P(T \leq t)}{P(T > t)} \quad . \tag{1}$$

Martingale characterization defines the rate in terms of the process $A(t)$ which is supposed to be H-predictable and such that the process $M(t)$ defined as

$$M(t) = I(T \leq t) - A(t)$$

is an H-adapted martingale. It turns out that for process $M(t)$ to be martingale, $A(t)$ should have the form

$$A(t) = \int_0^t I(u \leq T) \lambda(u) du \quad .$$

where $\lambda(u)$ is given by (1). The family H in this case is generated by the indicator process $X_t = I(t \geq T)$. If the stopping time T is correlated with some random variable $z(\omega)$, then the traditional approach defines the intensity in terms of conditional probabilities

$$\lambda(t,z) = \frac{\frac{d}{dt} P(T \leq t \mid z)}{P(T > t \mid z)} \quad . \tag{2}$$

Martingale characterization shows that the process $M^z(t)$

$$M^z(t) = I(T \leq t) - \int_0^t I(u \leq T) \lambda(u,z) du$$

is a martingale with respect to the family of σ-algebras $\mathbf{H}^z$, generated in Ω by the indicator process $I(t \geq T)$ and random variable z. In the case of a discontinuous conditional distribution function for T, formula (2) should be corrected. The notion of cumulated intensity $\Lambda(t,z)$ is more appropriate in this case. The formula for it is

$$\Lambda(t,z) = \int_0^t \frac{dP(T \leq u \mid z)}{P(T \geq u \mid z)} \tag{3}$$

and martingale characterization is respectively (Liptzer and Shiryaev 1978; Jacod 1975):

$$M^z(t) = I(T \le t) - \int_0^t I(u \le T) d\Lambda(u,z) \quad . \tag{4}$$

The situation is not so clear however if one has the random processes, say Y_t, correlated with stopping time T. Assume for instance that Y_t simulates the changes of the physiological variables of some patient in hospital and T is the time of death. It is clear that in this case the process Y_t should terminate at time T, so observing Y_t one can tell about the alive/dead state of the patient, that is, can observe the death time.

It is clear also that the state or the whole history of the physiological variable influences the chances of occurring death. The question is how can one specify the random intensity in terms of conditional probabilities in order to establish the correspondence between intuitive traditional and martingale definitions of the random intensities.

The idea to use $\Lambda(t,Y)$ in the form

$$d\Lambda(t,Y) = \frac{dP(T \le t \mid H_t^y)}{P(T \ge t \mid H_t^y)}$$

where σ-algebra H_t^y is generated in Ω by the process Y_u up to time t fails because H_t^y contains the event $\{T \le t\}$. Taking H_{t-}^y instead of H_t^y seems to improve the situation, however the event $\{T \ge t\}$ is measurable with respect to H_{t-}^y.

So to find the proper formula for random intensity one needs to get the probabilistic representation result for H^y-predictable compensator. The representation result will be used in calculating the new version of the Cameron and Martin result (Cameron and Martin 1944, 1949).

3. REPRESENTATION OF COMPENSATORS

We will demonstrate the result and the ideas of proof on a simple particular case. The generalization on a more general situation is straightforward.

Assume that stopping time $T(\omega)$ and the Wiener process $W_t(\omega)$ are defined on some probability space (Ω,H,P). Define

$$Y_t = W_t - I(T \le t)W_t \quad .$$

Let $\mathbf{H}^y = (H_t^y)_{t\ge0}$, $\mathbf{H} = (H_t^w)_{t\ge0}$ where

$$H_t^y = \sigma\{Y_u, u \le t\}$$

$$H_t^w = \sigma\{w_u, u \le t\} \ .$$

The following statement is true.

Theorem 1. $\mathbf{H}^y$*-predictable compensator* $A^y(t)$ *of the process* $I(T \le t)$ *have the following representation*

$$A^y(t) = \int_0^t I(u \le T) \frac{d_u P(T \le u \mid H_u^w)}{P(T \ge u \mid H_u^w)} \ . \tag{5}$$

The proof of this theorem can be found in (Yashin 1985).

Remark. It turns out that even if the σ-algebra H_t^y has the more general structure than in the example above, for instance, $H_t^y = H_t^x \vee H_t^w$, the result of the probabilistic representation of the H^y-predictable compensator of the process $X(t) = I(T \le t)$ given by formula (5) is true.

Generalization of formula (5) for the compensator plays the key role in developing a new approach to the solution of the Cameron and Martin problem (Cameron and Martin 1944, 1949).

4. THE CAMERON AND MARTIN RESULT

The well-known Cameron and Martin formula (Cameron and Martin 1944, 1949; Liptzer and Shiryaev 1978) gives a way of calculating the mathematical expectation of the exponent which is the functional of a Wiener process. More precisely, let $(\Omega, H, \mathbf{P})$ be the basic probability space, $\mathbf{H} = (H_u)_{u \ge 0}$ be the nondecreasing right-continuous family of σ-algebras, and H_0 is completed by the events of $\mathbf{P}$-probability zero from $H = H_\infty$. Denote by W_u n-dimensional $\mathbf{H}$-adapted Wiener process and $Q(u)$ a symmetric non-negative definite matrix whose elements $q_{i,j}(u), i,j = 1,2,\dots,n$ satisfy for some t the condition

$$\int_0^t \sum_{i,j=1}^n |q_{i,j}(u)| du < \infty. \tag{6}$$

The following result is known as a Cameron-Martin formula.

Theorem 2. *Let (6) be true. Then*

$$\mathbf{E} \exp[-\int_0^t (W_u, Q(u) W_u) du] = \exp[\frac{1}{2} \int_0^t Sp\, \Gamma(u) du] \ . \tag{7}$$

where $(W_u, Q(u)W_u)$ *is the scalar product equal to* $W_u^*, Q(u)W_u$, *and* $\Gamma(u)$ *is a symmetric nonpositive definite matrix, being a unique solution of the Ricatti matrix equation*

$$\frac{d\Gamma(u)}{du} = 2Q(u) - \Gamma^2(u); \tag{8}$$

$\Gamma(t) = 0$ *is a zero matrix.*

The proof of this formula given in (Liptzer and Shiryaev 1978) uses the property of likelihood ratio for diffusion type processes. The idea of using this approach comes from Novikov (Novikov 1972). Using this idea Myers in (Myers 1981) developed this approach and found the formula for averaging the exponent when, instead of a Wiener process, there is a process satisfying a linear stochastic differential equation driven by a Wiener process. His result may be formulated as follows.

Theorem 3. *Let* $Y(t)$ *be an* m*-dimensional diffusion process of the form*

$$dY(t) = a(t)Y(t)dt + b(t)dW_t,$$

with deterministic initial condition $Y(0)$. *Assume that matrix* $Q(u)$ *has the properties described above. Then the next formula is true:*

$$\mathbf{E}\exp[-\int_0^t Y^*(u)Q(u)Y(u)du] = \tag{9}$$

$$\exp[Y^*(0)\Gamma(0)Y(0) + Sp\int_0^t b(u)b^*(u)\Gamma(u)du]$$

where $\Gamma(u)$ *is the solution of matrix Ricatti equation*

$$\frac{d\Gamma(u)}{du} = Q(u) - (\Gamma(u) + \Gamma^*(u))a(u) - \tag{10}$$

$$\frac{1}{2}(\Gamma(u) + \Gamma^*(u))b(u)b^*(u)(\Gamma(u) + \Gamma^*(u)),$$

with the terminal condition $\Gamma(t) = 0$.

These results have direct implementation to survival analysis: any exponent on the left-hand sides of (7) and (9) can be considered as a conditional survival function in some life cycle problem (Myers 1981; Woodbury and Manton 1977; Yashin 1983). The stochastic process in the exponent is interpreted in terms of spontaneously changing factors that influence mortality or failure rate.

Such interpretation was used in some biomedical models. The quadratic dependence of risk from some risk factors was confirmed by the results of numerous physiological and medical studies (Woodbury and Manton 1977). The results are also applicable to the reliability analysis.

The way of proving the Cameron-Martin formula and its generalizations given in (Cameron and Martin 1944, 1949; Liptzer and Shiryaev 1978) does not use an interpretation and unfortunately does not provide any physical or demographical sense to the variables $\Gamma(u)$ that appear on the right-hand side of the formulas (7) and (9). Moreover, the form of the boundary conditions for equation (8) and (10) on the right-hand side complicate the computing of the Cameron-Martin formula when one needs to calculate it on-line for many time moments t. These difficulties grow when there are some additional on-line observations correlated with the influential factors.

5. NEW APPROACH

Fortunately there is the straightforward method that allows avoidance of these complications. The approach uses the innovative transformations random intensities or compensators of a point process. Usage of this "martingale" techniques allows to get a more general formula for averaging exponents which might be a more complex functional of a random process from a wider class.

The following general statement gives the principal new solution of the Cameron and Martin problem.

Theorem 4. *Let* $Y(u)$ *be an arbitrary* **H**-*adapted random process and* $\lambda(Y,u)$ *is some non-negative* $\mathbf{H}^y$*-adaptive function such that for some* $t \geq 0$

$$\mathbf{E}\int_0^t \lambda(Y,u)du < \infty \tag{11}$$

Then

$$\mathbf{E}\exp[-\int_0^t \lambda(Y,u)du] = \exp[-\int_0^t \mathbf{E}\,[\lambda(Y,u)\,|\,\mathrm{T} > u]du] \tag{12}$$

where T *is the stopping time associated with the process* $Y(u)$ *as follows:*

$$\mathbf{P}(\mathrm{T} > t \mid H_t^y) = \exp[-\int_0^t \lambda(Y,u)du] \tag{13}$$

and $H_t^y = \bigcap_{u>t} \sigma\{Y(v), v \leq u\}$ *is* σ*-algebra generated by the history of the process* $Y(u)$ *up to time* t, $\mathbf{H}^y = (H_t^y)_{t\geq 0}$.

The proof of this statement based on the idea of "innovation", widely used in martingale approach to filtration and stochastic control problems (Liptzer and Shiryaev

1978; Yashin 1970; Bremaud 1980) is given in (Yashin 1984).

Another form of this idea appeared and was explored in the demographical studies of population heterogeneity dynamics (Yashin 1983; Vaupel, Manton, and Stallard 1979; Vaupel and Yashin 1983). Differences among the individuals or units in these studies were described in terms of a random heterogeneity factor called "frailty". This factor is responsible for individuals' susceptibility to death and can change over time in accordance with the changes of some external variables, influencing the individuals' chances to die (or to have failure for some unit if one deals with the reliability studies).

When the influence of the external factors on the failure rate may be represented in terms of a function which is a quadratic form of the diffusion type Gaussian process, the result of Theorem 4 may be developed as follows:

Theorem 5. *Let the m-dimensional* **H**-*adapted process* $Y(u)$ *satisfy the linear stochastic differential equation*

$$dY(t) = [a_0(t) + a_1(t)Y(u)]dt + b(t)dW_t, \quad Y(0) = Y_0,$$

where Y_0 *is the Gaussian random variable with mean* m_0 *and variance* γ_0. *Denote by* $Q(u)$ *a symmetric non-negative definite matrix whose elements satisfy condition (6). Then the next formula is true*

$$\mathbf{E}\exp[-\int_0^t (Y^*(u)Q(u)Y(u))du] = \exp[-\int_0^t (m_u^* Q(u)m_u + Sp(Q(u)\gamma_u))du]. \tag{14}$$

The processes m_u *and* γ_u *are the solutions of the following ordinary differential equations:*

$$\frac{dm_t}{dt} = a_0(t) + a_1(t)m_t - 2\gamma_t Q(t)m_t \tag{15}$$

$$\frac{d\gamma_t}{dt} = a_1(t)\gamma_t + \gamma_t a_1^*(t) + b(t)b^*(t) - 2\gamma_t Q(t)\gamma_t \tag{16}$$

with the initial conditions m_0 *and* γ_0, *respectively.*

The proof of this theorem is based on the Gaussian property of the conditional distribution function $\mathbf{P}(Y(t) \le x \mid T > t)$ and is given in (Yashin 1984). This situation recalls the well-known generalization of the Kalman filter scheme (Liptzer and Shiryaev 1978; Liptzer 1975; Yashin 1980; Yashin 1982).

Note that a similar approach to the averaging of the survival function was studied in (Woodbury and Manton 1977) under the assumption that the conditional Gaussian property take place.

Assume that $\mathbf{Z}_t$ is a finite state continuous time Markov process with vector initial probabilities $p_1,...,p_K$ and intensity matrix

$$\mathbf{R}(t) = ||r_{i,j}(t)||, \quad i,j = \overline{1,K} \quad t \geq 0.$$

with bounded elements for any $t \geq 0$. The process $\mathbf{Z}_t$ can be interpreted as a formal description of the individuals' transitions from one state to another in the multistate population model. Denote $H_t^z = \sigma\{\mathbf{Z}_u, u \leq t\}$. The following statement is the direct corollary of Theorem 4 (Yashin 1984).

Theorem 6. *Let the process* $\mathbf{Z}_t$ *be associated with the death time* T *as follows:*

$$\mathbf{P}(\mathrm{T} > t \mid H_t^z) = \exp[-\int_0^t \lambda(\mathbf{Z}_u, u)du].$$

Then the next formula is true:

$$\mathbf{E}\exp[-\int_0^t \lambda(\mathbf{Z}_u, u)du] = \exp[-\int_0^t \sum_{i=1}^{i=K} \lambda(i,u)\pi_i(u)du]$$

where the $\pi_i(t)$ *are the solutions of the following system of the ordinary differential equations:*

$$\frac{d\pi_j(t)}{dt} = \sum_{i=1}^{i=K} \pi_i(t)r_{i,j}(t) - \pi_j(t)[\lambda(j,t) - \sum_{i=1}^{i=K} \lambda(i,t)\pi_i(t)],$$

with $\pi_j(0) = p_j$.

The variables $\pi_j(t)$, $j = \overline{1,K}$ can be interpreted as the proportions of the individuals in different groups at time t.

6. FURTHER GENERALIZATIONS

Suppose one needs to analyze the survival problem for a certain population. The duration of life for any individual in the cohort is the functional of the two-component process $Z(t) = X(t),Y(t)$.

Assume that data which are available consist of the results of measurements of component $X(t)$ at some fixed times for the population cohort consisting of n individuals.

Let $X_i(t_1),...,X_i(t_k)$ be data-related to the i-th individual. Assume that both measured and unmeasured processes influence the mortality rate and this impact is specified as a quadratic form from both $X(t)$ and $Y(t)$, that is

$$\mu(t,X(t),Y(t)) = (X'(t)Y'(t))\begin{pmatrix} Q_{11}(t), & Q_{12}(t) \\ Q_{21}(t), & Q_{22}(t) \end{pmatrix}\begin{pmatrix} X(t) \\ Y(t) \end{pmatrix} + \mu_0(t)$$

where $Q_{11}(t)$, $Q_{22}(t)$ are positive-definite symmetric matrices and

$$Q'_{12}(t) = Q_{21}(t) \quad .$$

Note that one can always find the vector-function F and function G, such that the mortality rate $\mu(t,X,Y)$ can be represented in the form

$$\mu(t,X,Y) = (Y - F)'Q_{22}(t)(Y - F) + G$$

where F and G are the functions of t and X

$$F(t,X) = Q_{22}^{-1}(t)Q_{21}(t)X$$

$$G(t,X) = X'Q_{11}(t)X - X'Q_{12}(t)Q_{22}^{-1}(t)Q_{21}(t)X + \mu_0(t) \quad .$$

Assume that the problem is to estimate the elements of matrix Q on the base of data $X_i(t_1 \wedge T_i),...,X_i(t_k \wedge T_i)$, $i = \overline{1,n}$, where T_i are the observed death times and

$$Q = \begin{pmatrix} Q_{11} & Q_{12} \\ Q_{21} & Q_{22} \end{pmatrix} \quad .$$

Note that some parameters specifying the evolution of the process $Y(t)$ can also be known.

Assume that processes $X(t)$ and $Y(t)$ are the solutions of the following linear stochastic differential equations

$$d\begin{pmatrix} Y(t) \\ X(t) \end{pmatrix} = \left[\begin{pmatrix} a_{01}(t) \\ a_{02}(t) \end{pmatrix} + \begin{pmatrix} a_{11}(t) & a_{12}(t) \\ a_{22}(t) & a_{22}(t) \end{pmatrix}\begin{pmatrix} Y(t) \\ X(t) \end{pmatrix}\right]dt + \begin{pmatrix} b(t) \\ B(t) \end{pmatrix} d\begin{pmatrix} W_{1t} \\ W_{2t} \end{pmatrix}$$

where Q_{1t} and W_{2t} are vector-valued Wiener processes, independent on initial values $X(0),Y(0)$, and $b(t),B(t)$ are the matrices having the respective dimensions.

To avoid complications, we will omit index i related to some particular individual in the notations related to $X_i(t)$.

Let $\hat{x}(t)$ denote the vector $X(t_1),X(t_2),...,X(t_j(t))$, where

$$t_j(t) = sup\{t_m : t_m < t\} \quad .$$

Define the conditional survival function $S(t,\hat{x})$ with the help of equality

$$S(t,\hat{x}) = P(T > t \mid \hat{x}(t))$$

and let

$$\bar{\mu}(t,\hat{x}(t)) = -\frac{\partial}{\partial t} \ln S(t,\hat{x}) \quad .$$

The problem is to find the form of $\bar{\mu}(t,\hat{x}(t))$.

The following theorem about the form of $\bar{\mu}(t,\hat{x}(t))$ is true.

Theorem 7. Let the processes $X(t)$ and $Y(t)$ be defined as above. Then $\bar{\mu}(t,\hat{x}(t))$ can be represented in the form

$$\bar{\mu}(t,\hat{x}(t)) = (m'(t) - F'(t,\hat{x}))Q(t)(m(t) - F(t,\hat{x})) + Sp\,(Q(t)\gamma(t)) + \mu_0(t)$$

where $m(t) = \begin{pmatrix} m_1(t) \\ m_2(t) \end{pmatrix}$, $\gamma(t) = \begin{pmatrix} \gamma_{11}(t) & \gamma_{12}(t) \\ \gamma_{21}(t) & \gamma_{22}(t) \end{pmatrix}$ on the intervals $t_j \le t < t_{j+1}$ satisfy the equations .

$$\frac{dm(t)}{dt} = a_0(t) + a(t)m(t) - 2m(t)Q(t)\gamma(t)$$

$$\frac{d\gamma(t)}{dt} = a(t)\gamma(t) + \gamma(t)a^*(t) + b(t)b^*(t) - 2\gamma(t)Q(t)\gamma(t)$$

where

$$a_0(t) = \begin{pmatrix} a_{01}(t) \\ a_{02}(t) \end{pmatrix} \quad ,$$

$$a(t) = \begin{pmatrix} a_{11}(t)\,, & a_{12}(t) \\ a_{21}(t)\,, & a_{22}(t) \end{pmatrix}$$

$$m(t) = \begin{pmatrix} m_1(t) \\ m_2(t) \end{pmatrix}$$

$$\gamma(t) = \begin{pmatrix} \gamma_{11}(t)\,, & \gamma_{12}(t) \\ \gamma_{21}(t)\,, & \gamma_{22}(t) \end{pmatrix} \quad .$$

At time t_j, $j = 1,\ldots,k$, the initial values for these equations are

$$m_1(t_j) = m_1(tj-) + \gamma_{12}(tj-)\gamma_{22}^{-1}(tj-)(X(t_j) - m_2(tj-))$$

$$m_2(t_j) = X(t_j)$$

$$\gamma_{11}(t_j) = \gamma_{11}(tj-) - \gamma_{12}(tj-)\gamma_{22}^{-1}(tj-)\gamma_{21}(t_j)$$

$$\gamma_{22}(t_j) = 0$$

$$\gamma_{12}(t_j) = \gamma_{21}(t_j) = 0 \ .$$

The proof of this theorem can be done using for instance the approach developed in (Yashin, Manton, and Vaupel 1983).

Example. The relevance of the results such as formula (5) becomes evident from the following example. Assume that Wiener process W_t and stopping time T are interrelated and the random intensity of occurrence T is $\lambda(t)W_t^2$. Formula (5) gives immediately the form of the conditional survival function

$$S(t \mid W_0^t) = P(T > t \mid H_t) = e^{-\int_0^t \lambda(u) W_u^2 du} \ .$$

In survival analysis the stopping time T is associated with the death or failure time and the research is often focused on the properties of the survival function $S(t) = P(T > t)$ (Elandt-Johnson and Johnson 1979a; Nelson 1982; Lawless 1982; Cox and Oakes 1984). The straightforward way of its calculation is the averaging of the conditional survival function $S(t \mid W_0^t)$. It turns out that (see for instance, Yashin (1984))

$$E e^{-\int_0^t \lambda(u) W_u^2 du} = e^{-\int_0^t \lambda(u) E(W_u^2 \mid T > u) du} \ .$$

For the Wiener process the conditional mathematical expectation on the right-hand side of this formula can be easily calculated (the condition $W_0 = 0$ is used there)

$$E(W_t^2 \mid T > t) = \gamma(t)$$

where $\gamma(t)$ is the solution of the differential equation

$$\dot{\gamma}(t) = 1 - 2\lambda(t)\gamma^2(t) \ , \quad \gamma(0) = 0 \ .$$

When $\lambda(t)$ is constantly equal, say, to $\frac{1}{2}$ the straightforward calculations lead to the formula

$$E e^{-\frac{1}{2}\int_0^t W_u^2 du} = \frac{1}{\sqrt{cht}}$$

which coincides with the result based on the Cameron and Martin formula (Liptzer and Shiryaev 1978).

7. CONCLUSION

Formula (5) can be generalized in more complex cases including the sequence of observed stopping times and semimartingale as an influential stochastic process. It can be useful in the field of survival analysis, reliability theory and risk analysis. It shows which particular conditional distribution functions should be used in specification of the random intensities. The specification of the influential process and the measurement schemas provide the particular forms for the distributions and the intensities. Some examples, related to the biomedical and demographical applications are discussed in (Yashin, Manton, and Vaupel 1983; Yashin and Manton 1984; Yashin 1984).

REFERENCES

Anderson, P.K. and Gill, R. (1981). Cox Regression Model for Counting Processes: A Large Sample Study. RR-81-6. Statistical Research Unit, Danish Medical Research Council, Danish Social Science Research Council, Copenhagen.

Bailey, K. (1983). The Asymptotic Joint Distribution of Regression and Survival Parameter Estimates in the Cox Regression Model. The Annals of Statistics 11(1):39-48.

Barlow, R.E. and Proschan, F. (1975). Statistical Theory of Reliability and Life Testing. Holt, Rinehart, and Winston, Inc., New York.

Bremaud, P. (1980). Point Processes and Queues. Springer Verlag, New York, Heidelberg, and Berlin.

Cameron, R.H. and Martin, W.T. (1944). The Wiener Measure of Hilbert Neighborhoods in the Space of Real Continuous Functions. Journal of Mathematical Physics 23:195-209.

Cameron, R.H. and Martin, W.T. (1949). Transformation of Wiener Integrals by Nonlinear Transformation. Transactions of American Mathematical Society 66:253-283.

Cox, D.R. (1972). Regression Models and Life Tables. Journal of Royal Statistical Society B 34:187-220.

Cox, D.R. (1975). Partial Likelihood. Biometrika A 62:269-276.

Cox, D.R. and Oakes, D. (1984). Analysis of Survival Data. Chapman and Hall, London.

Dellacherie, C. (1972). Capacities et Processus Stochastiques (Capacities and Stochastic Processes). Springer Verlag, Berlin and New York.

Elandt-Johnson, R.C. and Johnson, N.L. (1979). Survival Models and Data Analysis. John Wiley and Sons, New York.

Gill, R.D. (1984). Understanding Cox's Regression Model: A Martingale Approach. Journal of American Statistical Association 79(386):441-447 (June).

Jacod, J. (1975). Multivariate Point Processes: Predictable Projection, Radon-Nicodim Derivatives, Representation of Martingales. Zeitschrift für Wahrscheinlichkeitstheorie und Verw. Gebiete 31:235-253.

Jacod, J. (1979). Calcules Stochastique et Probleme de Martingales. Lectures Notes in Mathematics, Vol. 714. Springer Verlag, Heidelberg.

Lawless, J.F. (1982). Statistical Models and Methods for Lifetime Data. John Wiley and Sons, New York.

Liptzer, R.S. (1975). Gaussian Martingales and Generalization of the Kalman-Bucy Filter. Theory of Probability and Applications 20(2):34-45 (in Russian).

Liptzer, R.S and Shiryaev, A.N. (1978). Statistics of Random Processes. Springer Verlag, Berlin and New York.

Myers, L.E. (1981). Survival Functions Induced by Stochastic Covariate Processes. Journal of Applied Probability 18:523-529.

Nelson, W. (1982). Applied Life Data Analysis. John Wiley and Sons, New York.

Novikov, A.A. (1972). On Parameters Estimation of Diffusion Processes. Studia Science Mathematics (7):201-209.

Prentice, R.L. and Self, S.G. (1983). Asymptotic Distribution Theory for Cox-type Regression Models with General Relative Risk Form.

Vaupel, J.W., Manton, K., and Stallard, E. (1979). The Impact of Heterogeneity in Individual Frailty on the Dynamics of Mortality. Demography 16:439-454.

Vaupel, J.W. and Yashin, A.I. (1983). The Deviant Dynamics of Death in Heterogeneous Populations. RR-83-1. International Institute for Applied Systems Analysis, Laxenburg, Austria. An abridged version is in Nancy Tuma (Ed.), Sociological Methodology 1985. Jossey-Bass, San Francisco.

Woodbury, M.A. and Manton, K.G. (1977). A Random Walk Model of Human Mortality and Aging. Theoretical Population Biology 11(1):37-48.

Yashin, A.I. (1970). Filtering of Jumping Processes. Automatic and Remote Control 5:52-58.

Yashin, A.I. (1980). Conditional Gaussian Estimation of Characteristics of the Dynamic Stochastic Systems. Automatic and Remote Control (5):57-67.

Yashin, A.I. (1982). A New Proof and New Results in Conditional Gaussian Estimation Procedures. Pages 205-207 in Proceedings of the 6th European Meeting on Cybernetics and Systems Research, April 1982. North-Holland Publishing Company.

Yashin, A.I. (1983). Chances of Survival in a Chaotic Environment. WP-83-100. International Institute for Applied Systems Analysis, Laxenburg, Austria.

Yashin, A.I. (1984). Dynamics in Survival Analysis: Conditional Gaussian Property Versus Cameron-Martin Formula. WP-84-107. International Institute for Applied Systems Analysis, Laxenburg, Austria.

Yashin, A.I. (1985). On the Notion of Random Intensity. WP-85-13. International Institute for Applied Systems Analysis, Laxenburg, Austria.

Yashin, A.I., Manton, K.G., and Vaupel, J.W. (1983). Mortality and Aging in a Heterogeneous Population: A Stochastic Process Model with Observed and Unobserved Variables. WP-83-81. International Institute for Applied Systems Analysis, Laxenburg, Austria.

Yashin, A.I. and Manton, K.G. (1984). Evaluating the Effects of Observed and Unobserved Diffusion Processes in Survival Analysis of Longitudinal Data. CP-84-58. International Institute for Applied Systems Analysis, Laxenburg, Austria.

MARKOV DECISION PROCESSES WITH BOTH CONTINUOUS AND IMPULSIVE CONTROL

A.A. Yushkevich
Moscow Institute of Transport Engineering
Moscow, USSR

1. INTRODUCTION

In this paper we present some progress in the theory of Markov decision processes (MDPs) with both continuous and impulsive actions. This is based on analogues of Bellman's dynamic programming optimality equation. We also discuss a modification of some basic concepts of general stochastic process theory that appears useful in a formal treatment of impulsive control problems.

Historically one may trace two paths in the development of the theory of continuous time MDPs. The first, which is based on pioneering work by Bellman and Howard, deals with actions (we shall call them *controls*) which influence infinitesimal characteristics, i.e., the generator of the process and the reward rate. Here the problem of the existence of "good" policies is treated following Blackwell and Strauch's approach from discrete time dynamic programming, and optimality equations lie at the heart of the investigations. This path is described, for example, by Miller, Kakumanu, Plisca, Doshi, Yushkevich and Fainberg (for exact references see [1]).

Another approach is due to De Leve [2]. According to De Leve, decisions are made at isolated moments and produce immediate changes in both state and reward (we shall call such actions *impulsive controls*); during the time intervals between decisions the system is governed by a given Markov process. These concepts were not widely recognised by specialists in MDPs at first, but have been given a new lease on life in publications by the Netherlands school.

Schouten [3] and Hordijk and Schouten [4] have recently begun a systematic study of MDPs involving actions of both kinds. These authors consider a model with a deterministic drift between jumps, and with random jumps influenced by both continuous and impulsive controls. In the works cited above, particular attention is paid to discrete time approximation of optimal policies, and the related weak convergence problems on a functional space differing from Skorohod's space in that three different situations may occur at discontinuity points t: $x(t)$, $x(t-)$ and $x(t+)$. This is because Hordijk and Schouten permit an impulse to follow immediately after a jump generated by continuous controls, but forbid instantaneous repetitions of impulses.

Impulsive controls have also been studied in connection with stochastic differential equations, in some cases together with continuous controls (Bensoussan and Lions [5], Kushner [6]). In general these authors did not exclude immediate repetition of impulses (i.e., coincidence of stopping times), but such phenomena were treated in an informal way. An essential tool in [5] and related publications is provided by quasi-variational inequallities (QVIs), which represent a substitute for Bellman's optimality equation.

An approach to Hordijk–Schouten MDP models based on appropriate QVIs was previously initiated. Here we describe and develop this approach.

Examples of MDPs with impulsive controls and deterministic drift arising in inventory, storage or queueing problems can be found in [3]. Deterministic drift also occurs when a non-homogeneous jump model is reduced to a homogeneous model by including time as a space variable. A familiar example of an impulsive control problem with only one permitted decision is the optimal stopping problem. Of course, this can also be combined with continuous control (see Shiryaev [7], Krylov [8]).

Finally, we should mention that another treatment of (continuously) controlled stochastic processes without a diffusion component but with controlled drift and jumps is given by Davis [9] and Vermes [10]. Processes of this type with a concentration of small jumps have been studied by Pragarauskas [11] as a particular case of controlled, possibly degenerated diffusion with jumps.

2. MODIFICATION OF SOME STOCHASTIC THEORY CONCEPTS

In a Markovian model it is natural to desire that in any current state the set of possible decisions should not depend on the history of the system. There are no physical reasons to forbid jumps influenced by impulsive actions which bring the system into a state where impulsive controls are available. Therefore we should permit any finite number of successive impulsive decisions to coincide in time, so that at some random moment t there will be several state positions x_t^0, x_t^1, x_t^2 ,... . The information available after reaching x_t^0 is less than after reaching x_t^1, etc., and so any fixed t will be associated with various σ-fields $\mathbf{F}_t^0 \subseteq \mathbf{F}_t^1 \subseteq \cdots$. We shall kill the process after any decomposition of $x(t)$ into countably many states x_t^n, and possibly also on other occasions. We shall need stopping times ϑ with respect to the enlarged family of σ-fields $\{\mathbf{F}_t^n\}$. We must obtain some measurability properties analogous to the usual ones. We must also be able to identify all "split" time periods t and for this purpose we shall require that these periods be numbered in their natural order (maybe after some classification by "size"). And we must require that all x_s^m are observable at split random periods s, i.e., they are measurable with respect to σ-fields $\mathbf{F}_t^n$ if $s < t$, because this measurability does not follow automatically from the fact that x_t^n is adapted to $\mathbf{F}_t^n$ in the usual sense. So we arrive at the following definitions.

Denote by M the set of all pairs $T = (t, n)$, $t \in R_+$, $n \in Z$, called *moments*. We shall define ordering in M by setting $(s, m) < (t, n)$ if $s < t$ or $s = t$, $m < n$. For $T = (t, n)$ with $n \geq 1$ define $T- = (t, n-1)$. Suppose that a measurable state space $(X, \mathbf{X})$ and a measurable base space $(\Omega, \mathbf{F})$ are given. For any function ξ defined on some subset $D(\xi)$ of Ω we write $\{\xi \in B\}$ instead of $\{\omega: \omega \in D(\xi), \xi(\omega) \in B\}$, and we call ξ an $\mathbf{F}$-measurable variable if $D(\xi) \in \mathbf{F}$ and ξ is measurable with respect to $\mathbf{F} \cap D(\xi)$.

Definition 1. A *T-process* $\{x_T\}$ on $(\Omega, \mathbf{F})$ with values in $(X, \mathbf{X})$ is a map $x(T, \omega) = x_T(\omega) = x_t^n(\omega)$ from some $D \subset M \times \Omega$ into X with the following properties:

(1) if $(t, n, \omega) \in D$, $n \geq 1$, then $(t, m, \omega) \in D$ for $m = 0, 1, \ldots, n$;

(2) if $(t, 0, \omega) \notin D$, then $(u, 0, \omega) \notin D$ for any $u > t$;

(3) if $(t, n, \omega) \in D$ for all $n \in Z_+$, then $(u, 0, \omega) \notin D$ for any $u > t$;

(4) $(0, 0, , \omega) \in D$ for any $\omega \in \Omega$;

(5) $x_T(\omega)$ is $\mathbf{F}$-measurable for any $T \in M$;

(6) there exists a function d from $X \times X$ into $(0, \infty)$ such that for any $\varepsilon > 0$ and $\omega \in \Omega$ the set

$$\{T = (t, n): n \geq 1, (T, \omega) \in D, d(x_{T-}(\omega), x_T(\omega)) > \varepsilon\}$$

is empty or all its elements form a finite or countable ordered sequence $T_1^\varepsilon(\omega) < T_2^\varepsilon(\omega) < \ldots$, and T_i^ε and $x(T_i^\varepsilon(\omega), \omega)$ are $\mathbf{F}$-measurable variables.

Definition 2. A class $\mathbf{N}_T$ of events which are *observable* by means of $\{x_T\}$ up to a moment T, $T \in M$, is a minimal σ-field in Ω such that (i) variables x_S with $S \leq T$, (ii) sets $\{T_i^\varepsilon \leq T\}$, $\varepsilon > 0$, $i = 1, 2, \ldots$, and (iii) restrictions of variables T_i^ε and $x(T_i^\varepsilon)$ on these sets, are all measurable with respect to $\mathbf{N}_T$.

As usual a filter $\{\mathbf{F}_T\}$ is understood to be a family of σ-fields $\mathbf{F}_T$, $T \in M$, such that $\mathbf{F}_T \subset \mathbf{F}$ and $\mathbf{F}_S \subset F_T$ if $S < T$.

Definition 3. A T-process $\{x_T\}$ is said to be *adapted* to a filter $\{\mathbf{F}_T\}$ if $\mathbf{N}_T \subset \mathbf{F}_T$, $T \in M$. It is said to be *progressively measurable* with respect to $\{\mathbf{F}_T\}$ if for any $T \in M$, $i = 1, 2, \ldots$, $B \in \mathbf{X}$, $\varepsilon > 0$, we have

$$\{(S, \omega): x(S, \omega) \in B, S \leq T\} \in \mathbf{B}_T \times \mathbf{F}_T \quad ,$$

$$\{(S,\omega): S = T_i^{\varepsilon}(\omega), S \le T\} \in \mathbf{B}_T \times \mathbf{F}_T \quad ,$$

where $\mathbf{B}_T$ denotes a Borel σ-field on $\{S: S \le T, S \in M\}$.

Definition 4. A function ϑ from Ω into $M \cup \{\infty\}$ is said to be a ***stopping time*** of a T-process $\{x_T\}$ with respect to a filter $\{\mathbf{F}_T\}$, if $\{\vartheta \le T\} \in \mathbf{F}_T$ for any $T \in M$, and $x_{\vartheta(\omega)}(\omega)$ is defined for any $\omega \in \{\vartheta \ne \infty\}$.

Many of the usual properties of stochastic processes remain valid with slight modifications for T-processes. Using essentially the same arguments as in [12] one can prove that (i) if $(X, \mathbf{X})$, is a Borel space then a progressively measurable T-process is an adapted process; (ii) if $(X, \mathbf{X})$ is a metric space with a Borel measurability structure, if $\{x_T\}$ is a T-process adapted to $\{\mathbf{F}_T\}$, and if $x_t^0(\omega)$ is left-continuous in t for any $\omega \in \Omega$, then $\{x_T\}$ is progressively measurable with respect to $\{\mathbf{F}_T\}$; (iii) if $\{x_T\}$ is adapted to $\{\mathbf{F}_T\}$, then all variables T, $T \in M$, and T_i^{ε}, $\varepsilon > 0$, $i = 1,2 \cdots$ (equal to $+\infty$ if x_T does not reach them) are stopping times with respect to $\{\mathbf{F}_T\}$; (iv) if $\{x_T\}$ is adapted and progressively measurable with respect to $\{\mathbf{F}_T\}$ then $x_{\vartheta(\omega)}(\omega)$ is $\mathbf{F}_{\vartheta}$-measurable, where $A \in \mathbf{F}_{\vartheta}$ if $\{A, \vartheta \le T\} \in \mathbf{F}_T$ for all T; and so on.

3. A MARKOV DECISION PROCESS

We shall construct an MDP with deterministic drift and non-accumulating impulses analogous to that defined in [3,4], but allowing immediate repetition of impulses. In contrast to those papers and also to [1] we now define policies in terms of T-processes and stopping times in a way similar to that adopted in stochastic differential equations. In the next definition we use terminology introduced by Gihman and Skorohod [13].

Definition 5. A *controlled object* Y is a collection $(X, f, A, A^1, A^2, A(x), q)$ of the following Borel-measurable elements:

(1) a state space X;

(2) a drift function $f(s, x, t)$, $x \in X$, $s \le t \in R_+$, with $f(t, x, t) = x$ and an obvious semigroup property;

(3) an action space A divided into a set of impulsive controls A^1 and a control set $A^2 = A \setminus A^1$;

(4) a non-empty set of constraints $A(x)$, $x \in X$, defined as the x-section of a set $C \in \mathbf{B}(X \times A)$;

(5) a jump density $\lambda(t, x, a) \ge 0$, $t \in R_+$, $a \in A(x)$, $x \in X$, bounded for $a \in A^2$, and equal to $+\infty$ for $a \in A^1$;

(6) a jump distribution $q(t, x, a, dy)$, i.e., a probability measure on $X \setminus \{x\}$, $t \in R_+$, $a \in A(x)$, $x \in X$.

With any controlled object Y we associate a T-process $\{x_T\}$ in the following way. Denote by Ξ the set of all *skeletons* ξ, i.e., finite or countable collections of successive pairs $t_k x_k \in R_+ \times X$ of the form

$$\xi = t_0 x_0 t_1 x_1 t_2 x_2 \ldots,\ 0 = t_0 \le t_1 \le t_2 \le \ldots,\ x_{k+1} \ne f(t_k, x_k, t_{k+1}) \quad .$$

Ξ is a Borel space, and we consider Ξ to be an image of a base space $(\Omega, \mathbf{F})$ obtained by means of a measurable map $\xi = \xi(\omega)$ (for example, one may let $\Xi = \Omega$).

Any skeleton $\xi(\omega)$ defines a trajectory $x(\cdot, \omega)$ by drift: if $t \in (t_k, t_{k+1})$ or if t_{k+1} is not defined and $t \in (t_k, +\infty)$, then $x_t^0 = f(t_k, x_k, t)$; if i is the smallest number with $t_i = t$, then for any $j \ge i$ such that $t_j = t$ we have $x_t^{j-i}(\omega) = x_j$ in the case $t = 0$, or $x_t^{j-i+1}(\omega) = x_j$ and $x_t^0(\omega) = f(t_{i-1}, x_{i-1}, t)$ in the case $t > 0$; at other moments x_T is not defined. Definition 1 is satisfied here with $d = 1$; variables T_i^ε do not depend on ε for $\varepsilon < 1$, and we denote them by T_i, $i = 1, 2, \ldots$, setting $T_0 = (0, 0)$. It is evident that the first coordinate of T_i equals t_i.

Definition 6. For a controlled object Y and the T-process $\{x_T\}$ associated with it, a *policy* π is a collection $(\{a_T\}, \vartheta_1, \vartheta_2, \ldots)$, where

(1) ϑ_i, $i \ge 1$ are stopping times of $\{x_T\}$, with respect to $\{\mathbf{N}_T\}$ such that $\vartheta_i < \vartheta_{i+1}$ or $\vartheta_i = \vartheta_{i+1} = \infty$;

(2) $\{a_T\}$ is a T-process with values in A with the same domain D as $\{x_T\}$; it is progressively measurable with respect to $\{\mathbf{N}_T\}$, has $a_T(\omega) \in A(x_T(\omega))$, and is such that $a_T(\omega) \in A^1$ if T equals any of $\vartheta_i(\omega) < \infty$ and $a_T(\omega) \in A^2$ for all other $(T, \omega) \in D$.

As usual, if the correspondence $x \to A(x)$, $x \in X$, admits a measurable selection, then policies do exist. The following analytic representation of a policy (cf. [14] for an analogous decomposition of stopping times in jump processes) demonstrates that chosen decisions are functions of histories $t_0 x_0 \ldots t_m x_m t$ in the same sense as in dynamic programming. Let $\vartheta_i = (\tau_i, \nu_i)$ if $\vartheta_i < \infty$, and $\tau_i = +\infty$ if $\vartheta_i = +\infty$. Let $i_m(\omega, \pi)$ denote the number of moments $\vartheta_i < T_m$.

LEMMA 1. *For any policy π there are measurable functions $b(t_0 x_0 \ldots t_m x_m t)$ with values in $A(x_m)$, $i(t_0 x_0 \ldots t_m x_m)$ with values in $(0, 1, 2, \ldots, \infty)$, and $g(t_0 x_0 \ldots t_m x_m)$ with values in $[t_m, +\infty]$ (where $m = 0, 1, 2, \ldots$, elements $t_0 x_0 \ldots t_m x_m$ are the same as in skeleton ξ, and $t \ge t_m$), such that, setting $t_k x_k = t_k x_k(\omega)$, $T_k = T_k(\omega)$, we have*

(1) $a_T(\omega) = b(t_0 x_0 \ldots t_m x_m t)$ *for* $(T, \omega) \in D$, *where* $m = \max\{k : T_k \leq T\}$, *and* t *is the first coordinate of* T;

(2) $i_m(\omega, \pi) = i(t_0 x_0 \ldots t_m x_m)$ *if* $t_m(\omega)$ *is defined;*

(3) $\tau_i(\omega) = g(t_0 x_0 \ldots t_m x_m)$ *for* $i = i(t_0 x_0 \ldots t_m x_m)$, *if* $t_m(\omega)$ *is defined,* $i(t_0 x_0 \ldots t_m x_m)$ *is finite, and either* $t_{m+1}(\omega) \geq g(t_0 x_0 \ldots t_m x_m)$ *or* $t_{m+1}(\omega)$ *is not defined; in that case* ν_i *equals the second coordinate of* T_m *if* $\tau_i = t_m$, *and equals* 0 *if* $t_m < \tau_i < \infty$.

Using Lemma 1, induction in m and the Ionescu–Tulcea-theorem, we construct a probability measure P_x^π on Ω corresponding to any initial state x and policy π. P_x^π is a unique measure on the σ-field $\mathbf{N} = \lim_{T \to \infty} \mathbf{N}_T$ which satisfies the following conditions: for all $m \geq 0$ (almost surely for conditional probabilities) we have

$$P_x^\pi\{x_0 = x\} = 1 \quad ,$$

$$P_x^\pi\{t_{m+1} \in [t_m, t] \mid t_0 x_0 \cdots t_m x_m\} =$$

$$= \begin{cases} 1 - \exp[-\int_{t_m}^{t} \lambda(s, x(s), a(s))ds & \text{if } t_m \leq t < g(t_0 x_0 \cdots t_m x_m) \; , \\ 1 & \text{if } g(t_0 x_0 \cdots t_m x_m) \leq t \quad , \end{cases}$$

$$P_x^\pi\{x_{m+1} \in dy \mid t_0 x_0 \cdots t_m x_m t_{m+1}\} = q(t_{m+1}, x(t_{m+1}), a(t_{m+1}), dy) \quad ,$$

where for brevity $x(s) = f(t_m, x_m, s)$, $a(s) = b(t_0 x_0 \cdots t_m x_m s)$. By E_x^π we denote expectation with respect to P_x^π.

The T-process $\{x_T\}$ constructed on the probability space $(\Omega, \mathbf{N}, P_x^\pi)$ is a *controlled stochastic T-process* (with given initial state and policy).

A policy π is said to be a *Markov* policy if functions b and g corresponding to π (see Lemma 1) can be expressed in terms of a measurable function (called a *selector*) $\varphi(t, x)$ on $R_+ \times X$ taking values in sets $A(x)$:

(1) $b(t_0 x_0 \cdots t_m x_m t) = \varphi(t, f(t_m, x_m, t))$;

(2) $g(t_0 x_0 \cdots t_m x_m) = g_\varphi(t_m, x_m)$,

where

$$g_\varphi(t, x) = \begin{cases} +\infty \text{ if } \varphi(u, f(t, x, u)) \in A^2 \text{ for all } u \geq t \quad , \\ \min\{u : u \geq t \; , \; \varphi(u, f(t, x, u)) \in A^1\} \quad \text{otherwise} \end{cases}$$

and the minimum is necessarily attained. (If the last condition is not satisfied, then at the time inf $\{u: \cdots\}$ when the first impulsive action after t occurs, the distribution of that action is not defined.) If $\varphi(t,x) = \varphi(x)$, then the policy is said to be *stationary*. Let Π denote the set of all policies in Y.

Definition 7. The *cost structure* of a controlled object Y is given by

(i) a horizon $H \in (0,\infty]$;

(ii) a measurable real reward function $r(t,x,a)$, $t \in [0,H] \cap R_+$, $a \in A(x)$, $x \in X$;

(iii) if $H < \infty$, a measurable real final reward function $R(x)$, $x \in X$.

A controlled object Y with a corresponding T-process $\{x_T\}$, a collection of policies π and corresponding measures P_x^π, and a cost structure together form a *Markov decision process*.

For a given horizon H, a policy π is said to be *admissible*, if for any $x \in X$ with P_x^π – probability 1, the variables t_m (or equivalently τ_i) have no limit point on the segment $[0,H]$ for $H < \infty$ or on the interval $[0,\infty)$ for $H = \infty$. For an admissible policy, the value x_t^0 is almost surely defined for any $t \geq 0$ in the case $H = \infty$, and for some $t > H$ in the case $H < \infty$. The set of all admissible policies will be denoted by Π_0.

On the whole we shall deal with an *expected total reward criterion*

$$v^\pi(x) = \mathrm{E}_x^\pi\Big[\int_0^H r(t,x_t^0,a_t^0)dt + \sum_{\infty \neq \tau_i < H} r(\tau_i, x_{\vartheta_i}, a_{\vartheta_i}) + R(\bar{x}_H)1(H<\infty)\Big] \quad ,$$

where $\bar{x}_t = x_t^n$ with $n = \max\{m: x_t^m \text{ is defined}\}$. Assuming that $v^\pi(x)$ is well-defined for all $\pi \in \Pi_0$, $x \in X$, we introduce a *value function*

$$v(x) = \sup_{\pi \in \Pi_0} v^\pi(x) \, , \; x \in X \quad .$$

If v is finite, then a policy $\pi \in \Pi_0$ is said to be *optimal* (ε-optimal) if $v^\pi = v \, (v^\pi \geq v - \varepsilon)$.

4. OPTIMALITY CONDITIONS IN THE CASE OF A FINITE HORIZON

In addition to a given Markov decision process Z we consider a collection of MDPs Z_t obtained from Z by replacing the control period $[0,H]$ by $[t,H]$. Let $v_t^\pi(x) = v^\pi(t,x)$ and $v_t(x) = v(t,x)$ be the criterion and value functions for Z_t. We shall say that Z is *upper* (or *lower*) *bounded*, if in the process $\tilde{Z}$ obtained from Z by replacing r and R by r^+ and R^+ (r^- and R^-) the criterion function $\tilde{v}^\pi(t,x)$ is uniformly bounded from above (below).

For bounded measurable functions F on X, the following operators T_t, T_t^1, T_t^2 are defined for $t \in [0,H]$, $x \in X$:

$$T_t F(x,a) = \begin{cases} r(t,x,a) + \lambda(t,x,a) \int_X [F(y) - F(x)] q(t,x,a,dy), & a \in A^2(x), \\ r(t,x,a) + \int_X F(y) q(t,x,a,dy), & a \in A^1(x), \end{cases}$$

$$T_t^1 F(x) = \sup_{A^1(x)} T_t F(x,a), \quad T_t^2 F(x) = \sup_{A^2(x)} T_t F(x,a) .$$

Here T_t^1 has the form of a Bellman operator in discrete time dynamic programming, and T_t^2 has the same form for the continuous time case.

For brevity we shall write

$$-D_u \Phi(u) \geq \Psi(u), \quad u \in 1$$

instead of

$$\Phi(s) - \Phi(t) \geq \int_0^t \Psi(u) du \text{ for all } s < t \in 1$$

even in the case when Φ has no derivative. The equality $D_u \Phi(u) = \Psi(u)$, $u \in 1$ has an analogous meaning.

THEOREM 1 [1]. *Suppose that all constraint sets $A^2(x)$, $x \in X$, are non-empty, that the horizon H is finite, and that MDP Z is upper (or lower) bounded. If a selector $\varphi(t,x)$ generates a Markov policy $\pi \in \Pi_0$, and if φ and a bounded measurable real function $v(t,x) = v_t(x)$, $t \in [0,H]$, $x \in X$, satisfy the following conditions for any $t \in [0,H]$, $x \in X$:*

$$v_t(x) \geq T_t^1 v_t(x) \quad ; \tag{1}$$

$$-D_u v_u(f(t,x,u)) \geq T_u^2 v_u(f(t,x,u)), \quad u \in [t,H] \quad , \tag{2}$$

$$v_t(x) = T_t v_t(x, \varphi(t,x)) \text{ if } \varphi(t,x) \in A^1 \quad , \tag{3}$$

$$-D_u v_u(f(t,x,u)) = T_u v_u[f(t,x,u), \varphi(u, f(t,x,u))], \quad u \in [t, g_\varphi(t,x)] \quad , \tag{4}$$

$$v_H(x) \geq R(x), \quad v_H(x) = R(x) \text{ if } \varphi(H,x) \in A^2 \quad , \tag{5}$$

then v_0 is the value of Z and π is an optimal policy in Z.

In fact conditions (1)–(5) are necessary and sufficient for v_t to be the value of Z_t for all $t \in [0,H]$ and for selector φ to be an *optimal synthesis*, i.e., to produce an optimal policy π_t in any MDP Z_t (provided that $\pi_t \in \Pi_0$); we omit the exact formulations here (but see [1]). The proofs use arguments from dynamic programming, the Markov property, and measurable selection.

Inequalities (1)–(2), with an equality sign in one of them at any point (t, x) implied by (3)–(4), form a set of *quasi-variational inequalities* (QVIs) for the model under consideration. The question arises as to whether the QVIs are valid without the assumption that an optimal synthesis exists. We shall give a partial answer to this question in Section 5.

5. COUNTABLE HOMOGENEOUS NEGATIVE AND DISCOUNTED CASES

Here we shall make the following assumptions:

Assumptions 1.

(i) X is countable and $f(s, x, t) = x$;

(ii) $\lambda(t, x, a) = \lambda(x, a)$, $q(t, x, a, dy) = q(x, a, dy)$, $r(t, x, a) = r(x, a)e^{-\beta t}$ with constant $\beta \geq 0$, $H = \infty$;

(iii) Π_0 is non-empty;

(iv) criterion function $\tilde{v}^{\pi}$ in $\tilde{Z}$, obtained from Z by replacing r by r^+, is bounded from above;

(v) $-\infty < v(x) \leq K < +\infty$, $x \in X$.

In the homogeneous case $v(t, x) = v(x)e^{-\beta t}$, so that ${}_t v(t, x) = \beta v(t, x)$, and in the absence of drift (2) becomes $\beta v_u \geq T_u^2 v_u$. Operators T_t, T_t^1, T_t^2 from Section 4 are connected with the operators

$$TF(x, a) = \begin{cases} r(x, a) + \lambda(x, a) \sum\limits_y [F(y) - F(x)] q(x, a, y), & a \in A^2(x) \\ r(x, a) + \sum\limits_y F(y) q(x, a, y), & a \in A^1(x) \end{cases},$$

by the formula $T_t F(x, a) = e^{-\beta t} T(e^{\beta t} F)(x, a)$. Thus inequalities (1)–(2) reduce to $v \geq T^1 v$, $\beta v \geq T^2 v$. The denumerability of X allows us to avoid measurability difficulties, and dynamic programming arguments for both discrete and continuous time parameters may be used to obtain the QVIs.

THEOREM 2. *Under Assumptions 1, we have*

$$v \geq T^1 v, \quad \beta v \geq T^2 v, \quad (v - T^1 v)(\beta v - T^2 v) = 0 \quad .$$

By induction over T_m it is possible to extend to the present case a defect formula which has been known implicitly for a long time but was stated explicitly in [15] for discrete and in [16] for continuous time.

THEOREM 3. *Under Assumptions 1, for any $\pi \in \Pi_0$ we have*

$$v(x) - v^{\pi}(x) = \kappa +$$

$$+ E_x^{\pi}\{\int_0^{\infty} e^{-\beta t}[\beta v(x_t^0) - Tv(x_t^0, a_t^0)]dt + \sum_i e^{-\beta\tau_i}[v(x_{\vartheta_i}) - Tv(x_{\vartheta_i}, a_{\vartheta_i})]\} \tag{6}$$

where $\kappa \geq 0$, but if $E_x^{\pi}\{ \cdots \}$ is finite then

$$\kappa = \lim_{T \to \infty} E_x e^{-\beta t} v(x_t^0) \quad . \tag{7}$$

According to Theorem 2, the terms in square brackets in (6) are non-negative. If $\beta > 0$ or $v \leq 0$ we have $\kappa \leq 0$ (from (7)), and therefore $\kappa = 0$ (any policy is *equalizing*). If a Markov policy π is generated by a selector $\varphi(t, x)$ which satisfies

$$Tv(x, \varphi(t, x)) = \begin{cases} v(x) \text{ if } \varphi(t, x) \in A^1 , \\ \beta v(x) \text{ if } \varphi(t, x) \in A^2 , \end{cases}$$

(i.e., π is a *conservation* policy), then $E_x^{\pi}\{ \cdots \} = 0$ in Theorem 3. By selecting a conserving π if the constraint set $A(x)$ is finite, or a nearly conserving π in other cases, one can obtain an optimal or ε-optimal policy (of course, it is necessary to check that $\pi \in \Pi_0$). If

$$r(x, a) \leq \delta < 0 \text{ for } a \in A^1(x) , x \in X \tag{8}$$

then a policy $\pi \notin \Pi_0$ may provide a reward $v^{\pi} = -\infty$ because it may provoke infinitely many impulses in a finite time. Using this remark and an analogue of (6) for $\pi \notin \Pi_0$, it can be shown that a nearly conserving policy must belong to Π_0. This leads to the following result.

COROLLARY 1. *Suppose that either $\Pi = \Pi_0$ or (8) is valid.*

(1) *If* (i) *$A(x)$ is finite,* (ii) *$\beta > 0$ or $v \leq 0$, then a stationary optimal policy exists.*

(2) *If* (i) *$A^1(x)$ is finite,* (ii) *$\beta > 0$, then for any $\varepsilon > 0$ a stationary ε-optimal policy exists.*

(3) *If* (i) *$A^1(x)$ is finite,* (ii) *$v \leq 0$, then for any $\varepsilon > 0$ a Markov ε-optimal policy exists.*

A policy π will be said to be a *tracking* Markovian policy (a tracking stationary policy), if π differs from a Markov (stationary) policy only in that the decision may depend on i at stopping times ϑ_i.

COROLLARY 2. *Assume that* (8) *holds, and let* $\varepsilon > 0$. *In the case* $\beta > 0$ *(or* $v \leq 0$*) a tracking stationary (or tracking Markovian)* ε*-optimal policy exists.*

6. FINITE HOMOGENEOUS CASE

The existence of stationary optimal policies can be proved for finite homogeneous models. We shall make the following assumptions:

Assumptions 2.

(i) X and A are finite, $f(s, x, t) = x$;

(ii) the same as (ii) in Assumptions 1.

Let X^1 denote the set of those states x in which only impulsive actions are admitted: $A(x) = A^{1(x)}$. A non-empty subset $V \subseteq X^1$ is said to be *closed* if $q(x, a, V) = 1$ for any $a \in A(x)$, $x \in V$. Starting at $x \in V$, with probability 1 the system will never leave V under any arbitrarily chosen policy, so that x_T will be defined (only) for infinitely many $T = (0, n)$. In this case Π_0 is empty. We shall therefore introduce the following condition:

Condition 1. The set X^1 contains no cosed subsets.

Denote by Φ the set of all selectors $\varphi(t, x)$ that do not depend on t. According to Section 3, every stationary policy π is associated with a selector $\varphi \in \Phi$. If X is finite (or countable) and if there is no drift, then the converse is also true; and in this case ϑ_i equals the i-th successive time at which x_T enters the set $X^1_\varphi = \{x : \varphi(x) \in A^1\}$. Any selector $\varphi \in \Phi$ generates a Markov chain C_φ on the state space X^1_φ with transition probabilities $p_\varphi(x, y) = q(x, \varphi(x), y)$ and rewards $r_\varphi(x) = r(x, \varphi(x))$; this chain represents the alternation of states x^n_0, $n = 0,1, \ldots$ in the original T-process $\{x_T\}$ under policy π; the chain is terminated as soon as x^n_0 leaves X^1_φ. For any ergodic class E of states in the chain C_φ there exists a unique stationary distribution $p_\varphi(x)$, $x \in E$. By the law of large numbers there exists an average expected reward in C_φ and for any $x \in E$ this is equal to

$$g_\varphi(E) = \lim_{n \to \infty} \frac{1}{n+1} \mathrm{E}^\pi_x \sum_{m=0}^{n} r_\varphi(x^m_0) = \sum_E r_\varphi(y) p_\varphi(y) \quad .$$

If $g_\varphi(E)$ is positive, then starting in E it is possible to receive an arbitrarily large reward in the T-process $\{x_T\}$ without increasing t; we can therefore say that a selector $\varphi \in \Phi$ with $g_\varphi(E) > 0$ for some ergodic class E *immediately provides an infinite reward.* We shall now introduce a second condition:

Condition 2. there are no selectors which immediately provide an infinite reward in Z.

THEOREM 4 [1]. *Suppose that Assumptions* 2 *are satisfied, and that* $\beta > 0$. *Then the value* v *is finite and a stationary optimal policy exists if and only if Conditions* 1 *and* 2 *are satisfied.*

In the case when $\beta = 0$, we shall denote by $Z_H(R)$ the MDP obtained from Z by fixing a finite horizon H and a final reward function R; let v_H^R be the value of $Z_H(R)$. We say that a selector $\varphi \in \Phi$ and real functions g and h on X form a *canonical triple* in Z with $\beta = 0$, if φ is optimal in $Z_H(h)$ for any $H \geq 0$, and also

$$v_H^h(x) = g(x)H + h(x) \,, \; x \in X \,, \; H \geq 0 \quad .$$

THEOREM 5 [1]. *Under Assumptions* 2, *Conditions* 1 *and* 2 *are necessary and sufficient for the existence of a canonical triple.*

In both a discrete time MDP [17] and a continuous time MDP without impulsive actions [18], a policy φ from a canonical triple is optimal for the average criterion

$$w^\pi(x) = \lim_{H \to \infty} \frac{1}{H} \mathrm{E}_x^\pi \Big[\int_0^H r(x_t^0, a_t^0) dt + \sum_{\tau_i \leq H} r(x_{\vartheta_i}, a_{\vartheta_i}) \Big] \quad .$$

REFERENCES

1. A.A. Yushkevich. Continuous time Markov decision processes with interventions. *Stochastics*, 9(1983)235–274.

2. G. De Leve. Generalized Markov decision processes. Mathematical Centre, Amsterdam, 1964.

3. F.A. Van der Duyn Schouten. Markov decision processes with a continuous time parameters. Thesis, Amsterdam, 1979.

4. A. Hordijk and F.A. Van der Duyn Schouten . Discretization and weak convergence in Markov decision drift processes. *Mathematical Operations Research*, 9(1984)112–141.

5. A. Bensoussan and J.-L. Lions. Contrôle impulsionnel et contrôle continu: methode des inequations quasi-variationnelles non-linear. *Comptes Rendus de l'Academie des Sciences Series* A, 278(1974)675–679.

6. H.J. Kushner. *Probability Methods for Approximations in Stochastic Control and for Elliptic Equations*. Academic Press, New York, 1977.

7. A.N. Shiryaev. Statistical sequential analysis. *Translations of Math. Monographs*, Vol. 18, Providence, R.I., 1973.

8. N.V. Krylov. *Controlled Diffusion Processes.* Springer-Verlag, 1980.

9. M.H.A Davis. Piecewise deterministic Markov processes: a general class of non-diffusion stochastic models. *Journal of the Royal Statistical Society (B)* (1984).

10. D. Vermes. Optimal control of piecewise deterministic Markov processes. *Stochastics* (1984).

11. G. Pragarauskas. Some problems in optimal control theory concerning the solutions of stochastic integral equations. Thesis, Vilnius, 1983.

12. P.A. Meyer. *Probabilitiés et potentiel.* Hermann, Paris, 1966.

13. I.I. Gihman and A.V. Skorohod. *Controlled Stochastic Processes.* Springer-Verlag, 1979.

14. M. Kohlmann and R.A. Rishel. A variational inequality for a partially observed stopping time problem. In *Stochastic Control Theory and Stochastic Differential Systems*, Lecture Notes in Control and Information Sciences, Vol. 16, Springer-Verlag, 1979, pp. 472–480.

15. A.A. Yüshkevich and R.Ya. Chitashvili. Controlled random sequences and Markov chains (in Russian). *Uspehki Matematicheskie Nauk*, 37(1982)213–242.

16. A.A. Yushkevich. Controlled jump Markov processes (in Russian). *Doklady Academii Nauk SSSR*, 233(1977)304–307.

17. E.B. Dynkin and A.A. Yushkevich. *Controlled Markov Processes*, Springer-Verlag, 1979.

18. A.A. Yushkevich and E.A. Fainberg. On homogeneous controlled Markov models with continuous time and a finite or countable state space (in Russian). *Teoriya Veroyatnosti i ee Primeneniya*, 24(1979)155–160.

Section II
Stochastic Extremal Problems

STOCHASTIC PROGRAMMING METHODS: CONVERGENCE AND NON-ASYMPTOTIC ESTIMATION OF THE CONVERGENCE RATE

Ya.I. Al'ber and S.V. Shil'man
Gorky University, Gorky, USSR

Iterative stochastic optimization algorithms are widely used to solve problems of estimation, adaptation, learning, etc. It is clearly important to study the convergence, rate of convergence, stability and other properties of such procedures. For some time, the question of convergence has received a good deal of attention while the rate of convergence has received relatively little. The few estimates of convergence rates given in the literature are generally asymptotic in character. In addition, they have been obtained only for those optimization problems which satisfy the condition of strong convexity or the condition of strong pseudo-differentiability (see the work of Chung, Venter, Burkholder, Ermoliev, Tsypkin, Polyak, etc.).

Again, when using iterative methods it is necessary to estimate the smallest number of iterations required to obtain a given accuracy. It is clear that asymptotic estimates of the convergence rate are not suitable for this purpose. In addition, such estimates do not generally suggest any course of action regarding the iterative process as a whole.

This paper summarizes results obtained recently by the authors on the non-asymptotic (valid from the first step) analytical estimation of the convergence rate of stochastic iterative algorithms. We study both singular and non-singular problems, including those subject to the conditions of strong and uniform pseudo-differentiability.

We shall consider a function $f(x)$, $x \in R^l$, and two types of minimization problems. The first involves the finding of minimum points $x^* = \arg\min_{x \in X \subset R^l} f(x)$ while the second is concerned only with the lower edge of the function value $f^* = \inf_{x \in X \subset R^l} f(x)$. This distinction is particularly important when studying the singular cases. For problems of the first type we shall estimate the mean square of the error $\lambda_n = \mathrm{E}[\rho^2(x_n, X^*)]$, where $\{x_n\}$ is a random iterative sequence, E is the mathematical expectation, and $\rho(\cdot)$ is the distance from the point x_n to the set of minimum points X^*, which is assumed to be convex and closed. For problems of the second type we formulate the problem to investigate the mean values of the difference $\lambda_n = E[f(x_n) - f^*]$.

We shall consider iterative stochastic minimization algorithms of the form

$$x_{n+1} = \pi_X[x_n - \alpha_n S_n(x_n)] \, , \; n = 1, \ldots, \qquad (1)$$

where $S_n(x_n)$ is the vector of random observations at the n-th iteration at a point x_n, $\alpha_n \geq 0$ is the given numerical sequence, and $\pi_x(\cdot)$ denotes the projection operator for the closed convex set X. If $x = R^l$, then $\pi_X(\cdot) = I$ where I is the unit matrix.

Formula (1) describes various conditional and unconstrained stochastic optimization processes. In what follows we consider only the most well-known of these, although this approach may also be applied to a much wider range of methods and problems.

(A) *The Robbins-Monro algorithm.* In this case the vector $S_n(x_n)$ is given by

$$S_n(x_n) = f'(x_n) + \eta_n \quad ,$$

where $f'(x)$ is the (sub)gradient of the function $f(x)$ and η_n is a sequence of independent random vectors such that $E[\eta_n] = 0$, $E[\|\eta_n\|^2] \leq \sigma^2 < \infty$.

(B) *The Kiefer–Wolfowitz algorithm.* In this case the vector $S_n(x_n)$ has components

$$\frac{1}{2c_n} [f(x_n + c_n e_i) - f(x_n - c_n e_i) + \eta_{n,i}] \, , \; i = \overline{1,l} \quad .$$

at each point x_n. Here $\{e_i\}$ is the orthonormal basis in R^l, $\eta_n = [\eta_{n,1}, \ldots, \eta_{n,l}]^T$ is a vector characterizing the error in calculating difference of the functions, the set $\{\eta_n\}$ consists of random vectors which are independent for each value of n and such that $E[\eta_n] = 0$, $E[\|\eta_n\|^2] \leq \sigma_1^2 < \infty$, and $c_n \geq 0$ is a given sequence. For simplicity we shall take $c_n = c(n + n_0)^{-r}$, $c > 0$, $n_0 \geq 0$, $r > 0$.

(C) *The random search algorithms.* In this case the vector $S_n(x_n)$ is defined by the equality

$$S_n(x_n) = \frac{1}{2c_n} [f(x_n + c_n u_n) - f(x_n - c_n u_n) + \eta_n] u_n \quad ,$$

where η_n, c_n are defined as in the Kiefer–Wolfowitz algorithm and u_n is a sequence of independent vectors which are uniformly distributed over a sphere of unit radius centered at zero. The sets $\{\eta_n\}$ and $\{u_n\}$ are mutually independent.

Our aim is to give for each algorithm a worst-case estimate of the accuracy λ_n for a sufficiently wide class of functions $f(x)$. The choice of this class has an important influence on the estimate of the rate of decrease of λ_n to zero.

Non-asymptotic estimates of the convergence rate are derived from studies of the behavior of the solutions of the recursive numerical inequalities

$$\lambda_{n+1} \le (1+\beta_n)\lambda_n - \rho_n \Psi(\lambda_n) + \gamma_n \quad , \tag{2}$$

where $\beta_n \ge 0$, $\rho_n \ge 0$, $\gamma_n \ge 0$ $\sum_{n=1}^{\infty} \beta_n < \infty$, $\sum_{n=1}^{\infty} \rho_n = \infty$, and $\Psi(\lambda)$ is a convex, strictly increasing function for $\lambda \ge 0$, $\Psi(0) = 0$.

The function $\Psi(\lambda)$ characterizes the degree of singularity of the optimization problem, with $\Psi(\lambda) = \vartheta\lambda$, $\vartheta > 0$, corresponding to the non-singular case and $\Psi(\lambda) = \vartheta\lambda^p$, $p > 1$, to the singular case. However, particularly striking results are obtained when $\Psi(\lambda) = \vartheta\lambda^p$, $(1+\beta_n)^{-1}\rho_n \ge b(n+n_0)^{-t}$, $b > 0$, $0 < t \le 1$, $\gamma \le d(n+n_0)^{-s}$, $d > 0$, $s > t$. The corresponding statements are presented below. In this case we have $\bar{c} \ge \prod_1^{\infty}(1+\beta_n)$.

LEMMA 1. *Let* $\lambda_n \ge 0$, $n = 1, \ldots,$ *satisfy the inequality*

$$\lambda_{n+1} \le (1+\beta_n)\lambda_n - \rho_n\lambda_n + d(n+_0)^{-s} \quad ,$$

where $s > 1$, $t = 1$, $n_0 \ge 0$, $n_0 = \text{const}$ *and let*

$$u(x, C) = C\left(\frac{2+n_0}{x+n_0}\right)^{ab}, \quad v(x) = \left(\frac{d}{b}c_0 + d\right)\left(\frac{1}{x+n_0-1}\right)^{s-1}$$

$$A = \lambda_1\left(\frac{n_0+1}{n_0+2}\right)^{ab}, \quad C_1 = \left(\frac{d}{b}c_0 + d\right)\left(\frac{1}{n_0+1}\right) \quad .$$

Then $\lim_{n\to\infty} \lambda_n = 0$ *and the following statements are true*:

1. *If* $b > s-1$ *and the arbitrary parameter* c_0 $(c_0 > 1)$ *is chosen such that* $ab > s-1$, $a = c_0(c_0-1)^{-1}$, *then*

$$\lambda_n \le \bar{C}u(n, C), \text{ if } 2 \le n \le \bar{x} \quad ,$$
$$\lambda_n \le \bar{C}v(n), \text{ if } n > \bar{x} \quad , \tag{3}$$

where $C = \max\,[A, C_1]$ and $\bar{x}$ *is the single root of the equation* $u(x, C) = v(x)$ *on the interval* $(2, \infty)$.

If for $A \le C_1$ *we have* $ab \ge (s-1)(2+n_0)(1+n_0)^{-1}$, *then the estimate* (3) *holds for all* $n \ge 2$.

2. *If* $b > s-1$ *and* $ab \le s-1$, *or if* $b \le s-1$, *then* $\lambda_n \le \bar{C}u(n, C)$, $n \ge 1$.

LEMMA 2. *Let* $\lambda_n \geq 0$, $n = 1, \ldots,$ *satisfy the inequality*

$$\lambda_{n+1} \leq (1+\beta_n)\lambda_n - \rho_n \lambda_n + d(n+n_0)^{-s} \quad ,$$

where $0 < t < 1$, $s > t$, $n_0 \geq 0$, $n_0 = \text{const}$, *and let*

$$v(x) = \left(\frac{d}{b} c_0 + d\right) \left(\frac{1}{x+n_0-1}\right)^{s-t}, \quad C_1 = \left(\frac{d}{b} c_0 + d\right)(n_0+1)^{-(s-t)}$$

$$u(x, C) = C \exp\{-\frac{ab}{1-t}[(x+n_0)^{1-t} - (2+n_0)^{1-t}]\}$$

$$A = \lambda_1 \exp\{-\frac{ab}{1-t}[(2+n_0)^{1-t} - (1+n_0)^{1-t}]\} \quad .$$

Then $\lim_{n\to\infty} \lambda_n = 0$ *and we have the following estimates:*

$$\begin{aligned} &\lambda_n \leq \bar{C}u(n, C), \text{ if } 2 \leq n \leq \bar{x} \\ &\lambda_n \leq \bar{C}v(n), \text{ if } n > \bar{x} \quad , \end{aligned} \tag{4}$$

where $C = \max[A, C_1]$ *and* $\bar{x}$ *is the single root of the equation* $u(x, C) = v(x)$ *on the interval* $(2, \infty)$. *If for* $A \leq C_1$ *and* $c_0 > 1$ *we have* $ab \geq (s-t)(2+n_0) \times (1+n_0)^{-1}$, *then* (4) *holds for all* $n \geq 2$.

LEMMA 3. *Let* $\lambda_n \geq 0$, $n = 1, \ldots,$ *satisfy the inequality*

$$\lambda_{n+1} \leq (1+\beta_n)\lambda_n - \rho_n \lambda_n^p + d(n+n_0)^{-s} \quad ,$$

where $p > 1$, $s > 1$, $t = 1$, $n_0 \geq 0$, $n_0 = \text{const}$, *and let*

$$u(x, C) = C[1 + (p-1)C^{p-1}ab \ln\left(\frac{x+n_0}{2+n_0}\right)]^{-\frac{1}{p-1}}$$

$$C_1 = \left[\left(\frac{d}{b} c_0\right)^{1/p} + d\right](n_0+1)^{-\frac{s-1}{p}}$$

$$A = \lambda_1\left[1 + (p-1)\lambda_1^{p-1}ab \ln\left(\frac{2+n_0}{1+n_0}\right)\right]^{-\frac{1}{p-1}} \quad .$$

Then $\lim_{n\to\infty} \lambda_n = 0$, *and if* $c_0 > 1$ *is chosen to satisfy the condition*

$$[(\frac{d}{b} c_0)^{1/p} + d]^{p-1}ab \leq \frac{s-1}{p} \quad ,$$

then

$$\lambda_n \le \bar{C}u(n, C), \; C = \max[A, C_1]$$

for all $n \ge 1$.

LEMMA 4. *Let* $\lambda_n \ge 0$, $n = 1, \ldots,$ *satisfy the inequality*

$$\lambda_{n+1} \le (1+\beta_n)\lambda_n - \rho_n \lambda_n^p + d(n+n_0)^{-s} \quad ,$$

where $p > 1$, $s > t$, $0 < t < 1$, $n_0 \ge 0$, $n_0 = \text{const}$, *and let*

$$v(x) = [\frac{d}{b} c_0)^{1/p} + d] \left(\frac{1}{x+n_0-1}\right)^{\frac{s-t}{p}}$$

$$C_1 = [(\frac{d}{b} c_0)^{1/p} + d](n_0+1)^{-\frac{s-t}{p}}$$

$$u(x, C) = C\left\{1 + C^{p-1}ab\left(\frac{p-1}{1-t}\right)\left[(x+n_0)^{1-t} - (2+n_0)^{1-t}\right]\right\}^{-\frac{1}{p-1}}$$

$$A = \lambda_1\{1 + \lambda_1^{p-1}ab((\frac{p-1}{1-t})[(2+n_0)^{1-t} - (1+n_0)^{1-t}]\}^{-\frac{1}{p-1}} \quad .$$

Then $\lim_{n\to\infty} \lambda_n = 0$ *and the following statements are true:*

1. *If* $\frac{s-t}{p} < \frac{1-t}{p-1}$, *then*

$$\lambda_n \le \bar{C}u(n, C), \text{ if } 2 \le n \le \bar{x}$$
$$\tag{5}$$
$$\lambda_n \le \bar{C}v(n), \text{ if } n > \bar{x} \quad ,$$

where $C = \max[A, C_1]$ *and* $\bar{x}$ *is the single root of the equation* $u(x, C) = v(x)$ *on the interval* $(2, \infty)$.

If $A \le C_1$ *and* $c_0 > 1$ *is chosen to satisfy the condition*

$$ab[(\frac{d}{b} c_0)^{1/p} + d]^{p-1} \ge \frac{(2+n_0)^t}{(1+n_0)^\kappa} \frac{s-t}{p}$$

$$\kappa = 1 - \frac{(s-t)(p-1)}{p} \quad ,$$

then (5) *is valid for all* $n \ge 2$.

2. *If* $\frac{s-t}{p} \ge \frac{1-t}{p-1}$ *and* $c_0 > 1$ *is chosen to satisfy the condition*

$$ab[(\frac{d}{b} c_0)^{1/p} + d]^{p-1} \le \frac{s-t}{p}$$

then $\lambda_n \le \bar{C}u(n, C)$ *for all* $n \ge 1$.

In minimization problems of the first type we define classes of functions using the following conditions:

1. $(f'(x), x - x^*) \ge \Psi(\|x - x^*\|^2)$, $x \in X$;

2. $\|f'(x)\|^2 \le \sigma_2 + \tau \|x - x^*\|^2$, $\sigma_2 \ge 0$, $\tau \ge 0$, $x \in X$.

When considering search methods it is also assumed that $f'(x)$ satisfies the local Lipschitz–Gelder condition:

3. $\|f'(x) - f'(y)\| \le L\|x - y\|^\mu$, $L > 0$, $0 < \mu \le 1$, $\|x - y\| \le c$.

Condition 1 describes the structure of functions $f(x)$, condition 2 the order of growth of $f'(x)$ at infinity and condition 3 the smoothness of functions $f(x)$. We note that strongly convex functions satisfy condition 1 wth $\Psi(\lambda) = \vartheta\lambda$, $\vartheta > 0$, (the non-singular case), and uniformly convex functions satisfy this condition, for example, with $\Psi(\lambda) = \vartheta\lambda^p$, $p > 1$, even if $\lambda \le N$ (for a uniformly convex function with a power singularity). Arbitrary convex functions (with a non-regular singularity) do not in general satisfy a condition of type 1.

When analyzing minimization problems of the second type we shall take $X = R^l$ and define classes of functions using the following conditions:

1'. $\|f'(x)\|^2 \ge \Psi[f(x) - f^*]$

and there exists an $N > 0$ such that $\Psi(\lambda) = \vartheta\lambda^m$, $\vartheta > 0$, $1 \le m \le 2$, on the segment $[0, N]$;

2'. $\|f'(x) - f'(y)\| \le L\|x - y\|^\mu$, $L > 0$, $0 < \mu \le 1$, (the Lipschitz–Gelder condition defining the class of function $C^{1,\mu}$).

Instead of 1' we may take an alternative condition:

1." $E[\|f'(x_n)\|^2] \ge \vartheta\,[E(f(x_n) - f^*)]^m$, $1 \le m \le 2$,

where x_n is the sequence generated by algorithm (1).

We have stated that, for convex functions condition 1" is satisfied if $\Psi(\lambda) = \vartheta\lambda^2$ for algorithms (A)–(C). In the Robbins–Monro algorithm it is also necessary to have algorithm $\sum_{n=1}^{\infty} \alpha_n = \infty$, $\sum_{n=1}^{\infty} \alpha_n^2 < \infty$ and in the search methods

$$\sum_{n=1}^{\infty} \alpha_n = \infty, \quad \sum_{n=1}^{\infty} \alpha_n C_n^\mu < \infty, \quad \sum_{n=1}^{\infty} \alpha_n^2 c_n^{-2} < \infty \quad .$$

These assumpstions are commonly made in theorems on convergence with probability 1. Thus, condition 1" makes it possible to treat convex problems with a non-regular singularity. Strongly convex functions of the class $C^{1,1}$ satisfy condition 1 when $\Psi(\lambda) = \vartheta\lambda$, $\vartheta > 0$.

We shall now briefly describe the results concerning convergence and the convergence rate which can be derived from Lemmas 1–4. The classes of function introduced above allow us to find recursive inequalities of form (2) for each of the algorithms (A)–(C) [1–3]. In particular, taking $a_n = a(n+n_0)^{-t}$, $a>0$, $0<t\le 1$ for the Robbins–Monro algorithm we obtain $\rho_n(1+\beta_n)^{-1} \ge b(n+n_0)^{-t}$. In this case we have

$$b(a,\vartheta)>0\ ,\ \gamma_n \le d(n+n_0)^{-2t}\ ,\ d(a,\sigma^2,\sigma_2)\ge 0\quad ,$$

for minimization problems of the first type and

$$b(a,\vartheta,\mu)>0\ ,\ \gamma_n \le d(n+n_0)^{-(1+\mu)t}\ ,\ d(a,\sigma^2,\mu)\ge 0\quad .$$

for minimization problems of the second type.

Turning to Kiefer–Wolfowitz and random search algorithms, $\rho_n(1+\beta_n)^{-1}$ has a form similar to that given above, $\gamma_n \le d(n+n_0)^{-s}$, while

$$s=\min\{2(t-r)\ ,\ t+2(2p-1)^{-1}p\,\mu\tau\}$$

for minimization problems of the first type and

$$s=\min\{t+2\mu\tau\ ,\ (1+\mu)(t-\tau)\}$$

for minimization problems of the second type.

All of this makes it possible to study the effects of the structure $p(m)$ and the smoothness of the function μ, and the values of steps t and r, on the convergence of the procedure and the estimates of its convergence rate. Depending on values of these parameters, the estimates may decrease according to some power term (Lemmas 1,2,4) or some logarithmic term (Lemma 3). As a rule, the form of the estimates is different at the initial stage of the search and at large n. For example, the estimates may depend on an exponential expression at small n and on some power term for $n\to\infty$ (Lemma 2). It can happen that the estimates are of the same form in both cases but depend on power terms (Lemmas 1,4). This means that the results obtained in the initial section of the search and as $n\to\infty$ are essentially different. In the case when the uncertainty of the *a priori* information on the position of the extremum is greater than the uncertainty due to noise from estimations, it is reasonable to use a non-decreasing step parameter at the initial stage of the search. Then if $p(m)$ is fixed, the convergence rate increases as t decreases and at fixed t it decreases as p or m increases.

Asymptotic estimates for $p=1$ $(m=1)$ coincide over orders with the results obtained by Chung [4], although they differ slightly by majorant constants. Our estimates differ from Venter estimates [4] over the order by a factor a, $0<a<1$, but they contain the majorant constant which is absent in Venter estimates. For $p>1$ $(1<m\le 2)$ all of the results given above are new, including statements on the

convergence of the process.

For $n \to \infty$, many of our estimates are better. They allow us to find the parameters which produce the fastest decrease to zero in the estimates as $n \to \infty$. We shall now give some of these parameters.

1. For minimization problems of the first type and the Robbins–Monro algorithm we have $t = t^* = p/(2p-1)$ and $E[\|x_n - x^*\|^2] \le 0[(\frac{1}{n})^{\frac{1}{2p-1}}]$. For Kiefer–Wolfowitz and random search algorithms

$$t = t^* = (2p + p\mu - 1)/(1+\mu)(2p-1), \quad r = r^* = (2p-1)t^*/2(2p+p\mu-1) \quad ,$$

and

$$E[\|x_n - x^*\|^2] \le 0\,[(\frac{1}{n})^{\mu/(1+\mu)(2p-1)}] \quad .$$

In particular, for $p = 1$ we obtain $t^* = 1$, $r^* = 1/2(1+\mu)$, and $E[\|x_n - x^*\|^2] \le 0$ $[n^{-\mu/(1+\mu)}]$. If $p = \mu = 1$, then $t^* = 1$, $r^* = 1/4$, and $E[\|x_n - x^*\|^2] \le 0\,(n^{-1/2})$. This coincides with the result obtained by Dupač for the Kiefer–Wolfowitz algorithm.

2. For minimization problems of the second type and the Robbins–Monro algorithm we have $t = t^* = m/(m + m\mu - \mu)$ and

$$E[f(x_n) - f^*] \le 0\,[(\frac{1}{n})^{\mu/(m+\mu m-\mu)}] \quad .$$

If $\mu = 1$ we have $t^* = m/(2m-1)$ and

$$E[f(x_n) - f^*] \le 0\,[(\frac{1}{n})^{1/(2m-1)}] \quad .$$

If $m = 2$, we have $t^* = 2/(2+\mu)$ and

$$E[f(x_n) - f^*] \le 0\,[(\frac{1}{n})^{\mu/(2+m)}] \quad .$$

In particular, if $\mu = 1$ and $m = 2$ we have $t^* = 2/3$ and

$$E[f(x_n) - f^*] \le 0\,[(\frac{1}{n})^{1/3}] \quad ;$$

as μ decreases the asymptotically optimal rate falls and $t^* \to 1$.

If $\mu = 1$ and $m = 1$ we have $t^* = 1$ and

$$E[f(x_n) - f^*] \le O\,(\frac{1}{n}) \quad .$$

For the Kiefer–Wolfowitz and random search algorithms we have

$$t = t^* = m(1+3\mu)/(m+3m\mu+2m\mu^2-2\mu^2)\ ,\ r = r^* = \mu t^*/(1+3\mu)\quad .$$

In this case

$$E[f(x_n) - f^*] \le O\,[(\frac{1}{n})^{2\mu^2/(m+3m\mu+2m\mu^2-2\mu^2)}]\quad .$$

If $\mu = 1$ we have $t^* = 2m/(3m-1)$, $r^* = m/2(3m-1)$, and

$$E[f(x_n) - f^*] \le O\,[(\frac{1}{n})^{1/(3m-1)}]\quad .$$

If $m = 2$ we have $t^* = (1+3\mu)/(1+3\mu+\mu^2)$, $r^* = \mu/(1+3\mu+\mu^2)$, and

$$E[f(x_n) - f^*] \le O\,[(\frac{1}{n})^{\mu^2/(1+3\mu+\mu^2)}]\quad .$$

In particular, if $\mu = 1$ and $m = 2$ we have $t^* = 4/5$, $r^* = 1/5$, and

$$E[f(x_n) - f^*] \le O\,[(\frac{1}{n})^{1/5}]\quad ;$$

and $m = 1$ we have $t^* = 1$, $r^* = 1/4$, and

$$E[f(x_n) - f^*] \le O\,(\frac{1}{n})\quad .$$

If $t \ne t^*$ the rate may fall sharply. Thus if $t = 1$, $m = 2$, then

$$E[f(x_n) - f^*] \le O\,(\frac{1}{\ln n})\quad .$$

REFERENCES

1. Ya.I. Al'ber and S.V. Shil'man. Non-asymptotic estimates of the convergence rate of stochastic iterative algorithms. *Avtomatika i Telemekhanika*, 1(1980)41–54.

2. Ya.I. Al'ber and S.V. Shil'man. On methods of stochastic optimization. *Doklady Akademii Nauk SSSR*, 255(1980)265–269.

3. Ya.I. Al'ber and S.V. Shil'man. Optimal parameters and non-asymptotic estimates of the convergence rate of stochastic algorithms in problems of criterial optimization. *Avtomatika i Telemekhanika*, 10(1984)96–106.

4. M. Vazan. *Stochastic Approximation*. Mir, Moscow, 1972.

SOLUTION OF A STOCHASTIC PROGRAMMING PROBLEM CONCERNING THE DISTRIBUTION OF WATER RESOURCES

I.A. Aleksandrov[1], V.P. Bulatov[1], S.B. Ognivtsev[2] and F.I. Yereshko[2]
[1]Siberian Energy Institute, Irkutsk, USSR
[2]VASHNIL Cybernetics Institute, Moscow, USSR

1. INTRODUCTION

Many optimization problems with economic criteria can be solved using the following linear programming formulation:

$$\min \{cx : Ax \leq b \ , \ x \geq 0\} \quad . \tag{1}$$

This expression is correct for deterministic values of the parameters, but requires additional explanation if some or all components of the matrix A or the vectors b and c are random values. Substitution of these by average values may mean that the model is no longer an adequate representation of the initial problem. Game criteria should be used only when this discrepancy results in a penalty large enough to reduce the effect of minimization of the linear form to zero.

In [1] the stochastic linear programming problem is given as follows:

$$\min \{cx : \mathrm{P}(Ax \leq b) \geq p \ , \ 0 < p \leq 1 \ , \ x \geq 0\} \quad . \tag{2}$$

It is possible to consider the (linear) mathematical expectation $\overline{cx} = \mathrm{E}(cx)$ or its variance $\mathrm{E}(cx - \overline{cx})^2$ instead of cx. The conditions under which the constraints $Ax \leq b$ are satisfied can be given as in (2) or for each line separately. In the latter case p is replaced by the set p_i, $i = \overline{1,m}$, thus allowing the comparative values of individual inequalities to be taken into account. Problems of type (2) can often be reduced to deterministic problems, in particular to convex programming problems. Here we shall use this approach to solve the problem of water resource distribution which arises when planning the allocation of agricultural production.

2. PROBLEM STATEMENT

We shall assume that there are K climatic zones in which I types of agricultural output can be produced using J technologies and L sources of water.

We shall use the following notation:

x_{ijk} – area allocated for cultivation of crop i using technology j in region k

a_{ijk}, u_{ijk}, v_{ijk}, c_{ijk} – crop yielding, labor-intensiveness, water consumption rate and land price associated with area x_{ijk}

y_k – labor supply in region k

z_{kl} – amount of water taken from source l in region k

d_{kl} – discounted cost of water intake z_{kl}

w_l – total water available (flow) from source l

b_{i0} – given total volume of crop i produced

b_{ik} – given minimum volume of crop i produced in region k

t_i – sequence coefficient for crop i

f_k – amount of land available in region k.

Now we shall formulate a deterministic linear programming problem:

$$\min \Big(\sum_{i,j,k} c_{ijk} x_{ijk} + \sum_{k,l} d_{kl} z_{kl} \Big) \tag{3}$$

$$\sum_{j,k} a_{ijk} x_{ijk} \geq b_{i0} \, , \; i \in I \tag{4}$$

$$\sum_{j} a_{ijk} x_{ijk} \geq b_{ik} \, , \; i \in I \, , \; k \in K \tag{5}$$

$$\sum_{\alpha \in I} x_{\alpha jk} - t_i x_{ijk} \geq 0 \, , \; i \in I \, , \; j \in J \, , \; k \in K \tag{6}$$

$$\sum_{i,j} u_{ijk} x_{ijk} \leq y_k \, , \; k \in K \tag{7}$$

$$\sum_{i,j} x_{ijk} \leq f_k \, , \; k \in K \tag{8}$$

$$\sum_{i,j} v_{ijk} x_{ijk} - \sum_{l} z_{kl} \leq 0 \, , \; k \in K \tag{9}$$

$$\sum_{k} z_{kl} \leq w_l \, , \; l \in L \tag{10}$$

$$x_{ijk} \geq 0 \, ; \; z_{kl} \geq 0 \, ; \; i \in I \, , \; j \in J \, , \; k \in K \, , \; l \in L \quad . \tag{11}$$

The unknown variables here are x_{ijk} and z_{kl} $(i \in I,\ j \in J,\ k \in K,\ l \in L)$. Constraints (4) and (5) fix the minimum volumes of agricultural output, constraint (6) ensures that crop sequence conditions are satisfied, while constraints (7)–(10) relate demands for labor, land and water resources to their given maximum values y_k, f_k and w_l.

Now assume that the matrix a and the vectors y and w are no longer determinate. Let the elements of the matrix a and the components of the vector y be independent, normally distributed random variables with mathematical expectations $\overline{a_{ijk}}$, $\bar{y}_k$ and variances σ_{ijk}^2, γ_k^2, and the components of the vector w be independent random variables with a gamma distribution described by parameters $\bar{w}_l$ and w_l^2.

Furthermore, let $p_{i0} \geq 0.5$, $p_{ik} \geq 0.5$, q and r be given probabilities.

In this case constraints (4) and (5) may be replaced by the following inequalities:

$$\mathrm{P}(\sum_{j,k} a_{ijk} x_{ijk} \geq b_{i0}) \geq p_{i0}\ ,\ i \in I$$

$$\mathrm{P}(\sum_{j} a_{ijk} x_{ijk} \geq b_{ik}) \geq p_{ik}\ ,\ i \in I\ ,\ k \in K \quad ,$$

which may be reformulated as follows [1]:

$$\Phi^{-1}(p_{i0}) \sqrt{\sum_{j,k} \sigma_{ijk}^2 x_{ijk}^2} \leq \sum_{j,k} \overline{a_{ijk}} x_{ijk} - b_{i0}\ ,\ i \in I \tag{12}$$

$$\Phi^{-1}(p_{ik}) \sqrt{\sum_{j} \sigma_{ijk}^2 x_{ijk}^2} \leq \sum_{j} \overline{a_{ijk}} x_{ijk} - b_{ik}\ ,\ i \in I\ ,\ k \in K \quad , \tag{13}$$

where $\Phi^{-1}(p)$ is a quantile of the normal distribution.

Constraints (7) and (10) may also be replaced by the probabilistic inequalities

$$\mathrm{P}(\sum_{i,j} u_{ijk} x_{ijk} \leq y_k\ ,\ k \in K) \geq q$$

$$\mathrm{P}(\sum_{k} z_{kl} \leq w_l\ ,\ l \in L) \geq r$$

which may be reduced to the form

$$\ln q - \sum_{k} \ln\,[1 - \Phi(y_k)] \leq 0 \tag{14}$$

$$\ln r - \sum_{l} \ln\,[1 - \Gamma(\alpha_l\ ,\beta_l)] \leq 0 \quad , \tag{15}$$

where $\Gamma(\alpha_l\ ,\beta_l)$ is an incomplete gamma function with parameters

$$\alpha_l = w_l \left[\frac{\bar{w}_l}{w_l^2} \right] , \quad \beta_l = \left[\frac{\bar{w}_l^2}{w_l^2} \right] .$$

In this case inequalities (7) and (10) still hold but the values y_k, w_l on the right-hand sides should be interpreted as the desired values of corresponding components of the random vectors y and w.

The left-hand sides of inequalities (12)–(15) with $p_{i0} \geq 0.5$, $p_{ik} \geq 0.5$ are convex functions. This therefore leads to a convex programming problem determined by conditions (3), (6)–(15).

3. SOLUTION PROGRAMS

Two programs, MODEL and CONE, have been developed to solve problems of the above type.

MODEL makes it possible to write down the problem conditions in a compact form, to input the initial data with the required comments and names, and to specify print formats by means of a specially developed procedure language. Using the recorded data and problem conditions, MODEL forms a matrix of coefficients for linear constraints, a vector of right-hand sides, a (row) vector of objective function coefficients, and upper and lower bounds on the variables, as well as calculating an initial approximation of the problem solution. For nonlinear constraints the program constructs an information table including the type of constraint, a one-dimensional index for each random variable, and the nature of the distribution law and its parameters (the mathematical expectation and variance for the normal distribution, α and β for the gamma distribution, etc.). When the solution has been found, MODEL decodes the results and prints them out, giving additional calculations, aggregating tables, etc., as necessary.

CONE is designed to solve the general convex programming problem:

$$\min \{\varphi(x): x \in R\} \quad , \tag{16}$$

where $\varphi(x)$ is a convex scalar function and $R \in E^n$ is a convex set, int $R \neq \phi$.

Before describing the support cone method [2] used in CONE we must first introduce some definitions.

1. A direction s ($\|s\| \leq c < \infty$) at a point $x \notin R$ is said to be ***admissible*** if there exists a $\lambda > 0$ such that $x + \lambda s \in R$. The set of admissible directions will be denoted by $S(x)$.

2. The set R^k will be called a *support set* with respect to the convex set R at its boundary point $\bar{x}^k$ if $\bar{x}^k$ is also a boundary point of R^k and $R^k \supset R$.

Suppose that the set $R^0 = \{x: \underline{x} \le x \le \bar{x}\}$ which contains the minimum point x^* of $\varphi(x)$ on R is known.

Let $x^1 = \text{argmin}\,\{\varphi(x): x \in R^0\}$. If $x^1 \in R$, x^1 is the solution of problem (16). Otherwise, determine $s^1 \in S^1(x^1)$ and find the intersection point $\bar{x}^1$ of the ray $x = x^1 + \lambda^1 s^1$, $\lambda^1 > 0$, with the boundary of R, and the convex closed support set R^1 corresponding to $\bar{x}^1$.

Determine

$$x^2 = \text{argmin}\,\{\varphi(x): x \in R^1 \cap R^0\} \quad .$$

Then, in a similar way, we construct a convex set R^2 which is the support set with respect to R at the point $\bar{x}^2 = x^2 + \lambda^2 s^2 \in R_G$, where R_G is the set of boundary points of R; we then obtain the following approximation:

$$x^3 = \text{argmin}\,\{\varphi(x): x \in \bigcap_{j=1}^{2} R^j \cap R^0\} \quad .$$

If $x^3 \in \bigcap_{j=1}^{2} R_G^j$ then

$$x^4 = \text{argmin}\,\{\varphi(x): x \in \bigcap_{j=1}^{3} R^j \cap R^0\} \quad .$$

If $x^3 \notin R_G^1$ then

$$x^4 = \text{argmin}\,\{\varphi(x): x \in \bigcap_{j=2}^{3} R^j \cap R^0\} \quad .$$

Now we shall write down the general step. Let

$$x^k = \text{argmin}\,\{\varphi(x): x \in \bigcap_{j \in J_{k-1}} R^j \cap R^{k-1}\} \quad .$$

Suppose that $x^k \in \text{int}\, R^j \;\forall j \in J_{k-1}^1 \subset J_{k-1}$ (the set J_{k-1}^1 can be empty).

Determine

$$\bigcap_{j \in J_k} R^j = \bigcap_{j \in J_{k-1}/J_{k-1}^1} R^j \cap R^{k-1} \cap R^0 \quad ,$$

where R^j is a convex closed support set with respect to R at the point $\bar{x}^j = x^j + \lambda^j s^j$ and find

$$x^{k+1} = \text{argmin}\,\{\varphi(x): x \in \bigcap_{j \in J_k} R^j \cap R^k\} \quad ,$$

where R^k is a convex closed support set with respect to R at the point $\bar{x}^k$.

Thus, the significance of the support sets R^j is checked at each step of the iterative process. If any such sets do not affect the problem solution, i.e., the minimum point belongs to the interior of their intersection they are not included in the next problem description.

The following theorem holds:

THEOREM 1. *Let $\varphi(x)$ be strictly convex. Then*

1. $\lim_{k \to \infty} x^k = \bar{x} \in R_G$;
2. $\lim_{k \to \infty} \varphi(x^k) = \varphi(x^*)$.

In CONE we have $\varphi(x) = c^T x$, $R = \{x\colon g_j(x) \le 0 \, , \, j = \overline{1,m}\}$, and the sets R^k are constructed as the intersections of half-spaces which are support sets with respect to R, i.e.,

$$R^k = \{x : A^k x \le b^k\} \quad ,$$

where A^k is a non-singular $n \times n$ matrix.

Hence, the solution of problem (16) is reduced to the sequential solution of a system of algebraic equations $A^k x = b^k$, the matrices of which differ from step to step only in one row and one column. If the inverse matrix $(A^{k-1})^{-1}$ is known, the inversion of matrix A^k obviously presents no difficulty. Moreover, it is quite unnecessary to invert matrices of large dimensions as the number of active constraints of the general type is, as a rule, considerably less than n.

This method gives a two-sided estimate of the error in the approximate solution at each step. By virtue of the above construction we have

$$\varphi(x^k) < \varphi(x^*) < \varphi(\bar{x}^k) \quad ,$$

i.e., the iterative process terminates when

$$|\varphi(x^k) - \varphi(\bar{x}^k)| < \varepsilon \quad .$$

4. REMARKS AND CONCLUSIONS

The model described above does not include all the constraints present in the original (dealing with cattle breeding, fodder production, etc.), but these are not necessary for our purposes. The original model involved about 350 variables with 225 linear constraints of the general type and 1–10 nonlinear constraints. Calculations were performed for approximately 100 variants. The computation time per variant was, on the average, 15–17 minutes using a BESM-6 computer. The number of active constraints in the optimal solution varied from 135 to 150 and the number of iterations from 320 to 420, i.e., the number of iterations exceeded the number of active constraints by a factor of 2–3.

Our calculations have shown that the support cone method used for the solution of convex programming problems works reliably and with high accuracy. It may be possibile to increase the speed of the method, especially for a series of calculations with small changes in the initial parameters. The program CONE is currently being modified along these lines.

The calculations have also shown that the approach used to solve stochastic linear programming problems offers substantially more scope for the analysis of the models under consideration than variant calculations using the deterministic models. At the same time the reduction of the initial stochastic model to a convex deterministic model is often a non-trivial problem and may lead to a model which is no longer an adequate representation of the phenomenon studied. However, even in this case the information obtained can be useful in the analysis of initial models.

REFERENCES

1. D.B. Judin. *Mathematical Methods for Control under Conditions of Insufficient Information*. Sovietskie Radio, Moscow, 1974 (in Russian).
2. V.P. Bulatov. *Imbedding Methods in Optimization Problems*. Nauka, Novosibirsk, 1977 (in Russian).

LIMIT THEOREMS FOR PROCESSES GENERATED BY STOCHASTIC OPTIMIZATION ALGORITHMS

V.V. Anisimov
Kiev State University, Kiev, USSR

Attempts to solve certain optimization, computation and estimation problems have led to studies of the convergence of recursive stochastic algorithms. These arise in connection with stochastic approximation problems (Robbins–Monro and Kiefer–Wolfowitz procedures), random search methods, adaptive control procedures, problems involving the recursive estimation of parameters, and so on. Sequential iterations of such algorithms generate special classes of step processes in the Skorohod space **D** which under sufficiently general assumptions converge to Markov diffusion processes.

A new technique for investigating the convergence of such processes is suggested. It uses and develops the results of Gihman and Skorohod [1,2] concerning the convergence of triangular arrays of random variables to a solution of a stochastic differential equation (SDE).

The main features of this approach are outlined below. Consider a recursive vector procedure

$$x_{k+1} = x_k + \alpha_k \gamma_k \, , \; k \geq 0 \quad , \tag{1}$$

where the α_k are real numbers, the γ_k are random vectors in R^m whose distribution may depend on $\bar{x}_k = (x_0, \ldots, x_k)$ as on a parameter, and x_0 is an initial value. Let $k \geq 0$ be an increasing sequence of σ-algebras such that x_0 is $\mathbf{F}_0$-measurable and γ_k is $\mathbf{F}_{k+1}$-measurable for every k. Hence $\sigma_k = \sigma(x_0, \ldots, x_k) \subset \mathbf{F}_k$, $k > 0$. We study the convergence of the processes

$$\eta_n(t) = B_n x_k \quad \text{for} \quad \frac{k}{n} \leq t < \frac{k+1}{n} \, , \; t \geq 0 \quad , \tag{2}$$

where B_n is a normalizing factor. Put $\eta_{nk} = B_n x_k$, $k \geq 0$. Then (1) can be rewritten [1] in the form

$$\eta_{nk+1} = \eta_{nk} + a_{nk}(\bar{\eta}_{nk})\frac{1}{n} + b_{nk})(\bar{\eta}_{nk})\xi_{nk} \, , \; k \geq 0 \quad , \tag{3}$$

where $a_{nk}(\bar{y}_k)$, $b_{nk}(\bar{y}_k)$ are a σ_k-measurable vector and matrix, respectively, and ξ_{rk} is a σ_k-difference martingale determined as follows. Let

$$a_k(\bar{x}_k) = \mathrm{E}[\gamma_k / \bar{x}_k] \quad ,$$

$$\mathrm{D}_k^2(\bar{x}_k) = \mathrm{E}[(\gamma_k - a_k(\bar{x}_k))(\gamma_k - a_k(\bar{x}_k))^* / \bar{x}_k] \quad .$$

Then

$$\begin{aligned} a_{nk}(\bar{y}_k) &= nB_n \alpha_k a_k(B_n^{-1}\bar{y}_k) \quad , \\ b_{nk}(\bar{y}_k) &= \sqrt{n} B_n \alpha_k \mathrm{D}_k(B_n^{-1}\bar{y}_k) \quad , \end{aligned} \tag{4}$$

and

$$\xi_{nk} = \frac{1}{\sqrt{n}} \mathrm{D}_k(\bar{x}_k)^{-1}(\gamma_k - a_k(\bar{x}_k)) \quad ,$$

so that by construction

$$\mathrm{E}[\xi_{nk} / \sigma_k] = 0 \quad , \quad \mathrm{E}[\xi_{nk}\xi_{nk}^* / \sigma_k] = \frac{1}{n} \quad .$$

(Here and elsewhere a^* denotes the transpose of the row vector a, and (a, b) represents the scalar product $|a|^2 = (a, a)$, $|A|^2 = \sup_{|x|=1} (Ax, Ax)$, and $q\bar{x}_k = (qx_0, \ldots, qx_k)$, $k \geq 0$).

Note that for any fixed k we generally have $\eta_{nk} \xrightarrow{P} 0$, $n \to \infty$ and the coefficients are unbounded. Thus the standard technique which assumes convergence of the initial values and regular coefficients is not applicable. For this reason the investigation of (1), (2) is divided into the following stages:

1. Conditions are found under which the measures corresponding to $\eta_n(t)$ are weakly compact in $\mathbf{D}$ on each segment $[\delta, T]$, $\delta > 0$.

2. Conditions for step processes (sums of the ξ_{nk} converging to a square-integrable martingale $\psi(t)$) are found.

3. Under certain assumptions about the coefficients (4) (bounded growth rate, convergence to sufficiently smooth functions $a(t, y)$, $b(t, y)$), we prove convergence by subsequences in $\mathbf{D}$ on $[\delta, T]$ of the processes $\eta_n(t)$ to the process $\eta(t)$. This is the solution to an SDE of the form

$$d\eta(t) = a(t, \eta(t))dt + b(t, \eta(t))d\psi(t), \quad t \geq \delta \quad , \tag{5}$$

and $\eta(\delta)$ is a proper initial value whose distribution is generally speaking unknown.

4. Using the properties of the class of equations (5), we take the limit as $\delta \to 0$, thus getting rid of the unknown value $\eta(\delta)$ and establishing the uniqueness of the representation of $\eta(t)$ on $(0, T]$, $T > 0$.

This approach was first applied in [3] to procedures of the Robbins–Monro type with weakly dependent disturbances in the observations, and was extended in [4,5] to more general stochastic optimization algorithms.

We shall now state a general theorem which is useful when investigating the convergence of such processes. We will assume for simplicity that the coefficients $a_{nk}(\bar{y}_k)$, $b_{nk}(\bar{y}_k)$ depend uniformly with respect to k on only a finite number of arguments, i.e., there exists an integer N such that for any $k \geq N$ we have $a_{nk}(\bar{y}_k) = a_{nk}(y_{k-N}, y_{k-N+1}, \ldots, y_k)$ (the same is true for $b_{nk}(\cdot)$). Write $\|\bar{y}_k\| = \max_{0 \leq i \leq k} |y_i|$, $k \geq 0$. We shall consider a normalizing factor of the form $B_n = n\beta$, $\beta > 0$.

THEOREM 1. *Let the following conditions hold for any* $\delta > 0$:

1. $\overline{\lim}_{k \to \infty} k^{2\beta} \mathrm{E} |x_k|^2 < K < \infty$;

2. *For any* $n > 0$, $n\delta \leq k \leq nT$, *we have*

$$|a_{nk}(\bar{y}_k)| + |b_{nk}(\bar{y}_k)| \leq c_\delta (1 + \|\bar{y}_k\|) \quad ;$$

3. *The finite-dimensional distributions of the processes* $\sum_{k=0}^{[nt]} \xi_{nk}$, $0 \leq t \leq T$, *converge weakly to those of a square-integrable martingale* $\psi(t)$ *with independent increments;*

4. *Functions* $a(t, y)$, $b(t, y)$ *exist such that for any* $L > 0$ *we have*

$$\max_{n\delta \leq k \leq nT} \sup_{\|\bar{y}_k\| < L} \{ |a_{nk}(\bar{y}_k) - a(\frac{k}{n}, y_k)| +$$

$$+ |b_{nk}(\bar{y}_k) - b(\frac{k}{n}, y_k)| \} \underset{n \to \infty}{\longrightarrow} 0 \quad ;$$

5. *Functions* $a(t, y)$ $b(t, y)$ *satisfy the conditions* $L > 0$ *for any* $\delta \leq t \leq T$ *and* $\max(|y|, |z|) \leq L$, *and*

$$|a(t, y) - a(t, z)| + |b(t, y) - b(t, z)| \leq c_\delta^{(2)} \cdot |y - z|$$

$$|a(t, y) - a(s, y)| + |b(t, y) - b(s, y)| \leq \rho(|t - s|), \; t - s \to 0 \quad .$$

6. *Solutions of* (5) *possess the following property: if* $\eta(t, \delta, \xi)$, $t \geq \delta$, *is the solution of* (5) *on* $[\delta, T]$ *with an initial value* ξ *which is independent of increments in* $\psi(t)$ *for* $t > \delta$, *then for sequences* $\delta_m \downarrow 0$ *and* ξ_{δ_m} *such that*

$$\overline{\lim_{m\to\infty}}\ \delta_m^{2\beta}\mathrm{E}|\xi_{\delta_m}|^2 < K$$

for any $t > 0$, *the distribution of* $\eta(t, \delta_m, \xi_{\delta_m})$ *converges weakly to the distribution of some random variable* η_t *as* $m \to \infty$ *(independent of the choice of* δ_m *and* ξ_{δ_m}*).*

Then the measures generated by $\eta_n(t)$ *converge weakly in* **D** *on any segment* $[a, T]$, $a > 0$, *to the measure generated by the process* $\eta(t)$, $t > 0$. *Here, on any segment* $[\delta, T]$, $\delta > 0$, $\eta(t)$ *is the solution to the SDE* (5) *with initial value* η_δ. *Thus for any* $t > 0$ *the distribution of* $\eta(t)$ *and* η_t *coincide.*

Proof. For arbitrary $\delta > 0$, consider $\eta_n(t)$, $t \geq \delta$. It can easily be shown from Condition 1 that

$$\overline{\lim_{n\to\infty}}\ \mathrm{E}\,|\eta_n(\delta)|^2 = \overline{\lim_{n\to\infty}}\ \frac{n^{2\beta}}{[n\delta]^{2\beta}}\cdot[n\delta]^{2\beta}\mathrm{E}\,|x_{[n\delta]}|^2 \leq \delta^{-2\beta}K \quad . \tag{6}$$

Then by Chebyshev's inequality the sequence $\eta_n(\delta)$ is weakly compact in R^m. Together with Condition 2, this implies (see [1,2]) weak convergence of the measures generated by $\eta_n(t)$ on $[\delta, T]$.

Choose an arbitrary subsequence $n_k \to \infty$. Then a subsequence (denoted again by n_k) can be selected from it such that the sequence $\eta_{nk}(\delta)$ converges weakly to a proper random value ξ_δ. It then follows from Conditions 3–5 and results given in [1,2] that the measures generated by $\eta_{nk}(t)$ on $[\delta, T]$ converge weakly to the measure generated by the process $\eta(t, \delta, \xi_\delta)$ – the solution to (5). The inequality (6) implies that $\delta^{2\beta}\ \mathrm{E}|\xi_\delta|^2 \leq K$. Now choose a sequence $\delta_m \downarrow 0 (\delta_1 = \delta)$ and using the diagonal method select a subsequence of n_k (again denoted by n_k) such that for any δ_m, $\eta_{nk}(\delta_m)$ converges weakly to a proper value ξ_{δ_m}. Thus from (6) $\delta_m^{2\beta}\ \mathrm{E}\,|\xi_{\delta_m}|^2 \leq K$. Then by construction the distributions of $\eta(t, \delta_m, \xi_{\delta_m})$ coincide for $t \geq \delta$ as the limit of $\eta_{nk}(t)$, and by Condition 6 the distribution of $\eta(t, \delta_m, \xi\delta_m)$ for any $m \geq 1$ coincides with that of η_t. Since subsequence n_k is arbitrary, we may conclude that the weak limit of $\eta_n(t)$ exists, is unique and coincides with η_t for any $t > 0$. This completes the proof of the theorem.

Note that verification of Conditions 1, 3, 6 can be rather complicated for concrete examples.

The results given in [3–5] on the convergence of procedures of the stochastic approximation type can also be obtained in the framework of Theorem 1. In the one-dimensional case, the coefficients given in [5] were linear:

$$a(t, y) = -\frac{\alpha}{t}(ay + bt^{-\beta}),\ b(t, y) = \frac{\alpha}{\sqrt{t}}(cy + qt^{-\beta}) \quad , \tag{7}$$

with constants defined through the parameters of the procedure and where $\psi(t) = w(t)$ is a standard Wiener process. In particular, for $c = 0$ we have

$$\eta(t) = c_1 t^{-\beta} + c_2 t^{-\alpha a} w(t^{2\alpha a - 2\beta}) \, , \; t > 0 \quad , \tag{8}$$

where $\alpha a > \beta$. The results obtained generalize those of [6] concerning the convergence of Robbins–Monro processes.

The investigation of algorithms which are inhomogeneous in time and affected by random disturbances generated by a random environment leads to models with more complicated random coefficients which may not satisfy Condition 4 for uniform convergence of the coefficients. In such cases one can use the results given in [7], where a theorem on convergence to the solution of (5) under conditions of an integral type on $a_{nk}(\cdot)$, $b_{nk}(\cdot)$ is proved using the techniques described in [8,9]. We shall state here a modification of Theorem 1 in which Condition 4 is replaced by a weaker condition emerging from [7].

Suppose that a random sequence z_{nk}, $k \geq 0$, taking values in a measurable space $(Z, \mathbf{B}_Z)$ and adapted to an increasing sequence $\mathbf{F}_{nk}$ of σ-algebras is given. Let the η_{nk} be defined by the inequalities

$$\eta_{nk+1} = \eta_{nk} + a_{nk}(\eta_{nk}, z_{nk})\frac{1}{n} + b_{nk}(\eta_{nk}, z_{nk})\xi_{nk} \, , \; k \geq 0 \quad , \tag{9}$$

where $a_{nk}(y, z)$, $b_{nk}(y, z)$ are $\mathbf{B}_{R^m} \times \mathbf{B}_Z$-measurable functions and ξ_{nk} is an $\mathbf{F}_{nk}$-difference martingale with $\mathrm{E}[\xi_{nk}\xi_{nk}^* / \mathbf{F}_{nk}] = \frac{1}{n}$, $k \geq 0$.

Here we assume for simplicity that the parameters of the algorithm at the κ-th step depend only on the k-th iteration η_{nk} and on the state of the random environment z_{nk}. Define

$$\varphi_{nk}(j) = \sup_{A,B \in \mathbf{B}_Z} |\mathrm{P}\{z_{nk} \in A \, , \, z_{nk+j} \in B\} -$$

$$- \mathrm{P}\{z_{nk} \in A\}\mathrm{P}\{z_{nk+j} \in B\}| \, , \; k \geq 0 \, , \; j > 0 \quad , \tag{10}$$

$$a_{nk}(y) = \mathrm{E}a_{nk}(y, z_{nk}) \, , \; b_{nk}^2(y) = \mathrm{E}b_{nk}^2(y, z_{nk}) \quad .$$

THEOREM 2. *Let Conditions* 1, 3, 6 *of Theorem* 1 *hold with* $\psi(t) = w(t)$, *and the remaining conditions be replaced by the following:*

2'. *For any* $n > 0$, $n\delta \leq k < nT$, $z \in Z$, *we have*

$$|a_{nk}(y, z)| + |b_{nk}(y, z)| \leq C_\delta(1 + |y|)$$

and for any $L > 0$ *there exists a function* $q_{L,\delta}(u) \geq 0$ *such that* $q_{L,\delta}(+0) = 0$

and for $\max(|x|, |y|) < L$, $n\delta \le k \le nT$, *we have*

$$|a_{nk}(y, z) - a_{nk}(x, z)| + |b_{nk}(y, z) - b_{nk}(x, z)| \le q_{L,\delta}(|x - y|) \quad ; \tag{11}$$

4'. *Functions* $a(t, y)$, $b(t, y)$ *exist such that for any* $y \in R^m$, $\delta > 0$, *we have*

$$\max_{n\delta \le k \le nT} \{ \Big| \frac{1}{n} \sum_{j=[n\delta]}^{k} a_{nj}(y) - \int_{\delta}^{\frac{k}{n}} a(t, y) dt \Big| +$$

$$+ \Big| \frac{1}{n} \sum_{j=[n\delta]}^{k} b_{nj}^2(y) - \int_{\delta}^{\frac{k}{n}} b^2(t, y) dt \Big| \} \underset{n \to \infty}{\longrightarrow} 0 \quad ;$$

5'. *The functions* $a(t, y)$, $b(t, y)$ *satisfy Condition 5 and*

$$|a(t, y)| + |b(t, y)| \le C_\delta (1 + |y|), \ \delta \le t \le T, \ \delta > 0 \quad ;$$

7'. $\lim_{j \to \infty} \overline{\lim_{n \to \infty}} \sup_{k > 0} \varphi_{nk}(j) = 0.$

Then the statement made in Theorem 1 holds with $\psi(t) = w(t)$ in (5).

We shall now consider applications of Theorem 2 to concrete stochastic optimization algorithms. Let a procedure (which for simplicity we shall assume to be one-dimensional) of the Robbins–Monro type in a Markov random environment be defined by the formula

$$x_{k+1} = x_k - \alpha_k (f(x_k, z_k) + \xi_k(x_k, z_k)), \ k \ge 0 \quad . \tag{11}$$

Here the $\xi_k(x, z)$ are independent families of random variables whose distributions depend measurably on the parameters (x, z), and z_k is a z-valued Markov chain. Set

$$a_k(x, z) = \mathrm{E}\xi_k(x, z), \ b_k^2(x, z) = \mathbf{D}\xi_k(x, z) \quad . \tag{12}$$

Put $B_n = n^\beta$. Then in (9)

$$a_{nk}(y, z) = -n\alpha_k (n^\beta f(yn^{-\beta}, z) + n^\beta a_k(yn^{-\beta}, z)) \quad ,$$

$$b_{nk}(y, z) = -n\alpha_k \cdot n^{\beta - 1/2} b_k(yn^{-\beta}, z) \quad , \tag{13}$$

$$\xi_{nk} = \frac{1}{\sqrt{n}} b_k^{-1}(x_k, z_k)(\xi_k(x_k, z_k) - a_k(x_k, z_k)) \quad .$$

Suppose that the chain z_k is homogeneous and ergodic, i.e., for any A, $B \in \mathbf{B}_z$ we have

$$\mathrm{P}\{z_k \in B / z_0 \in A\} \underset{k \to \infty}{\longrightarrow} \pi(B) \quad , \tag{14}$$

where $\pi(B)$ is the stationary measure on Z.

We shall write

$$f(y) = \int_Z f(y,z)\pi(dz), \quad a_k(y) = \int_Z a_k(y,z)\pi(dz) \quad ,$$

$$b_k^2(y) = \int_Z b_k^2(y,z)\pi(dz) \quad .$$

THEOREM 3. *Let* (14) *hold, and assume that* $k\alpha_k \to \alpha > 0$, $k \to \infty$ *and positive constants* λ, C, N_1, N_2 *exist such that for all* $z \in Z$ *we have*

1. $xf(x,z) \geq \lambda x^2$, $f(x,z)^2 \leq C(1+x^2)$, $-\infty < x < +\infty$;
2. $\overline{\lim}_{n\to\infty} \sup_{y,z} (n^{\beta} |a_n(yn^{-\beta},z)| - N_1|y|) < \infty$;
3. $\overline{\lim}_{n\to\infty} \sup_{y,z} (n^{\beta-1/2} |b_n(yn^{-\beta},z)| - N_2|y|) < \infty$;
4. *For all* $z \in Z$, $k \geq 0$ *and any* $L > 0$, *we have*

$$|f(x,z) - f(u,z)| + |a_k(x,z) - a_k(u,z)| + |b_k(x,z) - b_k(u,z)| \leq$$

$$\leq C_L\ |x-u| \quad ,$$

where $\max(|x|,|u|) < L$;

5. $2\alpha\lambda > \alpha^2 N_2^2 + 2\alpha N_1 + 2\beta$;
6. $f'(0) = r > 0$ *and constants* a, b, σ *exist such that for any* $y \in R$, $\delta > 0$, *we have*

$$\max_{n\delta\leq k\leq nT} \Big\{ \Big| \sum_{j=[n\delta]}^{k} \frac{n^{\beta}}{j} a_j(yn^{-\beta}) - \int_{\delta}^{k/n} t^{-1}(ay+bt^{-\beta})dt \Big| +$$

$$+ \Big| \sum_{j=[n\delta]}^{k} \frac{n^{2\beta}}{j^2} b_j^2(yn^{-\beta}) - \int_{\delta}^{k/n} t^{-2\beta-1}\sigma^2 dt \Big| \Big\} \to 0 \quad ;$$

7. *For any* $L > 0$ *we have*

$$\lim_{R\to\infty} \overline{\lim}_{n\to\infty} = \sup_{k>0} \sup_{|y|<L, z\in Z} \mathrm{E}\xi_k^2(yn^{-\beta},z)\chi(|\xi_k(yn^{-\beta},z)| > R) = 0 \quad ;$$

Then the measures generated in **D** *by the sequence* $n^{\beta}x_{[n+t]}$ *converge weakly on any segment* $[\delta, T]$, $\delta > 0$, *to the measure generated by the process*

$$\eta(t) = -\frac{1}{t^{\beta}} \frac{ab}{\alpha(r+a)-\beta} - \frac{1}{t^{\alpha(r+a)}} \frac{a\sigma}{\sqrt{2\alpha(r+a)-2\beta}} w(t^{2\alpha(r+a)-2\beta}) \quad . \qquad (15)$$

Proof. Condition 1 of Theorem 1 follows in a standard way from Conditions 1–3,5 above (see [10], p. 169). Using (13) and Conditions 2–4 above, one can immediately verify

Condition 2' of Theorem 2. Condition 3 of Theorem 1 (with $\psi(t) = w(t)$) follows from Condition 7 of the Lindeberg type and the formula for ξ_{nk}. Furthermore, Condition 6 above evidently implies Condition 4 of Theorem 2 with

$$a(t,y) = -\alpha t^{-1}((r+a)y + bt^{-\beta}),\ b(t,y) = \alpha\sigma t^{-\beta-1/2} \quad . \tag{16}$$

These functions automatically satisfy Condition 5', while Condition 7' of Theorem 2 follows from (14). Thus we only have to verify Condition 6 of Theorem 1. Consider equation (5) with coefficients given by (16). Solving this equation on $[\delta, T]$ with the initial value ξ_δ using the formulae in [11, p. 37], we obtain

$$\eta(t,\delta,\xi_\delta) = \eta_\delta\left(\frac{\delta}{t}\right)^{\alpha(r+a)} - \int_\delta^t \left(\frac{s}{t}\right)^{\alpha(r+a)} \cdot \frac{\alpha}{s^{\beta+1}}\,(b\,ds + \sigma\sqrt{s}\,dw(s)) \quad . \tag{17}$$

Furthermore, it is clear from Condition 1 above that $r \geq \lambda$ and from Conditions 2 and 6 that $|a| \leq N_1$. Then from Condition 5 we have $\alpha r > \alpha\,|a| + \beta$, and hence $\alpha(r+a) > \beta$. Condition 1 of Theorem 1 implies that $\mathrm{E}\eta_\delta^2 \leq K\delta^{-2\beta}$ (see (6)). Hence $\mathrm{E}(\eta_\delta(\frac{\delta}{t})^{\alpha(r+a)})^2 \leq Kt^{-\alpha(r+a)} \cdot \delta^{2(\alpha(r+a)-\beta)} \to 0,\ \delta \to 0$. It can easily be seen that the integral in (17) has a limit in the quadratic mean for $\delta \to 0$ which is equivalent to the process $\eta(t)$ (see (15)). Thus in the quadratic mean $\eta(t,\delta,\xi_\delta) \to \eta(t),\ \delta \to 0$, i.e., Condition 6 of Theorem 1 is fulfilled and Theorem 3 holds.

Remark 1. Condition 6 is fulfilled, for example, if for some fixed $N > 0$

$$\lim_{n\to\infty} \frac{n^\beta}{N} \sum_{k=n+1}^{n+N} a_k(yn^{-\beta}) = ay + b \quad ,$$

$$\lim_{n\to\infty} \frac{n^{2\beta-1}}{N} \sum_{k=n+1}^{n+N} b_k^2(yn^{-\beta}) = \sigma^2 \quad ,$$

which takes in the case of circularly altering parameters.

Remark 2. In Condition 6 of Theorem 3 we confined ourselves for the sake of clarity to the linear functions $a(t,y)$, $b(t,y)$. The case $b(t,y) = cy + q$ was studied in [5], with the limiting distributions being non-Gaussian. In the general case it can be shown that under the assumptions of Theorem 3 the coefficients in Condition 6 must be of the form $t^{-1-\beta}a(yt^\beta)$ and $t^{-1/2-\beta}b(yt^\beta)$. Taking into account the function $f(x)$, the coefficients in (5) will be

$$a(t,y) = -\frac{\alpha}{t}\left(\frac{ny+1}{t^\beta}a(yt^\beta)\right) \quad ,$$

$$b(t, y) = -\alpha t^{-1/2-\beta} b(yt^{\beta}) \quad . \tag{18}$$

We shall assume that the functions $a(y)$ and $b(y)$ are such that a solution to equation (5) exists and is unique. When studying this equation we introduce a new process $\zeta(u) = e^{u\beta}\eta(e^u)$, $u > -\infty$, thus reducing the problem to the solution of the equation

$$d\zeta(u) = ((\beta - \alpha r)\zeta(u) - \alpha a(\zeta(u)))du -$$
$$- \alpha b(\zeta(u))dw(u) \quad , \quad u > -\infty \tag{19}$$

for which, from (6) $E|\zeta(u)|^2 \leq K$, $u > -\infty$. If the function $b(u)$ is nondegenerate in every bounded domain, then using Conditions 2, 3, 5 it is possible to verify that the coefficients of equation (19) satisfy the ergodicity condition and the condition for the existence of a stationary solution in [12]. Thus in this case Condition 6 of Theorem 1 is satisfied (since $\delta_m \downarrow 0$ and the time parameter t is replaced by e^u, the initial moments in (19) will be $u_m = \ln \delta_m$, $u_m \to -\infty$) and the stationary solution to equation (19) acts as a limiting solution. (Note that such models with homogeneous coefficients and a non-random environment were investigated in the multivariate case in [13].) This leads to a theorem generalizing Theorem 3:

THEOREM 4. *Let the conditions of Theorem 3 hold, taking the integrals of* $t^{-1-\beta}a(yt^{\beta})$, $t^{-1/2-\beta}b(yt^{\beta})$ *in Condition 6, and the function* $b(y)$ *be nondegenerate in every bounded domain. Then the measures corresponding to* $n^{\beta}x_{[nt]}$ *converge weakly in* **D** *on any segment* $[\delta, T]$ *to the measure corresponding to the solution* $\eta(t)$ *of equation* (5) *with coefficients given by* (18) *with* $\psi(t) = w(t)$, *and initial value* $\eta_{\delta} = \delta^{-\beta}\zeta$. *Here* ζ *is the stationary solution to* (19).

Theorems 2–3 show that an algorithm operating in a random environment is equivalent to the same algorithm with its parameters averaged over the environment state space and over time, provided that the environment satisfies the mixing condition and the coefficients satisfy Conditon 2'.

REFERENCES

1. I.I. Gihman. Limit theorems for triangular arrays of random variables (in Russian). In *Theor. Slutchaynyh Prosessov*, Vol. 2, 1974, p. 37–47.
2. I.I. Gihman and A.V. Skorohod. *The Theory of Random Processes*, Vol. 3, Wiley, New York, 1978.

3. V.V. Anisimov and Z.P. Anisimova. On the convergence of random processes generated by a procedure of the Robbins–Monro type (in Russian). In *Analytical Methods in Probability Theory*, p. 104–108. Naukova Dumka, Kiev, 1979.

4. V.V. Anisimov. On the convergence of multivariate inhomogeneous stochastic approximation procedures (in Russian). *Doklady Akademii Nauk Ukrainskoj SSR*, 5(1981) 82–85.

5. V.V. Anisimov. Limit theorems for random processes generated by stochastic approximation procedures (in Russian). In *Vychislit. i Pricl. Matematika*, Vol. 46, p. 32–41, Kiev, 1982.

6. D. McLeish. Functional and random limit theorems for the Robbins–Monro process. *Journal of Applied Probability*, 13(1)(1976)148–154.

7. V.V. Anisimov and A.P. Yurachkovsky. A limit theorem for stochastic difference schemes with random coefficients (in Russian). In *Teoriya Veroyatnosti i Matematicheskaya Statistika*, Vol. 33, 1985.

8. I.I. Gihman and A.V. Skorohod. *Stochastic Differential Equations and their Applications*. Naukova Dumka, Kiev, 1982 (in Russian).

9. R.Sh. Liptser and A.N. Shyryaev. Weak convergence of a sequence of semimartingales to a diffusion-type process (in Russian). In *Matematecheskie Sbornik*, 121(2)(1983)176–204.

10. M.B. Nevelson and R.Z. Hasminskiy. *Stochastic Approximation and Recursive Estimation*. Nauka, Moscow, 1972 (in Russian).

11. I.I. Gihman and A.V. Skorohod. *Stochastic Differential Equations*. Naukova Dumka, Kiev, 1968 (in Russian).

12. R.Z. Hasminskiy. *Stability of Systems of Differential Equations with Randomly Perturbed Parameters*. Nauka, Moscow, 1969 (in Russian).

13. R.Z. Hasminskiy. Convergence of processes generated by multivariate stochastic optimization procedures (in Russian). In *Teoriya Veroyatnosti i Matematicheskaya Statistika*, Vol. 33, 1985.

ON THE STRUCTURE OF OPTIMALITY CRITERIA IN STOCHASTIC OPTIMIZATION MODELS

V.I. Arkin and S.A. Smolyak
Central Economic–Mathematical Institute, Moscow, USSR

1. INTRODUCTION

This paper is concerned with the problem of estimating the efficiency of economic actions under uncertainty, and its correct formulation. Optimization problems in economics are often reduced to the choice of the best of a finite or sometimes infinite number of given *alternatives*, i.e., different economic actions or decisions. Here it is very important both to justify the optimality criterion and to estimate the efficiency achieved as a result of choosing the best alternative. In the deterministic case each alternative is characterized by some *profit* (income, efficiency, gain) and a rational choice can be made using the criterion of profit maximization. In this case if two alternatives both have the same profit value then one cannot be preferred to the other; however, this does not exclude the choice of one as the best on the basis of some other criterion. This of course presupposes that a method for evaluating profit in the deterministic case has been established.

In practice the profit resulting from some real economic action actually depends on the conditions under which the action takes place, which are generally not known in advance. The methods by which this uncertainty regarding the profit of various economic actions is taken into account should be uniform so that decisions taken in different economic sectors will be consistent, i.e., local decisions should be consistent with the global optimum. We shall call the criterion which takes into account information about the possible values of the profit ξ of alternative ξ (the alternatives and their profits will be denoted by the same symbol) *the expected profit* and denote it by $E(\xi)$. Let us consider the structure of this criterion. To do this we should first formalize the concept of uncertainty. There are two methods by which this could be done.

The first is to formalize the causes and effects of the uncertainty. The cause of the uncertainty in the profit ξ is incomplete information about the conditions under which the alternative ξ will be implemented. Thus the uncertainty may be treated as a combination of three elements:

- a known function $\xi(s)$ expressing the dependence of the profit ξ on external conditions s;
- the set S of possible external conditions s;
- information I about "the degree of possibility" (e.g., the probability) of specific conditions $s \in S$ occurring.

Thus the uncertain profit ξ is considered to be a function $\xi(s)$ on the "information probability" space $\{S, I\}$ and the expected profit is a functional of this function.

The disadvantage of this method is the possible dependence of the structure of the functional on the set of possible conditions S. For example, let $S = \{s_1, \ldots, s_n\}$ and the information I comprise known probabilities of occurrence p_i for each condition s_i. It is obvious that the functional

$$E(\xi) = [\sum_i \xi(s_i) 2^{p_i}] / \sum_i 2^{p_i} \tag{1}$$

possesses "good" properties but on *this* probability space only. If we replace any condition s_i by two other conditions with the same total probability but each having the same profit as the original, the value of the criterion will change.

To avoid such situations the expected profit criteria for different "information probability" spaces should be linked. Such an approach is developed in [1–3] in connection with probability uncertainty (see Section 2). Below we shall consider another method of formalizing the uncertainty concept which leads to more general criteria but which limits itself to describing the effects of uncertainty only, ignoring its causes. Here the uncertainty is characterized by two elements: the set $X = X(\xi)$ of possible profit values and information I on "the degree of possibility" (e.g., probability distribution) of each value $x \in X(\xi)$ of the profit. These elements must determine the value of the expected profit $E(\xi)$.

This method allows us to compare alternatives from different "information probability" spaces. Alternatives with equal values of expected profit will be called *equally profitable*, e.g., alternatives with the same profit probability distribution are equally profitable. We shall consider only alternatives for which the set X is bounded (*finite alternatives*).

We shall first discuss the main features of the proposed method in order to clarify the structure of the expected profit criterion and its results under conditions of *probability uncertainty*.

2. PROBABILITY UNCERTAINTY

We shall say that an alternative ξ is *stochastic* and its profit is characterized by *probability uncertainty* if this profit is a random variable with a known probability distribution function $P = P_\xi(t) = P\{\xi \leq t\}$. The limiting case of stochastic alternatives is deterministic alternatives (ξ = const.), which corresponds to degenerate probability distributions. Finite alternatives correspond to finite (i.e., contained in a bounded interval) distributions. The expected profit $E(\xi)$ is now considered to be a functional of P_ξ. The conventional functional of this type is the *mean* (denoted by an overbar): $E(\xi) = \bar{\xi}$. It has many "good" properties but a lot of economists have criticized this criterion because it does not take into account the scatter in the profit values, e.g., they do not agree that alternative ξ, which has zero profit, and alternative η, which has a profit of one million or a loss of one million with equal probability, are equally profitable. Thus they repeatedly propose the criteria:

$$E(\xi) = \bar{\xi} - kD(\xi) \quad \text{or} \quad E(\xi) = \bar{\xi} - k\sqrt{D(\xi)} \quad , \tag{2}$$

where $D(\xi) = \overline{(\xi - \bar{\xi})^2}$, $k > 0$. These proposals are discussed in [4]. Consider the alternatives

$$\xi = 0$$

$$\eta : P\{\eta = 0\} = p \; , \; P\{\eta = A\} = 1 - p \quad .$$

The values of $A > 0$ and $0 < p < 1$ can be chosen in such way that $E(\eta) < E(\xi)$ for both criteria in spite of the fact that η is preferable to ξ. It is shown in [5] that functions of the mean profit value or any of its central high-order moments or other "dispersion indexes" are equally inappropriate.

A better criterion can be constructed by formulating certain reasonable suggestions about criterion properties as axioms. Various combinations of these axioms lead to optimality criteria of different structures. In the deterministic case this approach leads to a criterion consistent with traditional economic representations [2,3]. Thus we can hope that this approach can be adapted to the non-deterministic case.

Let us now consider such "reasonable" (from the economic point of view) axioms separately. They can be divided into three groups.

The first group contains statements of a general character; some examples follow.

Consistency between estimates of the efficiency of deterministic and stochastic alternatives is guaranteed by the first axiom:

C1. $E(\xi) = \xi$ if ξ is deterministic.

The stability of expected profit to small changes in the technical or economic parameters of the alternatives is guaranteed by axiom C2:

C2. The functional $E(\xi)$ is continuous.

In order for this statement to be rigorous we shall suppose that the weak convergence topology generated by the Lévy metric is introduced into the space of probability distributions.

Sometimes one alternative is obviously preferable to others for economic reasons. For example, ξ is obviously preferable to η if $\xi \geq 2$, $\eta \leq 1$, or if ξ and η can only take the values 0 or 1, but the value 1 is more probable under alternative ξ. We would expect the criterion of expected profit in these cases to suggest the choice of the obviously preferable alternative. Thus we introduce the ***obvious preference relation*** (>>) as follows:

$$\xi \gg \eta \text{ if } P_\xi \neq P_\eta ,\ P_\xi(t) \leq P_\eta(t) \text{ for all } t \quad . \tag{3}$$

This is the well-known stochastic prevalence relation.

Consistency between the expected profit criterion and the obvious preference relation is expressed by the axiom of ***monotonicity***:

C3. If $\xi \gg \eta$ then $E(\xi) > E(\eta)$.

This axiom is not satisfied by criterion (2) above. Note that there is a natural desire to extend C3 to guarantee $E(\xi) > E(\eta)$ whenever $\xi(s) \geq \eta(s)$ and $\xi(s) > \eta(s)$ for any s. Such (supermonotonic) criteria can exist in any probability space (e.g., criterion (1)), but any attempt to coordinate these criteria in different probability spaces fails. As far as we are concerned, supermonotonic criteria do not exist! For example, if $\xi(s) \geq \eta(s)$, but $P\{\xi(s) > \eta(s)\} = 0$, then $P_\xi = P_\eta$ and therefore ξ and η are equally profitable.

The axioms of the second group reflect natural economic interpretations of the "convexity" of the set of equally profitable alternatives. Let the expected profit of alternatives ξ and η be not greater than some e. We would like alternative ζ, which has technical and economic parameters between those of ξ and η, also to have an expected profit less than e. This will be formalized later in the weak invariance axioms C4, C6. Let (ξ, η) and (ξ_1, η_1) be two pairs of equally profitable alternatives. Consider the pair formed by taking values midway between the first and second elements of these pairs. We would like the resulting alternatives also to be equally profitable. This will be formalized later in the strong invariance axioms C5, C7. To enable such formalization to take place, however, it is necessary to introduce operations which can be used to find the "middle" of distribution functions. We shall introduce two such operations – *mixing* $_{\mathrm{o}}$ and *averaging* ($\Box$).

The *mixture* $\xi \circ \eta = \eta \circ \xi$ of alternatives ξ and η is defined as the alternative ζ obtained by choosing ξ and η with equal probability. It corresponds to the following *mixture* of distribution functions:

$$P_\zeta(t) = (P_\xi \circ P_\eta)(t) = (P_\xi(t) + P_\eta(t))/2 \quad . \tag{4}$$

The *average* $\xi \Box \eta = \eta \Box \xi$ is defined as the alternative ζ associated with a profit which is the mean of profits ξ and η. It corresponds to the following *composition* of distribution functions:

$$P_\zeta(t) = (P_\xi \Box P_\eta)(t) = \int_{-\infty}^{\infty} P_\xi(2t - s)dP_\eta(s) \quad . \tag{5}$$

We can now formulate all of the axioms in this group.

Weak invariance with respect to averaging:

C4. If $E(\xi) = E(\eta) = e$ then $E(\xi \Box \eta) = e$.

Strong invariance with respect to averaging:

C5. If $E(\xi) = E(\eta)$ then $E(\xi \Box \zeta) = E(\eta \Box \zeta)$ for all ζ.

Weak invariance with respect to mixing:

C6. If $E(\xi) = E(\eta) = e$ then $E(\xi \circ \eta) = e$.

Strong invariance with respect to mixing:

C7. If $E(\xi) = E(\eta)$ then $E(\xi \circ \zeta) = E(\eta \circ \zeta)$ for all ζ.

The third group of axioms includes the property of "additivity", which makes it possible to describe the effect of several alternatives implemented simultaneously by summing the individual effects of these alternatives. However, this property cannot be written $E(\xi + \eta) = E(\xi) + E(\eta)$. Such an equation makes no mathematical or economic sense. Firstly, the operation of addition is defined for random variables, not for their distribution functions. Moreover, the distribution function of a sum of two random variables cannot be expressed in terms of the distribution functions of the individual variables. Secondly, this axiom does not take into account the synergic (assuming the economic system to be closed) effects which can arise when two actions take place simultaneously.

However, there are two special cases in which the above axiom does make mathematical and economic sense.

1. Let alternatives ξ and η be independent (from the economic point of view) and their profits be independent random variables. Now consider the simultaneous implementation of ξ and η as a new alternative $\zeta = \xi * \eta = \eta * \xi$, with a distribution function

$P_\xi = P_\zeta * P_\eta$ (the convolution of P_ξ and P_η). The following *I*-additivity axiom says that the expected profit of a number of independent alternatives implemented simultaneously is equal to the sum of the profits of the individual alternatives:

C8. $E(\xi * \eta) = E(\xi) + E(\eta)$.

This property reflects the real economic situation quite adequately, since if the implementation of one alternative affects the efficiency of another then their efficiency is estimated ***jointly*** in practical calculations.

In the special case where $\eta = e$ = constant is a deterministic alternative, C8 becomes ***invariant to translation*** (i.e., to changes in the profit origin):

C9. $E(\xi * e) = E(\xi) + e$ for all e.

2. Suppose that alternative ξ can be repeated, producing the same effect each time. This corresponds to the *scale* operation $\zeta = k \otimes \xi$ with distribution function $P_\zeta(t) = P_\xi(t/k)$. For such alternatives, which are said to have a limited dependence on each other, the additivity demand is transformed into an axiom of ***homogeneity***:

C10. $E(k \otimes \xi) = kE(\xi)$.

Suppose that C10 is valid for all k including $k \le 0$. Suppose also that $0 \otimes k = 0$, and if $k < 0$ consider an alternative $k \otimes \xi$ which compensates for the effect of alternative $|k| \otimes \xi$. The distribution function in this case is $1 - P_\xi(t/k - 0)$. Condition C10 allows us to take "compensating" economic actions into account in the efficiency calculations, and to calculate the expected profit of repeating an action by multiplying the number of times it is repeated by the specific expected profit, as is usually done in economic calculations.

The structures of criteria which satisfy certain combinations of the above axioms are given in Table 1.

We shall now clarify some of the statements made in Table 1.

Statement 1 is almost trivial. Form the sequence $\xi_1 = \xi$, $\xi_{n+1} = \xi_n \square \xi_n (n \ge 1)$. From C4 we deduce that $E(\xi_\eta) = E(\xi)$ for any n. Further, for $n \to \infty$ the distributions of the ξ_n converge to a degenerate distribution concentrated at the point $\bar{\xi}$. Hence and from C1 and C2 we deduce that $E(\xi) = \bar{\xi}$. Thus condition C4 is sufficiently strong and there is no need to replace it by the stronger C5.

Statement 2 is well-known and widely cited in mathematical texts, e.g., [6,7], although in other formulations.

Statements 3, 7, 8 are proved in [5]; the other statements are simple corollaries of these three. We shall now prove some of them.

Proof of Statement 4. Let axioms C1–C3, C7, C10 be satisfied. From C2 there must exist a continuous monotonic function $u(x)$ such that $E(\xi) = z$ is the root of the

Table 1 The structure of criteria $E(\xi) = z$ which satisfy certain combinations of axioms.

No.	Axioms	Criteria Structure
1.	C1, C2, C4	$z = \bar{\xi}$
2.	C1–C3, C7	The root of the equation $u(z) = \overline{u(\xi)}$, where $u(x)$ (the "utility function") is continuous and strictly increasing.
3.	C1–C3, C7, C8	As above but with $u(x) = \exp(cx)$ for $c \neq 0$ or $u(x) = x$.
4.	C1–C3, C7, C10	As above but with $u(x) = \|x\|^p \operatorname{sign}(x)$, $p > 0$.
5.	C1–C3, C7, C8, C10	$z = \bar{\xi}$
6.	C1–C3, C7, C9, C10	$z = \bar{\xi}$
7.	C1–C3, C6	The root of the equation $\overline{u(z, \xi)} = 0$, where $u(x, t)$ (the compared utility function) is continuous on x, strictly increasing on t, vanishing for $t = x$ and such that $u(x, t)/u(x, s) > u(y, t)/u(y, s)$ for any $s < y < x < t$.
8.	C1–C3, C6, C8	As above but with $u(x, t) = \exp(a(x-t)) - \exp(c(x-t))$ for $a \geq 0 \geq c$, $a \neq c$, or $u(x, t) = x - t$.
9.	C1–C3, C6, C8, C10	$z = \bar{\xi}$
10.	C1–C3, C6, C10	The root of the equation $\overline{u(z, \xi)} = 0$, where $u(x, t)$ is continuous on x and may be represented by the following formula: $u(x, t) = \begin{cases} v(t/x)/v(1+1/x) & \text{for } x \neq 0 \\ \|t\|^p \operatorname{sign}(t) & \text{for } x = 0. \end{cases}$ In addition $p > 0$, $v(1) = 0$, and $v(y)$ is a continuous, strictly increasing function on y.
11.	C1–C3, C6, C9	The root of the equation $\overline{u(\xi - z)} = 0$, where $u(x)$ is a continuous, strictly increasing function on x, $u(0) = 0$.
12.	C1–C3, C6, C9, C10	As above but with $u(x) = \|x\|^p \operatorname{sign}(x)$, $p > 0$.

equation $u(z) = \overline{u(\xi)}$.

Without loss of generality we can take $u(0) = 0$, $u(1) = 1$. Choose any $x > 0$, $y < 0$ and assign them the probabilities $P\{\xi = x\} = u(y)/(u(y) - u(x))$, $P\{\xi = y\} = u(x)/(u(x) - u(y))$. It is not difficult to show that $\overline{u(\xi)} = 0$, and thus $E(\xi) = 0$. Therefore, using C10 we have $E(k \otimes \xi) = 0$, and thus $u(kx)P\{\xi = x\} + u(ky)P\{\xi = y\} = 0$. From this we can deduce that $u(kx)/u(x) = u(ky)/u(y)$. This means that both parts of this expression are equal to the same constant, independent of x and y. The value of this constant can be found by making the substitution $x = 1$, when we obtain $u(kx)/u(x) = u(k)$ for all $x \neq 0$. The continuous solution of this equation has the form $u(x) = |x|^p \operatorname{sign}(x)$, $p > 0$.

Proof of Statements 11 and 12. From Statement 7 and C9 there must exist a "compared utility function" $u(x, t)$ such that $\overline{u(z, \xi)} = 0$ implies $\overline{u(z+e, \xi+e)} = 0$ for any e. Making the substitutions $e = -z$ and $u(0, x) = u(x)$ leads to Statement 11. The proof of Statement 12 is now completely analogous to that of Statement 4.

Consideration of Table 1 shows that weak combinations of axioms lead to wide classes of criteria, while strong combinations lead to unique mean criteria. For practical purposes we recommend the use of only one- or two-parameter criteria from Statements 3, 4, 8, 12. All of these revert to the mean criterion for specific parameter values, while for other values they take into account the scatter in the profit. The most interesting criterion is the invariant to mixing, I-additive criterion from Statement 3, proposed by Massé [4]:

$$\mathrm{E}(\xi) = \psi(c)/c \ , \quad \text{where } \psi(c) = \ln\left(\overline{\exp(c\xi)}\right) \ . \tag{6}$$

Iff ξ has a normal distribution then $\mathrm{E}(\xi)$ has the form $\mathrm{E}(\xi) = \bar{\xi} + 0.5cD(\xi)$. The parameter c may therefore be treated as a special norm which takes into account the scatter in the profit. Other properties of criteria from Statements 3 and 8 are analyzed in [5].

3. UNCERTAINTY OF GENERAL FORM

The axiomatic approach can be adapted for use in situations when the profit distribution function ξ is not known precisely. Suppose that the available information only enables us to describe the class H_ξ of distribution functions which contains the unknown distribution function P_ξ. In such situations we will say that the profit ξ is characterized by *uncertainty of general form* and consider the expected profit to be a functional of the corresponding class H_ξ. Thus we will sometimes use the expressions: "expected profit of class ..." and "classes ... are equally profitable". Uncertainty of general form becomes probability uncertainty if the class H contains only one distribution function. We shall demonstrate this by means of two examples.

1. Let profit ξ be a non-random variable and the only available information be $a \le \xi \le b$. Then H_ξ is $L(a, b)$, i.e., the class of all degenerate distributions on the interval $[a, b]$.

2. Let $\xi = \wedge(x_1, \ldots, x_n)$ be a "lottery" which guarantees gains $x_1, \ldots, x_2, \ldots, x_n$ with unknown probabilities. Then H_ξ contains all distributions $(p_1, \ldots, p_n)$ on the set $\{x_1, \ldots, x_n\}$. If we know in addition that gain x_1 is most probable, then H_ξ contains only distributions for which $0 \le p_2, \ldots, p_n < p_1 \le 1$.

The formulations of all axioms remain unchanged under these conditions if the definitions of the corresponding concepts, operations and relations are extended to classes of distribution functions. This can be done using the same natural approach.

An alternative ξ is said to be ***deterministic*** if the class H_ξ contains a unique distribution concentrated at the point ξ. Now the axiom C1 has some meaning.

The distance between two classes is determined using the ***Hausdorff–Lévy metric***, i.e., the supremum of the Lévy distance between an element of one class and the nearest element of the other. This metric allows us to define convergence for sequences of classes and makes sense of axiom C2.

The ***obvious preference relation*** may be introduced by the rule: $\xi \gg \eta$, if every distribution $P \in H_\xi$ dominates (as defined by (3)) every distribution $Q \in H_\eta$. This allows us to describe the monotonicity property by axiom C3. Many economists consider this formulation to be very weak. They would like to introduce into the space of classes an ordering relation that corresponds to the relation >> (i.e., $P >> Q \Longrightarrow \{P\} \quad \{Q\}$) and satisfies one of the following conditions:

(M) $H' \quad H'' \Longrightarrow H \cup H' \quad H \cup H''$ for any H,

(M') $P \gg Q \Longrightarrow \{P\} \quad \{P, Q\} \quad \{Q\}$,

$H' \quad H'' , H \cap H' = H \cap H'' = \phi \Longrightarrow H \cup H' \quad H \cup H''$,

(M") $P \gg Q \gg R \gg T \Longrightarrow \{P, Q, T\} \quad \{P, R, T\}$.

It is found that such orderings do not in fact exist! If M is satisfied, then for $P \gg Q \gg T$ we have:

$$\{P\} \quad \{Q\} \Longrightarrow \{P,T\} \cup \{P\} \quad \{P,T\} \cup \{Q\} \Longrightarrow \{P,T\} \quad \{P,T,Q\} \quad ,$$

$$\{Q\} \quad \{T\} \Longrightarrow \{P,T\} \cup \{Q\} \quad \{P,T\} \cup \{T\} \Longrightarrow \{P,T,Q\} \quad \{P,T\} \quad ,$$

which is impossible. □

If M' is satisfied, then for $P \gg Q >> T$ we have [8]:

$$\{P\} \quad \{Q\} \Longrightarrow \{P\} \quad \{P,Q\} \Longrightarrow \{T\} \cup \{P\} \quad \{T\} \cup \{P,Q\} \Longrightarrow \{P,T\} \quad \{P,T,Q\} \quad ,$$

$$\{Q\} \quad \{T\} \Longrightarrow \{Q,T\} \quad \{T\} \Longrightarrow \{P\} \cup \{Q,T\} \quad \{P\} \cup \{T\} \Longrightarrow \{P,T,Q\} \quad \{P,T\} \quad ,$$

which is impossible. □

If M" is satisfied, then for $R \downarrow T$ we have $\{P,Q,T\} \quad \{P,T\}$. Similarly, for $P \gg R \gg Q \gg T$, $R \uparrow P$, it follows from M" that $\{P,T\} \quad \{P,Q,T\}$, which is impossible. □

Other similar examples of the impossibility of extending ordering relations on any set to a power set are discussed in [8,9] and in other papers in the same issue of the journal.

To preserve the formulation of axioms C4–C10 we shall define the average (mixture, convolution, etc.) of classes H_ξ and H_η as the class of various averages (mixtures, convolutions, etc.) of their elements $P \in H_\xi$ and $Q \in H_\eta$, and define the scale operation by the formula: $k \otimes H = \{k \otimes P / P \in H\}$.

Let $\overline{v(\xi)}$ denote the supremum and $\overline{v(\xi)}$ the infimum of possible values of the mean of the random value $v(\xi)$ for distributions $\mathrm{P}_\xi \in H_\xi$:

$$\overline{v(\xi)} = \sup_{P \in H_\xi} \int v(t) \mathrm{d}P(t) , \quad \overline{v(\xi)} = \inf_{P \in H_\xi} \int v(t) \mathrm{d}P(t) \quad .$$

Let us consider the structure of criteria which satisfy C1–C3,C5. Form the sequence $H_1 = H_\xi$, $H_{n+1} = H_n \square H_n (n \geq 1)$. Under the assumed metric $H_n (n \rightarrow \infty)$ tends to the class $L(m, M)$ of all degenerate distributions concentrated at $[m, M]$, where $m = \bar{\xi}$, $M = \bar{\xi}$. Just as it follows from C5 that all of the H_n are equally profitable, so $\mathrm{E}(\xi)$ is equal to the expected profit $f(m, M)$ of class $L(m, M)$.

Let $F(m, M)$ denote the class which contains a unique distribution concentrated at the point $f(m, M)$. It follows from C1 that $F(m, M)$ and H_ξ are equally profitable. Further, the function f is continuous (C1, C2) and $f(x, x) = x$. Hence and from C3 it follows that

$$m \leq f\left(m, M\right) \leq M \quad . \tag{7}$$

Making use of the averaging operation we have:

$$L(m, M) \square L(m_1, M_1) = L\left(\frac{m + m_1}{2}, \frac{M + M_1}{2}\right) \quad .$$

We shall now replace $L(m, M)$ and $L(m_1, M_1)$ by the equally profitable classes $F(m, M)$ and $F(m_1, M_1)$. Then from C5 we obtain:

$$\frac{f(m, M) + f(m_1, M_1)}{2} = f\left(\frac{m + m_1}{2}, \frac{M + M_1}{2}\right) \quad . \tag{8}$$

It is not difficult to prove that every continuous solution (7)–(8) can be written in the form:

$$f(m, M) = \lambda M + (1 - \lambda) m , \quad (0 \leq \lambda \leq 1) \quad . \tag{9}$$

Thus axioms C1–C3, C5 are satisfied only for criteria such that

$$\mathrm{E}(\xi) = \lambda \bar{\xi} + (1 - \lambda) \bar{\xi} , \quad (0 \leq \lambda \leq 1) \quad . \tag{10}$$

For classes of degenerate distributions these criteria are known as *Hurwicz's "optimism–pessimism" criteria* [10], and are justified in [1,3].

A wider class of criteria, which are also designed for profit maximization, is justified in [11].

Parameter λ from (10) may be viewed as a ***norm which takes into account the uncertainty in the mean of the profit.*** It has a different interpretation to norm c from criterion (6) – it allows us to take into account the dispersion of the profit around its known mean.

Criterion (10) is continuous, monotonic, homogeneous, I-additive and invariant to averaging and translation. This suggests the use of this criterion in practical estimations of the efficiency of economic actions. The centralized regulation of parameter λ allows us to adjust local decisions made under uncertainty with national economic interests.

Criteria with the following structure satisfy axioms C1–C4:

$$E(\xi) = f(\underline{\xi}, \overline{\xi}) \quad , \tag{11}$$

where $f(m, M)$ is the solution of equations

$$f(m, M) = x + \lambda(x) = m + (M - m)\lambda(x) \quad . \tag{12}$$

Here $\lambda(x)$ is any continuous non-increasing function on $(-\infty, +\infty)$ such that $0 \leq \lambda(x) \leq 1$ and $x + \lambda(x)$ is a non-decreasing function.

For $\lambda(x)$ = const. this criterion reverts to (10), while in other cases it is non-homogeneous and non-I-additive.

We cannot establish the structure of criteria satisfying C1–C3, C7 or C6. However, between these there are criteria which take into account the scatter in the profit, e.g., $E(\xi) = z$, where z is a root of the equation $u(z) = \lambda \overline{u(\xi)} + (1-\lambda)\overline{u(\xi)}$, $(0 \leq \lambda \leq 1)$, and $u(x)$ is a continuous monotonic function. However, if $u(x)$ is non-linear, such criteria are not I-additive, although with appropriate $u(x)$ they may be homogeneous or invariant to translation.

Consideration of Table 1 suggests that we may hope to construct "good" parametric criteria, which take into account the scatter of the profit around the mean and the uncertainty in the mean itself, under combinations of axioms C1–C3, C8, C7 or C6. However, the structure of such criteria and even their existence are uncertain. Let us try to combine (6) and (10). This leads to the following criterion:

$$E(\xi) = \lambda \ln(\overline{\exp(c\xi)})/c + (1-\lambda)\ln(\overline{\exp(c\xi)})/c \quad ,$$

which is monotonic and I-additive, but for $0 < \lambda < 1$ is non-invariant to mixing.

REFERENCES

1. S.A. Smolyak. On rules for comparing some variants of economic actions under uncertainty. In *Studies in Stochastic Control Theory and Mathematical Economics*. Central Economics–Mathematical Institute (CEMI), Moscow, 1981 (in Russian).

2. R.M. Merkin and S.A. Smolyak. *Taking Probability Factors into Account in Capital Construction Planning*. Moscow Civil Engineering Institute (MCEI), Moscow, 1982 (in Russian).

3. S.A. Smolyak and B.P. Titarenko. *Mathematical Models and the Economics of Construction: A Summary of Lectures*. Moscow Civil Engineering Institute (MCEI), Moscow, 1983 (in Russian).

4. P. Massé. *Le choix des investissement. Critères et méthodes*. Dunod, Paris, 1968.

5. S.A. Smolyak. *Taking into Account the Effect of Dispersion when Evaluating Economic Efficiency Under Uncertainty. Models and Methods of Stochastic Optimization*. Central Economics–Mathematical Institute (CEMI), Moscow, 1983 (in Russian).

6. J.M. Grandmont. Continuity properties of a von Neumann–Morgenstern utility function. *Economic Theory*, 4(1972).

7. M.H. De Groot. *Optimal Statistical Decisions*. McGraw Hill, 1970.

8. S. Barberà and P.K. Pattanaik. Extending an order on a set to the power set. Some remarks on Kannai and Peleg's approach. *Journal of Economic Theory*, 32(1984).

9. Y. Kannai and B. Peleg. A note on the extension of an order on a set to the power set. *J.Econ. Theory*, 32(1984).

10. L. Hurwicz. *Optimality Criteria for Decision Making under Ignorance*. Cowles Commission Papers, No. 370, 1951.

11. K.J. Arrow and L. Hurwicz. *An Optimality Criterion for Decision-Making under Ignorance. Uncertainty and Expectation in Economics*. Basil Blackwell and Mott, Oxford, 1972.

STRONG LAWS FOR A CLASS OF PATH-DEPENDENT STOCHASTIC PROCESSES WITH APPLICATIONS

W. Brian Arthur
Stanford University
Stanford, CA 94305
USA

Yu. M. Ermoliev
Glushkov Institute
of Cybernetics
Kiev, USSR

Yu. M. Kaniovski
Glushkov Institute
of Cybernetics
Kiev, USSR

1. *INTRODUCTION*

In many simple sequential processes (rolls of a die say), outcomes at each time may be labelled by category or type (the die turns up 1, or 2, or 3, etc.), with type i having fixed probability $q(i)$, and $\sum_i q(i) = 1$. The strong law of large numbers then tells us that over time the proportion of outcomes of each type must converge to the probability for that type.

We consider an important generalization of such processes, wherein the probability $q(i)$ is no longer fixed, but becomes itself a function of the proportions at each moment. This is the case, for example, where new firms in a growing industry each in turn make a locational choice between N possible cities, but where the probability that a given city is chosen next for location depends on the number of firms already located there. Transitions in the proportions of the industry in the various cities now depend upon the path these proportions follow. We seek strong laws for processes of this path-dependent type.

It is convenient to formulate such path-dependent processes as generalized urn schemes of the Polya kind. Consider an urn of infinite capacity that contains balls of N possible colors or types. Let the vector $X_n = (X_n^1, X_n^2, \ldots, X_n^N)$ describe the proportions of balls of type 1 to N respectively, at time n; and let $\{q_n\}_{n=1}^{\infty}$ be a sequence of Borel functions from the N-dimensional unit simplex S into itself. One ball is added to the urn at each time n; it is of type i with probability $q_n^i(X_n)$. Starting with an *initial vector* of balls $b_1 = (b_1^1, b_1^2, \ldots b_1^N)$ the process is iterated to yield $X_1, X_2, X_3, \ldots$. We investigate conditions under which X_n converges to a limit random vector X, and the support set of X, under different specifications of the urn functions q_n. In general, we find that where q_n possesses a limit function q and where the process converges, it converges to a limit which belongs to a subset of the fixed points of q.

The literature on this problem is small. In a recent elegant paper, Hill, Lane and Sudderth [1] analyze the special case where $N=2$ and the urn functions q_n are stationary. Blum and Brennan [2] present strong laws for a related problem (with $N=2$) where additions to each category are not restricted to 0 or 1. In this paper we extend our own previous results [3] for the general N-dimensional, time-varying case. We use, for the most part, Lyapunov techniques and stochastic approximation methods. We pay special attention to unstable points (fixed points of q that are not in the support of X); and to convergence to the vertices of the simplex (where a single color dominates). We also present examples of path-dependent processes in economic theory, optimization theory, and chemical kinetics, for which this N-dimensional, non-stationary, path-dependent process is a natural model.

Non-stationary functions arise even in simple urn schemes. Consider *Example 1.1. A Sampled Urn.* (a) An urn contains red and white balls. Sample at random r balls. If m, where $0 \leq m \leq r$, or more are white, replace the sample and add a white. Otherwise add a red. (b) As before, but if m or more are white, replace the sample and add a *red*. Otherwise add a *white*. In (a) the probability that a white is added is

$$q_n^w = \sum_{k=m}^{r} H(k;\, n,\, n_w,\, r)$$

where H is the Hypergeometric distribution parametrized by n, r, and n_w, the number of white balls at time n. In this sampled urn scheme the urn function (path-dependent on n_w) is non-stationary: the Hypergeometric varies with n.

As a simple N-dimensional urn example consider *Example 1.2.* An urn contains balls of N colors. Choose one ball. If it is of type j replace it and add a ball of type i with probability $q(i, j)$, where $\sum_{i=1}^{N} q(i, j) = 1$, for all j. For example, when $N=3$, we might have the rule: Choose one ball, replace it and add a ball which is one of the two possible other colors with equal probability one half.

Notice that the well-known basic Polya scheme [1] (sample one ball and replace it together with a ball of the same color) is a special case both of 1.1 (a) (where r and m are 1) and of 1.2 (where $q(i, i)=1$ and $N=2$). For this scheme, the proportion of white balls, n_w/n, converges almost surely to a random limit variable that has a beta distribution with parameters dependent on the initial urn composition. This case however, is singular. When r in 1.1 (a) is greater than 2, our results below show that n_w/n converges to a random variable with support $\{0, 1\}$ only. In 1.1 (b) they show that it converges to a single interior point $\{p\}$. The process of 1.2 also converges, as we will show later.

The general scheme above covers other path-dependent processes.

Example 1.3. A Position-Dependent Random Walk. Consider a simple one-dimensional random walk, where $Y_i = \pm 1$, with the position at n given by partial sum $S_n = \sum_{i=1}^{n} Y_i$, but with position-dependent transition probabilities $P(Y_i=+1) = p_n(S_n)$. If we add a white ball to the urn when $Y_i = +1$, a red ball when $Y_i = -1$ (starting from an empty urn), the position of the random walk, S_n, is given by $(2X_n-1)n$, where X_n is the proportion of white balls in the total n. We can then treat the limiting behavior of the random walk within our present framework.

The general N-dimensional time-varying urn process described above does not always converge. Theorem 3.1 establishes a test for convergence, expressed in terms of the existence of a limit function for $\{q_n\}$ and of an appropriate Lyapunov Function. Theorem 6.1 shows more general conditions, for the particular case where the q_n functions are separable. In general, continuity of the q_n functions is not required for convergence. Where the process does converge and the q_n functions are continuous, the support of the limit vector lies within the set $\{X : q(X) = X,\ q(X) = \lim_{n\to\infty} q_n(X),\ X\varepsilon S\}$, that is, within the set of fixed points of the limit function q. (A slight modification is required for non-continuous urn functions.) However, not all fixed points of q are in the support. Theorems 5.1 and 5.2 show that certain fixed points can be classed as *stable* and *unstable*, with stable fixed points in the support, but unstable ones excluded. We pay particular attention in theorems 4.1 and 6.4 to conditions under which the vertices of the simplex are in the support, that is conditions under which the process tends to single-color dominance. In a final section applications in economic theory, optimization theory, and chemical kinetics are outlined.

2. *Preliminaries*

The general process starts at time 1 with a vector $b_1 = (b_1^1, b_1^2, \ldots b_1^N)$ of balls in the urn, with total $\gamma = \sum_i b_1^i$. Balls are added indefinitely, according to the urn probability functions q_n. At time n, define the random variable

$$\beta^i_n(x) = \begin{cases} 1 & \text{with probability } q^i_n(x) \\ 0 & \text{with probability } 1-q^i_n(x), \end{cases} \qquad i = 1, \ldots, N .$$

Then additions of i-type balls to the urn follow the dynamics

$$b^i_{n+1} = b^i_n + \beta^i_n(X_n) \qquad i = 1, \ldots, N .$$

Thus the evolution of the proportion of i-types, $X^i_n = b^i_n/(\gamma+n-1)$, is described by

$$X^i_{n+1} = X^i_n - \frac{1}{\gamma+n}\,[X^i_n - \beta^i_n(X_n)] \qquad n = 1, 2, \ldots \tag{1}$$

with

$$X^i_1 = b^i_1/\gamma$$

We can rewrite (1) in the form

$$X^i_{n+1} = X^i_n - \frac{1}{\gamma+n}\,[X^i_n - q^i_n(X_n)] + \frac{1}{\gamma+n}\,\eta^i_n(X_n) \tag{2}$$

$$X^i_1 = b^i_1/\gamma ,$$

where

$$\eta^i_n(X_n) = \beta^i_n(X_n) - q^i_n(X_n)$$

Noting that the conditional expectation of η^i_n with respect to X_n is zero,

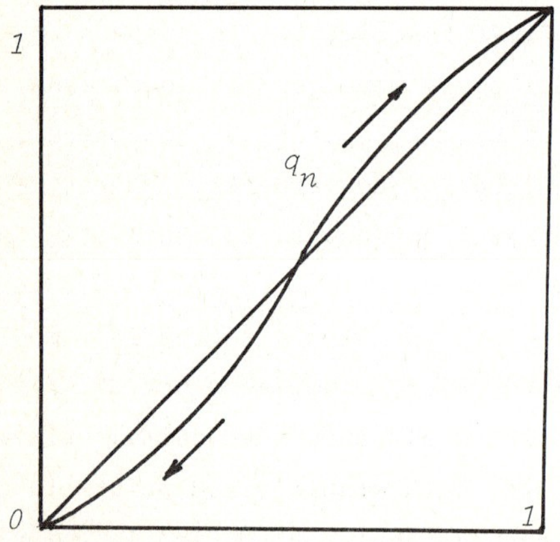

Figure 1

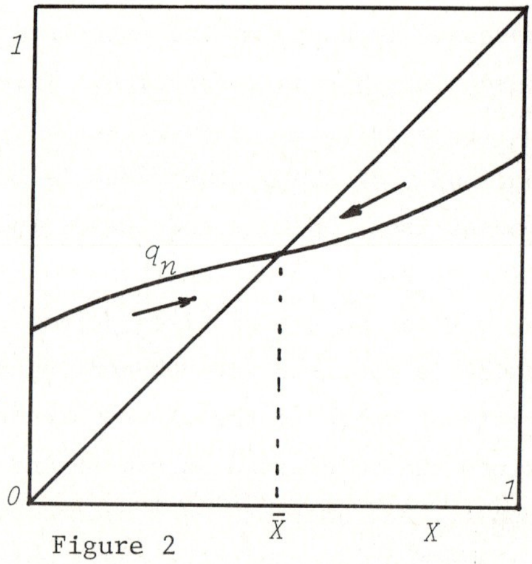

Figure 2

we can derive the expected motion of X_{n+1} as

$$E\{X^i_{n+1}|X_n\} = X^i_n - \frac{1}{\gamma+n}\,(X^i_n - q^i_n(X_n)) \tag{3}$$

Thus we see that motion tends to be directed by the term $q_n(X_n) - X_n$. In Figure 1, for example, this tendency is toward 0 or 1. In Figure 2 it is toward $\bar{X}$.

3. *A Convergence Test*

We begin with a convergence theorem that is a stochastic analog of the Lyapunov asymptotic stability theorem for deterministic systems. It serves as a very general test for convergence in the N-dimensional case. We denote the N-dimensional unit simplex by S, and use $||\cdot||$ to denote the Euclidean norm.

Theorem 3.1. Given continuous urn functions $\{q_n\}$, suppose there exists a Borel function $q: S \to S$, constants $\{a_n\}$, and a (Lyapunov) function $v: S \to R$ such that:

(a) $\sup_{x \in S} ||q_n(x) - q(x)|| \leq a_n \, , \quad \sum_{n=1}^{\infty} a_n/n < \infty$

(b) The set

$B = \{x : q(x) = x, \, x \in S\}$

contains a finite number of connected components

(c) (i) v is twice differentiable

(ii) $v(x) \geq 0 \, , \qquad x \in S$

(iii) $\langle q(x) - x \, , \, v_x(x) \rangle < 0 \, , \qquad x \in S \setminus U(B)$

where $U(B)$ is an open neighborhood of B.

Then $\{x_n\}$ converges to a point of B or to the border of a connected component.

Proof. The theorem follows from stochastic approximation results of Nevelson and Hasminskii [4], Ch. 2. Applied to our problem, we can summarize the argument as follows. Note first that $v(X_n)$ eventually becomes a non-negative supermartingale on S. On the set $S \setminus U(B)$, v has expected increment always less than some $-\delta$; hence the process must exit this set in finite time. It thus enters $U(B)$ infinitely often. Next, the cumulated perturbations $\sum_{n=0}^{t} \frac{1}{(n+\gamma)} \eta_n$ form a martingale and converge; thus after sufficient time the process cannot cumulate sufficient perturbation to counter expected motion and exit an ε-neighborhood surrounding $U(B)$. Now the B-components are separated by finite distances. Hence the process converges to a single component of B or its border. Finally, since expected motions within B are zero, and cumulated perturbations converge, the process cannot visit distinct points inside B infinitely often. Thus $\{X_n\}$ converges to a point of B, or to the border of a connected component. □

For the most general cases, an appropriate Lyapunov function may be difficult to find. For the special case $N = 2$, an appropriate Lyapunov function is simply the norm

$$v(x) = \int_0^x (t - q(t)) \, dt + a$$

providing q is differentiable. (For this reason a norm can be used in place of Lyapunov functions in the two-color case in [1]. We can also construct a Lyapunov function in the case $N>2$, providing the q^i are differentiable and symmetric in the sense that

$$\partial q^i/\partial x^k = \partial q^k/\partial x^i\ , \qquad x\varepsilon S\ .$$

Remark 3.1. In the case of example 1.2 it is easily shown that

$$q_n(x) = Qx$$

where the matrix $Q = (q(i,j))$.

We can take $v(x) = \langle (I-Q)x,\ x\rangle$ for

$$B = \{x : (I-Q)x = 0\}$$

where I is the identity matrix.

The theorem then tells us that the scheme converges to a fixed point $\bar{x} = Q\bar{x}$.

4. *Convergence to the Vertices*

We next establish conditions under which the urn may converge to single-color dominance, that is, conditions under which X_n may converge to a vertex of the simplex S. Without loss of generality we take the vertex to be $(0, 0, \ldots, 1)$.

Theorem 4.1. Given the process characterized by initial urn vector b_1 and $\{q_n\}$. Let $\tilde{b}^i_n = b^i_1$, for $i = 1, \ldots, N-1$; $\tilde{b}^N_n = b^N_1 + n-1$.

If (a) $$\sum_{i=1}^{N-1} q^i_n\left(\frac{\tilde{b}_n}{\gamma+n-1}\right) < 1\ , \quad n \geq 1$$

and (b) $$\sum_{n=1}^{\infty} \sum_{i=1}^{N-1} q^i_n\left(\frac{\tilde{b}_n}{\gamma+n-1}\right) < \infty$$

then

$$P\{\bigcap_{i=1}^{N-1} [\lim_{n\to\infty} X^i_n = 0]\} > 0\ .$$

Proof. Let

$$A_n = \{\omega | b^i_n = b^i_1\ ,\ i = 1, \ldots, N-1;\ b^N_n = b^N_1 + n-1\}$$

then

$$P\{\bigcap_{i=1}^{N-1} [\lim_{n\to\infty} X^i_n = 0]\} \geq P\{\bigcap_{n=1}^{\infty} A_n\}$$

$$= \prod_{n=1}^{\infty} \{1 - \sum_{i=1}^{N-1} q^i_n\left(\frac{\tilde{b}_n}{\gamma+n-1}\right)\} > 0\ ,$$

the inequality following from standard results on the convergence of infinite products. The theorem then follows. □

Notice that this theorem is independent of the previous one--no special conditions are imposed on the $\{q_n\}$ beyond the condition (a) that the vertex is reachable from the starting point, and the condition (b) that $q^N(X)$ approaches 1 sufficiently fast as X approaches the vertex.

5. *Stable and Unstable Fixed Points*

We now wish to show that convergence is restricted to only certain subsets of the fixed points of q. We will find useful a lemma of Hill, Lane, and Sudderth [1] extended to the N-dimensional case.

Lemma 5.1. Suppose $\{X_n\}$ and $\{Y_n\}$ are generalized urn processes with the same initial urn compositions and with urn functions $\{f_n\}$ and $\{g_n\}$ respectively. Suppose all urn functions map the interior of S into itself, and suppose f_n and g_n agree a.e. in a neighborhood U of the point Θ. Then $\{X_n\}$ converges to θ with positive probability if and only if $\{Y_n\}$ does.

Proof. The argument follows that in [1]. In essence, if $\{X_n\}$ converges to θ, the process must be contained within U from some stage k onward and must therefore have reached some point a within U at that stage. Since the q_n map the interior of the simplex into the interior, a must also be reachable by $\{Y_n\}$ at stage k with positive probability, and once in this state a at k the two processes become identical. Hence $\{Y_n\}$ converges to Θ with positive probability, and the lemma is proved. □

We now consider fixed points Θ of q of two special types. Given Θ and a neighborhood U of Θ, we will say that Θ is a *stable point* if there exists a symmetric positive-definite matrix C such that

$$\langle C[x-q(x)],\ x-\theta\rangle > 0\ , \qquad x \neq \theta \qquad x\varepsilon U \cap S\ . \tag{4}$$

Similarly we will call θ an *unstable point* if θ is such that

$$\langle C[x-q(x)],\ x-\theta\rangle < 0\ , \quad \text{for} \quad x \neq \theta \qquad x\varepsilon U \cap S\ . \tag{5}$$

Notice that we impose no requirement that q is continuous within U.

In the $N = 2$ case, stable points are those where q downcrosses the diagonal, unstable ones are where q upcrosses the diagonal. In N-dimensions downcrossing and upcrossing are inappropriate: the Lyapunov criterion (4) tests whether expected motion is locally always toward θ, the Lyapunov criterion (5) tests whether it is locally always away from θ.

We now show that the process converges to stable points with positive probability:

Theorem 5.1. Let θ be a stable point in the interior of S. Given a process with transition functions $\{q_n\}$ which map the interior of S into itself, and which converge in the sense that

$$\sup_{x \in U \subset S} || q_n(x) - q(x) || \leqq a_n , \qquad \sum_{n=1}^{\infty} a_n/n < \infty$$

then

$$P\{X_n \to \theta\} > 0.$$

Proof. Construct the functions $\{\tilde{q}_n\}$ and $\{\tilde{q}\}$ which are identical to $\{q_n\}$ and $\{q\}$ respectively within the neighborhood U, and are equal to θ outside it. Let $\{Y_n\}$ be the urn scheme corresponding to $\{\tilde{q}_n\}$, with initial state identical to that of the X-scheme. It is clear that $\{\tilde{q}_n\}$ converges to $\tilde{q}$, in the sense given above, and that θ is the unique solution of $\tilde{q}(y) = y$. Now introduce the function

$$v(y) = \langle C(y-\theta), y - \theta \rangle$$

using the fact that θ is a stable point to select C, a positive-definite symmetric matrix. It is easy to check that v is a Lyapunov function, as specified in Thm. 3.1. It follows from Thm. 3.1 (the discontinuity in $\tilde{q}$ does not affect the argument) that $\{Y_n\}$ converges to θ with probability 1. Finally, $\{X_n\}$ and $\{Y_n\}$ as a pair fulfill the conditions of Lemma 5.1. Therefore $\{X_n\}$ converges to θ with positive probability, and the theorem is proved. □

Remark 5.1. If the Lyapunov criterion (4) holds over the interior of S, so that θ is the only stable point, then, by Thm. 3.1 or as shown in [3], $\{X_n\}$ converges to θ with probability 1.

We now wish to establish that, given an additional Hölder condition, convergence to unstable points has probability zero. We adapt a stochastic approximation result of Nevelson and Hasminskii [4] (Chapt. 5) in the lemma that follows. Consider the process

$$z_{n+1} = z_n - a_n F_n(z_n) + \beta_n \gamma_n(z_n, \omega) \tag{6}$$

where $F_n : R^N \to R^N$, where γ_n is a random vector,

where F_n converges uniformly to F, and where $\sum_{n=1}^{\infty} a_n^2 < \infty$, $\sum_{n=1}^{\infty} \beta_n^2 < \infty$.

Lemma 5.2. Given the process described by (6), such that:

(a) If $B = \{z : F(z) = 0\}$, and $\tilde{B}$ is a subset of B such that, for $\tilde{z} \in \tilde{B}$ and z in a neighborhood of $\tilde{z}$, there exists a symmetric positive-definite matrix C such that $\langle C(z-\tilde{z}), F(z)\rangle < 0$;

(b) $\{\gamma_n\}$ has bounded fourth moments, and there exist positive constants a_1 and a_2 such that

$$a_1 \leqq Tr\, D(n,\tilde{z}) \leqq a_2$$

where $D(n,z)$ is the matrix $(E\,[\gamma_n^i(z,\omega) \times \gamma_n^j(z,\omega)]$;

(c) $|F(z)|^2 + |Tr[D(n,z) - D(n,\tilde{z})]| \leqq k\ |z-\tilde{z}|^{\mu}$

for some k and some $\mu \in (0,1]$.

Then $P\{z_n \to \tilde{z} \in \tilde{B}\} = 0$.

Proof. See [4]. Proof involves constructing a Lyapunov function w, infinite on $\tilde{B}$, and such that $w(z_n)$ becomes a non-negative supermartingale. $\{z_n\}$ then cannot converge to any $\tilde{z}\in\tilde{B}$.

We now apply this lemma to our urn scheme $\{X_n\}$, assuming as before that $\{q_n\}$ converges to some function q.

Theorem 5.2. Suppose θ is a non-vertex unstable point with a neighborhood U such that:

$$||q(x) - q(\theta)|| \leqq k||x-\theta||^{\mu} \text{ for } x\in U, \text{ and for some } k, \text{ and } \mu \in (0,1].$$

Then $P\{x_n \to \theta\} = 0$.

Proof. Using the previous lemma, and the dynamic equation (2), we identify z_n with X_n, F_n with $(X-q_n(X))$, γ_n with η_n, and $\tilde{z}$ with θ. Then condition (a) of the lemma is fulfilled and we need only check (b) and (c). Now η_n and q_n are bounded and η_n has a fourth moment. It is easy to see that the diffusion matrix $D(n, X) = (E[\eta_n^i(X) \times \eta_n^j(X)]$ approaches a limiting matrix $D(X)$ uniformly for $x \in U$. We also have

$$E[\eta_n^i(X)^2] = q_n^i(X) \quad (1 - q_n^i(X))$$

and since $q(\theta) = \theta$, we have $[D(\theta)]_{ii} = \theta_i(1-\theta_i)$.

Finally, since $Tr\ D(\theta) = \sum_{i=1}^{N} \theta_i(1-\theta_i)$, we have $Tr\ D(\theta)$, given θ non-vertex, bounded above and below. Then all requirements of Lemma 5.2 are fulfilled and the theorem follows. □

Remark 5.2. If θ is the sole non-vertex fixed point, if it is unstable, if q is continuous, and if the process converges, then it must converge to one of the vertices.

6. *Separable Urn Functions*

Until now we have used Lyapunov techniques to prove or rule out convergence to points in the simplex. For a certain restricted class of urn functions we can dispense with Lyapunov techniques and instead use martingale methods, the restrictions allowing us to sharpen our results. We will say that the urn function q is *separable* if

$$q(x) = (q^1(x^1),\ q^2(x^2),\ \ldots,\ q^{N-1}(x^{N-1}),\ q^N(x) = 1 - \sum_{i-1}^{N-1} q^i(x^i))$$

where the indices are of course arbitrarily determined. Note that this restricted class always includes the important case where $N=2$. We further impose a requirement that the urn function does not cross the diagonal "too often." That is, we suppose that for each open interval $J \in [0,1]$ and $i = 1, \ldots, N-1$, there exists a subinterval $J_1 \subset J$ such that $x^i - q^i(x^i) \leqq 0$, or $x^i - q^i(x^i) \geqq 0$ for $x^i \in J_1$. The theorems in this section assume separable urn functions that fulfill this condition. For reasons of brevity, we state the theorems that follow in terms of a stationary urn function q. All proofs extend rather simply to the non-stationary case, providing $\{q_n\}$ converges to q in the sense given in Theorem 3.1 and providing that $\{q_n\}$ fulfills the above subinterval condition (with the same subinterval for all q_n) for n greater than some time n_1.

We begin by establishing convergence to the fixed points of q.

Theorem 6.1. Given a continuous (and separable) urn function q, $\{X_n\}$ converges with probability one to a random variable X which has support in the set of fixed points of q.

Proof. Let B_n be the σ-field generated by $X_1, X_2, \ldots, X_n$. Using the dynamical system (2), consider, for index i:

$$\mu_n^i = \sum_{t=1}^{n} \eta_t^i(X_t)\,(\gamma+t)^{-1}$$

Since

$$E(\eta_t^i \mid B_t) = 0, \text{ and } |\eta_t^i| \leqq 2,$$

the pair μ_n^i, B_n, for $n \geqq 1$ define a martingale, with $E|\mu_n^i|^2 <$ constant. It follows that there exists a $\mu^i < \infty$ such that $\mu_n^i \to \mu^i$ with probability one. From (2) we thus obtain the convergence:

$$X_{n+1}^i - X_1^i + \sum_{t=1}^{n} [X_t^i - q^i(X_t^i)]\,(\gamma+t)^{-1} \to \mu^i \tag{6}$$

for all events ω in $\tilde{\Omega}_i$, the set where μ_n^i converges. (Note that $P\{\tilde{\Omega}_i\} = 1$.)

Now, to establish the convergence of X_n^i on $\tilde{\Omega}_i$ suppose the contrary, that is,

$$\underline{\lim}_{n\to\infty} X_n^i < \overline{\lim}_{n\to\infty} X_n^i .$$

Under our specified condition, we may now choose a subinterval J_1 of $(\underline{\lim}\, X_n^i, \overline{\lim}\, X_n^i)$ within which (without loss of generality) $x^i - q(x^i) \geqq 0$. Choose within this a further subinterval (a^i, b^i). There must exist times $m_k, n_k, m_k < n_k, k = 1, 2, \ldots,$ such that

$$X_{m_k}^i < a^i,\; X_{n_k}^i > b^i \quad \text{and} \quad a^i \leqq X_n^i \leqq b^i \quad \text{for } m_k < n < n_k .$$

Summing (2) between m_k and n_k we have

$$\mu^i_{n_k} - \mu^i_{m_k} = X^i_{n_k} - X^i_{m_k} + \sum_{t=m_{k+1}}^{n_{k-1}} [X^i_t - q^i(X^i_t)]\ (\gamma+t)^{-1}$$

$$\geq a^i - b^i ,$$

which, for k large enough, contradicts (6). Convergence for index i with probability one to a point X^i is established.

Now suppose X^i_n fails to converge to a fixed point of q. That is that $X^i - q^i(X^i) = \beta > 0$. From the argument above, the quantity

$$\sum_{t=\tau}^{\infty} [X^i_t - q^i(X^i_t)]\ (\gamma + t)^{-1} \to 0 \tag{7}$$

with probability one, as τ goes to infinity. Since X^i_n converges to X^i, it eventually lies within a neighborhood U of X^i where, by continuity of q^i, $X^i_t - q^i(X^i_t) > \beta/2$. But then the summation in (7) becomes infinite, which contradicts (7). Thus X^i_n converges to a fixed point of q^i.

A similar argument holds for other indices $j\ (\neq N)$: X^j_n converges to a fixed point of q^j, on the set $\tilde{\Omega}_j$. We have $P\{\bigcap_1^{N-1} \tilde{\Omega}_j\} = 1$. Therefore the residual, X^N_n, is constrained to converge, with probability 1, to a fixed point $q^N = X^N$. The theorem is proved. □

Remark 6.2. Note that continuity of q is required only for the fixed-point property, and not for the overall convergence of the process.

As before, we wish to narrow the set of points to which the process may converge. We call the interior fixed point θ a *downcrossing* point of the function q, if for all indices $i = 1$ through $N - 1$ in some neighborhood U of θ:

$$x^i < q^i(x^i) \quad \text{where} \quad x^i < \theta^i$$

$$x^i > q^i(x^i) \quad \text{where} \quad x^i > \theta^i .$$

(It is easy to check that it follows that $x^N < q^N$ where $x^N < \theta^N$, and $x^N > q^N$ where $x^N > \theta^N$, so that the term downcrossing is consistent.) Upcrossing can be defined analogously.

Theorem 6.2. If q: *Int* $S \to$ *Int* S, then the process converges to downcrossing points θ with positive probability.

Proof. Let θ be a downcrossing point. Then the function

$$\sum_{i=1}^{N-1} \langle x^i - q^i(x^i),\ x^i - \theta^i \rangle + \langle x^N - q^N(x^N),\ x^N - \theta^N \rangle$$

is positive where $x \neq \theta$ in a neighborhood U of θ. Hence θ qualifies as a stable point and, by theorem 5.1,

$$P\{X_n \to \theta\} > 0 .$$

Remark 6.3. The restriction that q should map the interior of S into the interior of S ensures that the neighborhood of θ is reachable from any starting conditions. This is a stronger condition than normally required in practice.

Corollary 6.1. If θ is the only fixed point of q continuous on S, and if $q^i > 0$ at $x^i = 0$, for all $i = 1$ to $N - 1$, then θ is a downcrossing point and convergence to θ follows with probability 1.

Theorem 6.3. If for any single index i, q^i upcrosses the diagonal at θ, and the upcrossing satisfies the Hölder condition of Thm. 5.2, then $P\{X_n \to \theta\} = 0$.

Proof. Follows from Theorem 5.2. □

Finally, we give a useful condition for convergence to the vertices. We will say that q possess the strong S-property, if it has a single interior fixed point θ, which is a point of upcrossing for each index i, and where each upcrossing satisfies the Hölder condition (see Thm. 5.2).

Theorem 6.4. Suppose q is continuous and satisfies the strong S-property. Then the process converges to one of the vertices with probability one.

Proof. Consider index i. The function q^i, it is easy to show, must have fixed points $\{0, \theta^i, 1\}$. By Theorem 5.2 convergence to θ^i has probability zero. In combination over all indices, the only other fixed points are vertices. □

7. *Conclusion*

To summarize, we can conclude that where a limiting urn function exists and where a suitable Lyapunov function can be found (we have shown several), the process in N dimensions converges. If the limiting urn function is continuous, only fixed points of this urn function belong to the support of the limiting random variable. Where expected motion is *toward* a reachable fixed point, it is in the support: where it is *away* from a fixed point, it is not in the support. In the special case of separable urn function, we may talk about "upcrossing" and "downcrossing" in N dimensions, with results that become extensions of the two-dimensional case. And where the strong S-property is fulfilled (see also [3]), the process must converge to a vertex.

8. *Applications*

8.1 Economic Allocation. Economic agents, drawn from a large pool, each demand a single unit of a durable good that comes in N different types or brands. The agents, in general, are heterogeneous and they choose in random sequence. Where there are increasing or decreasing supply costs; or where agents' preferences are endogenous (their tastes are influenced by the purchases of others); or where agents gain information on the products by learning of other agents' use of them; then the probability that the nth agent to choose purchases brand i depends upon the market-share proportions of the N brands at his time of purchase. Market-share dynamics for this type of allocation problem are thus path-dependent and we may enquire as to the limiting market share outcome as the market expands to an indefinitely large size. For the case where agents choose between competing *technologies*, rather than goods, see [6]. This market-share problem becomes more complex [7] when sellers of goods (or technologies) can strategically price to gain market share; but the overall structure remains the same.

8.2. Industrial Location. As outlined in the introduction, firms in a growing industry may each make a locational choice between N cities in random sequence. Choice will be influenced both by internal firm needs and by economies of agglomeration--returns from locating where other firms of the industry have established themselves. We might inquire as to whether cities eventually share the industry, or whether the industry coalesces and agglomerates in a single city (in a vertex solution). For analysis of this locational problem see [8].

8.3. Chemical Kinetics. Consider the dual autocatalytic chemical reaction:

$$S + 2W \rightarrow 3W + E$$

$$S + 2R \rightarrow 3R + F$$

A single substrate molecule S is converted into either W or R form (with waste molecules E and F) according to whether it encounters two W-molecules before two R-molecules. Given initial concentrations, we may inquire as to the final proportions of chemical products. Notice that this example is equivalent to Example 1.1(a) above; if we think of the process as "sampling" the next three W or R molecules encountered and adding one to W or R according as 2 out of the 3 molecules sampled are W or R. More general N-dimensional kinetics can be similarly modeled.

8.4 Stochastic Optimization. In stochastic optimization methods based on the Kiefer-Wolfowitz procedure or its modern variants, an approximation to the solution is iteratively updated as:

$$X_{n+1} = X_n - \rho_n [Y_n(X_n, \omega)] \tag{7}$$

where X_n is an N-dimensional vector in R^N; the step-size ρ_n satisfies

$$\sum_n \rho_n = \infty \quad , \qquad \sum_n \rho_n^2 < \infty \quad ;$$

and Y_n is a random vector, serving as an estimate for or approximation to the gradient of the function to be minimized. Often it is computationally expedient to calculate only the sign of Y_n. This gives the Fabian procedure [9]:

$$X_{n+1} = X_n - \rho_n \; sgn \; [Y_n(X_n), \omega] \quad . \tag{8}$$

We leave it to the reader to show that (8) can be put in the form of (2). Thus convergence of the Fabian algorithm to a local minimum can now be established.

References

[1] Hill, Bruce M., Lane, David, and Sudderth, William (1980). A Strong Law for Some Generalized Urn Processes. *Ann. Prob.*, 214-216.

[2] Blum, J.R. and Brennan, M. (1980). On the Strong Law of Large Numbers for Dependent Random Variables. *Israeli J. Math.* 37, 241-245.

[3[Arthur, W.B., Ermoliev, Yu. M., and Kaniovski, Yu. M. (1983). A Generalized Urn Problem and Its Applications. *Kibernetika* 19, 49-57 (in Russian). Translated in *Cybernetics* 19, 61-71.

[4] Nevelson, M.B., and Hasminskii, R.Z. (1972). *Stochastic Approximation and recursive estimation.* American Math. Society Translations of Math. Monographs, Vol. 47. Providence.

[5] Ljung, L. (1978). Strong Convergence of a Stochastic Approximation Algorithm. *Annals of Stat.* 6, 680-696.

[6] Arthur, W.B. (1983). Competing Technologies and Lock-In By Historical Small Events: The Dynamics of Allocation under Increasing Returns. Committee for Economic Policy Research, Paper No. 43. Stanford Univ.

[7] Hanson, W.A. (1985). Bandwagons and Orphans: Dynamic Pricing of Competing Systems Subject to Decreasing Costs. Ph.D. dissertation. Stanford Univ. Forthcoming.

[8] Arthur, W.B. (1984). Industry Location and the Economies of Agglomeration: Why a Silicon Valley? Stanford Univ., mimeo.

[9] Fabian, V. (1965). Stochastic Approximation of Constrained Minima. In: *Transactions of the 4th Prague Conference on Information Theory, Statistical Decision Functions, and Random Processes.* Prague.

THE GENERALIZED EXTREMUM IN THE CLASS OF DISCONTINUOUS FUNCTIONS AND FINITELY ADDITIVE INTEGRATION

V.D. Batuhtin and A.G. Chentsov
Institute of Mathematics and Mechanics
Sverdlovsk, USSR

In this paper we consider an approach to the solution of extremum problems involving discontinuous functions, defined on finite-dimensional spaces. Using the concept of the approximate gradient introduced in [1,2], necessary and sufficient conditions for the extremum of discontinuous functions are obtained, a numerical method is constructed and some classical theorems of analysis are generalized. Integration with respect to a finitely additive measure on a semi-algebra of sets is defined. We give statements concerning conditions for the universal integrability of bounded functions, the integrability of functions defined on the Cartesian product, and the relations between measurable spaces.

Definition 1. We shall say that the minimum of the bounded function $f(x)$ is reached at a point $x^* \in R_n$, if there exists a neighborhood $V(x^*)$ of the point x^* such that for all $x \in V(x^*)$ the inequality $\underline{f}(x^*) \leq \underline{f}(x)$ holds, where $\underline{f}(x)$ is the lower limit of the function f at the point x.

Problem. Let a bounded measurable (Lebesque) function of n variables $f(x) = f(x_1, x_2, \ldots, x_n)$ be defined on the closed set $D \subset R_n$. Find the minima of this function.

To solve the problem we proceed as follows. We extend the function $f(x)$, defined on D, onto the whole space R_n. This extension may be achieved in a number of different ways, but should be such that none of the new minimum points appears on D. The simplest such extension has the form

$$f^*(x) = \begin{cases} f(x), & x \in D \\ \omega = \text{const}, & x \notin D \end{cases},$$

where $\omega = \sup_{x \in D} f(x)$. The extension of $f(x)$ allows us, firstly, to analyze the minima of the function not only at inner points of D, but also on the boundary, and secondly, the extension of $f(x)$ in fact removes the constraints in extremum problems. Further, we consider the extension of the function f onto R_n, denoting it by f^*.

For an arbitrary point $x^* \in D$ set

$$\Omega(x^*;\tau) = \{x : \|x - x^*\| \le \tau\}$$

$$\Omega = \{s : s = x - x^*,\ x \in \Omega(x^*;\tau)\} \quad .$$

Let $p(s)$ be the distribution density of the continuous random variables $\xi_i = s_i$, $i \in \overline{1,n}$, and let it satisfy the following conditions: $p(s) = 0$, if $s \notin \Omega$: $p(s) \ge 0$ if $s \in \Omega$;

$$\int_\Omega p(s)\mu(ds) = 1\,,\ \mathrm{E}[(s_i - \mathrm{E}[s_i])(s_j - \mathrm{E}[s_j])] = 0\,,\ i \ne j \quad ;$$

$$i,j \in \overline{1,n}\,,\ \mathrm{E}[s_i] = 0\,,\ \mathrm{E}[s_i^2] \ne 0 \quad .$$

Let us replace the function $f(x)$ on $\Omega(x^*;\tau)$ by the linear function

$$\varphi(x^* + s;\Omega,p,f) = a_0(x^*;\Omega;p;f) + \langle a(x^*;\Omega;p;f), s\rangle \quad ,$$

where $a(x^*;\Omega;p;f) = (a_1(x^*;\Omega;p;f), \ldots, a_n(x^*;\Omega;p;f))$ and the coefficients a_0, a_i are defined by the minimum condition of the functional

$$\gamma(x^*;a_0;a;\Omega;p) = \int_\Omega [f(x^* + s) - a_0 - \langle a, s\rangle]^2 \times p(s)\mu(ds) \quad .$$

These coefficients may be represented as follows:

$$a_0(x^*;\Omega;p;f) = \int_\Omega f(x^* + s)p(s)\mu(ds) \quad ,$$

$$a_i(x^*;\Omega;p;f) = (\mathrm{E}^{-1}[s_i^2])^{-1} \int_\Omega s_i f(x^* + s)p(s)\mu(ds)\,,\ i \in \overline{1,n} \quad .$$

Definition 2. A point $x^* \in D$ is said to be an *approximate stationary point* of the function $f(x)$ on D for fixed $\{\Omega, p\}$ if

$$a_i(x^*;\Omega;p;f) = 0\,,\ i \in \overline{1,n} \quad .$$

Definition 3. A point $x^* \in D$ is said to be an *approximate local minimum* of the function $f(x)$ on D for fixed $\{\Omega, p\}$ if there exists a neighborhood $V(x^*)$ of the point x^* such that

$$\langle a(x;\Omega;p;f),\ x - x^*\rangle \ge 0$$

for each point $x \in V(x^*)$.

Definition 4. A point $x^* \in D$ is said to be a *generalized stationary point* of the function $f(x)$ on D if there exist sequences $\Omega^{(k)}$, $p^{(k)}$, $k = 1,2,\ldots$, $\mu(\Omega^{(k)}) > 0$ for each k and a sequence $x^{(k)}$ of approximate stationary points for $\Omega^{(k)}$, $p^{(k)}$ such that $\mu(\Omega^{(k)}) \to 0$, $x^{(k)} \to x^*$ as $k \to \infty$.

Definition 5. Let the points $x^{(k)}$ in Definition 4 be approximate minima. Then we call the point $x^* = \lim x^{(k)}$ as $k \to \infty$ the *generalized local minimum*.

It turns out that for a wide class of discontinuous functions the extremum points and the generalized extremum points are the same. This allows us to replace the problem of finding the extremum of the function by the problem of finding its approximate extremum. The approach described above permits us to obtain necessary and sufficient extremum conditions for discontinuous functions from one side, and, using these conditions, to construct numerical methods for finding extrema. The approximate gradient $a(x;\Omega;p;f)$, not the actual gradient of the function $f(x)$, serves as the basis of these numerical methods.

We shall now give some statements which illustrate the above points.

THEOREM 1. *Let p be a continuously differentiable function on R_n. Then the functions $a_0(x;\Omega;p;f)$ and $a_i(x;\Omega;p;f)$, $i \in \overline{1,n}$, are also continuously differntiable on D.*

In what follows we assume that the function p is continuously differentiable (it should be noted that this function may be chosen by the experimenter).

THEOREM 2. *Let a function $f: R_1 \to R_1$ be piecewise-monotonic with a finite number of semiintervals of monotonicity and discontinuities of the first type. Then the point $x^* \in [a,b] \subset R_1$ is a minimum point of the function f if and only if there exist sequences $\tau^{(k)}$, $p^{(k)}$, where $\tau^{(k)} > 0$ for each k, $\tau^{(k)} \to 0$ as $k \to \infty$, and a sequence of approximate minimum points $x^{(k)}$ for these $\tau^{(k)}$, $p^{(k)}$ such that $x^{(k)} \to x^*$.*

We say that $f \in G$ if $\lim a_0(x;\Omega;p;f) = f_*(x)$ as $\mu(\Omega) \to 0$ for all $x \in R_n$, and $a_0(x;\Omega;p;f)$ converges to $f(x)$ μ-almost everywhere on R_n. It should be noted that if the function f is discontinuous, then f_* is also discontinuous in the general case.

Let us define the iterative procedure by the equality

$$x^{(k+1)} = x^{(k)} - \alpha^{(k)} a(x^{(k)};\Omega^{(k)};p^{(k)};f) \quad , \tag{i}$$

where $\mu(\Omega^{(k)}) \to 0$ as $k \to \infty$, and the step length $\alpha^{(k)}$ is defined by the following condition:

$$\min_{\alpha \geq 0} h^{(k)}(\alpha) \leq h^{(k-1)}(\alpha^{(k)}) \leq \min_{\alpha \geq 0} h^{(k)}(\alpha) + \delta^{(k)}$$

$$h^{(k)}(\alpha) = f(x^{(k)} - \alpha a(x^{(k)}; \Omega^{(k)}; p^{(k)}; f)) \tag{ii}$$

$$\delta^{(k)} \geq 0, \sum_{k=0}^{\infty} \delta^{(k)} = \delta < \infty .$$

Define

$$E(x^{(0)}; \delta) = \{x : f_*(x) \leq f_*(x^{(0)}) + \delta\}$$

$$E^{\infty}(\delta) = \{x \in E(x^{(0)}; \delta) : x = \lim_{k \Rightarrow \infty} x^{(k)}, x^{(k)} \in E(x^{(0)}; \delta) ,$$

$$\lim_{k \to \infty} a_i(x^{(k)}; \Omega^{(k)}; p^{(k)}; f) = 0 , i \in \overline{1,n}\} .$$

THEOREM 3. *Let a bounded measurable function $f \in G$, defined on R_n, satisfy the condition*

$$|a(x^*; \Omega; p; f) - a(x_*; \Omega; p; f)| \leq L |x^* - x_*| ,$$

where L is the Lipschitz constant, which for fixed p is the same for all $x_*, x^* \in R_n$, but in general depends on Ω. Then the iterative procedure (i), (ii) ensures that the sequence $f_*(x^{(k)})$ is monotonically decreasing and

$$\lim_{k \to \infty} [f_*(x^{(k)}) - f_*(x^{(k+1)})] = 0 ,$$

regardless of the choice of $x^{(0)}$. If, in addition, L is the same for all k, $\mu(\Omega^{(k)}) > 0$, then

$$\lim_{k \to \infty} a_i(x^{(k)}; \Omega^{(k)}; p^{(k)}; f) = 0 , i \in \overline{1,n}$$

and since the set $E(x^{(0)}; \delta)$ is assumed to be bounded,

$$\lim_{k \to \infty} \rho(x^{(k)}, E^{\infty}(\delta)) = 0 .$$

It should be noted that many statements analogous to those derived in the classical analysis of smooth functions can be proved for discontinuous functions using approximate gradients. For example, results which are extensions of the theorems of Rolle, Fermat, Lagrange, and Cauchy turn out to hold for the class of discontinuous functions.

THEOREM 4 (mean value theorem). *Let $f \in G$. Then*

$$f_*(c) - f_*(b) = \langle a(x(\Omega); \Omega; p; f) + \beta(x(\Omega), \Omega), c - b \rangle ,$$

$$x(\Omega) \in [b, c], |\beta(x(\Omega), \Omega)| \to 0 \text{ as } \mu(\Omega) \to 0$$

If $(f \mid K) \in C$, where $K \subset R_n$ is compact, then $f(c) - f(b) =$

$<a(x(\Omega);\Omega;p;f) + \beta(\Omega), c - b>$, $|\beta(\Omega)| \to 0$ as $\mu(\Omega) \to 0$.

The use of constructions based on the concept of approximate extrema and, in particular, the use of these constructions in the optimization of functionals on infinite-dimensional spaces, requires some means of integrating with respect to abstract (and, possibly, finite additive) measures, thus allowing the easy estimation of integrals by finite sums.

Consider one class of discontinuous functions. Let M be a non-empty set, $\mathbf{L}$ be a semialgebra ([3, p. 46]) of subsets of M, $B_0(M, \mathbf{L})$ be the linear span of the set $\{\chi_L : L \in \mathbf{L}\}$ of characteristic functions χ_L of the sets $L \in \mathbf{L}$, and $B(M, \mathbf{L})$ be the closure of $B_0(M, \mathbf{L})$ with respect to the sup-norm $\|\cdot\|$ of the space $\mathbb{B}(M)$ of all real bounded functions on M. A set-valued function μ operating from $\mathbf{L}$ to R_1 is said to be a *finite additive measure* (FAM) on $\mathbf{L}$, if $\mu(L) = \sum_{i=1}^{m} \mu(L_i)$ for all $L \in \mathbf{L}$ and all finite subdivisions $(L_1 \in \mathbf{L}, \ldots, L_m \in \mathbf{L})$ of the set L. If μ is non-negative on $\mathbf{L}$ it is said to be a positive FAM, or FAM_+, on $\mathbf{L}$. Any FAM that may be represented as the difference of positive FAMs we call a FAM with bounded variation, or FAMBV. If μ is a FAM and $g \in B_0(M, \mathbf{L})$, then the elementary integral is defined by the finite sum ([4]; [5, p. 15]). If μ is a FAMBV, $g \in B(M, \mathbf{L})$, then the μ-integral g is defined [5, p. 18] as $c = \int_M g(x)\mu(dx)$; if $(g_1 \in B_0(E, \mathbf{L}), g_2 \in B_0(M, \mathbf{L}), \ldots)$, $\|g_i - g\| \to 0$ is satisfied for all g_i, then this implies that the μ-integrals of functions g_i converge to c. If μ is a FAMBV on $\mathbf{L}$, then $\int d\mu$ denotes the functional on $B(M, \mathbf{L})$ which associates each bounded function $g \in B(M, \mathbf{L})$ with its μ-integral.

The mapping $\mu \mapsto \int d\mu$ is ([5, p. 18]) an isometric isomorphism of the space $B^*(M, \mathbf{L})$, of the topologically conjugated space $(B(M, \mathbf{L}) \|\cdot\|)$, and of the space of finite additive measures with bounded variation on $\mathbf{L}$ with the (strong) norm variation (see [5, p. 17]). The indefinite integral is introduced in a standard way ([5, p. 19]) by the integration of the "reductions" $g\chi_L$ of the functions $g \in B(M, \mathbf{L})$ as described above (indefinite μ-integrals at functions $g \in B(M, \mathbf{L})$ are FAMBVs on $\mathbf{L}$ if μ is a FAMBV on $\mathbf{L}$). Elementary indefinite integrals are defined for any finite additive measure and bounded function from $B_0(M, \mathbf{L})$ by means of finite sums (see [5]). In addition, an elementary indefinite integral with respect to a FAMBV uniformly on $\mathbf{L}$ approximates the indefinite integral if the integrands in the elementary indefinite integral approximate the integrand of this indefinite integral with respect to $\|\cdot\|$ (see [5, p. 20]). Following [5], we introduce for all $g \in B(M)$ the sets $B^0_{\leq}(M, \mathbf{L}, g)$ and $B^0_{\geq}(M, \mathbf{L}, g)$ of all minorants and majorants, respectively, of g contained in $B_0(M, \mathbf{L})$. Here and elsewhere ordering is assumed to be pointwise. If μ is a FAM_+ and $g \in \mathbb{B}(M)$, then the set of all elementary indefinite μ-integral functions $s \in B^0_{\leq}(M, \mathbf{L}, g)$ $(s \in B^0_{\geq}(M, \mathbf{L}, g))$ is bounded above (below) in an ordered full linear lattice of all FAMBVs on $\mathbf{L}$. The least upper bound and the greatest lower bound of the above sets are properly defined and are called the upper and lower finite additive Darboux indefinite integrals of bounded function g with respect to the FAM_+ μ. If the Darboux indefinite integrals of g are the same for all FAM_+, then bounded function g is said to be *universally integrable* [5, p. 45]).

THEOREM 5 ([5, p. 45]). *The set of all universally integrable bounded functions is equal to $B(M, \mathbf{L})$, and the Darboux integral for these bounded functions is the indefinite integral constructed by extending the elementary integral from $B_0(M, \mathbf{L})$ to $B(M, \mathbf{L})$ by taking the limit with respect to $\|\cdot\|$.*

Let $X \neq \phi$, $Y \neq \phi$, and $\mathbf{X}$ and $\mathbf{Y}$ be the semialgebras of subsets of X and Y respectively. Let μ be a FAMBV on $\mathbf{X}$ and ν be a real-valued function on $X \times \mathbf{Y}$ such that [4], [6]:

(a) $\nu(\cdot, H) \in B(X, \mathbf{X}) \quad \forall H \in \mathbf{Y}$;

(b) $\nu(x, \cdot)$ $\forall x \in X$ s a FAMBV on $\mathbf{Y}$

(c) $x \mapsto \nu(x, \cdot)$ is a (strongly) bounded function with values in the space of FAMBVs on $\mathbf{Y}$ with norm variation.

The semialgebra $\mathbf{X}\{\times\}\mathbf{Y}$ for the subsets of $X \times Y$ is defined as the family of all "rectangles" $U \times V$, where $U \in \mathbf{X}$ and $V \in \mathbf{Y}$. We define the set-valued function $\mu \otimes \nu$ on $\mathbf{X}\{\times\}\mathbf{Y}$, taking the value $(\mu \otimes \nu)(K \times L)$ of this for all $K \in \mathbf{X}$, $L \in \mathbf{Y}$ as the μ-integral $\nu(\cdot, L)$ on the set K [4,6]; $\mu \otimes \nu$ is a FAMBV.

THEOREM 6 ([4,6]). *Let $g \in B(X \times Y, \mathbf{X}\{\times\}\mathbf{Y})$. Then $\int\int g(x, y)\nu(x, dy)\mu(dx)$ is properly defined and coincides with $\int g(z)(\mu \otimes \nu)dz$):*

$$\int_U \int_V g(x, y)\nu(x, dy)\mu(dx) = \int_{U \times V} g(z)(\mu \otimes \nu)(dz)$$

$$U \in \mathbf{X}, V \in \mathbf{Y}) \quad .$$

If $\nu(x, \cdot) = \text{const}$ we have the following corollary [4,6]:

$$\int\int g(x, y)\mu(dx)\nu(dy) = \int\int g(x, y)\nu(dy)\mu(dx)$$

Another corollary: let X be a strongly bounded set of FAMBVs on $\mathbf{Y}$ and $\mathbf{X}$ be a semialgebra for subsets of X which are locally closed ([7, p. 42]) in terms the weak* topology induced in X from the set $\mathbf{Y}$ (i.e., the family of all intersections $G \cap F$, where G is open and F is closed. Let $\nu(x, L) = x(L)$ for all $x \in X$ and $L \in \mathbf{Y}$. Then ν satisfies (a), (b), (c) and, since μ is a FAMBV on $\mathbf{X}$,

$$\int\int g(x, y)x(dy)\mu(dx) = \int g(z)(\mu \otimes \nu)(dz)$$

on $\mathbf{X}\{\times\}\mathbf{Y}$. We shall consider the relations between $(B(X,\mathbf{X}), B^*(X,\mathbf{X}))$ and $(B(Y,\mathbf{Y}), B^*(Y,\mathbf{Y}))$. Let us define the product in $\mathbf{X}$ and $\mathbf{Y}$ as the intersection, we consider, as in [8], the homomorphisms φ acting from $\mathbf{X}$ into $\mathbf{Y}$ such that if $L \in \mathbf{X}$ and $(L_1 \in \mathbf{X}, \ldots, L_m \in \mathbf{X})$ is finite subdivision of $X \setminus L$, then $(g(L_1), \ldots, g(L_m))$ is subdivision of $Y \setminus g(L)$. Such homomorphisms are said to be *decomposition homomorphisms* [8]. Each decomposition homomorphism φ defines operators T_1, T_2 $(T_i = T_i(\varphi))$, where T_1 acts from $B(X,\mathbf{X})$ into $B(Y,\mathbf{Y})$, and T_2 operates from the space of FAMBVs on $\mathbf{Y}$ into the space of FAMBVs on $\mathbf{X}$ [8]. The operator T_1 on $B_0(X,\mathbf{X})$ satisfies the condition

$$T_1\left(\sum_{i=1}^{m} \alpha_i \chi_{L_i}^{x}\right) = \sum_{i=1}^{m} \alpha_i \chi_{\varphi(L_i)}^{Y}$$

and is continuously extendable on $B(X,\mathbf{X})$ in terms of the sup norms $B(X)$ and $B(Y)$. Here $\chi_L^X = \chi_L$, $L \in \mathbf{X}$, is a characteristic function defined on the set X (χ_L^x is defined analogously). The operator T_2 is defined by the condition:

$$(T_2(\mu))(L) = \mu(\varphi(L)) = (\mu \cdot \varphi)(L) \quad ,$$

where μ is a FAMBV on $\mathbf{Y}$ and $L \in \mathbf{X}$. The following theorem [8] is analogous to the formula for change of variables (see [9, p. 200]). However, the theorem does not use the concept of mappings which are measurable with respect to $\mathbf{X}$, $\mathbf{Y}$ on the sets X, Y.

THEOREM 7 [8]. *If μ is a FAMBV on $\mathbf{Y}$ and $g \in B(X,\mathbf{X})$*, then

$$\int g(x)(T_2(\mu))(dx) = \int (T_1(g))(y)\mu(dy) \quad .$$

REFERENCES

1. V.D. Batuhtin and L.A. Maiboroda. *Optimization of Discontinuous Functions.* Nauka, Moscow, 1984 (in Russian).

2. V.D. Batuhtin and L.A. Maiboroda. On a formalization for extremal problems. *Doklady Akademii Nauk SSSR*, 250(1)(1980)11–14.

3. G. Neve. *Mathematical Foundations of Probability Theory*. Mir, Moscow, 1969 (in Russian).

4. A.G. Chentsov. On integration on additive functions of sets. In *Some Problems in the Optimization of Discontinuous Functions*. UNC AN SSSR, Sverdlovsk, 1984, pp. 94–108.

5. A.G. Chentsov. *Finite Additive Integration of Bounded Functions*. UNC AN SSSR, Sverdlovsk, 1984.

6. A.G. Chentsov. *Integration on Additive Functions of Sets*. S.M. Kirov UPI, Sverdlovsk, 1983.

7. R.A. Aleksandryan and A.A. Mirzakchanyan. *General Topology*. Moscow, 1979.

8. A.G. Chentsov. *On the Relations Between Measurable Spaces*. IMM UNC AN SSSR, Sverdlovsk, 1984.

9. N. Dunford and J.T. Schwartz. *Linear Operators: General Theory*. Moscow, 1962.

CONVEX MULTIVALUED MAPPINGS AND STOCHASTIC MODELS OF THE DYNAMICS OF ECONOMIC SYSTEMS

N.N. Bordunov
V. Glushkov Institute of Cybernetics, Kiev, USSR

This paper gives a brief description of some results obtained by using the theory of convex multivalued mappings [1] to investigate a reasonably general stochastic model of the dynamics of economic systems. All proofs are given in [2]; see also [3].

We shall first give some definitions from [1]. Let L and L' be Banach spaces. A mapping a of the space L into the set of all subsets of L' is said to be *convex* if

$$\text{gr } a = \{(l, l') \mid l' \in a(l)\}$$

is a convex subset of the cartesian product $L \times L'$. We shall write

$$\text{dom } a = \{l \mid a(l) \neq \phi\}$$

$$w(l, l'^*) = \inf \{\langle l'^*, l' \rangle \mid l' \in a(l)\} \quad ,$$

where l'^* is from the conjugate space L'^*. Here and elsewhere we shall suppose that $\inf \phi = +\infty$. Obviously, if l'^* is fixed, then $w(l, l'^*)$ is a convex function of l. The subdifferential of w at the point $\bar{l} \in L$ will be denoted by $a^*(\bar{l}, l'^*)$. Thus, if l is fixed, we can consider a mapping $a^*(l, \cdot)$ which associates a set $a^*(l, l'^*) \subset L^*$ (possibly the empty set) with every $l'^* \in L'^*$. This mapping is said to be *conjugate* to a at the point l.

Now we shall construct a stochastic optimization model of the dynamics of economic systems. Necessary and sufficient conditions for maxima and minima are formulated below. A similar model was investigated in [4].

Let (Ω, F, P) be the basic probability space, $\{F_t\}_{t=0}^{T}$ be a non-decreasing sequence of σ-subalgebras of F, and T be a fixed natural number. We suppose that F is complete with respect to P and the inclusion $A \in F$ holds for all $A \in F_t$, $0 \leq t \leq T$, $P(A) = 0$. The finite-dimensional normed space R^n is assumed to be endowed with a Borel σ-algebra $\mathbf{B}^n$. Let S_ε^n denote an open ball with radius ε and centre 0 in R^n, and $L_{\infty,t}$ denote Banach spaces of F_t-measurable, R^n-valued, bounded random variables x_t with the norm

$$\text{ess sup } \{ \|x_t(\omega)\|_{R^n} \mid \omega \in \Omega \} \quad .$$

In addition, let $L_{1,t,*}$ denote Banach spaces of F_t-measurable, $(R^n)^*$-valued random variables x_t^* with finite mathematical expectations; in $L_{1,t,*}$ the norm is

$$\mathrm{E} \|x_t^*(\omega)\|_{(R^n)^*} \quad ,$$

where E represents mathematical expectation.

Let $\bar{x}_0 \in L_{\infty,0}$ be fixed. We shall take $\bar{x}_0$ as the initial state of model. Let random variables $f_1^*, \ldots, f_T^*$, $f_t \in L_{1,t,*}$, and multivalued mappings $a_t \colon R^n \times \Omega \to R^n$, $1 \le t \le T$, be given. We shall assume that the following conditions are satisfied:

(C1) Mappings $a_t(\cdot, \omega) \colon R^n \to R^n$ are convex multivalued mappings for $1 \le t \le T$, $\omega \in \Omega$; graphs gr $a_t(\cdot, \omega)$ are non-empty closed subsets of $R^n \times R^n$.

(C2) Mapping a_t is measurable (see, for example, [5]) with respect to $\mathbf{B}^n \otimes F_t$.

(C3) If $U_{t-1} \subset R^n$, $1 \le t \le T$, is a bounded set, there exists a bounded set $U_t \subset R^n$ (possibly dependent on U_{t-1}) such that $a_t(x, \omega) \subset U_t$ for all $(x, \omega) \in U_{t-1} \times \Omega$.

(C4) There exist R^n-valued, F_t-measurable random variables $\tilde{x}_t$ and positive numbers ε_t, $1 \le t \le T$, such that

$$(\tilde{x}_{t-1}(\omega), \tilde{x}_t(\omega)) + S^n_{\varepsilon_t} \times S^n_{\varepsilon_t} \subset \text{gr } a_t(\cdot, \omega)$$

$$\tilde{x}_1(\omega) + S^n_{\varepsilon_1} \subset a_1(\bar{x}_0(\omega), \omega)$$

P-a.s. for $1 < t \le T$.

(C5) The sets dom $a_t(\cdot, \omega)$ are closed for $1 < t \le T$ and all $\omega \in \Omega$; the multivalued mapping $D_t \colon \Omega \to R^n$, where $D_t(\omega) = \text{dom } a_t(\cdot, \omega)$, is F_t-measurable.

(C6) For all $(x, \omega) \in R^n \times \Omega$, $1 \le t < T$, we have

$$a_t(x, \omega) \subset \text{dom } a_{t+1}(\cdot, \omega) \quad .$$

Consider the following extremum problem: find random variables $x_1, \ldots, x_T$ such that

$$\sum_{t=1}^{T} \mathrm{E} f_t^* x_t \to \min \tag{1}$$

$$x_t(\omega) \in a_t(x_{t-1}(\omega), \omega) \ \text{(a.s.)}, \ 1 \le t \le T \tag{2}$$

$$x_0(\omega) = \bar{x}_0(\omega) \tag{3}$$

$$\mathrm{E} \{ x_t \mid F_t \} = x_t \ \text{(a.s.)}, \ 1 \le t \le T \tag{4}$$

where $x_t^* x_t = \langle x_t^*(\omega), x_t(\omega)\rangle_{R^n}$.

To simplify the notation we shall no longer indicate every dependence on ω; we shall write, for example, $x_t \in a_t(x_{t-1}, \omega)$ (a.s.) instead of (2).

A reasonably detailed economic interpretation of this model can be found in [2,6]; of course, this interpretation is very similar to that given in [4]. Comments on conditions (C1)–(C6) are given in [2,7]. Condition (C6) is very important here; using (C6) we can deduce extremum conditions for (1)–(4) (see Theorem 2 below). The idea of using conditions of type (C6) was first proposed in [8].

Note. If we replace (1) by the criterion

$$\sum_{t=1}^{T} \mathrm{E} f_t(x_t(\omega), \omega) \quad ,$$

where the $f_t : R^n \times \Omega \to R^1$ are $\mathbf{B}^n \otimes F_t$-measurable functions which are convex with respect to x (where ω is fixed), then the conclusions and results obtained in [2] require only weak and obvious modifications.

Now we shall introduce Bellmann's function for (1)–(4). Let

$$B_t(x_t) = \inf \{ \sum_{\tau=t}^{T} \mathrm{E} f_\tau^* x_\tau \mid x_\tau \in a_\tau(x_{\tau-1}, \omega) \text{ (a.s.)} \quad ,$$

$$\mathrm{E}\{x_\tau \mid F_\tau\} = x_\tau \text{ (a.s.)}, \; t < \tau \le T\}, \; 1 \le t < T, \; B_T(x_T) = \mathrm{E} f_T^* x_T \quad ,$$

assuming that $f_0^* = 0$ (a.s.). It is easy to verify that the B_t are proper convex functions, $B_t : L_{\infty,t} \to R^1 \cup \{+\infty\}$.

Let $\chi_t^* \in (L_{\infty,t})^*$. It is known (see [9]) that $\chi_t^* = \chi_t^{*a} + \chi_t^{*s}$, where χ_t^{*a} and χ_t^{*s} are the absolutely continuous and singular components of χ_t^*, respectively; this representation is unique. Absolutely continuous functionals from $(L_{\infty,t})^*$ possess the following property: for every $\chi_t^{*a} \in (L_{\infty,t})^*$ there exists one and only one random variable $x_t^* \in L_{1,t}$, such that for all $x_t \in L_{\infty,t}$

$$\langle \chi_t^{*a}, x_t \rangle = \mathrm{E}\, x_t^* x_t \quad .$$

THEOREM 1. *Let* $1 \le t \le T$, $\bar{x}_t \in \operatorname{dom} B_t = \{x_t \in L_{\infty,t} \mid B_t(x_t) \ne +\infty\}$, *and* $\chi_t^* \in \partial B_t(\bar{x}_t)$. *Then*

$$\chi_t^{*a} \in \partial B_t(\bar{x}_t)$$

$$\langle \chi_t^{*s}, x_t - \bar{x}_t \rangle \le 0$$

for all $x_t \in \operatorname{dom} B_t$.

The essence of Theorem 1 (but not its proof) was prompted by Corollary 1B of Theorem 1 in [10]. Theorem 1 plays a central role in the proof of Theorem 2, which states solution conditions for problem (1)–(4).

THEOREM 2. *The sequence* $\bar{x}_1, \ldots, \bar{x}_T$ *is a solution of* (1)–(4) *if and only if there exist random variables* $\bar{x}_t^*$, $\bar{g}_t^*$, $1 < t \leq T$, *such that* $\bar{x}_t^*, \bar{g}_t^* \in L_{1,t}$, *and the following relations hold with probability* 1:

$$\bar{x}_t^* + \bar{g}_{t+1}^* \in f_t^* + a_{t+1}^*(\bar{x}_t, \bar{x}_{t+1}^*, \omega) , \; 1 \leq t < T \tag{5}$$

$$\bar{x}_T^* = f_T^* \quad , \tag{6}$$

$$\bar{x}_t^* \bar{x}_t = w_t(\bar{x}_{t-1}, \bar{x}_t^*, \omega) , \; 1 \leq t \leq T \tag{7}$$

$$\mathrm{E}\,\{\bar{g}_{t+1}^* \mid F_t\} \cdot (x - \bar{x}_t) \geq 0 , \; \forall x \in \operatorname{dom} a_{t+1}(\cdot, \omega) , \; 1 \leq t < T \quad . \tag{8}$$

The proof of Theorem 2 allows us to see the connection between the conjugate variables $\bar{x}_1^*, \ldots, \bar{x}_T^*$ and the subgradients of the Bellmann function. If $F_0 = F_1 = \cdots = F_T = F = \{\phi, \Omega\}$ (the deterministic case), then $\bar{x}_t^* \in \partial B_t(\bar{x}_t)$ [11]. The idea of applying dynamic programming methods to the problem (1)–(4) was originally suggested in [11]; see also [4,12,13].

It should be mentioned that Theorem 2 can be deduced from the main theorem given in [2]; however, this deduction is nontrivial and not so clear and simple as the proof given in [2].

It is interesting to note that there is an analogy between (5)–(8) and Pontryagin's maximum principle. Set $\bar{p}_t^* = \bar{x}_t^* + \bar{g}_{t+1}^*$, $1 \leq t < T$, (in which case $\bar{p}_t^* \in F_{t+1}$) and let $\bar{p}_T^* = f_T^*$. It is not very difficult to verify that

$$\bar{p}_t^* \in f_t^* + a_{t+1}^*(\bar{x}_t, \mathrm{E}\,\{\bar{p}_{t+1}^* \mid F_{t+1}\}, \omega) \text{ (p–a.s.)} , \; 1 \leq t < T \tag{9}$$

and by definition

$$\bar{p}_T^* = f_T^* \text{ (a.s.)} \quad . \tag{10}$$

In addition,

$$\mathrm{E}\,\{\bar{p}_t^* \mid F_t\} \cdot \bar{x}_T = w_t(\bar{x}_{t-1}, \mathrm{E}\,\{\bar{p}_t^* \mid F_t\}, \omega) \text{ (p–a.s.)} , \; 1 \leq t \leq T \quad . \tag{11}$$

Now it is obvious that (9) is analogous to the conjugate system of equations, (10) is the condition of transversality, and (11) is analogous to the maximum principle. These analogies confirm that the theory of convex multivalued mappings is useful in investigating stochastic convex models of the dynamics of economic systems.

REFERENCES

1. B.N. Pschenichny. Convex multivalued mappings and their conjugates (in Russian). *Kibernetika*, (3)(1972)94–102.

2. N.N. Bordunov. On a general stochastic model of economic dynamics (in Russian). In *Models and Methods of Stochastic Optimization*, CEMI, Moscow, 1983, pp. 27–60.

3. N.N. Bordunov. Dynamic programming and optimality conditions for the control of random convex mappings (in Russian). *Kibernetika*, (3)(1984)51–55.

4. E.B. Dynkin. Stochastic concave dynamic programming (in Russian). *Matematicheskie Sbornik*, 87(4)(1972)490–503.

5. R.T. Rockafellar. Measurable dependence of convex sets and functions on parameters. *Journal of Mathematical Analysis and Applications*, 28(1969)4–25.

6. N.N. Bordunov. Optimality conditions for trajectories of a stochastic Neumann-type model of economic dynamics (in Russian). *Kibernetika*, (1)(1981)143–144.

7. N.N. Bordunov. Optimality conditions for the many-step problem of controlling random convex mappings (in Russian). *Kibernetika*, (1)(1983)40–45.

8. R.T. Rockafellar and R. Wets. Stochastic convex programming: relatively complete recourse and induced feasibility. *SIAM Journal on Control and Optimization*, 14(1976)574–589.

9. K. Yosida and E. Hewitt. Finitely additive measures. *Transactions of the American Mathematical Society*, 72(1952)46–66.

10. R.T. Rockafellar. Integrals which are convex functionals. II. *Pacific Journal of Mathematics*, 39(1971)439–469.

11. V.V. Beresnev. On optimal programs for systems of discrete inclusions over an infinite time (in Russian). *Kibernetika*, (3)(1978)93–99.

12. R.T. Rockafellar and R. Wets. Non-anticipativity and L^1-martingales in stochastic optimization problems. *Mathematical Programming Study*, 6(1976)170–187.

13. I.V. Evstigneev. Measurable selection and dynamic programming. *Mathematics of Operations Research*, 1(1976)267272.

STABILITY IN STOCHASTIC PROGRAMMING – PROBABILISTIC CONSTRAINTS

Jitka Dupačová
Charles University, Prague

1. INTRODUCTION

When solving stochastic programs, a complete knowledge of the distribution of random coefficients is usually taken for granted. In many real-life situations, however, this assumption is not justified and the results should at least be supplemented by suitable stability studies.

The stability of the optimal solution of stochastic programs with respect to the distribution and its parameters can be studied to some extent using the methods of parametric programming and techniques developed for studying the stability of nonlinear programming problems (Armacost and Fiacco 1974, Garstka 1974). These methods can be complemented by suitable statistical approaches capable of dealing with the statistical character of the original problem.

The stability of the optimal solution of stochastic programs with recourse has been studied, e.g., by Dupačová (1983, 1984, 1985).

In this paper, the methods outlined above will be applied to stochastic programs with probabilistic constraints. We shall consider the following model:

Let z be a random vector on (Z, B_z), $Z \subset R^l$; $c: R^n \rightarrow R^l$ be a given function; $g_j: R^n \times R^l \rightarrow R^{m_j}$, $1 \le j \le k$, be given Borel mappings; and $X \subset R^n$ be a given nonempty convex set. The problem may then be stated as follows:

$$\begin{aligned} &\text{maximize} \quad c(x) \\ &\text{subject to} \quad \underset{x \in X}{\mathrm{P}_F} \{ g_j(x; z) \ge 0 \} \ge \alpha_j, \; 1 \le j \le k \quad , \end{aligned} \tag{1}$$

where F denotes a given probability distribution on (Z, B_z). The optimal solution, the Lagrange function and the Kuhn–Tucker points of (1) will be denoted by $x(F)$, $L(x; F)$ and $w(F)$, respectively, to indicate their dependence on the chosen distribution F.

As a starting point, consider the following deterministic nonlinear program, which depends on a vector parameter y:

Let $Y \subset R^q$ be an open set and $c: R^n \times Y \to R^1$, $h: R^n \times Y \to R^{m+p}$ be given, continuously differentiable functions. For a fixed $y \in Y$, the problem is to

$$\text{maximize} \quad c(x;y)$$

$$\text{subject to } h_i(x;y) \geq 0, \; 1 \leq i \leq m \qquad \mathrm{M}(y)$$
$$h_i(x;y) = 0, \; m+1 \leq i \leq m+p \quad .$$

The corresponding Lagrange function has the form:

$$L(x,u,v;y) = c(x;y) + \sum_{i=1}^{m} u_i h_i(x;y) + \sum_{i=1}^{p} v_i h_{m+i}(x;y)$$

and the Kuhn–Tucker point of M(y) will be denoted by $w(y) = [x(y), u(y), v(y)] \in R^n \times R^m_+ \times R^p$. Knowledge of the first- and second-order Kuhn–Tucker conditions and of the linear independence condition and the strict complementarity conditions (Fiacco 1976, Robinson 1980) will be assumed throughout the text.

THEOREM 1. *Let $y^0 \in Y$ and let $w(y^0)$ be a Kuhn–Tucker point of M(y^0) for which the Kuhn–Tucker conditions of the first and second order, the linear independence condition and the strict complementarity conditions hold. Let c and h_i, $1 \leq i \leq m+p$, be twice continuously differentiable with respect to x on a neighborhood of $[x(y^0); y^0]$ and continuous derivatives*

$$\frac{\partial^2 c(x;y)}{\partial y_k \partial x_j}, \quad \frac{\partial^2 h_i(x;y)}{\partial y_k \partial x_j} \quad \text{exist for all } 1 \leq k \leq q, \; 1 \leq j \leq n, \; 1 \leq i \leq m+p$$

Then the following statements are true:

(a) *For $y \in 0_\varepsilon(y^0)$, there exists a unique, continuously differentiable function $w(y) = [x(y), u(y), v(y)]$ satisfying the Kuhn–Tucker conditions of the first and second order, the linear independence condition and the strict complementarity conditions for M(y).*

(b) *Let the index set $I(y) \subset \{1, \ldots, m\}$ contain the indices of the active inequality constraints $h_i(x(y);y) = 0$, $i \in I(y)$, and define*

$$w_I(y) = [x(y), u_i(y), i \in I(y), v(y)] \quad ,$$

$$\nabla_x h_I(x;y) = [\nabla_x h_i(x;y), i \in I(y), \nabla_x h_r(x;y), m+1 \leq r \leq m+p]$$

$$\nabla_y h_I(x;y) = [\nabla_y h_i(x;y), i \in I(y), \nabla_y h_r(x;y), m+1 \leq r \leq m+p] \quad .$$

Further, set

$$D(y) = \begin{bmatrix} \nabla^2_{xx} L(w\,;y) & \nabla_x h_I(x\,;y) \\ [\nabla_x h_I(x\,;y)]^{\mathrm{T}} & 0 \end{bmatrix}_{[w(y);y]} \tag{2}$$

$$B(y) = [\nabla^2_{yx} L(w\,;y), \nabla_y h_I(x\,;y)]^{\mathrm{T}}_{[w(y);y]} \quad . \tag{3}$$

Then for $y \in 0_\varepsilon(y^0)$, *we have*

$$\frac{\partial w_I(y)}{\partial y} = -D^{-1}(y)B(y) \quad . \tag{4}$$

The statements of Theorem 1 represent modifications of results obtained by Fiacco (1976) and Robinson (1974). The assumptions can be weakened using, e.g., results derived by Robinson (1980); for application of these more general results see Dupačová (1986).

As we shall see later, the parameter vector y may correspond to the parameters of the underlying distribution F (see Section 2), to the contamination parameter (see Section 3) or, finally, to the probability levels α_i, $1 \le i \le k$. The results can be extended without difficulty to the case when the objective function in (1) also depends on the parameter.

To provide some motivation, let us first consider a few examples.

Example 1: The cattle-feed problem (Van de Panne and Popp 1963). The problem is to find the amounts x_j of ingredient j which lead to the cheapest final mixture which still satisfies certain nutritional requirements. The protein content (as a weight percentage per ton), a_j, of each of the four constituents is assumed to be a normally distributed random variable with mean μ_j and variance σ_j^2, $1 \le j \le 4$. In addition to deterministic linear constraints, one probabilistic constraint

$$\mathrm{P}\{\sum_{j=1}^{4} a_j x_j \ge p\} \ge 1 - \alpha \tag{5}$$

is constructed.

Assuming a normal distribution, (5) can be rewritten as follows:

$$\sum_{j=1}^{4} \mu_j x_j + \Phi^{-1}(\alpha)\,(\sum_{j=1}^{4} \sigma_j^2 x_j^2)^{1/2} \ge p \quad ,$$

where $\Phi^{-1}(\alpha)$ denotes the α-quantile of the $N(0,1)$ distribution. The parameters μ_j, σ_j^2, $1 \le j \le 4$, are estimated by sampling, and in practical applications the estimates are used instead of the true parameter values. The problem of the stability of the optimal solution with respect to parameter values was solved, i.e., derivatives of the optimal solution with respect to the parameter values were obtained (Armacost and Fiacco 1974).

Now that we are familiar with the statistical background of the parameters we shall try to complement the deterministic stability results with some statistical ones (see Section 2).

Example 2. In simple stochastic models of water reservoir design based on individual probabilistic constraints, the variables take the form of monthly inflows r_i whose marginal distributions F_i are assumed to be known. Usually, a log-normal distribution is used; its parameters are estimated on the basis of observations of the monthly inflows over a relatively long period of time. However, in particular months, specific deviations from the assumed distribution may appear: in spring, the distribution may be relatively close to the normal distribution. Under these circumstances, we can accept the hypothesis that the true marginal distributions are mixtures of the assumed log-normal and normal distributions. We are interested in describing the changes in the original optimal decision arising from the influence of the second distribution.

Example 3. The STABIL model (Prékopa et al. 1980) was applied to the fourth Five-Year Plan for the electrical energy sector in Hungary. In addition to numerous deterministic linear constraints, this model also contains one joint probabilistic constraint:

$$\mathrm{P}_F\{\sum_{j=1}^{n} a_{ij}x_j \geq z_i \ , 1 \geq i \geq 4\} \geq p \quad .$$

The four right-hand sides z_i, $1 \leq i \leq 4$ were taken to be stochastic and the joint distribution of these random variables was assumed to be normal. Due to the lack of reliable data, some of the correlations could not be obtained with sufficient precision. For this reason two alternative correlation matrices were considered and the numerical results compared.

Alternatively, instead of considering the separate normal distributions $N(\mu, \sum_1)$ and $N(\mu, \sum_2)$, we could construct a mixture

$$(1-t)N(\mu, \textstyle\sum_1) + tN(\mu, \textstyle\sum_2) \tag{6}$$

which in principle allows us to study the changes in the optimal solution for $0 \leq t \leq 1$; (6) corresponds to the ***gross error*** or ***contamination model.***

2. ESTIMATED PARAMETERS

Assume now that the parameter vector y in $M(y)$ is connected with some statistical assumptions about the distribution F of random coefficients in a stochastic program. In particular, let y be the parameter identifying the distribution F, which is known to belong to a parametric family of distributions $\{F_y, y \in Y\}$. $M(y)$ is the corresponding program:

$$\text{maximize} \quad c(x)$$

$$\text{subject to} \quad h_j(x\,;y) := \mathrm{P}_{F_y}\{g_j(x\,;z) \geq 0\} - \alpha_j \geq 0\,,\ 1 \leq j \leq k \quad , \tag{7}$$

$$x \in X \quad .$$

Our aim is to solve program (7) for the true parameter vector, say $\eta \in Y$. However, our decision can only be based on an estimate, say y^N, of η. This means that we are actually solving the substitute program $\mathrm{M}(y^N)$ instead of $\mathrm{M}(\eta)$. Assuming that the distribution of the estimate y^N is asymptotically normal, the deterministic stability results of Theorem 1 can then be complemented by statistical stability results.

THEOREM 2. *Let y^N be an estimate of the true parameter vector η, based on a sample of size N and with an asymptotically normal distribution*

$$\sqrt{N}\,(y^N - \eta) \sim N(0, \textstyle\sum)$$

and a known variance matrix $\sum$. Let the assumptions of Theorem 1 be satisfied for $M(\eta)$. Then the distribution of the optimal solution $x(y^N)$ of $M(y^N)$ is asymptotically normal

$$\sqrt{N}\,(x(y^N) - x(\eta)) \sim N(0, V) \tag{8}$$

with a variance matrix

$$V = \left[\frac{\partial x(\eta)}{\partial y}\right] \sum \left[\frac{\partial x(\eta)}{\partial y}\right]^{\mathrm{T}} \quad ,$$

where $(\frac{\partial x(\eta)}{\partial y})$ is the $(n\,, q)$ submatrix of (4).

Proof. Under the assumptions of Theorem 1, $x(y)$ is continuously differentiable on a neighborhood of $x(\eta)$. Using the normality assumption and the δ-method (Rao 1973, p. 388), we immediately get the desired result. ▪

The application of Theorem 2 to Example 1 is straightforward.

Remark 1. All elements of $(\frac{\partial x}{\partial y})$ are continuous on a neighborhood of η, so that the asymptotic distribution (8) can be replaced by

$$N(0, (\frac{\partial x(y^N)}{\partial y}) \sum (\frac{\partial x(y^N)}{\partial y})^{\mathrm{T}}) \quad ;$$

see Rao (1973).

A similar theorem for stochastic programs with recourse was proved in detail and applied to a stochastic program with simple recourse under special assumptions on the parametric family of distributions by Dupačová (1984).

3. CONTAMINATED DISTRIBUTIONS

The local behaviour of the optimal solution $x(F)$ of program (1) with respect to small changes in the distribution F can be studied by t-contamination of F by a suitably chosen distribution G, i.e., instead of F, consider distributions of the form

$$F_t = (1-t)F + tG \, , \, 0 \le t \le 1 \quad . \tag{9}$$

The original stability problem is thus transformed to a simple problem linearly perturbed by a scalar parameter t. In principle, it is possible to calculate the trajectory of the optimal solution $x(F_t)$, $0 \le t \le 1$; for a suitable method see Gfrerer et al. (1983). We shall aim to obtain the Gâteaux differential $dx(F;G-F)$ of the optimal solution of (2) in the direction $G-F$; for parallel results concerning stochastic programs with recourse, see Dupačová (1985). To get explicit results, it is necessary to check the differentiability and regularity assumptions of Theorem 1 and to compute matrices $B(0)$, $D(0)$ corresponding to the contamination parameter $t = 0$.

For the sake of simplicity we shall put $X = R^n$ in (1), thus considering only the probabilistic constraints, and we shall concentrate on the special case

$$g_i(x;z) = \sum_{j=1}^{n} a_{ij}x_j - z_i \, , \, 1 \le i \le m \quad ,$$

or

$$g(x;z) = Ax - z$$

which corresponds to Examples 2 and 3. The rows of matrix $A = (a_{ij})$ will be denoted by a^i. Similar results can be proved for $g_i(x;z) = g_i(x) - z_i$, $1 \le i \le m$, and $g(x;z) = g(x) - z$. In the constrained case, e.g., for X polyhedral, explicit results can be obtained using the same approach. However, these formulas are rather cumbersome.

THEOREM 3. *Consider the program*

$$\begin{aligned} &\text{maximize} \quad c(x) \\ &\text{subject to} \quad P_F\{\sum_{j=1}^{n} a_{ij}x_j \ge z_i\} \ge \alpha_i \, , \, 1 \le i \le m \quad , \end{aligned} \tag{10}$$

where it is assumed that:

(i) *$\alpha_i \in (0,1)$, $1 \le i \le m$, and the matrix $A(m,n)$ are given.*

(ii) *$c: R^n \to R^1$ is twice continuously differentiable.*

(iii) *z is a random vector on (Z, B_z), $Z \subset R^m$, whose distribution F is absolutely continuous.*

(iv) *(10) has a Kuhn–Tucker point $w(F) = [x(F), u(F)]$ such that the strict complementarity conditions are fulfilled.*

(v) *Marginal densities f_i, $1 \le i \le m$, are continuously differentiable in a neighborhood of $X(F) := Ax(F)$ and*

$$f_i\left(\sum_{j=1}^{n} a_{ij} x_j(F)\right) > 0 \,,\; i \in I(F) \quad .$$

The rank of $A_I = (a_{ij})_{\substack{i \in I(F) \\ 1 \le j \le n}}$ equals the cardinality of $I(F)$.

(vi) *For all $l \in R^n$, $l \ne 0$, for which $A_I l = 0$, the inequality $l^{\mathrm{T}} \nabla^2_{xx} c(x(F)) l < 0$ holds and the matrix*

$$L = \nabla^2_{xx} L(w(F);F) = \nabla^2_{xx} c(x(F)) + \sum_{i \in I(F)} u_i(F) f_i'(a^i x(F)) a^{i\mathrm{T}} a^i \tag{11}$$

is nonsingular.

(vii) *The marginal distribution functions G_i of the distribution G on (Z, B_z) are twice continuously differentiable on a neighborhood of $X(F)$.*

Then

(a) *There is a neighborhood $O(w(F)) \subset R^n \times R^m_+$, a real number $t_0 > 0$ and a continuous function $w: \langle 0, t_0) \to O(w(F))$, $w(0) = w(F)$ such that for any $t \in \langle 0, t_0)$, we have that $w(t) = [x(t), u(t)]$ is the Kuhn–Tucker point of the problem*

maximize $c(x)$

$$\text{subject to } \mathrm{P}_{F_i}\left\{\sum_{j=1}^{n} a_{ij} x_j \ge z_i\right\} \ge \alpha_i \,,\; 1 \le i \le m \quad ,$$

for which the second-order sufficient condition, the linear independence condition and the strict complementarity conditions are all satisfied.

(b) *The Gâteaux differential $\mathrm{d}x(F; G-F)$ of the isolated local maximizer $x(F)$ of (10) in the direction $G - F$ is given by*

$$\mathrm{d}x(F; G-F) = -L^{-1} A_I^{\mathrm{T}} (A_I L^{-1} A_I^{\mathrm{T}})^{-1} f_I^{-1} (G_I - \alpha_I) \quad , \tag{13}$$

where L is given by (13) *and*

$$f_I \;= \operatorname{diag}\,\{f_i(\sum_{j=1}^{n} a_{ij}x_j(F)),\ i \in I(F)\}\quad , \tag{14}$$

$$G_I \;= [G_i(\sum_{j=1}^{n} a_{ij}x_j(F)),\ i \in I(F)]\quad , \tag{15}$$

$$\alpha_I \;= [\alpha_i\ ,\ i \in I(F)]\quad .$$

Proof. From assumptions (v), (vi), the condition of linear independence and the second-order sufficient conditions are satisfied for (10) so that the general results of Theorem 1 can be applied. For t sufficiently small, we have

$$D(t) = \begin{pmatrix} \nabla^2_{xx} L(w(t);F_t) & [\nabla_x h_I(x(t);F_t)] \\ [\nabla_x h_I(x(t);F_t)]^T & 0 \end{pmatrix}$$

where

$$\nabla_x h_I(x(t);F_t) = ([(1-t)f_i(a^i x(t)) + tG_i'(a^i x(t))]a_i^{iT} i \in I(F))$$

and

$$B(t) = \begin{pmatrix} \sum_{i \in I(F)} u_i(t)[G'_i(a^i x(t)) - f_i(a^i x(t))]a^{iT} \\ G_i(a^i x(t)) - F_i(a^i x(t))\ ,\ i \in I(F) \end{pmatrix} .$$

For t = 0, formula (13) follows from (4) by inversion of the block matrix $D(0)$. ▪.

We have an alternative result for a linear objective function:

THEOREM 4. *Consider the program*

$$\begin{aligned} &\text{maximize} \quad c^T x \\ &\text{subject to} \quad P_F\{\sum_{j=1}^{n} a_{ij}x_j \ge z_i\} \ge \alpha_i\ ,\ 1 \le i \le m \end{aligned} \tag{17}$$

and let assumptions (i), (iii) *of Theorem 3 be satisfied. Assume in addition that the optimal solution* $x(F)$ *of* (17) *is unique and nondegenerate with* $f_i(\sum_{j=1}^{n} a_{ij}x_j(F)) > 0$, $i \in I(F)$, *and that the marginal distribution functions* G_i *of the distribution* G *on* (Z, B_z) *are continuously differentiable on a neighborhood of* $X(F) := Ax(F)$.

Then the Gâteaux differential $dx(F;G-F)$ *of* $x(F)$ *in the direction* $G-F$ *is given by*

$$dx(F;G-F) = -(f_I \cdot A_I)^{-1}[G_I - \alpha_I] \quad ,$$

where f_I , G_I , α_I *are given by* (14), (15), (16), *respectively*.

Proof. Let $b_i = F_i^{-1}(\alpha_i)$ denote the α_i-quantile of the marginal distribution function F_i. Using the properties of the linear program

$$\max \{c^T x : Ax \geq b\}$$

and the assumptions of the theorem, we have that A_I is a nonsingular (n, n) matrix and $x(F)$ is the unique solution of the system

$$F_i(a^i x) - \alpha_i = 0 \, , \, i \in I(F) \quad .$$

Define $\Psi_i(x;t) = (1-t)F_i(a^i x) + tG_i(a^i x) - \alpha_i$, $i \in I(F)$, and apply the implicit function theorem to the system

$$\Psi_i(x;t) = 0 \, , \, i \in I(F) \quad .$$

It is clear that

$$D(t) = [\nabla_x \Psi_i \, , i \in I(F)]^T = [((1-t)f_i(a^i x) + tG_i'(a^i x))a^i \, , \, i \in I(F)] \quad ,$$

$$B(t) = [\frac{d}{dt}\Psi_i \, , i \in I(F)] = [G_i(a^i x) - F_i(a^i x) \, , \, i \in I(F)] \quad ,$$

so that

$$dx(F;G-F) = -D(0)^{-1}B(0) = -(A_I^{-1}f_I^{-1}) \cdot [G_I - \alpha_I] \quad . \blacksquare$$

The case of one joint probabilistic constraint is treated in the following theorem:

THEOREM 5. *Consider the program*

$$\begin{aligned} &\text{maximize} \quad c(x) \\ &\text{subject to} \quad P_F\{Ax \geq z\} \geq \alpha \end{aligned} \tag{18}$$

and let assumptions (i), (ii), (iii) *of Theorem 3 be satisfied. Assume in addition that*

(iv') *There is a Kuhn–Tucker point* $w(F) = [x(F), u(F)]$ *for* (18) *such that* $u(F) > 0$ *and the second-order sufficient condition is fulfilled.*

(v') *The distribution functions F and G are twice continuously differentiable in a neighborhood of $X(F)$: $= Ax(F)$ and*

$$A^{\mathrm{T}}\nabla_X F(X(F)) \neq 0 \quad .$$

Then

(a) *There is a neighborhood $0(w(F)) \subset R^n \times R^1_+$, a real number $t_0 > 0$ and a continuous function w: $<0, t_0) \rightarrow 0(w(F))$, $w(0) = w(F)$ such that for any $t \in <0, t_0)$ we have that $w(t) = [x(t), u(t)]$, $u(t) > 0$, is the Kuhn–Tucker point of the problem*

maximize $c(x)$

subject to $\mathrm{P}_{F_t}\{Ax \geq z\} \geq \alpha$,

for which the second-order sufficient condition is satisfied.

(b) *The Gâteaux differential* $\mathrm{d}w(F; G-F)$ *is given by*

$$\mathrm{d}w(F; G-F) = \begin{pmatrix} \dfrac{\mathrm{d}x(0)}{\mathrm{d}t} \\ \dfrac{\mathrm{d}u(0)}{\mathrm{d}t} \end{pmatrix} =$$

$$= \begin{pmatrix} -L^{-1}A^{\mathrm{T}}\{u(F)[\nabla_X G(X(F)) - \dfrac{l(G)}{l(F)}\nabla_X F(X(F))] + \dfrac{1}{l(F)}\nabla_X F(X(F))[G(X(F)) - \alpha]\} \\ u(F)[1 - \dfrac{l(G)}{l(F)}] + \dfrac{1}{l(F)}[G(X(F)) - \alpha] \end{pmatrix} ,$$

where

$$L = \nabla^2_{xx} L(w(F); F) = \nabla^2_{xx} c(x(F)) + u(F)A^{\mathrm{T}}\nabla^2_{XX} F(X(F))A \quad ,$$

$$l(G) = \nabla_X F(X(F))^{\mathrm{T}} A L^{-1} A^{\mathrm{T}} \nabla_X G(X(F)) \quad ,$$

$$l(F) = \nabla_X F(X(F))^{\mathrm{T}} A L^{-1} A^{\mathrm{T}} \nabla_X F(X(F)) \quad .$$

Proof. The result can again be obtained by application of the implicit function theorem, this time to the system of equations

$$\nabla_x c(x) + u\, \nabla_x h(x; t) = 0$$

$$h(x; t) = 0$$

with

$$h(x; t) = (1-t)F(Ax) + tG(Ax) - \alpha \quad . \blacksquare$$

Having solved the original problem (18), we know $x(F)$ and now have to compute $u(F)$, L^{-1} and to evaluate $G(Ax(F)) - \alpha$, $\nabla_X G(Ax(F))$, $\nabla_X F(Ax(F))$ to obtain the

Gâteaux differential.

For a given $x(F)$, $u(F)$, F and G, the Gâteaux differential depends on the difference between the values of the distribution functions $F(Ax(F))$, $G(Ax(F))$ and on the relative differences between their gradients, as measured by $\frac{l(G)}{l(F)}$ and $\nabla_X G(Ax(F)) - \frac{l(G)}{l(F)} \nabla_X F(Ax(F))$. The assumptions of Example 3 make it realistic to consider numerical evaluation of the Gâteaux differential.

REFERENCES

Armacost, R.L. and Fiacco, A.V. (1974). Computational experience in sensitivity analysis for nonlinear programming. *Mathematical Programming* 6:301–326.

Dupačová, J. (1983). Stability in stochastic programming with recourse. *Acta Univ. Carol., Math. et Phys.* 24:23–34.

Dupačová, J. (1984). Stability in stochastic programming with recourse – Estimated parameters. *Mathematical Programming* 28:72–83.

Dupačová, J. (1985). Stability in stochastic programming with recourse – contaminated distributions. *Mathematical Programming Studies* (forthcoming).

Dupačová, J. (1986). On some connections betwen parametric and stochastic programming. *Proceedings of the International Conference Parametric Programming and Related Topics*, Plane 1985, Akademie Verlag, Berlin.

Fiacco, A.V. (1976). Sensitivity analysis for nonlinear programming using penalty methods. *Mathematical Programming* 10:287–311.

Garstka, S.J. (1974). Distribution functions in stochastic programs with recourse: a parametric analysis. *Mathematical Programming* 6:339–351.

Gfrerer, H., Guddat, J., and Wacker, H. (1983). A globally convergent algorithm based on imbedding and parametric optimization. *Computing* 30:225–252.

Van de Panne, C. and Popp, W. (1963). Minimum-cost cattle feed under probabilistic protein constraints. *Management Science* 9:405–430.

Prékopa, A., Deák, I., Ganczer, S. and Patyi, K. (1980). The STABIL stochastic programming model and its experimental application to the electrical energy sector of the Hungarian economy. In M. Dempster (Ed.), *Stochastic Programming*. Academic Press, London, pp. 369–385.

Rao, C.R. (1973). *Linear Statistical Inference and its Applications*. Wiley, New York.

Robinson, S.M. (1974). Perturbed Kuhn–Tucker points and rates of convergence for a class of nonlinear programming algorithms. *Mathematical Programming* 7:1–16.

Robinson, S.M. (1980). Strongly regular generalized equations. *Mathematics of Operations Research* 5:43–62.

DUALITY IN IMPROPER MATHEMATICAL PROGRAMMING PROBLEMS UNDER UNCERTAINTY

I.I. Eremin and A.A. Vatolin
Institute of Mathematics and Mechanics
Sverdlovsk, USSR

This paper deals with linear programming problems with random or interval data. The problems under consideration are not assumed to be solvable. To deal with such problems we use the approach described in [1]. In the case of uncertain (interval) data, we shall concentrate on the development of various approaches for the "elimination" of uncertainty.

Consider the linear programming problem

$$\max \{(c, x): b - Ax \in K, x \in G\} \tag{1}$$

and its dual

$$\min \{(b, u): A^{T}u - c \in G^*, u \in K^*\} \quad . \tag{2}$$

Here c, $x \in R^n$, b, $u \in R^m$; the notation $b - Ax \in K$, $x \in G$ means that each component of the vector $b - Ax$ satisfies one of the relations: ≤ 0, ≥ 0, $= 0$, and that each component of x is either arbitrary or satisfies one of the inequalities: ≥ 0, ≤ 0. Thus, K and G are closed convex cones, and K^*, G^* are their duals, i.e., $K^* = \{u : (u, u^*) \geq 0, \forall u^* \in K\}$. Note that the arguments put forward below are also valid if K and G are arbitrary closed convex cones.

If all elements of vectors c, x and matrix $A = [a_{ji}]_{m,n}$ are random variables defined on the probability space (Ω, Σ, P), we may consider the following stochastic program:

$$\max \{E(\sum_{i=1}^{n} c_i x_i): b - E(Ax) \in K, x \in C\} \quad , \tag{3}$$

where $E(\cdot)$ denotes mathematical expectation; $E(Ax)$ is a vector with components $E(\sum_{i=1}^{n} a_{ji} x_i)$, $j = 1, \ldots, m$; C is a cone consisting of all $x = x(\omega) = [x_1(\omega), \ldots, x_n(\omega)]$ such that $x(\omega) \in G$ for almost all $\omega \in \Omega$. All the random variables considered are assumed to be square-integrable.

It is well-known (see [2]) that in this case problem (3) can be formulated as a linear programming problem:

$$\max \{(c, x): b - Ax \in K, x \in C\} \tag{4}$$

with a Hilbert space H^n of vectors $x = x(\omega) = [x_1(\omega), \ldots, x_n(\omega)]^T$. Here Ax denotes a vector with components $(a_{j.}, x)$; $(\cdot, \cdot)$ denotes an inner product in the Hilbert space H^n. We do not distinguish between random variables which coincide almost everywhere. The problem dual to (4) can be written

$$\min \{(b, u): A^T u - c \in C^*, u \in K^*\} \quad , \tag{5}$$

where $b, u \in R^m$.

Problems (4) and (5) are treated using the approach described in [1]. More specifically, problems D and $D^\#$ are associated with the dual linear programming problems (4) and (5) according to the same transformation scheme π, making no assumptions about solvability (i.e., whether they are solvable and have coincident optimal values). Problems D and $D^\#$ are connected by duality relations; they play the role of approximating problems for (4) and (5).

To formulate problems D and $D^\#$ we arbitrarily partition matrix A into submatrices A_j, $j = 0, 1, \ldots, m_0$, and B_i, $i = 0, 1, \ldots, n_0$, as follows:

$$A = \begin{bmatrix} A_0 \\ A_1 \\ \cdot \\ \cdot \\ \cdot \\ A_{m_0} \end{bmatrix} = [B_0, B_1, \ldots, B_{n_0}] \quad .$$

This determines a partitioning of vectors b, u, c, x:

$$b^T = [b^0, b^1, \ldots, b^{m_0}] \quad , \quad u^T = [u^0, u^1, \ldots, u^{m_0}] \quad ,$$

$$c = [c^0, c^1, \ldots, c^{n_0}] \quad , \quad x^T = [x^0, x^1, \ldots, x^{n_0}]$$

and also of cones (into direct products):

$$K = K_0 \times \cdots \times K_{m_0} \quad , \quad K^* = K_0^* \times \cdots \times K_{m_0}^* \quad ,$$

$$C = C_0 \times \cdots \times C_{n_0} \quad , \quad C^* = C_0^* \times \cdots \times C_{n_0}^* \quad .$$

If necessary, we can assign the value ϕ to some of the submatrices.

Let $\|\cdot\|^i$ be the norm in Hilbert space of the vector x^i. Let $\|\cdot\|_j$ be arbitrary norms in spaces of the same dimension as the vectors u^j, $j = 0, 1, \ldots, m_0$, and $\|\cdot\|_j^*$ be norms dual to them. By $\|u^j - K_j\|_j$ and $\|x^i - C_i^*\|^i$ we shall denote the distances between u^j and K_j and also between x^i and C_i^* in the metrics induced by the norms $\|\cdot\|_j$ and $\|\cdot\|^i$, respectively.

We introduce parameters $R_j > 0$, $j = 1, \ldots, m_0$, and $r_i > 0$, $i = 1, \ldots, n_0$, and formulate the problems

$$\sup \{(c, x) - \sum_{j=1}^{m_0} R_j \|(b^j - A_j x) - K_j\|_j :$$

$$b^0 - A_0 x \in K_0,\ x \in C,\ \|x^i\|^i \le r_i \quad (i = 1, \ldots, n_0)\} , \tag{D}$$

$$\inf \{(b, u) + \sum_{i=1}^{n_0} r_i \|(B_i^{\mathrm{T}} u - c^i) - C_i^*\|^i :$$

$$B_0^{\mathrm{T}} u - c^0 \in C_0^*,\ u \in K^*,\ \|u^j\|_j^* \le R_j \quad (j = 1, \ldots, m_0)\} . \tag{D$^\#$}$$

In what follows it will be assumed that the feasible sets M and $M^\#$ of the problems D and $D^\#$ are non-empty (this condition is fulfilled, e.g., for $A_0 = \phi$, $B_0 = \phi$). Let f and g be optimal values and $\tilde{M}$ and $\tilde{M}^\#$ be optimal sets of the problems D and $D^\#$, respectively. We can now formulate a statement concerning the duality relating problems D and $D^\#$:

THEOREM 1. *The following statements hold:*

1. *The inequalities* $-\infty < f \le g < +\infty$ *hold.*

2. *Let at least one of the following assumptions be satisfied:*

   (a) $A_0 = \phi$, $B_0 = \phi$;

   (b) $B_0 = \phi$ *and* $\exists\, x$, $b^0 - A_0 x \in \operatorname{int} K_0$.

Then $\tilde{M} \ne \phi$, $\tilde{M}^\# \ne \phi$, $f = g$.

3. *If* $B_0 = \phi$, *then* $f = g$ *and* $\tilde{M} \ne \phi$.

4. *If* $A_0 = \phi$ *or the second assumption from* (b) *holds, then* $f = g$ *and* $\tilde{M}^\# \ne \phi$.

Now consider the following system of linear equations and inequalities over R^n:

$$Ax = b,\ Qx \le p,\ x \ge 0 \tag{6}$$

and the linear programming problem

$$\max \{(c, x) : Ax = b,\ Qx \le p,\ x \ge 0\} , \tag{7}$$

where the data ($A = [a_{ji}]_{m,n}$, $b = [b_1, \ldots, b_m]^{\mathrm{T}}$, $Q = [q_{ji}]_{l,n}$, $p = [p_1, \ldots, p_m]^{\mathrm{T}}$,

$c = [c_1, \ldots, c_n]^T$) takes the form not of specific values, but of intervals:

$$c' \le c \le c'' , A' \le A \le A'' , b' \le b \le b'' , Q' \le Q \le Q'' , p' \le p \le p'' \quad .$$

Here $c' = [c_1', \ldots, c_n']^T$, $c'' = [c_1'', \ldots, c_n'']^T$, etc. are given (and fixed) matrices and vectors satisfying the conditions

$$c' \le c'' , A' \le A'' , b' \le b'' , Q' \le Q'' , p' \le p''$$

and matrix inequalities are regarded as component-wise inequalities.

Let $\mu = m(n+1) + l(n+1) + n$. The vector

$$T = [a_{11}, \ldots, a_{1n}, b_1, a_{21}, \ldots, a_{2n}, b_2, a_{31}, \ldots, b_m, q_{11}, \ldots, p_l, c_1, \ldots, c_n]$$

$$\equiv [t_1, \ldots, t_\mu]$$

is formed by the elements of matrices $[A, b]$, $[Q, p]$ and vector $\bar{c}$. Denote by $T_1, \ldots, T_\mu$ the range of variation of the components of vector T, i.e., $T_1 = \{t_1 : a_{11}' \le t_1 \le a_{11}''\}, \ldots, T_\mu = \{t_\mu : c_n' < t_\mu \le c_n''\}$. Set $\nu = \mu - n$, $\tau = T_1 \times \cdots \times T_\mu \subset R^\mu$, $\tau_0 = T_1 \times \cdots \times T_\nu \subset R^\nu$.

Now we shall outline the methods which we will use to "remove" uncertainty [3].

Let $\Omega = \{\omega_1, \ldots, \omega_\mu\}$, $\Omega_0 = \{\kappa_1, \ldots, \kappa_\nu\}$ be arbitrary ordered sets consisting of the quantors $\exists$ and $\forall$. Let $K = \{r_1, \ldots, r_\mu\}$ and $K_0 = \{k_1, \ldots, k_\nu\}$ be arbitrary permutations of the sets $\{1, \ldots, \mu\}$ and $\{1, \ldots, \nu\}$, respectively. Then the solution set of system (6) with interval data is defined as follows:

$$M(\tau_0, \Omega_0, K_0) = \{x : \kappa_1 t_{k_1} \in T_{k_1}, \ldots, \kappa_\nu t_{k_\nu} \in T_{k_\nu} ,$$

$$Ax = b , Qx \le p , x \ge 0\} \quad .$$

For problem (7) we shall consider two ways of "removing" uncertainty. Set

$$M_1(\tau, \Omega, K, \vartheta) = \{x : \omega_1 t_{r_1} \in T_{r_1}, \ldots, \omega_\mu t_{r_\mu} \in T_{r_\mu} ,$$

$$(c, x) \ge \vartheta , Ax = b , Qx \le p , x \ge 0\} ,$$

$$M_2(\tau, \Omega, K, \vartheta) = \{x : \omega_1 t_{r_1} \in T_{r_1}, \ldots, \omega_\mu t_{r_\mu} \in T_{r_\mu} ,$$

$$(c, x) = \vartheta , Ax = b , Qx \le p , x \ge 0\} \quad .$$

We can now define the optimal value and the optimal and feasible sets of problem (7) as either (a) the value

$$v_1 = \max \{\vartheta : M_1(\tau, \Omega, K, \vartheta) \ne \phi\}$$

and the sets $M_1(\tau, \Omega, K, v_1)$ and $\bigcup_{\vartheta \in R} M_1(\tau, \Omega, K, \vartheta)$, respectively, or (b) the value

$$v_2 = \max \{\vartheta : M_2(\tau, \Omega, K, \vartheta) \neq \phi\}$$

and the sets $M_2(\tau, \Omega, K, v_2)$ and $\bigcup_{\vartheta \in R} M_2(\tau, \Omega, K, \vartheta)$, respectively.

The problems with optimal values and optimal and feasible sets thus defined we denote by $L_1(\tau, \Omega, K)$ and $L_2(\tau, \Omega, K)$, respectively.

The parameters t_{k_i}, t_{τ_i} with $\kappa_i = \exists$, $\omega_i = \exists$ can be interpreted as controllable parameters; the parameters t_{k_j}, t_{τ_j} with $\kappa_j = \forall$, $\omega_j = \forall$ can be interpreted as uncertain parameters. For example, if we put

$$m = n = l = 1, \ \Omega_0 = \{\exists, \exists, \forall, \forall\}, \ K_0 = \{1, 4, 3, 2\} ,$$

then

$$T = [a, b, q, p, c] \in R^5 ,$$

$$M(\tau_0, \Omega_0, K_0) = \{x : \exists a \in T_1, \exists p \in T_4, \forall q \in T_3, \forall b \in T_2,$$
$$ax = b, \ qx \leq p, \ x \geq 0\} ,$$

so that $x \in M(\tau_0, \Omega_0, K_0)$ iff we can find values of controllable parameters $a \in T_1$, $p \in T_4$ such that x satisfies (6) for any $q \in T_3$, $b \in T_2$.

In general, the model described above represents a multistep decision process under uncertainty. Note that insolvability of problem (7) for some or even for all $T \in \tau$ does not necessarily imply that the problems $L_1(\tau, \Omega, K)$ or $L_2(\tau, \Omega, K)$ are insoluble.

If $\omega_i = \exists$, $i = 1, \ldots, \mu$, then $L_1(\tau, \Omega, K)$ represents a generalized linear programming problem as defined by Dantzig and Wolfe [4]. On the other hand, if $\omega_i = \forall$, $i = 1, \ldots, \mu$, then $L_1(\tau, \Omega, K)$ represents an "inexact linear programming" problem as defined by Soyster [5]. For these problems see, e.g., [6–9].

To formulate Theorem 2 we must introduce some new notation. Put $y = [y_1, \ldots, y_{n+1}]^T$. For $j \in N_m = \{1, \ldots, m\}$, $k \in N_l = \{1, \ldots, l\}$ we set

$$a_{j,n+1} = -b_j , \quad a'_{j,n+1} = -b''_j , \quad a''_{j,n+1} = -b'_j ,$$

$$q_{k,n+1} = -p_k , \quad q'_{k,n+1} = -p''_k , \quad q''_{k,n+1} = -p'_k .$$

Fix arbitrary $j \in N_m$. The indices of the parameters

$$t_{(j-1)(n+1)+1} \equiv a_{j1}, \ldots, t_{j(n+1)} \equiv b_j \equiv -a_{j,n+1}$$

form the set

$$N(j) = \{i : (j-1)(n+1) + 1 \leq i \leq j(n+1)\} .$$

Put

$$\{i: k_i \in N(j)\} = \{i_1, \ldots, i_{n+1}\}, i_1 < i_2 < \cdots < i_{n+1} \quad ,$$

$$N'(j) = \{i_s : s \le n , \kappa_{i_s} = \exists , \kappa_{i_{s+1}} = \forall\} = \{l_1, \ldots, l_{n_j}\} \quad ,$$

$$l_1 < l_2 < \cdots < l_{n_j} \quad ,$$

$$Z(j,\gamma) = \{i : i = k_\lambda - (j-1)(n+1) \quad , \quad k_\lambda \in N(j) , \lambda > l_\gamma\} , \gamma = 1, \ldots, n_j \quad .$$

Thus, for each $j \in N = \{j: N'(j) \ne \phi\}$ we have a collection of sets $Z(j,\gamma)$, $\gamma = 1, \ldots, n_j$. The elements of the matrices $\bar{\bar{A}} = [\bar{\bar{a}}_{ji}]_{m,n+1}$, $\bar{A} = [\bar{a}_{ji}]_{m,n+1}$, $\bar{Q} = [\bar{q}_{ji}]_{l,n+1}$ are defined as follows:

$$\bar{\bar{a}}_{ji} = a''_{ji} , \bar{a}_{ji} = a'_{ji} , \text{ if } (j-1)(n+1) + i = k_s , \kappa_s = \exists \quad ;$$

$$\bar{\bar{a}}_{ji} = a'_{ji} , \bar{a}_{ji} = a''_{ji} , \text{ if } (j-1)(n+1) + i = k_s , \kappa_s = \forall \quad ;$$

$$\bar{q}_{ji} = q'_{ji} , \text{ if } m(n+1) + (j-1)(n+1) + i = k_s , \kappa_s = \exists \quad ;$$

$$\bar{q}_{ji} = q''_{ji} , \text{ if } m(n+1) + (j-1)(n+1) + i = k_s , \kappa_s = \forall \quad .$$

THEOREM 2. *The equality $M(\tau_0, \Omega_0, K_0) = M'$ holds, where*

$$M' = \{x = [y_1, \ldots, y_n]^T : \bar{A}y \le 0 , \bar{\bar{A}}y \ge 0 , \bar{Q}y \le 0 \quad ,$$

$$\sum_{i \in Z(j,\gamma)} (\bar{a}_{ji} - \bar{\bar{a}}_{ji})\bar{\bar{y}}_i \le 0 , \gamma = 1, \ldots, n_j , j \in N , y \ge 0 , y_{n+1} = 1\} \quad . \tag{8}$$

From Theorem 2 we can also obtain, for each $x \in M(\tau_0, \Omega_0, K_0)$ and for each combination of uncertain (and controllable) parameters at every step of the above-mentioned multistep process (numbered from 1 to k), a representation, analogous to (8), of the set of (feasible) values of the controllable parameters which should be chosen at the $(k+1)$-th step.

Results concerning problems $L_1(\tau, \Omega, K)$ and $L_2(\tau, \Omega, K)$ can also be obtained from Theorem 2. Set $\bar{c} = [\bar{c}_1, \ldots, \bar{c}_n]^T$, $\bar{\bar{c}} = [\bar{\bar{c}}_1, \ldots, \bar{\bar{c}}_n]^T$, where $\bar{c}_i = c'_i$, $\bar{\bar{c}}_i = c''_i$, if $\nu + i = r_\lambda$, $\omega_\lambda = \exists$, and $\bar{c}_i = c''_i$, $\bar{\bar{c}}_i = c'_i$, if $\nu + i = r_\lambda$, $\omega_\lambda = \forall$. It will be convenient to use the following concordance law for the sets Ω_0, K_0, Ω, K:

$$(k_\lambda = r_i , k_\sigma = z_j , i < j) \Longrightarrow (\lambda < \sigma , \kappa_\lambda = \omega_i , \kappa_\sigma = \omega_j) \tag{9}$$

COROLLARY 1. *Let* (9) *be satisfied. Then problem $L_1(\tau, \Omega, K)$ is equivalent to (i.e., has the same optimal value and optimal and feasible sets as) the problem*

$$\max\{(\bar{\bar{c}}, x) : x \in M'\} \quad .$$

We put

$$\{i : \nu < r_i \le \mu\} = \{i_1, \ldots, i_n\}, \; i_1 < i_2 < \cdots < i_n \quad ,$$

$$\{i_s : s < n \, , \, \omega_{i_s} = \exists \, , \, \omega_{i_{s+1}} = \forall\} = \{l_1, \ldots, l_{n'}\} \quad ,$$

$$l_1 < l_2 < \cdots < l_{n'} \quad ,$$

$$Z(\gamma) = \{i : i \ge 1, \, \nu + i = z_\lambda \quad , \, \lambda > l_\gamma\}, \; \gamma = 1, \ldots, n' \quad .$$

COROLLARY 2. *Let* (9) *be satisfied. Then problem* $L_2(\tau, \Omega, K)$ *is equivalent to the problem*

$$\max\{(\bar{\bar{c}}, x) : (\bar{\bar{c}} - \bar{c}, x) \ge 0 \, , \sum_{i \in Z(\gamma)} (\bar{c}_i - \bar{\bar{c}}_i) x_i \le 0 \, , \, \gamma = 1, \ldots, n' \, , \, x \in M'\} \quad .$$

REFERENCES

1. I.I. Eremin, V.D. Mazurov and N.N. Astafiev. *Improper Problems in Linear and Convex Programming*. Nauka, Moscow, 1983 (in Russian).

2. D.B. Judin. *Mathematical Methods of Control under Uncertainty*. Sov. Radio, Moscow, 1974 (in Russian).

3. A.A. Vatolin. On the approximation of inconsistent systems of linear equations and inequalities. In: *Methods for Approximating Improper Mathematical Programming Problems*. Urals Scientific Center, Sverdlovsk, 1984, pp. 39–54 (in Russian).

4. G.B. Dantzig. *Linear Programming and Extensions*. Princeton University Press, Princeton, New Jersey, 1963.

5. L.A. Soyster. Convex programming with set-inclusive constraints and applications to inexact linear programming. *Operational Research*, 2(1973) 1154–1157.

6. L.G. Pliskin. *Bilinear Models of Production Optimization*. Sov. Radio, Moscow, 1979 (in Russian).

7. S.G. Timohin and A.B. Shapkin. On linear programming problems under uncertainty (in Russian). *Ekonomiki i Matimaticheskie Metodi*, 17(1981) 955–963.

8. J.D. Thuente. Duality theory for generalized linear programs with computational methods. *Operational Research,* 28(1980)1005–1011.

9. C. Singh. Convex programming with set-inclusive constraints and its applications to generalized linear and fractional programming. *Journal of Optimization Theory and Applications,*, 38(1982) 33–42.

EQUILIBRIUM STATES OF MONOTONIC OPERATORS AND EQUILIBRIUM TRAJECTORIES IN STOCHASTIC ECONOMIC MODELS

I.V. Evstigneev
Central Mathematical–Economical Institute
Moscow, USSR

In the present note we consider a stochastic model of a developing economy which generalizes deterministic models proposed by Gale (see, e.g., [1]) and Polterovich [2]. The central result is an existence theorem for equilibrium trajectories. The proof of the theorem is based on the monotonic operators method [3].

We start with a description of the model. Let $s_0, s_1, \ldots, s_N$ be a random process with values in a measurable space (S, I) (s_t may be interpreted as the state of the environment at time t). Let a sequence of natural numbers $n_0, \ldots, n_N$ be given, where n_t represents the number of different goods available in the economy at time t. Denote by L_t, $t = 0, \ldots, N$, the set of non-negative measurable functions $x(s^t)$ of $s^t = (s_0, \ldots, s_t)$ which take values in n_t-space R^{n_t} and satisfy the condition $\|x\|_1 = E|x(s^t)| < \infty$, where $|x| = |(x^1, \ldots, x^{n_t})| = |x^1| + \cdots + |x^{n_t}|$ and $E(\cdot)$ represents mathematical expectation. A function $x \in L_t$ is said to belong to the set X_t if $\|x\|_\infty = \text{ess sup } |x(s^t)| < \infty$. We identify functions which coincide almost surely (a.s.).

Suppose that sets

$$Q_t \subset X_{t-1} \times X_t \ (t \geq 1), \ K_t \subset X_t \ \ (t \geq 0)$$

and multivalued mappings

$$L_t \ni p \mapsto C_t(p) \subset K_t \ (t \geq 0) \, ; \, L_{t-1} \times L_t \ni l \mapsto Z_t(l) \subset Q_t \ (t \geq 1)$$

are given. The set $C_t(p)$ (consumption set) contains the most preferred consumption vectors $c \in K_t$ given the vector prices of p at time t. Pairs of vectors $(x, y) \in Q_t$ are interpreted as technological processes: x is the input at time $t-1$ and y is the output at time t. The set $Z_t(p, q)$ (production set) consists of the most preferred technological proceses $(x, y) \in Q_t$ given the vector of prices p at time $t-1$ and the vector of prices q at time t.

A sequence $\{x_t, y_t, c_t, p_t\}_0^N$ is said to be an *equilibrium trajectory* if the following conditions are satisfied:

(E1) $p_t \in L_t$, $(0 \le t \le N)$;

(E2) $y_0 \in X_0$, $(x_{t-1}, y_t) \in Z_t(p_{t-1}, p_t)$ $(1 \le t \le N)$, $x_N \in X_N$;

(E3) $c_t \in C_t(p_t)$ $(0 \le t \le N)$;

(E4) $y_t - x_t - c_t \ge 0$, $p_t(y_t - x_t - c_t) = 0$ $(0 \le t \le N)$, $p_N x_N = 0$ (a.s.).

The equilibrium trajectory is thus a sequence of inputs x_t, outputs y_t, consumption vectors c_t and prices p_t. From (E1)–(E4), this sequence possesses the following properties:

(a) The state of the economy $e_t = (x_t, y_t, c_t, p_t)$ depends only on the past and present states of the environment $s_0, \dots, s_t$ and does not depend on the future $s_{t+1}, s_{t+2}, \dots$.

(b) Given the system of prices $\{p_t\}$, the technological processes (x_{t-1}, y_t) the consumption vectors c_t are the most highly preferred alternatives (see (E2) and (E3)).

(c) Demand and supply are balanced, and the cost of over-supplied goods equals zero (see (E4)).

Let us now fix a vector $\bar{y}_0 \in X_0$ (an initial vector). We shall assume that the following requirements are fulfilled:

(A1) For all t , p and l, the sets Q_t, K_t, $Z_t(l)$ and $C_t(p)$ are convex, closed relative to a.s. convergence and bounded in the norm $\|\cdot\|_\infty$ (uniformly in t , p and l).

(A2) The mappings $l \mapsto Z_t(l)$ and $p \mapsto C_t(p)$ are closed in the following sense. If $z^k \in Z_t(l^k)$, $c^k \in C_t(p^k)$, $z \in Q_t$, $c \in K_t$ and $\|l^k - l\|_1 \mapsto 0$, $\|p^k - p\|_1 \mapsto 0$, $E(z^k - z)\xi \to 0$, $E(c^k - c)\eta \to 0$ for all integrable ξ and η, then $z \in Z_t(l)$ and $c \in C_t(p)$.

(A3) There exists a constant H such that

$$pc \le H \text{ (a.s.)} \tag{1}$$

and

$$E[(z' \circ l - z \circ l) \mid s^t] \le H \text{ (a.s.)} \tag{2}$$

for all $p \in L_t$, $c \in C_t(p)$, $z' \in Q_{t+1}$, $l \in L_t \times L_{t+1}$ and $z \in Z_{t+1}(l)$, where

$$z \circ l = qy - px \text{ , } z = (x, y) \text{ , } l = (p, q) \text{ .}$$

(A4) There exist $(\mathring{x}_{t-1}, \mathring{y}_t) \in Q_t$ $(t = 1, \ldots, N)$, $\bar{y}_0 \in X_0$, $x_N \in X_N$ and a real number $\delta > 0$ such that $y \geq \mathring{y}_t$ for $(x, y) \in Q_t$ and

$$\mathring{y}_t \geq \mathring{x}_t + \delta e \ (t = 0, 1, \ldots, N), \ \bar{y}_0 \geq \mathring{y}_0 \ [e = (1, 1, \ldots, 1)] \quad . \tag{3}$$

(A5) If $z \in Z_t(l)$, $z' \in Z_t(l')$, then $\mathrm{E}(z - z') \circ (l - l') \geq 0$. (The multivalued operator $l \mapsto Z_t(l)$ is monotonic with respect to the bilinear form $\mathrm{E}(z \circ l)$.)

(A6) For each $p \in L_t$, it is possible to find a function

$$\gamma(s^N, r) = \gamma_t^p(s^N, r) \geq 0 \ (r \geq 0)$$

which possesses the following properties:

(i) $\gamma(s^N, r)$ is measurable in (s^N, r) and non-decreasing in r;

(ii) $\gamma(s^N, r) > 0$ for $r > 0$;

(iii) for all $c \in C_t(p)$, p' and $c' \in C_t(p')$, we have

$$\mathrm{E}(c - c')(p - p') \leq -\mathrm{E}\gamma(|p - p'|) \quad .$$

(The multivalued operator $p \mapsto -C_t(p)$ is strictly monotonic with respect to the bilinear form $\mathrm{E}cp$.)

(A7) The σ-algebra I is countably generated.

THEOREM 1. *Under assumptions* (A1)–(A7), *there exists an equilibrium trajectory* $\{x_t, y_t, c_t, p_t\}_0^N$ *with* $y_0 = \bar{y}_0$.

Remarks. It is assumed in (1) that the cost of the most preferred consumption vectors is bounded. According to inequality (2), the expected profit $\mathrm{E}[(qy' - px') \mid s^t]$ for every technological process $(x', y') \in Q_{t+1}$ is not much greater than the expected profit for $(x, y) \in Z_{t+1}(p, q)$. It is assumed in (A4) that the outputs of all technological processes $(x, y) \in Q_t$ are not less than some minimal output $\mathring{y}_t$. Moreover, the program of minimal outputs $\mathring{y}_0, \mathring{y}_1, \ldots, \mathring{y}_N$ can be realized in such a way that at each time t, a positive amount δ of every good is consumed (see (3)). A deterministic variant of (A6) is discussed in [2].

We shall now outline the proof of Theorem 1. Let $L_\infty(t)$ denote the space of measurable functions $x(s^t)$ with values in R^{n_t} such that $\|x(s^t)\|_\infty < \infty$. Consider the spaces $L_\infty = L_\infty(0) \times \cdots \times L_\infty(N)$, $L = L_0 \times \cdots \times L_N$ and the multivalued operator $p \mapsto D(p) = F(p) - C(p) \subset L_\infty$, where $p = (p_0, \ldots, p_N) \in L$, $C(p) = C_0(p_0) \times \cdots \times C_N(p_N)$ and

$$F(p) = \{(\bar{y}_0 - x_0, y_1 - x_1, \ldots, y_{N-1} - x_{N-1}, y_N): (x_{t-1}, y_t) \in Z_t(p_{t-1}, p_t) \quad ,$$

$$t = 1,2, \ldots, N\} \quad .$$

A pair of random vectors (p, d) ($p \in L$, $d \in L_\infty$) is called an ***equilibrium state*** of the operator $p \mapsto D(p)$ if $d \in D(p)$, $d \geq 0$ and $pd = 0$ (a.s.). The existence of equilibrium trajectories with $y_0 = \bar{y}_0$ is equivalent to the existence of equilibrium states for the operator $p \mapsto D(p)$. In order to establish the latter assertion, we note that $D(p)$ possesses the following properties:

(D1) The set $D(p)$ is convex and bounded in L_∞ uniformly with respect to $p \in L$.

(D2) If $d^k \in D(p^k)$, $\|p^k - p\|_1 \to 0$ and $\mathrm{E} d^k b \to \mathrm{E} db$ for all $b \in L$, then $d \in D(p)$.

(D3) We have $\mathrm{E}(d - d')(p - p') \geq \mathrm{E}\tilde{\gamma}^p(|p - p'|)$ $(d \in D(p), d' \in D(p'))$, where $\tilde{\gamma}^p \geq 0$ satisfies (i) and (ii) in (A6).

(D4) If $(d_0, \ldots, d_N) \in D(p_0, \ldots, p_N)$, then

$$|p_t| \leq A \cdot \sum_{i-t}^{N} \mathrm{E}(p_i d_i \mid s^t) + B \quad ,$$

where $A > 0$ and $B > 0$ are constants.

Conditions (D1) and (D2) follow from (A1) and (A2); (D3) follows from (A5) and (A6); (D4) is a consequence of (A3) and (A4).

In order to prove Theorem 1, we use the following result (cf. [4, Theorem 4]).

THEOREM 2. *If an operator $p \mapsto D(p)$ $(p \in L$, $D(p) \subseteq L_\infty)$ possesses the properties* (D1)–(D4) *and I is countably generated, then $D(p)$ has an equilibrium state.*

We shall sketch the proof of Theorem 2. Let $I_k \subseteq I$ $(k = 1,2, \ldots)$ be an increasing sequence of finite algebras such that

$$I = \bigvee_{k=1}^{\infty} I_k$$

(see (A7)). Let $\sigma_k: (S, I_k) \to (S, I)$ be measurable mappings which generate I_k $(k = 1,2, \ldots)$. We define $s_k^t = (\sigma_k(s_0), \ldots, \sigma_k(s_t))$,

$$\mathrm{E}_k h = (\mathrm{E}(h_0 \mid s_k^0), \ldots, \mathrm{E}(h_N \mid s_k^N))$$

for $h = (h_0, \ldots, h_N) \in L_\infty$ and $D_k(p) = \mathrm{E}_k D(p)$, $L_\infty^{(k)} = \mathrm{E}_k L_\infty$.

LEMMA 1. *There exist $p^k \in L_\infty^{(k)}$ and $d^k \in L_\infty^{(k)}$ such that*

$$d^k \in D_k(p^k), \; d^k \geq 0, \; p^k d^k = 0, \; p^k \geq 0 \text{ (a.s.)} \quad . \tag{4}$$

In order to prove this statement, we use (D1), (D2), (D4) and the following lemma (a consequence of Kakutani's fixed point theorem), which is applied to the finite-dimensional space $L_\infty^{(k)}$.

LEMMA 2. *Let $V(q)$ be a compact convex subset of R^m depending on $q \in R^m$, $q \geq 0$. Suppose that $\{(v, q): v \in V(q)\}$ is closed, $\cup \{V(q),\ q \in W\}$ is bounded for each bounded $W \subset R^m$ and there exist constants $b > 0$, $a_1 > 0, \ldots, a_m > 0$ such that $|q| \leq a_1 v_1 q_1 + \ldots + a_m v_m q_m + b$ for $(v_1, \ldots, v_m) \in V(q)$ and $q = (q_1, \ldots, q_m) \geq 0$. Then there exist $q \geq 0$ and $0 \leq v \in V(q)$ such that $vq = 0$.*

Thus, by virtue of Lemma 1, there exist p^k, $d^k \in L_\infty^{(k)}$ $(k = 1,2, \ldots)$ with properties (4). Using (4) and (D4), we deduce that $\sup_k \|p^k\|_\infty < \infty$. It is proved in [5, Lemma 4] that this is sufficient for the existence of an equilibrium state with the operator $p \mapsto D(p)$, provided that $D(p)$ possesses the properties (D1)–(D3).

REFERENCES

1. D. Gale. On optimal development in a multi-sector economy. *Review of Economic Studies*, 34(1)(1967)1–18.

2. V.M. Polterovich. Equilibrium trajectories of economic growth. In *Methods of Functional Analysis in Mathematical Economics*, Nauka, Moscow, 1978, pp. 56–97.

3. M.M. Vainberg. *The Variational Method and the Method of Monotone Operators in the Theory of Nonlinear Equations*. Nauka, Moscow, 1972.

4. I.V. Evstigneev. An optimality principle and an equilibrium theorem for controlled random fields on a directed graph. *Doklady Akademii Nauk SSSR*, 274(4)(1984)782–786.

5. I.V. Evstigneev and P.K. Katyshev. Equilibrium trajectories in stochastic models of economic dynamics. *Teoriya Veroyatnosti i ee Primeneniya*, 27(1)(1982)120–128.

FINITE HORIZON APPROXIMATES OF INFINITE HORIZON STOCHASTIC PROGRAMS

Sjur D. Flåm, Christian Michelsen Institute, 5036 Fantoft, Norway
Roger J.-B. Wets, IIASA, 2361 Laxenburg, Austria

ABSTRACT

This paper deals with approximation schemes for infinite horizon, discrete time, stochastic optimization problems. We construct finite horizon approximates that yield upper and lower estimates and whose optimal solutions converge to long-term optimal trajectories. The results extend those of [3] from the deterministic case to the stochastic.

1. INTRODUCTION

We are concerned with open ended stochastic problems of the following type: At each stage $t \in \{1,2,\ldots\}$ a decision $x_t \in R^n_+$ must be chosen under uncertainty. Decisions are required to be adapted to increasing information. Formally, $x_t : \Omega \to R^n_+$ must be B_t-measurable where $B_t \subseteq B_{t+1}$ are sigma-fields included in B and (Ω, B, μ) is a given probability space. We shall find it convenient to assume that B is generated by countably many atoms. Then without further loss we take Ω to be countable and B to equal the power set. Let $\xi_t : \Omega \to R^m$ be a stochastic vector corresponding to the random factors that affect cost and constraints in period t. The cost incured in that period, denoted by

$$f_t(\xi_t, x_{t-1}, x_t),$$

is finite unless some implicit constraint is violated in which case it equals $+\infty$. Future costs are discounted at rate $\alpha \varepsilon (0,1)$ and the performance criterion is the expected accumulated present value of all future costs. Thus we are led to consider the following infinite horizon problem:

$$P : \text{minimize} \quad \limsup_{T\to\infty} E \sum_{t=1}^{T} \alpha^{t-1} f_t(\xi_t, x_{t-1}, x_t)$$

over all sequences $x = (x_t)_{t=1}^{\infty}$ such that $x_t : \Omega \to R_+^n$ has finite expectation and is B_t-measurable. The initial point $x_0 \in R_+^n$ is known.

The open-endedness of P is (among other things) a major deterrent for efficient computation. In order to mitigate this situation the original problem must be replaced with more tractable finite horizon versions intended to provide good approximations. Here we shall design two approximation schemes that yields lower and upper estimates of the optimal value function. In this way we are able to bracket the optimal value. Moreover, these two approximation schemes allow us to approach the limit. Specifically, as we extend the planning horizon towards infinity, the optimal values and the optimal solutions of the finite time problems cluster to those of P. We shall resolve these problems of stability by making an appeal to the theory of epi-convergence.

The organization of the paper is now outlined. Section 2 introduces the basic assumptions needed to obtain existence and convergence results. In Section 3 we formulate the finite horizon approximates that furnish lower and upper bounds for the optimal value of problem P. Section 4 provides statements and proofs about the existence of optimal solutions, and finally in Section 5 convergence of finite horizon approximates is established.

2. ASSUMPTIONS

The <u>essential objective function</u> F of P is defined by

$$F(x) = \begin{cases} \limsup_{T\to\infty} E \sum_{t=1}^{T} \alpha^{t-1} f_t(\xi_t, x_{t-1}, x_t) \\ \quad \text{if } x_t : \Omega \to R_+^n \text{ is } B_t\text{-measurable with a finite expectation,} \\ +\infty \quad \text{otherwise.} \end{cases}$$

Three basic assumptions are imposed in order to derive our results:

ASSUMPTION 2.1 (Problem P is proper convex). <u>We assume that</u>

(i) <u>for all</u> $t \geq 1$ <u>and</u> $\xi \in R^m$, $f_t(\xi, \bullet)$ <u>is lower semicontinuous convex;</u>

(ii) for some sequence $\tilde{x} = (\tilde{x}_t)_{t=0}^{\infty}$ we have $\tilde{F}(x) < +\infty$.

We take $\tilde{x} = 0$ which we can do without loss of generality.

ASSUMPTION 2.2 (The dampening effect of discounting is sufficiently strong). We assume that

(i) with $\eta \in (0, \alpha]$, the discounted expectation $\bar{\xi}(\eta) := (1-\eta)\, E\, [\Sigma_{t=1}^{\infty} \eta^{t-1} \xi_t]$ is well defined (it suffices to assume that $\bar{\xi}(\alpha)$ exists);

(ii) no trajectories except those with $\limsup_{t\to\infty} (E|x_t|)^{1/t} < +\infty$ are of interest;

(iii) there exists $h : R^{m+2n} \to R \cup \{+\infty\}$ lower semicontinuous convex such that for all $t \geq 1$, we have $h \leq f_t$, $h(\bullet, 0, 0)$ is continuous at $\bar{\xi}(\eta)$ for every $\eta \in (0, \alpha]$, and to every non-zero $z \in R^n_+$, $r' \geq r$ with $\alpha = \frac{1}{1+r}$ there corresponds $\lambda > 0$ such that

$$h(\bar{\xi}(\eta), 0, 0) < h(\bar{\xi}(\eta), \lambda z, \lambda(1+r')z). \tag{2.1}$$

Essentially assumption 2.2 limits growth: (i) makes the procedure of exponential smoothing of the noise process well defined, (ii) excludes super-linear growth, and finally, (iii) shows that at very high stock levels it does not pay, even under certainty, to pile up resources at a rate $r' \geq r$. Thus it is not worthwhile to counteract the effect of discounting when the resource endowment is sufficiently rich. This means that the own interest rate of the resources eventually becomes inferior to the rate r of impatience.

In mathematical terms assumption 2.2 (iii) makes questions about compactness easier to handle, the reason being that unboundedness is related to directions at infinity. Indeed, the last condition of 2.2 (iii) tells that it is not profitable to embark on a trajectory the asymptotic direction of which has a certain form.

Denote by $\mathrm{rc}h(\bar{\xi}(\eta), \bullet)$ the recession function of $h(\bar{\xi}(\eta), \bullet)$, cf [6]. Then (2.1) requires that

$$\mathrm{rc}\, h(\bar{\xi}(\eta), \eta z, z) > 0 \tag{2.2}$$

for all nonzero $z \in R^n_+$ and all $\eta \in (0, \alpha]$. (2.2) is a stochastic form of a condition introduced by Grinold [5].

ASSUMPTION 2.3 (Sustainability of tail-stationary trajectories)

If $F(x) < +\infty$, then $\limsup_{T\to\infty} \limsup_{T'\to\infty} E \sum_{t=T}^{T'} \alpha^{t-1} f_t(\xi_t, x_{T-1}, x_{T-1}) \leq 0.$

3. FINITE HORIZON APPROXIMATES

This section introduces two approximation schemes for P. We begin with

Approximates from below motivated as follows:

Define $h_T : R^{m+2n} \to R \cup \{+\infty\}$ to be the largest lower semicontinuous convex function majorized by the f_t for all $t \geq T$, i.e. $h_T = \text{cl co} (\inf_{t \geq T} f_t)$.

Denote by E_T the conditional expectation $E(\,|B_T)$. Suppose for the time being that for any feasible trajectory $x = (x_t)_{t=1}^{\infty}$, the average

$$z_T := (1-\alpha) E_T \sum_{t=T}^{\infty} \alpha^{t-T} x_t$$

of the tail $x_T, x_{T+1}, \ldots$ is well defined. Indeed, in Section 4 we state that this results from the growth conditions 2.2 already imposed on P. Since

$$(1-\alpha) E_T \sum_{t=T}^{\infty} \alpha^{t-T} (x_{t-1}, x_t) = ((1-\alpha) x_{T-1} + \alpha z_T, z_T),$$

the convexity and the lower semicontinuity of h_T implies that

$$\frac{\alpha^{T-1}}{1-\alpha} h_T(\bar{\xi}_T, (1-\alpha) x_{T-1} + \alpha z_T, z_T) \leq \limsup_{T'\to\infty} E_T \sum_{t=T}^{T'} \alpha^{t-1} f_t(\xi_t, x_{t-1}, x_t)$$

where $\bar{\xi}_T := (1-\alpha) E_T \sum_{t=T}^{\infty} \alpha^{t-T} \xi_t$.

Thus we are led to consider the following finite horizon problem

P_T : find a trajectory $(x_t)_{t=1}^{T}$ with $x_t : \Omega \to R_+^n$ integrable and B_t-measurable, which minimizes

$$E\{\sum_{t=1}^{T-1}\alpha^{t-1}f_t(\xi_t,x_{t-1},x_t)+\frac{\alpha^{T-1}}{1-\alpha}h_T(\bar{\xi}_T,(1-\alpha)x_{T-1}+\alpha x_T,x_T)\}.$$

Denote by $V(x_0) := \inf_x F(x) = \inf(P)$ and $V_T(x_0) := \inf(P_T)$ the optimal values of P and P_T, respectively. Then the observations that motivated the formulation of problem P_T, T = 1, ... yields in a straightforward manner:

PROPOSITION 3.1. (i) Suppose $F(x) < +\infty$ and $z_T = (1-\alpha)E_T\sum_{t=T}^{\infty}\alpha^{t-T}x_t < +\infty$ a.s. Then $x_1, \ldots, x_{T-1}, z_T$ is feasible for P_T. Moreover,

$$E\{\sum_{t=1}^{T-1}\alpha^{t-1}f_t(\xi_t,x_{t-1},x_t)+\frac{\alpha^{T-1}}{1-\alpha}h_T(\bar{\xi}_T,(1-\alpha)x_{T-1}+\alpha z_T,z_T)\}\le F(x).$$

Hence $V_T(x_0)\le V(x_0)$ for all $T\ge 1$.

(ii) Suppose $x_1, \ldots, x_T, x_{T+1}$ is a feasible solution for P_{T+1}. Then, with

$x'_T = (1-\alpha)x_T+\alpha E_T x_{T+1}$, the sequence $x_1, \ldots, x_{T-1}, x'_T$ is a feasible solution of P_T, since

$$h_T(\bar{\xi}_T,(1-\alpha)x_{T-1}+\alpha x'_T,x'_T)\le E_T\{(1-\alpha)f_T(\xi_T,x_{T-1},x_T)+\alpha h_{T+1}(\bar{\xi}_{T+1},x'_T,x_{T+1})\}$$

from which it also follows that $V_T(x_0)\le V_{T+1}(x_0)$.

Thus the process of averaging the tail generates a sequence of optimal values $\{V_T(x_0),\ T=1,\ldots\}$ which is monotone increasing and bounded from above by $V(x_0)$. That we actually have convergence is proved in section 4 where we also demonstrate that optimal solutions of P_T, $T\ge 1$ cluster to those of P. We now turn to

Approximates from above.

Suppose $x=(x_t)_{t=1}^{\infty}$ verify $x_t = x_{T-1}$ for all $t\ge T$. Then

$$F(x)=E\sum_{t=1}^{T-1}\alpha^{t-1}f_t(\xi_t,x_{t-1},x_t)+\limsup_{T'\to\infty}E\sum_{t=T}^{T'}\alpha^{t-1}f_t(\xi_t,x_{T-1},x_{T-1})$$

and we are led to consider the following finite time problem:

$$P^T: \text{minimize } E \sum_{t=1}^{T-1} \alpha^{t-1} f_t(\xi_t, x_{t-1}, x_t) + \limsup_{T'\to\infty} E \sum_{t=T}^{T'} \alpha^{t-1} f_t(\xi_t, x_{T-1}, x_{T-1})$$

over all $x = (x_t)_{t=1}^{T-1}$ such that $x_t : \Omega \to R^n_+$ is B_t-measurable with finite expectation.

4. EXISTENCE OF OPTIMAL SOLUTIONS

This section shows that problems P, P_T and P^T all admit optimal solutions when restricted to appropriate decision spaces. In order to state this some notations are needed. For each integer $T \geq 1$, denote by $L_1(T)$ the space of all $x = (x_t)_{t=1}^{T}$ such that

$$x_t : \Omega \to R^n \text{ is integrable and } \| x \| := E \; \Sigma_{t=1}^{T} \alpha^{t-1} | x_t | < +\infty$$

where $| \bullet |$ is the l_1-norm in R^n. When $T = +\infty$, we write simply L_1 in place of $L_1(\infty)$. These are all Banach spaces when equipped with the norm $\| \bullet \|$ and elements which are equal a.e. are regarded as identical. The justification for confining the decision space of P to L_1 is provided by the following lemma proven in [4].

LEMMA 4.1. <u>Assumption</u> 2.1, 2.2 <u>and</u> $\| x \| = E \sum_{t=1}^{\infty} \alpha^{t-1} | x_t | = +\infty$ <u>imply</u> $F(x) = +\infty$.

We may also prove

THEOREM 4.2 <u>Under assumption</u> 2.1 <u>and</u> 2.2 $F : L_1 \to R \cup \{\pm \infty\}$ <u>is a proper convex lower semicontinuous function and</u>

<u>for any real</u> β, <u>the level set</u>

$$\text{lev}_\beta F = \{ x \in L_1 \mid F(x) \leq \beta \}$$

<u>is compact. Hence optimal solutions of P exist</u>.

Entirely parallell results obtain for problems P_T and P^T, the functions

$$E\{\sum_{t=1}^{T-1}\alpha^{t-1}f_t(\xi_t,x_{t-1},x_t)+\frac{\alpha^{T-1}}{1-\alpha}h(\xi_T,(1-\alpha)x_{T-1}+\alpha x_T,x_T)\},$$

$$E\sum_{t=1}^{T-1}\alpha^{t-1}f_t(\xi_t,x_{t-1},x_t)+\limsup_{T'\to\infty}E\sum_{t=T}^{T'}\alpha^{t-1}f_t(\xi_t,x_{T-1},x_{T-1}),$$

and the spaces $L_1(T)$, $L_1(T-1)$, respectively.

In any case the proof can be outlined as follows. The lower semicontinuity essentially follows by appealing to Fatou's lemma. (Recall that the functions f_t, $t \geq 1$ are uniformly bounded below by the proper convex function h). Suppose without loss of generality that $0 \in K := \mathrm{lev}_\beta F$. then the halfspace

$$\{x \mid E\sum_{t=1}^{\infty}\alpha^{t-1}(1,\ldots\ldots,1)\bullet x_t \leq \rho\}$$

intersects K in a weakly closed neighbourhood of 0 for any $\rho > 0$. Let $\{x^\nu, \nu = 1,2,\ldots\}$ be any sequence contained in this relative neighbourhood of 0. Since Ω is countable we may, by a standard diagonal argument, extract a subsequence x^ν, $\nu \in N_1$ such that $\{x_t^\nu(\omega)\}_{\nu\in N_1}$ converges for each t and ω. This subsequence converges weakly and by Schur's theorem it also converges in the norm [2]. Now apply the growth condition to establish that K is norm-bounded, hence compact [3].

5. CONVERGENCE OF FINITE HORIZON APPROXIMATES

Recall that the problems P, P_T and P^T are defined on different spaces. In order to establish convergence it is necessary somehow to conceive of all these problems as being defined on the same space, namely L_1. We begin with the approximates from below. Motivated by the construction that led us to the problem P_T we assign to

$$F_T(x) := \begin{cases} E\{\sum_{t=1}^{T-1}\alpha^{t-1}f_t(\xi_t,x_{t-1},x_t)+\frac{\alpha^{T-1}}{1-\alpha}h_T(\xi_T,(1-\alpha)x_{T-1}+\alpha z_T,z_T)\} \\ \text{if } x\in L_1,\ x_t\in R_+^n \text{ is } B_t\text{-measurable and } z_T=(1-\alpha)E_T\sum_{t=T}^{\infty}\alpha^{t-T}x_t, \\ +\infty \text{ otherwise} \end{cases}$$

the role of being the essential objective function. Indeed, to minimize F_T over L_1

corresponds to problem P_T in the following way: If $F_T(x) < +\infty$, then $(x_1, \ldots, x_{T-1}, z_T)$ is a feasible for P_T and provides the same value. Conversely, if $(x)_{t=1}^{T}$ is feasible for P_T and $x_t := x_T$ for all $t > T$, then $F_T(x) < +\infty$, in fact $F_T(x)$ coincides with the value of the criterion of P_T. Thus $V_T(x_0) = \inf F_T$

We now turn to approximates from above. Here the function

$$F^T(x) := \begin{cases} F(x) \text{ if } x_t = x_{T-1} & \text{for all } t \geq T, \\ +\infty \text{ otherwise} & \end{cases}$$

acts as the essential objective of problem P^T. There is of course a one-one correspondence between feasible solutions of P^T and long-term trajectories $x \in L_1$ such that $F^T(x) < +\infty$. Therefore, the optimal value function $V^T(x_0) := \inf(P^T)$ satisfies $V^T(x_0) = \inf F^T$.

We now address the major issue whether the sequences F_T, F^T, $T = 1, \ldots$ converge to F in an apporiate sense.

DEFINITION 5.1. A sequence $G^T : L_1 \to R \cup \{\pm\infty\}$, $T = 1,2, \ldots$ epi-converges to $G : L_1 \to R \cup \{\pm\infty\}$, and we write $G = \text{epi-}\lim_{T\to\infty} G^T$ if

(a) for every sequence $x^T \to x$ we have $\liminf_{T\to\infty} G^T(x^T) \geq G(x)$; and

(b) for every x there exists a sequence $x^T \to x$ such that $\limsup_{T\to\infty} G^T(x^T) \leq G(x)$.

Epi-convergence does not in general imply pointwise convergence nor does it follow from the latter. The two types of convergence coincide, however, if the sequence $(G^T)_{T=1}^{\infty}$ is monotone increasing. This fact is utilized in the first statement of the following

PROPOSITION 5.2. (i) Under assumptions 2.1, 2.2, $F_T(x) \uparrow F(x)$ for all x which implies that $F = \text{epi-}\lim_{T\to\infty} F_T$

(ii) Under assumptions 2.1 – 2.3, $F = \text{epi-}\lim_{T\to\infty} F^T$

PROOF (i). Since $F_1 \leq F_2 \leq \ldots \leq F$, it suffices for the monotone convergence to prove that

$$\liminf_{T\to\infty} \frac{\alpha^{T-1}}{1-\alpha} Eh_T(\xi_T, (1-\alpha)x_{T-1} + \alpha z_T, z_T) \geq 0$$

when $x \in (L_1)_+$ and $z_T = (1-\alpha) E_T \sum_{t=T}^{\infty} \alpha^{t-T} x_t$. This would follow from

$$\liminf_{T\to\infty} \alpha^{T-1} h(E(\xi_T, z_{T-1}, z_T)) \geq 0$$ since $z_{T-1} = (1-\alpha)x_{T-1} + \alpha E_{T-1} z_T$. But observe that $\alpha^{T-1} E(\xi_T, z_{T-1}, z_T) \to (0,0,0)$. Hence

$$\liminf_{T\to\infty} \alpha^{T-1} h(E(\xi_T, z_{T-1}, z_T)) =$$

$$\liminf_{T\to\infty} \alpha^{T-1} h(\alpha^{-T+1}(\alpha^{T-1} E(\xi_T, z_{T-1}, z_T))) = (rch)(0,0,0) = 0 .$$

This completes the proof of (i).

(ii) First observe that $F^T(x) \geq F(x)$ for any $x \in L_1$. Hence $x^T \to x$ implies by the lower semicontinuity of F that

$$\liminf_{T\to\infty} F^T(x^T) \geq \liminf_{T\to\infty} F(x^T) \geq F(x)$$

This takes care of (a) in definition 5.1. To prove (b) define for given $x \in L_1$ such that $F(x) < +\infty$ the sequence $(x^T)_{T=1}^{\infty}$ in L_1 by

$x_t^T = x_t$ for $t = 1, \ldots.. T-1$, and $x_t^T = x_{T-1}$ for $t \geq T$. Then $x^T \to x$ and

$$\limsup_{T\to\infty} F^T(x^T) =$$

$$\limsup_{T\to\infty} \{E \sum_{t=1}^{T-1} \alpha^{t-1} f_t(\xi_t, x_{t-1}, x_t) + \limsup_{T'\to\infty} E \sum_{t=T}^{T'} \alpha^{t-1} f_t(\xi_t, x_{T-1}, x_{T-1})\}$$

$$\leq F(x) + \limsup_{T\to\infty} \limsup_{T'\to\infty} E \sum_{t=T}^{T'} \alpha^{t-1} f_t(\xi_t, x_{T-1}, x_{T-1})$$

$\leq F(x)$ by assumption 2.3. This completes the proof. □

The epi-convergence is essentially a one-sided version of uniform convergence on compacta, one having the same implications regarding the convergence of minima, see [7]. This is recorded in the following theorem that states the major result of this paper.

THEOREM 5.3. (i) Under assumptions 2.1, 2.2

$$V_T(x_0) \uparrow V(x_0)$$

Moreover, problems P and P_T admit optimal solutions, and given any sequence $(x_t^T)_{t=1}^T$, $T = 1, 2, \ldots\ldots$ of optimal solutions of P_T, we may find a subsequence N' such that

$$\lim_{T\varepsilon N'} x_t^T = x_t \text{ a. s. for all } t \geq 1, \tag{5.1}$$

where $x = (x_t)_{t=1}^{\infty}$ solves the long term problem P. Finally, if $x = (x_t)_{t=1}^{\infty}$ solves P, then there exists a sequence of real numbers $\varepsilon_T \downarrow 0$ and ε_T-optimal solutions $(x_t^T)_{t=1}^T$ of P_T such that

$$x_t = \lim_{T\to\infty} x_t^T \text{ a.s. for all } t \geq 1 .$$

(ii) Under assumption 2.1 – 2.3

$$V^T(x_0) \downarrow V(x_0)$$

Moreover, problems P and P^T admit optimal solutions, and given any sequence $(x_t^T)_{t=1}^{T-1}$, $T = 1, 2, \ldots\ldots$

of optimal solutions of P^T, we may extract a subsequence N' such that

$$\lim_{T\varepsilon N'} x_t^T = x_t \text{ a.s. for all } t \geq 1 ,$$

where $x = (x_t)_{t=1}^{\infty}$ solves the long term problem P. Finally, if $x = (x_t)_{t=1}^{\infty}$ solves P, then there exists a sequence of real numbers $\varepsilon_T \downarrow 0$ and ε_T-optimal solutions $(x_t^T)_{t=1}^{T-1}$ of P^T such that

$$x_t = \lim_{T\to\infty} x_t^T \text{ a.s. } \underline{\text{for all}}\ t \geq 1.$$

Proof (i). From $F_T \leq F_{T+1} \leq \ldots\ldots \leq F$ it follows that $V_T(x_0) = \inf F_T$ is monotone increasing and bounded above by $V(x_0) = \inf F$. By the corresponding analog of Theorem 4.2 we may find an optimal $(x_t^T)_{t=1}^T$ for P_T.

Then $x^T := (x_1^T, x_2^T, \ldots\ldots, x_t^T, x_T^T, x_T^T, \ldots)$ is a minimum of F_T. Since

$$V_T(x_0) = F_T(x^T) \leq F_T(0) \leq F(0) < +\infty$$

we may by theorem 4.2 extract a convergent subsequence with limit x. From this subsequence, we may extract a further subsequence x^T, $T \in N'$ converging almost surely. This takes care of (5.1). Moreover, the inequalities

$$V(x_0) \geq \lim_{T\to\infty} V_T(x_0) \geq \liminf_{T\in N'} F_T(x^T) \geq F(x) \geq V(x_0)$$

implies that $V_T(x_0) \uparrow V(x_0)$ and x is optimal. Since epi-convergence implies $\limsup_{T\to\infty} (\arg\min F_T) \subseteq \arg\min F$

and whenever $\inf F_T \to \inf F$

$$\arg\min F = \bigcap_{\varepsilon>0} \liminf_{T\to\infty} (\varepsilon - \arg\min F_T)$$

see [1], the last statement of (i) now follows from the fact that a mean convergent sequence contains a subsequence which converges almost surely.

The proof of (ii) is entirely similar. We are allowed to identify P and P^T with minimizing F and F^T, respectively. Since $F = \text{epi-}\lim_{T\to\infty} F^T$ it follows, see [1], that

$$\limsup_{T\to\infty} (\inf F^T = V^T(x_0)) \leq \inf F = V(x_0)$$

which together with $V^1(x_0) \geq V^2(x_0) \geq \ldots\ldots \geq V(x_0)$ gives us $V^T(x_0) \downarrow V(x_0)$.

The remainder of the proof is identical and is therefore omitted. □

REFERENCES

1. Attouch, H. (1984). Variational convergence for functions and operators. Pitman Research Notes, London.
2. Diestel, J. (1984). Sequences and series in Banach Spaces. Springer Verlag, New York.
3. Flåm, S.D., and Wets, R.J.-B. (1984). Existence results and finite horizon approximations of inifinite horizon optimization problems. Rep.No. 842555-1, CMI, Bergen, Norway.
4. Flåm, S.D., and Wets, R.J.-B. (1984). Infinite horizon discrete time stochastic Bolza type problems: Existence results. Rep.No. 842650-1, Bergen, Norway.
5. Grinold, R. (1983). Finite horizon approximations of infinite horizon programs. Mathemathical Programming, 25, pp. 64-82.
6. Rockafellar, R.T. (1970). Convex Analysis. Princeton University Press, Princeton.
7. Rockafellar, R.T., and Wets, R.J.-B. (1984). Variational systems, an introduction. Rep.No. 8405, CEREMADE, Univ. Paris-Dauphine.

STOCHASTIC OPTIMIZATION TECHNIQUES FOR FINDING OPTIMAL SUBMEASURES

A. Gaivoronski
International Institute for Applied Systems Analysis and
V. Glushkov Institute of Cybernetics, Kiev, USSR.

1. INTRODUCTION

Optimality conditions based on duality relations were studied in [1,2] for the following optimization problem.

Find the positive Borel measure H such that

$$\Psi^0(H) = \max \tag{1}$$

with respect to constraints

$$\Psi^i(H) \le 0 \quad , \quad i = 1:m \quad , \tag{2}$$

$$H^l(A) \le H(A) \le H^u(A) \tag{3}$$

for all Borel $A \subset Y \subset R^n$

$$H(Y) = 1 \tag{4}$$

where Y— some subset of Euclidean space R^n, $\Psi^i(H)$— function which depends on the measure H, usually some kind of directional differentiability and convexity is assumed. H^u and H^l are some positive Borel measures. Stochastic optimization methods for solving (1)-(4) in case when functions $\Psi^i(H)$ are linear with respect to H were developed in [1]. In this paper such methods are developed for nonlinear functions $\Psi^i(H)$ and for arbitrary finite measures. Interest for such a problem is originated from statistics where it appears in finite population sampling [3,4].

Another application of the problem (1)-(4) are approximation schemes for stochastic optimization [5,6]. In section 2 the characterization of solutions for quite general classes of measures is obtained. The conseptual algorithm for solving nonlinear problems is proposed in the Section 3. Missing proofs and results of some numerical experiments can be found in [2].

2. CHARACTERIZATION OF THE OPTIMAL SOLUTIONS

We shall consider subset Y of Euclidean space R^n and some σ-field Ξ on it. We shall assume that all measures specified below are defined on this σ-field.

In this section, the representation of measures H, which are the solution of the following problem, will be developed:

$$\max \ \Psi(H) \tag{5}$$

subject to constraint

$$H^l \leq H \leq H^u \tag{6}$$

$$H(Y) = b \tag{7}$$

The constraint (6) means that $H^l(E) \leq H(E) \leq H^u(E)$ for any $E \subset \Xi$. Define $H^\Delta = H^u - H^l$. In what follows the spaces $L_1(Y,\Xi,H^\Delta)$ and $L_\infty(Y,\Xi,H^\Delta)$ play an important role, where $L_1(Y,\Xi,H^\Delta)$ is the space of all H^Δ-measurable functions $g(y)$ defined on Y and such that $\int_Y |g(y)| dH^\Delta < \infty$, $L_\infty(Y,\Xi,H^\Delta)$ is the space of all H^Δ-measurable and H^Δ-essentially bounded functions $g(y)$, defined on Y. In what follows we shall denote by $||\cdot||_\infty$ the norm in the space $L_\infty(Y,\Xi,H^\Delta)$, i.e.

$$||g(y)|| = H^\Delta - \operatorname{ess\,sup}_{y \in Y} |g(y)|$$

Let us denote by G the set of all measures, satisfying (6):

$$G = \{H: \quad H^u \leq H \leq H^l\}$$

and by G_b the set of all measures, satisfying in addition (7):

$$G_b = \{H: H \in G, \ H(Y) = b\}$$

Suppose that $f(y)$ is some function defined on Y, c-some number and define the following sets

$$Z^+(c,f) = \{y: y \in Y, f(y) > c\}$$

$$Z^{-}(c,f) = \{y : y \in Y, f(y) < c\}$$

$$Z^{0}(c,f) = \{y : y \in Y, f(y) = c\}$$

In notations below we shall substitute in this definition instead of f various particular functions. Take

$$c^{*} = \inf\{c : H^{\Delta}(Z^{+}(c,g)) \le b - H^{l}(Y)\}$$

and define, as usual, by H^{-}, H^{+} and $|H|$ positive, negative and total variation of the measure H.

We shall first consider the problem in which function $\Psi(H)$ is linear:

$$\max_{H \in G_b} \int g(y)dH \tag{8}$$

and describe the set of all solutions of (10). The following result is generalization of Lemma 1 from [1].

THEOREM 1. Suppose that the following conditions are satisfied:

1. $H^{l}(Y) \le b, |H^{l}|(Y) < \infty, H^{u}(Y) \ge b$

2. For any $E \in \Xi, H^{\Delta}(E) > 0$ exists $E_1 \in \Xi$, $E_1 \subset E$, such that either E_1 is H^{Δ}-atom or $0 < H^{\Delta}(E_1) < \infty$

3. $g(y) \in L_1(Y,\Xi,H^{\Delta})$, $\int_Y |g(y)| d|H^{l}| < \infty$

4. If $c^{*} = 0$ then $H^{\Delta}(Y \setminus Z^{-}(0,g)) \ge b - H^{l}(Y)$

Then the solution of problem (8) exists and any such solution has the following representation:

(i) $H^{*}(A) = H^{u}(A)$ for any $A \in \Xi$, $A \subset Z^{+}(c^{*},g)$

(ii) $H^{*}(A) = H^{l}(A)$ for any $A \in \Xi$, $A \subset Z^{-}(c^{*},g)$

(iii) $H^{u}(A) \ge H^{*}(A) \ge H^{l}(A)$ for any $A \in \Xi$, $A \subset Z^{0}(c^{*},g)$ and $H^{*}(Z^{0}(c^{*},g)) = b - H^{l}(Y) - H^{\Delta}(Z^{+}(c^{*},g))$

Conversely, any measure defined by (i)-(iii) is the solution of the problem (8).

Note that if measure H^{Δ} has bounded variation conditions 2 and 4 are satisfied automatically. For such measures the structure of solutions can be studied using general duality theory [7].

Let us now consider in more detail the set G_b. If the measure H^l has finite variation we have the following representation for arbitrary $H \in G_b$:

$$H = H^l + (H - H^l)$$

where measure $H - H^l$ is finite, positive and continuous with respect to measure H^Δ. If H^Δ is σ-finite we can use Radon-Nycodym theorem [8] and for arbitrary $H \in G_b$ obtain the following representation:

$$H(E) = H^l(E) + \int_E h_H(y)dH^\Delta \qquad \forall E \in \Xi \tag{9}$$

where $h_H \in L_1(Y,\Xi,H^\Delta)$ and this representation is unique. For arbitrary $E \in \Xi$ we have:

$$0 \le \int_E h_H(y)dH^\Delta \le H^\Delta(E)$$

and therefore $0 \le h_H(y) \le 1$ H^Δ-everywhere. Consider now the set $K_b \subset L_1(Y,\Xi,H^\Delta)$:

$$K_b = \{h:\ 0 \le h(y) \le 1,\ \int_Y h(y)dH^\Delta = b - H^l(Y)\} \tag{10}$$

Each function from this set defines measure H_h from G_b:

$$H_h(E) = H^l(E) + \int_E h(y)dH^\Delta,\ E \in \Xi \tag{11}$$

Therefore (9), (11) defines isomorphism between sets G_b and K_b such that the problem (5)-(7) is equivalent to the following one:

$$\max \overline{\Psi}(h) \tag{12}$$

subject to constraints

$$0 \le h(y) \le 1 \tag{13}$$

$$\int_Y h(y)dH^\Delta = b - H^l(Y) \tag{14}$$

where the function $\bar{\Psi}(h) = \Psi(H_h)$. Optimal values of problems (12)-(14) and (5)-(7) are the same and each solution of (12)-(14) defines solution of (5)-(7) through (11) and vice versa. This equivalence together with certain convexity assumptions lead to solution representation for problems (5)-(7) similar to theorem 1:

THEOREM 2. Suppose that the following assumptions are satisfied:

1. Measures H^l and H^u have bounded variation,

$$H^l(Y) \le b, \quad H^u(Y) \ge b$$

2. $\Psi(H)$ is concave and finite for $H \in G_{b,\varepsilon} = G_b + G_\varepsilon$

where $G_\varepsilon = \{H_\varepsilon : |H_\varepsilon|(Y) \le \varepsilon, \ H_\varepsilon \text{ is } H^\Delta\text{-continuous}\}$ for some $\varepsilon > 0$. Then

1) For each $H_1 \in G_b$ exists $g(y, H_1) \in L_\infty(Y, \Xi, H^\Delta)$ such that

$$\Psi(H_2) - \Psi(H_1) \le \int_Y g(y, H_1)\, d(H_2 - H_1) \tag{15}$$

for all $H_2 \in G_b$

2) The solution H^* of problem (5)-(7) exists.

3) For any $E \in \Xi$ and any optimal solution H^* of the problem (7)-(9) we have the following representation:

$$H^*(E) = \begin{cases} H^u(E) & \text{for} \quad E \subset Z^+(c^*, g(y, H^*)) \\ H^l(E) & \text{for} \quad E \subset Z^-(c^*, g(y, H^*)) \\ H^l(E) \le H(E) \le H^u(E) & \text{for} \quad E \subset Z^0(c^*, g(y, H^*)) \end{cases} \tag{16}$$

where

$$c^* = \inf\{c : H^\Delta(Z^+(c, g(y, H^*)) \le b - H^l(Y)\}$$

and

$$g(y, H^*) \in L_\infty(Y, \Xi, H^\Delta), \quad \Psi(H) - \Psi(H^*) \le \int_Y g(y, H^*)\, d(H - H^*)$$

for all $H \in G_b$. Conversely, if for some $H_1 \in G_b$ exists $g(y, H_1) \in L_\infty(Y, \Xi, H^\Delta)$ such that (15) is fulfilled and H_1 can be represented according to (16) then H_1 is the optimal solution of the problem (5)-(7).

Proof. The previous argument shows that under assumptions of the theorem problem (5)-(7) is equivalent to the problem (12)-(14) and there is isomorphism between set $G_{\delta,\varepsilon}$ as defined in condition 2, and the following set $K_{\delta,\varepsilon} \in L_1(Y,\Xi,H^\Delta)$:

$$K_{\delta,\varepsilon} = K_\delta + K_\varepsilon,$$

$$K_\varepsilon = \{h:\ h \in L_1(Y,\Xi,H^\Delta), \int_Y |h(y)| dH^\Delta \le \varepsilon\}$$

Function $\bar{\Psi}(h)$ from (12) is concave on the set $K_{\delta,\varepsilon}$, which is ε-vicinity of K_δ in $L_1(Y,\Xi,H^\Delta)$. Therefore for each $L \in K_\delta$ exists subdifferential of concave function $\bar{\Psi}(h)$ [9, 10], which in this case is linear continuous functional $\bar{g} \in L_1^*(Y,\Xi,H^\Delta)$ such that

$$\bar{\Psi}(h_1) - \bar{\Psi}(h) \le \bar{g}(h_1 - h)$$

Taking into account representation of $L_1^*(Y,\Xi,H^\Delta)$ [8] we get:

$$\bar{\Psi}(h_1) - \bar{\Psi}(h) \le \int_Y \bar{g}(y,h)(h_1(y) - h(y)) dH^\Delta \tag{17}$$

where

$$\bar{g}(y,h) \in L_\infty(Y,\Xi,H^\Delta)$$

which together with (9) implies

$$\Psi(H_1) - \Psi(H) \le \int_Y g(y,H) d(H_1 - H)$$

for all $H, H_1 \in G_\delta$ where $g(y,H) = \bar{g}(y,h_H)$. Thus, (15) is proved. Note that we may consider function $g(y,H_1)$ from (15) (possibly non-unique) as subdifferential of the function $\Psi(H)$ at point H_1.

Now observe that the set K_δ is weakly sequentially compact in $L_1(Y,\Xi,H^\Delta)$ because $H^\Delta(Y) < \infty$ and

$$\lim_{H^\Delta(E) \to 0} \int_E h(y) dH^\Delta = 0$$

uniformly for $h \in K_b$ (see [8, p.294]). Let us prove that it is also weakly closed. Consider the sequence $h^s(y)$, $h^s \in K_b$ and

$$\int_Y g(y)h^s(y)dH^\Delta \to \int_Y g(y)h(y)dH^\Delta$$

for some $h \in L_1(Y,\Xi,H^\Delta)$ and all $g \in L_\infty(Y,\Xi,H^\Delta)$. In particular, we have

$$\int_E h^s(y)dH^\Delta \to \int_E h(y)dH^\Delta$$

for all $E \in \Xi$ because the indicator function of the set $E \in \Xi$ clearly belongs to $L_\infty(Y,\Xi,H^\Delta)$. This gives $0 \le h(y) \le 1$ $\quad H^\Delta$-everywhere. Taking $g(y) \equiv 1$ we have also

$$\int_Y h^s(y)dH^\Delta \to \int_Y h(y)dH^\Delta$$

which gives

$$\int_Y h(y)dH^\Delta = b - H^l(Y)$$

Thus, $h \in K_b$ and K_b is weakly closed.

It follows from (17) that for any sequence $h^s \in K_b$, $h^s \to h$ weakly, $h \in K_b$ we have

$$\overline{\lim_{s \to \infty}} \bar{\Psi}(h^s) \le \bar{\Psi}(h)$$

This together with sequential compactness and closeness of K_b implies existence of h^* such that

$$\bar{\bar{\Psi}}(h^*) = \sup_{h \in K_b} \bar{\bar{\Psi}}(h)$$

Thus, solution of the problem (5)-(7) exists.

The general results of convex analysis [9] now imply that under assumption 2 of the theorem for any solution H^* of the problem (5)-(7) exists subdifferential $g(y,H^*)$ of the function $\Psi(H)$ at point H^* such that

$$\int_Y g(y,H^*)d(H-H^*) \le 0, \quad \Psi(H)-\Psi(H^*) \le \int g(y,H^*)d(H-H^*)$$

for all $H \in G_b$ or, in other words H^* is one of the solutions of the following problem:

$$\max_{H \in G_b} \int_Y g(y,H^*)dH \tag{18}$$

This problem is exactly of the type (8) and its solutions are characterized by the Theorem 1. Conversely, if for some $H^* \in G_0$ exists subdifferential $g(y,H^*)$ such that H^* is the solution of the problem (18) then H^* is the optimal solution of the original problem. Proof is now completed by using theorem 1. Similar results were obtained for a special kind of function $\Psi(H)$, atomless probability measure H^u and $H^l \equiv 0$ in [4].

Theorem 2 shows that solutions of the problem (5)-(7) can be viewed as indicator functions of some sets. Therefore many problems involving selection of optimal set [13] can be reformulated as problems of finding optimal measures.

3. STOCHASTIC OPTIMIZATION METHOD

Using the results of the previous section we can construct numerical methods for solving problem (5)-(7). From now on we shall assume that function $\Psi(H)$ is concave and finite on some vicinity of the set G and possess certain differentiability properties:

$$\Psi(H_1+\alpha(H_2-H_1)) = \Psi(H_1)+\alpha\int_Y g(y,H_1)d(H_2-H_1)+o(\alpha) \tag{19}$$

where $o(\alpha)/\alpha \to 0$ as $\alpha \to 0$ for all $H_1,H_2 \in G$. This means that subdifferential $g(y,H_1)$ from (15) is unique for all interior points of G and we can assume that $g(y,H^*)$ from (16) satisfies also (19).

Consider now the mapping $\Gamma(c,f)$ from $R\times L_\infty(Y,\Xi,H^\Delta)$ to G: if $H = \Gamma(c,f)$ then

$$H(E) = \begin{cases} H^u(E) & \text{for } E \subset Z^+(c,f) \\ H^l(E) & \text{for } E \subset Y\setminus Z^+(c,f) \end{cases} \tag{20}$$

for any $E \in \Xi$.

First of all we shall give an informal description of the algorithm. Suppose that some $H^s \in G$ is the current approximation to the solution of the problem (5)-(7). According to (19) local behavior of $\Psi(H)$ around H^s is approximated by linear form:

$$\Psi(H^s+\alpha(H-H^s)) = \Psi(H^s)+\alpha\int g(y,H^s)d(H-H^s)+o(\alpha)$$

and if H_*^s is the solution of the problem

$$\max_{H\in G_0} \int_Y g(y,H^s)dH \tag{21}$$

then direction $H_*^s-H^s$ will be the ascent direction at point H^s. Therefore we can take as the next approximation to the optimal solution

$$H^{s+1} = H^s+\alpha(H_*^s-H^s) \tag{22}$$

for some $\alpha > 0$. Consider now the problem of finding H_*^s or suitable approximation to it.

Suppose that we know the function $g(y, H^s)$ exactly. Then, according to theorem 1, all the possible H_*^s are fully described by pair $(c^*, g(y, H^s))$, where c^* is the solution of the problem

$$\inf_c c \tag{23}$$

$$H^u(Z^+(c, g(y, H^s))) + H^l(Y \setminus Z^+(c, g(y, H^s))) \leq b \tag{24}$$

Observe now that function

$$W_c^s(c) = H^u(Z^+(c, g(y, H^s))) + H^l(Y \setminus Z^+(c, g(y, H^s))) - b \tag{25}$$

is nonincreasing and therefore solving (22)-(23) is equivalent to solving

$$\max_c W^s(c),\ W^s(c) = \int_T^c W_c^s(t)dt$$

for some T and $W_c^s(c)$ can be considered as subgradient of the function $W^s(c)$. Therefore we can use subgradient method for finding c^*:

$$c^{k+1} = c^k + \rho_k W_c^s(c^k) \tag{26}$$

However, computation of $W_c^s(c^k)$ according to (25) involves multidimensional integration over complex regions and this may be too complicated from the computational point of view. In this situation stochastic quasigradient methods [12] can be used. In such methods the statistical estimate ξ^k of W_c^k is implemented in (26) instead of W_c^k.

Once c^* is determined the measure $\Gamma(c^*, g(y, H^s))$ defined in (20), may be a reasonable approximation to the solution H_*^s of the problem (21) and can be used in algorithm (22). However, precise estimation of c^* from (26) requires infinite number of iterations and to make algorithm implementable, it is necessary to avoid this. It appears that under certain assumptions about stepsizes in (22) and (26), we may take in (26) $k = s$ and perform only one iteration in (26) per iteration in (22) using as approximation to H_*^s the measure $\bar{H}^s = \Gamma(c^s, g(y, H^s))$. Thus, along with sequence H^s we obtain also the sequence of numbers c^s. Note now that although $\bar{H}^s$ is quite simple, measure H^s would be excessively complex even for small s. However, H^s is only needed for getting gradient $g(y, H^s)$ and in particular cases some approximation $f(s, y)$ to $g(y, H^s)$ can be obtained using only $\bar{H}^s$ in the sort of updating formula similar to (22).

Once sequence $f(s, y)$ with property $|f(s, y) - g(y, H^s)| \to 0$ is obtained together with sequence c^s: $V^s(c^s) - \max_c V^s(c) \to 0$, the optimal solution of problem (5)-(7) is defined by Theorem 2 through accumulation points of these sequences. The structure of optimal solution is close to (20).

Now we shall define the algorithm for solving (5)-(7) formally.

1. At the beginning select initial approximation to solution H^0, function $f(0,y)$ and number c^0.

2. Suppose that at the step number s we get measure H^s, function $f(s,y)$ and number c^s. Then on the next step we do the following:

2a. Pair $(c^s, f(s,y))$ defines measure $\bar{H}^s$ according to (20):

$$\bar{H}^s = \Gamma(c^s, f(s,y))$$

New approximation to solution is obtained in the following way:

$$H^{s+1} = (1-\alpha_s)H^s + \alpha_s \bar{H}^s \tag{27}$$

2b. Now number c^{s+1} is obtained:

$$c^{s+1} = c^s + \rho_s \xi^s \tag{28}$$

where

$$E(\xi^s / c^0, \cdots, c^s) = V_c^s(c^s)$$

$$V_c^s = H^u(Z^+(c, f(s, f(s,y)))) + H^l(Y \setminus Z^+(c, f(s,y))) - b$$

$$V^s(c) = \int_{T^c} V_c^s(t)dt$$

i.e., the function $V^s(c)$ is defined similarly to $W^s(c)$ with the difference that $f(s,y)$ is used instead of $g(y, H^s)$.

2c. New function $f(s+1,y)$ is obtained in such a way as to approximate $g(y, H^{s+1})$. The precise way of achieving this can be specified only after considering particular ways of dependence $g(y.H)$ on H. One quite general case is considered in the next section. Here we shall only assume that

$$\| f(s,y) - g(y, H^s) \|_\infty \to 0 \tag{29}$$

as $s \to \infty$. The method of achieving this in particular situation will be described in the next section.

Let us now investigate convergence of algorithm (27)-(28). In all statements concerning convergence of measures from the set G we shall use the weak-L_1 convergence, used already in the proof of Theorem 2:

$$H^k \to H \text{ iff } \int_Y g(y)dH^k \to \int_Y g(y)dH$$

for all $g \in L_\infty(Y,\Xi,H^\Delta)$, and topology, induced by this convergence will be used without further reference.

We shall assume that random variables $\xi^1,\cdots,\xi^s,\ldots$ are defined on some probability space, therefore $c^s, H^s, \bar{H}^s$ from (27)-(28) depend on event ω of this space. For simplicity of notations this dependence will be omitted in formulas. Convergence, boundedness, etc. will be considered almost everywhere with respect to this probability space. It should be stressed that we are primarily interested in convergence properties of the sequences c^s and $f(s,y)$. The following theorem gives results in this direction.

THEOREM 3. Suppose that the following assumptions are satisfied:

1. Measures H^l and H^u have bounded variation, $H^l(Y) \le b$, $H^u(Y) \ge b$
2. $\Psi(H)$ is finite concave function for $H \in G + G_\varepsilon$ where

   $$G_\varepsilon = \{H_\varepsilon\colon |H_\varepsilon|(Y) \le \varepsilon,\ H_\varepsilon \text{ is } H^\Delta\text{--continuous}\}$$

   for some $\varepsilon > 0$, and satisfies (19) for $H_1, H_2 \in G$.
3. $\|g(y,H^k) - g(y,H)\|_\infty \to 0$ if $H^k \to H$;

   $\|g(y,H^s) - g(y,H^{s+1})\|_\infty \le \delta_s \to 0$ as $s \to \infty$.
4. $f(s,y) \in L_\infty(Y,\Xi,H^\Delta)$, $\|g(y,H^s) - f(s,y)\|_\infty \le \Delta_s \to 0$ as $s \to \infty$

   $\sup_s \|f(s,y)\|_\infty = \bar{f} < \infty$
5. $\rho_s \to 0,\ \sum_{s=0}^{\infty} \alpha_s = \infty,\ \sum_{s=0}^{\infty} \rho_s^2 < \infty,\ \alpha_s/\rho_s \to 0,\ \delta_s/\rho_s \to 0$

   $E((\xi^s - V_c^s(c^s))^2 / c^0,\cdots,c^s) \le M_1$
6. One of the following conditions is satisfied:

   (i) $\dfrac{\alpha_{s+1}}{\rho_{s+1}} \le \dfrac{\alpha_s}{\rho_s}$

   (ii) $\alpha_s > 0$ and $\dfrac{1}{\alpha_s}\left[\dfrac{\alpha_{s-1}}{\rho_{s-1}} - \dfrac{\alpha_s}{\rho_s}\right] \to 0$

   (iii) $\sum_{s=0}^{\infty} \left| \dfrac{\alpha_s}{\rho_s} - \dfrac{\alpha_{s+1}}{\rho_{s+1}} \right| < \infty$

Then

1) $\Psi(H^s) \to \max_{H \in G} \Psi(H)$, $H^s(Y) \to b$ and all accumulation points of the sequence H^s belong to the set $\Phi = \{H\colon H \in G, \Psi(H) = \max_{H \in G} \Psi(H)\}$

2) For any convergent subsequence $c^{s_k} \to c^*$ exists measure $H^* \in \Phi$ such that

$$H^*(A) = \begin{cases} H^+(A) & \text{for } A \subset Z^+(c^*, g(y, H^*)) \\ H^-(A) & \text{for } A \subset Z^-(c^*, g(y, H^*)) \\ H^-(A) \le H(A) \le H^+(A) & \text{for } A \subset Z^0(c^*, g(y, H^*)) \end{cases}$$

and $\| f(s_l, y) - g(y, H^*) \|_\infty \to 0$ where s_l is some subsequence of the sequence s_k.

Condition 4 of the theorem means that it is possible to use approximations to gradient $g(y, H)$ and it is necessary that precision of these approximations increase as $s \to \infty$. Condition 6 is necessary to assure $H^s(Y) \to b$ although $\bar{H}^s(Y)$ from (27) may not be equal to b. In case if $\bar{H}^s(Y) = b$, i.e. $H^\Delta(Z^0(c^s, f(s, y))) = 0$ starting from some s, condition 6 is not necessary.

Theorem 3 means that if $H^\Delta(Z^0(c, f(s, y))) = 0$ starting from some s then the measure $\bar{H}^s = \Gamma(c^s, f(s, y))$ defined in (20) is good approximation for the optimal solution if s is large. If this is not the case then $f(s, y)$ can still be used for constructing optimal solution, but more careful choice of c is needed.

REFERENCES

1. Yu. Ermoliev, and A. Gaivoronski (1984), Duality relations and numerical methods for optimization problems on the space of probability measures with constraints on probability densities. Working Paper WP-84-46, Laxenburg, Austria: International Institute for Applied Systems Analysis.

2. A. Gaivoronski, (1985) Stochastic Optimization Techniques for finding optimal submeasures. Working Paper WP-85-28, Laxenburg, Austria: International Institute for Applied Systems Analysis.

3. H.P. Wynn (1977), Optimum designs for finite population sampling. S.S. Gupta and D.S. Moore, eds., in: Statistical Decision and Related Topics II. Academic Press, New York.

4. H. P Wynn. Optimum submeasures with application to finite population sampling. Private communication.

5. J. Birge, and R. Wets (1983), Designing approximation schemes for stochastic optimization problems, in particular for stochastic problems with recourse. Working paper WP-83-114, Laxenburg, Austria: International Institute for Applied Systems Analysis.

6. P. Kall, Karl Frauendorfer, and A. Ruszczyński (1984), Approximation techniques in stochastic programming. Working paper, Institute of Operations Research, University of Zurich.

7. W.K. Klein Haneveld (1984), Abstract LP duality and bounds on variables. Discussion paper 84-13-OR, University of Groningen.

8. N. Dunford, and J.T. Schwartz (1957), Linear Operators. Part I: General Theory Interscience Publ. Inc. New York.

9. R.T. Rockafellar (1970), Convex Analysis. Princeton University Press, Princeton.

10. F.H. Clarke (1983), Optimization and nonsmooth analysis, John Wiley & Sons, New York.

11. R.J. T. Morris (1979), Optimal constrained selection of a measurable set, *J. Math. Anal. Appl.* 70:546-562.

12. Yu. Ermoliev (1976), Methods of stochastic programming (in Russian). Nauka, Moscow.

13. A. Gaivoronski (1978), Nonstationary problem of stochastic programming with varying constraints, in: Yr. Ermoliev, I. Kovalenko, eds., Mathematical Methods of Operations Research and Reliability Theory. Institute of Cybernetics Press, Kiev, 1978.

14. Yu. Ermoliev, and A. Gaivoronski, Simultaneous nonstationary optimization estimation and approximation procedures. *Stochastics*, to appear.

15. J. Kiefer, and J. Wolfowitz (1959), Optimum designs in regression problems. Annals of *Mathematicsl Statistics* 30:271-294.

16. P. Whittle (1973), Some general points in the theory of optimal experimental design. Journal of the *Royal Statistical Society*, Series B 35:123-150.

17. V. Fedorov (1972), Theory of Optimal Experiments. Academic Press, New York, 1972.

STRONG CONSISTENCY THEOREMS RELATED TO STOCHASTIC QUASI-NEWTON METHODS

L.Gerencsér
Computer and Automation Institute of the Hungarian Academy of Sciences
H-1053, Budapest, Pf. 63.

1. INTRODUCTION

An important task in the theory of recursive identification is the design of stochastic quasi-Newton methods. In the simplest setting the problem is to solve the nonlinear algebraic equation $f(\theta)=0$ under the assumption that we can observe $f(\theta) + \dot{w}_t$ where $\dot{w}_t$ is a continuous time white noise. We want to find a stochastic approximation process of the form

$$d\theta_t = -t^{-1}H_t(f(\theta_t)dt + dw_t) \qquad t \geq 1,$$

where H_t is an estimation of the inverse Jacobian $f_\theta^{-1}(\theta_t)$ converging to $H^* = (G^*)^{-1} = f_\theta^{-1}(\theta^*)$ in some probabilistic sense. Having in mind real-time control applications it is not allowed to make extra measurement in order to estimate H_t, the estimation process must be based on the observation process $f(\theta_t) + \dot{w}_t$. The first results which conform to this aspect are due to Lai and Robbins (1981). Their paper considers the scalar case only. A stochastic quasi-Newton method for the vector-case is given in Gerencsér (1984a). Both paper assume some knowledge of the position of the derivative matrix G^*, so that the resulting adaptive stochastic approximation process is apriori stable. The behaviour of the proposed processes is unknown if the initial guess $H = G^{-1}$ is such that the stochastic approximation process

$$d\theta_t = -t^{-1}H(f(\theta_t)dt + dw_t) \tag{1.1}$$

is unstable.

In this paper we solve a less ambitious problem; we consider the case when f is linear and we show that a stochastic approximation process (1.1) is generally rich enough to estimate G^*. The observation process is assumed to have the form

$$dy_t = (G^*\Theta_t + b^*)dt + dw_t$$

with $y_1 = 0$.

Let us introduce the notations $\psi_t^T = (\Theta_t^T, 1)$, $K^* = (G^*, b^*)$, then the observation process can be written in the form

$$dy_t = K^*\psi_t dt + dw_t.$$

The least square estimation of K^* is given by

$$K_t = \left(\int_1^t dy_s \psi_s^T\right)\left(\int_1^t \psi_s \psi_s^T ds\right)^{-1}.$$

This is obtained by splitting the matrix estimation problem into n independent vector-estimation problem.

To prove strong consistency of K_t we shall apply a linear transformation so that the original estimation problem falls apart into three separate estimation problems, in each of which the location of the eigenvalues of the matrix HG^* is conveniently specified. In one case we can apply Ljung's theorem (Ljung 1977) to prove strong consistency. In two cases we shall use the condition for strong consistency given by Lai and Wei (1982) for discrete time processes. This condition has been extended by Novikov to more general processes. (Novikov (1984)). A continuous time version which we shall use later will be stated in the third section.

2. SOME USEFUL TRANSFORMATIONS

A simple linear transformation in the parameter space (G,b) will move Θ^* into 0. Therefore we shall assume in the future that $\Theta^* = 0$.

We apply a standard transformation to Θ_t: first we normalize it by multiplying it by $t^{1/2}$ and then introduce a new time-scale $t = e^s$. Thus we get

$$x_s = e^{s/2}\Theta_{e^s}.$$

The process x_s satisfies the stochastic differential equation

$$dx_s = \bar{A}s_s ds - Hd\tilde{w}_s \qquad x_o(0) = \theta_1(1) \tag{2.1}$$

where $\tilde{w}_s$ is a new Wiener process defined by

$$\tilde{w}_s = \int_1^{e^s} t^{-1/2} dw_t$$

and the transition matrix $\bar{A}$ is defined by

$$\bar{A} = -HG^* + I/2.$$

The behaviour of the regressor process is determined by the matrix $\bar{A}$ in (2.1). We will apply a linear transformation to this equation so as to obtain a simpler problem. Let T be an nxn real nonsingular matrix and define a new vector $\bar{x} = Tx$. Then $\bar{x}$ will satisfy

$$d\bar{x}_s = T\bar{A}T^{-1}\bar{x}_s ds + THd\tilde{w}_s \qquad \bar{x}_o = Tx_o.$$

The observation process will be transformed into

$$d(THy_t) = (THG^*T^{-1}T\theta_t + THb^*)dt + THdw_t. \tag{2.2}$$

As $\bar{A} = -HG^* + I/2$ we observe that if $T\bar{A}T^{-1}$ is quasi-diagonal or block-diagonal then the unknown matrix THG^*T^{-1} will be also blockdiagonal. The regressor process in the observation model (2.2) is $(\bar{\theta}_t^T,1)$ with $\bar{\theta}_t = T\theta_t$. The relation $\bar{x}_s = e^{s/2}\bar{\theta}_{e^s/2}$ remains valid. Thus the problem is split into independent subproblems.

A convenient form of $\bar{A}$ which can be obtained by a real-valued similarity transformation is its second normal form. In this from $T\bar{A}T^{-1}$ will be blockiagonal and the characteristic polynomial (which is equal to the minimal polynomial) of each block is an elementary divisor of the matrix $\bar{A}$. Elementary divisors are now regarded over the field of real numbers, that is we have powers of first and second order real polynomials (for details see e.g. Gantmacher (1962). As a first step we shall group these blocks and form three blocks $\bar{A}_-,\bar{A}_+$, and $\bar{A}_0$ such that the characteristic values of these matrices are in the left half plane, in the right half plane and on the imaginary axis, respectively. We shall consider these three cases separately.

The coefficient matrix of the normal equation is

$$\tilde{R}_t = \int_1^t \psi_r \psi_r^T dr.$$

Let $R_s = \tilde{R}_{e^s}$. In the next two section we summarize some auxiliary results.

3. CONTINUOUS-TIME STOCHASTIC REGRESSION

In this section we state the result of Lai and Wei for continuous-time observation process in the form that we shall need later. We shall formulate the problem in the case when the unknown parameter γ is a vector in R^m. Assume that we have an observation process

$$\dot{y}_t = \psi_t^T \gamma^* + \dot{w}_t \qquad t \geq 0$$

or equivalently

$$dy_t = \psi_t^T \gamma^* dt + dw_t \qquad y_o = 0 \qquad t \geq 0$$

where ψ_t is an adapted process in $\mathbb{R}^m$ such that $E \int_0^t \psi_s^T \psi_s ds < \infty$ for every finite t, and w_t is tandard Wiener process in $\mathbb{R}^1$. Let

$$R_t = \int_0^t \psi_s \psi_s^T ds ,$$

Then the least square estimation γ_t of γ^* is obtained from the normal equation

$$R_t \gamma_t = \int_0^t \psi_s dy_s .$$

Let $\lambda_{min}(R_t)$, $\lambda_{max}(R_t)$ denote the minimal and the maximal eigenvalues of R_t, respectively. Then we have the following

Theorem 3.1. A sufficient condition for the strong consistency of the least square estimation method is that

$$\lim_{t\to\infty} \lambda_{min}(R_t) = \infty \qquad \text{w.p.1.} \tag{3.1}$$

and

$$\lim_{t\to\infty} \lambda_{min}^{-1}(R_t)(\log\lambda_{max}(R_t)) = 0 \qquad \text{w.p.1.} \tag{3.2}$$

Under these conditions

$$|\gamma_t - \gamma^*| = O(\lambda_{min}^{-1}(R_t)\cdot \log\lambda_{max}(R_t))^{1/2} \qquad \text{w.p.1.} \tag{3.3}$$

Remark. It is easy to see that conditions (3.1), (3,2) are invariant under a coordinate transformation in the regression model. If the transformation matrix is T then R_t is replaced by $\bar{R}_t = TR_tT^T$. From the variantional property of λ_{max} it is easy to see that

$$\lambda_{max}(\bar{R}_t) \leq C\, \lambda_{max}(R_t)$$

where C depends only on T. The same argument for the inverse matrices gives

$$\lambda_{min}(\bar{R}_t) \geq C\, \lambda_{min}(R_t)$$

with some other $C > 0$. From these two inequalities follows that if (3.1), (3.2) are satisfied for R_t then they also hold for $\bar{R}_t$.

4. A CLASS OF STOCHASTIC PROCESSES

We shall need some results which were developed in connection with the off-line identification of continuous-time systems. They were stated with a sketch of proof in Gerencsér (1984a). First we define a class of stochastic processes which is very close to the class $\mathcal{A}$ introduced by Krylov (Krylov (1977)). Our aim is to give an idea on some technical details.

Definition. A stochastic process ξ_t is in class M if for every $m \geq 1$ we have for all $t \geq 0$

$$E|\xi_t|^m < K_m < \infty.$$

We say that ξ_t is M-continuous in t if it is in M and for $h > 0$ $t \geq 0$

$$E|\xi_{t+h} - \xi_t|^m < K_m h^{r(m)}$$

where $r(m)\to\infty$ for $m\to\infty$.

We formulate two theorems which state that the property of beeing in M is preserved under certain linear operations.

Lemma 4.1. Let A be a stable nxn matrix i.e. $\mathrm{Re}\lambda_i(A) < 0$ for i = 1,n and consider the linear filter

$$\dot{x}_t = Ax_t + u_t$$

with initial condition x_o beeing in M, and assume that u_t is measurable and is in M. Then x_t is in M, and it is M-continuous in t.

An immediate consequence of this lemma is the following

Lemma 4.2. Assume that he conditions of the previous lemma hold and consider

$$\dot{x}_t = Ax_t + g_t u_t$$

where g_t, $t \geq 0$ is a positive differentiable function such that $\lim_{t\to\infty} \dot{g}_t/g_t = 0$. Then x_t/g_t is in M and it is M-continuous in t.

Our third lemma is the following

Lemma 4.3..Let g_t be a positive locally integrable function and let

$$G_t = \int_0^t g_s ds.$$

Furthermore let u_s be a measurable process in M. Then

$$x_t = G_t^{-1} \int_0^t g_s u_s ds$$

is in M and is M-continuous in t.

We need a strong law of large numbers:

Theorem 4.4. Let ξ_t $t \geq 0$ be measurable and M continuous in t. Then for every $\varepsilon > 0$ we have

$$\lim_{t\to\infty} t^{-\varepsilon}\xi_t = 0 \qquad \text{w.p.1.}$$

We shall also need the following large deviation theorem.

Theorem 4.5. Let Q_s, $s \geq 0$ be a locally integrable symmetric, positive semi-definite nxn matrix function such that $\mathrm{tr}Q_s > \delta > 0$ for all s, and let w_s be an n-dimensional standard Wiener process. Then

$$\lim_{s\to\infty}\left(\int_0^s w_u^T Q_u w_u du\right)/s^{2-\varepsilon} = \infty \qquad \text{w.p.1.}$$

For the proofs see Gerencsér (1984a) except for Theorem 4.5.

In the next section we state our strong consistency theorems.

5. STORNG CONSISTENCY THEOREMS

5.1. $\bar{A}$ is stable

By this we mean the case when $\mathrm{Re}\lambda_i(\bar{A}) < 0$ $i = 1,n$, or $\bar{A} = \bar{A}_-$

It is easy to see that in this case the conditions of Theorem 3.1 are violated. However we have the following

Theorem 5.1. If the matrix $-HG^* + I/2$ is stable, then the least square estimation of G^*, b^* based on the observation process

$$dy_t = (G^*\theta_t + b^*)dt + dw_t \qquad t \geq 1$$

converge to their true values with probability 1.

The idea of the proof is that we consider the recursive form of the least square estimation method and then this recursion is transformed by changing the time scale into a recursion, which can be analyzed by Ljung's covergence theorem (Ljung 1977)). The aasociated oridnary differential equation is

$$\dot{G}(s) = (G^* - G(s))\Sigma_{xx}(H)\alpha^{-1}(s) \tag{5.1}$$

$$\dot{\alpha}(s) = \Sigma_{xx}(H) - \alpha(s) \tag{5.2}$$

where $\Sigma_{xx}(H)$ is the solution of the Lyapunov-equation

$$\bar{A}\Sigma + \Sigma\bar{A}^T + HH^T = 0.$$

As the associated ordinary differential equation (4.1), (4.2) is globally asymtotically stable, the proposition follows from Ljung's theorem.

5.2. $\bar{A}$ is unstable

By this we mean that $\mathrm{Re}\lambda_i(\bar{A}) > 0$ for $i = 1,n$. We shall apply the results of section 3 to prove strong consistency. Before formulating our theorem we remark that we can transform $\bar{A}$ by a real transformation into a blockiagonal matrix, all the blocks of which are simple matrices (i.e. the minimal polynomial of any block coincides with the characteristic polynomial). Therefore we can restrict ourselves to the consideration of simple matrices.

Theorem 5.2. Let $\bar{A}$ be a simple matrix the eigenvalues of which have positive real parts. Then the matrix R_s satisfies the conditions obtained in Theorem 3.1. Moreover if G_t denotes the least square estimate of G^* up to time t in the original time scale then with some $c > 0$ we have

$$\| G_t - G^* \| = O(t^{-c}) \qquad \text{w.p.1.}$$

This theorem was first stated in Gerencsér (1984b).

5.3. $A = \bar{A}_0$.

We say that a matrix $\bar{A}$ has simple structure if it is similar to a diagonal matrix. If $\bar{A}$ is real and simple then $\bar{A}$ is similar over the real field to a blockdiagonal matrix all the blocks of which are real numbers or 2x2 real matrices of the form

$$\begin{pmatrix} a & b \\ -b & a \end{pmatrix}.$$

If all the eigenvalues of $\bar{A}$ are purely imaginary then all diagonal elements are equal to 0. (See Gantmacher (1966)). We have

Theorem 5.3. Assume that $\bar{A}$ has simple structure. The the matrix R_s satisfies the conditions obtained in Theorem 3.1. Moreover have for all $\varepsilon > 0$

$$\| G_t - G^* \| = O((\log t)^{-1/2+\varepsilon}) \qquad \text{w.p.1.}$$

6. DISCUSSION

The extension of our results to adaptive procedure is of interest. This has been done for the case when $\bar{A}$ is restricted to be stable. However if the initial value for $\bar{A}$ is unstable the complete analysis is yet to be done.

7. REFERENCES

Gantmacher, F.R. (1966). Theory of Matrices (In Russian), Nauka, Moscow, 1966.

Gerencsér, L. (1984a). Off-line identification of continuous-time linear systems. In K. Sarkadi, P.K. Sen and P. Révész (Eds.), Goodness of fit. Akadémiai Kiadó and North-Holland, to appear.

Gerencsér, L. (1984b). Stable and unstable stochastic approximation processes as stochastic regressors. Report on Workshop in Multidimensional Analysis, Information Theory and Asymptotic Methods, Stanford University, (1984b), pp. 17-20.

Gerencsér, L. (1985). Strong Consistency and asymptotic efficiency of stochastic quasi-Newton methods-theory and applications. Preprints of the 7-th IFAC Sympósium on Identification and System Parameter Estimation, 1985, York, U.K.

Th. Kailath, Linear Systems, Prentice Hall, Englewood Cliffs, N.J., 1980.

Krylov, N.V. (1977). Controlled Diffusion Processes. (In Russian). Nauka, Moscow,

Lai, T.L. and Robbins H. (1981). Consistency and asymptotic efficiency of slope estimate in stochastic approximation schemes, Z. Wahrsch. Verw. Gebiete, 56: 329-366.

Lai, T.L. and Wei, C.Z. (1982). Least squares estimates in stochastic regression models with applications to identification and control of dynamic systems, Ann. Statist. 10:154 - 166.

Ljung, L. (1977). Analysis of recursive stochastic algorithms. IEEE Trans. Automat. Control 22 : 551-575.

Ljung, L. and Söderström, T. (1984), Theory and Practice of Recursive Identification. MIT Press, Cambridge, Massachuessets.

Novikov A.A. (1984). Consistency of least squares estimates in regression models with martingale errors. To appear.

Stochastic Gradient Methods for Optimizing Electrical Transportation Networks

M.Goursat, J.P.Quadrat & M.Viot INRIA

This work has been supported by EDF Etudes et Recherches , Service Etude de réseaux, Département Methodes d'Optimisation.

We have to determine the best capacities for an electrical transmission network in order to minimize the sum of the investment cost and the expectation of the optimal generation cost for a given power demand and a set of generation facilities.

The physical laws governing the system are the two Kirchoff laws in the d.c. appcroximation .

The system is perturbed by the stochastic nature of the demand and the possible breakdowns of the generation units.

The main difficulty is the large size of the problem.

We consider here a simplified case.In the first part we present a stochastic subgradient algorithm for the french aggregated network modelled by the first Kirchoff law only .We compare the results with the previous method used at EDF.In the following part we derive some heuristic algorithms for the integer value case .The last part of the paper is devoted to the two Kirchoff laws case.

1.INTRODUCTION

Let be given a set of generation facilities and the power demand in an electrical network with a given geometry.Our purpose is to determine the transmission line capacities which minimize the sum of the investment cost and the generation cost.

Some characteristic points of the problem are the following :

i) The system is subjected to some uncertainties namely the breakdowns of the generation units and of the transmission lines , the stochastic nature of the demand.

ii)The optimization problem is dynamic since one must determine every year which new transmission capacities should be added to the existing ones.

iii)We have two kinds of control variables :

-the capacities which do not depend of the uncertainties,

-the generation outputs which are functions of the perturbations.

iv)The system is governed by the two Kirchoff laws in d.c. approximation.

v)The line capacities must be multiple of a given value.

Here we are interested by the methodology and we will consider a simplified approach of the problem .The main simplifications made are:

i) In the first two parts we only take into account the first Kirchoff law.

ii) The demand is deterministic.The only stochastic variables are the possible breakdowns of the generation capacities.We neglect the breakdowns of the lines and the reliability problem of the transmission.

iii) In the first part the capacities may take continuous values.

iv) We consider the one time period case.

The previous method used by EDF (Electricite de France) was based on linear programming algorithms for a simplified formulation of a very large deterministic problem based on a bundle (e.g. 500) of perturbation realizations .This method was too much computer time consuming and could not be extended to the general model.

The paper is organized as follows.

In the first part we present the simplified mathematical model using only the first Kirchoff law and we descibe the optimization problem.

In the second one we present the stochastic gradient method and give some numerical results in the continuous value case for comparison with the linear programming type method used previously at EDF.

In the third part we study the integer value case.We give three heuristic algorithms based on the stochastic gradient and the corresponding numerical results.

In the last part the best method is applied to the more realistic case of the two Kirchoff laws .We give some numerical results for a simplified network.

2 . MATHEMATICAL MODEL AND OPTIMIZATION PROBLEM

2.1. The directed graph associated with the generation-transmission system

The nodes are $i \in N$,they represent the regions .A fictitious node representing the source of generation and the sink of consumption is denoted o.

The arcs are belonging to $T \cup D \cup G$ where:

T is the set of the transmission lines ,

D is the set of arcs representing a demand ,

G is the set of arcs representing the generation.

The node-arc incidence matrix is A(N,TGD) defined by :

$A_{ij}=1$ if i is the terminal node of the arc j,

$A_{ij}=-1$ if i is the initial node of the arc j,

$A_{ij}=0$ otherwise.

2.2. System situation

The stochastic variables , i.e. the availability of the generation units of the lines and the level of the demand , are supposed to be known functions of $\omega \in \Omega$ which is called the system situation. The variables and the data are associated with arcs of the three sorts previously described. In a situation ω we have :

-for a generation arc $j \in G$:

$q_j(\omega)$ is the chosen power output,

Q_j is the invested power generation capacity ,

$C_j(Q_j)$ the investment cost,

$\xi_j(Q_j,\omega)$ the maximal power output ,

$c_j(q_j)$ the generation cost .

$0 \leq q_j(\omega) \leq \xi_j(Q_j,\omega) \leq Q_j$ $\forall \omega, \forall j \in G$ and $\xi_j = Q_j$ or 0 (breakdown).

-for a transmission arc $j \in T$:

$q_j(\omega)$ is the power flow in the line,

Q_j is the installed line capacity ,

$C_j(Q_j)$ the investment cost,

$\xi_j(Q_j,\omega)$ the available line capacity .

$0 \le |q_j(\omega)| \le \xi_j(Q_j,\omega) \le Q_j \ \forall\omega, \forall j \in T$ where q_j is positive or negative according to its direction in the arc.

In our case $\xi_j = Q_j$ and $c_j(q_j)=0$ because we neglect the breakdowns of the lines and the transportation cost.

-for a demand arc $j \in D$:

$q_j(\omega)$ is the supplied part of the demand $\xi_j(Q_j,\omega)$,

Q_j being the maximal demand ,

$\xi_j(Q_j,\omega)$ is a random variable on $[0,Q_j]$ with a known distribution,

$c_j(\xi_j - q_j)$ for $q_j \le \xi_j$ a penalty cost associated to the shortage of production or transportation capabilities.

The demand is obviously time varying denoted by $\xi_j(s)$ where s denotes here the time.In order to get a stationary model we define the load curve that is:

$t_j(x) = \int_0^S 1_{[x,\infty]}(\xi_j(s))\, ds$ is the part of the period $[0,S]$ with a demand greater than x and $F_j(x)=t_j(x)/S$ is decreasing and $0 \le F_j(x) \le 1$. Thus $1-F(x)$ can be interpreted as the distribution of a random variable $\xi_j(Q_j,\omega)$ on $[0,Q_j]$.

2.3 Formulation of the optimization problem

With the previous notations we have :

(i) $A\,q=0$ (first Kirchoff law)

(ii) $|q_j| \le Qj \quad j \in T$

(iii) $0 \le q_j \le \xi_j \quad \forall j \in G \cup D$

-for $j \in G$:

$\xi_j = 0$ with probability $1-\delta_j$,

$\xi_j = Q_j$ with probability δ_j.

δ_j is the failure rate of the unit j.

-for $j \in D$ ξ_j is a random variable generated with the distribution given by the load graph.

Problem : minimize the sum of the investment cost and the optimal generation cost:

$$\text{Min}\,[C_j(Q_j)+\mathbb{E}\Phi(Q,\omega)\ \mid \mathbf{Q}_j\leq Q_j\leq \mathbb{Q}_j\ ].$$

$\mathbf{Q}_j$ is the existing network, $\mathbb{Q}_j$ is a technological bound for every route and $\Phi(Q,\omega)$ is the optimal generation cost i.e. the solution of the following optimization problem:

$$\Phi(Q,\omega)=\text{Min}\,[\Sigma_j c_j q_j(\omega) \mid q_j(\omega)\ ,j\in G\cup T\cup D\ \ (i)\ (ii)\ (iii)].$$

3.THE CONTINUOUS VALUES CASE

3.1 The stochastic gradient method

Consider a function

$f : R^n \times \Omega$ ---------->R^+

x,ω $\qquad\qquad$ $f(x,\omega)$

on the probabity space$(\Omega,\mathbb{B},p)$.

We want to minimize the function:

$\mathbf{f}(x)=\mathbb{E}f(x)=\int f(x,\omega)p(d\omega)$

For this purpose we use the well-known stochastic gradient algorithm (B.T.Polyak[2] ,H.Robbins- S.Monro[1],B.T.Polyak -Y.Z.Tsypkin[3] , H.J.Kushner- D.S.Clark[4]) that is:

$$x_{n+1}=x_n-a_n D_x f(x_n,\omega_n)$$

with a_n such that:

$$a_n\in R^+\ ,a_n\text{-->}0,\ \ \Sigma_n a_n=+\infty,\ \Sigma_n a_n^{\ 2}<+\infty.$$

More precisely, consider the following set of assumptions :

A1: ω------>$f(x,\omega)$ is $L1(\Omega,\mathbb{B},p)$ $\forall x\in R^n$;

A2: x------>$f(x,\omega)$ is convex ,continuous $\forall\omega$;

A3: $\exists\delta>0 : \forall g\in Df(x,\omega)\ \ |g|<\delta\ ,\ \forall x\ ,\forall\omega$;

A4 :$C\subset R^n$ convex compact set ;

A5 :$a_n\in R^+,\ \ a_n>0\ \ ,\ \Sigma_n a_n=+\infty$;

A6: f continuous $\exists k>0 \quad f(x)-f^* \geq kd^2(x)$ where:

$f^*=\min [\, f(x) \mid x \in C]$;

$P=\{x \in C,\ f(x)=f^*\}$;

$d(x)=dist(x,P)$;

$\mathbb{P}_C$=projection onto C ;

$g(x,\omega) \in Df(x,\omega)$ measurable selection of the subgradient

We define a sequence of random variables (x_n) by:

$x_{n+1}=\mathbb{P}_C(x_n-a_n.g(x_n,\omega_n))$;

x_0 given , ω_n independent realization of the random variable of distribution p .

Denoting $z_n=\mathbb{E}d^2(x_n)$ we have the following result:

Theorem 1

Under the assumptions A1 to A6 we have:

$\lim_n z_n=0.$

Moreover ,if we take $a_n=(r.A)/(n.A+B)$ with:

$A=k^2/\delta^2$, $B=1/z_0$, $r=1/k$,

we have the following speed of convergence

$(nA+B)z_n \leq 1$, $\forall n$.

Theorem 2

Under the previous assumptions and with a_n such that :

$\Sigma_n a_n^2 < +\infty$,

we have the estimation:

$P(\sup_{n>m} d^2(x_n) \geq \varepsilon) \leq (z_m+\Sigma_{n\geq m} a_n^2 \delta^2)/\varepsilon.$

In the case of Theorem 1 this estimation becomes

$P(\sup_{n>m} d^2(x_n) \geq \varepsilon) \leq K(m)/\varepsilon(m.k^2/\delta^2+B).$

3.2 Solving the continuous value capacities optimization problem

Our problem is:

$\mathrm{Min}\ \Sigma_j C_j(Q_j)+\mathbb{E}\Phi(Q,\omega),$

with Φ solution of:

P_ω: $\Phi(Q,\omega)=\text{Min}\ \Sigma_j c_j q_j(\omega)$.

$Q \dashrightarrow \Phi(Q,\omega)$ is convex , non differentiable ;

P_ω is a flow problem . We use the flow algorithm of [5] to solve it. The multipliers associated with the constrained variables $q_j \le Q_j$ give us an elementof $D_Q\Phi$. By adding C_i we get immediatly a subgradient for our problem.We apply the previous stochastic subgradient algorithm.

3.3 Numerical results

Our case is the 400 kV french aggregated network with 48 nodes, 82 lines and 185 power plants.

In order to compare with the results of [6] (linear programming method on a bundle of trajectories) we make the two following simplifications :

-the production cost is zero but we take the available plants according to the increasing cost up to the satisfaction of the demand if possible , if not we add a new plant with a shortage cost of generation;

-we take a constant demand corresponding to the peak hours.

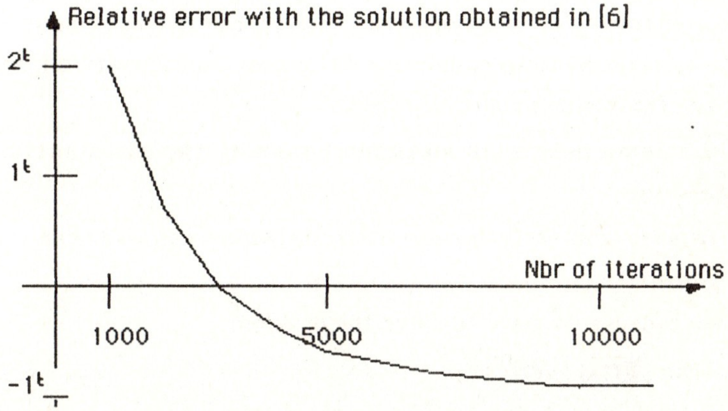

figure 1.Convergence of the stochastic gradient method

On Figure 1 we have represented the evolution of the criterion with the number of iterations of the stochastic gradient algorithm : let SOL(n) be the solution obtained with n iterations for the algorithm. The criterion corresponding to any solution is the global mean cost obtained on a fixed bundle of 15000 different realizations . We represent here the variation of this cost with regard to the cost of the solution given by [].

The same precision is obtained after less than 3000 iterations. The computer time on IBM 370/168 needed by [6] to compute his solution is 15mn . The cost of the stochastic gradient on the same computer is 1mn for 3000 iterations .

4. THE INTEGER VALUE CASE

4.1 Some remarks

We are here in a more realistic case. The formulation of the problem is the same excepted for Q_j which is replaced by $Q.u_j$ where the optimization variable is u_j a positive integer, Q being a given value which represents the capacity of a standard line unit.

The corresponding deterministic problem (i.e. card(Ω)= 1) is NP complete. We have to solve now a large scale system with mixed variables .To solve it -the first possibility is to use the classical methods of branch and bound , but with this approach the computer time will be absolutely prohibitive in the stochastic case -we have to content ourselves with good heuristic methods.

The last remark concerns the use of the continuous solution . The values for this solution belongs to [0,3000] and the unit line is Q=700.The rounded continuous solution has a cost approximatively equal to two times the optimal continuous one.

4.2 Some heuristic methods based on the stochastic gradient

(a) Stochastic gradient calculated at the nearest integer point

We suppose we have to solve the problem:

$\text{Min}\,[\,\mathbb{E}f(x) \mid x \in \mathbb{N}^m]$,

and we consider the following algorithm:

$x_{n+1} = x_n - a.D_x f([x_n], \omega_n)$ $a \in R$,$a>0$ fixed ,

$[x_n]$=nearest integer of x_n.

$[x_n]$ cannot converges but moves on some recurrent points. We suppose that these points belong to the hypercube $[0,1]^m$ (for simplicity we take u=0 or 1).

We denote:

p_i^1=the visit frequency of $[x_n]_i$ in 1,

p_i^0=the visit frequency of $[x_n]_i$ in 0.

We take the solution $[x_n]^*$ given by the maximum frequencies :

$[x_n]^*_i = 1$ if $p_i^1 > p_i^0$

$[x_n]^*_i = 0$ if $p_i^1 \leq p_i^0$

The algorithm can be slighty improved by the following way : after n iterations the components of $[x_n]$ are classified in the order of the decreasing frequencies $q_i = \sup(p_i^1, p_i^0)$. The r first components being fixed to $\mathrm{argmax}_k(p_i^k)$ .We reiterate the algorithm . We have obtained the best of our heuristic methods with 3 steps and n=2000.

(b) Using the stochastic gradient for changing the probability of the optimization variable

Consider the problem:

$$\mathrm{Min}\ [\ \mathbb{E}f(x) \mid x \in \{0,1\}^m]\quad ,f \text{ convex.}$$

x is generated with the distribution $P = \Pi_i P_i$ with P_i such that:

$P_i(0) = 1 - p_i$,

$P_i(1) = p_i$.

We will adapt p_i by a stochastic gradient method in order to get the convergence to p_i^* optimal in the sense of the minimal gradient :

$p_i^* = 0$ if $-g_i(0) \leq g_i(1)$,

$p_i^* = 1$ if $-g_i(0) > g_i(1)$;

where g denotes the gradient:

$g(x,\omega) = D_x f(x,\omega)$.

The algorithm is:

$$p_i^{n+1} = \mathbb{P}_m\{p_i^n - a^n . g_i(x^n, \omega^n)[\chi_{\{1\}}(x^n_i)/p_i^n + \chi_{\{0\}}(x^n_i)/(1-p_i^n)]\}$$

$$a_n > 0\ ,\ \Sigma_n a_n = +\infty\ ,\ \Sigma_n a_n^2 < +\infty\ ;$$

where $\mathbb{P}_m$ is the projection onto $[0,1]^m$ and $\chi_A(x)$ the characteristic function of A.

In the case of dimension 1 it is easy to see that the algorithm converges to the minimum gradient point; but this result is false for larger dimension cases.

The main problem with such an algorithm is that we have a denominator equal to zero at the convergence point. We can remove this problem by a slight modification : we suppose that $g_i^{1}(p)=C_i$ (investment cost); this choice is realistic if the continuous minimum point is smaller than 1 ; if not we will have a large shortage cost in 0 and the gradient at 0 will be greater than the one at 1 and it is not a trouble to take C_i for this point.

The modified algorithm is :

$$p_i^{n+1}=\mathbb{P}\{p_i^{n}-a^n[g_i(x^n,\omega^n)+C_i]\chi_{\{o\}}(x_i^{n})\}.$$

(c)The penalization method

We penalize the integer constraints in a non differentiable way. Starting with the continuous solution ,the penalization term is increased gradually and the solution of the penalized problem is computed by a stochastic gradient algorithm .

We have to solve :

$$\mathrm{Min}[\,\mathbb{E}f(x,\omega)\mid x\in\mathbb{N}\,].$$

Consider the piecewise linear function, continuous ,≥0 and

$$\Psi_{b,c}(x)= 0\ ,\ D\Psi(x^+)=b,\ D\Psi(x^-)=c \text{ if } x\in\mathbb{N}\ .$$

The penalized problem is :

$$\mathrm{Min}_x H_{b,c}(x)\ ,$$

with :

$$H_{b,c}(x)=\mathbb{E}f(x,\omega)+\Psi_{b,c}(x).$$

We want to apply the stochastic gradient to this problem but we have lost the convexity.The procedure is the following :

we start with $H_{o,o}(x)$. To solve H_{b^n,c^n} , where b^n and c^n increase with the iterations , we use the stochastic gradient algorithm.

In our investment problem b_n---> b very large and c_n--->c , where c is the investment cost for the line .

The figure 2 gives the limit penalization term and the limit global cost.

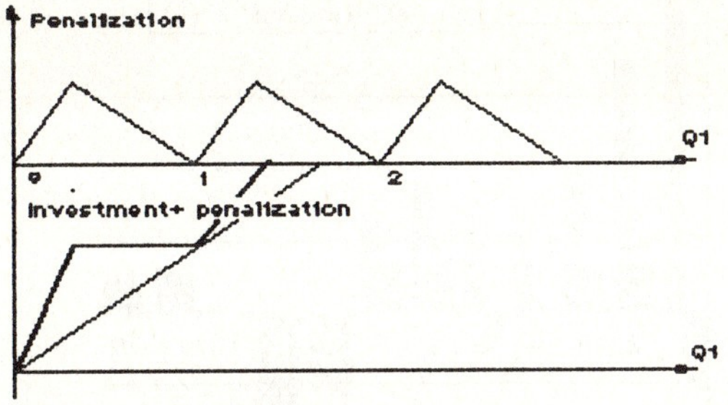

figure 2 :Penalization function

4.3 Numerical results

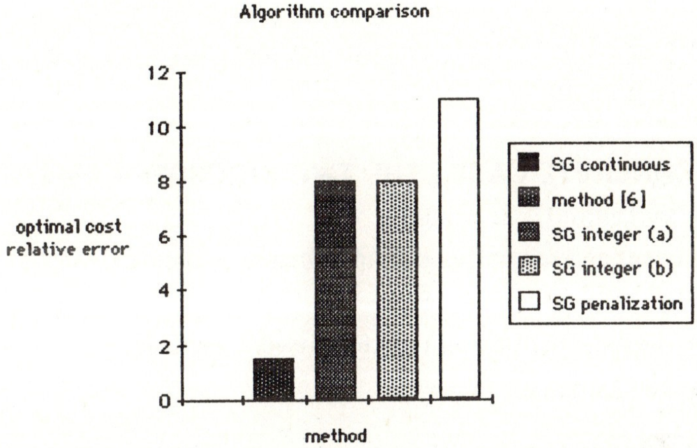

figure 3:Comparison of the precision of the result obtained by the different algorithms based on a bundle of 12000 realizations of the perturbations.

The datas are the same as in 3.3. The best results are given by the algorithms(a) and (b) : the criterion is about 8 percent above the cost given by the best continuous solution . The difference of 11 percent for the penalization algorithm (c) is due to the fact that this solution was obtained with only 3000 iterations and can be improved but the main default of this method is probably that the results depend on the adjustment of the evolution of the penalization term. From the point of view of the robustness the algorithm (a) is the best. . The best solution with (a) is obtained with 6000 iterations i.e. 2mn of computer time of IBM 370/168.

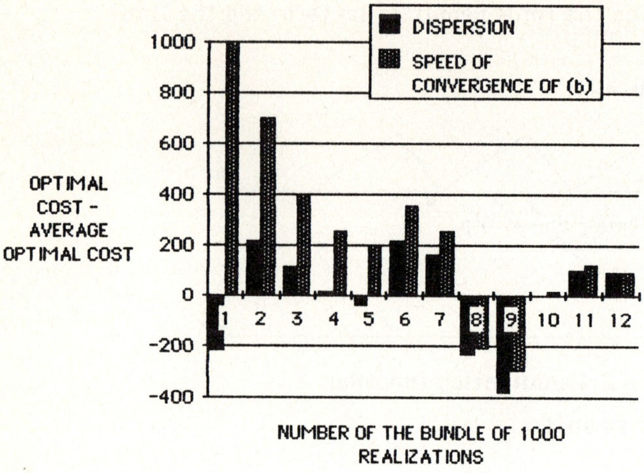

figure4: Dispersion of 1000-average-optimal-cost and speed of convergence of algorithm (b).

5.THE CONCRETE CASE : THE TWO KIRCHOFF LAWS

5.1 The mathematical model

The electrical approximation considered here is the d.c. approximation with the two Kirchoff laws.

Consider the existing liaison l of the network we denote:

$y_l^0 \geq 0$ the admittance,

$X_l^0 = 1/y_l^0$ the reactance (ohm),

Q_l^0 the capacity (MW) .

The additive (reinforcing) line is denoted z_l, X_l, Q_l .

Thus the global new line becomes $y_l = y_l^0 + z_l$.

We have the technical constraint to avoid the destruction of the line :

$X_l Q_l = \alpha_l$ (α_l=constant).

More precisely, in the integer case we have:

$Q_l = 0, q_l, 2q_l$ .... (q_l=capacity of a standard new unit line)

$X_l = +\infty, r_l, r_l/2, r_l/3$

$\alpha_l = q_l r_l$.

The problem is now:

$\mathrm{Min}\,[C.y+\mathbb{E}\Phi(y,\omega)\mid y\geq y^0]$,

with Φ solution of :

$\Phi(y,\omega)=\mathrm{Min}_p\ \Sigma_i c_i p_i$ subject to the constraints (i) to (iv).

(i) $AYv=p-d$ is the 1st Kirchoff law,where p denotes the generation and d the consumption ,A is the incident matrix nodes arcs associated only to the transportation arcs,

(ii) $Sv=0$ is 2nd Kirchoff law where S denotes the incident matrix basis of cycles arcs,

(iii) $|v|\leq\mathbf{v}$ is the constraint on transportation,

(iv) $0\leq p\leq\mathbf{p}(\omega)$ is the generation constraints.

$\Sigma_i c_i p_i$ represents the sum of the generation cost and the shortage cost.

$Y=\mathrm{diag}(y_i)$

At $\mathbf{v}=0$ appears a discontinuity of the cost indeed:

-*case 1* there is no existing line for the liaison l : $y_1^0=0$

$\mathbf{v}_1=\alpha_1$ if $y_1>0$ (a line is installed)

$\mathbf{v}_1=0$ if not

-*case 2* there exists a line $y_1^0>0$:

we compare α_1 and $X_1^0 Q_1^0$

(i) $\alpha_1>X_1^0Q_1^0$ implies $\mathbf{v}_1=X_1^0Q_1^0$

(ii)$\alpha_1<X_1^0Q_1^0$ $\quad\mathbf{v}_1=\alpha_1$ if $z_1>0$

$\mathbf{v}_1=X_1^0Q_1^0$ if $z_1=0$

in fact , thanks to technological improvements we are in the case (i) and so we have a discontinuity only in the case of the setting up of a new line.

The gradient is:

$g_1=C_1+D_y\Phi$ with

$D_y\Phi=[u_{r(1)}-u_{s(1)}].v_1^*$

where v_l^* is the value of the transit for the optimum and u_i are the Lagrange multipliers associated with the components i of AYv=p-d .

r(l) and s(l) are the two nodes linked by the arc l.

The u_i are obtained in the computation of Φ.

5.2 Numerical experiments

We use the same methods than the ones employed in the part 2. In the integer case we only use the algorithm (a) (gradient computed at the nearest integer point) .To avoid the difficulty od the discontinuity at the point 0 we take the gradient at 0^+.

In order to save a part of computer time we consider here a smaller network which is an aggregated version of the previous 400kV french network: 15 nodes, 26 lines and 140 power plants.

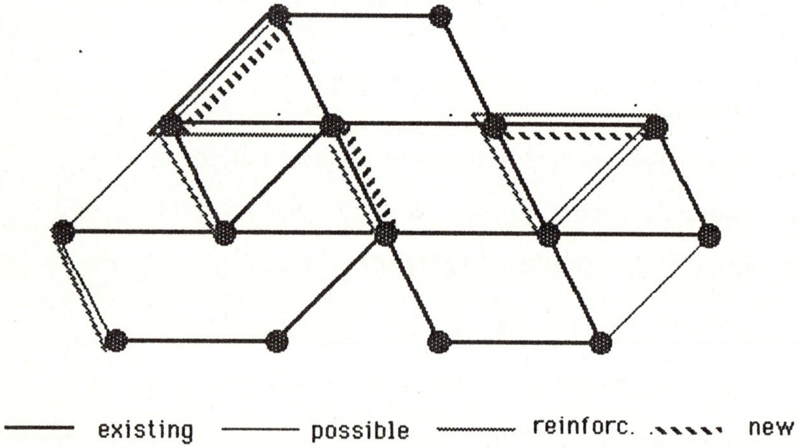

figure 5:The network

(i) The continuous case:

we use the gradient with a constant step starting from two different points :-a network *high* or over-equiped - a network *low* or underequiped . The evolution of the network during the iterations shows that we have the same convergence point and that it is possible to have departures from 0 and returns to 0 for a given liaison (there is no problem with the discontinuity). The Figures 5 gives the existing networks and the lines reinforced.

(ii) The integer case :

we compute the integer solution with (a) .We get the new arcs of the figure 5 with two lines reinforced and one new liaison. The difference between the integer cost and the continuous one is of order of the investment cost of a line unit. Systematic exploration has shown also that the solution obtained is good.The exemple solved is too small to formulate definitive conclusion.

CONCLUSION

It appears that the stochastic gradient method is

-very easy to implement

-very efficient

-easy to extend in the integer case

The main interest of this study is to show that this method may be applied in a domain which is not classical for it that is : large NP complete problems .In these cases it seems to give good heuristics.

REFERENCES

[1]**Robbins,H. and Monro,S.** (1951) A stochastic approximation method Ann.Math.Statist.,22,400-407.

[2] **Polyak,B.T.** (1976) Convergence and convergence rate of iterative stochastic algorithms. Automatica i Telemekhanika,12,83-94.

[3] **Polyak,B.T. and Tsypkin, Y.Z.** (1973). Pseudogradient adaptation and training algorithms. Automatica i Telemekhanica,3,

[4] **Kushner,H.J. and Clark,D.S.** (1976). Stochastic approximation methods for constrained and unconstrained systems. Springer Verlag.

[5] **Maurras,J.F.** (1972) Flot optimal sur un graphe avec multiplicateurs.EDF Publication DER Serie C,(1)

[6] **Dodu,J.C.** (1978) Modele dynamique d'optimisation a long terme d'un reseau de transport d'energie electrique. French Journal RAIRO,12,(2).

[7] **Goursat,M.,Quadrat,J.P. and Viot,M.** (1984) Report of EDF contract. INRIA.

[8]**Dodu J.C., Goursat,M.,Hertz A.,Quadrat,J.P. and Viot,M.** (1981) Méthodes de gradient stochastique pour l'optimisation des investissements dans un réseau électrique.EDF Bulletin des Etudes et Recherches ,serie C , N°2 , pp133-164 .

ON THE FUNCTIONAL DEPENDENCE BETWEEN THE AVAILABLE INFORMATION AND THE CHOSEN OPTIMALITY PRINCIPLE

V.I. Ivanenko and V.A. Labkovsky
V. Glushkov Institute of Cybernetics, Kiev, USSR

We shall use the term ***decision problem*** to describe any ordered triple $Z = (\Theta_Z, \mathbf{D}_Z, L_Z)$ consisting of arbitrary non-empty sets Θ_Z, $\mathbf{D}_Z$ and a real bounded function L_Z: $\Theta_Z \times \mathbf{D}_Z \rightarrow \mathbb{R}$. The class of all decision problems is denoted by $\mathbf{Z}$. When we interpret a decision problem as the problem of finding a decision $d \in \mathbf{D}_Z$ which "minimizes" the loss $L_Z(\vartheta, d)$ depending on a parameter $\vartheta \in \Theta_Z$, we need an exact definition of the term "minimization" as used here. In other words, we need an optimality principle which associates a ranking on $\mathbf{D}_Z$ with any $Z \in \mathbf{Z}$. Intuitively it is clear that the choice of the optimality principle is somehow connected with our information about the behavior of ϑ. There has been considerable research on the classification of "information situations" and the corresponding optimality principles [1–3]. The finite set of natural restrictions on the optimization principle for certain information situations then allows us to identify a unique optimality principle which satisfies these restrictions.

This paper describes a different approach to the problem. The general form of information about ϑ, or as we shall express it, uncertainty on Θ, may be specified for a sufficiently wide class of optimality principles in such a way that any uncertainty will generate one and only one optimality principle in this class and each such principle will be generated by some particular uncertainty.

More formally, we fix one arbitrary non-empty set Θ and assume that all the decision problems under consideration belong to the class $\mathbf{Z} \in (\Theta) = \{Z \in dlbZ\colon \Theta_Z = \Theta\}$.

Definition. The term *optimality principle* is used to describe any mapping γ defined on $\mathbf{Z}(\Theta)$ which associates a certain real bounded function $L_Z^* = \gamma(Z)$, L_Z^*: $\mathbf{D}_Z \rightarrow \mathbb{R}$, with each decision problem Z. An optimality principle is said to belong to a class Γ if it satisfies the following conditions:

(C1) If $Z_i = (\Theta, \mathbf{D}_i, L_i)$, $d_i \in \mathbf{D}_i$ $(i = 1,2)$, and $L_1(\vartheta, d_1) \leq L_2(\vartheta, d_2)$ $(\vartheta \in \Theta)$, then $L^*_{Z_1}(d_1) \leq L^*_{Z_2}(d_2)$;

(C2) If $Z_i = (\Theta, \mathbf{D}, L_i)$ $(i = 1,2)$, $a, b \in \mathbf{R}$, $a \geq 0$, and $L_1(\vartheta, d) = aL_2(\vartheta, d) + b$ $(\vartheta \in \Theta, d \in \mathbf{D})$, then $L^*_{Z_1}(d) = aL^*_{Z_2}(d) + b$ $(d \in \mathbf{D})$;

(C3) If $Z = (\Theta, \mathbf{D}, L)$, $d_1, d_2, d_3 \in \mathbf{D}$, and

$$L(\vartheta, d_1) + L(\vartheta, d_2) = 2L(\vartheta, d_3), \quad (\vartheta \in \Theta) \tag{1}$$

then

$$L^*_Z(d_1) + L^*_Z(d_2) \geq 2L^*_Z(d_3) \quad . \tag{2}$$

Here $L^*_Z(d)$ should be interpreted as an *a priori* estimate of the loss associated with decision d.

The purpose of the first two conditions should be clear, so we shall consider only (C3) in detail.

Let us compare the two decision problems $Z = (\Theta, \mathbf{D}, L)$ and $\tilde{Z} = (\tilde{\Theta}, \tilde{\mathbf{D}}, \tilde{L})$, where $\tilde{\Theta} = \Theta \times \Theta$, $\tilde{\mathbf{D}} = \mathbf{D} \times \mathbf{D}$

$$\tilde{L}((\vartheta_1, \vartheta_2), (d_1, d_2)) = L(\vartheta_1, d_1) + L(\vartheta_2, d_2) \quad .$$

Interpreting $\tilde{Z}$ as a two-stage decision problem under the conditions described above, it is natural to assume that

$$L^*_{\tilde{Z}}(d_1, d_2) = L^*_Z(d_1) + L^*_Z(d_2) \quad .$$

Inequality (2) thus implies that it is better to choose d_3 twice than to choose d_1 first and then d_2. What is the reason for this? We have from (1) that

$$[L(\vartheta_1, d_3) + L(\vartheta_2, d_3)] - [L(\vartheta_1, d_1) + L(\vartheta_2, d_2)] =$$

$$= [L(\vartheta_2, d_1) + L(\vartheta_1, d_2)] - [L(\vartheta_1, d_3) + L(\vartheta_2, d_3)] \quad .$$

This may be interpreted as follows. For any pair $\{\vartheta_1, \vartheta_2\}$, the loss l associated with the choice (d_3, d_3) does not depend on the order of ϑ_1, ϑ_2. However, if (d_1, d_2) are chosen then the associated loss *does* depend on the order of ϑ_1, ϑ_2, being $l + \Delta$ in one case and $l - \Delta$ in the other. Thus C3 leads us to a guaranteed result.

Now we need to introduce some additional notation. Let $PF(\Theta)$ be the set of all finite-additive probabilities on Θ, i.e.,

$$PF(\Theta) = \{\psi \in (2^{\Theta} \to [0,1]) : \psi(\Theta) = 1, \psi(A \cup B) =$$

$$= \psi(A) + \psi(B \setminus A) \ \forall A, B \subset \Theta\} \quad .$$

In addition let M be the set of all bounded real functions defined on Θ, $\mathbf{Q}$ be the set of all finite partitions of Θ into non-intersection subsets, and $\geq$ be a natural ordering relation in $\mathbf{Q}$ such that

$$Q_1 \_ Q_2 \Longleftrightarrow \forall q_1 \in Q_1 \quad q_2 \in Q_2 \ (q_1 \subset q_2) \quad .$$

It is clear that for any $\varepsilon > 0$ there exists a $Q_\varepsilon \in \mathbf{Q}$ such that

$$\left| \sum_{q \in Q} f(\vartheta_q')\psi(q) - \sum_{q \in Q} f(\vartheta_q'')\psi(q) \right| < \varepsilon \quad ,$$

where $Q \_ Q_\varepsilon$, ϑ_q', $\vartheta_q'' \in q$, $f \in M$, $\psi \in \mathrm{PF}(\Theta)$, and therefore the limit of integral sums on the ordered set $(\mathbf{Q}, \_)$ exists for all $f \in M$, $\psi \in PF(\Theta)$. Let this be denoted by $\int f(\vartheta)\psi(d\tilde{\vartheta})$.

Definition 2. A non-empty subset $\Psi \subset PF(\Theta)$ represents some uncertainty on Θ.

THEOREM 1. *It is possible to associate some uncertainty* $\Psi \in \mathrm{PF}(\Theta)$ *with each* $\gamma \in \Gamma$ *such that for* $Z \in \mathbf{Z}(\Theta)$, $L_Z^* = \gamma(Z)$ *we have*

$$L_Z^*(d) = \sup_{\psi \in \Psi} \int L_Z(\vartheta, d)\psi(d\tilde{\vartheta}), \quad d \in \mathbf{D}_Z \quad . \tag{3}$$

The converse is also true: if Ψ represents some uncertainty on Θ and the mapping $\gamma: Z \mapsto L_Z^* \ (Z \in \mathbf{Z}(\Theta))$ is given by formula (3), then $\gamma \in \Gamma$.

Definition 3. The ordered pair $S = (Z_S, \Psi_S)$, where Z_S is a decision problem and Ψ_S is some uncertainty on Θ_{Z_S}, is called a *decision system*. The class of all decision systems will be denoted by $\mathbf{S}$ and we shall write Θ_S, $\mathbf{D}_S$, L_S instead of Θ_{Z_S}, $\mathbf{D}_{Z_S}$, L_{Z_S} respectively.

Interpreting a decision system as a mathematical model of a situation in which a decision is to be made, we will associate with any decision system S a function

$$L_S^*: \mathbf{D}_S \to \mathbf{R}, \ L_S^*(d) = \sup_{\psi \in \Psi_S} \int L_S(\vartheta, d)\psi(d\tilde{\vartheta})$$

and a value

$$\rho(S) = \inf_{d \in \mathbf{D}_S} L_S^*(d) \quad ,$$

which will be called the "risk" by analogy with the Bayesian case (Ψ_S consists of a single element).

Now let the decision system be given and let it be possible to observe the value of some function $\eta: \Theta_S \to Y$ before taking the decision. Clearly we need only $\eta^{-1}(y) \subset \Theta_S$

and not the value $y \in Y$ in order to make a decision. We can therefore assume that the result of any observation is a corresponding subset of the set Θ_S and introduce the following definition:

Definition 4. Any function $C\colon \Theta_S \to 2^{\Theta_S}$ such that (i) $\vartheta \in C(\vartheta)$ for all $\vartheta \in \Theta_S$ and (ii) sets $C(\vartheta)(\vartheta \in \Theta_S)$ either do not intersect at all or coincide completely, is said to be an observation scheme on Θ_S. The set of all such observation schemes will be denoted by $\mathbf{C}(\Theta_S)$, and C_1 is said to be sufficient for C_2 if $C_1(\vartheta) \subset C_2(\vartheta)$ $(\vartheta \in \Theta_S)$.

Let $S \in \mathbf{S}$, $C \in \mathbf{C}(\Theta_S)$, $S^C \in \mathbf{S}$, $\Theta_{S^C} = \Theta_S$, $\Delta_{S^C} = (C(\Theta_S) \to \Delta_S)$, and $L_{S^C}(\vartheta, \delta) = L_S(\vartheta, \delta(C(\vartheta)))$ $(\vartheta \in \Theta_S, \delta \in \Delta_{S^C})$, i.e., S^C is another decision system, where Δ_{S^C} is the set of all functions δ which associate a certain decision with each observation.

The difference $\rho(S) - \rho(S^C)$ will be written INF (C/S) and will be called the *informativity* of the observation scheme C with respect to the decision system S. The reason for this term is that the informativity of a stochastic experiment in a Bayesian decision system (see [4,5]) is given by such a difference, which in this case is unique. (Note: it may be multiplied by a positive real number.)

THEOREM 2. *Let* $S \in \mathbf{S}$; C, C_*, $C^* \in \mathbf{C}(\Theta_S)$; $C_*(\vartheta) = \Theta$, and $C^*(\vartheta) = \{\vartheta\}$ $(\vartheta \in \Theta_S)$. *Then*

$$0 = \mathrm{INF}\,(C_*/S) \le \mathrm{INF}\,(C/S) \le \mathrm{INF}\,(C^*/S) =$$

$$= \rho(S) - \sup_{\psi \in \Psi_S} \int \Big(\inf_{d \in \mathbf{D}_S} L_S(\vartheta, d) \psi(d\tilde{\vartheta}) \quad .$$

Therefore the right-hand side of this chain of inequalities is the maximum possible informativity of the observation scheme with regard to S, or, in other words, that part of the *a priori* risk which is introduced by observation.

THEOREM 3. *Let* $\mathbf{S}(\Theta) = \{S \in \mathbf{S}\colon \Theta_S = \Theta\}$ and C_1 $C_2 \in \mathbf{C}(\Theta)$. *Then to obtain* INF $(C_1/S) \ge$ INF (C_2/S) $\forall S \in \mathbf{S}(\Theta)$ *it is necessary and sufficient that* C_1 *is sufficient for* C_2.

An observation system is said to be *optimal* for decision system S if $C \in \mathbf{C}(\Theta_S)$ and for any other observation system $C_1 \in \mathbf{C}(\Theta_S)$ the following relations hold:

$$\mathrm{INF}\,(C/S) \ge \mathrm{INF}\,(C_1/S) \quad .$$

$$\mathrm{Card}\,(C(\Theta_S)) \le \mathrm{Card}\,(C_1(\Theta_S)) \quad ,$$

where Card (A) is the cardinality of set A.

THEOREM 4. *There is an optimal observation system for any decision system and each observation system is optimal for some decision system.*

REFERENCES

1. N.N.Moiseev (Ed.). *The State-of-the-Art of Operations Research Theory*. Nauka, Moscow, 1979 (in Russian).

2. R.D. Luce and H. Raiffa. *Games and Decisions*. Inostrannaya Literatura, Moscow, 1960 (in Russian).

3. R.I. Truhaev. *Decision-Making Models under Conditions of Uncertainty*. Nauka, Moscow, 1981 (in Russian).

4. V.I. Ivanenko and V.A. Labkovsky. The uncertainty function in Bayesian systems (in Russian). *Doklady Akademii Nauk SSSR*, 248(2)(1979)307–309.

5. V.A. Labkovsky. The information content of stochastic experiments (in Russian). *Kibernetika*, (4)(1982)90–93.

UNCERTAINTY IN STOCHASTIC PROGRAMMING

Vlasta Kaňková, Institute of Information Theory and Automation Czechoslovak Academy of Sciences, Pod vodárenskou věží 4 182 08 Praha 8-Libeň Czechoslovakia

1. INTRODUCTION

As it is well known many optimization problems with random parameters arising in practice can be treated as optimization problems with respect to the expectation of some random function. In this case both the optimum solution and the optimum value will depend on the distribution function of the random parameters occurring in the considered problem. Consequently, the distribution function can be viewed as a parameter (important for the stability) of the original problem.

It is well known from the literature, e.g. Kall (1976), Dupačová (1976), Tsybakov (1981), and Kaňková (1974, 1978), that in many cases a small variations of the distribution function evokes only a small changes of the optimum value. This property was employed e.g. for finding estimates on the optimum solution of problems with unknown distribution function or for the construction of approximation methods for solving two-stage stochastic programming problems.

In the present paper we shall investigate stability of the above problems. First we present a simple example showing that a small pertubation of the distribution function may cause a large deviation of the optimum value even if the original optimalized function is bounded and Lipschitz. Then we present some general conditions guaranteeing stability of the considered problems in the class of all distribution functions. Finally, some specific cases are discussed.

Let $(\Omega, \mathcal{S}, P)$ be a probability space,

$\mathcal{K} \subset E_n$ be a non-empty compact set,

$\xi = \xi(\omega)$ be an s-dimensional random vector defined on $(\Omega, \mathcal{S}, P)$,

$F(z)$ be the distribution function of ξ,

$g(x,z)$ be a continuous bounded function defined on $\mathcal{K} \times E_s$.

(E_k denotes a k-dimensional Euclidean space.)

Under this conditions $g(x, \xi)$ for every $x \in \mathcal{K}$ is a random variable defined on $(\Omega, \mathcal{S}, P)$. So we can consider $Eg(x, \xi)$ for every $x \in \mathcal{K}$, where E is the operator of mathematical expectation. We can formulate the stochastic optimization problem in which the optimum is sought with respect to the mathematical expectation as to find

$$\max_{x \in \mathcal{K}} Eg(x, \xi). \tag{1}$$

Further, we shall denote by $\mathcal{F}_s$ the space of s-dimensional distribution functions - this is the space of all functions satisfying the necessary and sufficient conditions to be distribution functions.

In this paper we shall study the stability of (1) with respect to some topologies in $\mathcal{F}_s$. Especially, we shall consider the topology given by the Kolmogorov metric

$$\rho_{\mathcal{F}}(G_1, G_2) = \sup_{z \in E_s} |G_1(z) - G_2(z)|$$

for all $G_1(z), G_2(z) \in \mathcal{F}_s$.

Remark. Kall and Stoyan (1982) deal with a similar problem with respect to the L_2 metric.

At the end of this part we shall introduce a simple example showing that the stability does not hold generally in the space $(\mathcal{F}_s, \rho_{\mathcal{F}})$. More precisely we shall find $G_N(z), G(z) \in \mathcal{F}_s$, $N = 1,2,\dots$ a and bounded continuous and Lipschitz functions $g_N(x,z)$, $N = 1,2,\dots$ defined on $\mathcal{K} \times E_s$ such that for some $K \in E_1$, $K > 0$

$$\sup_{z \in E_s} |G_N(z) - G(z)| \underset{(N\to\infty)}{\to} 0$$

$$\left| \max_{x \in \mathcal{K}} \int_{E_s} g_N(x,z)\, dG_N(z) - \max_{x \in \mathcal{K}} \int g_N(x,z)\, dG(z) \right| \geqq K \tag{2}$$

simultanously.

To obtain (2) it is enough to set s = 1,

$$G(z) = z \quad \text{for } z \in \langle 0,1\rangle ,$$
$$= 0 \quad \text{for } z < 0 ,$$
$$= 1 \quad \text{for } z > 1 ,$$

$$G_N(z) = Nz/N+1 \quad \text{for } z \in \langle 0,(N+1)/N\rangle ,$$
$$= 0 \quad \text{for } z < 0$$
$$= 1 \quad \text{for } z > (N+1)/N$$

$$g_N(x,z) = 1 \quad \text{for } z < 1 , x \in \mathcal{K} ,$$
$$= (2N^2+2N)z + 1 - 2N^2 - 2N \quad \text{for } z \in \langle 1,2\rangle , x \in \mathcal{K} ,$$
$$= 2N^2+2N+1 \quad \text{for } z > 2 , x \in \mathcal{K} ,$$

where $\mathcal{K} \in E_n$ is an arbitrary non-empty compact set.

Surely, in this case it is easy to see

$$\int_{E_s} g_N(x,z)\, dG(z) = 1 , \quad \int_{E_s} g_N(x,z)\, dG_N(z) = 2$$

for all N = 1,2, ... ,$x \in \mathcal{K}$

and simultanously

$$\sup_{z \in E_s} |G_N(z) - G(z)| = \frac{1}{N+1} \quad \text{for all } N = 1,2 \ldots$$

2. SOME AUXILIARY ASSERTIONS

Let 1. $f(z)$ be a real valued uniformly continuous bounded function defined on E_s,

then for an arbitrary $\varepsilon > 0$ there exists δ_ε such that

$$\rho_s(z,z') < \delta_\varepsilon \Rightarrow |f(z) - f(z')| < \varepsilon/2 \tag{3}$$

where ρ_s denotes the Euclidean norm in E_s .

Further, according to 1 there exists $a,b \in E_s$ $a = (a_1,\ldots,a_s)$, $b = (b_1,\ldots,b_s)$ such that

$$\int_{E_s - I\langle a-\delta_\varepsilon, b+\delta_\varepsilon\rangle} M \, dF(z) < \varepsilon/4 \qquad (4)$$

for $I_{\langle a,b\rangle} = \{z \in E_s, z = (z_1,\ldots,z_s) : a_i \leq z_i \leq b_i, \; i=1,2,\ldots n\}$

and $M \in E_1$ fulfilling the condition $|f(z)| \leq M$, $z \in E_s$.

If now we define the points $z_{i,j}$, z^k, $i=1,2,\ldots,N_i$ $i = 1,2,\ldots,s$, $k = 1,2,\ldots,N$ by

$$a_i = z_{i,1}, \; b_i > z_{i,N-1}, \; b_i < z_{i,N_i}, \; z_{ij} < z_{i,j+1}$$
$$|z_{i,j+1} - z_{i,j}| < \delta_\varepsilon/s, \; j=1,2,\ldots,N_i-1, \; j=1,2,\ldots,s,$$
$$z^k = (z_1^k,\ldots,z_s^k) \text{ where } z_i^k = z_{i,j} \quad \text{for some } j = \{1,2,\ldots,N_i\}$$

$j = 1,2,\ldots,s$ then we can, for every $z \in E_s$, define the discrete function $F_N(z)$ such that

2. the discontinuity points of $F_N(z)$ can be only the points z^k,

3. $F(z^k) = \lim_{\substack{z \to z^k \\ z_i \searrow z_i^k}} F_N(z)$.

We shall present the following lemma.

Lemma 1. If conditions 1,2,3 are fulfilled then

$$\left| \int_{E_s} f(z)\, dF(z) - \int_{E_s} f(z)\, dF_N(z) \right| < \varepsilon .$$

Proof. If we define the function $f_N(z)$ by

$$f_N(z) = f(z^k) \Leftrightarrow z = (z_1,\ldots,z_s), \quad z_i \in (z^k_{i,j-1}, z^k_{i,j} \rangle$$
$$j = 2,\ldots,N, \quad z^k = (z^k_{1,j1},\ldots,z^k_{s,js})$$

it is easy to see that

$$|f_N(z) - f(z)| < \varepsilon/2 \qquad (5)$$

and also

$$\int_{I_{\langle a,b\rangle}} f_N(z)\, dF(z) = \int_{I_{\langle a,b\rangle}} f(z)\, dF_N(z) \, .$$

But from this we get

$$\left| \int_{I_{\langle a,b\rangle}} f(z)\, dF_N(z) - \int_{I_{\langle a,b\rangle}} f(z)\, dF(z) \right| \leqq \int_{I_{\langle a,b\rangle}} | f(z) - f_N(z) | dF(z) < \varepsilon/2 \qquad (6)$$

But as from (4) it follows that

$$\left| \int_{E_s - I_{\langle a,b\rangle}} f(z)\, dF(z) \right| < \varepsilon/4 \quad \text{and} \left| \int_{E_s - I_{\langle a,b\rangle}} f(z)\, dF_N(z) \right| < \varepsilon/4$$

Using the triangular inequality the assertion of the Lemma follows from (4),(6) and(7).

Remark. Under rather stronger conditions a similar assertion was proved in Kaňková (1980), where it was used for the approximative method of the two-stage stochastic nonlinear programming problems.

If further the functions $\underline{F}(z)$, $\bar{F}(z)$ are defined by

$$\begin{aligned} \underline{F}(z) &= F(z) - P\{\xi_i \in \langle z_i - \delta_\varepsilon / s, z_i), \; i = 1,2,\ldots,s\} \\ \bar{F}(z) &= F(z) + P\{\xi_i \in \langle z_i, z_i + \delta_\varepsilon / s), \; i = 1,2,\ldots,s\} \end{aligned} \qquad (8)$$

for δ_ε given by (3),

then we can formulate the following assertion.

Lemma 2. Let condition 1 be fulfilled and let $\varepsilon > 0$ be arbitrary. If the functions $\underline{F}(z)$, $\bar{F}(z)$ are defined by (8) then

$$\left| \int_{E_s} f(z)\, dG(z) - \int_{E_s} f(z)\, dF(z) \right| < 4\varepsilon$$

for every distribution function $G(z) \in \langle \underline{F}(z), \bar{F}(z) \rangle$, $z \in E_s$.

Proof. We can take without loss of generality the case $s = 2$. If we define the points $z'_{i,j}$, z'^k, $j = 1,2,\ldots,N_i$, $i = 1,2$ $k = 1,2,\ldots,N$ such that $a_i = z'_{i,1}$, $z'_{i,N_i-1} \leqq b_i$, $b_i \geqq z'_{i,N_i}$,

$$|z'_{i,j+1} - z'_{i,j}| = \delta_\varepsilon / s, \quad z_{i,j} - z'_{i,j} = \delta_\varepsilon / 2s, \quad z'_{i,j} - z_{i,j} = \delta/2$$

$$z'^k = (z_1^k, \ldots, z_s^k) \text{ where } z'^k_i = z'_{i,j} \quad \text{for some } j \in \{1,2,\ldots,N_i\}, \; i = 1,2 \, .$$

We can easily see that

$$F(z_{1,i}, z_{2,j}) \leq G(z'_{1,i+1}, z'_{2,j+1}) \leq F(z_{1,i+1}; z_{2,j+1})$$

for an arbitrary distribution function $G(z) \in \langle \underline{F(z)}, \bar{F}(z) \rangle$, $z \in E_s$, $i = 1,2,\dots,N_1$, $j = 1,2,\dots,N_2$.

According to Kolmogorov's theorem there exist two-dimensional discrete random vectors $Y = (Y_1, Y_2)$, $V = (V_1, V_2)$, $U = (U_1, U_2)$ defined on (Ω, S, P) and having jump points only $(z_{1,i}, z_{2,j})$, $(z_{1,i}, z_{2,j})$, $(z'_{1,i}, z'_{2,j})$, $(z'_{1,i}, z'_{2j})$ respectively $i = 1,2,\dots,N_1$, $j = 1,2,\dots,N_2$ such that

$$P\{Y_1 = z_{1,i}, Y_2 = z_{2,j}\} = P\{\xi_1 \in (z_{1,i-1}, z_i\rangle, \xi_2 \in (z_{2,j-1}, z_{2,j}\rangle\} = p_{i,j},$$

$$P\{V_1 = z_{1,i}, V_2 = z_{2,j}\} = p_{i,j} - p'_{i,j}, \tag{9}$$

$$P\{V_1 = z'_{1,i}, V_2 = z'_{2,j}\} = p'_{i,j},$$

$$P\{U_1 = z'_{1,i}, U_2 = z'_{2,j}\} = P\{\xi_1 \in (z'_{1,i-1}, z'_{1,i}\rangle, \xi_2 \in (z'_{2,j-1}, z'_{2,j}\rangle\}$$

$$= p_{i-1,j-1} - p'_{i-1,j-1} + p'_{i,j}$$

where $\xi = (\xi_1, \xi_2)$ is a random vector having the distribution function $G(z)$. Further in (9) we set $z_{1,0} = z_{2,0} = z'_{1,0} = z'_{2,0} = -\infty$ and, of course, in this case the intervals of the type $(\cdot, \cdot)$ instead of $\langle \cdot, \cdot)$.

Using the assertion of Lemma 1 we get by the triangular inequality

$$\left| \int_{E_s} f(z)\, dG(z) - \int_{E_s} f(z)\, dF(z) \right| \leq 4\varepsilon$$

Remark. Taking $s = 1$ we get

$$\underline{F(z)} = F(z - \delta_\varepsilon), \quad \bar{F}(z) = F(z + \delta_\varepsilon) \tag{10}$$

Of course the relation (10) can be utilized in case of the stochastically independent components of the random vector or in case of a separable function $f(z)$.

3. MAIN RESULTS

Theorem 1. Let $\varepsilon > 0$ be arbitrary. If

1. $g(x,z)$ is a real valued bounded continuous function defined on $E_n \times E_s$,

2. $g(x,z)$ is for every $x \in \mathcal{K}$ a Lipschitz function of $z \in E_s$ with the Lipschitz constant L not depending on x,

3. the functions $\underline{F}(z)$, $\bar{F}(z)$ are defined by (8) for $\delta_\varepsilon = \varepsilon / 2L$ then for every distribution function $G(z) \in \langle \underline{F}(z), \bar{F}(z) \rangle$, $z \in E_s$

$$\left| \max_{x \in \mathcal{K}} \int_{E_s} g(x,z)\, dG(z) - \max_{x \in \mathcal{K}} \int_{E_s} g(x,z)\, dF(z) \right| < 4\varepsilon$$

Proof. As it is easy to see that under the assumptions 3 relation (3) is fulfilled, hence the assertion of Theorem 1 follows immediately from Lemma 2.

Theorem 2. Let the assumptions of Theorem 1 be fulfilled. Let, further, there exists the probability density $\varphi(z)$ corresponding to the distribution function $F(z)$. If $G(z)$ is an arbitrary distribution function for which

$$\rho_{\mathcal{F}}(F,G) < \eta , \quad \eta \leq [\inf_{z \in I_{\langle a,b \rangle}} \varphi(z)] \, (\delta_\varepsilon / 2)^s$$

where $I_{\langle a,b \rangle}$ is given by 4 , then

$$\left| \max_{x \in \mathcal{K}} \int_{E_s} g(x,z)\, dG(z) - \max_{x \in \mathcal{K}} \int_{E_s} g(x,z)\, dF(z) \right| < 4\varepsilon .$$

Theorem 2 presents the sufficient stability conditions for the problem (1) with respect to the metric space $(\mathcal{F}_s, \rho_{\mathcal{F}})$. Because the Lévy metric is stronger then the Kolmogorov one we get also sufficient conditions of stability in the modified Hampell's sense.

4. APPLICATIONS

It is well known from the literature that the results similar to those presented in this paper can be utilized in a few directions. We can, for example, construct a solution approximative method of two-stage stochastic problems Kall (1976), Kall and Stoyan (1982). Further we can find the estimation of the optimal solution and the optimal value in the case of the unknown probability laws. However till now, this was practically possible only on the independent random sample basis. Using this paper results under some conditions we can get the estimates on the basis of dependent sequence too. But as this problem is rather extensive we shall omit it.

In this paper we shall consider the case of unknown location parameter. We shall assume without loss of generality s=1. If $Y \subset E_1$ is a non-empty set and if

$$F(z) = F_o(z-y) \tag{11}$$

where $F_o \in \mathcal{F}_1$, $y \in Y$ then y presents the location parameter in the class of the distribution function $\mathcal{F}_y$ given by (11). We shall consider the problem (1) under the assumption $F \in \mathcal{F}_y$. We shall assume that y is inknown and that we can find an estimation of the optimal value, setting some statistic estimation $\hat{y}$ instead of the theoretical value y.

If we note

$$E_y g(x,z) = \int_{E_s} g(x,z)\, dF_o(z-y) = \int_{E_s} g(x,z)\, dF_y(z),$$

the aim of this part is to construct upper bounds on

$$p\left\{ \left| \max_{x \in \mathcal{K}} E_{\hat{y}} g(x,z) - \max_{x \in \mathcal{K}} E_y\, g(x,z) \right| > \varepsilon \right\}.$$

The next theorem follows from Theorem 1.

Theorem 3. Let Y be a compact set. Let the assumptions of Theorem 1 be fulfilled. Let the distribution function $F \in \mathcal{F}_y$, $F = F_{y_o}$ for an y_o. If $\hat{y}_N$ is a statistical estimation y_o for which

$$P\left\{ \left| \hat{y}_N - y_o \right| \right\} \geq 1-\varepsilon$$

then

$$P\left\{ \left| \max_{x \in \mathcal{K}} E_y\, g(x,z) - \max_{x \in \mathcal{K}} E\, g_{\hat{y}_N}(x,z) \right| < 4\delta\, L \right\} \geq 1-\varepsilon.$$

References

Billingsley, P. (1977). Convergence of Probability Measures. Wiley, New York.

Birge, J. and Wets, R.J.-B. (1983). Designing Approximation Schemes for Stochastic Optimization Problems, in Particular for Stochastic Programs with Recourse. WP-83-111. International Institute for Applied Systems Analysis, Laxenburg, Austria.

Dupačová, J. (1976). Experience in stochastic programming models. IXth International Symposium on Mathematical Programming, Budapest.

Huber, P.J. (1981). Robust Statistics. Wiley, New York.

Kall, P. (1976). Stochastic Linear Programming. Springer, Berlin - Heildeberg - New York.

Kall, P. and Stoyan, D. (1982). Solving stochastic programming problems with recourse including error bounds. Math. Operationsforsch. Statist., Ser. Optimization 13(3): 431-447.

Kaňková, V. (1974). Optimum solution of a stochastic optimization problem with unknown parameters. In J. Kožešník and L. Kubát (Ed.), Trans. of the Seventh. Prague Conference, Prague 1974. Academia, Prague 1978.

Kaňková, V. (1978). An approximative solution of a stochastic optimization problem. In J. Kožešník and M. Driml (Ed.), Trans. of the Eighth Prague Conference, Prague 1978. Academia, Prague 1978.

Kaňková, V. (1974). Approximative solution of problems of two--stage stochastic nonlinear programming. In Czech . Ekonomicko-matematický obzor 16 (1): 64-76.

Tsybakov, G.B. (1981). Error bounds for the method of minimization of empirical risk. Problemy Peredači Informacii, 17 (1): 50-61.

STOCHASTIC PROGRAMMING MODELS FOR SAFETY STOCK ALLOCATION

Kelle Péter
Computer and Automation Institute
Hungarian Academy of Sciences
Budapest, P.O.Box 63
H-1502

1. THE PRODUCTION-INVENTORY SYSTEM

In the production line considered the subsequent phases of processing form a multi-stage production-inventory system with internal stocking. The raw material is processed successively at the N facilities before reaching the costumer having a stationary final product demand. There is a final product store I_{N+1} and an internal store I_i before each stage of processing $(i=1,2,\dots,N)$ including raw material store I_1. For each store a safety stock M_i $(i=1,\dots,N+1)$ is planned as initial stock for a production cycle to ensure the continuous supply for the whole production line. It is necessary because of the uncertainties in demand and in production which may often be disturbed by random factors such as machine failures, faulty products, breakdowns, etc. In this case it is a great difficulty to provide for continuous production with reasonable in-process inventories. Stochastic programming models are formulated for the allocation of the safety stocks on an optimal way defined later.

The uncertainty in material requirements planning systems was considered and different buffering policies were given by Berry and Whybark (1977), Whybark and Williams (1976), Miller (1979) and New (1975). The effect of random demand in internal stocking was analysed recently in the papers of Schmidt and Nahmias (1981), Lambrecht et. al. (1982) and De Bodt and Graves (1982). The optimal safety stock policy for a continuous deterministic production process and Poisson demand was derived in a recursive form by Axsäter and Lundell (1983). We consider a simple multi-stage batch production system but both demand and production process may have random factors.

2. THE STOCHASTIC PROGRAMMING MODELS

The models of inventory allocation are particularly important in case of considerable random influence in production and demand. The papers referred to in the first chapter consider only the effect of random demand.

For production managers it is a great problem to provide the continuous supply of customers and production stages with a reasonable law level of raw material, in-process and final product stocks. We intend to allocate the safety stocks for the system of these stores in such a way that they should jointly provide a high service level of the whole production line. The service level is measured by the probability of non-interruption in supply of processing and in demand satisfaction.

In the first stochastic programming model a constrained investment capacity K has to be allocated in safety stock among the different stores in such a way that it should provide for the maximal service level of production and supply. On production level i there is no interruption during the production cycle if the following inequality holds for all $0 \leq t \leq T$:

$$g_t^i \leq M_i + g_t^{i-1} ,$$

where g_t^i denotes the cumulative amount processed in the period $[0,t]$ on level i. Here g_t^1 means the delivery of raw material and g_t^{N+1} means the external demand until time t. If a unit of safety stock on level i has the investment cost d_i, this model can be formulated in the following way:

$$\text{maximize} \quad P\left(g_t^i \leq M_i + g_t^{i-1} \ ,\ 0 \leq t \leq T \ ,\ i=1,2,\ldots,N+1\right)$$

$$\text{subject to} \quad \sum_{i=1}^{N+1} d_i M_i \leq K \ .$$

In the second stochastic programming model a prescribed high service level $1-\varepsilon$ of the production-supply system should be ensured with minimal investment in safety stocks:

$$\text{minimize} \quad \sum_{i=1}^{N+1} d_i M_i$$

$$\text{subject to} \quad P\left(g_t^i \leq M_i + g_t^{i-1}\ ,\ 0 \leq t \leq T,\ i=1,2,\ldots,N+1\right) \geq 1-\varepsilon .$$

The first model is a typical allocation model, which can be used for safety stock planning in a static case when the total investment is prescribed. In case of considerable changes in demand, production or raw material supply system the second model should be preferred to safety stock planning. The service level should be fixed using the results of the static model applied to the previous planning periods. On the joint application and analysis of the above two models a decision support system of safety stock planning can be built.

3. SOLUTION OF THE MODELS

Exact and approximate solutions of the above stochastic programming models have their own importance in practice. The approximations are often satisfactory and the exact solution method is used only for the estimation of the failure. The first step in the solution of the stochastic programming problem is to derive the probability of the continuous supply in connection with two consecutive levels and given initial stock. Here we can describe it only for the most important models of delivery, batch processing and demand.

3.1. The raw material stock depends on the lead-time of delivery and on the processing at the first level realized usually in fixed batch sizes with certain periodicity. Let n_1 denote the number of these batches in the production cycle $[0,T]$ considered. The setup times $t_1^{(1)} < t_2^{(1)} < \ldots < t_{n_1}^{(1)}$ are uniformly distributed, but owing to random disturbances in production, such as machine failures, break downs, faulty material etc., they are random.

For the raw material store we have a random lead-time L, characterized by the known distribution function $H(L)$. The random setup times of the first processing level $t_1^{(1)} < t_2^{(1)} < \dots < t_{n_1}^{(1)}$ are assumed to correspond to a random sample of the uniform distribution on $[0,T]$ arranged in increasing order. The batch sizes are fixed and uniform in Model I. of batch processing. The total amount processed in $[0,T]$ is Q. By choosing appropriate units of time and amount we may assume further in this paper that $T=1$ and $Q=1$ for the sake of simple notations. The initial stock $k/n_1 \leq M_1 < (k+1)/n_1$ ensures the continuous raw material supply with probability

$$P\left(t_k^{(1)} \leq L\right) = n_1 \binom{n_1-1}{k-1} \int_0^{\infty} \int_0^{L} t^{k-1} (1-t)^{n_1-k} \, dt \, dH(L) .$$

3.2. The in-process stocks are characterized by the processing of two consecutive levels. The input (delivery) of the internal store I_i is the amount processed on level $i-1$ and the demand is represented by the scheduled processing of level i. There are usually fixed batch sizes on both levels. The number of batches processed in $[0,T]$ on level i is denoted by n_i. For the optimal batch sizes usually the equation $n_i = kn_{i-1}$ holds with an integer k. The consequence of the random influences in processing is that the processing time of the different batches with the same size may be different, the setup times $t_1^{(i)} < t_2^{(i)} < \dots < t_{n_i}^{(i)}$ of level i are random.

For the internal stores both input and demand can be approximated by Model I. of batch processing. Often not the whole amount of a batch can be immediately processed on the next stage because of quality problems which have to be corrected. Afterwards the repaired quantity is added to the next batch. In this case the random batch size model of Prékopa (1965) can be applied to approximate the input process. Here the batch size has a deterministic fraction v $(0 \leq v \leq 1)$ and the fraction $(1-v)$ is randomly subdivided among the batches. The subdivision happens by $n-1$ random points which are uniformly distributed random points arranged in increasing order: $q_1 < q_2 < \dots < q_{n-1}$.

The cumulative amount processed during time $[0,t]$ $(0 \leq t \leq 1)$ can be expressed by

$$g_t = F_n(t,v) = \begin{cases} 0 & \text{if } 0 \leq t \leq t_1 \\ v\frac{k}{n} + (1-v)\,q_k & \text{if } t_k < t \leq t_{k+1},\ k=1,\dots,n-1 \\ 1 & \text{if } t_n < t \leq 1 \end{cases}$$

This Model II. of batch processing is the generalized version of Model I. which represents the special case $v=1$.

If levels $i-1$ and i are both processing according to Model I. the probability of continuous supply for given initial stock M_i can be expressed using the result of Gnedenko (1951) in the case when the number of batches is equal to n at both levels

$$P\left(g_t^i - g_t^{i-1} \leq M_i,\ 0 \leq t \leq 1\right) = P\left(F_n^i(t,1) - F_n^{i-1}(t,1) \leq M_i,\ 0 \leq t \leq T\right) = 1 - \frac{\binom{2n}{n-c}}{\binom{2n}{n}}.$$

Here the constant c is the smallest integer for which $c \geq Mn$. In the case of $n_i = k\,n_{i-1}$ we get the following probability using the result of Koroliuk (1955):

$$P\left(F_{n_i}^i(t,1) - F_{n_{i-1}}^{i-1}(t,1) \leq M_i,\ 0 \leq t \leq 1\right) =$$

$$= 1 - \frac{c}{\binom{n_i+n_{i-1}}{n_i}} \sum_{\substack{m=c \\ m \equiv c \,(\mathrm{mod}\,(k+1))}}^{n_i - n_{i-1} - \left[\frac{c}{k}\right]} \frac{1}{m} \binom{m}{\frac{m-c}{k+1}} \binom{n_i+n_{i-1}-m}{n_i - \frac{m-c}{k+1}}$$

where the constant c is the smallest integer for which $c \geq M_i n_i^2 / n_{i-1}$. For the general case of arbitrary integers n_i, n_{i-1} exact probability is not known. Here an approximation can be derived based on the asymptotic distribution of Smirnov (1944) in the form

$$P\left(F^{i}_{n_i}(t,1) - F^{i-1}_{n_{i-1}}(t,1) \leq M_i,\ 0\leq t\leq 1\right) \approx 1 - \exp\left[-\frac{2n_i n_{i-1}}{n_i+n_{i-1}} M_i^2\right],$$

which must be close to the exact value if n_i, n_{i-1} are great enough (greater than 10).

For the generalized Model II. a similar approximation has been published by Prékopa (1965)

$$P\left(F^{i}_{n_i}(t,v_i) - F^{i-1}_{n_{i-1}}(t,v_{i-1}) \leq M_i,\ 0\leq t\leq 1\right) \approx$$

$$\approx 1 - \exp\left[\frac{2n_i n_{i-1} M_i^2}{n_i+n_{i-1}+n_{i-1}(1-v_i)^2+n_i(1-v_{i-1})^2}\right]$$

3.3. <u>The final product store</u> has a stationary external demand, but at the time of stock planning the intensity of demand cannot be exactly forecasted. We consider a continuous demand with random intensity α. The final processing level is the input. It can be described also by the model of batch production with fixed batch sizes and random setup times. However, a part of each batch may be faulty product, which is realized only at the final quality control. In this case random batch sizes have to be considered from the point of view of demand satisfaction. Random batch sizes may occur also on other processing levels.

For the final product store we assume a continuous demand with random intensity α which has a known distribution function $G(\alpha)$. The setup times of the final processing level are uniformly distributed as in Model I. and II. but the random batch sizes are not necessarily uniform distributed. Using the statistical data of the final quality control we fit appropriate distributions. Let $H_k(x)$ denote the distribution function characterizing the cumulative amount of perfect final products of the first k batches, $k=1,2,\dots,n_N$. This Model III. of batch processing is a generalized version of Model II:

$$g_t^N = F_n^{\beta}(t) = \begin{cases} 0 & \text{if } 0 \le t \le t_1 \\ \sum_{i=1}^{k} \frac{\beta_i}{n} & \text{if } t_k \le t \le t_{k+1} \quad (k=1,\dots,n,\ t_{n+1} \equiv 1) \end{cases}$$

We have derived the following exact distribution for fixed α (>0), when the random vector $\underline{\beta} = (\beta_1, \dots, \beta_n)$ is interchangeable (see Kelle (1984 a))

$$P(\alpha t - F_n^{\beta}(t) \le M,\ 0 \le t \le 1) = p(M,\alpha) =$$

$$= 1 - \left(1 - \frac{M}{\alpha}\right)^n - \frac{M}{\alpha} \sum_{k=1}^{n} \binom{n}{k} \int_0^{n(\alpha - M)} \left(\frac{M + \frac{x}{n}}{\alpha}\right)^{k-1} \left(1 - \frac{M + \frac{x}{n}}{\alpha}\right)^{n-k} dH_k(x)$$

where $H_k(x)$ denotes the distribution function of $\sum_{i=1}^{k} \beta_i$ $(k=1,\dots,n)$.

This result can be applied also for Model II. of batch processing to derive the exact distribution of the service level for a continuous demand (see Kelle (1980)). By the total probability of continuous supply for random demand intensity α can be expressed in the form

$$P(\alpha t - F_{n_N}^{\beta}(t) \le M_{N+1},\ 0 \le t \le 1) = \int_0^{\infty} p(M_{N+1}, \alpha | \alpha = x)\, dG(x).$$

If α has normal distribution with parameters m and s $(s < 1/\sqrt{n})$ the following approximation has been derived based also on an asymptotic distribution (Kelle (1981)):

$$P(\alpha t - F_n(t) \le M,\ 0 \le t \le 1) \approx 1 - \exp\left[-2nM(M+1-m-nMs^2)\right].$$

3.4. The <u>solution of the stochastic programming models</u> formulated in chapter 2 can be reduced to deterministic nonlinear programming problems for independent joint constraint or objective function. This formulation is simple using the exact distributions described above. The consideration of stochastic dependence among the subsequent processing levels leads to a lot of difficulties in solution and in data requirement. Most of these problems can be solved only by using simulation technique and the statistical data of previous periods (see e.g. Prékopa and Kelle (1978)).

The approximate distributions described above enable us to construct a very simple, effective solution method based on Lagrange multiplier method. The algorithm is detailed in the paper Kelle (1984 b). The error of the approximate solution has been analysed compared with the exact solution in the cases when the letter was available, too. For the most part, especially when the number of batches is above 20 during the production cycle, the approximation is satisfactory. The relative error is less than 5 %. In other cases the approximation can serve as starting point for the solution of the nonlinear programming problem defined by the exact distributions.

4. APPLICATION

The stochastic programming models formulated for the optimal allocation of the investment in raw material, in-process and final product safety stocks could be reduced to simple deterministic problems in some important cases. These results have made it possible for us to solve these problems effectively and have given an efficient tool for decision makers in planning the safety stock allocation. The above models and methods have been tested on real production processes and have partly been introduced in practice in a rolling mill in Hungary.

REFERENCES

Axsäter, S. and Lundell, P. (1983). In-Process Safety Stocks, Working Paper, Linköping Institute of Technology.

Berry, W.L. and Whybark, D.C. (1977). Buffering Against Uncertainty in Material Requirement Systems, Discussion Paper No. 82, School of Business, Indiana University.

De Bodt, M.A. and Graves, S.C. (1982). Continuous-Review Policies for a Multi-Echelon Inventory System with Stochastic Demand, Working Paper 82-18, Katolicke Univ. Leuven.

Gnedenko, G. (1951). On the Maximal Difference of Two Empirical Distribution Function, Dokladi Akad. Nauk SSSR, 80 525-528. (in Russian)

Kelle, P. (1980). Reliability-Type Inventory Models for Random Delivery Process, in: Chikán A. ed. : Proc. First Int. Symp. on Inventories, Budapest, 1980. II. 385-395.

Kelle, P. (1981). Chance Constrained Inventory Models and their Application, Methods of OR 44. Athenäum/Hain, 607-616.

Kelle, P. (1984 a). On the Safety Stock Problem for Random Delivery Process, European Journal of OR. 17 1984 191-200.

Kelle, P. (1984 b). Safety Stock Planning in a Multi-Stage Production-Inventory System, Proc. of III.Int.Working Sem. on Production Economics, Igls, Austria, 1984. February to appear .

Koroliuk, D. (1955). On the Difference of Two Independent Empirical Distribution Functions, Isvestia Akad. Nauk SSSR, Ser. Math. 19, 81-96. (in Russian).

Lambrecht, M.R., Muckstadt, J.A. and Luyten, R. (1982). Protective Stocks in Multi-Stage Production Systems, Onderzocksrapport No.8215, Katholicke Univ. Leuven.

Miller, J.G. (1979). Hedging the Master Schedule, in: Ritzman et.al. ed. : Disaggregation: Problems in Manufacturing and Service Organizations, Martinus Nijhoff Publ., Boston.

New, C. (1975). Safety Stocks for Requirements Planning, Production and Inventory Management, 2nd quarter, 1975.

Prékopa, A. (1965). Reliability Equation for an Inventory Problem and its Asymptotic Solution, in: Prékopa, A. ed. : Coll. on Appl. of Math. to Economics, Akadémia, Budapest, 1965. 317-327.

Prékopa, A. and Kelle, P. (1978). Reliability Type Inventory Models Based on Stochastic Programming, Math. Programming Study 9, 43-58.

Schmidt, C.P. and Nachmias, S. (1981). Optimal Policy for a Two Stage Assembly System Under Random Demand, Working Paper, Univ. of Santa Clara, California.

Smirnov, N.V. (1944). Approximation of Random Amount by Empirical Data, Uspechi Math. Nauk 179-206.

Whybark, D.C. and Williams, J.C. (1976). Material Requirements Planning Under Uncertainty, Decision Sciences, 7.

DIRECT AVERAGING AND PERTURBED TEST FUNCTION METHODS FOR WEAK CONVERGENCE

Harold J. Kushner
Lefschetz Center for Dynamical Systems
Division of Applied Mathematics
Brown University, Providence, RI 02912, USA

ABSTRACT

Perturbed test function methods have been very useful for proving weak convergence of a sequence $\{x^{\varepsilon}(\cdot)\}$ of processes to a diffusion $x(\cdot)$, under quite weak conditions. The basic ideas are reviewed here. Typically one constructs two small perturbations, the first being a sum (or integral) and the second a double sum (or double integral). If the noise is 'state-dependent' and in other tricky cases, it can sometimes be difficult to verify the conditions on the second perturbation. We also discuss a method which 'averages' only the first perturbation, and which is often quite easy to use - it does not require construction of the second perturbation. The conditions are often quite easy to verify - even in the complicated 'state-dependent' noise case.

1. INTRODUCTION

Typically, in control and communication theory models, one is given a nonlinear system, with possibly discontinuous dynamical terms and a noise process which (loosely speaking) is 'wide bandwidth' and might or might not depend on the state of the system. A main problem is to find a diffusion or jump-diffusion process whose statistics are close to those of the physical system. For this, weak convergence methods are very useful. The particular methods which seem to be most useful and versatile are various versions of the perturbed test function method - pioneered by Kurtz (1969,1975), Papanicolaou, Stroock and Varadhan (1976), Blankenship and Papanicolaou (1978) and in Kushner (1979,1980a,b,1984). The book (Kushner 1984) presents a comprehensive development of several such methods and illustrates their use in numerous concrete applications. The emphasis in this paper is on methods which we have found to be most useful in the areas of application of greatest interest to us, but these same techniques are of general use whenever weak convergence methods are called for.

Applications of typical forms involve models of the type

$$X_{n+1}^{\varepsilon} = X_n^{\varepsilon} + \varepsilon G_{\varepsilon}(X_n^{\varepsilon},\xi_n^{\varepsilon}) + \sqrt{\varepsilon}F_{\varepsilon}(X_n^{\varepsilon},\xi_n^{\varepsilon}), \quad X_n^{\varepsilon} \in R^r, \tag{1.1}$$

where $EF_{\varepsilon}(x,\xi) \equiv 0$, G_{ε} or F_{ε} might be discontinuous and $\{\xi_n^{\varepsilon}\}$ might be 'state dependent'. For example, digital phase locked loops (Kushner and Huang Hai (1982)) or adaptive quantizers. A continuous parameter system of interest is the phase locked loop of Figure 1, where the box marked "limiter" is a normalized 'sign' function and $n^{\varepsilon}(\cdot)$ is a 'band-pass wide bandwidth' noise process. $\theta(\cdot)$ is the phase to be tracked (Kushner (1984), Kushner and Ju (1982), Lindsey and Simon (1973)).

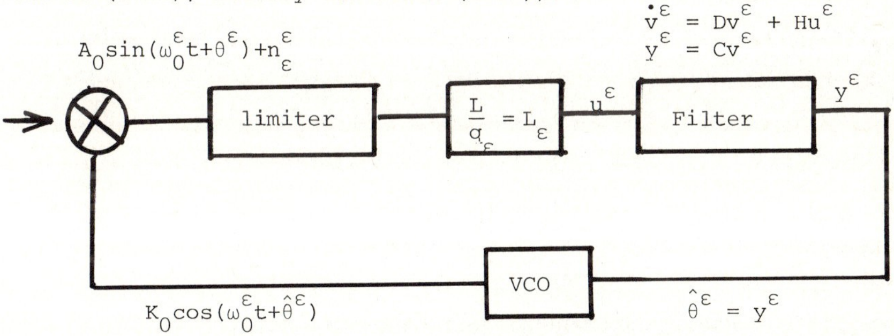

We discuss the general approach for two particular methods. The aim is mainly expository and, to this end, the regularity conditions and the conditions on the noise processes are much stronger than needed, and much of the discussion will be heuristic, and we concentrate on the problem where the limit process is a diffusion. Also, for notational convenience, we deal only with the discrete parameter problem. The continuous parameter case is treated in an almost identical manner. Fuller developments and extensions can be found in Kushner (1984).

Let $D^r[0,\infty)$ denote the Skorohod space of R^r-valued functions which are continuous on the right and have left hand limits and endowed with the Skorohod topology, Billingsley (1978). Let $\mathcal{F}_n^{\varepsilon}$ denote the minimal σ-algebra measuring $\{X_i^{\varepsilon}, i \le n\}$, define $x^{\varepsilon}(\cdot)$ by $x^{\varepsilon}(t) = X_n^{\varepsilon}$ on $[n\varepsilon,n\varepsilon+\varepsilon)$ and let $f(\cdot)$ be constant on each interval $[n\varepsilon,n\varepsilon+\varepsilon)$, with $f(n\varepsilon)$ being $\mathcal{F}_n^{\varepsilon}$ measurable. Define the operator $\hat{A}^{\varepsilon}$ by $\hat{A}^{\varepsilon}f(t) = \hat{A}^{\varepsilon}f(n\varepsilon)$ on $[n\varepsilon,n\varepsilon+\varepsilon)$

and

$$\hat{A}^{\varepsilon} f(n\varepsilon) = [E_n^{\varepsilon} f(n\varepsilon+\varepsilon) - f(n\varepsilon)]/\varepsilon, \tag{1.2}$$

where E_n^{ε} denotes expectation, conditioned on $\mathscr{F}_n^{\varepsilon}$. Then

$$f(t) - f(0) - \varepsilon\sum_0^{[t/\varepsilon]-1} \hat{A}^{\varepsilon} f(i\varepsilon) \equiv M_f^{\varepsilon}(t) \tag{1.3}$$

is a $\{\mathscr{F}_{[t/\varepsilon]}^{\varepsilon}\}$ martingale.

Let A denote an elliptic operator such that the associated martingale problem in $D^r[0,\infty)$ has a unique solution for each initial condition. Equivalently, let the Itô equation $x(\cdot)$ whose differential generator is A have a unique solution (in the sense of the multivariate distributions) for each initial condition.

The first method to be discussed (Section 2) is the second order perturbed test function method, which was used in one way or another in [3-10]. In this method, one chooses a nice real valued test function (on R^r), adds small perturbations $f_1^{\varepsilon}(\cdot) + f_2^{\varepsilon}(\cdot)$ to it, and tries to show that

$$\hat{A}^{\varepsilon}[f(x^{\varepsilon}(t)) + f_1^{\varepsilon}(t) + f_2^{\varepsilon}(t)] \sim A\, f(x^{\varepsilon}(t))$$

in an appropriate sense. Then, under a tightness condition (which can also be conveniently proved by a perturbed test function method) we have $x^{\varepsilon}(\cdot) \Rightarrow x(\cdot)$ (the arrow '$\Rightarrow$' denotes weak convergence) in $D^r[0,\infty)$.

There is a simple procedure for obtaining the perturbations $f_1^{\varepsilon}(\cdot)$ and $f_2^{\varepsilon}(\cdot)$, and the form of the operator A appears automatically in the course of the construction of the $f_i^{\varepsilon}(\cdot)$. The numerous applications in the references attest to the usefulness of the technique. The first perturbation $f_1^{\varepsilon}(\cdot)$, usually represented as a simple integral or sum, is straightforward to get, and verification of the required conditions is often quite easy. The second perturbation $f_2^{\varepsilon}(\cdot)$, usually represented as a double integral or double sum, is also often straightforward to get, but verification of the required conditions is harder.

The second method to be discussed (Sections 3,4) called the 'perturbed test function-direct averaging method' uses only a first order perturbed test function - (uses $f_1^{\varepsilon}(\cdot)$ only), and then performs an 'averaging' on $f(x^{\varepsilon}(\cdot)) + f_1^{\varepsilon}(\cdot)$ in order to get the result. The form of the operator A also comes directly out of the constructions. For this method, we require weaker regularity conditions, and the conditions are often much easier to verify than those on $f_2^\varepsilon(\cdot)$. The method is actually a development of the averaging method used in a stochastic approximation problem in Kushner and Shwartz (1984).

2. THE SECOND ORDER PERTURBED TEST FUNCTION.

Let $\hat{C}_0(R^r)$ denote the set of real valued continuous functions on R^r with compact support, and the sup norm topology. The basic perturbed test function result is the following.

Theorem 1. *For each* $f(\cdot)$ *in a dense set in* $\hat{C}_0(R^r)$ *and* $T < \infty$, *let there be* $f^\varepsilon(\cdot)$ *satisfying*

$$f^\varepsilon(\cdot) \text{ is constant on each interval } [n\varepsilon,\ n\varepsilon+\varepsilon) \tag{2.1}$$

$$f^\varepsilon(n\varepsilon) \text{ is } \mathcal{F}_n^\varepsilon\text{-measurable} \tag{2.2}$$

$$\sup_{\substack{\varepsilon,n\\ \varepsilon n\le T}} E|f^\varepsilon(n\varepsilon)| < \infty, \quad \sup_{\substack{\varepsilon,n\\ \varepsilon n\le T}} E|\hat{A}^\varepsilon f(n\varepsilon)| < \infty, \tag{2.3}$$

$$E|f^\varepsilon(n\varepsilon) - f(x^\varepsilon(t))| \to 0 \tag{2.4}$$

$$E|\hat{A}^\varepsilon f^\varepsilon(n\varepsilon) - Af(x^\varepsilon(t))| \to 0, \text{ as } \varepsilon \to 0 \text{ and } n\varepsilon = t \le T. \tag{2.5}$$

Let $\{x^\varepsilon(\cdot)\}$ *be tight in* $D^r[0,\infty)$ *and* $X_0^\varepsilon \Rightarrow x_0$. *Then* $x^\varepsilon(\cdot) \Rightarrow x(\cdot)$, *the unique solution to the martingale problem (or unique solution to the Itô equation) with operator* A *and initial condition* x_0.

The typical construction of the $f^\varepsilon(\cdot)$ and A will be given below. It is usually easier to prove the tightness if the $\{x^\varepsilon(\cdot)\}$ are bounded. Then, in applications, Theorem 1 is applied to the truncated $x^\varepsilon(\cdot)$ processes, and then the uniqueness assumption and a simple piecing-together argument are used to obtain the desired result. For truncated processes, we have the tightness theorem (a special case of Kushner (1984), Theorem 3.4), and based on a 'perturbed' form of Aldous' result; (see Kurtz (1981), Theorem 2.7):

Theorem 2. *Let* C_2 *be a dense set in* $\hat{C}_0(R^r)$ *which contains the square of each function in it. Let* $\{x^\varepsilon(\cdot)\}$ *be truncated or, more generally, let* $\sup_{\varepsilon>0} P\{\sup_{t\le T}|x^\varepsilon(t)| \ge N\} \to 0$ *as* $N \to \infty$, *for each* $T < \infty$. *For each* $f(\cdot) \in C_2$ *and* $T < \infty$, *let there be* $f^\varepsilon(\cdot)$ *satisfying (2.1) to (2.3) and*

$$\lim_{\varepsilon} P\{\sup_{t\le T}|f^{\varepsilon}(t) - f(x^{\varepsilon}(t))| > \alpha\} = 0, \quad \underline{\text{each}} \quad \alpha > 0, \tag{2.6}$$

$$\{\hat{A}^{\varepsilon}f^{\varepsilon}(t),\ \varepsilon > 0,\ n\varepsilon \le T\} \ \underline{\text{is uniformly integrable}}. \tag{2.7}$$

Then $\{x^{\varepsilon}(\cdot)\}$ is tight in $D^{r}[0,\infty)$.

Remark. Typically $f^{\varepsilon}(\cdot)$ is a 'first order' perturbed test function; i.e., $f^{\varepsilon}(t) = f(x^{\varepsilon}(t)) + f_1^{\varepsilon}(t)$, where the $f_1^{\varepsilon}(\cdot)$ is the same as that used to get the first perturbation for the $f^{\varepsilon}(\cdot)$ in Theorem 1. See the typical construction below.

Typical construction of the perturbed test function for the discrete parameter case. For simplicity, we use the scalar system

$$X_{n+1}^{\varepsilon} = X_n^{\varepsilon} + \varepsilon G(X_n^{\varepsilon}) + \sqrt{\varepsilon}\, F(X_n^{\varepsilon},\xi_n), \tag{2.8}$$

where $E\, F(x,\xi_n) \equiv 0$, $\{\xi_n\}$ is bounded, stationary and (sufficiently) strongly mixing, and the $G(\cdot)$ and $F(\cdot)$ are smooth and bounded. The aim is an illustration of the formal technique. Extensions to more general cases are in Kushner (1984).

Let C_1 be the subset of $\hat{C}_0(R^r)$ whose partial derivatives up to third order are bounded and continuous. We get $f^{\varepsilon}(\cdot)$ in the form $f^{\varepsilon}(n\varepsilon) = f(X_n^{\varepsilon}) + f_1^{\varepsilon}(n\varepsilon) + f_2^{\varepsilon}(n\varepsilon)$. Note that

$$\varepsilon\, \hat{A}^{\varepsilon} f(X_n^{\varepsilon}) = \varepsilon f_x(X_n^{\varepsilon}) G(X_n^{\varepsilon}) + \varepsilon \frac{f_{xx}(X_n^{\varepsilon})}{2} E_n^{\varepsilon} F^2(X_n^{\varepsilon},\xi_n) + \sqrt{\varepsilon}\, f_x(X_n^{\varepsilon}) E_n^{\varepsilon} F(X_n^{\varepsilon},\xi_n) + o(\varepsilon). \tag{2.9}$$

The $f_x G$ term in (2.9) could be part of an operation Af for an appropriate A, but the two middle terms cannot. Fix $T < \infty$. Define $f_1^{\varepsilon}(t)$ by $f_1^{\varepsilon}(t) = f_1^{\varepsilon}(x^{\varepsilon}(n\varepsilon),n\varepsilon)$ on $[n\varepsilon,n\varepsilon+\varepsilon)$, where

$$\begin{aligned} f_1^{\varepsilon}(x,n\varepsilon) &= \sqrt{\varepsilon} \sum_{j=n}^{[T/\varepsilon]} f_x(x) E_n^{\varepsilon} F(x,\xi_j) \\ &\quad + \varepsilon \sum_{j=n}^{[T/\varepsilon]} \frac{f_{xx}(x)}{2} E_n^{\varepsilon}[F^2(x,\xi_j) - EF^2(x,\xi_j)]. \end{aligned} \tag{2.10}$$

By the (sufficiently) strong mixing condition, the sums are bounded. Thus $f_1^{\varepsilon}(n\varepsilon) = O(\sqrt{\varepsilon})$. The tightness Theorem 2 can be used with the test function $f(x^{\varepsilon}(\cdot)) + f_1^{\varepsilon}(\cdot)$.

To see what has been accomplished, note that

$$\varepsilon\hat{A}[f(X_n^\varepsilon) + f_1^\varepsilon(n\varepsilon)] = \varepsilon[f_x(X_n^\varepsilon)G(X_n^\varepsilon) + \frac{f_{xx}(X_n^\varepsilon)EF^2(X_{n}^\varepsilon,\xi_n)}{2}] + o(\varepsilon)$$

$$+ \varepsilon q(X_n^\varepsilon, n\varepsilon), \tag{2.11}$$

where

$$q(x,n\varepsilon) = \sum_{n+1}^{[T/\varepsilon]} E_n^\varepsilon[f_x(x)F(x,\xi_j)]_x F(x,\xi_n).$$

The use of the first perturbation $f_1^\varepsilon(\cdot)$ has allowed us to average the second term on the right of (2.9), and to 'partially' average the third term. We must continue the procedure - one more step - in order to average the $\varepsilon q(X_n^\varepsilon, n\varepsilon)$ term.

Define $f_2^\varepsilon(n\varepsilon) = f_2^\varepsilon(X_n^\varepsilon, n\varepsilon)$, where

$$f_2^\varepsilon(x,n\varepsilon) = \sum_{n}^{[T/\varepsilon]} E_n^\varepsilon[q(x,j\varepsilon) - Eq(x,j\varepsilon)]. \tag{2.12}$$

Define A by

$$Af(x) = f_x(x)G(x) + \frac{1}{2}f_{xx}(x)EF^2(x,\xi) + \lim_{\varepsilon\to 0} Eq(x,n\varepsilon). \tag{2.13}$$

Then

$$\varepsilon\, \hat{A}^\varepsilon f^\varepsilon(n\varepsilon) = \varepsilon\, Af(X_n^\varepsilon) + o(\varepsilon) \tag{2.14}$$

and Theorem 1 can be applied.

The method is widely and readily applicable, as attested to by the applications in the cited references. But the second perutrbation is sometimes troublesome to work with - particularly when the noise $\{\xi_j\}$ is 'state-dependent', or the dynamics are not smooth. In the next two sections an alternative will be discussed.

3. AN AVERAGED PERTURBED TEST FUNCTION METHOD. I.

We work only with the relatively simple first order perturbed test function. Tightness is handled as in the previous section. Again, for simplicity in the exposition, we stay with simple scalar cases. We first do a simple case, where the limit process $x(\cdot)$ satisfies an ordinary differential equation, and the noise is not state dependent. This case can be treated by many other methods - but it provides a convenient vehicle for our discussion of the basic ideas. In order to avoid more technicalities than needed to illustrate the idea, we always let $\{\xi_n\}$ be bounded and the functions $G(x,\xi)$, $F(x,\xi)$ have compact x-support - uniformly in ξ. See Kushner (1984) for the general case. In the arguments below, whenever convenient, we use the Skorohod imbedding method to put all the processes $\{x^\varepsilon(\cdot)\}$ on the same probability space, and turn weak convergence into w.p.1 convergence in the metric of $D^r[0,\infty)$. We do this often without specific mention and without altering the notation.

The system is

$$X_{n+1}^\varepsilon = X_n^\varepsilon + \varepsilon\, G(X_n^\varepsilon,\xi_n). \tag{3.1}$$

If the $G(x,\xi)$ were not bounded, then we would require

$$\{\sup_{|x|\leq N} |G(x,\xi_j)|,\ j<\infty\} \text{ uniformly integrable for each } N.$$

We assume either (3.2) or (3.3) for each bounded random variable X, and either (3.4) or (3.5). Below, n_ε is some sequence such that $n_\varepsilon \to \infty$ and $\varepsilon n_\varepsilon = \delta_\varepsilon \to 0$ as $\varepsilon \to 0$, and $\overline{G}(\cdot)$ is continuous.

$$E \sup_{|Y|\leq\delta} |G(X,\xi_j) - G(X+Y,\xi_j)| \overset{\delta}{\to} 0 \tag{3.2}$$

$$\lim_{\varepsilon,n,\delta} \frac{1}{n_\varepsilon} \sum_n^{n+n_\varepsilon-1} E \sup_{|Y|\leq\delta} |G(X,\xi_j) - G(X+Y,\xi_j)| = 0 \tag{3.3}$$

$$\frac{1}{n_\varepsilon} \sum_n^{n+n_\varepsilon-1} G(x,\xi_j) \overset{P}{\to} \overline{G}(x), \text{ each } x, \tag{3.4}$$

$$\frac{1}{n_\varepsilon} \sum_n^{n+n_\varepsilon-1} E_n^\varepsilon G(x,\xi_j) \overset{P}{\to} \overline{G}(x), \text{ each } x, \text{ as } \varepsilon \to 0 \text{ and } n \to \infty. \tag{3.5}$$

<u>Theorem 3.</u> <u>Let</u> $\dot{x} = \overline{G}(x)$ <u>have a unique solution</u> $x(\cdot)$ <u>for each initial condition, and let</u> $X_0^\varepsilon \Rightarrow x_0$. <u>Then</u> $x^\varepsilon(\cdot) \Rightarrow x(\cdot)$ <u>in</u> $D[0,\infty)$, <u>where</u> $x(0) = x_0$.

<u>Outline of proof.</u> Fix $T < \infty$, and divide $[0,T]$ into intervals $[\ell\delta_\varepsilon, \ell\delta_\varepsilon+\delta_\varepsilon)$, where $\delta_\varepsilon = \varepsilon n_\varepsilon \to 0$. Define $\tilde{G}^\varepsilon(\cdot)$ and $\hat{G}^\varepsilon(\cdot)$ by: for $s \in [\ell\delta_\varepsilon, \ell\delta_\varepsilon+\delta_\varepsilon)$,

$$\tilde{G}^\varepsilon(s) = \frac{1}{n_\varepsilon} \sum_{\ell n_\varepsilon}^{\ell n_\varepsilon+n_\varepsilon-1} E_{\ell n_\varepsilon}^\varepsilon f_x(X_j^\varepsilon) G(X_j^\varepsilon,\xi_j)$$

$$\hat{G}^\varepsilon(t) = \int_0^t \tilde{G}^\varepsilon(s)\,ds.$$

The set $\{x^{\varepsilon}(\cdot), \hat{G}^{\varepsilon}(\cdot)\}$ is tight in $D^2[0,\infty)$ and all limits are continuous. Choose and fix a weakly convergent subsequence, also indexed by ε, and with limit denoted by $(x(\cdot), \hat{G}(\cdot))$. The limit will not depend on this subsequence. We will show that

$$\hat{G}(t) = \int_0^t f_x(x(s))\bar{G}(x(s))ds,$$

and for each k, s, t, and $s_i < t$ and bounded continuous $h(\cdot)$ and smooth $f(\cdot)$,

$$Eh(x(s_j), j \leq k)[f(x(t+s)) - f(x(s)) - (\hat{G}(t+s) - \hat{G}(t))] = 0.$$

This will yield the desired result, since it is equivalent to showing that $x(\cdot)$ solves the martingale problem for operator $\bar{G}(x)\partial/\partial x$.

We proceed as follows. We have

$$Eh(x^{\varepsilon}(s_j), j \leq k)[f(x^{\varepsilon}(t+s)) - f(x^{\varepsilon}(t)) - \sum_{\varepsilon j=t}^{t+s} \varepsilon f_x(X_j^{\varepsilon})G(X_j^{\varepsilon},\xi_j)] \to 0$$

$$Eh(x^{\varepsilon}(s_j), j \leq k)[f(x^{\varepsilon}(t+s)) - f(x^{\varepsilon}(t)) - \sum_{\ell\delta_{\varepsilon}=t}^{t+s} \delta_{\varepsilon}\, \tilde{G}^{\varepsilon}(\ell\delta_{\varepsilon})] \to 0$$

$$Eh(x(s_j), j \leq k)[f(x^{\varepsilon}(t+s)) - f(x^{\varepsilon}(t)) - \int_t^{t+s} \tilde{G}^{\varepsilon}(u)du] \overset{\varepsilon}{\to} 0.$$

Since $x^{\varepsilon}(\cdot) \to x(\cdot)$, we need only prove that, for each s,

$$\tilde{G}^{\varepsilon}(s) \overset{P}{\to} f_x(x(s))\bar{G}(x(s)).$$

To do this it is enough to show either (3.8a) or (3.8b) for s fixed in $[\ell_{\varepsilon}\delta_{\varepsilon}, \ell_{\varepsilon}\delta_{\varepsilon}+\delta_{\varepsilon})$ and $m_{\varepsilon} = \ell_{\varepsilon}n_{\varepsilon}$:

$$\frac{1}{n_{\varepsilon}} \sum_{m_{\varepsilon}}^{m_{\varepsilon}+n_{\varepsilon}-1} f_x^{\varepsilon}(X_j)G(X_j^{\varepsilon},\xi_j) \overset{P}{\to} f_x(x(s))\bar{G}(x(s)), \tag{3.8a}$$

$$\frac{1}{n_{\varepsilon}} \sum_{m_{\varepsilon}}^{m_{\varepsilon}+n_{\varepsilon}-1} E_{m_{\varepsilon}}^{\varepsilon} f_x^{\varepsilon}(X_j^{\varepsilon})G(X_j^{\varepsilon},\xi_j) \overset{P}{\to} f_x(x(s))\bar{G}(x(s)). \tag{3.8b}$$

But either (3.8a or b) is guaranteed by the 'ergodic' assumptions (3.4, 3.5) and the smoothness assumptions (3.2,3.3). Q.E.D.

4. AVERAGED PERTURBED TEST FUNCTION METHOD. II STATE DEPENDENT NOISE

We continue with the case where the limit process $x(\cdot)$ satisfies an ordinary differential equation. Write the noise as ξ_j^ε. The state dependence is modelled by assuming that $\{X_n^\varepsilon, \xi_{n-1}^\varepsilon\}$ is a homogeneous Markov process. If $X_n^\varepsilon \equiv x$, then we have a Markov chain $\{\xi_n^\varepsilon(x)\}$. The essential assumption is that there is a 'limit' process with a unique invariant measure, and even this can be weakened (see comments below). In many applications $\{\xi_j^\varepsilon\}$ only depends on ε via its dependence on the $\{X_j^\varepsilon\}$. Then, if $\{X_j^\varepsilon\}$ is replaced by a constant x, the corresponding process $\{\xi_j^\varepsilon(x)\}$ does not depend on ε.

We assume the following. Condition (4.2) is used only to simplify the development. See Kushner (1984), Chapter 5. For each x, define $P^\varepsilon(\xi,1,\cdot|x) = P\{\xi_n^\varepsilon \in \cdot | X_n^\varepsilon = x, \xi_{n-1}^\varepsilon = \xi\}$, and define the n-step $P^\varepsilon(\xi,n,\cdot|x)$ by convolution. Define $P^\varepsilon(x,\xi,\alpha,\beta,\cdot) = P\{(X_{\alpha+\beta}^\varepsilon, \xi_{\alpha+\beta-1}^\varepsilon) \in \cdot | X_\alpha^\varepsilon = x, \xi_{\alpha-1}^\varepsilon = \xi\}$.

$$X_{n+1}^\varepsilon = X_n^\varepsilon + \varepsilon\, G(X_n^\varepsilon, \xi_n^\varepsilon) \tag{4.1}$$

$\{\xi_j^\varepsilon\}$ is bounded, $G(\cdot,\cdot)$ is bounded (4.2)

There is a transition function $P(\xi,\ell,\cdot|x)$ such that for bounded and continuous $f(\cdot)$, (4.3)

$\int f(\xi')P(\xi,1,d\xi'|x)$ is (x,ξ)-continuous

$\int f(\xi')P^\varepsilon(\xi,1,d\xi'|x) \to \int f(\xi')P(\xi,1,d\xi'|x)$ uniformly on compact (x,ξ) sets. (4.4)

for each x, $P(\xi,1,\cdot|x)$ has a unique invariant measure $P^x(\cdot)$ and $\{P^x(\cdot),\ x \in \text{compact}\}$ is tight. (4.5)

$\int G(x,\xi')P^\varepsilon(\xi,1,d\xi'|x) \to \int G(x,\xi')P(\xi,1,d\xi'|x)$, continuous, uniformly on compact (x,ξ)-sets. (4.6)

A c-step smoothing can be used in (4.6) in lieu of the one-step smoothing.

Theorem 4. Let $\dot{x} = \overline{G}(x) = \int G(x,\xi)P^x(d\xi)$ have a unique solution for each $x(0)$, and let $x_0^\varepsilon \Rightarrow x_0$. Then $x^\varepsilon(\cdot) \Rightarrow x(\cdot)$, satisfying this equation for $x(0) = x_0$.

Remark. If the $P^x(\cdot)$ is not unique, we get the limit equation

$$\dot{x} \quad \{\int G(x,\xi)P^x(dt), \text{ all invariant } P^x\}. \tag{4.7}$$

Outline of proof. As in Theorem 3, and with the same notation,

$$Eh(x^\varepsilon(s_j), j \le k)[f(x^\varepsilon(t+s)) - f(x^\varepsilon(t)) - \sum_{\ell\delta_\varepsilon=t}^{t+s} \delta_\varepsilon \tilde{G}^\varepsilon(\ell\delta_\varepsilon)] \overset{\xi}{\to} 0$$

where

$$\tilde{G}^\varepsilon(\ell\delta_\varepsilon) = \frac{1}{n_\varepsilon} \sum_{j=\ell n_\varepsilon}^{\ell n_\varepsilon+n_\varepsilon-1} E^\varepsilon_{\ell n_\varepsilon} f_x(X^\varepsilon_j)G(X^\varepsilon_j,\xi^\varepsilon_j). \tag{4.8}$$

The set $\{x^\varepsilon(\cdot), \hat{G}^\varepsilon(\cdot)\}$ is tight and all limits are continuous. Extract and fix a weakly convergent subsequence, also indexed by ε, and with limit $(x(\cdot),\hat{G}(\cdot))$. The result will not depend on the subsequence. We need only show that as $\ell n_\varepsilon \to s$, $\tilde{G}^\varepsilon(\ell n_\varepsilon) \to f_x(x(s))\bar{G}(x(s))$.

Define the measure

$$Q(\omega,\ell,\varepsilon,\cdot) = \frac{1}{n_\varepsilon} \sum_{\ell n_\varepsilon}^{\ell n_\varepsilon+n_\varepsilon-1} P\{\xi^\varepsilon_j \in \cdot | X^\varepsilon_{\ell n_\varepsilon}, \xi^\varepsilon_{\ell n_\varepsilon}-1\}.$$

A consequence of the 'smoothing' assumption (4.6) is that we can replace the X^ε_j in (4.8) by $X^\varepsilon_{\ell n_\varepsilon}$ without altering the limits. Thus $(\ell_\varepsilon n_\varepsilon = m_\varepsilon$, $s \in [\ell_\varepsilon\delta_\varepsilon, \ell_\varepsilon\delta_\varepsilon+\delta_\varepsilon))$

$$\tilde{G}^\varepsilon(s), \int Q(\omega,\ell_\varepsilon,\varepsilon,d\xi)f_x(X^\varepsilon_{m_\varepsilon})G(X^\varepsilon_{m_\varepsilon},\xi)$$

have the same limits.

The set $\{Q(\omega,\ell,\varepsilon,\cdot), \omega,\ell,\varepsilon\}$ is tight. We work with each ω (not in some null set) - in order to get the limits of $\tilde{G}^\varepsilon(s)$. (The ω in Q only indexes the $\xi^\varepsilon_{m_\varepsilon-1}, X^\varepsilon_{m_\varepsilon}$.) Fix s and ω. Extract a weakly convergent subsequence of $^\varepsilon\{Q(\omega,\ell^\varepsilon_\varepsilon,\varepsilon,\cdot)\}$ with limit $\bar{P}^\omega$. It will be shown that $\bar{P}^\omega = P^{x(s)}$. Define, for bounded and continuous $g(\cdot)$,

$$\lim_\varepsilon \int P^\varepsilon(\xi,1,d\xi'|x)g(\xi') = \bar{g}(x,\xi).$$

By the weak convergence and Skorohod imbedding

$$\sup_{m_\varepsilon+n_\varepsilon \ge j \ge m_\varepsilon} |X^\varepsilon_j - X^\varepsilon_{m_\varepsilon}| \to 0.$$

Also,

$$\int \overline{P}^{\omega}(dz)g(z) = \lim_{\varepsilon} \int Q(\omega,\ell_{\varepsilon},\varepsilon,d\xi)g(\xi)$$

$$= \lim_{\varepsilon} \frac{1}{n_{\varepsilon}} \sum_{m_{\varepsilon}+1}^{m_{\varepsilon}+n_{\varepsilon}} \int P^{\varepsilon}(X^{\varepsilon}_{m_{\varepsilon}},\xi^{\varepsilon}_{m_{\varepsilon}-1},m_{\varepsilon},j-m_{\varepsilon},d\xi)g(\xi)$$

$$= \lim_{\varepsilon} \int Q(\omega,\ell_{\varepsilon},\varepsilon,d\tilde{\xi})P(\tilde{\xi},1,d\xi|X^{\varepsilon}_{m_{\varepsilon}})g(\xi) \tag{4.8}$$

$$= \int \overline{P}^{\omega}(d\tilde{\xi})\overline{g}(x(s),\tilde{\xi})$$

$$= \int \overline{P}^{\omega}(d\tilde{\xi})P(\tilde{\xi},1,d\xi|x(s))g(\xi).$$

Then $\overline{P}^{\omega}$ is an invariant measure for $P(\tilde{\xi},1,d\xi|x(s))$ for each ω-irrespective of the chosen subsequence, and $\overline{P}^{\omega} = P^{x(s)}$.

Now that $\lim_{\varepsilon} Q(\omega,\ell_{\varepsilon},\varepsilon,\cdot)$ is characterized, it is a simple matter to show that

$$\lim_{\varepsilon} \tilde{G}^{\varepsilon}(\omega,s)$$

$$= \lim_{\varepsilon} \frac{1}{n_{\varepsilon}} \sum_{m_{\varepsilon}+1}^{m_{\varepsilon}+n_{\varepsilon}} \int P^{\varepsilon}(X^{\varepsilon}_{m_{\varepsilon}},\xi^{\varepsilon}_{m_{\varepsilon}-1},j-m_{\varepsilon}-1,d\tilde{\xi},dx)$$

$$\cdot \int P^{\varepsilon}(\tilde{\xi},1,d\xi|x)f_{x}(x)G(x,\xi)$$

$$= \int P^{x(s)}(d\xi)f_{x}(x(s))G(x(s),\xi),$$

as desired. Q.E.D.

5. AVERAGED PERTURBED TEST FUNCTION METHOD III.

We now treat the next level, where the limit process is a diffusion $x(\cdot)$ for whose operator A the martingale problem has a unique solution for each initial condition. We use

$$X^{\varepsilon}_{n+1} = X^{\varepsilon}_{n} + \varepsilon\, G(X^{\varepsilon}_{n},\xi^{\varepsilon}_{n}) + \sqrt{\varepsilon}\, F(X^{\varepsilon}_{n},\xi^{\varepsilon}_{n}), \tag{5.1}$$

$$EF(x,\xi) = 0.$$

The $f^{\varepsilon}(\cdot)$ below are typically <u>first order</u> perturbed test functions. We use the following conditions. For each $f(\cdot)$ in a dense set in $\hat{C}_0(R^r)$ there are $f^{\varepsilon}(\cdot)$ satisfying (2.1) to (2.4) and

$$E\left|\frac{1}{n_{\varepsilon}} \sum_{j=n}^{n+n_{\varepsilon}-1} E^{\varepsilon}_{n}(\hat{A}^{\varepsilon}f^{\varepsilon}(j\varepsilon) - Af(X^{\varepsilon}_{j}))\right| \to 0 \tag{5.2}$$

as $n \to \infty$ and $\varepsilon \to 0$.

Theorem 5. *If* $\{x^\varepsilon(\cdot)\}$ *is tight in* $D^r[0,\infty)$ *and* $X_0 \Rightarrow x_0$, *then* $x^\varepsilon(\cdot) \Rightarrow x(\cdot)$, *the solution to the martingale problem for operator* A *and initial condition* x_0.

Remark. Tightness can be obtained by a first order perturbed test function and Theorem 2, exactly as for the case of Section 2. If necessary, we truncate the $x^\varepsilon(\cdot)$ and use a piecing-together argument.

Outline of Proof. Using the notation of Sections 3 and 4, we have $(\delta_\varepsilon/n_\varepsilon = \varepsilon)$

$$Eh(x^\varepsilon(s_i), i \le k)[f(x^\varepsilon(t+s)) - f(x^\varepsilon(t)) - \varepsilon \sum_{\varepsilon j=t}^{t+s} \hat{A}^\varepsilon f^\varepsilon(\varepsilon j)] \overset{\varepsilon}{\to} 0,$$

$$Eh(x^\varepsilon(s_i), i \le k)[f(x^\varepsilon(t+s)) - f(x^\varepsilon(t))$$
$$- \sum_{\ell\delta_\varepsilon=t}^{t+s} \delta_\varepsilon \frac{1}{n_\varepsilon} \sum_{j=\ell n_\varepsilon}^{\ell n_\varepsilon+n_\varepsilon-1} E^\varepsilon_{\ell n_\varepsilon} \hat{A}^\varepsilon f^\varepsilon(\varepsilon j)] \overset{\varepsilon}{\to} 0,$$

$$Eh(x^\varepsilon(s_i), i \le k)[f(x^\varepsilon(t+s))-f(x^\varepsilon(t)) - \varepsilon \sum_{j\varepsilon=t}^{t+s} Af(x^\varepsilon(j\varepsilon))] \overset{\varepsilon}{\to} 0$$

Choose and fix a weakly convergent subsequence of $\{x^\varepsilon(\cdot)\}$, indeed by ε also, and with limit $x(\cdot)$. Then

$$Eh(x(s_i), i \le k)[f(x(t+s))-f(x(t)) - \int_t^{t+s} Af(x(u))du] = 0 \qquad (5.3)$$

Owing to the arbitrariness of $f(\cdot)$, $h(\cdot)$, t, s, k, and $s_i \le t$, $x(\cdot)$ solves the martingale problem for operator A and initial condition x_0 and we are done.

Sufficient conditions for Theorem 5 are often quite easy to get. Apart from tightness, mainly (5.2) need be verified for the appropriate $f^\varepsilon(\cdot)$. For $f^\varepsilon(\cdot)$, we use the <u>first order</u> perturbed test function $f^\varepsilon(\cdot) = f(x^\varepsilon(\cdot)) + f_1^\varepsilon(\cdot)$, where $f_1^\varepsilon(\cdot)$ is defined in Section 2. In order to illustrate the flavor, we state conditions for one simple case. We use state-independent noise $\{\xi_j\}$. Let $G(\cdot,\xi)$, $F(\cdot,\xi)$ and $F_x(\cdot,\xi)$ be continuous in x, uniformly in (x,ξ) in each compact set. If ξ_j^ε does not depend on ε, then the continuity can be for each ξ. (This correction should be added to condition A5.8.2 in Kushner (1984). The continuity of G can be replaced by conditions of the type (3.2) or (3.3), and the smoothness of $F(\cdot,\xi)$ can also be weakened, since only conditional expectations of this function appear in the calculations.

We also use the following, where $\bar{G}(\cdot)$, $\Phi(\cdot)$ are continuous, and the operators A^{ℓ} and A_0 are defined below, and T is arbitrary. The convergence is for each x as $\varepsilon \to 0$ and $n \to \infty$, except where otherwise noted. Also, $f(\cdot) \in \hat{C}_0(R^r)$.

$$\frac{1}{n_\varepsilon} \sum_{j=n}^{n+n_\varepsilon-1} E_n^\varepsilon G(x,\xi_j^\varepsilon) \to \bar{G}(x) \tag{5.4}$$

$$\frac{1}{n_\varepsilon} \sum_{j=n}^{n+n_\varepsilon-1} E_n^\varepsilon F(x,\xi_j^\varepsilon)F'(x,\xi_j^\varepsilon) \to \Phi(x) \tag{5.5}$$

$$\sum_{j=n+M}^{T/\varepsilon} \sup_{|x|\le N} |E_n^\varepsilon(f_x'(x)F(x,\xi_j^\varepsilon))_x| \to 0, \text{ each } N, \text{ as } M\to\infty,\ \varepsilon\to 0,\ n\to\infty \tag{5.6}$$

$$\frac{1}{n_\varepsilon} \sum_{j=n}^{n+n_\varepsilon-1} E_n^\varepsilon(F'(x,\xi_{j+\ell}^\varepsilon)f_x(x))_x' F(x,\xi_j^\varepsilon) \to A^\ell f(x) \tag{5.7}$$

$$\sum_1^\infty A^\ell f(x) \equiv A_0 f(x), \text{ uniform convergence.} \tag{5.8}$$

The following sets are uniformly integrable

$$\begin{gathered}\{\sup_{|x|\le N} |F(x,\xi_j^\varepsilon)|^2,\quad \sup_{|x|\le N} |G(x,\xi_j^\varepsilon)|^2\} \\ \{\sup_{|x|\le N} |\textstyle\sum_n^{T/\varepsilon} E_n^\varepsilon F(x,\xi_j^\varepsilon)|^2,\quad \sup_{|x|\le N} |\textstyle\sum_n^{T/\varepsilon} E_n^\varepsilon F_x(x,\xi_j^\varepsilon)|^2,\ n \le T/\varepsilon,\ \varepsilon > 0\}.\end{gathered} \tag{5.9}$$

Define A by

$$Af(x) = f_x'(x)\bar{G}(x) + \frac{1}{2}\,\text{trace}\, f_{xx}(x)\bar{\Phi}(x) + A_0 f(x). \tag{5.10}$$

The $A_0 f(x)$ is essentially the centering term in the construction of the second perturbation in Section 2.

Theorem 6. Let $X_0^\varepsilon \Rightarrow x_0$. Under the above conditions and the uniqueness to the solution to the martingale problem for operator A under each initial condition, $\{x^\varepsilon(\cdot)\}$ is tight in $D^r[0,\infty)$ and the limit is the solution to the martingale problem for operator A and initial condition $x(0) = x_0$.

Remark. There are extensions to state-dependent noise and to the continuous parmaeter case. The above conditions are different from those used in Theorem 5.9, Kushner (1984). We also note that (5.8.6) in Kushner (1984) (used in Theorem 5.9) should be interpreted to read

$$\lim_{n_\Delta,\varepsilon} P\{\sup_{|x-x'|<\Delta} |\textstyle\sum_n^{T/\varepsilon} {}_n^\varepsilon(F(x,\xi_j^\varepsilon)-F(x',\xi_j^\varepsilon)| > \delta\} = 0, \text{ each } \delta > 0.$$

Outline of proof. (Truncate (5.1) if necessary.) Use $f^\varepsilon(n\varepsilon) = f(X_n^\varepsilon) + f_1^\varepsilon(X_n^\varepsilon, n\varepsilon)$, where

$$f_1^\varepsilon(x,n\varepsilon) = \sqrt{\varepsilon} \sum_n^{T/\varepsilon} E_n^\varepsilon f_x'(x) F(x,\xi_j^\varepsilon).$$

By hypothesis, $f_1^\varepsilon(\cdot) \to 0$ and $\{\hat{A}f^\varepsilon(t),\ t \le T,\ \varepsilon > 0\}$ is uniformly integrable. Hence $\{x^\varepsilon(\cdot)\}$ is tight. Also,

$$\varepsilon\, \hat{A}^\varepsilon f^\varepsilon(n\varepsilon) = o(\varepsilon) + \varepsilon\, E_n^\varepsilon L(X_n^\varepsilon, n\varepsilon),$$

where

$$L(x,n\varepsilon) = f_x'(x) G(x,\xi_n^\varepsilon) + F'(x,\xi_n^\varepsilon) f_{xx}(x) F(x,\xi_n^\varepsilon)/2 + \hat{L}(x,n\varepsilon),$$

$$\hat{L}(x,n\varepsilon) = \sum_{n+1}^{T/\varepsilon} [E_{n+1}^\varepsilon f_x'(x) F(x,\xi_j^\varepsilon)]_x F(x,\xi_n^\varepsilon).$$

Now choose a weakly convergent subsequence, indexed also by ε and with limit denoted by $x(\cdot)$. Possibly excluding some countable set of t, s, s_i values, we have

$$Eh(x^\varepsilon(s_i),\ i \le k)[f^\varepsilon(t+s) - f^\varepsilon(t) -$$

$$- \sum_{\ell n_\varepsilon = t}^{t+s} \delta_\varepsilon \frac{1}{n_\varepsilon} \sum_{j=\ell n_\varepsilon}^{\ell n_\varepsilon + n_\varepsilon - 1} E_{\ell n_\varepsilon}^\varepsilon \{f_x'(X_j^\varepsilon) G(X_j^\varepsilon, \xi_j^\varepsilon) + F'(X_j^\varepsilon,\xi_j^\varepsilon) f_{xx}(X_j^\varepsilon) F(X_j^\varepsilon,\xi_j^\varepsilon)/2$$

$$+ \hat{L}(X_j^\varepsilon, j\varepsilon)] \overset{\varepsilon}{\to} 0.$$

Now average as in Section 3, using the hypotheses and the weak convergence.

REFERENCES

Billingsley, P. (1968). Convergence of Probability Measures. Wiley, New York.

Blankenship, G. and Papanicolaou, G.C. (1978). Stability and control of stochastic systems with wide band noise disturbances. SIAM Journal Appl. Math., 34: 437-476.

Kurtz, T.G. (1969). Extensions of Trotter's operator semigroup approximation theorems. Journal of Functional Analysis, 3:111-132.

Kurtz, T.G. (1975). Semigroups of conditioned shifts and approximations of Markov processes. Ann. Prob., 4:618-642.

Kurtz, T.G. (1981). Approximation of Population Processes. Vol. 36 in CBMS-NSF Regional Conference Series in Applied Mathematics, Society for Ind. and Applied Mathematics, Philadelphia, PA.

Kushner, H.J. (1979). Jump-diffusion approximations for ordinary differential equations with wideband random right hand sides. SIAM J. on Control and Optimization, 17:729-744.

Kushner, H.J. (1980). Diffusion approximations to output processes of nonlinear systems with wide-band inputs, and applications. IEEE Trans. on Inf. Theory, IT-26, 715-725.

Kushner, H.J. (1980). A martingale method for the convergence of a sequence of processes to a jump-diffusion process. Z. Wahr, 53:207-219.

Kushner, H.J. (1984). Approximation and Weak Convergence Methods for Random Processes; with Applications to Stochastic Systems Theory. MIT Press, Cambridge, Mass., USA.

Kushner, H.J. and Hai Huang (1981). Averaging methods for the asymptotic analysis of learning and adaptive systems with small adjustment rate. SIAM J. on Control and Optimization, 19:635-650.

Kushner, H.J. and Huang Hai (1982). Diffusion approximations for the analysis of digital phase locked loops. IEEE Trans. on Inf. Theory, IT-28, 384-390.

Kushner, H.J. and Ju, W.T.Y. (1982). Diffusion approximations for nonlinear phase locked loop-type systems with wide band inputs. Journal of Mathematical Analysis and Applications, 86:518-541.

Kushner, H.J. and Shwartz, A. (1984). An invariant measure approach to the convergence of stochastic approximations with state-dependent noise. SIAM Journal on Control and Optimization.

Lindsey, W.C. and Simon, M.K. (1973). Telecommunication Systems Engineering. Prentice-Hall, Englewood Cliffs, New Jersey, USA.

ACKNOWLEDGEMENTS

This research was partially supported by the Office of Naval Research under Grant No. N00014-83-K-0542 and by the Army Research Office under Grant No. DAAG-29-84-K-0082.

ON THE APPROXIMATION OF STOCHASTIC CONVEX PROGRAMMING PROBLEMS

R. Lepp
Institute of Applied Mathematics
Academy of Sciences of the Estonian SSR
Tallin

1. INTRODUCTION AND PROBLEM STATEMENT

In certain classes of stochastic programming problems we look for a solution in the form of a measurable vector-valued function of a random parameter. For example, many of the stochastic programming problems studied in [1] and static formulations of two-stage stochastic programming problems (see [2]) are treated in this way. As pointed out in [3], static formulation of two-stage stochastic programming problems allows us to construct an elegant duality theory [2] and is more computationally tractable (in some cases it can be solved by a sequence of finite-dimensional "discretizations").

In this paper the convex two-stage stochastic programming problem in the space $R^r \times L^\infty$ is replaced by its finite-dimensional analogue. Conditions under which this substitution is justified are given, i.e., conditions which guarantee the convergence of the solutions of the approximate problems to the solution of the original problem.

Consider the following stochastic programming problem:

$$\min_{x,y(\xi)} \{f_1(x) + \int_\Xi f_2(\xi, x, y(\xi))\mu(d\xi) \mid x \in C_1, g_{ij} \le 0, j \in J_1 , \tag{1}$$

and almost everywhere (a.e.) $y(\xi) \in C_2$, $g_{2j}(\xi, x, y(\xi)) \le 0$, $j \in J_2\} = f^*$.

Here $x \in R^r$, $y(\xi) \in L^\infty(\Xi, \sum, \mu; R^m)$, $\Xi \subset R^s$, μ is the probability measure induced by the random vector ξ, $\mu(\Xi) = 1$, $\sum$ is a Borel σ-algebra, and J_1, J_2 are finite sets of indices. It is well-known (see [2]) that if the set C_2 is bounded, then problem (1) is equivalent to the dynamic formulation of a stochastic programming problem with recourse (a two-stage stochastic programming problem).

Assume that the sequence of discrete probability measures

$$\mu_n = \{\mu_{in}, i = 1, \ldots, n\}, \mu_{in} > 0, \sum_{i=1}^n \mu_{in} = 1, n \in N = \{1,2, \ldots\} ,$$

converges weakly to the probability measure μ, i.e.,

$$\lim_{n \to \infty} \sum_{i=1}^{n} h(\xi_{in})\mu_{in} = \int h(\xi)\mu(d\xi) \tag{2}$$

for every bounded continuous function $h(\xi)$, where ξ_{in} , $i = 1, \dots, n$, are points in Ξ.

Using the approximation (2) in problem (1), we obtain the following extremum problem in Euclidean space $R^r \times l_n^\infty$ with norm $\|y_n\|_n = \max_{1 \le i \le n} |y_{in}|$:

$$\min_{x, y_n} \{f_1(x) + \sum_{i=1}^{h} f_2(\xi_{in}, x, y_{in})\mu_{in} \mid x \in C_1, g_{1j}(x) \le 0, j \in J_1 ;$$

$$y_n \equiv (y_{1n}, \dots, y_{nn}) \in C_2, g_{2j}(\xi_{in}, x, y_{in}) \le 0, j \in J_2, i = 1, \dots, n\} = f_n^* . \tag{1n}$$

Using the general theory for the approximate minimization of functionals [4], we define a system of linear connection operators between the spaces L^∞ and l_n^∞ as follows:

$$p_n : L^\infty \to l_n^\infty$$

$$p_n y(\xi) = (\mu(A_{in})^{-1} \int_{A_{in}} y(\xi)\mu(d\xi), i = 1, \dots, n) , \tag{3}$$

where $\bigcup_{i=1}^{n} A_{in} = \Xi$, $A_{in} \cap A_{jn} = \phi$ if $i \ne j$, and diam $A_{in} \to 0$. We say that a sequence of spaces $l_n^\infty (n \in N)$ is a *discrete approximation* of the space L^∞ if

$$\|p_n y\|_n \to \|y\| \quad \text{a.s.} \quad n \to \infty . \tag{4}$$

Sequences of connection operators p_n , $p_n' (n \in N)$ are said to be *equivalent* if

$$\|p_n y - p_n' y\|_n \to 0 \quad \text{a.s.} \quad n \to \infty .$$

A sequence $y_n (n \in N)$, $y_n \in l_n^\infty$, is said to *converge discretely* to $y \in L^\infty$ (writing $y_n \to y$) if

$$\|y_n - p_n y\|_n \to 0 \quad \text{a.s.} \quad n \to \infty .$$

A sequence $y_n (n \in N)$, $y_n \in l_n^\infty$, is said to *display weak discrete convergence* to $y \in L^\infty$ (writing $y_n \to y$) if $y_n (n \in N)$ and y, considered as elements of dual spaces of $l_n^1 (n \in N)$ and L^1, respectively, satisfy the condition $<z_n , y_n> \to <z , y>$ for every discretely converging sequence of elements z_n ($n \in N$), $z_n \in l_n^1$, $z \in L^1$. Here l_n^1 is the Euclidean space with norm

$$\|z_n\|_n = \sum_{i=1}^{n} |z_{in}| \mu_{in} \quad .$$

We shall let the pair $\{f, (f_n)\}$ denote a functional f with a region of definition $D(f) \subset L^\infty$ and a sequence of functionals f_n ($n \in N$) with a region of definition $D(f_n) \subset l_n^\infty$ ($n \in N$).

The pair $\{f, (f_n)\}$ is *discretely lower(upper)-semicontinuous* if for $y_n \to y$ we have $f(y) \le \lim_{n\to\infty} \inf f_n(y_n)$ ($f(y) \ge \lim_{n\to\infty} \sup f_n(y_n)$).

If the elements display weak discrete convergence rather than discrete convergence, then the pair $\{f, (f_n)\}$ is said to be *weakly discretely lower(upper)-semicontinuous*.

2. CONDITIONS FOR CONVERGENCE OF DISCRETE APPROXIMATIONS

In order to guarantee that the solutions of problems (1n) converge to the solution of problem (1), we must impose comparatively strict restrictions on the measure μ. Suppose that

(R1)the support Ξ of the measure μ is bounded in R^s;

(R2)the probability measure μ has a Riemann-integrable density $\varphi(\xi)$, i.e., for every $A \in \sum$ we have

$$\mu(A) = \int_A \varphi(\xi) d\xi \quad ,$$

where $\varphi(\xi)$ is a Rieman-integrable non-negative function with $\int \varphi(\xi) d\xi = 1$.

We should perhaps explain the point of restriction (R2). If the function $\varphi(\xi)$ is only Lebesgue-integrable, then on changing the values of $\varphi(\xi)$ on a set of measure zero it can happen that the sum on the left-hand side of formula (2) is equal to zero at every point $\xi_{1n}, \xi_{2n}, \ldots, \xi_{nn}$ (here $\mu_{in} = \varphi(\xi_{in}) h_{in}$), but that the value of the integral on the right-hand side of formula (2) does not change. Hence, if we want to replace the integral by a sum of the form (2), the class of Lebesgue-integrable functions is too wide to guarantee convergence of the solutions of problems (1n) to the original problem (1).

Before continuing further we must introduce the notion of *compatibility of approximations*.

We say that the compatibility condition is fulfilled if $z_n \to z$, $y_n \to y$, where $z_n \in l_n^1$, $z \in L^1$, $y_n \in l_n^\infty$, $y \in L^\infty$, implies that $\langle z_n, y_n \rangle \to \langle z, y \rangle$.

If (R1) and (R2) are satisfied, then we can guarantee that the compatibility condition holds for our approximations. Let $z_n \to z$, $y_n \to y$, i.e.,

$$\sum_{i=1}^{n} |z_{in} - q_n z(\xi)|\, \mu_{in} \to 0 \quad a.s. \quad n \to \infty \quad ,$$

$$\max_{1 \le i \le n} |y_{in} - p_n y(\xi)| \to 0 \quad a.s. \quad n \to \infty \quad ,$$

where the connection operator q_n is also of the form (3). Restrictions (R1) and (R2) on the measure μ justify the use of an equivalent connection system $p_n'(n \in N)$ of the following form (cf. [5]):

$$p_n' y(\xi) = (y(\xi_{in}),\, i = 1,2, \ldots, n) \quad . \tag{3'}$$

Take also an equivalent connection system $q_n'(n \in N)$ in the same form (3'). Then we see immediately that

$$\left| \sum_{i=1}^{n} (z_{in}, y_{in})\mu_{in} - \sum_{i=1}^{n} (z(\xi_{in}), y(\xi_{in}))\mu_{in} \right| \to 0 \quad a.s. \quad n \to \infty \quad ,$$

i.e., the compatibility condition is fulfilled for our approximations.

Define

$$F(x, y) = \int f_2(\xi, x, y(\xi))\mu(d\xi)$$

and

$$F_n(x, y_n) = \sum_{i=1}^{n} f_2(\xi_{in}, x, y_{in})\mu_{in} \quad .$$

We shall impose the following restrictions on the function $f_2(\xi, x, y)$:

(R3) the function $f_2(\xi, x, y)$ is continuous in (ξ, x, y) and convex and differentiable in y for all (ξ, x); the function $f_{2y}(\xi, x, y)$ is continuous in (ξ, x, y).

LEMMA 1. *Let a sequence of discrete probability measures μ_n converge weakly to the probability measure μ, and the restrictions* (R1)–(R3) *be satisfied. Then the pair $\{F(x, y), (F_n(x, y_n))\}$ is discretely upper-semicontinuous.*

Proof. Consider the inequality

$$\sum_{i=1}^{n} f_2(\xi_{in}, x, y_{in})\mu_{in} - \int f_2(\xi, x, y(\xi))\mu(d\xi) \le$$

$$\le \sum_{i=1}^{n} (f_{2y}'(\xi_{in}, x, y_{in}),\, y_{in} - p_n' y(\xi))\mu_{in} +$$

$$+ \sum_{i=1}^{n} f_2(\xi_{in}, x, p_n' y(\xi))\mu_{in} - \int f_2(\xi, x, y(\xi))\mu(d\xi) \quad .$$

In view of restrictions (R1) and (R2) we are able to use a connection operator p_n' of form (3') for the function $y(\xi)$, since in essence the restriction (R2) contracts the measure μ to the algebra $\sum_0$ generated by sets A which satisfy

$$\mu(A^0) = \mu(A) = \mu(\bar{A}) \quad ,$$

where A^0 and $\bar{A}$ denote the interior and the closure of the set A, respectively. From this restriction we then infer that the function $y(\xi)$ will be Riemann-integrable (for details see [6]). Then from the restrictions on $f_{2y}'(\xi, x, y)$ and the discrete convergence $y_n \to y(\xi)$ we have

$$\sum_{i=1}^{n} (f_{2y}'(\xi_{in}, x, y_{in}), y_{in} - p_n' y(\xi))\mu_{in} \leq$$

$$\leq \max_{1 \leq i \leq n} |y_{in} - y(\xi_{in})| \sum_{i=1}^{n} |f_2' y(\xi_{in}, x, y_{in})| \mu_{in} \leq \varepsilon/2$$

for $n \geq n_1$.

The weak convergence of the discrete probability measures μ_n together with the restrictions (R1)–(R3) guarantee that the inequality

$$\left| \sum_{i=1}^{n} f_2(\xi_{in}, x, p_n' y(\xi))\mu_{in} - \int f_2(\xi, x, y(\xi))\mu(d\xi) \right| \leq \varepsilon/2 \quad .$$

will be satisfied for $n \geq n_2$. This completes the proof.

We shall say that the sequence $y_n (n \in N)$, $y_n \in l_n^\infty$, is *weakly discrete compact* if there exist $N' \subset N$ and $y(\xi) \in L^\infty$ such that $y_n \to y(n \in N)$.

LEMMA 2. *Every bounded sequence $y_n (n \in N)$, $y_n \in l_n^\infty$, is weakly discretely compact.*

Proof. Consider the sequence $(q_n^* y_n) \in L^\infty$, where $q_n^* \in \mathbf{L}(l_n^\infty, L^\infty)$ is adjoint to the operator $q_n \in \mathbf{L}(L^1, l_n^1)$. Let $\|y_n\|_n \leq$ const $(n \in N)$. Since condition (4) is fulfilled (diam $A_{in} \to 0$) for operators q_n of form (3), we have

$$\|q_n^*\| = \|q_n\| \leq \text{const}(n \in N) \quad .$$

Thus, the sequence $q_n^* y_n (n \in N)$ is bounded in L^∞. It is well-known that in L^∞ every bounded sequence is compact in the sense of weak "star" topology (i.e., in the topology imposed on L^∞ by L^1). Hence we can extract a subsequence $q_n^* y (n \in N')$ from the sequence $q_n^* y (n \in N)$ such that $\langle z, q_n^* y_n \rangle \to \langle z, \bar{y} \rangle$ $(n \in N')$ for every $z \in L^1$. By the same token

$$<q_n z , y_n> \to <z , \bar{y}>(n \in N') \ \forall z \in L^1 \quad .$$

The criterion concerning the weak discrete convergence of functionals [5] then guarantees that $y_n \to y(n \in N')$, i.e., the sequence $y_n (n \in N)$ is weakly discretely compact. This completes the proof.

We are now able to formulate and prove the main result of this paper, which concerns the discrete stability of convex stochastic programming problems with recourse.

We shall impose the following restrictions on the functions $g_{2j}(\xi, x, y)$, $j \in J_2$:

(R4)The functions $g_{2j}(\xi, x, y)$, $j \in J_2$, are continuous in (ξ, x, y), and convex and differentiable in y for all (ξ, x); the functions $g'_{2jy}(\xi, x, y)$, $j \in J_2$, are continuous in (ξ, x, y) and for a fixed x

$$\sup_{\xi \in \Xi} \{ |y| \ | g_{2j}(\xi, x, y) \le 0 \} \le \alpha < \infty , \ j \in J_2 \quad .$$

(R5)The functions $f_1(x)$, $g_{1j}(x)$, $j \in J_1$, are convex and differentiable.

Note that if the sets of constraints in problems (1) and (1n) $(n \in N)$ are not empty then the conditions (R1)–(R5) guarantee the existence of $(x^*, y^*, (\xi)) \in R^r \times L^\infty$ and $(x_n^*, y_n^*) \in R^r \times l_n^\infty$ $(n \in N)$ corresponding to the minimum in problems (1) and (1n) $(n \in N)$, respectively [7].

THEOREM 1. *Let the restrictions* (R1)–(R5) *be satisfied, the sets* C_1 *and* C_2 *be bounded, and the sequence of discrete probability measures* $\mu_n (n \in N)$ *converge weakly to the probability measure* μ. *If the constraint sets of problems* (1) *and* (1n) $(n \subset N)$ *are not empty, then*

$$F_n(x_n^*, y_n^*) \to F(x^*, y^*) \quad a.s. \quad n \to \infty$$

and we can extract a subsequence from the sequence of solutions (x_n^*, y_n^*) *to problems* (1n) $(n \in N)$ *which displays weak discrete convergence to the solution of problem* (1).

Proof. The contraction of the measure μ implied by (R2) allows us to use the equivalent connection system of operators $p'_n (n \in N)$ of form (3'). Thus $p'_n y^*(\xi)$ is an admissible point for the problems (1n) $(n \in N)$. From Lemma 1 the pair $\{F(x, y), \{F_n(x, y_n))\}$ is discretely upper-semicontinuous and hence for $n \ge n_1$ we have

$$\lim_{n \to \infty} \sup F_n(x_n^*, p_n y^*) \le F(x^*, y^*) \quad ,$$

where p_n is the connection operator from (3).

We shall now prove the converse. By Lemma 2, the sequence (x_n^*, y_n^*) $(n \in N)$ is weakly discretely compact, i.e.,

$$(x_n^*, y_n^*) \to (\bar{x}, \bar{y}(\xi))(n \in N') \quad .$$

We shall show that the point $(\bar{x}, \bar{y}(\xi)) \in R^r \times L^\infty$ is admissible. Assuming the converse, let there exist a sphere $S \in \sum$ with positive measure $\mu(S) > 0$ such that for every $\xi \in S$ and some $k \in J_2$ we have

$$g_{2k}(\xi, \bar{x}, \bar{y}(\xi)) \geq \delta > 0 \quad .$$

Then we also have $\int \chi_S(\xi) g_{2k}(\xi, \bar{x}, \bar{y}(\xi))\mu(d\xi) \geq \delta\mu(S)$, where

$$\chi_s(\xi) = \begin{cases} 1, & \text{if } \xi \in S \quad , \\ 0 & , \text{ if } \xi \notin S \end{cases} \quad .$$

Since the points (x_n^*, y_n^*) $(n \in N')$ are admissible for problems (1n) $(n \in N')$, we have

$$\sum_{i=1}^{n} \chi_s(\xi_{in}) g_{2k}(\xi_{in}, x_n^*, y_{in}^*)\mu_{in} \leq 0 \quad .$$

Then

$$\delta\mu(S) \leq \int \chi_s(\xi) g_{2k}(\xi, \bar{x}, \bar{y}(\xi))\mu(d\xi) \leq$$

$$\leq \int \chi_s(\xi) g_{2k}(\xi, \bar{x}, \bar{y}(\xi))\mu(d\xi) - \sum_{i=1}^{n} \chi_s(\xi_{in}) g_{2k}(\xi_{in}, \bar{x}, \bar{y}(\xi_{in}))\mu_{in} +$$

$$+ \sum_{i=1}^{n} (\chi_s(\xi_{in}) g'_{2xy}(\xi_{in}, \bar{x}, \bar{y}(\xi_{in})) , \bar{y}(\xi_{in}) - y_{in}^*)\mu_{in} \quad .$$

Restrictions (R1), (R2) and (R4) guarantee that for $n \geq n_2$ both terms in the last sum are less than $\delta\mu(s)/4$. Thus the weak discrete limit point $(\bar{x}, \bar{y}(\xi))$ of the sequence $(x_n^*, y_n^*)(n \in N')$ is also an admissible point for problem (1).

We can also show that if the constraints (R1)–(R3) are satisfied then the pair $\{F(x, y) , (F_n(x, y_n))\}$ is weakly discretely lower-semicontinuous. Assuming the contrary, we can easily show that

$$\lim_{n\to\infty} \sup (F(\bar{x}, \bar{y}) - F_n(x_n^*, y_n^*)) \leq 0 \quad .$$

This completes the proof.

Remark. The discrete approximation scheme for linear multistage programming problems in reflexive spaces L^p proposed in [8] may diverge in the absence of restriction (R2) for the reasons described in Sectin 2 of this paper.

REFERENCES

1. Yu.M. Ermoliev. *Methods of Stochastic Programming*. Nauka, Moscow, 1976 (in Russian).

2. R.T. Rockafellar and R.J.-B. Wets. Stochastic convex programming: Basic duality. *Pacific Journal of Mathematics*, 62(1976) 173–195.

3. P. Olsen. Multistage stochastic programming with recourse as mathematical programming problem in an L_p space. *SIAM Journal on Control and Optimization*, 14(1976) 528–537.

4. J.W. Daniel. On the approximate minimization of functionals. *Mathematics and Computing*, 23(107) (1969) 573–581.

5. G. Vainikko. Approximative methods for nonlinear equations. *Nonlinear Analysis*, 2(1978) 647–687.

6. G. Vainikko. On the convergence of the quadrature formulae method for integral equations with discontinuous kernels (in Russian). *Sibirski Matematicheski Zhurnal*, 12(1971) 40–53.

7. B.T. Poljak. Semicontinuity of integral functionals and theorems for the existence of solutions in extremum problems (in Russian). *Matematicheski Sbornik*, 78(120) (1969) 65–84.

8. P. Olsen. Discretizations of multistage stochastic programming problems. *Mathematical Programming Study*, 6(1976) 111–124.

EXTREMAL PROBLEMS WITH PROBABILITY MEASURES, FUNCTIONALLY CLOSED PREORDERS AND STRONG STOCHASTIC DOMINANCE

V.L. Levin
Central Institute of Economics and Mathematics
Moscow, USSR

This work is devoted to extremal problems with probability measures on topological spaces* and their applications in some aspects of decision making.

Let X be a completely regular topological space and $\mathbf{B}(X)$ the σ-algebra of its Borel subsets. We shall use $C(X)$ to denote the vector space of continuous real-valued functions on X and $C^b(X)$ to denote the vector subspace of $C(X)$ consisting of bounded functions.

Let $V_+(X)$ denote the set of finite non-negative interiorly regular Borel measures on X, i.e., the set of countably additive functions σ: $\mathbf{B}(X) \rightarrow R^1_+$ satisfying the condition

$$\sigma B = \sup \{\sigma K\colon K \subset B \text{ , } K \text{ is compact } \} \quad \forall B \in \mathbf{B}(X) \quad .$$

Let

$$V(X) = V_+(X) - V_+(X)$$

$$V_0(X) = \{\rho \in V(X)\colon \rho X = 0\}$$

$$M(X) = \{\sigma \in V_+(X)\colon \sigma X = 1\} \quad .$$

The elements of $M(X)$ are called *probability measures on X*.

If $\sigma \in V(X)$ and a function φ: $x \rightarrow R^1$ is bounded and σ-measurable, the finite integral

$$\sigma(\varphi) \triangleq \int_X \varphi(x)\sigma(\mathrm{d}x)$$

is said to be *defined*. If $\sigma \in V_+(X)$, then the integral $\sigma(\varphi)$, (which is finite or equal to $+\infty$) is *well-defined* for any σ-measurable function φ: $X \rightarrow R^1 \cup \{+\infty\}$ which is bounded

*All topological spaces considered here are assumed to be Hausdorff.

below. Recall that a function φ is said to be *σ-measurable* if its Lebesgue sets

$$\{x \in X: \varphi(x) \le \alpha\}, \ \alpha \in R^1$$

belong to the completion of $\mathbf{B}(X)$ by the measure σ.

Any measure $\mu \in V(X \times X)$ may be associated with the pair of marginal measures $P_1\mu, P_2\mu \in V(X)$,

$$(P_1\mu)B = \mu(B \times X), \ (P_2\mu)B = \mu(X \times B) \quad \forall B \in \mathbf{B}(X) \quad .$$

Clearly, P_1, P_2 are linear operators $V(X \times X) \to V(X)$ and

$$P_i V_+(X \times X) = V_+(X), \ P_i M(X \times X) = M(X), \ i = 1,2 \quad .$$

Let $\sigma_1, \sigma_2 \in M(X)$ and let a function $c: X \times X \to R^1 \cup \{+\infty\}$ be bounded below, universally measurable (i.e., μ-measurable for any $\mu \in V_+(X \times X)$) and satisfy the triangle inequality

$$c(x, y) \le c(x, z) + c(z, y) \quad \forall x, y, z \in X \quad .$$

Take $\rho = \sigma_1 - \sigma_2$, so that $\rho \in V_0(X)$, and consider a pair of extremal problems (the mass translocation problem and the dual problem) which require us to find the values

$$\mathbf{A}(c, \rho) = \inf\{\mu(c): \mu \in V_+(X \times X), \ (P_1 - P_2)\mu = \rho\}$$

$$\mathbf{B}(c, \rho) = \sup\{\rho(u): u \in C^b(X), \ u(x) - u(y) \le c(x, y) \quad \forall x, y \in X\} \quad ,$$

respectively.

The duality problem is to describe the class of functions $c(x, y)$ for which the duality relation

$$\mathbf{A}(c, \rho) = \mathbf{B}(c, \rho) \quad \forall \rho \in V_0(X) \tag{1}$$

holds. For compact X the duality problem was completely solved in [1]. In this case (1) holds if and only if the function

$$\bar{c}(x, y) = \begin{cases} c(x, y) & \text{for } x \ne y \\ 0 & \text{for } x = y \end{cases}$$

is lower-semicontinuous (l.s.c.) on $X \times X$. It was also shown in [1] that if c is l.s.c. and $c(x, x) = 0 \ \forall x \in X$, then the representation

$$c(x, y) = \sup_{u \in Q} [u(x) - u(y)] \quad \forall x, y \in X \tag{2}$$

holds with $Q \subset C^b(X)$.

This paper is concerned with the duality problem for non-compact spaces.*

Definition 1. We shall say that X belongs to the class $\mathbf{L}$ if it is homeomorphic to a universally measurable subset of some compact space.

It is known that Polish spaces (separable metrizable spaces that may be metrized in such a way that they become complete) are homeomorphic to Borel (G_δ) subsets of a metrizable compact space. Therefore, Polish spaces belong to the class $\mathbf{L}$. Locally compact spaces and σ-compact spaces clearly belong to $\mathbf{L}$ as well. It is not difficult to show that $X, Y \in \mathbf{L} \Rightarrow X \times Y \in \mathbf{L}$. Thus, the class $\mathbf{L}$ is sufficiently wide.

Let βX denote the Stone-Cech compactification of X. The next lemma proves to be useful when dealing with spaces from $\mathbf{L}$:

LEMMA 1 ([2]). *The following assertions are equivalent:*

(1) $X \in \mathbf{L}$

(2) *X is universally measurable in βX.*

THEOREM 1. *Let $X \in \mathbf{L}$, $\sigma_1, \sigma_2 \in M(X)$. Suppose that a function c may be represented in the form* (2), *where $Q \subset C_b(X)$. Then $\mathbf{A}(c, \sigma_1 - \sigma_2) = \mathbf{B}(c, \sigma_1 - \sigma_2)$ and a measure $\mu \in M(X \times X)$ exists such that $P_1\mu = \sigma_1$, $P_2\mu = \sigma_2$, $\mu(c) = \mathbf{A}(c, \sigma_1 - \sigma_2)$.*

How wide is the class of functions that may be represented in the form (2)? If function $c(x, y)$ is continuous in x for every y, satisfies the triangle inequality and the equality $c(x, x) = 0 \; \forall x \in X$, and is either bounded or non-negative, then it can be represented in the required form. In the first case one can take $Q = \{u_z : z \in X\}$, $u_z(\cdot) = c(\cdot, z)$ and in the second $Q = \{u_{z,n} : z \in X, n = 1,2, \ldots\}$, where $u_{z,n}(\cdot) = \min[c(\cdot, z), n]$.

COROLLARY 1. *If $x \in \mathbf{L}$ and d is a continuous metric on X, then the assertion of the theorem is true for $c = d$.*

This result was first proved in [3] for metric compact spaces. In this case $C(X)$ is a Banach space with respect to the sup-norm, $V(X) = C(X)^*$ is the dual Banach space, and the function $d_1(\sigma_1, \sigma_2) = \mathbf{A}(d, \sigma_1 - \sigma_2)$ is a metric on $M(X)$ topologizing the weak* convergence of probability measures (the Kantorovich–Rubinstein metric).

Theorem 1 was completely proved for compact X in [1]; the case of continuous c was investigated in [4].

THEOREM 2. *Let $X \in \mathbf{L}$, a function $c: X \times X \to R^1 \cup \{+\infty\}$ be universally measurable and satisfy the triangle inequality. Suppose that a bounded universally measurable function $v(x)$ exists such that $v(x) - v(y) \le c(x, y) \; \forall x, y \in X$.*

*In [1] the duality problem was also solved for functions c failing to satisfy the triangle inequality; however, this goes beyond the scope of this work.

Then for the duality relation (1) *to be true it is necessary and sufficient that* c *can be represented in the form*

$$c(x,y) = \sup_{u \in Q} [u(x) - u(y)] \quad \forall x, y \in X, x \neq y \quad ,$$

where $Q \subset C^b(X)$.

For the proofs of Theorems 1 and 2 see [2].

Let us now look at applications. In decision problems one often has to compare two different alternatives or states (vectors of consumer goods, different modes of economic development, technological projects, etc.). The fact that pairs of states can be compared implies that the state space is endowed with a preference binary relation satisfying some intuitively acceptable conditions (axioms). This type of situation is generally formalized using the notion of preorder, i.e., with the help of a reflexive and transitive binary relation.

A preordering relation $\precsim$ on a set X is said to be *linear* (the terms "complete" and "connected" are also sometimes used), if every pair of elements of X is comparable, i.e., at least one of the relations $x \precsim y$ or $y \precsim x$ is satisfied for any $x, y \in X$.

A preordering relation $\precsim$ on a topological space X is said to be *closed* if its graph

$$\mathrm{gr}(\precsim) \triangleq \{(x,y) \colon x \precsim y\}$$

is closed in $X \times X$.

Any function $u : X \rightarrow R^1$ satisfying the conditions

$$x \precsim y \Rightarrow u(x) \leq u(y) \quad , \tag{3}$$

$$x \prec y \Rightarrow u(x) < u(y) \quad , \tag{4}$$

where $x \prec y \overset{\Delta}{\Longleftrightarrow} x \precsim y, y \not\precsim x$, is called a *utility function* of the preordering relation .

A function $u(x)$ which satisfies (3) is said to be *isotonic* with respect to $\precsim$.

If $\precsim$ is a linear preordering relation, then the pair of conditions (3), (4) is evidently equivalent to the single condition:

$$x \precsim y \Longleftrightarrow u(x) \leq u(y) \quad .$$

One of the fundamental results in mathematical economics and general decision theory is the Debreu theorem [5], which asserts the existence of a continuous utility function for any closed linear preordering relation on a separable metrizable space. It is not difficult to show that the assumptions that the space is metrizable and separable, and that the preordering relation is closed, cannot be omitted. The assumption that the preordering relation is linear is used in all existing proofs of the Debreu theorem, but it is unknown whether it is necessary for the theorem to be true. It should be noted that this assumption is sufficiently restrictive for both mathematical and economics purposes.

In [6] (see also [7]) I prove the existence of a continuous utility function for an arbitrary closed preordering relation on a separable locally compact metrizable space. The main step in the proof is to establish the representation

$$\mathrm{gr}(\precsim) = \{(x, y):\ u(x) \le u(y) \quad \forall u \in Q\} \quad ,$$

where Q is a non-empty set in $C^b(X)$. Here we basically use the duality theorem for a mass translocation problem on a compact space.

Definition 2. A preordering relation $\precsim$ on a completely regular space X is said to be functionally closed if its graph can be represented in form (5) with $Q \subset C^b(X)$.*

Thus, every closed preordering relation on a separable metrizable locally compact space is functionally closed.

THEOREM 3. *Let $\precsim$ be a preordering relation on a completely regular space x. Then the following assertions are equivalent:*

(1) *$\precsim$ is functionally closed;*

(2) *$\precsim$ is a restriction of some closed preordering relation on βX;*

(3) *(the extension theorem). For any compact set $F \subset X$ and function $v \in C^b(F)$ which is isotonic with respect to the restriction of $\precsim$ to F, there exists an isotonic function $u \in C^b(X)$ which coincides with v on F and satisfies the equalities*

$$\max u(X) = \max v(F) \ , \ \min u(X) = \min v(F) \quad ;$$

(4) *(the separation theorem) For any compact sets F_1, F_0, in X such that $(F_1 \times F_0) \cap \mathrm{gr}(\precsim) = \phi$, there exists a continuous isotonic function $u: X \to [0,1]$ which equals 1 on F_1 and 0 on F_0.*

Proof. (1) $\Rightarrow$ (2). Every function $u \in C^b(X)$ may be uniquely extended to X with preservation of continuity. Then $\precsim$ is a restriction to X of a preordering relation $\precsim_1$ defined on βX by

$$x \precsim_1 y \overset{\Delta}{\Longleftrightarrow} u(x) \le u(y) \quad \forall u \in Q \quad .$$

*IF gr ($\precsim$) can be represented in form (5) with $Q \subset C(X)$, then $\precsim$ is functionally closed because Q can be replaced in (5) by

$$Q_1 = \{\frac{u}{1+|u|}: u \in Q\} \subset C^b(X) \quad .$$

(2) $\Longrightarrow$ (3). Passing from X to βX, we reduce the extension theorem to a similar assertion for a compact space. The extension theorem is proved for a compact space in [8] (see also [6]).

(3) $\Longrightarrow$ (4). Let $F = F_1 \cup F_0$ and take a function $v \in C^b(F)$ which equals 1 on F_1 and 0 on F_0. It is isotonic and by assumption may be extended to an isotonic function $u \in C^b(X)$, $u(X) \subset [0,1]$.

(4) $\Longrightarrow$ (1). Let Q denote the set of all isotonic functions in $C^b(X)$. Suppose that $(x, y) \notin \mathrm{gr}(\precsim)$. Take the singleton sets $F_1 = \{x\}$, $F_0 = \{y\}$ and find an isotonic function $u \in C^b(X)$, $0 \le u \le 1$, $u(x) = 1$, $u(y) = 0$. By virtue of the arbitrariness of the pair $(x, y) \notin \mathrm{gr}(\precsim)$, this implies representation (5).

THEOREM 4. *Let $\precsim$ be a preordering relation on a separable metrizable space X. Then the following assertions are equivalent:*

(1) *representation* (5) *holds with a countable set* $Q \subset C^b(X)$;

(2) $\precsim$ *is a restriction to x* of a closed preordering relation $\quad_1$ *on* X_1, *where* X_1 *is some metrizable compactification X.*

If these equivalent assertions are true, the preordering relation permits a continuous utility function.

Proof. (1) $\Longrightarrow$ (2). Let $Q = \{u_1, u_2, \dots\} \subset C^b(X)$. We may suppose without loss of generality that $u_k(X) \subset C[0,1]$, $k = 1, 2, \dots$. Due to the metrizability and separability of X there exists a countable family of continuous functions $\varphi_k : X \to [0,1]$, $k = 1, 2, \dots$ separating points in X. Denote by Y the topological product of the countable family of segments [0,1] and consider a mapping $f : X \to Y \times Y$, where

$$f(x) = ((u_k(x))_{k=1}^{\infty}, (\varphi_k(x))_{k=1}^{\infty}) \quad .$$

The mapping f is a continuous embedding of X into $Y \times Y$. Let X_1 denote the closure of X in metrizable compact space $Y \times Y$ and consider the preordering relation $\precsim_1$ on X_1:

$$((a_k), (b_k)) \precsim_1 ((a_k'), (b_k')) \overset{\Delta}{\Longleftrightarrow} a_k \le a_k', \; k = 1, 2, \dots \quad .$$

Clearly, $\precsim_1$ has the desired property.

(2) $\Longrightarrow$ (1). Since $\precsim_1$ is a closed preordering relation on a compact space X_1, it is functionally closed, i.e.,

$$x \precsim_1 y \Longleftrightarrow u(x) \le u(y) \quad \forall u \in Q_1 \quad ,$$

where $Q \subset C(X_1)$. Further, by virtue of the fact that X_1 is metrizable, the Banach space $C(X_1)$ is separable; hence

$$x \precsim_1 y \iff u_k(x) \le u_k(y)\,,\ k = 1,2,\dots\ ,$$

where the sequence $\{u_1, u_2, \dots\}$ is dense in Q_1. Then representation (5) holds for $\mathrm{gr}\,(\precsim)$ with $Q = \{u_k \mid X\}_{k=1}^{\infty}$.

Finally, if the equivalent conditions (1), (2) are satisfied and $Q = \{u_k\}_{k=1}^{\infty}$, then

$$u_0(x) = \sum_{k=1}^{\infty} \frac{u_k(x)}{2^k(1+|u_k(x)|)}$$

is a continuous utility function.

THEOREM 5. *Let $\precsim$ be a preordering relation on a metric space (X, d). Consider the function*

$$c(x, y) = \begin{cases} 0\ , & \text{if } x \precsim y \\ \omega(d(x, y))\ , & \text{otherwise} \end{cases}\ ,$$

where $\omega\colon R^1_+ \to R^1_+$ is an increasing continuous function, $\omega(0) = 0$. Then the following assertions are equivalent:

(1) *representation* (5) *holds with $Q \subset C(X)$ and*

$$|u(x) - u(y)| \le \omega(d(x, y)) \quad \forall u \in Q\,,\ x\,,\ y \in X\ ;$$

(2) $c_*(x, y) > 0 \quad \forall (x, y) \notin \mathrm{gr}\,(\precsim)$, where

$$c_*(x, y) = \lim_{n\to\infty} \inf\,\{c(x, z_1) + c(z_1, z_2) + \dots + c(z_n, y)\colon z_1, \dots, z_n \in X\}\ .$$

If X is separable, and either of the equivalent assertions (1), (2) holds (in which case both assertions hold), then $\precsim$ has a continuous utility function u_0 satisfying the inequality

$$|u_0(x) - u_0(y)| \le \omega(d(x, y)) \quad \forall x\,,\ y \in X\ .$$

Note that in certain cases assertion (2) may be verified directly.

Proof. It is easy to see that c_* satisfies the triangle inequality and $c_*(x, y) \le \omega(d(x, y))$; therefore c_* is continuous as a function of two variables.

(1) $\Longrightarrow$ (2). From representation (5) and the definition of c we have

$$u(x) - u(y) \le c(x, y) \quad \forall u \in Q\,,\ x\,,\ y \in X\ ,$$

and hence

$$u(x) - u(y) \le c_*(x, y) \quad \forall u \in Q\,,\ x\,,\ y \in X\ .$$

If $(x, y) \notin \mathrm{gr}\,(\precsim)$, then, by virtue of (5), a function $u \in Q$ exists such that $u(x) - u(y) > 0$. We thus obtain

$$c_*(x, y) \geq u(x) - u(y) > 0 \quad .$$

(2) $\Longrightarrow$ (1). Take $Q = \{u_z : z \in X\}$, where $u_z(\cdot) = c_*(\cdot, z)$. We then have

$$u_z(x) - u_z(y) = c_*(x, z) - c_*(y, z) \leq c_*(x, y) \leq \omega(d(x, y)) \quad \forall u_z \in Q,\ x, y \in X \quad .$$

Further, if $(x, y) \in \mathrm{gr}\,(\precsim)$, then

$$u_z(x) - u_z(y) = c_*(x, z) - c_*(y, z) \leq c_*(x, y) = 0 \quad ,$$

and therefore $u_z(x) \leq u_z(y)\ \forall u_z \in Q$.

If $(x, y) \notin \mathrm{gr}\,(\precsim)$, then

$$u_y(x) - u_y(y) = c_*(x, y) > 0 \quad ,$$

and hence representation (5) holds.

The equivalence of assertions (1) and (2) is thus established.

Now let a sequence (x_k) be dense in X and the assertions (1), (2) be satisfied. Using the density of (x_k) in X and the continuity of c_*, we obtain

$$\begin{aligned} \mathrm{gr}\,(\precsim) &= \{(x, y) : c_*(x, z) \leq c_*(y, z) \quad \forall z \in X\} \\ &= \{(x, y) : c_*(x, x_k) \leq c_*(y, x_k),\ k = 1, 2, \ldots\} \\ &= \left\{(x, y) : \frac{c_*(x, x_k)}{1 + c_*(x, x_k)} \leq \frac{c_*(y, x_k)}{1 + c_*(y, x_k)},\ k = 1, 2, \ldots\right\} \quad , \end{aligned}$$

i.e., representation (5) holds with the countable set

$$Q = \left\{\frac{c_*(\cdot, x_k)}{1 + c_*(\cdot, x_k)} : k = 1, 2, \ldots\right\} \subset C^b(X) \quad .$$

Then

$$u_0(X) = \sum_{k=1}^{\infty} \frac{c_*(x, x_k)}{2^k (1 + c_*(x, x_k))}$$

is the required utility function.

Let (X, d) be a separable metric space with a bounded metric, and $\mathbf{F}(X)$ be the space of closed sets in X with the Hausdorff metric

$$d_H(A, B) \triangleq \max\,\{\inf\,\{\alpha > 0 : A^\alpha \supset B\},\ \inf\,\{\alpha > 0 : B^\alpha \supset A\}\} \quad ,$$

where

$$A^\alpha \triangleq \{x \in X: \text{dist}\,(x, A) < \alpha\} \quad \forall A \in \mathbf{F}(X),\ \alpha > 0 \quad ,$$

$$\text{dist}\,(x, A) \triangleq \inf\,\{d(x, y): y \in A\} \quad .$$

THEOREM 6. *There exists a function φ on $\mathbf{F}(X)$ with the following properties:*

(a) $|\varphi(A) - \varphi(B)| \le d_H(A, B)$;

(b) *if $A \subset B$ and $A \ne B$, then $\varphi(A) < \varphi(B)$;*

(c) $\varphi(A \cup B) \le \varphi(A) + \varphi(B)$.

Proof. Since X is a separable metric space, there exists a countable family of sets $A_n \in \mathbf{F}(X)$, $n = 1,2, \ldots,$ such that every $A \in \mathbf{F}(X)$ may be represented in the form

$$A = \cap\,\{A_n : n \in N(A)\} \quad ,$$

where

$$N(A) \triangleq \{n : A_n \supset A\} \quad .$$

Let

$$\varphi(A) = \sum_{n=1}^{\infty} \frac{1}{2^n} \inf\,\{\alpha > 0: A_n^\alpha \supset A\} \quad \forall A \in \mathbf{F}(X)$$

and check that the function φ has the required properties.

Let A, $B \in \mathbf{F}(X)$. If $A_n^\alpha \supset B$, then $A_n^{\alpha + d_H(A,B)} \supset A$ and hence

$$\inf\,\{\alpha > 0: A_n^\alpha \supset A\} \le \inf\,\{\alpha > 0: A_n^\alpha \supset B\} + d_H(A, B) \quad .$$

Replacing A and B, we obtain

$$\inf\,\{\alpha > 0: A_n^\alpha \supset B\} \le \inf\,\{\alpha > 0: A_n^\alpha \supset A\} + d_H(A, B) \quad .$$

Thus,

$$|\inf\,\{\alpha > 0: A_n^\alpha \supset B\} - \inf\,\{\alpha > 0: A_n^\alpha \supset A\}| \le d_H(A, B) \quad ,$$

Then

$$|\varphi(A) - \varphi(B)| \le \sum_{n=1}^{\infty} \frac{1}{2^n} |\inf\,\{\alpha: A_n^\alpha \supset A\} - \inf\,\{\alpha: A_n^\alpha \supset B\}| \le d_H(A, B) \quad .$$

This proves property (a).

If $A \subset B$ and $A \ne B$, then

$$\inf\,\{\alpha > 0: A_n^\alpha \supset A\} \le \inf\,\{\alpha > 0: A_n^\alpha \supset B\},\ n = 1,2,\ldots$$

and there exists an $n_0 \in N(A)\setminus N(B)$. We obtain

$$\inf\{\alpha > 0: A_{n_0}^{\alpha} \supset A\} = 0$$

$$\inf\{\alpha > 0: A_{n_0}^{\alpha} \supset B\} \geq \operatorname{dist}(x_0, A_{n_0}) > 0 \quad ,$$

where $x_0 \in B\setminus A_{n_0}$. This implies (b).

Finally, if $A_n^{\alpha} \supset A$, $A_n^{\beta} \supset B$, we have

$$A_n^{\alpha+\beta} \supset A_n^{\max(\alpha,\beta)} \supset A \cup B \quad .$$

Then

$$\inf\{\alpha > 0: A_n^{\alpha} \supset A\} + \inf\{\beta > 0: A_n^{\beta} \supset B\} \geq \inf\{\gamma > 0: A_n^{\gamma} \supset A \cup B\}$$

which implies (c), thus completing the proof.

COROLLARY 2. *Let $\precsim$ be a preordering relation on X. Suppose that a mapping $a: X \to \mathbf{F}(X)$,*

$$a(x) \triangleq \{y \in X: y \precsim x\}$$

is continuous with respect to the Hausdorff metric d_H on $\mathbf{F}(X)$. Then representation (5) *holds with countable $Q = \{u_n\} \subset C^b(X)$, where*

$$u_n(x) = \inf\{\alpha > 0: A_n^{\alpha} \supset a(x)\}\ ,\ n = 1,2,\ldots \quad ,$$

and $u_0(x) = \varphi(a(x))$ is a continuous utility function for $\precsim$.

Any closed preordering relation $\precsim$ on a completely regular space X may be associated with a preordering relation $\precsim_{\bullet}$ on $M(X)$:

$$\sigma_1 \precsim_{\bullet} \sigma_2 \overset{\Delta}{\Longleftrightarrow} \sigma_1(u) \leq \sigma_2(u) \quad \forall u \in H(\precsim) \quad ,$$

where $H(\precsim)$ is the cone of isotonic functions in $C^b(X)$. We call the preordering relation $\precsim_{\bullet}$ *strong stochastic dominance.*

If $(X, \precsim)$ is a real line segment with natural order, $\precsim_{\bullet}$ coincides with the usual stochastic dominance of probability measures $\precsim_{SD}$:

$$\sigma_1 \precsim_{SD} \sigma_2 \overset{\Delta}{\Longleftrightarrow} \sigma_1\{y: y \precsim x\} \geq \sigma_2\{y: y \precsim x\} \quad \forall x \in X \quad .$$

This has been studied in connection with problems of rational behavior under risk (see, for example, [9–12] and the works cited therein). Note that for sets $X \subset R^2$ with natural order, strong stochastic dominance $\precsim_{\bullet}$ does not coincide with but is strictly stronger than $\precsim_{SD}$ (an example is given in [8]).

THEOREM 7. *Let $X \in \mathbf{L}$ and $\precsim$ be a functionally closed preordering relation on X, $\sigma_1, \sigma_2 \in M(X)$. Then the following assertions are equivalent:*

(1) $\sigma_1 \precsim. \sigma_2$;

(2) *there exists a measure $\mu \in M(X \times X)$ such that $P_1\mu = \sigma_1$, $P_2\mu = \sigma_2$, and μ gr $(\precsim)) = 1$.*

For compact X Theorem 7 is proved in [8,13].

Proof. Consider the mass translocation problem on X with

$$c(x, y) = \begin{cases} 0 , & \text{if } (x, y) \in \text{gr}(\precsim) \\ +\infty , & \text{otherwise} \end{cases} .$$

This function satifies the triangle inequality and the equality $c(x, x) = 0 \ \forall x \in X$ because the preordering relation is transitive and reflexive. In addition, it can be represented in form (2), since $\precsim$ is functionally closed. Then by Theorem 1 the duality relation $\mathbf{A}(c, \sigma_1 - \sigma_2) = \mathbf{B}(c, \sigma_1 - \sigma_2)$ holds and a measure $\mu \in M(X \times X)$ exists such that $P_1\mu = \sigma_1$, $P_2\mu = \sigma_2$, $\mu(c) = \mathbf{A}(c, \sigma_1 - \sigma_2)$. Further, the condition $\sigma_1 \precsim. \sigma_2$ is evidently equivalent to the equality $\mathbf{B}(c, \sigma_1 - \sigma_2) = 0$ It remains only to note that the equality $\mu(c) = 0$ may be rewritten as $\mu(\text{gr}(\precsim)) = 0$.

A number of other characterizations of strong stochastic dominance may be derived from Theorem 7; the case of compact X is treated in [8].

Finally, I shall list some open problems.

Let X be a completely regular space.

(1) Does there exist an l.s.c. function $c: X \times X \to R^1 \cup \{+\infty\}$, bounded below, that satisfies the triangle inequality and the equality $c(x, x) = 0 \ \forall x \in X$ but cannot be represented in form (2)?

(2) Does there exist a closed preordering relation on X that is not functionally closed?

An affirmative answer to the second question would imply an affirmative answer to the first.

(3) Let X be a separable metrizable space. Does there exist a functionally closed preordering relation $\precsim$ on X such that gr $(\precsim)$ does not admit representation (5) with countable $Q \subset C^b(X)$?

REFERENCES

1. V.L. Levin and A.A. Milyutin. Mass translocation problem with discontinuous cost function and mass statement of duality relation for convex extremal problems (in Russian). *Uspekhi Matematicheskie Nauk*, 34(3)(1979)3–68; English translation in *Russian Mathematical Surveys*, 34(1979).

2. V.L. Levin. Mass translocation problem in topological space and probability measures with given marginals on product of two spaces (in Russian). *Doklady Academii Nauk SSSR*, 276(5)(1984)1059–1064; English translation in *Soviet Mathematics Doklady* (1984).

3. L.V. Kantorovich and G.S. Rubinstein. On a space of absolutely additive functions (in Russian). *Vestnik Leningradskogo Universiteta, Matematika, Mekhanika, Astronomiya*, 7(1958)52–59.

4. V.L. Levin. Monge–Kantorovich mass translocation problem (in Russian). In *Methods of Functional Analysis in Mathematical Economics*, Nauka, Moscow, 1978, pp. 23–55.

5. G. Debreu. Representation of a preference ordering by a numerical function. In *Decision Processes*, Wiley, New York, 1954, pp. 159–165.

6. V.L. Levin. Measurable utility theorems for closed and lexicographic preference relations (in Russian). *Doklady Academii Nauk SSSR*, 270(3)(1983)542–546; English translation in *Soviet Mathematics Doklady*, 27(3)(1983)639–643.

7. V.L. Levin. Continuous utility theorem for closed preorders on σ-compact metrizable space (in Russian). *Doklady Academii Nauk SSSR*, 273(4)(1983)800–804; English translation in *Soviet Mathematics Doklady* (1983).

8. V.L. Levin. Mass translocation problem, strong stochastic dominance and probability measures with given marginals on a product of two compact spaces (in Russian). Preprint, Central Institute of Economics and Mathematics, Moscow, 1983.

9. G.A. Whitmore and M.C. Findlay (Eds.). *Stochastic Dominance: An Approach to Decision Making under Risk*. Lexington, Massachusetts, 1977.

10. P.C. Fishburn. Stochastic dominance: theory and applications. In *The Role and Effectiveness of Theories of Decision in Practice*, London, 1975, pp. 60–72.

11. A.W. Marshall and I. Olkin. *Inequalities: Theory of Majorization and Its Applications*. Academic Press, 1979.

12. A.Ya. Kiruta, A.M. Rubinov and E.B. Yanovskaya. *Optimal Choice of Distributions in Complicated Social Economic Problems* (in Russian). Nauka, Leningrad, 1980.

13. V.L. Levin. Some applications of duality for mass translocation problems. Closed preference relations and Choquet theory (in Russian). *Doklady Academii Nauk SSSR*, 260(2)(1981)284–288; English translation in *Soviet Mathematics Doklady*, 24 (1981).

EXPECTED VALUE VERSUS PROBABILITY OF RUIN STRATEGIES

L.C. MacLean and W.T. Ziemba
Dalhousie University, Halifax, Canada
University of British Columbia, Vancouver, Canada

1. INTRODUCTION

In this paper we consider a discrete time control problem involving the accumulation of risky capital. Capital accumulation problems have occupied a prominent rôle in the literature of economics, finance and gambling. The general formulation involves investment decisions at each point in time with the return on investment being uncertain. The resulting stream of capital is a discrete time stochastic process depending upon the investment strategy. The problem is to choose a strategy which is best according to specified criteria.

The choice of criteria for evaluating strategies varies depending upon the discipline. The usual objective in models of capital growth in economics/finance is the maximization of the expected value of a function of capital (Hankasson 1979, Mirman 1979). The gambling/probability models emphasize the probability of ruin (Ferguson 1965, Gottlieb 1984) in specifying their objectives. Our purpose is to simultaneously evaluate a variety of properties of the capital accumulation process as the decision rule changes. It may be that optimizing according to one criterion at the expense of another is not satisfactory. In section 2 the process and properties of interest are defined. Then those properties are analytically evaluated in section 3 for the class of proportional decision rules. As well, an index trading off expectation and risk is defined, somewhat analogous to the mean-variance ratio in static portfolio theory. Finally, in section 4 some numerical examples are considered.

2. CAPITAL ACCUMULATION

The fundamental process we will study is generated by a stochastic return on investment. To develop the process, consider the following definitions:

1) (Ω,B,P) - a probability space with (Ω^t,B^t,P^t) the corresponding product space

2) y_0 - the initial amount of capital available

3) $Y_t(\omega^t)$ - capital available at the beginning of period t, given the history $\omega^t \varepsilon \Omega^t$.

4) $X_{it}(\omega\ )$ - capital investment in opportunity i, $i = 1,\dots,n$, in period t given the history $\omega^t \varepsilon \Omega^t$.

5) $K_i(\omega)$ - the return on a unit of capital invested in opportunity i, $i = 1,\dots,n$, given the event $\omega \varepsilon \Omega$. It is assumed the return function is favourable, that is, $E_\omega K_i(\omega) > 0$ for at least one i, $i = 1,\dots,n$.

So the return function we have defined here is linear and stationary (independent of t). The investment decision at time t can be written in terms of investment fractions $p_t^{(\omega^t)} = (p_{it}^{(\omega^t)},\dots,p_{nt}^{(\omega^t)})$, $\sum_i p_{it}(\omega^t) \leqq 1$, so that $X_{it}(\omega^t) = p_{it}(\omega^t)Y_t(\omega^t)$, $t = 0,1,\dots$. Then the accumulated capital at time t, given the investment policy $p(\omega^t) = (p_1(\omega_1),\dots,p_t(\omega^t))$, is

$$Y_t(p(\omega^t),\omega^t) = y_0 \Pi_{s=1}^{t} (1 + \textstyle\sum_i K_i(\omega_s)p_{is-1}(\omega^{s-1})) \quad . \tag{2.1}$$

For the stochastic process $\{Y_t(p)\}_{t=1}^{\infty}$ we are interested in the following properties:

1) $\mu_t(p) = E\ Y_t(p)$ - mean accumulation to time t

2) $\phi_t(p) = E \log(Y_t(p)^{1/t})$ - mean growth rate to time t

3) $\eta(p) = E\ \tau_{\{Y(p) \geqq U\}}$ - mean first passage time to the set $[U,\infty)$

4) $\gamma_t(p) = \text{Prob}[Y_t(p) \geqq b_t]$ - probability of specified accumulation at time t

5) $\alpha(p) = \text{Prob}[Y_t(p) \geqq b_t, t=1,\dots]$ - probability accumulation is above a specified path

6) $\beta(p) = \text{Prob}[\tau_{\{Y(p) \geqq U\}} < \tau_{\{Y(p) \leqq L\}}]$ - probability of reaching U before L.

Of these properties 1) - 3) are expected values whereas 4) - 6) consider the risk of achieving desired goals. The objective is to study the behaviour of these properties as functions of the policy p, a task which is made easier by restricting our attention to the class of fixed fraction strategies

$$S = \{p(\omega^\infty) | p_{it}(\omega^t) = p_i, \ \omega^t \in \Omega^t, \ t=1,2,\ldots, \ i=1,\ldots,n\} \quad . \qquad (2.2)$$

A partial justification for this restriction is given in the next theorem.

Theorem 1.

Given the stochastic process $Y(p)$ defined by (2.1) we have:

(i) If $\phi(p^*) = \sup_p\{\lim_{t\to\infty} \phi_t(p) | \alpha(p) \geqq \alpha\}$, then there is a $p' \in S$ with $\mu(p') = \mu(p^*)$.

(ii) If $\eta(p^*) = \inf_p\{\eta(p) | \beta(p) \geqq \beta\}$, then there is a $p'' \in S$ with with $\mu(p'') = \mu(p)$.

(iii) If $\phi(p^*) = \sup_p \lim \phi_t(p)$ and $p = r \cdot p^*$, $0 < r \leqq 1$, then for any $\varepsilon > 0$ there exist $T(\varepsilon)$ and $a > 1$ such that $\alpha(p) > 1 - \varepsilon$ for $b_t = 0 \ (t < T(\varepsilon))$, $b_t = y_0 a^t (t \geqq T(\varepsilon))$.

Since the results in this theorem are variations on results in the literature no proof will be given. For reference (i) is from MacLean and Ziemba (1985), (ii) from Gottlieb (1984) and (iii) from Harkansson (1979). The point to be made is that whether we are considering mean growth rate, first passage times or probability of ruin the class of fixed fraction strategies is acceptable for a stationary linear return function.

3. Computation of Measures

Suppose then we have a fixed fraction strategy $p \in S$ and $Z_t(p,\omega^t) = \log Y_t(p,\omega^t)$. From (2.1) $Z_t(p,\omega^t) = z_0 + \sum_{s=1}^t \log(1 + \sum_i K_i(\omega_s)p_i) = Z_{t-1}(p,\omega^{t-1}) + J(p,\omega_t)$ where $J(p,\omega) = \log(1 + \sum_i K_i(\omega)p_i)$ is a stationary jump process. That is, Z_t, $t = 1,\ldots,$ is a random walk and we will evaluate the various measures of interest with this process.

First consider the mean growth rate $\eta(p) = \lim_{t\to\infty} E \log(Y_t(p)^{1/2}) = \lim_{t\to\infty} 1/t \ E \ Z_t(p) = E \log(1 + \sum_i K_i(\omega)p_i)$.

For the remainder of this section we assume that Ω is a finite set, and therefore the jump process $J(p)$ has a discrete distribution with

$\text{Prob}[J(p,\omega) = J_i] = \pi_i(p)$, $i = 1,\ldots,N$. This is a reasonable assumption since in practice we would in most cases use a discrete approximation to any continuous distribution.

So consider the measure $\beta_{z_0}(p) = \text{Prob}[\tau\{Y(p) \geqq U\} < \tau\{Y(p) \leqq L\}$, starting from $y_0] = \text{Prob}[\tau\{Z(p) \geqq \log U\} < \tau\{Z(p) \leqq \log L\}$, starting from $z_0]$. Letting $\beta_z(p) = \text{Prob}[\tau\{Z(p) \geqq \log U\} < \tau\{Z(p) \leqq \log L\}$, starting from $z]$ we have the fundamental recursion

$$\beta_{z_0}(p) = \sum_{i=1}^{N} \beta_{z_0+J_i}(p)\, \pi_i(p) \quad . \tag{3.1}$$

We need to solve (3.1) for $\beta_z(p)$ subject to the boundary conditions

$$\beta_z(p) = 0 \; (z \leqq \log L) \; ; \qquad \beta(p) = 1 \; (z \geqq \log U) \quad . \tag{3.2}$$

It is easy to see that given the roots $\theta_1,\ldots,\theta_s$ to the equation

$$\sum_k \pi_k(p)\theta^k = 1 \tag{3.3}$$

then $\beta_z(p) = \sum A_k \theta_k^z$ is a solution to (3.1), where the A_k are chosen to satisfy the boundary conditions. If we take the minimum and maximum jumps, $J_m(p) < 0$ and $J_M(p) > 0$ respectively, then with the positive roots to (3.2) as $\theta_1 = 1$ and $\theta_2 = \theta$ we can solve a simple system so that only the extreme boundary conditions are satisfied as equalities ($\beta_{\log L} - J_m(p) = 0$, $\beta_{\log U} = 1$ for upper bound; $\beta_{\log L} = 0$, $\beta_{\log U + J}(p_m) = 1$ for lower bound). We have then

$$\frac{\theta^{z_0} - \theta^{\log L}}{\theta^{\log U + J_M(p)} - \theta^{\log L}} \leqq \beta_{z_0}(P) \leqq \frac{\theta^{z_0} - \theta^{\log L + J_m(p)}}{\theta^{\log U} - \theta^{\log L + J_m(p)}} \quad . \tag{3.4}$$

Next consider the measure $\eta_{z_0}(p) = E\, \tau\{Y(p) \geqq U\} U \{Y(p) \leqq L\}$, starting from z_0 . As before we have a recursion equation

$$\eta_{z_0}(p) = \sum \eta_{z_0+J_i}(p)\pi_i(p) + 1 \tag{3.5}$$

subject to the boundary conditions

$$\eta_z(p) = (z \leqq \log L \text{ or } z \geqq \log U) \quad . \tag{3.6}$$

Now one solution to (3.5) is given by

$$\eta_z^*(p) = \frac{-Z}{E\,J(p)} \quad . \tag{3.7}$$

Any other solution can be written as

$$\eta_z(p) = \eta_z^*(p) + \Delta_z(p) \quad .$$

From (3.5) we have that Δ_z satisfies the system

$$\Delta_z(p) = \sum \Delta_{z+J_i}(p)\pi_i(p) \tag{3.8}$$

Proceeding as before we have solutions of (3.8) of the form

$$\Delta_z(p) = \sum A_k\theta_k^z$$

and then solutions of (3.5) are given by

$$\eta_z(p) = \sum A_k\theta_k^z - \frac{Z}{E\,J(p)}$$

where the A_k are chosen to satisfy the boundary conditions (3.6). Again solving the boundary conditions as inequalities

$$(\eta_{\log L + J_m} = 0,\ \eta_{\log u + J_M} = 0;\ \eta_{\log L} = 0,\ \eta_{\log u} = 0) \ .$$

we have

$$\left[\frac{\log U - \log L}{E\,J(p)}\right]\left(\frac{\theta^{z_0} - \theta^{\log L}}{\theta^{\log U} - \theta^{\log L}}\right) - \frac{Z_0 - \log L}{E\,J(p)} \leq \eta_{z_0}(p) \leq$$

$$\left[\frac{(\log U + J_M(p) - (\log L + J_m(p)}{E\,J(p)}\right]\left(\frac{\theta^{z_0} - \theta^{\log L + J_m(p)}}{\theta^{\log U + J_M(p)} - \theta^{\log L + J_m(p)}}\right) -$$

$$\frac{Z_0 - (\log L + J_m(p)}{E\,J(p)} \quad . \tag{3.10}$$

If we consider the process $Z_t(p,\omega^t) = z_0 + \sum_{s=1}^{t} J(p,\omega_s)$, we can rewrite this as $Z_t(p) = z_0 + \sum_i n_iJ_i(p)$, where n_i = number of $J_i(p)$

jumps in t trials, has a multinomial distribution with $E(n_i) = t\,\pi_i$ and $\sigma^2(n_i) = Var(n_i) = t\,\pi_i\,(1-\pi_i)$. Then $Z_t(p)$ has an approximate normal distribution $Z_t(p) \sim N(z_0 + t \sum_i \pi_i J_i(i-p),\ t \sum_i \pi_i (1-\pi_i) J_i^2(p))$. With this approximation we have

$$\gamma_t(p) \simeq 1 - F\left(\frac{\beta_t - z_0 - t \sum \pi_i J_i(p)}{\sqrt{t \sum \pi_i (1-\pi_i) J_i^2(p)}} \right) \tag{3.11}$$

where F is the cumulative for the standard normal.

Finally consider the quantity $\alpha(p)$. We have (with $b_t' = \log b_t$)

$$\alpha(p) = Pr[Z_t(p) \geqq b_t',\ t=1,\ldots] =$$

$$\cdot\ \Pi_{t=1}^{\infty} Pr[Z_t(p) \geqq b_t' | Z_s(p) \geqq b_s',\ s=1,\ldots,t-1] = \Pi_{t=1}^{\infty} \alpha_t(p)\ .$$

Then

$$\alpha_t = \sum_{z_t \geqq b_t'} P_r[Z_t = z_t | Z_s \geqq b_s',\ s=1,\ldots,t-1\}$$

$$= \sum_{z_t \geqq b_t'} [\sum_{z_t - J_i \geqq b_{t-1}'} Pr[Z_{t-1} = z_t - J_i | Z_s \geqq b_s',\ s=1,\ldots,t-1] \pi_i]$$

$$= \sum_{z_t \geqq b_t'} [\sum_{z_t - J_i \geqq b_{t-1}'} \frac{Pr[Z_{t-1} = z_t - J_i | Z_s \geqq b_s',\ s=1,\ldots,t-2]}{Pr[Z_{t-1} \geqq b_{t-1}' | Z_s \geqq b_s',\ s=1,\ldots,t-2]} \cdot \Pi_i]$$

which we will write as

$$\alpha_t(p) = \sum_{z_t \geqq b_t'} [\sum_{J_i \leqq z_t - b_{t-1}'} \frac{g_{t-1}(z_t - J_i(p))}{\alpha_{t-1}(p)} \cdot \pi_i]\ . \tag{3.12}$$

In this latter expression g_{t-1} is the conditional distribution for Z_{t-1} which is known from the previous stage. So we have a sequential computational procedure requiring only the distribution from the previous stage and the jump probabilities.

Bringing together the various measures we have the following:

1) $\mu(p) = \lim_{t\to\infty} E\, Y_t(p) = \infty$.

2) $\phi(p) = \lim_{t\to\infty} E \log(Y_t(p)^{1/2}) = E\, J(p)$.

3) $$\eta(p) = \frac{(\log U - \log L)(\theta^{z_0} - \theta^{\log L})(\theta^{\log U + J_M} - \theta^{\log L + J_m}) + (\log U + J_M - \log L - J_m)(\theta^{z_0} - \theta^{\log L + J_m})(\theta^{\log U} - \theta^{\log L})}{2EJ\cdot(\theta^{\log U} - \theta^{\log L})(\theta^{\log U + J_M} - \theta^{\log L + J_m})} - \frac{z_0 - \log L - \frac{1}{2} J_m}{E\, J} .$$

4) $$Y_t(p) = 1 - F\left(\frac{b_t' - z_0 - t\, E\, J(p)}{t\, \delta^2(J(p)}\right) .$$

5) $$\alpha(p) = \Pi_{t=1}^{\infty} \alpha_t(p), \qquad \alpha_t(p) = \sum_{z_t \geq b_t'} [\sum_{J_i \leq z_t - b_{t-1}'} \frac{g_{t-1}(z_t - J_i(p)}{\alpha_{t-1}(p)} \cdot \pi_i] .$$

6) $$\beta(p) = \frac{(\theta^{Z_0} - \theta^{\log L})(\theta^{\log U} - \theta^{\log L + J_m}) + (\theta^{z_0} - \theta^{\log L + J_m})(\theta^{\log U + J_M} - \theta^{\log L})}{2(\theta^{\log U + J_M} - \theta^{\log L})(\theta^{\log U} - \theta^{\log L + J_m})} .$$

Since the objective is to consider various measures simultaneously in evaluating a decision rule, this task would be simplified by an index combining measures. For example, consider the trade-off index for $\phi(p)$ and $\beta(p)$ given by

$$I(p) = \frac{\nabla\phi(p) \cdot \beta(p^*)}{\nabla\beta(p) \cdot \phi(p^{**})} \quad . \tag{3.13}$$

In this definition ∇ is the differential and p^*, p^{**} the optimal decision rules for β and ϕ respectively.

4. EXAMPLES

In this final section we will look at simple examples of some of the measures being discussed. For this purpose we will take $\Omega = \{0,1\}$ and consider a single investment opportunity with return $K_0 = 1$ with probability π and $K_1 = -1$ with probability $1 - \pi$.

Figure 1a gives the effect on $\alpha(p)$ of reducing the investment proportion p. The strategy which maximizes growth rate $\phi(p)$ is called the KELLY fraction. We see that halving this fraction increases $\beta_{z_0}(p)$ -- the probability of reaching \$2,500 before falling to \$0, \$500 and \$1,000 respectively, for various wealth levels z_0, when $\pi = .625$.

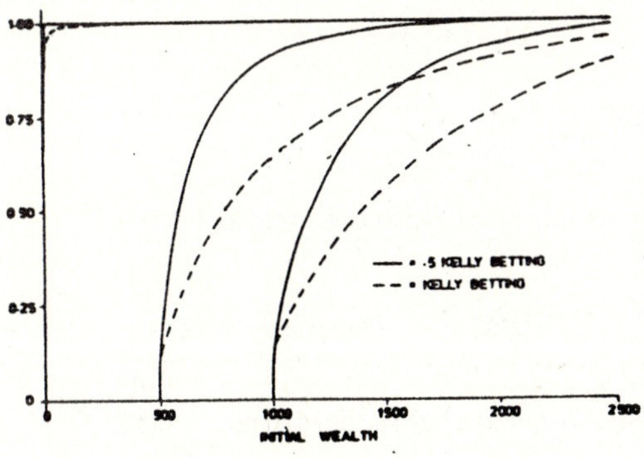

FIGURE 1 $\beta_{z_0}(p)$ for $\pi = .625$

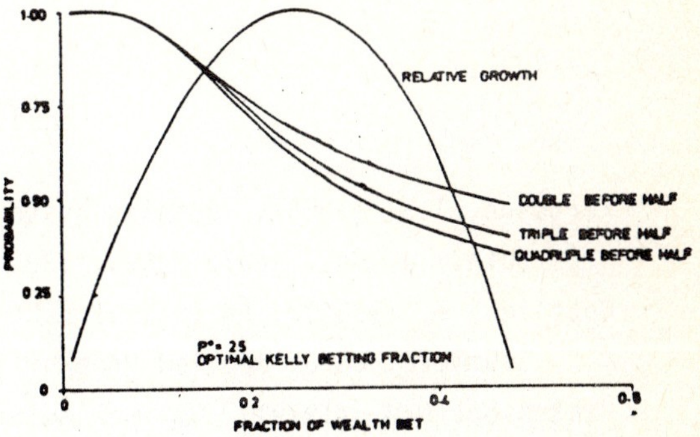

FIGURE 1b $\phi(p), \beta(p)$ for $\pi = .625$

In Figure 1b we have a comparison of $\phi(p)$ -- growth rate, and $\beta(p)$ -- probability of A before B . For $p \leqq p^* = .25$ the growth rate increases while the security decreases with p . A similar pattern is displayed in Figure 2a for $\phi(p)$ and $\beta(p)$ in this case with $\pi = .52$. Note that as the advantage decreases (EK from .25 to .02) the growth and security are less and the range of acceptable investment fractions is smaller.

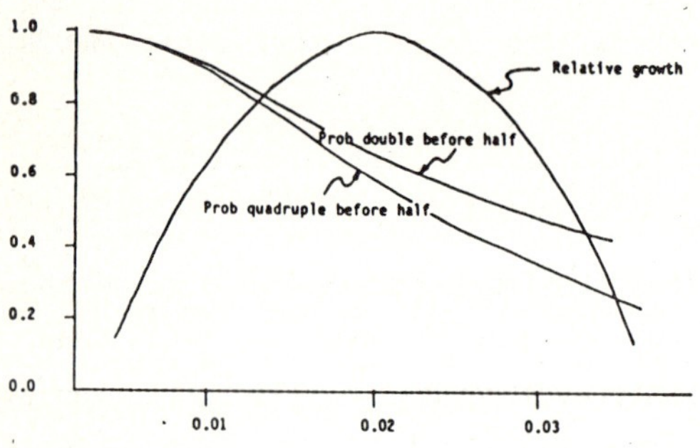

FIGURE 2a $\phi(p)$ and $\beta(p)$ for $\pi = .52$

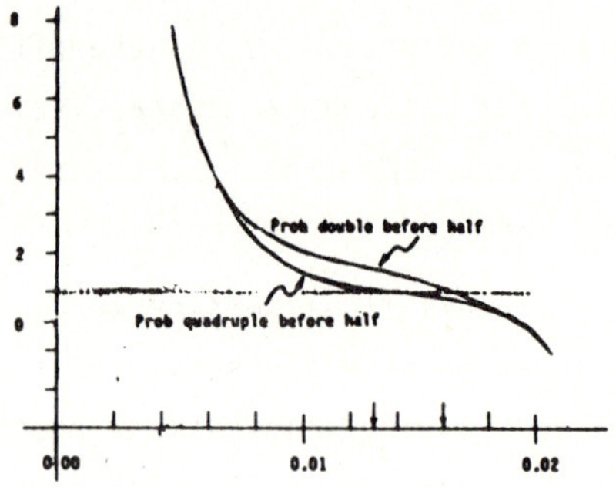

FIGURE 2b $I(p)$ for $\pi = .52$

To facilitate the comparison of ϕ and β we have presented the trade-off index in Figure 2b. The equilibrium values where the rates of change of ϕ and β are equal are highlighted. These values are at $p = .016$ and $p = .013$ for double before half and quadruple before half respectively (80% and 65% of the optimal growth value $p = .02$).

REFERENCES

Ferguson, T.S. (1965). Betting systems which minimize the probability of ruin. J. Soc. Appl. Math., 13.

Gottlieb, G. (1984). An optimal betting strategy for repeated games. Submitted to J. Applied Probability.

Hakansson, N.H. (1979). Optimal multi-period portfolio policies. TIMS Studies in the Management Sciences, 11. North-Holland.

MacLean, L. and Ziemba, W.T. (1985). Optimality of Proportional Investment. Research Report, School of Business Administration, Dalhousie University.

Mirman, L. (1979). One sector economic growth and uncertainty: A survey. In: Stochastic Programming. M.A.H. Dempster, Ed. Academic Press.

CONTROLLED RANDOM SEARCH PROCEDURES FOR GLOBAL OPTIMIZATION

K. Marti
HSBw Munich, FB LRT
Werner-Heisenberg-Weg 39
8-8014 Neubiberg/Munich
W-Germany

1. Introduction

Solving optimization problems arising from engineering and economics, as e.g. parameter- or process - optimization problems,

minimize $F(x)$ s.t. $x \in D$, (1)

where D is a measurable subset of $\mathbb{R}^d$ and F is a measurable real function defined (at least) on D, one meets often the following situation:

I) One should find the global minimum F^* and/or a global minimum point x^* of (1). Hence, most of the deterministic programming procedures, which are based on local improvements of the objective function $F(x)$, will fail.

II) Concerning the objective function $F(x)$ one has a black-box-situation, i.e. there is only few a priori information about F, especially there is no (complete) knowledge about the direct functional relationship between the control or input vector $x \in D$ and its function value $y = F(x)$. Hence - besides the limited a priori information about F - only by evaluating F numerically or by experiments at certain points $z_1, z_2, \ldots$ of $\mathbb{R}^d$ one gets further information about F.

Consequently, engineers use in these situations usually a certain search procedure for finding the global minimum F^* and an optimal solution x^* of (1), see e.g. Box'EVOP- method (1957) and the random search methods as first proposed by Anderson (1953),

Brooks (1959) and Karnopp (1963). More recent descriptions of random search procedures were given by Schwefel (1977), Rappl (1978), Marti (1980 a-c), Müller et.al. (1983), Zielinski and Neumann (1983).

In the random search method considered in this paper the sequence $X_o(\omega), X_1(\omega), \ldots, X_n(\omega), \ldots$ of random iterates is constructed according to the following recurrence scheme:

$$X_{n+1}(\omega) = \begin{cases} z_{n+1}, \text{ if } z_{n+1} \in D \text{ and } F(z_{n+1}) < F(X_n(\omega)) \\ X_n(\omega), \text{ else,} \end{cases} \tag{2}$$

$n=0,1,\ldots,$ where $X_o(\omega)=x_o \in D$ is a given starting-point in D and $z_1, z_2, \ldots, z_n, \ldots$ are realizations of a sequence of random d-vectors

$$Z_1(\omega), Z_2(\omega), \ldots, Z_n(\omega), \ldots$$

having conditional distributions

$$\begin{aligned} &P(Z_{n+1}(\omega) \in B \mid X_o=x_o, X_1=x_1, \ldots, X_n=x_n,\ Z_1=z_1, \ldots, Z_n=z_n) = \\ &= P(Z_{n+1}(\omega) \in B \mid X_n=x_n) = \\ &= \pi_n(x_n, B) \end{aligned} \tag{3}$$

for each Borel subset B of $\mathbb{R}^d$. Here

$$\pi_n(x_n, \cdot),\ n=0,1,2,\ldots$$

is a sequence of transition probability measures which may be selected by the engineer. In many concrete cases Z_{n+1} has a d-dimensional normal distribution with mean μ_n and covariance matrix Λ_n, i.e.

$$\pi_n(x_n, \cdot) = N(\mu_n, \Lambda_n),\ n=0,1,\ldots,$$

where $\mu_n=\mu_n(x_n)$ and $\Lambda_n=\Lambda_n(x_n)$ are certain functions of the last state (n, x_n).

Let the area of success $G_F(x)$ at a point $x \in \mathbb{R}^d$ be defined by

$$G_F(x) = \{y \in D: F(y) < F(x)\}. \tag{4}$$

At an iteration point x_n by the random search procedure (2) a d-vector z_{n+1} is generated randomly according to the transition probability distribution $\pi_n(x_n, \cdot)$ and from x_n we move to $x_{n+1} = z_{n+1}$ provided that $z_{n+1} \in G_F(x_n)$. Otherwise we stay at $x_{n+1}=x_n$

and generate a new random point z_{n+2} according to the distribution $\pi_{n+1}(x_{n+1},\cdot) = \pi_{n+1}(x_n,\cdot)$.

We observe that $X_{n+1} \in G_F(x_n)$ implies $X_t \in G_F(x_n)$ for all $t>n$. Let the set D_ε of ε-optimal solutions of our global minimization problem (1) be defined by

$$D_\varepsilon = \{y \in D: F(y) \leq F^* + \varepsilon\}, \tag{5}$$

where $\varepsilon>0$ and F^* is given by

$$F^* = \inf\{F(x): x \in D\};$$

let $F^*>-\infty$. We observe that $X_n \in D_\varepsilon$ implies $X_t \in D_\varepsilon$ for all $t>n$, hence

$$P(X_n \in D_\varepsilon),\ n=1,1,\ldots$$

is a nonincreasing sequence for each fixed $\varepsilon>0$.

2. Convergence of the random search procedure (2)

Let $\alpha_n(D_\varepsilon)$ denote the minimal probability that at the nth iteration step $X_n \rightarrow X_{n+1}$ we reach the set D_ε from any point $X_n=x_n$ outside this set, i.e.

$$\alpha_n(D_\varepsilon) = \inf\{\pi_n(x_n,D_\varepsilon): x_n \in D \setminus D_\varepsilon\}. \tag{6}$$

According to Marti (1980) we have this

Theorem 2.1

a) If for an $\varepsilon>0$

$$\sum_{n=0}^{\infty} \alpha_n(D_\varepsilon) = +\infty, \tag{7}$$

then $\lim_{n\to\infty} P(X_n \in D_\varepsilon) = 1$ for every starting-point $x_0 \in D$.

b) Suppose that

$$\lim_{n\to\infty} P(X_n \in D_\varepsilon) = 1 \text{ for every } \varepsilon>0. \tag{8}$$

Then $\lim_{n\to\infty} F(X_n) = F^*$ w.p.1 (with probability one) for every starting-point $x_0 \in D$.

c) Assume that F is continuous and that the level sets D_ε are nonempty and compact for each $\varepsilon>0$. Then $\lim_{n\to\infty} F(X_n) = F^*$ implies that also $\lim_{n\to\infty} \mathrm{dist}(X_n,D^*) = 0$, where $\mathrm{dist}(X_n,D^*)$ denotes the distance between X_n and the set $D^*=D_0$ of global points x^* of D.

Example

If $\pi_n(x_n,\cdot) = \pi(\cdot)$ is a fixed probability measure, then $\lim_{n\to\infty} F(X_n) = F^*$ w.p.1 holds provided that

$\pi(\{y \in D: F(y) \le F^* + \varepsilon\}) > 0$ for each $\varepsilon > 0$.

This is true e.g. if D_ε has a nonzero Lebesgue measure for all $\varepsilon > 0$ and π has a probability density ϕ with $\phi(x) > 0$ almost everywhere.

Note

Further convergence results of this type were given by Oppel (1976), Rappl (1978), Solis and Wets (1981).

Knowing several (weak) conditions which guarantee the convergence w.p.1 of (X_n) to the global minimum F^* (to the set of global minimum points D^*, resp.), one should also have some information concerning the rate of convergence of (X_n) to F^*, D^*, respectively.

By Rappl's doctoral thesis (1984) we have now the following result. Of course, as in the deterministic optimization, in order to prove theorems about the speed of convergence, the optimization problem (1) must fulfill some additional regularity conditions.

Theorem 2.2

Suppose that $D^* \neq \emptyset$ and the transition probability measure $\pi(x_n,\cdot)$ is a d-dimensional normal distribution $N(\mu(x_n),\Lambda)$ with a fixed covariance matrix Λ.

a) Let D_ε be bounded for some $\varepsilon = \varepsilon_0 > 0$ and assume that F is convex in a certain neighbourhood of D^*. Then

$$\lim_{n\to\infty} n^\gamma (F(X_n) - F^*) = 0 \quad \text{w.p.1} \tag{9}$$

for each constant γ such that $0 < \gamma < \frac{1}{d}$ and every starting-point x_o.

b) Let D be compact, $D^* = \{x^*\}$, where $x^* \in \overset{o}{D}$(= interior of D), and suppose that F is continuous and twice continuously differentiable in a certain neigbourhood of x^*. Moreover, assume that F has a positive definite Hessian matrix at x^*. Then for each starting-point $x_o \in D$ it is

$$\lim_{n\to\infty} n^\gamma (F(X_n) - F^*) = 0 \text{ w.p.1 for each } 0 < \gamma < \frac{2}{d},$$

$$\lim_{n\to\infty} n^\gamma \|X_n - x^*\| = 0 \quad \text{w.p.1 for each } 0 < \gamma < \frac{1}{d}, \tag{10}$$

$$\limsup_{n\to\infty} n^{\frac{2}{d}} E(F(X_n)-F^*) \le \tau(x_o)<+\infty,$$

where $\tau(x_o)$ is a nonnegative finite constant depending on the starting-point $x_o \in D$ and "E" denotes the expectation operator.

c) Under the same assumptions as in (b) we also have for each starting-point $x_o \in D, x_o \neq x^*$,

$$\liminf_{n\to\infty} n^{\frac{2}{d}} E(F(X_n)-F^*) \ge h(x_o), \tag{11}$$

where $h(x_o)$ is a nonnegative constant depending on the starting point x_o. Furthermore for each $x_o \in D$, $x_o \neq x^*$, it is

$$\liminf_{n\to\infty} n^{\gamma}||X_n-x^*|| = +\infty \text{ for each } \gamma>\frac{2}{d}. \tag{12}$$

Note

a) Theorem 2.2 holds also for many non-normal classes of transition probability measures $\pi_n(x_n,\cdot)$, see Rappl (1984).

b) It turns out that under the assumptions of Theorem 2.2b the speed of convergence of (2) to the global minimum of (1) is exactly given by

$$E(F(X_n)-F^*) = O(n^{-\frac{2}{d}}). \tag{13}$$

c) The above convergence rates reflect the fact that in practice one observes that the speed of convergence may be very poor - especially near to the optimum of (1).

Hence, using random search procedures, a main problem is the control of the basic random search algorithm (2) such that the speed of convergence of (X_n) to F^*, D^*, respectively, is increased.

3. Controlled Random Search methods

A general procedure how to speed up the search routine (2) is described in Marti (1980 a-c).

By the following items (I)-(III) a sequential stochastic decision process is associated with the random search routine (2):

I. We observe that the conditional probability distribution $\pi_n(x_n,\cdot)$ of Z_{n+1} given $X_n=x_n$ depends in general on a certain (vector valued) parameter a, i.e.

$$\pi_n(x_n,\cdot) = \pi_n(a,x_n,\cdot),\ a \in A, \tag{15}$$

where A is the set of admissible parameters a. The idea, developed first in Marti (1980 a-c), is now to run the algorithm (2) not with a fixed parameter a, but to use an "optimal" control

$$a = a_n^*(x_n) \tag{16}$$

of a such that a certain criterion - to be explained in (II) - is maximized.

In the present paper $\pi_n(x_n,\cdot)$ is assumed to be a d-dimensional normal distribution with mean μ_n and covariance matrix Λ_n. Hence, in our case we have

$$a = (\mu,\Lambda) \in A = M \times \mathbb{Q}, \tag{17}$$

where $M \subset \mathbb{R}^d$ and $\mathbb{Q} = \{0\} \cup \{\Lambda : \Lambda$ is a symmetric, positive definite d×d matrix$\}$.

II. To each search step $X_n \to X_{n+1}$ there is associated a mean search gain

$$U_n(a_n,x_n) = E(u(x_n,X_{n+1})|X_n=x_n), \tag{18}$$

where the gain function $u(x_n,x_{n+1})$ is defined e.g. by

$$u(x_n,x_{n+1}) = \begin{cases} 1, & \text{if } x_{n+1} \in G_F(x_n) \\ 0, & \text{else} \end{cases} \tag{19.1}$$

$$u(x_n,x_{n+1}) = \begin{cases} F(x_n)-F(x_{n+1}), & \text{if } x_{n+1} \in G_F(x_n) \\ 0, & \text{else} \end{cases} \tag{19.2}$$

$$u(x_n,x_{n+1}) = \begin{cases} ||x_n-x_{n+1}||, & \text{if } x_{n+1} \in G_F(x_n) \\ 0, & \text{else.} \end{cases} \tag{19.3}$$

Hence, in the first case $U_n(x_n)$ is the probability of a search success, in the second case $U_n(x_n)$ is the mean improvement of the value of the objective function and in case (19.3) $U_n(x_n)$ is the mean step length of a successful iteration step $X_n \to X_{n+1}$.

III. Obviously, the convergence behaviour of the random search process (X_n) can be improved now by maximizing the mean total search gain

$$U_\infty = U_\infty(a_0,a_1,\ldots) = E \sum_{n=0}^{\infty} \rho^n u(X_n,X_{n+1})$$

subject to the controls $a_n = a_n(x_n) \in A$, n=0,1,..., where $\rho>0$ is a certain discount factor. This maximization can be done in principle by the methods of stochastic dynamic programming, see e.g. Müller and Nollau (1984).

4. Computation of optimal controls

In order to weaken the computational complexity, the infinite stage stochastic decision process defined in section 3 is replaced by the sequence of 1-stage decision problems

$$\max U_n(a_n,x_n) \text{ subject to } a_n \in A,$$

n=0,1,2,..., hence the "optimal" control $a_n^* = a^*(x_n)$ is defined as a solution of

$$\underset{a \in A}{\text{maximize}} \int_{y \in G_F(x)} u(x,y)\pi(a,x,dy). \tag{20}$$

In the following we consider the gain function (19.2), i.e.

$$u(x,y) = F(x) - F(y).$$

Since an exact analytical solution of (20) is not possible in general, we have to apply some approximations. Firstly, the area of success $G_F(x)$ is approximated according to

$$G_F(x) \approx \{y \in \mathbb{R}^d \colon \nabla F(x)'(y-x) + \tfrac{1}{2}(y-x)'\nabla^2 F(x)(y-x)<0\}, \tag{21}$$

where $\nabla F(x)$ denotes the gradient of F and $\nabla^2 F$ is the Hessian matrix of F at x. We assume that $\nabla^2 F(x)$ is regular and $\nabla F(x) \neq 0$. Defining then the vector $w \in \mathbb{R}^d$ by

$$y - x = w - \nabla^2 F(x)^{-1} \nabla F(x), \tag{22}$$

the quadratic inequality contained in (21) has the form

$$w' \frac{\nabla^2 F(x)}{r} w < 1,$$

where $r>0$ is defined by

$$r = \nabla F(x)' \nabla^2 F(x)^{-1} \nabla F(x).$$

By the Cholesky-decomposition of $\frac{\nabla^2 F(x)}{r}$ we can compute a matrix Γ such that

$$\frac{\nabla^2 F(x)}{r} = \Gamma\Gamma'. \tag{23}$$

Defining

$$v = \Gamma' w, \tag{24}$$

the approximation (21) of $G_F(x)$ can be represented according to (22) and (23) by

$$G_F(x) \approx \{x_N + \Gamma'^{-1} v : \ ||v|| < 1\}, \tag{25}$$

where $||\cdot||$ is the Euclidean norm and x_N is given by

$$x_N = x - \nabla^2 F(x)^{-1} \nabla F(x).$$

It is then easy to verify that by the same transformations the search gain $u(x,y) = F(x)-F(y)$ can be approximated by

$$u(x,y) \approx \frac{r}{2}(1-||v||^2). \tag{26}$$

By means of (25) and (26) the objective function $U(a,x)$, $a=(\mu,\Lambda)$, of (20) can be approximated by

$$\tilde{U}(q,Q) = \frac{r}{2} \int_{||v||<1} (1-||v||^2 f(q,Q,v)dv, \tag{27}$$

where $f(q,Q,v)$ is given by

$$f(q,Q,v) = \frac{1}{(2\pi)^{d/2}(\det Q)^{1/2}} \exp(-\frac{1}{2}(v-q)Q^{-1}(v-q)).$$

Here the d-vector q and the positive definite dxd matrix Q are given by

$$q = q(\mu) = \Gamma'(x_N-\mu), \tag{28}$$

$$Q = Q(\Lambda) = \Gamma'\Lambda\Gamma. \tag{29}$$

By the 1-1-transformations (28) and (29), the maximization problem (20) can be approximated by

$$\begin{array}{l} \text{maximize } \tilde{U}(q,Q), \\ q \in K \\ Q \in \mathbb{Q} \end{array} \tag{30}$$

where K and $\mathbb{Q}$ are defined by

$$K = \{\Gamma'(x_N-\mu): \ \mu \in M\},$$

$$\mathbb{Q} = \{0\} \cup \{Q: Q \text{ positive definite dxd matrix}\}$$

and M is a certain subset of $\mathbb{R}^d$.

By the preceding considerations we obtain now this result.

<u>Theorem 4.1</u>

Let q^*, Q^* be an optimal solution of (30) and define μ^*, Λ^* by

$$\mu^* = x_N + \Gamma'^{-1}q^*,$$
$$\Lambda^* = (\Gamma Q^{*-1}\Gamma')^{-1}. \qquad (31)$$

Then the 1-stage optimal control $a^*(x) = (\mu^*,\Lambda^*)$ is given approximatively by (31).

In order to determine q^* and Q^*, we suppose now that the feasible set M for the mean value μ is defined by

$$M = \{\mu \in \mathbb{R}^d: \gamma_1^2 r \le (\mu-x_N)'\nabla^2F(x)(\mu-x_N) \le \gamma_2^2 r\}, \qquad (32)$$

where $0<\gamma_1\le\gamma_2$ are arbitrary, but fixed constants. In this case $K=K(M)$ is given by

$$K = \{q \in \mathbb{R}^d: \gamma_1 \le ||q|| \le \gamma_2\},$$

where $||\cdot||$ denotes the Euclidean norm. Note that the important case $M = \{x\}$, i.e. $\mu = x$(= last iteration point), corresponds to the case $\gamma_1=\gamma_2=1$.

Assume now that M is given by (32). Since each $Q \in \mathbb{Q}$ has the form $Q=T\Delta T'$, where T is an orthogonal matrix and Δ is a diagonal matrix, the minimization problem (30) is equivalent to

$$\text{maximize } \tilde{U}(q,\Delta) \text{ s.t. } \gamma_1 \le ||q|| \le \gamma_2, \qquad (33)$$
$$\Delta \in \mathbb{Q},\ \Delta \text{ diagonal.}$$

By a further approximation, we find then that an optimal solution q^*, Q^* of (33) is given approximatively by this equations

$$q^* = k^*1,\ 1 = (1,\dots,1)',$$
$$Q^* = c^*I,\ I = \text{identity matrix}, \qquad (34)$$

where the parameters $k^* \in \mathbb{R}$ and $c^*>0$ are defined by a certain remaining maximization problem.

Now, (31), (34) and (23) yield

$$\mu^* = x_N + k^*\Gamma'^{-1}1,$$

$$\Lambda^* = (\Gamma \frac{1}{c^*}I\Gamma')^{-1} = c^*(\Gamma\Gamma')^{-1} = c^*\left(\frac{\nabla^2F(x)}{r}\right)^{-1}$$
$$= c^*r\ \nabla^2F(x)^{-1}.$$

Hence, we have this result.

<u>Theorem 4.2</u>

The 1-stage optimal control $a^*(x) = (\mu^*,\Lambda^*)$ of the random search procedure (2) is given approximatively by

$$\mu^* = x_N + k^*\Gamma'^{-1},$$
$$\Lambda^* = c^*(\nabla F(x)'\nabla^2 F(x)^{-1}\nabla F(x))\nabla^2 F(x)^{-1}, \qquad (35)$$

where $k^* \in \mathbb{R}$, $c^*>0$ are certain fixed parameters.

5. Convergence rates of controlled random search procedures

Assume that the random search procedure (2) has normal distributed search variates $Z_1(\omega), Z_2(\omega), \ldots, Z_n(\omega), \ldots$ controlled by means of the following control law

$$\mu^0(x) = x$$
$$\Lambda^0(x) = c(\nabla F(x)'\nabla^2 F(x)^{-1}\nabla F(x))\nabla^2 F(x)^{-1}, \qquad (36)$$

where $c>0$ is a fixed parameter. Applying Rappl's results (1984) to control (36), we obtain this

Theorem 5.1

Suppose that D is a compact, convex subset of $\mathbb{R}^d$ and let x^* be the unique optimal solution of (1). Let $x^* \in \mathring{D}$ (= interior of D) and assume that F is twice continuously differentiable in a certain neighbourhood of x^*. Moreover, suppose that $\nabla^2 F$ is positive definite at x^*. Then there is a constant $q>1$ duch that

$$q^n E(F(X_n)-F^*) \to 0 \text{ as } n \to \infty$$

and

$$q^n (F(X_n)-F^*) \to 0 \text{ as } n \to \infty, \text{ w.p.1} \qquad (37)$$

for all starting points contained in a certain neighbourhood of x^*.

Note

a) Comparing Theorem 2.2 and Theorem 5.1, we find that -at least locally- the convergence rate of (2) is increased very much by applying a suitable control, as e.g. the control (36).

b) However, the high convergence rate (37) holds only if the starting point x_0 is sufficiently close to x^*, while the low convergence rate found in Theorem 2.2 holds for arbitrary starting points $x_0 \in D$.

Hence, the question arises whether by a certain combination

of a controlled and uncontrolled random search procedure we also can guarantee a linear convergence rate for all starting points $x_0 \in D$.

Given an increasing sequence N of integers

$n_1 < n_2 < \ldots < n_k < n_{k+1} < \ldots,$

let the controls $a_n = (\mu_n, \Lambda_n)$ of the normal distributed search variates $Z_{n+1}(\omega)$, $n=0,1,2,\ldots$, be defined by

$$\mu_n = x_n$$

and (38)

$$\Lambda_n = \begin{cases} \Lambda^o(x_n), & \text{if } n \in N \\ R, & \text{if } n \in N, \end{cases}$$

where $\Lambda^o(x)$ is defined by (36) and R is a fixed positive definite dxd matrix.

Hence, according to (38), the search procedure is controlled only at the times $n_1, n_2, \ldots$.

Now, by Rappl (1984) we have this result.

Theorem 5.2

Suppose that D is a compact, convex subset of $\mathbb{R}^d$ and let $x^* \in \overset{o}{D}$ be the unique optimal solution of (1). Assume that F is twice continuously differentiable in a certain neighbourhood of x^* and let $\nabla^2 F(x^*)$ be positive definite. Define then

$$h_n = \max\{k: n_k \leq n\}.$$

Then for every starting point $x_o \in D$ there is a constant $\beta > 1$ such that

$$\beta^{h_n} E(F(X_n) - F^*) \to 0 \text{ as } n \to \infty$$

and (39)

$$\beta^{h_n}(F(X_n) - F^*) \to 0 \text{ as } n \to \infty, \text{ w.p. } 1,$$

provided that $\limsup_{n\to\infty} \frac{h_n}{n} < 1$.

Note

a) Hence, the linear convergence rate (39) can be obtained by a suitable control of the type (38) for each starting points $x_o \in D$.

b) If $n_k = \frac{k}{p}$ for some $p \in \mathbb{N}$, then $\beta^{h_n} = (\sqrt[p]{\beta})^n$.

6. Numerical realizations of optimal control laws

In order to realize the control laws obtained in (35), (36), (38), one has to compute the gradient $\nabla F(x)$ and the inverse Hessian matrix $\nabla^2 F(x)^{-1}$ of F at x. However, since the derivatives ∇F and $\nabla^2 F$ of F are not given in analytical form in practice, the gradient and the Hessian matrix of F must be approximated by means of the information obtained about F during the search process. Hence, for an approximative computation of ∇F and $\nabla^2 F$ we may use the sequence of sample points, iteration points and function values

$$x_o, F(x_o), z_1, F(z_1), x_1, z_2, F(z_2), x_2, \ldots$$

In order to define a recursive approximation procedure, for n=0,1,2,... let denote

g_n the approximation of $\nabla F(x_n)$,

B_n the approximation of $\nabla^2 F(x_n)$,

H_n the approximation of $\nabla^2 F(x_n)^{-1}$.

Proceeding recursively, we suppose at the n-th stage of the search process we know the approximations g_n, B_n and H_n of $\nabla F(x_n)$, $\nabla^2 F(x_n)$ and $\nabla^2 F(x_n)^{-1}$, respectively. Hence, we may compute - approximatively - the control $a_n = (\mu_n, \Lambda_n)$ according to one of the formulas (35),(36) or (38) by replacing $\nabla F(x_n)$ and $\nabla^2 F(x_n)^{-1}$ by g_n, H_n, respectively. The search process (2) yields then the sample point z_{n+1}, its function value $F(z_{n+1})$ and the next iteration point x_{n+1}. Now we have to perform the update

$$g_n \to g_{n+1},\ B_n \to B_{n+1} \text{ and } H_n \to H_{n+1} \tag{40}$$

by using the information x_n, $F(x_n)$, z_{n+1}, $F(z_{n+1})$, x_{n+1} about F.

a) Search failure at x_n

If $z_{n+1} \notin D$ or $F(z_{n+1}) \geq F(x_n)$, then $x_{n+1} = x_n$. Since in this case we stay at x_n, we may define the update (40) by

$g_{n+1} = g_n$,

$B_{n+1} = B_n$,

$H_{n+1} = H_n$.

b) Search success at x_n

In this case it is $z_{n+1} \in D$ and $F(z_{n+1}) < F(x_n)$, hence

$x_{n+1} = z_{n+1} \neq x_n$. By a quadratic approximation of F at x_{n+1} we find then

$$F(x_n) \approx F(x_{n+1}) + \nabla F(x_{n+1})'(x_n - x_{n+1}) + \\ + \frac{1}{2}(x_n - x_{n+1})'\nabla^2 F(x_{n+1})(x_n - x_{n+1})$$

and therefore

$$\nabla F(x_{n+1})'s_n - \frac{1}{2} s_n' \nabla^2 F(x_{n+1}) s_n \approx \Delta F_n, \tag{41}$$

where

$$s_n = x_{n+1} - x_n = z_{n+1} - x_n,$$

$$\Delta F_n = F(x_{n+1}) - F(x_n) = F(z_{n+1}) - F(x_n).$$

Now we have to define the new approximations g_{n+1} and B_{n+1} of $\nabla F(x_{n+1})$ and $\nabla^2 F(x_{n+1})$, respectively.

Because of (41), in order to define the update (40), we demand next to the following

Modified Quasi - Newton Condition

$$g_{n+1}'s_n - \frac{1}{2} s_n' B_{n+1} s_n = \Delta F_n \tag{42}$$

or

$$g_{n+1}'s_n - \frac{1}{2} s_n' B_{n+1} s_n < 0. \tag{43}$$

Note

i) In contrary to (42), the modified Quasi-Newton condition (43) uses only the information that the function value of F at x_{n+1} is less than that at x_n.

ii) If $\Delta F_n = F(x_{n+1}) - F(x_n) < 0$, then $- s_n = x_n - x_{n+1}$ is an ascent direction of F at x_{n+1}. Hence, since $\nabla F(x_{n+1})$ is the best ascent direction of F at x_{n+1}, $- s_n$ may be used to define the approximation g_{n+1} of $\nabla F(x_{n+1})$.

Since g_{n+1}, B_{n+1} is not completely determined by the modified Quasi-Newton condition (42) or (43), respectively, there are still many possibilities to define the update formulas (40).

Clearly, since B_n is an approximation to a symmetric matrix, we suppose that B_n is a symmetric matrix.

A) Additive rank-one-updates

In order to select a particular tuple (g_{n+1}, B_{n+1}) we may require that (g_{n+1}, B_{n+1}) is an optimal solution $(\bar{g},\bar{B})$ of the distance-minimization problem

$$\text{minimize } d_1(\bar{B},B) + d_2(\bar{g},g) \tag{44}$$
$$\text{s.t. } \bar{g}'s - \frac{1}{2}s'\bar{B}s = \Delta F,$$

where $B=B_n$, $g=g_n$, $\Delta F=\Delta F_n$ and d_1,d_2 are certain distance measures. We suppose here that d_1, d_2 are defined by

$$d_1(\bar{B},B) = \frac{1}{2}\sum_{i,j=1}^{d}(\bar{b}_{ij} - b_{ij})^2,$$
$$d_2(\bar{g},g) = \frac{1}{2}\sum_{j=1}^{d}(\bar{g}_j - g_j)^2, \tag{45}$$

where $\bar{b}_{ij}$, b_{ij} are the elements of $\bar{B}$ and B, resp., and $\bar{g}_j$, g_j denote the components of $\bar{g}$, g, respectively.

Note.

The minimization (44) is a generalization of the minimality principles characterizing some of the well known Quasi-Newton update formulas, see e.g. J.E. Dennis and R.B. Schnabel (1983).

Solving (45), we find that $\bar{g}$, $\bar{B}$ are given by

$$\bar{g} = g - \lambda s \tag{46}$$
$$\bar{B} = B + \frac{\lambda}{2}ss', \tag{47}$$

where the real parameter λ is given by

$$\lambda = \frac{g's - \frac{1}{2}s'Bs - \Delta F}{s's(1 + \frac{1}{4}s's)}. \tag{48}$$

If the distance functions d_1, d_2 are changed, then other update formulas may be generated. If. e.g. d_2 is replaced by $d_2(\bar{g},g) = \frac{1}{2}(\bar{g}-g)B^{-1}(\bar{g}-g)$, then $\bar{g} = g - \lambda Bs$.

Supposing now that B is positive definite, it is known that the matrix $\bar{B}$ defined by (47) is positive definite if and only if

$$1 + \frac{\lambda}{2}s'Hs > 0, \tag{49}$$

where $H = B^{-1}$ is our approximation to the inverse Hessian matrix $\nabla^2F(x)^{-1}$ of F at $x=x_n$. Hence, if $\bar{H} = \bar{B}^{-1}$ denotes the approximation of the inverse Hessian matrix of F at x_{n+1}, then by (47) and (49) the following update formula $H \to \bar{H}$ for the inverse Hessian matrix of F may be established:

$$\bar{H} = \begin{cases} H - \frac{1}{2}\frac{\lambda}{1+\frac{\lambda}{2}s'Hs} Hss'H, & \text{if (49) holds} \\ H, & \text{else.} \end{cases} \tag{50}$$

Updates in the case of a search failure.

If $z_{n+1} \notin D$ or $F(z_{n+1}) \geq F(x_n)$, then we stay at $x_{n+1} = x_n$ and we may define therefore $\bar{g}=g$, $\bar{B}=B$ and $\bar{H}=H$. However, also in the case of a search failure the tuple $(z_{n+1}, F(z_{n+1}))$ yields new information about F, provided only that $z_{n+1} \neq x_n$. Hence, replacing the modified Quasi-Newton condition (42) by

$$\bar{g}'s + \frac{1}{2} s'\bar{B}s = \Delta F,$$

where now $s = z_{n+1} - x_n$, $\Delta F = F(z_{n+1}) - F(x_n)$, we may derive by the above procedure also update formulas $g \to \bar{g}$, $B \to \bar{B}$, $H \to \bar{H}$ for defining improved approximations $\bar{g}, \bar{B}, \bar{H}$ of $\nabla F, \nabla^2 F$ and $\nabla^2 F^{-1}$, respectively, at $x_{n+1} = x_n$.

B) Multiplicative rank-one-updates

By (46) - (50) we have given a first concrete procedure for the realization of the optimal control laws (35), (36) and (38), respectively. Indeed, having e.g. the mean $\mu_n = x_n$ and the covariance matrix

$$\Lambda_n = c^*(g_n'H_n g_n)H_n, \tag{51}$$

the random variable Z_{n+1} may be defined by

$$Z_{n+1} = \mu_n + \Gamma_n Z^o_{n+1},$$

where Z^o_{n+1} is a normal distributed with mean zero and covariance matrix equal to the identity matrix and Γ_n is a dxd matrix such that

$$\Gamma_n \Gamma_n' = \Lambda_n. \tag{52}$$

Hence, at each iteration point x_n the (Cholesky-) decomposition (52) of Λ_n must be computed.

In order to omit this time consuming step, we still ask whether update formulas $\Gamma_n \to \Gamma_{n+1}$ for the Cholesky-factors Γ_n may be obtained.

Since $H_n = B_n^{-1}$ we suppose that B_n may be represented by

$T_n T_n' = B_n$.

Then Λ_n is given by

$$\Lambda_n = c^*((T_n^{-1} g_n)' T_n^{-1} g_n) T_n^{-1'} T_n^{-1}$$

and the factor Γ_n may be defined by

$$\Gamma_n = \sqrt{c^*}\,||T_n^{-1} g_n||\,T_n^{-1'}. \tag{53}$$

In order to define the update $T \to \bar{T}$, where $T=T_n$ and $\bar{T}=T_{n+1}$ with $T_{n+1}\,T_{n+1}'=B_{n+1}$, we require that T is changed only in the direction of $s=x_{n+1}-x_n$, hence we assume that

$$\bar{T} = (I + \frac{\gamma-1}{ss'}\,ss')T,$$

where γ is real parameter. Furthermore, the distance minimization problem (44) is then replaced by

$$\begin{aligned} &\text{minimize } d_1(\bar{T},T) + d_2(\bar{g},g) \\ &\text{s.t. } \bar{g}'s - \tfrac{1}{2}s'\bar{B}s = \Delta F, \end{aligned} \tag{54}$$

where now

$$\bar{B} = (I + \frac{\gamma-1}{s's}\,ss')B(I + \frac{\gamma-1}{s's}\,ss'). \tag{55}$$

If the distance functions d_1, d_2 are again defined by (45), then

$$d_1(\bar{T},T) = \frac{1}{2}\left(\frac{\gamma-1}{s's}\right)^2 ||T|| \;, \quad ||T|| = \sum_{i,j=1}^{d} \tau_{ij}^{\,2}, \tag{56}$$

where $T = (\tau_{ij})$. Hence, by (54) a particular tuple $(\bar{g},\gamma)$ is selected. Because of (55) and (56), the minimization problem (54) has the form

$$\begin{aligned} &\text{minimize } \frac{||T||}{2}\left(\frac{\gamma-1}{s's}\right)^2 + \frac{1}{2}||\bar{g}-g||^2 \\ &\text{s.t. } \bar{g}'s - \frac{\gamma^2}{2}\, s'Bs = \Delta F, \end{aligned} \tag{57}$$

hence, the tuple (g,1) is projected onto the parabola in $\mathbb{R}^{d+1}$ defined by the constraint of (57).

It is easy to verify that the optimal solution of (57) is given by

$$\bar{g} = g - \lambda s, \tag{58}$$

$$\gamma = \frac{||T||}{||T||-\lambda(s's)^2 s'Bs}, \tag{59}$$

where the parameter λ is a solution of the equation

$$g's - \lambda s's - \frac{1}{2} s'Bs \left(\frac{||T||}{||T||-\lambda(s's)^2 s'Bs}\right)^2 = \Delta F. \quad (60)$$

Supposing that $T = T_n$ is regular, we know that the matrix $\bar{T} = T_{n+1}$ defined by (53) is regular if and only if $\gamma \neq 0$. In this case we finally obtain

$$\bar{T}^{-1} = T^{-1}\left(I + \frac{1-\gamma}{\gamma} \frac{ss'}{s's}\right). \quad (61)$$

By (53), (58)-(61) we have now an update procedure omitting the Cholesky-decomposition (52) of Λ_n.

Note

a) Other update formulas may be gained by changing the objective function of (57).

b) Also in the case of a search failure, by a similar method update formulas may be derived.

References

Anderson, R.L. (1953). Recent advances in finding best operation conditions. J. Amer. Statist. Assoc. 48: 789-798.

Box, G.E.P. (1957). Evolutionary operations: A method for increasing industrial productivity. Applied Statistics 6: 81-101.

Brooks, S.H. (1959). A comparison of maximum - seeking methods. Operations Research 7: 430-457.

Dennis, J.E., Schnabel, R.B. (1983). Numerical methods for unconstrained optimization and nonlinear equations. Prentice-Hall, Englewood Cliffs.

Karnopp, D.C. (1963). Random search for optimization problems, Automatica 1: 111-121.

Marti, K. (1980a). Random search in optimization as a stochastic decision process (adaptive random search). Methods of Operations Research 36: 223-234.

Marti, K. (1980b). On accelerations of the convergence in random search methods. Methods of Operations Research 37: 391-406.

Marti, K. (1980c). Adaptive Zufallssuche in der Optimierung. ZAMM 60: T357-359

Müller, P.H. et. al. (1983). Optimierung mittels stochastischer Suchverfahren. Wissenschaftliche Zeitschrift der TU Dresden 32(1): 69-75.

Müller, P.H., Nollau, V. (1984). Steuerung stochastischer Prozesse. Akademie Berlag, Berlin.

Oppel, U.G. (1976). Random search and evolution. Trans. Symp. Appl. Math., Thessaloniki

Rappl, G. (1978). Theorie und Anwendungen von auf der Zufallssuche basierenden Evolutionsprozessen. Diplomarbeit, Uni München.

Rappl, G. (1984). Konvergenzraten von Random-Search-Verfahren zur globalen Optimierung. Dissertation, HSBw München.

Schwefel, H.-P. (1977). Numerische Optimierung von Computer-Modellen mittels der Evolutions-Strategie. Birkhäuser, Basel und Stuttgart.

Solis, F.J., Wets, R. (1981). Minimization by random search techniques. Mathematics of Operations Research 6(1), 19-30.

Zielinski, R., Neumann, P. (1983). Stochastische Verfahren zur Suche nach dem Minimum einer Funktion. Akademie Verlag, Berlin.

ON BAYESIAN METHODS IN NONDIFFERENTIAL AND STOCHASTIC PROGRAMMING

J.B. Mockus
Institute of Mathematics and Cybernetics
Academy of Sciences, Lithuanian SSR,
K. Pozelos 54, Vilnius, USSR

1. INTRODUCTION

The idea of the Bayesian approach is to consider the function to be minimized as a sample observation of a stochastic function. We can therefore define a search procedures which minimize the average deviation of the observation from the minimum. This differs from classical numerical analysis, where the maximum deviation is usually considered (Mockus 1972). The average deviation can be defined mathematically if the probability distribution is fixed. Since the main objections to the Bayesian approach are doubts about the existence and nature of *a priori* distributions, a system of simple and natural assumptions are introduced which provide for the existence of a unique family of finite-dimensional probability density functions (Katkauskaite and Zilinskas 1977; Zilinskas 1978) and define the *a priori* distributions on the space of functions to be minimized (Mockus 1984a,b).

2. BAYESIAN APPROACH TO GLOBAL OPTIMIZATION

Let

$$f = f(x) = f(x, \omega) \tag{1}$$

denote the objective function to be minimized, where f is a continuous function of x and a measurable function of ω, and

$$x \in A \subset R^m \ , \ \omega \in \Omega \quad . \tag{2}$$

Here A is a compact set and Ω is a set of indices corresponding to all continuous functions of x in A.

Assume that we can observe (calculate or define by a physical experiment) the values of $f(x_i)$ at the point x_i. The results of observations will be denoted by

$$y_i = f(x_i)\ ,\ i = 1, \ldots, N \quad , \tag{3}$$

where N is the total number of observations.

We shall denote the vector of observations by

$$z_n = (x_i\ , y_i\ , i = 1, \ldots, N) \tag{4}$$

and define the decision function as

$$d_n = d_n(z_n)\ ,\ n = 0, \ldots, N \quad , \tag{5}$$

where d_n is a mapping of $(A \times R)^n$ into A.

Let

$$d = (d_0, \ldots, d_N) \tag{6}$$

denote the sequence of decision functions, and assume that $d_n \in D_n\ ,\ n = 0, \ldots, N$, where D_n is the set of all measurable mappings of $(A \times R)^n$ into A and

$$d \in D = \mathop{X}_{n=0}^{N} D_n \quad .$$

Assume that the observation points are defined by the decision function d:

$$x_{i+1} = x_{i+1}(d) = d_i(z_i)\ ,\ i = 0, \ldots, N \quad . \tag{7}$$

Let δ denote the deviation from the minimum of $f(x)$ when a sequence of decision functions d is used:

$$\delta = \delta(d) = \delta(d\ , \omega) = f(x_{N+1}, \omega) - \min_{x \in A} f(x\ , \omega) \quad . \tag{8}$$

Here x_{N+1} is the point at which the final decision should be made (after all N observations have been completed).

The average deviation Δ can be expressed as a Lebesque integral

$$\Delta = E\{\delta(d)\} = \int_{\Omega} \delta(d\ , \omega) P(d\omega) \quad , \tag{9}$$

where P is a probability measure defined on $B \subset \Omega$.

From (8),

$$\Delta = \int_{\Omega} f(x_{N+1}(d), \omega) P(d\omega) - \int_{\Omega} \min_{x \in A} f(x\ , \omega) P(d\omega) \quad . \tag{10}$$

The decision function d' is called the Bayesian method by Mockus (1972) if

$$\int_{\Omega} f(x_{N+1}(d'),\omega)P(d\omega) = \inf_{d \in D} \int_{\Omega} f(x_{N+1}(d),\omega)P(d\omega) \quad . \tag{11}$$

Condition (11) minimizes the expected deviation (1) if

$$\left| \int_{\Omega} \min_{x \in A} f(x,\omega)P(d\omega) \right| < \infty \quad , \tag{12}$$

since the second integral in (10) does not depend on d.

Note that definition (11) is more general than one based on minimization of the expected deviation (10).

It is convenient to reduce condition (11) to recursive equations in a dynamic programming framework (Mockus 1969,1972).

It is not easy to solve recursive equations (Mockus 1972), and so some approximation is needed. The one-stage approximation (Mockus 1972), in which the next observation is considered to be the final observation, is both simple and natural. In this case

$$u_N(z_N) = \min_{x \in A} E\{f(x) \mid z_N\}$$

$$u_{n-1}(z_{n-1}) = \min_{x \in A} R_x^{n-1} \, , \; 1 \le n \le N \tag{13}$$

$$x_n \in \arg\min_{x \in A} R_x^{n-1} \quad , \tag{14}$$

where

$$R_x^{n-1} = E\{\inf_{v \in A} E\{f(v) \mid z_{n-1}, x, f(x)\} \mid z_{n-1}\} \quad . \tag{15}$$

Denote the conditional density of $f(v)$ with respect to z_{n-1} by $p_v(y \mid z_{n-1})$ and the conditional density of $f(v)$ with respect to vector (z_{n-1}, x, y') by $p_v^n(x, y')$.

Set

$$c_{n-1} = \min_{x \in A} \int_{-\infty}^{\infty} y p_v(y \mid z_{n-1}) dy$$

and

$$c_n(x, y') = \min_{x \in A} \int_{-\infty}^{\infty} y p_v^n(x, y') dy \quad ,$$

and let

$$c_n(x, y') = \min(c_{n-1}, y') \quad .$$

Then

$$R_x^{n-1} = \int_{-\infty}^{\infty} c_n(x, y) p_v(y \mid z_{n-1}) dy$$

and

$$R_x^{n-1} = \int_{-\infty}^{\infty} \min(c_n, y) p_v(y \mid z_{n-1}) dy$$

or

$$R_x^{n-1} = c_{n-1} - \int_{-\infty}^{c_{n-1}} (c_{n-1} - y) p_v(y \mid z_{n-1}) dy \quad .$$

Mockus (1984a) gives conditions which define a family of *a priori* distributions P on Borel subsets of Ω such that Bayesian methods converge to the global minimum of any continuous function

$$f(x) = f(x, \omega), \; x \in A \subset R^m, \; \omega \in \Omega \quad ,$$

where A is a compact set. It was shown that homogeneity of the *a priori* distribution, continuity of the sample functions $f(x)$ and independence of the partial differences of m-th order are sufficient for the *a priori* distribution P to be Gaussian with mean μ and variance σ_0^2. The covariance is defined as

$$\sigma_{jk} = \sigma_0^2 \Gamma_{i=1}^m (1 - |x_j^i - x_k^i| / 2), \; x_j, \; x_k \in [-1, 1] \quad .$$

In the Gaussian case the optimal point taking into account the next observation should minimize the *a posteriori* risk function

$$x_{n+1} \in \arg\min_{x \in A} R_x^n(\mu_n, \sigma_n) \quad , \tag{16}$$

where

$$R_x^n(\mu_n, \sigma_n) = 1/(\sqrt{2\pi}\sigma_n) \int_{-\infty}^{\infty} \min(y, c_n) exp(-1/2 \left(\frac{y - \mu_n}{\sigma_n} \right)^2) dy \quad . \tag{17}$$

Here μ_n, σ_n are the conditional expectation and conditional variance with respect to the observed (calculated) values of function $f(x_i)$, $i = 1, \ldots, n$, and

$$c_n = \min_{1 \le i \le n} f(x_i) - \varepsilon_n \quad , \tag{18}$$

where

$$\varepsilon_n = \sigma_n \left(\sqrt{2 \ln N} - \sqrt{2 \ln n} \right) \quad .$$

ε_n is introduced to take into account the influence of subsequent observations; the above expression for ε_n was derived under the assumption that $f(x_n)$, $n = 1, \ldots, N$, are independent. (Here N is the total number of observations and n is the number of the current observation.) Formula (17) corresponds to the Gaussian case with a linear loss function, assuming that

$$\min_{x \in A} \mu_n = \min_{1 \le i \le n} f(x_i) = c_n \quad . \tag{19}$$

If $N - n$ is large, (1),(2) can be written in the form

$$x_{n+1} \in \arg\max_{x \in A} \sigma_n / (\mu_n - c_n) \quad . \tag{20}$$

Expression (16) is still too complicated, because it is necessary to compute the inverse of the covariance matrix which defines the conditional expectation μ_n and the conditional variance σ_n. This is the inevitable cost of requiring the system to obey Kolmogorov's consistency conditions:

$$(1)\ P\{\omega : f(x_1, \omega) < y_1, \ldots, f(x_k, \omega) < y_k\} =$$

$$= P\{\omega : f(x_1, \omega) < y_1, \ldots, f(x_k, \omega) < y_k, f(x_{k+1}, \omega) < \infty, \ldots, f(x_{k+l}, \omega) < \infty\} \quad ,$$

$$k, l = 1, 2, \ldots$$

$$(2)\ P\{\omega : f(x_1, \omega) < y_1, \ldots, f(x_k, \omega) < y_k\} =$$

$$= P\{\omega : f(x_{s_1}, \omega) < y_{s_1}, \ldots, f(x_{s_k}, \omega) < y_{s_k}\}, \ k = 1, 2, \ldots$$

Relaxation of these conditions means that we will consider not one probabilistic model of function $f(x)$, but some sequence of probabilistic models P_n, updating them after each observation. This is apparently the only way to avoid the inversion of matrices of order n. Obviously in such a case μ_n and σ_n^2 will not represent the conditional expectation and conditional variance as usually defined and can be regarded only as approximations of these functions. In order to define μ_n and σ_n more precisely when Kolmogorov's consistency conditions are relaxed, some additional conditions must be satisfied. The most natural conditions in the one-step Bayesian case seem to be the following:

1. The probability measures P_n, $n = 1, \ldots, N$, are absolutely continuous and Gaussian.
2. The probability measures P_n, $n = 1, \ldots, N$, are consistent in the sense that

(a) the risk function R_x^n, $n = 1, \ldots, N$, defined by (17) is continuous;

(b) the observation points x_{n+1}, $n = 0, \ldots, N$, defined by (16) converge to the global minimum of any continuous function on a compact set A.

3. The parameters μ_n and σ_n are functions of x which are as simple as possible.

It follows from these conditions that

$$x_{n+1} \in \arg \max_{x \in A} \min_{1 \le i \le n} \sigma_i / (\mu_i - c_n) \quad , \tag{21}$$

where

$$\mu_i = f(x_i) \, , \; \sigma_i = || x - x_i || \quad . \tag{22}$$

It was shown by Mockus (1984a) that under some general assumptions the convergence condition 2(b) will be satisfied if

$$\mu_n \to f(x) \tag{23}$$

and

$$\sigma_n \to \begin{cases} 0 \, , & \text{if } \tau_x^n \to 0 \\ \alpha > 0 \, , & \text{if } \tau_x^n \to \beta > 0 \end{cases} \tag{24}$$

for $n \to \infty$.

Here

$$\tau_x^n = \min_{1 \le i \le n} || x - x_i || \quad .$$

We can see that even in the simplest case (21), a considerable amount of auxiliary calculation is necessary to find the optimal point for the next observation. Thus it is reasonable to use the global Bayesian method in the case of "expensive" observations, i.e., when the calculation of the function is sufficiently complicated. Otherwise it may be better to use less efficient but much simpler methods, such as Monte Carlo simulation.

It will be shown later that local Bayesian methods can also be very simple.

The use of Bayesian methods is certainly justified when the observations are "noisy" because these methods filter out the noise during the optimization process. In this case $\mu_i = h(x_i)$, where

$$h(x_i) = f(x_i) + \xi_i \, , \; E\{\xi_i\} = 0 \, , \; \text{var} \, \{\xi_i\} = \sigma^2 \quad .$$

In the presence of noise Bayesian methods converge with probability $P_0 = 1$, where P_0 is defined on the space of noise.

Global Bayesian methods have been used in real problems arising in optimal design and experimental planning (Mockus 1983, 1984a). They also proved to be superior in some respects to other global optimization algorithms in an "international competition" (Dixon and Szego 1978).

One good example of a set of test functions is the following family of two-dimensional functions with parameters a_{ij}, b_{ij}, c_{ij}, $d_{ij} \in (0,1)$:

$$f(x) = \Big(\Big(\sum_{i,j=1}^{I} (a_{ij} \sin(\pi j x^1) \sin(\pi j x^2) + b_{ij} \cos(\pi i x^1) \cos(\pi i x^2))\Big)^2 + \left[\sum_{i,j=1}^{I} (c_{ij} \sin(\pi i x^1) \sin(\pi j x^2) - d_{ij} \cos(\pi i x^1) \cos(\pi j x^2))\right)^2\right]^{1/2} \tag{25}$$

where the number of components $I = 7$.

This family of functions satisfies our conditions. It represents the stress function in an elastic square plate under a cross-sectional load. These functions were considered by Grishagin (1978) when testing different versions of the method of maximum likelihood; see Strongin (1978). A full account of the experimental conditions was published, so it was relatively easy to compare maximum-likelihood-type methods with other methods of global optimization, such as the LP-based uniform search method (Sobol 1969), two versions of the one-step Bayesian method (Mockus 1972) and the uniform random search (Monte Carlo) method.

In all cases a local optimization was performed after the termination of the global search, using an algorithm of Nelder and Mead type. Only one local search was carried out, from the best point found in global search. In view of (25), it was thought that to do the local optimization more than once would be too expensive if the derivatives cannot be calculated directly and must be estimated using function differences.

In addition to the various completely automatic searches a purely interactive optimization performed by an expert was included in the trial.

50 sample paths corresponding to the randomly distributed parameters a_{ij}, b_{ij}, c_{ij}, $d_{ij} \in (0,1)$ were considered.

The relation between the percentage of successes (in which the global minimum was found) and the total number of observations $N_t = N + N_L$ (where N_L is the number of observations in the local Nelder-Mead search) is shown in Table 1 for six methods.

The family of functions (25) has some limitations as a set of test functions. It cannot be generalized to the multidimensional case without losing its physical meaning.

A different family of functions was used for comparison of global optimization algorithms by Dixon and Szego (1978).

Table 1: The relation between the percentage of successful cases and the total number of observations

Iteration number	Method 1[a]	2[b]	3[c]	4[d]	5[e]	6[f]
60					48	
80	46	46		30	26	
90	60					
100		56		38		
105		62	56			
110						81
125	80	72				
135						92
140	88	86	68	44	68	
200			82		52	
240	96					84
340					92	
370			94			
400			100	78	94	

[a]Bayesian algorithm (21); [b]standard one-step Bayesian algorithm (1) with Gaussian *a priori* distributions; [c]Strongin's algorithm (1978); [d]uniform random search; [e]uniform deterministic search; [f]interactive optimization performed by an expert.

The best performance in this competition was produced by the method of De Biase and Frontini (1978). Unfortunately there is some ambiguity regarding the choice of starting points for the local search; the reasons for the success of the method also remain unclear.

The second best result was provided by the one-step Bayesian algorithm. However, this method is good only in the sense of the minimal number of observations regarded as reasonable if one observation is very expensive. Otherwise simpler approaches, such as Torn's method (1978) or even the Monte Carlo method are preferable. The adaptive Bayesian method (21) had not been developed at the time of the comparison.

3. BAYESIAN APPROACH TO LOCAL OPTIMIZATION

Stochastic approximation methods are often used to find a local minimum in the presence of noise. These methods are simple and convenient, and converge to the local minimum. However, the step length is not defined uniquely. It is well known that the efficiency of solution depends on the length of the step (Ermoliev 1976).

The most convenient way to solve the step length problem is to use the Bayesian approach. Naturally a different *a priori* distribution should be used in the local case. The step direction in the local Bayesian method is defined by a gradient estimate and the step length calculated using the condition for a minimum of the Bayesian risk function (Mockus 1984b). Once the step direction is fixed we only need to consider the average behaviour of function $f(x)$ and of some of its derivatives along the line of search. It is then sufficient to define the optimal step size for the case of a non-negative one-dimensional variable, as the generalization is straightforward.

When applying the Bayesian framework it is natural to begin by considering some family of functions which are convex on the average. A simple and convenient way to construct such a family is to suppose that there exist positive a, k_n such that

$$E\{f_x''\} = a(2+k_n)/2 \quad , \tag{26}$$

where

$$f_x'' = \frac{\partial^2 f(x,\omega)}{\partial x^2} \; , \; f_x' = \frac{\partial f(x,\omega)}{\partial x} \; , \; f_x = f(x,\omega) \; , \; x \geq 0 \; , \; \omega \in \Omega \quad .$$

To derive the average properties of f_x' and f_x from (26) we shall assume that

$$E\{\int_A f_x'' \mathrm{d}x\} = \int_A E\{f_x''\} \mathrm{d}x \tag{27}$$

and

$$E\{\int_A f_x' \mathrm{d}x\} = \int_A E\{f_x'\} \mathrm{d}x \quad .$$

Suppose also that only the sum

$$h(x_i) = f(x_i) + \xi_i \tag{28}$$

can be observed, and that the current point of observation is the point

$$x_n = 0 \quad . \tag{29}$$

We shall assume that the loss function is linear

$$\delta(x, x^*) = f(x) - f(x^*) \quad , \tag{30}$$

where

$$f(x^*) = \inf_{x \in A} f(x) \quad .$$

From (26)-(30) and some additional assumptions, it follows that

$$x_{n+1} = \begin{cases} x_n - \beta_n , & \text{if } h(x_n - \beta_n) \le h(x_n) + \varepsilon_n \\ x_n , & \text{if } h(x_n - \beta_n) > h(x_n) + \varepsilon_n \end{cases} \quad . \tag{31}$$

In the second case the gradient estimate is updated using the results of additional observations. In expression (31) we have

$$\beta_n = \frac{1}{2+k_n} \frac{\Delta_n}{\alpha_n} \quad , \tag{32}$$

where

$$\Delta_n = h(x_n + q_n) - h(x_n - q_n) , \; q_n = n^{-\nu} \tag{33}$$

$$\alpha_n = (nq_n^2)^{-1} \sum_{i=1}^{n} a(i) q_i^2 \quad . \tag{34}$$

Here

$$a(i) = h(x_i + q_i^2) - 2h(x_i) + h(x_i - q_i^2) \quad . \tag{35}$$

If the value of the above expression is less that ε_0 then $a(i) = \varepsilon_0$. Parameter k_n in (26), (32) is defined as

$$k_n = \begin{cases} 1 , & \text{if } n < n' \\ (n - n')^{1-\alpha-\nu} , & \text{if } n \ge \min(n', n_{\max}) \end{cases} \quad , \tag{36}$$

where

$$\alpha \ge 0 , \; \nu > 0 , \; \alpha + \nu < \frac{1}{2} , \; \nu - \alpha > 0 \tag{37}$$

and n' is the first value of n for which

$$|\Delta_n| \le q_n^2 \sigma / \alpha_n \quad . \tag{38}$$

Parameter ε_n in (31) can be expressed as

$$\varepsilon_n = k_\varepsilon \sigma / \pi , \; k_\varepsilon = 0.9 \quad . \tag{39}$$

Inequalities (37) follow from the usual convergence conditions. Formula (38) defines the point at which the average error of the iteration procedure (31) becomes less than its estimate calculated using the fixed point theorem, assuming that the function $f(x)$

is quadratic.

Formula (39) follows from the condition $E\{h(x_n - \beta_n)\} = E\{h(x_n)\} + \varepsilon_n$ assuming that $h(x_n - \beta_n)$ and $h(x_n)$ are independent and normal (Senkiene 1983). The generalization of algorithm (31)-(39) to the multidimensional case is straightforward and is given by Mockus (1984b).

Table 2 shows the results of computer simulation using a test function which is unimodal but not convex. The average deviation from the minimum over 20 random runs was calculated. The first row of Table 2 gives the results obtained using the local Bayesian method, the second the results produced by the classical stochastic approximation method, with parameters optimal as defined by Wasan (1969), for continuously differentiable functions.

Table 2: The average deviation in percent from the minimum obtained in a computer simulation.

Method	m=2	m=5	m=10	m=20
Local Bayesian	2.5	1.25	5.1	1.85
Stochastic Approximation	4.1	-	-	-

The computational algorithms and portable FORTRAN programs for the global and local Bayesian methods described here were developed by V. Tieshis, J. Valevichene and L. Zukauskaite, whose suggestions also helped to improve the efficiency of the methods described in this paper.

REFERENCES

De Biase, L. and F. Frontini, F. (1978). A stochastic method for global optimization: its structure and numerical performance. In L.C.W. Dixon and G.P. Szego (Eds.), "Towards Global Optimization". North-Holland, Amsterdam.

Dixon, L.C.W. and Szego, G.P. (1978). The global optimization problem: an introduction. In L.C.W. Dixon and G.P. Szego (Eds.), "Towards Global Optimization". North-Holland, Amsterdam.

Ermoliev, Y.M. (1976). Methods of Stochastic Programming. Nauka, Moscow (in Russian).

Grishagin, V.A. (1978). Operational characteristics of some global search algorithms. In L.A. Rastrigin (Ed.), "Problems of Random Search" (Vol. 7), Zinatne, Riga (in Russian).

Katkauskaite, A. and Zilinskas, A. (1977). On the construction of statistical models of functions under uncertainty. In "Theory of Optimal Decisions", Vol. 3, Institute of Mathematics and Cybernetics, Academy of Sciences, Lithuanian SSR, Vilnius (in Russian).

Mockus, J.B. (1972). On Bayesian methods for searching for extrema (in Russian). Avtomatika i Vytchislitelnaya Technika, 3.

Mockus, J.B. (1983). The Bayesian approach to global optimization. In G.I. Marchuk and L.N. Belykh (Eds.), "Mathematical Modelling in Immunology and Medicine". North-Holland, Amsterdam.

Mockus, J.B. (1984a). The Bayesian Approach to Global Optimization. Freie Universität Berlin, May, Preprint No. 176.

Mockus, J.B. (1984b). The Bayesian Approach to Local Optimization. Freie Universität Berlin, May, Preprint No. 175.

Senkiene, E. (1983). Statistical simulation of the distribution of minimal terms in the normal random variable sequence. In "Theory of Optimal Decisions", Vol. 9, Institute of Mathematics and Cybernetics, Academy of Sciences, Lithuanian SSR, Vilnius (in Russian).

Strongin, R.G. (1978). Numerical Methods for Multiextremal Problems. Nauka, Moscow (in Russian).

Torn, A.A. (1978). A search-clustering approach to global optimization. In L.C.W. Dixon and G.P. Szego (Eds.), "Towards Global Optimization". North-Holland, Amsterdam.

Wasan, M.T. (1969). Stochastic Approximation. Cambridge University Press, Cambridge.

Zilinskas, A.G. (1978). On statistical models for multimodal optimization. Math.Operat.Stat., Ser. Statistics, 9(2):

ON STOCHASTIC PROGRAMMING IN HILBERT SPACE

N.M. Novikova
Computing Center, USSR Academy of Sciences
Moscow, USSR

1. INTRODUCTION

Let us consider the problem of finding I^0 and $z \in Z^0$ which satisfy the following expression:

$$\inf_{z \in Z} \int_U w(z, u)\mu(z, du) \triangleq \int_U w(z^0, u)\mu(z^0, du) \triangleq I(z^0) \triangleq I^0 \quad , \tag{1}$$

where $Z \triangleq \{z \in Z^0 \subset L_2(Y) \mid G(z(y)) \leq 0 \quad \forall y \in Y\}$; $U \triangleq \{u \in U^0 \subset L_2(X) \mid H(u(x)) \leq 0 \quad \forall x \in X\}$ is $\mu(z, \cdot)$-measurable for any $z \in Z^0$; U^0, Z^0 are weakly compact sets; H and G are continuous functionals on U^0 and Z^0, respectively; $X \subset R^{|X|}$, $Y \subset R^{|Y|}$ are finite-dimensional sets; and $w(z, \cdot)$ is $\mu(z, \cdot)$-integrable for any $z \in Z^0$. Assume that the functional w may be represented in the form

$$w(z, u) \triangleq \varphi\left(\int_Y g(z(y))dy \,,\, \int_X h(u(x))dx \,,\, \int_{Y\times X} f(z(y), u(x))d(y, x)\right) \quad ,$$

where $\varphi(\cdot,\cdot,\cdot)$ is a function of three variables and $g(\cdot)$, $h(\cdot)$ and $f(\cdot,\cdot)$ are operators on Z^0, U^0 and $Z^0 \times U^0$, respectively. For simplicity, we shall require φ to be non-negative. The probability measure $\mu(z, \cdot)$ on U^0 is continuously dependent on z. If $w(z, \cdot)$ is integrable over a finitely additive measure then $\mu(z, \cdot)$ may be a quasi-measure (weak distribution).

In stochastic programming it is usually assumed that $\mu(z, \cdot)$ is unknown but that various observations of the random variable u may be made. Such observations seem to be meaningless in an infinite-dimensional space, so we shall suppose that only finite projections of $u(\cdot)$ may be used in a numerical search algorithm for (1). We fix the orthonormal basis $\xi = \{\xi_j\}$ in $L_2(X)$ and introduce a sequence of compatible measures $\mu_n(z, \cdot)$ on R^n, $n \in N = \{1, 2, \ldots\}$. It is required that $\mu_n(z, Q) = \mu(z, \{a\xi \mid [a]_n \in Q\})$ for any $\mu_n(z, \cdot)$-measurable set $Q \subseteq R^n$. Here and elsewhere $[\cdot]$ denotes the first n components of a vector and for any set $V \subset L_2(X)$ we write $[V]_n =$

$\{a^n \in R^n \mid \quad a\xi \in V: a^n = [a]_n\}$, where $a\xi = \sum_{j=1}^{\infty} a_j \xi_j$ denotes the scalar product of vectors (without brackets). We assume that for any integer n we may make independent observations of the random variable a^n distributed in accordance with $\mu_n(z,\cdot)$. The following representation holds for weak distributions by definition and for measures because of their continuity [1–3]:

$$I(z) \triangleq \int_U w(z,u)\mu(z,du) = \lim_{n\to\infty} \{ \int_{[U]_n} w(z,a^n\xi)\mu_n(z,da^n) \triangleq I_n(z)\}$$

for any $z \in Z^0$. Suppose that the functional $w(\cdot,u)$ is weakly lower-semicontinuous on Z^0 for any $u \in U^0$ and the set Z is weakly compact. Then the minimum in (1) is attainable, $z^0 \in Z$, and if $w(z,\cdot)$ is uniformly continuous on U^0 $\forall z$ then

$$I^0 = \min_{z\in Z} I(z) = \lim_{n\to\infty} \min_{z\in Z} I_n(z) \quad .$$

Indeed, since $|w(z,a\xi) - w(z,[a]_n\xi)| \le \varepsilon_n \downarrow 0$ and $w(z,\cdot) \ge 0$, then

$$\int_U w(z,u)\mu(z,du) = \inf_n \int_{[U]_n} \{\varepsilon_n + w(z,a^n\xi)\}\mu_n(z,da^n), \quad \varepsilon_n \mu_n(z,[U]_n) \downarrow 0$$

because $\mu(z,\cdot)$ is bounded, i.e., $I(z) = \inf_n I_n(z)$ $\forall z \in Z^0$. Hence the minimum with respect to z and the limit as $n \to\infty$ are interchangeable.

Now the numerical search for I^0 using formula (2) uses finite-dimensional vectors only. This requires us to combine stochastic gradient methods [1,4] with increases in n (*limiting optimization*). But in addition to the problem of coordinating n and the precision of minimization, other difficulties may arise in the numerical search for gradients in Hilbert space. It may be that $\mu_n(z,\cdot)$ is given only for z with a finite number of non-zero basis coefficients. In this case it is necessary to use finite-dimensional approximations of the set Z in the algorithm or to approximate the set Z^0 and to take the constraints on Z into account with the aid of penalty functions. In addition we propose the computation of approximate integrals in Euclidean space. The above considerations lead to the following algorithm, which links the Ritcs method and the combined penalty functiions/stochastic quasi-gradient method according to the scheme offered in [5].

Algorithm 1. Fix an orthonormal basis $\xi(\cdot) = \{\xi_j\}$ in $L_2(Y)$ and specify control sequences of numbers $l_t, n_t, C_t, R_t, S_t \uparrow \infty$ and $\vartheta_t, \beta_t \downarrow 0$. Choose an initial approximation $b^1 \in [Z^0]_{l_1}$ for b^0 such that $z^0 = b^0\xi$. Subsequent approximations $b^t \in [Z^0]_{l_t}$, $t \in N$, are determined by the iterative procedure

$$[b^{t+1}]_{l_t} \triangleq \pi^t \{b^t - \beta_t \operatorname{grad}_b \{\varphi(\frac{d(Y)}{R_t} \sum_{r=1}^{R_t} g(b^t\xi(y_r^t)) \quad ,$$

$$\frac{d(X)}{S_t}\sum_{s=1}^{S_t} h(a^t\zeta(x_s^t)),\ \frac{d(Y)d(X)}{R_tS_t}\sum_{r=1}^{R_t}\sum_{s=1}^{S_t} f(b^t\zeta(y_r^t),\ a^t\zeta(x_s^t))) +$$

$$+ c_t[\max[0, G(b^t\zeta(y_1^t))]]^2\}\},\ b_j^{t+1} \triangleq 0\ \ \forall j = l_t+1,\dots,l_{t+1}\ .$$

Here π^t is the projection operator onto the set $[Z^0]_{l_t}$; grad_b denotes the generalized gradient of the expression in braces with respect to b; for a^t we choose the next independent value of the random variable distributed in accordance with $\mu_{n_t}(b^t\zeta,\cdot)$ and satisfying the condition $H(a^t\zeta(x_s^t)) \le \vartheta_t\ \ \forall x_s^t,\ s = 1,2,\dots,S_t,\ \forall t \in N$; the sequences $\{y_r^t\}_{r=1}^{R_t}$ and $\{x_s^t\}_{s=1}^{S_t}$ for $t \in N$ are sequences of independent values of random variables equidistributed on Y and X, respectively. If the set Y (and/or X) is defined by its restrictions on Y^0 (and/or X^0), we use a random number generator on Y^0 (and/or X^0) and test whether the restrictions are satisfied. With regard to R_t, S_t, we shall assume that there exist $\alpha_t, \gamma_t \downarrow 0$ such that for fixed $P_0 < 1$ we have

$$\Pi(1-\alpha_t) \triangleq 1-\alpha > P_0^{1/3},\ \Pi(1-\gamma_t) \triangleq 1-\gamma > P_0^{1/3},\ (R_t\alpha_t)^{-1/2}\downarrow 0, \tag{3}$$

$$(S_t\gamma_t)^{-1/2}\downarrow 0\ .$$

We shall also assume a rather slow increase in the penalty constants $c_t \uparrow \infty$:

$$c_t\sum_{j>l_t}|b_j^0|^2 \to 0,\ \sum c_t^2\beta_t^2 < \infty,\ \sum\beta_t = +\infty,\ \beta_t \downarrow 0\ . \tag{4}$$

Here and elsewhere we omit the bounds on sums, etc., over $t \in N$.

Let us analyze the convergence of Algorithm 1. Set

$$\psi(b,a) \triangleq w(b\zeta, a\xi)\ ,$$

$$\psi_t(b,a\mid\{y_r^t\}_{r=1}^{R_t},\{x_s^t\}_{s=1}^{S_t}) \triangleq \varphi\Big(\frac{d(Y)}{R_t}\sum_{r=1}^{R_t} g(b^t\zeta(y_r^t)),\ \frac{d(X)}{S_t}\sum_{s=1}^{S_t} h(a^t\xi(x_s^t))\ ,$$

$$\frac{d(Y)d(X)}{R_tS_t}\sum_{r=1}^{R_t}\sum_{s=1}^{S_t} f(b^t\zeta(y_r^t),\ a^t\xi(x_s^t))\Big)\ ,$$

$$\chi(b) \triangleq \frac{1}{d(Y)}\int_Y \{\max[0, G(\zeta(y)b)]\}^2 dy\ ,\ \chi_t(b\mid y^t) \triangleq \{\max[G(b\zeta(y_1^t)),0]\}^2\ ,$$

$$d(Y) \triangleq \int_Y dy\ ,\ d(X) \triangleq \int_X dx\ ,\ d(Y)d(X) < \infty\ .$$

We require these functions to be bounded uniformly on finite-dimensional projections of sets Z^0 and U^0:

$$|g([b]_l\zeta(y))| \le K_g(l)\ ,$$

$$\chi_t([b]_l \mid y) \le K_1(l)\,,\ \|\mathrm{grad}_b\, \chi_t([b]_l \mid y)\| \le K_2(l)\,,\ \forall l \in N\,,\ \forall [b]_l \in [z^0]_l\,,\ \forall y \in Y$$

$$|f([b]_l \xi(y)\,,\ [a]_n \xi(x))| \le K_f(l\,,n) \quad ,$$

$$|h([a]_n \xi(x))| \le K_h(n)\,,\ |\psi([b]_l\,,\ [a]_n)| \le K_3(l\,,n)\,,\ \forall n \in N\,,\ \forall [a]_n \in [U^0]_n\,,\ \forall x \in X$$

$$\|\mathrm{grad}_b\, \psi_t([b]_l\,,[a]_n \mid \{y_r\}_{r=1}^R\,,\ \{x_s\}_{s=1}^S)\| \le K_4(l\,,n)\,,\ \forall y_r \in Y\,,\ \forall x_s \in X \quad .$$

We choose a rather slow increase in the dimensions l_t and n_t, coordinated with the indicated bounds. If the bounds do not depend on the approximation dimension then the following conditions are included in (3)–(4):

$$\sum C_t^2 \beta_t^2 \{K_1^2(l_t) + K_2^2(l_t)\} < \infty\,,\ C_t K_2(l_t) \sum_{j > l_t} (b_j^0)^2 \underset{t \to \infty}{\longrightarrow} 0 \quad , \tag{5}$$

$$\sum C_t \beta_t^2 K_2(l_t) K_3(l_t\,,n_t) < \infty\,,\ \sum \beta_t^2 \{K_3^2(l_t\,,n_t) + K_4^2(l_t\,,n_t)\} < \infty \quad , \tag{5'}$$

$$\rho_t \triangleq \frac{K_g(l_t)}{(R_t \alpha_t)^{1/2}} \downarrow 0\,,\ \sigma_t \triangleq \frac{K_h(n_t)}{(S_t \gamma_t)^{1/2}} \downarrow 0\,,\ \tau_t \triangleq \frac{K_f(l_t\,,n_t)}{(R_t S_t \gamma_t)^{1/2}} \downarrow 0 \quad . \tag{6}$$

We take a control sequence $\{\vartheta_t\}$ such that

$$\vartheta_t \ge L_H \sup_{a\xi \in U^0} \Big(\sum_{j > n_t} (a_j)^2 \Big)^{1/2}\,,\ t \in N \quad , \tag{7}$$

where L is the Lipschitz constant of the functional H. If we do not know whether $H \in \mathrm{Lip}\,(U^0)$ then choosing a^t we check the condition: $\forall a\xi \in U^0 \{a^t = [a]_{n_t}\}$ implies $\{H(a\xi(x_s^t)) \le 0\ \forall s = 1,2\,, \ldots, S_t\}$. Both this check and (7) are easy for simple sets U^0. However, it is more convenient to combine the correction of ϑ_t with the updating of other parameters of the algorithm.

THEOREM 1. *Let $I(\cdot)$ be a strongly convex continuous functional on Z^0, $G(\cdot)$ be a convex continuous functional on Z^0, $H(\cdot) \in \mathrm{Lip}\ (U^0)$ with constant L, $w(\cdot,\cdot) \ge 0$ be convex with respect to $z \in Z^0$ and continuous with respect to $z \in Z^0$ uniformly over $u \in U^0$, sets Z^0 and U^0 be weakly compact in L_2, $\varphi(\cdot,\cdot,\cdot) \in Lip\ (R^3)$, and $\mathrm{d}(Y)$, $\mathrm{d}(X) < \infty$. Then, under the assumptions made above, the sequence $z^t \triangleq b^t \xi$ generated by Algorithm 1 with control sequences satisfying (3–7) converges to z^0 with probability P_0^2 on the set of random sequence $\{\{y_r^t\}_{r=1}^{R_t}, \{x_s^t\}_{s=1}^{S_t}, a^t\}_{t \in N}$ in the strong metric of Hilbert space.*

In practice it is necessary to select the control parameters in Algorithm 1 according to the problem being solved since conditions (3–7) give the asymptotes only. Therefore we carry out some test computations with different initial ratios between β_t and c_t (the other parameters, apart from ϑ_t, are sufficiently large) until the coefficients b^t are relaively stable. After that we slowly increase n_t and l_t, watching the stability. Note that the assumption that z belongs to the Sobolev space simplifies the computations [5].

2. A PRIORI ESTIMATION OF I^0

The case considered above requires a great number of observations $a^t \in [U^0]_{n_t}$. However, these observations may be very complicated and an *a priori* estimate of I^0 may be necessary. In this case suppose that a finite number of the following integral inequalities are known for $\mu(z,\cdot)$-integrable w_t:

$$D_k(z) \le \int_U w_k(z,u)\mu(z,du) \le F_k(z) \quad \forall k = 1,2,\dots, m \quad . \tag{8}$$

Inequalities (8) are insufficient for single-valued determination of $\mu(z,\cdot)$ on U. The set of $\mu(z,\cdot)$ which are integrable under conditions (8) with respect to the functionals w, w_k will be denoted by $M(z)$, $z \in Z^0$. Let $\mu(z,U) = 1$ in (8) for $k = 1$. If the functionals are integrable with respect to finite-additive measures the latter are also included in $M(z)$.

An *a priori* estimate (upper bound) $\hat{I}^0$ of I^0 under (8) is given by

$$\hat{I}^0 = \inf_{z \in Z} \sup_{\mu(z,\cdot) \in M(z)} \int_U w(z,u)\mu(z,du) \triangleq \inf_z \hat{I}(z) \quad . \tag{9}$$

With the aim of finding the value $\hat{I}^0$ and the corresponding value $\hat{z} \in Z^0$, we shall transform the interior problem of maximization on the set of measures M to the simpler problem of maximization on U as described in [6, pp. 106–114] and [4, pp. 72–74] for Euclidian space.

THEOREM 2. *Suppose that U is weakly compact, the w_k are weakly continuous with respect to u and w is weakly upper semicontinuous with respect to u and non-negative. Then for any $z \in Z$ there is a measure $\mu(z,\cdot)$ in $M(z)$ which maximizes the integral in* (9) *and which is concentrated at no more than m points $u^1, \dots, u^m$.*

Under the hypotheses of Theorem 2, the problem of finding $\hat{I}^0$ and $\hat{z}$ in (9) is equivalent to solving

$$\inf_{z \in Z} \max_{\{u^1,\dots,u^m \in U \mid A(u^1,\dots,u^m_z) \neq \phi\}} \Phi(u^1,\dots,u^m,z) \quad , \tag{10}$$

where

$$\Phi(u^1,\dots,u^m,z) \triangleq \max_{p \in A(u^1,\dots,u^m,z)} \sum_{i=1}^m w(z,u^i)p^i \ , \ p \triangleq (p^1,\dots,p^m) \quad ,$$

$$A(u^1,\dots,u^m,z) \triangleq \{p \geq 0 \mid D_k(z) \leq \sum_{i=1}^{m} w_k(z,u^i)p^i \leq F_k(z) \; \forall k = 1,2,\dots,m\} \; .$$

If z does not determine whether the set A is non-empty, then a similar algorithm to that given above may be used to compute the solution of (10). This is based on a method given in [5] for finding a minimax in Hilbert space. For the associated minimax function the results of [7] may be used. We shall compute generalized gradients of the maximum function Φ with respect to finite-dimensional vectors b with the help of a formula from [8, Remark 1]. After some transformations we obtain

$$\frac{\partial \Phi(u^1,\dots,u^m,b\zeta)}{\partial b_j} = p \in$$

$$\operatorname{Arg} \max_{A(u^1,\dots,u^m,b\zeta)} \max_{\sum_{i=1}^{m} w(b\zeta,u^i)p^i} \tag{11}$$

$$\{\sum_{i=1}^{m} p^i \frac{\partial w(b\zeta,u^i)}{\partial b_j} + \max_{\eta \in M} \sum_{i=1}^{m} w(b\zeta,u^i)\eta^i\} \; ,$$

where $M = \{\eta = (\eta^1,\dots,\eta^m) \mid$ if $\sum_{i=1}^{m} w_k(b\zeta,u^i)p^i = F_k(b\zeta)$ (and/or $D_k(b\zeta)$) then $\sum_{i=1} \{w_k(b\zeta,u^i)\eta^i + \frac{\partial w(b\zeta,u^i)}{\partial bj} p^i\} \leq 0$ (and/or ≥ 0), $k = 1,2,\dots,m\}$. This simple form for the partial derivative of the associated maximum is due to linearity with respect to p. It is now sufficient to use the simplex method twice to find the partial derivative from (11). The dimensionality m of the simplex is usually small but we must regularize the problem as described in [9]. Formula (11) holds when the restrictions are regular and the setting mapping is continuous. If these conditions are not satisfied then difference formulae may be used for the gradients as well. Thus we use the following algorithm to find $\hat{I}^0$ and $\hat{z}$ in (9):

Algorithm 2. Choose an orthonormal basis ζ in $L_2(Y)$. Assume that for any $l \in N$ a sufficiently simple set B^l $[\hat{b}]_l$ is known, where $\hat{b}\zeta = \hat{z}$ in (9). Let $\omega \triangleq (J,b) = (\omega_0,\omega_1,\dots)$, where the first component J approximates the value $\hat{I}^0$ in (9). We know that $J \in [J_0,J^0]$ for any $J_0 \leq \inf_{z,M} I(z)$, $J^0 \geq \max_{z,m} I(z)$. Set $\hat{\omega} \triangleq (\hat{I}^0,\hat{b})$, $[\omega]_l = (J,[b]_l) = (\omega_0,\omega_1,\dots,\omega_l)$. Choose an orthonormal basis ξ in $L_2(X)$. Assume that for any $n \in N$ a sufficiently simple set A^n is known, where $[U]_n \subset A^n \subset [U^0]_n$, $A^n \subset A^{n+1}$. Choose a continuous measure ν on U^0. Let ν_n denote the cylindrical projection of ν onto $[U^0]_n$. In particular, if $A^{n+1} \cap R^n = A^n$ and $d(A^n) < \infty \; \forall n \in N$, then we may take $\nu_n(\cdot) = d_n(\cdot)[d_n(A^n)]^{-1}$, where $d_n(\cdot)$ is the Lebesgue measure on R^n. Take an initial approximation $\omega^1 = (J^1,b^1)$ for $(\hat{I}^0,\hat{b})$ and control sequences of numbers l_t, n_t, R_t, S_t, $C_t \uparrow \infty$; κ_t, $\beta_t \downarrow 0$; $t \in N$. Subsequent approximations $\omega^t = (J^t,b^t)$, $t \in N$, are determined with the help of the following iterative procedure:

$$[\omega^{t+1}]_{l_t} = \pi_t \{\omega^t - \beta_t \operatorname{grad}_\omega \{J^t + C_t \{[\min[0 ,$$

$$J^t - \max_{\{p \geq 0 \mid \forall k = \overline{1,m} D_k(6^t \zeta) - \kappa_t \leq \sum_{i=1}^{m} \psi_i^k(6^t, a^{it} \mid \{y_r^t\}_{r=1}^{R_t}, \{x_s^t\}_{s=1}^{S_t}) p_i \leq F_k(6^t \zeta) + \kappa_t\}} \sum_{i=1}^{m} p_i x$$

$$\psi_t(6^t, a^{it} \mid \{y_r^t\}_{r=1}^{R_t}, \{x_s^t\}_{s=1}^{S_t})]]^2 + \chi_t(6^t \mid y_1^t)\}\}\} , \omega_j^{t+1} = 0 \ \forall j = l_t + 1 , \ldots , l_{t+1} .$$

Here π_t is the projection operator onto the set $[J_0, J^0] \times B^{l_t}$; for a^{it} we take values of the random variable distributed on A^{n_t} in accordance with ν_{n_t} (in particular, equidistributed) such that $H(a \, \xi(x_s^t)) \leq 0 \ \forall s = 1, S_t$ for any a satisfying the conditions $[a]_n \in A^n \ \forall n > n_t$, $[a]_{n_t} = a^{it}$; $i = \overline{1,m}$. The functions ψ_t^k correspond to w_k for $k = \overline{1,m}$ in the same way that ψ_t corresponds to w. All the other notation remains the same.

For simplicity we took equal parameters ξ, A^n, ν_n for different u^k, $k = \overline{1,m}$ in Algorithm 2 but this is not obligatory. With regard to the penalty constants C_t, the possibility of choosing different values for different restrictions (first of all for $z \in Z$ and then for J) may prove useful in practice.

Algorithm 2 is based on an idea put forward in [6, p. 256]. In the case when $A(u^1, \ldots, u^m, z) \neq \phi$ this idea may be represented schematically in the following way:

$$\inf_{z \in Z} \max_{u^1, \ldots, u^m \in U} \Phi = \inf_{\substack{z,y \\ J = \max_{u^1,\ldots,u^m} \Phi}} J = \inf_{\substack{z,y \\ J \geq \Phi \ \forall u^1,\ldots,u^m \in U}} J =$$

$$\inf_{\substack{z,y \\ \int_U \cdots \int_U \{[J-\Phi]^-\}^2 \nu(du^1) \cdots \nu(du^m) = 0}} J =$$

$$= \lim_{C \to \infty} \inf_{Z \times [J_0, J^0]} \{J + C \int_U \cdots \int_U \{[J - \Phi]^-\}^2 \nu(du^1) \cdots \nu(du^m) .$$

The version for infinite-dimensional sets Z and U is given in [7] under the conditions $\nu(\{u \in U^0 \mid \|u - u^0\| \leq \varepsilon\} \geq \delta(\varepsilon) \ \forall \varepsilon > 0$, $u^0 \in U^0$. The possibility of non-computing integrals in penalty functions was investigated in [10] but for the finite-dimensional case only. There we suggested combining the penalty function method [6] with the stochastic quasi-gradient method [4]. The combined method constructed in this way is not included in [4] because of the necessity to increase the penalty constant to infinity. New techniques developed for this case extend the set of problems that can be solved by stochastic programming methods. Using this technique together with the Ritcs method for the case of Hilbert space we obtain the required results.

We shall now make the same assumptions concerning functions ψ^k, ψ_t^k and their generalized gradients as were made with regard to ψ, ψ_t and $\text{grad}_b\ \psi_t$. The correspnding parameters will be denoted by the superscript k, $k = \overline{1,m}$. In order to coordinate the control sequences of Algorithm 2 we shall replace conditions (5–7) by the following:

$$\sum \beta_t^2 C_t^2 \{[K_3^k(l_t, n_t)]^4 + [K_4^k(l_t, n_t)]^4\} < \infty, \ k = \overline{1,m} \tag{12}$$

$$C_t\{\rho_t + \sigma_t + \tau_t + \{K_1(l_t) + [K_3(l_t, n_t)]^2 + [K_4(l_t, n_t)]^2\} \sum_{j > l_t} (\hat{b} j)^2\} \to 0 \tag{13}$$

$$\kappa_t = \bar{\bar{o}}\ (\rho_t, \sigma_t, \tau_t) \ . \tag{14}$$

We shall now investigate the convergence properties of Algorithm 2.

THEOREM 3. *Let the assumptions made above and the hypotheses of Theorems 1 and 2 (except for $H \in Lip$) be satisfied. In addition, let the functional $\hat{I}(\cdot)$ be strongly convex; the functional $\Phi(u^1, \ldots, u^m, z)$ be non-empty for any $z \in Z^0$, $u^1, \ldots, u^m \in U$; vecotrs $(w_k(z, u^1), \ldots, w_k(z, u^m))$ be linearly independent for any*

$$k \in \{k = \overline{1,m} \mid \sum_{i=1}^{m} w_k(z, u^i)\hat{p}_i = D_k(z) \quad \forall \hat{p} \geq 0:$$

$$\sum_{i=1}^{m} w(z, u^i)\hat{p}_i = \Phi(u^1, \ldots, u^m, z)\} \cup$$

$$\cup \{k = \overline{1,m} \mid \sum_{i=1}^{m} w_k(z, u^i)\hat{p}_i = F_k(z) \quad \forall \hat{p} \geq 0: \sum_{i=1}^{m} w(z, u^i)\hat{p}_i = \Phi(u^1, \ldots, u^m, z)\} \ ,$$

and functions $w([b\zeta]_b, u)$, $w_k([b\zeta]_l, u)$, $k = \overline{1,m}$ be continuously differentiable with respect to b for any $l \in N$, $u \in U^0$. Then with probability P_0^{2m} the sequence $z^t = b^t\zeta$ generated by Algorithm 1 with control sequences satisfying (3–5) *and* (12–14) *strongly converges to $\hat{z}$ in $L_2(Y)$ and the sequence of numbers J^t converges to $\hat{I}^0$* (see (9)).

Remark 1. It is sufficient for the convergence of Algorithm 1 and 2 if the functionals I, $\hat{I}$ display strict uniform convexity, i.e., the assumption of strong convexity can be weakened. If the functionals to be minimized display other forms of convexity the algorithms may be used after special regularization.

Remark 2. Algorithms 1 and 2 may be simplified if for any t we use all previous pairs x^τ, y^τ and only take a few new elements in $\{x_s^t\}$, $\{y_r^t\}$, for example $S_t = R_t = t$, $x_s^t = x^s$, $y_r^t = y^r$, s, $r = \overline{1,t}$. With regard to vectors a^t, it is sensible to use more than one value for every t (at least for large t) and to take the corresponding averaged values for the approximation of integrals in the algorithms.

REFERENCES

1. A. Balakrishnan. *Introduction to Optimization Theory in Hilbert Space*. Mir, Moscow, 1974.
2. A.B. Skorohod. *Integration in Hilbert Space*. Nauka, Moscow, 1975.
3. J. Varga. *Optimal Control of Differential and Functional Equations*. Nauka, Moscow, 1977.
4. Yu.M. Ermoliev. *Methods of Stochastic Programming*. Nauka, Moscow, 1976.
5. N.M. Novikova. A stochastic quasigradient method for solving optimization problems in Hilbert space. *Journal of Computational Mathematics and Mathematical Physics*, 24(1984)348–362.
6. Yu.B. Germeier. *Introduction to the Theory of Operations Research*. Nauka, Moscow, 1971.
7. V.A. Gorelik. Maximum problems in Banach spaces. *Kibernetika*, (1)(1981)64–67.
8. L.I. Minchenko. Differentiability properties of maximum functions. *Journal of Computational Mathematics and Mathematical Physics*, 24(1984)210–217.
9. M. Klepikova. Private communication.
10. N. Novikova Stochastic quasigradient method for minimax problems. *Journal of Computational Mathematics and Mathematical Physics*, 17(1977)91–99.

REDUCTION OF RISK USING A DIFFERENTIATED APPROACH

I. Petersen,
Institute of Cybernetics
Tallin, USSR

1. INTRODUCTION

In this paper we consider the problem of subdividing the range of the random vector in a risk minimization problem such that the best piecewise-constant decision rule is obtained. This is a generalization of the optimal partition problems which arise in optimal location, standardization, piecewise approximation, and cluster analysis. Necessary and sufficient conditions for local optimality of the partition are obtained in the case of continuous density, and an iterative method for calculating such partitions is presented.

2. THE PROBLEM

Consider the decision problem

$$\min_x \{f(x, \xi) \mid x \in X\} \quad , \qquad (1)$$

where $X \subset R^n$, ξ is a random vector with range $\Theta \subset R^m$ and distribution $\mu(d\xi)$, and let $\sum$ be the σ-algebra of the μ-measurable Borel subsets of Θ. We shall assume that for almost all $\xi \in \Theta$ problem (1) has unique solution $x^*(\xi)$ which is a measurable function.

Let us interpret Θ as a set of individuals, X as the set of admissible decisions, and $f(x, \xi)$ as the loss incurred by the individual ξ when the decision x is taken. Then for every individual ξ the function $x^*(\xi)$ gives the decision associated with minimal loss. The decision $x^*(\xi)$ can therefore be called the ***individual optimal*** ($\dot{I}O$) ***solution*** and the mean value of the corresponding losses $I = \int_\Theta f(x^*(\xi), \xi)\mu(d\xi)$ the $\dot{I}O$***-risk*** of problem (1). If, however, only one decision is allowed for all the individuals, which is the same for all of them, then the risk is minimized if the decision is taken to be the solution x^* of the problem

$$\min_x \{\int_\Theta f(x, \xi)\mu(d\xi) \mid x \in X\} = I_1 \quad . \qquad (2)$$

Let us call x^* the ***mean optimal (MO) solution*** of problem (1) and I_1 its ***MO-risk***.

Let Π_k, where k is an integer, be the set of all partitions $\pi_k = \{\Theta_1, \ldots, \Theta_k\}$ of Θ such that $\Theta_i \in \sum$, $\cup \Theta_i = \Theta$, $\mu(\Theta_i \cap \Theta_j) = 0$ when $i \neq j$, and $\mu(\Theta_i) = p_i > 0$, $i = 1, \ldots, k$. The partition π_k is associated with k conditional measures induced by the measure μ on the subsets Θ_i: $\mu_i(A) = \mu(A)/p_i$, $A \subset \Theta_i$, $A \in \sum$. The decision $x_{\pi_k}(\xi)$ which is MO on Θ_i with respect to μ_i for $i = 1, \ldots, k$ we call the ***π_k-differentiated optimal (π_k-DO) solution*** of problem (1). Thus $x_{\pi_k}(\xi) = x_i^*$ for $\xi \in \Theta_i$ and $x_i^* = \arg\min_{x \in X} \int_{\Theta_i} f(x, \xi)\mu(d\xi)$. The π_k-DO risk for the whole range Θ is given by the sum

$$\sum_{i=1}^{k} \int_{\Theta_i} f(x_i^*, \xi)\mu(d\xi) = I_{\pi_k} \quad . \tag{3}$$

Finally, taking

$$\min_{\pi_k} \{I_{\pi_k} \mid \pi_k \in \Pi_k\} = I_{\pi_k^*} = I_k \tag{4}$$

we call π_k^* and I_k the ***k-DO solution*** and ***risk***, respectively, of problem (1).

The risk values defined above satisfy the inequalities

$$I_1 \geq I_{\pi_k} \geq I_k \geq I_{k+1} \geq I \quad . \tag{5}$$

Particular k-DO optimization problems (in which it is required to determine optimal partitions) have been considered in various (mostly discrete) formulations. In optimal location problems [1] $X = \Theta$ and $f(x, \xi) = \|x - \xi\|$. In cluser analysis [2,3] we also have $X = \Theta$, but $f(x, \xi) = \|x - \xi\|^2$. In standardization problems [4,5] $X = \Theta$, but $f(x, \xi)$ (the loss incurred by an individual with needs ξ which must be satisfied with standard x) may turn out to be a more complicated function of x and ξ. In problems of optimal piecewise approximation [6] of a function $y(\xi)$ by combinations of the coordinates of a vector function $f(\xi)$ which are linear on Θ_i, we use the function $f(x, \xi) = \|y(\xi) - x^T f(\xi)\|^2$, where x is the vector coefficient. The problem of finding k-DO solution also occurs in other fields. Some authors [7,8] have considered using piecewise-constant approximation of the decision function to obtain an approximate solution to the distribution problem in stochastic programming. In this case we use a ***differentiated optimal approach*** to find the regions where the decision function is most constant.

The aim of this paper is to investigate the k-DO problem in the continuous situation, i.e., in the case when the measure μ has continuous density $\mu(d\xi) = g(\xi)d\xi$. We shall give the necessary and sufficient conditions for local optimality and present an iterative process to calculate locally optimal solutions.

3. NECESSARY CONDITION FOR OPTIMALITY

THEOREM 1. *If* $\pi_k^* = \{\Theta_1^*, \ldots, \Theta_k^*\}$ *is a solution of the problem*

$$\min_{\pi_k \in \Pi_k} \sum_{i=1}^{k} \min_{x \in X} \int_{\Theta_i} f(x, \xi) g(\xi) d\xi = \sum_{i=1}^{k} \int_{\Theta_i^*} f(x_i^*, \xi) g(\xi) d\xi \quad , \tag{6}$$

and also $f(x, \xi)$ *and* $g(x, \xi)$ *are continuous in* ξ *on* Θ *for all* $x \in X$, ξ_0 *lies on the boundary* S_{ij}^* *between* Θ_i^* *and* Θ_j^*, *and* $g(\xi_0) > 0$, *then*

$$f(x_i^*, \xi_0) = f(x_j^*, \xi_0) \quad , \tag{7}$$

where

$$x_s^* = \arg\min_{x \in X} \int_{\Theta_s^*} f(x, \xi) g(\xi) d\xi, \ s = i, j \quad . \tag{8}$$

Conditions (7–8) determine the structure of the dividing surfaces of the optimal partition. In particular, when $X = \Theta$ and $f(x, \xi) = \varphi(\|x - \xi\|)$, where $\varphi(t)$ is strictly monotonic on $t > 0$, it follows from (7) that the regions Θ_i^* of an optimal partition are convex polyhedrons with their faces lying on the hyperplanes through the midpoints and perpendicular to the line segments connecting x_i^* and x_j^*.

The necessary conditions (7–8) lead to the following result:

THEOREM 2. *The optimal partition problem* (6) *is equivalent to*

$$\min \{I(x_1, \ldots, x_k) \mid x_1, \ldots, x_k \in X\} \quad , \tag{9}$$

where

$$I(x_1, \ldots, x_k) = \sum_{i=1}^{k} \int_{\Theta_i(x_1, \ldots, x_k)} f(x_i, \xi) \mu(d\xi) \quad , \tag{10}$$

$$\Theta_i(x_1, \ldots, x_k) = \{\xi \mid \xi \in \Theta, f(x_i, \xi) \le f(x_j, \xi), j \ne i\} \quad . \tag{11}$$

By means of this theorem the optimal partition problem is reduced to a finite-dimensional nonlinear programming problem.

4. AN ITERATIVE METHOD

Locally optimal partitions can be found using the following generalization of the "k-means" method: let $x_1^{(0)}, \ldots, x_k^{(0)}$ be some starting values for $x_1^*, \ldots, x_k^*$; having already found $x_1^{(s)}, \ldots, x_k^{(s)}$, construct $\Theta_i^s = \Theta_i(x_1^{(s)}, \ldots, x_k^{(s)})$, $i = 1, \ldots, k$, and take

$$x_i^{(s+1)} = \arg\min_{x \in X} \int_{\Theta_i^{(s)}} f(x, \xi) \mu(d\xi), \ i = 1, \ldots, k \quad . \tag{12}$$

The iteration process (12) is a quasi-gradient with respect to the functional (10) and a simple iteration process with respect to the operator $\bar{y} = T\bar{x}$, $\bar{x} = (x_1, \ldots, x_k)$, defined by (12). From the Banach fixed-point theorem we obtain the following result:

THEOREM 3. *Let the operator T defined by* (12) *map a subset $\bar{X}$ of the set $X \times \cdots \times X$ into itself, $f''_{x^2}(x, \xi)$ be positive definite for almost all $\xi \in \Theta$, and the inequality*

$$\sum_j \int_{S_{ij}} \frac{\|f'_x(y_i, \xi)\| (\|f'_x(x_i, \xi)\| + f'_x(x_j, \xi)\|}{\|f'_\xi(x_j, \xi) - f'_\xi(x_i, \xi)\|} g(\xi) dS \le \rho \left\| \int_{\Theta_i} f''_{x^2}(y_i, \xi) g(\xi) d\xi \right\| \quad (13)$$

hold for some ρ, $0 < \rho < 1$, in $\bar{X}$, and for $i = 1, \ldots, k$.

Then problem (9) *has a unique local solution in $\bar{X}$ and the iteration process* (12) *converges to this solution at rate ρ^s for arbitrary starting point $\bar{x}^{(0)} \in \bar{X}$.*

5. SUFFICIENT CONDITIONS FOR LOCAL OPTIMALITY

Let us consider the unconstrained minimum case $X = R^n$. Then problem (9), and therefore also the k-DO problem (6), becomes the problem of constrained minimization of the functional $I(x_1, \ldots, x_k)$ in $k \cdot n$-dimensional space. The Hessian of the functional I is a block matrix with $n \times n$ blocks:

$$\frac{\partial^2 y}{\partial x_i \, \partial x_j} = \int_{S_{ij}} \frac{f'_x(x_i, \xi) f'^T_x(x_j, \xi)}{\|f'_\xi(x_j, \xi) - f'_\xi(x_i, \xi)\|} g(\xi) dS \, , \; i \ne j \quad , \quad (14)$$

$$\frac{\partial^2 y}{\partial x_i^2} = \int_{\Theta_i} f''_{x^2}(x_i, \xi) g(\xi) d\xi - \sum_j \int_{S_{ij}} \frac{f'_x(x_i, \xi) f'^T_x(x_j, \xi)}{\|f'_\xi(x_i, \xi) - f'_\xi(x_j, \xi)\|} g(\xi) dS \quad . \quad (15)$$

THEOREM 4. *If $x_1^*, \ldots, x_k^*$ satisfy the necessary conditions for optimality* (7–8) *and at this point the block matrix with blocks* (14–15) *is positive definite, then the partition $\Theta_i^* = \Theta_i(x_1^*, \ldots, x_k^*)$, $i = 1, \ldots, k$, is locally k-DO optimal.*

For the matrix $\left(\frac{\partial^2 y}{\partial x_i \, \partial x_j}\right)$ to be positive definite it is, in particular, necessary for the diagonal blocks (15) to be positive definite. This leads to the conclusion that the optimal boundaries S_{ij} cannot go through points with positive μ measure.

In general, the functional $I(x_1, \ldots, x_k)$ is multi-extremal. However, if the Hessian is positive definite over the whole space $R^n \times \cdots \times R^n$, then the k-DO problem is unimodal. In the one-dimensional case with quadric losses this fact leads to the following result:

THEOREM 5. *If $X = \Theta = R^1$ and $f(x, \xi) = (x - \xi)^2$, then for the k-DO problem to be unimodal it is sufficient that the density $g(\xi)$ satisfies the condition*

$$g(y_1)(\xi_{12}-y_1)+g(y_2)(y_2-\xi_{12})<\int_{y_1}^{y_2} g(\xi)\mathrm{d}\xi \qquad (16)$$

for all $y_1, y_2 \in R^1$, $y_1 < y_2$ *where* ξ_{12} *is the centre of mass of* $g(\xi)$ between y_1 *and* y_2.

It can be shown from (16) that $g(\xi)$ is logarithmically concave [9].

REFERENCES

1. R.L. Frances, J.A. McGinns and J.A. White. Locational analysis. *European Journal of Operational Research*, 12(1983)220–252.
2. H. Späth. *Cluster Analysis Algorithms for Data Reduction and Classification of Objects* Ellis Harwood, 1980.
3. J. MacQueen. Some methods for classification and analysis of multivariate observations. In *Proceedings of the Vth Berkeley Symposium on Mathematical Statistics and Probability*. University of California Press, 1967, pp. 281–297.
4. C. Bongers. *Standardization: Mathematical Methods in Assortment Determination*. Martinus Nijhoff, Boston, 1980.
5. D.M. Komarov. *Mathematical Models of the Optimization of Demands to Standards*. Moscow, 1976 (in Russian).
6. R.E. Quandt. A new approach to estimating switching regression. *Journal of the American Statistical Association*, 67(1972)306–310.
7. P. Kall. Approximation to stochastic programs with complete fixed recourse. *Numerische Mathematik*, 22(1974)333–339.
8. W. Römisch. An approximation method in stochastic optimal control. *Lecture Notes in Control and Information Sciences, Vol. 22, Springer-Verlag, 1980, pp. 169–178.*
9. A. Prekopa. A class of stochastic decision problems. *Mathematische Operationsforschung und Statistik*, 3(1972)349–354.

A STOCHASTIC LAKE EUTROPHICATION MANAGEMENT MODEL

J. Pintér and L. Somlyódy
Research Centre for Water Resources Development (VITUKI)
H-1453 Budapest, P.O. Box 27, Hungary

1. INTRODUCTION

Due to the world-wide danger of environmental degradation, water quality management has attracted increasing attention in the last decades. As a consequence of this tendency, a large number of descriptive (simulation) and management (optimization) models have been developed and applied, cf. e.g. Dorfman et al. (1972), Marks (1974), Deininger (1977), Loucks et al. (1981), Somlyódy (1983), Somlyódy and Wets (1985), Somlyódy and van Straten (1985). These interrelated (sub)models frequently have much different temporal and/or spatial scaling (information structure): this renders difficult their proper combination. Moreover, even recognizing the inherent stochasticity of the studied problems (analyzed in details by descriptive submodels), most optimization (sub)models are formulated as deterministic: this fact again implies complicated methodological issues.

In this paper a stochastic programming model is presented for solving a lake eutrophication management problem. The main features of the investigated problem are outlined in Section 2. Principles of the solution methodology are summarized in Section 3, while Sections 4-5 describe the mathematical framework of the optimization model and the solution method (more details can be found in Somlyódy and Pintér (1984)). The conclusions summed up in Section 6 are of direct relevance to many similarly complicated stochastic optimization problems, such as e.g. reservoir system design, monitoring network operation planning, inventory control, mass-service systems etc.

2. LAKE EUTROPHICATION; MANAGEMENT ALTERNATIVES

Man-made eutrophication in lakes - caused primarily by increasing municipal and industrial waste water discharges and the intensive use of chemicals in agriculture - has been recently considered as one of the major water quality problems. The typical symptoms of eutrophication (algal blooms, water colouration, floating water plants, organic debris, unpleasant taste and odor) frequently lead to serious limitations of water use.

Both the main causes of artificial eutrophication (increasing amounts of nutrient load, reaching the lake) and the possible water quality control alternatives are physically originated from and connected to the region surrounding the lake. Their impact on lake water quality is the result of interdependent (physical, chemical and biological) in-lake processes. Thus, eutrophication management requires a complex analysis of the

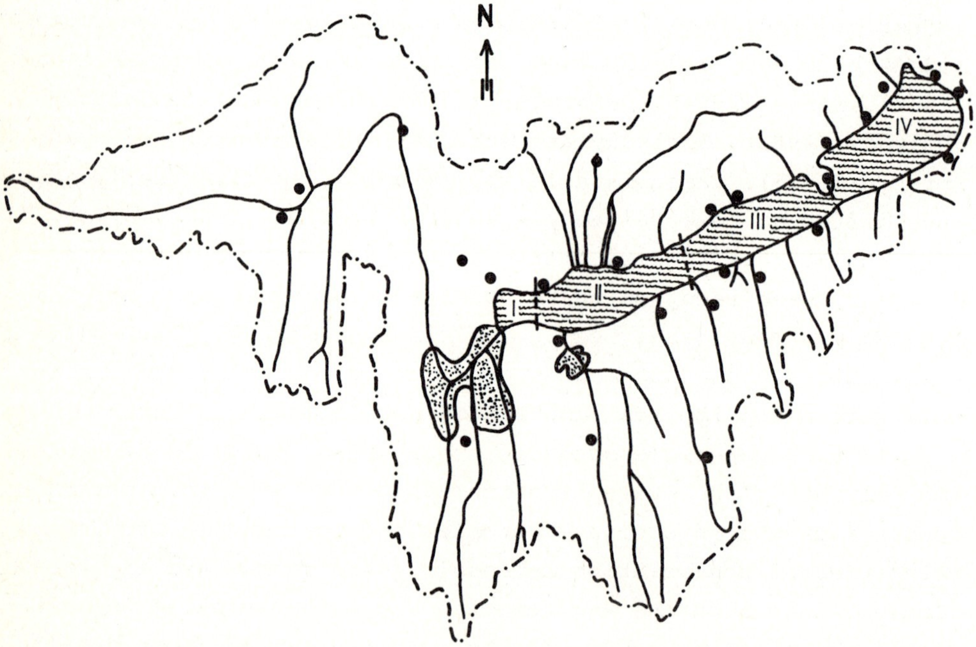

-·- boundary of the catchment
pre-reservoirs
● sewage discharges/treatment plants

FIGURE 1. Lake eutrophication management; water quality control alternatives

whole region, including all relevant natural phenomena/processes and human activities. As an example, consider the region of a lake (Figure 1), where the most important streamflows and possible water quality management options are also shown: for modelling reasons, the lake is subdivided into 4 basins, the water quality of which shows marked differences, with a well-defined longitudinal profile .

Because of the obvious stochasticity of the studied problem (caused primarily by hydrological and meteorological factors), even for a deterministic management option (e.g. a fixed config-

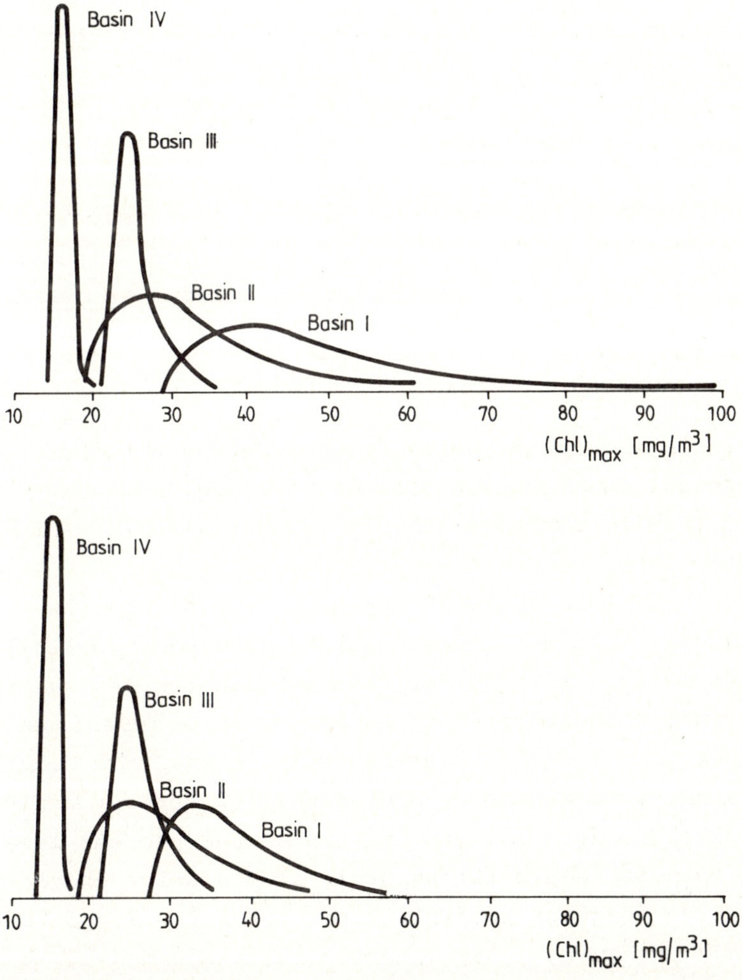

FIGURE 2. Water quality realizations vs. given treatment configurations

uration of treatment plants), the lake response will be stochastic. This fact can be illustrated e.g. by Figure 2, in which water quality density functions (based on 1000 Monte Carlo simulations) are displayed for a "cheap" and an "expensive" treatment configuration. (For each subbasin, the resulted water quality is characterized by $(Chl\text{-}a)_{max}$, the annual peak chlorophyll concentration.)

The relevance of random factors, influencing the impact of decision alternatives, implies that eutrophication management problems should be naturally formulated as stochastic programming models. Before doing this, in the next Section a short overview of the considered submodels and the applied methodology is presented.

3. DECOMPOSITION AND AGGREGATION

As already mentioned in the Introduction, most descriptive and optimization models have a different information structure: while the first model-type is applicable to analyze in details any single decision alternative, the second model-type is able to sequentially evaluate a set of alternatives, based on their aggregated characteristics. The coordinated, simultaneous use of these submodels is possible only by considering a number of related theoretical and technical issues, cf. e.g. Somlyódy (1982a, 1982b, 1983), Somlyódy and van Straten (1985) or Pintér (1980, 1983, 1984).

The present systems modelling approach is based on the principle of decomposition and aggregation (Somlyódy (1982b)). As a first step of this methodology, the studied complex system is decomposed into "homogeneous" parts which can be investigated in details: these parts form a hierarchical system. In the second step, the obtained information is aggregated for further analysis, while neglecting details, unnecessary on the present level of hierarchy. This procedure is obviously of iterative character, with adaptive feedback possibilities. As a result of this method, instead of having a huge, hardly tractable model, a coherent system of detailed and aggregated submodels is obtained.

The application of the outlined methodology for lake eutrophication management in shallow lakes is illustrated by Figure 3, in which the relevant submodels and their connections are displayed. On the top of this hierarchical structure a regional development model is situated, which summarizes the most important features of the considered man-made activities and natural processes (for more details see Somlyódy (1983) or Somlyódy and Pintér (1984)).

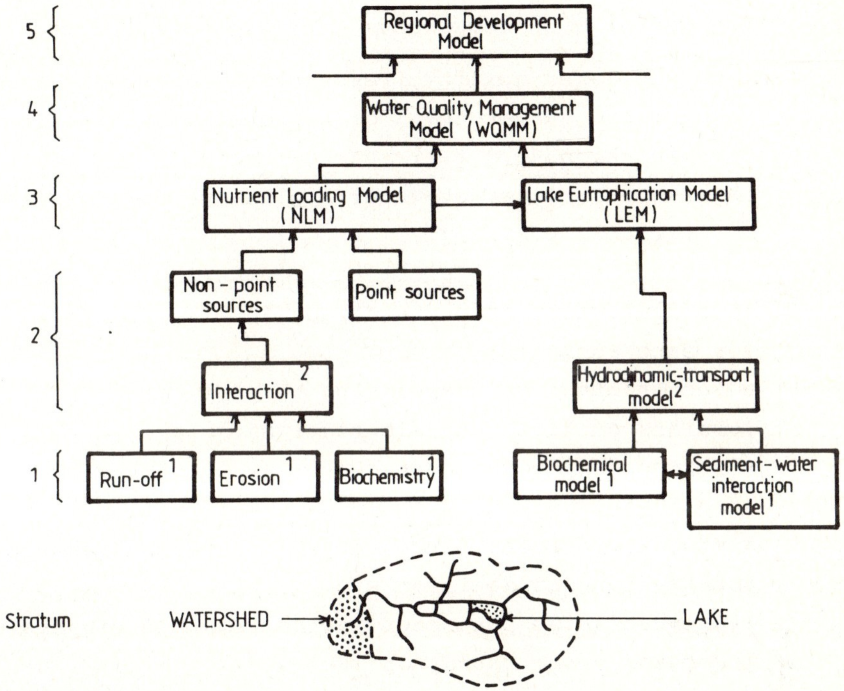

{ hierarchy levels

1 submodels for "homogeneous" segments (dotted areas)

2 integration of submodels

FIGURE 3. Decomposition and aggregation: submodels and their interrelations

4. LAKE EUTROPHICATION MANAGEMENT MODELS: A GENERAL FRAMEWORK

The following notations are introduced:

$\underline{x} \in R^n$ a particular feasible decision alternative (a water quality management configuration), represented as a vector of the n-dimensional real Euclidean space;

$\omega \in \Omega$	the elementary random events and their set which are considered in the stochastic model of the lake eutrophication management problem;
$(\Omega, \mathcal{A}, P)$	the underlying probability space;
$C^{(i)}(\underline{x},\omega)$	resulted water quality for a given decision and a random outcome, in lake segment (subbasin) i, i=1,.. ...,I);
$\underline{C}(\underline{x},\omega)$	The vector with components $C^{(i)}(\underline{x},\omega)$ i=1,..., I;
$C_g^{(i)}, C_g^{(i)}(\omega)$	target water quality, respectively given by a deterministic target level or a probability distribution function (p.d.f.);
$\underline{C}_g, \underline{C}_g(\omega)$	the respective target water quality vector (or vector variable p.d.f.);
F	a real-valued loss-functional, measuring the undesirable deviations from the target levels, specified for each subbasin;
$\underline{A}(\underline{x}):R^n \to R^m$	a vector-function, expressing technological constraints on treatment alternatives;
$\underline{b} \in R^m$	right-hand side constant vector of technological constraints;
$\underline{c}(\underline{x})$	a vector-function, representing financial (resource) implications of decision $\underline{x}$;
k	right-hand side parameter in the resource constraint;
$\underline{\ell}, \underline{u}$	lower and upper bounds on $\underline{x}$.

Applying the above notations, the following general model--type can be formulated for lake eutrophication management:

$$\min F \{\underline{C}(\underline{x},\omega), \underline{C}_g(\omega)\} \tag{1}$$

$$\underline{A}(\underline{x}) \leq \underline{b} \tag{2}$$

$$c(\underline{x}) \leq k \tag{3}$$

$$\underline{\ell} \leq \underline{x} \leq \underline{u} \ . \tag{4}$$

The technological and resource constraints (2)-(3) often can be reasonably well approximated by piecewise linear, convex functions (more details on this are given e.g. in Somlyódy and Pintér (1984), Somlyódy and van Straten (1985)). Consequently, (2) and (3) can be substituted by linear constraints of the form

$$A\ \underline{x} \leq \underline{b} \tag{5}$$

$$\underline{c}\ \underline{x} \leq k \tag{6}$$

with respective (m x n)-matrix A and n-vector $\underline{c}$.

The loss-functional F can be defined in a number of ways. First, it is often assumed that a proper weighting of the sub-basins can be expressed by some scalar factors w_i such that

$$w_i \geq 0 \quad i=1, \ldots, I \quad \text{and} \quad \sum_{i=1}^{I} w_i = 1.$$

This technique originates from multiobjective programming theory, cf. e.g. the survey of Hwang and Masud (1980). Thus, F is considered in the specialized form

$$F = \sum_{i=1}^{I} w_i F_i, \text{ where } F_i = F_i \{C^{(i)}(\underline{x},\omega),\ C_g^{(i)}(\underline{x},\omega)\} . \tag{7}$$

Again, the loss-functionals F_i can be defined in many different forms. We emphasize that, generally speaking, these formulations are not equivalent and reflect more or less different considerations on the stochastic factors. (Principles of formulating stochastic programming models with many examples are given e.g. in the works of Ermoliev (1976), Yudin (1979) or Kall and Prékopa, eds. (1980)). In Somlyódy (1983), the following objective components were considered:

$$F_i\{C^{(i)}(\underline{x},\omega),\ C_g^{(i)}\} = E(C^{(i)}(\underline{x},\omega) - C_g^{(i)})^w \tag{8}$$

and

$$F_i\{C^{(i)}(\underline{x},\omega), C_g^{(i)}\} = E(C^{(i)}(\underline{x},\omega) - C_g^{(i)}) + D\ (C^{(i)}(\underline{x},\omega) - C_g^{(i)}).w \tag{9}$$

where E and D^2 respectively denote the (existing) mathematical expectation and variance of the figuring random variables, $(w > 0)$.

In Somlyódy and Pintér (1984), we investigated objective functions with the following components:

$$F_i\ \{C^{(i)}(\underline{x},\omega),\ C_g^{(i)}\} = E(C^{(i)}(\underline{x},\omega) - C_g^{(i)})_+^2 \tag{10}$$

$$F_i\ \{C^{(i)}(\underline{x},\omega),\ C_g^{(i)}(\omega)\} = \int_{-\infty}^{\infty} |H_{\underline{x}}^{(i)}(z) - H_g^{(i)}(\underline{z})|\ dz \tag{11}$$

$$F_i\{C^{(i)}(\underline{x},\omega),\ C_g^{(i)}(\omega)\} = \int_{-\infty}^{\infty} (H_{\underline{x}}^{(i)}(z) - H_g^{(i)}(z))^2 \, d\, H_g^{(i)}(z), \qquad (12)$$

where

$$H_{\underline{x}}^{(i)}(z) = P(C^{(i)}(\underline{x},\omega) < z) \qquad (13)$$

and

$$H_g^{(i)}(z) = P(C_g^{(i)}(\omega) < z) \qquad (14)$$

are p.d.f.'s of the resulted and target water qualities (random variables) per basin. Note that the objective function (11) is based on Sherman-statistic (Sherman (1957)), while (12) is based on omega-square statistic (cf. e.g. Ermakov (1975)). Both statistics give a characterization of the "overall" deviation of two p.d.f.'s in a more complete manner, than e.g. the frequently applied Kolmogorov-Smirnov-type statistics; they have also less inherent subjectivity, than e.g. the more common chi-square statistics. We also note that Somlyódy and Wets (1985) formulated another stochastic model for lake eutrophication management which leads to a two-stage stochastic programming problem; its numerical solution was accomplished by King (1985).

5. SOLUTION METHOD

It is apparent from the model variants presented above that for arbitrary decision alternative, the result of that particular decision depends also on a number of random factors. This functional relations are far too complex to be treated in an explicit analytical form. Therefore the solution method is based on a careful combination of optimization and Monte Carlo simulation: the sequentially generated, improving decisions are evaluated with gradually increasing accuracy (for details on the theoretical background see Pintér (1983, 1984)).

As all considered versions of the optimization problem (1)-(4) are in principle deterministic nonlinear (or linear) programming problems (except the numerically very relevant fact that the objective function values can be estimated only via expensive simulation cycles and with some inherent inaccuracies), an arbitrary efficient nonlinear programming method could be applied to solve them. Considering the special structure (linear constraints) of the above models, the nonlinear optimization package, developed by Murtagh and Saunders (1977) could be applied advantageously.

6. SUMMARY OF RESULTS; CONCLUDING REMARKS

The numerical results for a number of runs with the different objective function types (8) - (12) were given in details elsewhere, cf. Somlyódy and Pintér (1984); therefore only a qualitative summary of these results is given here. Independently of the special form of the objective function, the results basically followed the same line: if the amount of available resources is relatively small, then only treatment plants figure in the optimal solution and the more costly reservoirs are absent. On the other hand, as the variance of the resulted water quality can be substantially decreased only via reservoirs, beyond a certain threshold value of available resources reservoirs begin to play an important role, gradually substituting treatment plants in the optimal solution, whenever this is possible. The basically analogous behaviour of the various models indicates their coherence. This point deserves attention as, in particular, it shows that properly aggregated, simpler models (such as the linear programming model variant, presented in Somlyódy (1983)) are capable to preserve the main features of the studied nonlinear, stochastic problem. The different stochastic programming versions, of course, yielded a fair amount of supplementary information and basically verified the findings of the LP model.

While following the mentioned general tendency, non-negligible differences could be also observed between the "optimal" solutions, depending on the form of the objective function. E.g. one of the sub-basins clearly had a dominating role in the composed objective function value (7), when using (8), (9) or (10) for F_i, while applying the statistical criteria (11), (12) this feature was much less evident. This clearly shows the importance of <u>proper</u> stochastic formulations of optimization problems (given a complex underlying random structure) - a fact frequently overlooked in practice, cf. e.g. the critical survey of Hogan et al. (1981).

The case study summarized above indicates that the field of stochastic optimization is full of challenging theoretical and practical (numerical) problems to be solved. Among these only some are enumerated here:

- in many cases the analyzed system can be studied only via costly and inherently inaccurate (simulation) methods: this fact motivates the investigation of approximation schemes (cf. e.g. Wets (1983)) as well as other (statistical) methods for reducing the necessary number of function evaluations (cf. e.g. Pintér (1983));
- complex systems can frequently be studied only by interactive, rather than automatic optimization procedures;
- multiobjective analysis and sensitivity studies may help to a large extent the identification of needed supplementary data and the correction of model structure and parameters.

ACKNOWLEDGEMENTS

This research was partially supported by the International Institute for Applied Systems Analysis (IIASA , Laxenburg,Austria), the National Water Authority (Budapest, Hungary) and the Hungarian Academy of Sciences (Budapest, Hungary). Fruitful co-operation with many colleagues from the above institutions is gratefully acknowledged. Special thanks are due to R. Wets, Z. Fortuna, A. King (IIASA) and I. Vályi (National Board for Technical Development in Hungary), for useful discussions which helped the numerical solution of some model variants.

REFERENCES

Deininger, R.A., ed. (1977). Systems Analysis for Water Quality Management (Proc. WHO Seminar, Budapest, 1975). University of Michigan Press, Ann Arbor, Michigan.

Dorfman, R., Jacoby, H.D. and Thomas, H.A., eds. (1972). Models for Managing Regional Water Quality. Harvard University Press, Cambridge, Massachusetts.

Ermakov, S.M. (1975). Monte Carlo Method and Related Topics. Nauka, Moscow (in Russian).

Ermoliev, Y.M. (1976). Stochastic Programming Methods. Nauka, Moscow (in Russian).

Hogan, A.J., Morris, J.G. and Thompson, H.E. (1981). Decision problems under risk and chance constrained programming: dilemmas in the transition. Management Science 27: 698-716.

Hwang, C.L. and Masud, A.S.M. (1979). Multiple Objective Decision Making - Methods and Applications. Springer, Berlin--Heidelberg.

Kall, P. and Prékopa,A.,eds.(1980). Recent Results in Stochastic Programming. Springer,New York- Berlin-Heidelberg.

King, A. (1985). An implementation of the Lagrangian finite-generation method, to appear in: Ermoliev, Y.M. and Wets,R., eds. Numerical Methods for Stochastic Programming.

Loucks, D.P., Stedinger, J.R. and Haith, D.A. (1981). Water Resources Systems Planning and Analysis. Prentice-Hall, Englewood Cliffs, New Jersey.

Marks, D.H. (1974) Models in water resources, 103-137. in: A Guide to Models in Governmental Planning and Operation. US EPA, Washington D.C.

Murtagh, B.A. and Saunders, M.A. (1977) MINOS: A Large-Scale Non-linear Programming System. User's Guide. Technical Report SOL 77-9, Stanford University, Stanford, California.

Pintér, J. (1980). Stochastic models for regional water quality management. Hidrological Bulletin 60: 364-373. (in Hungarian).

Pintér, J. (1983). A modified Bernstein-technique for estimating noise-perturbed function values. Calcolo (to appear).

Pintér, J. (1984). Convergence properties of stochastic optimization procedures. Mathematische Operationsforschung und Statistik, Series Optimization 15: 405-427.

Sherman, B. (1957). Percentiles of the ω_n statistic. Annals of Mathematical Statistics 28: 259-261.

Somlyódy, L. (1982a). Water quality modelling: A comparison of transport-oriented and ecology-oriented approaches. Ecological Modelling 17: 183-207.

Somlyódy, L. (1982b). Modeling a complex environmental system: The Lake Balaton Case Study. Mathematical Modelling 3: 481-502.

Somlyódy, L. (1983). Lake eutrophication management models, 207-250. in: Somlyódy, L., Herodek, S. and Fischer,J., eds. Eutrophication of Shallow Lakes: Modeling and Management. The Lake Balaton Case Study. CP-83-S3 IIASA, Laxenburg. Austria.

Somlyódy, L. and Pintér, J. (1984). Water Quality Management Models. Research Report 7614/3/36. Research Centre for Water Resources Development, Budapest, Hungary (in Hungarian).

Somlyódy, L. and van Straten, G. (1985). Modeling and Managing Shallow Lake Eutrophication, with Application to Lake Balaton (in press).

Somlyódy, L. and Wets, R. (1985). Stochastic optimization models for lake eutrophication management. Manuscript, IIASA, Laxenburg, Austria (to appear).

Wets, R. (1983). Stochastic programming: solution techniques and approximation schemes, 577-603. in: Bachem, A., Grötschel, M. and Korte B., eds. Mathematical Programming: The State-of-the-Art. Springer, New York-Berlin-Heidelberg.

Yudin, D.B. (1979). Stochastic Programming Models and Methods. Sovietskoye Radio, Moscow (in Russian).

A DYNAMIC MODEL OF MARKET BEHAVIOR

I.G. Pospelov
Computing Center, USSR Academy of Sciences,
Moscow, USSR

1. INTRODUCTION

Petrov and Pospelov have posed the problem of building an aggregate model of economic mechanisms based on micro-descriptions of the relations between individual economic agents [1]. This problem has been partially solved in [2]. Micro-descriptions of the process by which material wealth is produced have been constructed, and aggregated to give a macro-description of this process in terms of a production function. This approach allows relatively simple and well-structured models of market economies to be developed. However, a micro-description of the commodity market, i.e., of the process by which uniform prices are fixed as a result of exchanges between economic agents, is still lacking. This paper describes a new approach to the above problem and gives the first results achieved using this approach.

The problem of market description is traditionally treated as one of market equilibrium [3,4]. However, non-obvious *a priori* assumptions can be found in equilibrium models. Why is it that completely arbitrary interests are tolerated for some economic agents (consumers) while others (producers) are assumed to have identical interests (maximization of profit)? Why doesn't the price depend upon the volume of business? In any case, the dependence of usefulness on product quantity is non-linear.

Ideally, a micro-description of the market should explain rather than merely postulate such empirical phenomena as maximization of profit by producers and weak dependence of prices upon the volume of trade. To achieve this it is necessary to construct a dynamic model of market trading behavior. Equilibrium in such a model would be represented by either a partial solution or a limiting case. However, it is well-known that attempts to describe the approach to equilibrium come up against a number of fundamental difficulties [4]. These difficulties seem to be caused to a large extent by the interests of the parties involved being treated as fixed and unchanging. A simple example illustrating these points now follows.

Suppose that there are three traders (A,B,C) each of which holds some goods produced by the others in addition to their own. Assume that the first party (A) is in need of product A (its own) and product B; the second party (B) needs product B and product C, while the third party (C) is in need of product C and product A. Equilibrium exchange resulting in benefits to each party is possible. At the same time, however, any exchange between pairs of parties causes the situation of one of the parties to deteriorate. If a tripartite bargain cannot be struck, one of the parties is expected to become a dealer, acquiring products for further exchange rather than for consumption. If this event is described as a change in the utility function of the dealer, then the new consumption utility function cannot be defined arbitrarily – it must be in agreement with the interests of the other parties. In any case, the interests of professional dealers are fostered by market laws.

What are these interests? Because traders who ruin themselves are ousted from the market, it is clear that a party who is unable or unwilling to minimize the probability of ruin is expelled from the system before another trader who acts in such a way that the probability of ruin is minimized. Thus we can say that only those parties who strive (directly or indirectly) to minimize the probability of ruin will remain in the market.

2. THE MODEL AND ITS GENERAL PROPERTIES

One of the simplest possible models of intermediated trade is given below. It does not take into account the dealer's interests beyond the market or changes in behavior as a result of market competition. Our aim is to find out whether fear of ruin causes a dealer to perform the functions of an intermediary, and what value such a dealer places on the standard economic concepts of profit, price, etc.

Suppose a given product is exchanged for money. The producers sell the product to the only dealer in the market, who in his turn sells the product to n consumers (Fig. 1). A simple description of the behavior of producers and consumers now follows.

At each moment of time producer i has a fixed stock of product u_i^*; he is ready to sell any part of that product u, $0 \le u \le u_i^*$, for an amount of money not less than $U_i(u)$, where U_i is a given supply function. In a similar way, consumer j is always ready to buy any quantity of product $v \ge 0$ if the amount to be paid does not exceed $V_j(v)$, where V_j is a demand function. U_i, V_j are assumed to be smooth and monotonic, with U_i being strictly convex and V_j strictly concave (see Fig. 1).

$$V_i(0) = U_i(0) = 0\,,\; V_j(\infty) = V_j^* < \infty\,,\; U_i(u_i^*) = U_i^* < \infty \quad .$$

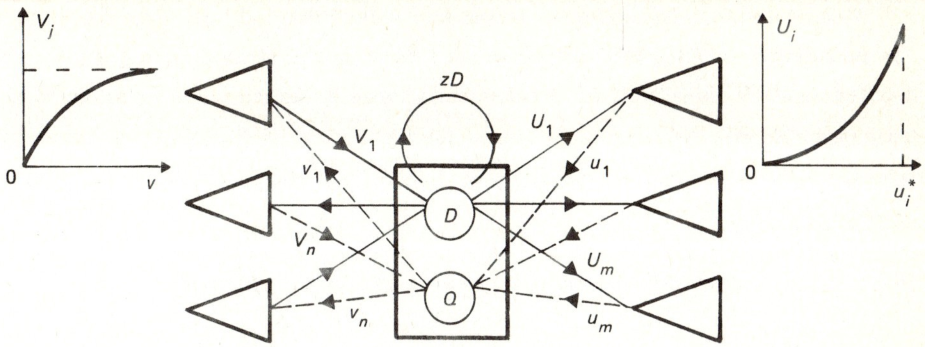

Fig. 1.

We shall make three assumptions regarding the behavior of the dealer:

1. Bargains between consumers and producers are made at random times. These times form Poisson distributions which are independent of each other. Under these conditions a dealer cannot always sell a newly purchased product immediately. Such products are accumulated as stock $Q > 0$.

2. The volume of business u_i, v_j within limits $0 \leq u_i \leq u_i^*$, $0 \leq v_j \leq Q(\tau)$, at the times τ when business can be performed is also determined by the dealer.

3. The dealer purchases on credit, so that he has debt D at any moment of time; the debt is charged with fixed interest r. D jumps in size on purchasing and decreases on selling.

As the behavior of the consumers and producers does not change, the state of the system is determined by Q and D, and the dealer's strategy is to choose functions $u_i(Q,D)$, $0 \leq u_i \leq u_i^*$, and $v_j(Q,D)$, $0 \leq v_j \leq Q(\tau)$, which tell him how much to buy or sell if at the time of bargaining he has stock Q and debt D. If strategy $\Omega = \{u_i, v_j\}$ has been chosen and the functions u_i, v_j are Borelean, then $\{Q(t), D(t)\}$ is a separable, stochastic, continuous, uniform Markov process (5):

$$dQ = \sum_{i=1}^{m} u_i(Q,D)d\xi_i - \sum_{j=1}^{n} v_j(Q,D)d\eta_j$$

(1)

$$dD = rDdt + \sum_{i=1}^{m} U_i(u_i(Q,D))d\xi_i - \sum_{j=1}^{n} V_j(v_j(Q,D))d\eta_j$$

Here ξ_i, η_j are independent Poisson processes with frequencies λ_i, μ_j, respectively.

Financial ruin is naturally connected with debts piling up over a certain level D^*. The probability of $D(t) > D^*$ for some $t \geq 0$ (with initial state $Q(0) = Q$ $D(0) = D$) is denoted as $\tilde{\omega}_\Omega(D^*, Q, D)$. If D^* is sufficiently large its actual value turns out to be of little importance.

Statement 1. If $D^* \to +\infty$, then $\tilde{\omega}_\Omega(D, Q, D) \to \omega_\Omega(Q, D)$ monotonically and uniformly on Ω, D and Q. Here

$$\omega_\Omega(Q, D) = \mathrm{P}\{\overline{\lim_{t \to +\infty}} D(t) = +\infty \mid Q(0) = Q \ , d(0) = D\} \quad , \tag{2}$$

where the right-hand side represents the probability of debts increasing beyond all bounds in trajectory (1), escaping from (Q, D).

From the mathematical point of view, it is the limiting value of ω_Ω that is most conveniently assumed to be the ruin probability; thus, ruin is treated merely as the unlimted growth of debts. A dealer is therefore assumed to choose strategies that minimize ω_Ω. It is important to note that if a dealer refuses to do business at all ($u_i = v_j = 0$, $\tilde{\Omega} = \{0, 0\}$), then the equation for D is of the form $dD/dt = rD$. Thus, from (2), $\omega_{\tilde{\Omega}} = 0$ when $D \leq 0$ and $\omega_{\tilde{\Omega}} = 1$ when $D > 0$. The danger of ruin in the form of an increase in debts therefore becomes an actual incentive to promote business activity when $D > 0$. If a dealer does not trade when $D > 0$, he will be ruined for sure; but if he still conducts his business he will probably manage to save himself and reduce $D \leq 0$.

The strategy $\hat{\Omega}$ is the optimal strategy if $\omega_{\hat{\Omega}} \leq \omega_\Omega$ for any Ω, Q, D. According to Bellman's principle, the optimal strategy should minimize the increment of $\omega_{\hat{\Omega}}$ at every transaction.

Statement 2. There exists an optimal strategy $\hat{\Omega} = \{\hat{u}_i, \hat{v}_j\}$ in a class of Borelean functions* u_i , v_j. $\omega_{\hat{\Omega}} = \hat{\omega}$ is the smallest lower semi-continuous function which satisfies the Bellman equation:

$$rD\frac{\partial \hat{\omega}}{\partial D} = \Lambda(\hat{\omega} - \sum_{j=1}^{n} \alpha_j \min_{0 \leq v \leq Q} \hat{\omega}(Q - v, D - V_j(v)) - \tag{3}$$

$$\sum_{i=1}^{m} \beta_i \min_{0 \leq u \leq u_i^*} \hat{\omega}(Q + u, D + U_i(u)))$$

$$\Lambda = \sum_{j=1}^{n} \mu_j + \sum_{i=1}^{m} \lambda_i \ ; \ \alpha_j = \mu_j / \Lambda \ ; \ \beta_i = \lambda_i / \Lambda \tag{4}$$

*This strategy is also optimal in the broader class where u_i, v_j are random quantities whose distribution is a Borelean function of t, Q, D.

and the conditions $0 \le \hat{\omega} \le 1$, $\lim_{D \to +\infty} \hat{\omega} = 1$ uniformly on Q. The values $\hat{u}_i(Q, D)$, $\hat{v}_j(Q, D)$ lead to minima on the right-hand side of (3).

3. STUDY OF THE BELLMAN EQUATION AND INTERPRETATION OF THE RESULTS

Two time constants $1/r$ and $1/\Lambda$ appear in equation (3). If the rate of interest is assumed to be ~ 10% per year and the frequency of transactions is $\Lambda \sim 100$ per year, then their ratio $\rho \sim 10^{-3}$ is a very small number. A small value of ρ means that in the period between transactions the debt increases rather slowly on the average, so that individual transactions do not have a decisive effect upon the fate of a dealer. The optimal behavior described above is intended for many transactions and displays features characteristic of large economic systems. Separation of the main part of equation (3), at $\rho \to 0$, makes it clear that the ruin probability differs essentially from 0 and 1 (by more than $O(\sqrt{\rho})$) only in a narrow band G (of relative width $O(\sqrt{\rho})$) in the vicinity of the curve $D = R(Q)$ (Fig. 2). The monotonic, non-negative, concave, bounded function R is the only solution of the equation

$$\rho R(Q) = \sum_{j=1}^{n} \alpha_j \max_{0 \le v \le Q} (V_j(v) + R(Q-v) - R(Q)) + \tag{5}$$

$$\sum_{i=1}^{m} \beta_i \max_{0 \le u \le u^*} (-U_i(u) + R(Q+u) - R(Q))$$

The upper bound of domain G is $\hat{\omega} = 1 - O(\sqrt{\rho})$ and its lower bound is $\hat{\omega} = O(\sqrt{\rho})$. Thus, $R(Q)$ is the maximum credit which, with a high degree of probability, will be repaid by a dealer if he has stock Q. In other words $R(Q)$ represents the dealer's solvency. The non-negative value $P = R(Q)$ gives the amount of credit assured reliably by the very position of the dealer in the market. This value is called the "price of the firm".

The unique and continuous functions $u_i^0(Q)$, $v_j^0(Q)$ which yield maxima on the right-hand side of (5) represent first approximations of optimal functions $\hat{u}_i(Q, D)$, $\hat{v}_j(Q, D)$. The summation from involving α_j on the right-hand side of (5) is apparently non-negative. From this we can deduce that a dealer will not sell all his stock Q at once ($v_j^0(Q) \ne Q$) if the price offered by the consumer is less than $M(Q) = R(Q) - R(0)$. By analogy, looking at the summation term involving β_i, we conclude that a dealer without stock will not buy stock $Q(u_i^0(0) \ne Q)$ if he is asked to pay more than $M(Q)$. Thus, the value $M(Q)$ acts as an exchange value, i.e., the monetary value of stock-in-trade formed by the market. Thus, the first solid economic result has been achieved: the monetary value of a dealer's assets – his position P and stock M – is set by the market and his solvency is equivalent to the value of his assets $R = M + P$. Replacing R by M on the right-hand side of equation (5), it follows that if

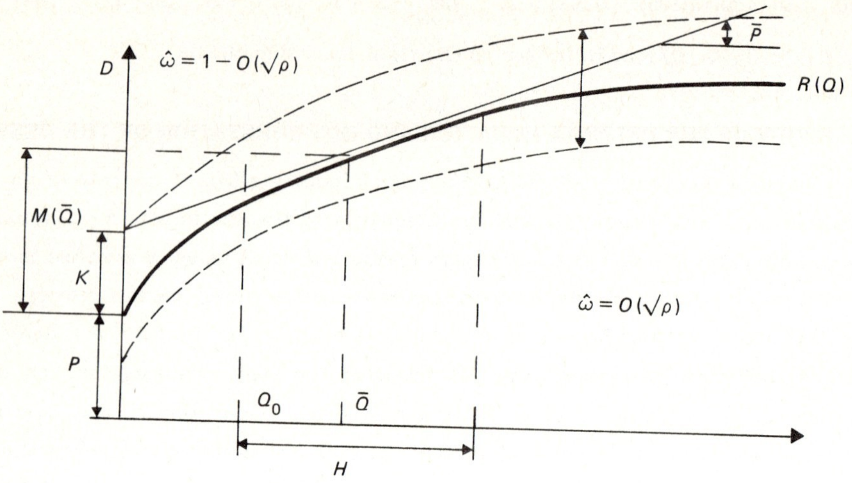

Fig. 2.

a dealer knows the exchange value $M(Q)$, his optimal strategy would be to maximize the profit he makes on every transaction, taking into account increments in the exchange value of his stock-in-trade. The right-hand side of (5) expresses the average amount of profit from trade, assuming stock Q. In the model under discussion, this profit is positive (a dealer sells at a price higher than the exchange value and buys at a lower price) and $D = R(Q)$ is exactly equal to the percentage charged for credit. This is the economic interpretation of equation (5).

There is also another way of describing optimal behavior. Let $\sum$ denote a set of strategies, with u_i, v_j being dependent only on Q. In particular, $\Omega_0 = \{u_i^0, v_j^0\} \in \sum$. It follows from (1) that the probability of a dealer having stock $Q \in A \subset R^1$ at time $t + t_0$ depends entirely on $Q(t_0)$ and not on D or t_0. Let $F^t(A \mid Q)$ denote this probability. It turns out that strategy Ω_0 yields the maximum of the functional

$$\mathrm{P}[\Omega] = \int_0^\infty dt\, e^{-rt} \int_{R'} F^t(Q' \mid Q)\Big[\sum_{j=1}^{n} \mu_j V_j(v_j(Q)) - \sum_{i=1}^{m} \lambda_i U_i(u_i(Q))\Big] ,$$

i.e., $\mathrm{P}[\Omega_0] \geq \mathrm{P}[\Omega]$ for all Ω at any initial Q. The value $\mu_j V_j(v_j(Q))$ is the average profit of a dealer trading with consumer j when his stock is Q; by analogy, $\lambda_i V_i(u_i(Q))$ is the average cost. Thus P is the average discounted profit and the dealer's optimal strategy, to a first aproximation, is to maximize this value.

The variation in stock associated with strategy Ω_0 is ergodic [5], and at time $\sim 1/\Lambda$ approaches a stationary and final distribution F_0. This means that a quasistationary state* is set by the market. Let us compare this with the classical market equilibrium. According to equilibrium theory [13], a dealer, starting from the product price, must set permanent** volumes of business, $u_i(p)$, $v_j(p)$, which maximize his profit:

$$V_j(v_j(p)) - pv_j(p)\, V_j(v) - pv \quad ; -U_i(u_i(p)) + pu_i(p) - U_i(u) + pu \tag{6}$$

and the price should be fixed at a level $\bar{p}$ which assures a balance between average purchases and sales:

$$\bar{w} = \sum_{j=1}^{m} \mu_j \bar{v}_j = \sum_{i=1}^{m} \lambda_i \bar{u}_i \; ; \; \bar{v} = v_j(\bar{p}) \; ; \; \bar{u} = u_i(\bar{p}) \quad . \tag{7}$$

It turns out that the equilibrium values define the average stock $\bar{Q} = \int QF_0(dQ)$ and characteristic domain H of stock variation ($F_0(H) = 1 - \rho$) in the final distribution, accurate to within $O(\rho^{2/3})$ (see Fig. 3):

$$\bar{\bar{Q}} = (\bar{\sigma}^2 \bar{a} / 8\bar{p}\, \rho)^{1/3} \; ; \; \bar{\sigma} = \sum_{j=1}^{n} \alpha_j \bar{v}_j^2 + \sum_{i=1}^{m} \beta_i \bar{u}_i^2 \quad ;$$

$$\bar{a} = \left(\sum_{i=1}^{m} \beta_i U_i(\bar{u}_i) - \sum_{j=1}^{n} \alpha_j V_j(v_j)\right)^{-1} \tag{8}$$

$$H = \{Q((\bar{w}/\Lambda) \lesssim Q \lesssim \bar{Q} \ln^{2/3}(1/\rho)\} \quad . \tag{9}$$

For $Q \in H$ one can find an explicit expression for the density f_0 of distribution F_0 and the optimal strategy v_j^0, u_i^0:

$$f_0 = A^2((Q/\bar{\bar{Q}}) - A_0)/A'(A_0) \tag{10}$$

$$v_j^0 = \bar{v}_j + (k/V_j''(\bar{v}_j))\psi(Q/\bar{Q}) \; ; \; u_i^0 = \bar{u}_i + (k/U_i''(\bar{u}_i))\psi(Q/\bar{Q}) \quad . \tag{11}$$

Here $k = (2\bar{\sigma}\bar{p}\bar{a}^2\rho)^{1/3}$, $\psi(x) = A'(A_0 - x)/A(A_0 - x)$. The latter is the bounded solu-

---

*Unlike the distribution of Q, the distribution of D "spreads out" rather than being set, although such "spreading" occurs quite slowly (its characteristic time is $1/\tau \gg 1/\Lambda$); in addition, the size of the debt has rather a feeble influence on the dealer's behavior. This is why we call state F_0 "quasistationary".

**Strategies with permanent v_j, $\sum_{j=1}^{n} v_j > 0$, are inadmissable in (1) since they contradict the condition $Q \geq 0$. This condition, which is very important in dynamics, cannot in principle be taken into account in statics; this is why the "equilibrium" structure described above is exclusively of an auxiliary character.

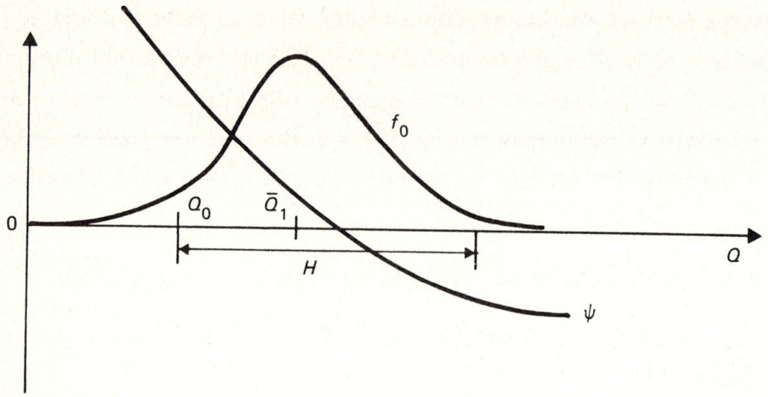

Fig. 3.

tion of the equation $A'' + xA = 0$, with A_0 representing its first root and A' its derivative.

It is important that in a typical situation the dealer does not require detailed information on supply and demand to determine his optimal strategy. It is only necessary to know the two values characterizing the market ($\bar{Q}$ and K) and two characterizing the trading partner ($\bar{v}_j$ and $V_j(\bar{v}_j)$ or $\bar{u}_i$ and $U_i(\bar{u}_i)$).

The most important thing is that the exchange value should be of the form $M = P + K + \bar{p}Q$ (to within an accuracy of $O(\rho^{1/3} \ln 1/\rho)$). Linear* dependence of the exchange value on the size of the stock means that there exists a price for product $\bar{p}$. From (6) this has the properties of an equilibrium price.

Let us now sum up the main points of our argument.

1. The dynamic description of the pairwise transaction mechanism (micro-description) given above can explain the most important features of market exchange: the existence of a price for a product, the fact that dealers strive to maximize their profit, and the existence of a dynamic equilibrium in the market.

2. Formal descriptions of economic concepts such as solvency, exchange value, trading profit, price of the firm, and active assets develop naturally from the model.

*Here K represents the additional value of a small amount of stock Q_0 which the dealer never actually sells (see (9)). Stock plays a specific role here. It is technologically necessary for regular trading and is present even in a deterministic sequence of purchases and sales. For this reason its market value K should be distinguished from the exchange value of the stock-in-trade. We could regard this stock as a special kind of asset (active assets). In this case, the exchange value is simply proportional to the amount of stock (in a typical domain H).

3. The behavior of a dealer minimizing his probability of ruin is rather simple, although not trivial, and can be described subjectively in a number of different ways: he could simply follow rule (11), or he could make a "correct" evaluation of stock-in-trade and maximize his current profit (10); he may also be more "far-sighted" and maximize his expected discounted profit.

REFERENCES

1. A.A. Petrov and I.G. Pospelov. Systems analysis of economic development, I–V (in Russian). *Technical Cybernetics*, 1979, No. 2, pp. 18–26; No. 3, pp. 28–37; No. 4, pp. 11–23; No. 5, pp. 15–31.

2. A.A. Shananin. Production and profit functions. *Vestnik of the USSR Academy of Sciences*, 258(1) (1981) 38–41.

3. I. Rosenmuller. *Cooperative Games and Markets*. Mir, Moscow, 1974.

4. H. Nikayado. *Convex Structures and Mathematical Economics*. Mir, Moscow, 1972.

5. J.L. Doub. *Stochastic Processes*. For.Lit., Moscow, 1956.

RECURSIVE STOCHASTIC GRADIENT PROCEDURES IN THE PRESENCE OF DEPENDENT NOISE

A.S. Poznyak
Institute of Control Sciences
Moscow, USSR

This paper is concerned with the asymptotic characteristics of recursive gradient procedures in the presence of dependent noise, which distorts observations of the gradient of the function to be optimized. A general characteristic is proposed for the correlation of elements of random sequences; this can be used to prove theorems of strong convergence and asymptotic normality for the class of algorithms under consideration. The relationship between this characteristic and existing correlation characteristics is defined and the conditions under which this procedure may be used are studied for specific cases. The maximum feasible rate of convergence for linear algorithms is shown to be attainable under conditions of Gaussian noise. We give a rule for nonlinear transformation of observations which makes the given procedure optimal in the presence of dependent noise. The optimal algorithm is shown to be one which corresponds to a recursive version of the maximum likelihood method.

1. INTRODUCTION AND STATEMENT OF THE PROBLEM

In recent years the problem of optimization in the presence of dependent noise has attracted increasing attention. Thus, Driml and Nedoma [1], Krasulina [2], Kultichitskii [3] and Borodin [4] have all carried out studies of convergence conditions and analyzed the rate of convergence of recursive optimization algorithms under dependent noise, making use of the "strong mixing" notion. Ljung [5], Kushner and Clark [6] and Korostelev [7] adopted the large deviations method and the theory of weak convergence to study these problems. Farden [8], Nemirovskii [9] and Solo [10] used the method of moments and results from martingale theory to analyze such procedures. A good survey of the state-of-the-art of recursive stochastic procedure theory is given by Tzypkin and Poznyak [11]. The properties of linear gradient optimization algorithms under dependent noise have been investigated by Poznyak and Tchickin [12]. In the present paper an approach proposed in [12] is described and then generalized to the class of nonlinear recursive algorithms. We prove the existence of optimal nonlinear transformations which lead to the maximum feasible rate of convergence.

We shall consider the problem of finding the point

$$c^* \stackrel{\Delta}{=} \arg\min_{c \in R^N} J(c) \quad , \tag{1}$$

assuming that at any point $c \in R^N$ the gradient $\nabla J(c)$ of the function to be optimized, distorted by noise ξ, is available for observation, i.e., we can observe

$$y = y(c, \xi) \stackrel{\Delta}{=} \nabla J(c) + \xi \quad . \tag{2}$$

Let the class of functions to be optimized satisfy the following condition:

(A1) $J(c) \geq J^* > -\infty \quad \forall c \in R^N$

and let there exist a unique point $c^* \in R^N$ at which $J(c^*) = J^*$.

In order to construct the sequence $\{c_n\}$ ($c_n \in R^N$, $n = 1,2, \ldots$) of estimates of the minimum point c^* of the function $J(c)$, we shall make use of the following recursive procedure:

$$\begin{aligned} c_n &= c_{n-1} - \Gamma_n \varphi_n(y_n) \,,\; c_0 = \bar{c} \in R^N \quad , \\ y_n &\stackrel{\Delta}{=} \nabla J(c_{n-1}) + \xi_n \,,\; n = 1,2, \ldots \quad , \end{aligned} \tag{3}$$

where $\{\Gamma_n\}$ is a sequence of $N \times N$ matrices which satisfy the condition

(B) $\lim_{n \to \infty} \gamma_n^{-1} \Gamma_n = \Gamma = \Gamma^T > 0 \,,\; 0 < \gamma_n \underset{n \to \infty}{\longrightarrow} 0 \,,\; \sum_{n=1}^{\infty} \gamma_n = \infty$

and $\{\xi_n\}$ is a sequence of dependent random vectors defined on the probability space $(\Omega, \mathbf{F}, P)$ with a given sequence of σ-algebras $\mathbf{F}_n$, $\mathbf{F}_n \subset \mathbf{F}_{n+1}$. Vectors ξ_n are measurable with respect to $\mathbf{F}_n$, and functions $\varphi_n(x)$, $x \in R^N$, are measurable with respect to the intersection of the Borelian σ-algebra defined on R^N and $\mathbf{F}_{n-1}$.

2. RECURSIVE LINEAR ALGORITHMS

Let us consider algorithm (3), assuming

$$\varphi_n(x) = x \,,\; n = 1,2, \ldots \quad . \tag{4}$$

Let the following conditions be satisfied:

(A2) There exist k, $K \in (0, \infty)$ such that for all $c \in R^N$ we have

$$(c - c^*)^T \Gamma \nabla J(c) \geq k \, \|c - c^*\|^2 \quad ,$$

$$\|\nabla J(c)\|^2 \leq K \, \|c - c^*\|^2 \quad ,$$

and

(C) $\mathrm{E}\{\zeta_n\} = 0$, $\|\zeta_n\|_{L_p} < \infty$ $(p \geq 2)$,

where

$$\|\zeta\|_{L_p} = \|(\|\zeta_1\|_p, \ldots, \|\zeta_N\|_p)\| \quad ,$$

$$\|\zeta_i\|_p \overset{\Delta}{=} (\mathrm{E}\{|\zeta_i|^p\})^{1/p} \, , \; \|\zeta_i\|_\infty \overset{\Delta}{=} \inf_{x \geq 0} \{x : \mathrm{P}\{|\zeta_i| > x\} = 0\} \quad .$$

We now introduce the following characteristics of noise dependence:

$$\psi_{k,m}^{(q)} \overset{\Delta}{=} \|\mathrm{E}\{\zeta_k / \mathbf{F}_m\}\|_{L_q} \, , \; k = m+1, \ldots; \; \frac{p}{p-1} \leq q \leq p \quad ,$$

$$\Theta_{t,m} \overset{\Delta}{=} \mathrm{E}\{\zeta_t \zeta_m^T\} \, , \; m, t = 1,2, \ldots, \quad \sigma_n^2 \overset{\Delta}{=} \mathrm{Sp}\, \Theta_{n,n} \quad . \tag{5}$$

Note that for independent sequences we have

$$\psi_{k,m}^{(q)} = 0 \; (k = m+1, \ldots) \, , \; \Theta_{t,m} = 0 \; (t \neq m) \quad .$$

Characteristic $\psi_{k,m}^{(q)}$ is new while $\Theta_{t,m}$ is the usual noise covariance matrix.

THEOREM 1 (on almost sure convergence). *Let assumptions* A1, A2, B, *and* C *be satisfied. If the series*

$$\sum_{t=1}^{n} \Gamma_t \zeta_t \tag{6}$$

converges (component-wise) almost surely as $n \to \infty$, *then for any* $c_0 = \bar{c} \in R^N$ *we have* $c_n \xrightarrow[n \to \infty]{a.s.} c^*$, *and in order for* (6) *to converge it is sufficient that*

$$r_n \overset{\Delta}{=} \sum_{m=n+1}^{\infty} \gamma_m \psi_{m,n}^{(2)} \xrightarrow[n \to \infty]{} 0 \, , \; \sum_{n=1}^{\infty} (\gamma_n^2 \sigma_n^2 + \gamma_n \sigma_n r_n) < \infty \quad . \tag{7}$$

THEOREM 2 (on mean-square convergence). *Let assumptions* A1, A2, B, *and* C *be satisfied. If*

$$\pi_n \sum_{t=1}^{n} \sum_{s=1}^{n} \pi_t^{-1} \Gamma_t \Theta_{t,s} \Gamma_s^T (\pi_s^{-1})^T \pi_n^T \xrightarrow[n \to \infty]{} 0 \quad , \tag{8}$$

where $\pi_n \overset{\Delta}{=} \prod_{t=1}^{n} (I - \Gamma_t \Gamma^{-1})$, *then for any* $\bar{c} \in R^N$ *we have* $\|c_n - c^*\|_{L_2} \underset{n\to\infty}{\longrightarrow} 0$, *and to meet condition* (8) *it is sufficient that*

$$\gamma_n \sigma_n^2 \underset{n\to\infty}{\longrightarrow} 0 \tag{9}$$

and, for some $q \in [p/p-1, p]$, *that*

$$\|\zeta_n\|_{L_q} \sum_{t=n+1}^{\infty} \gamma_t \psi_{t,n}^{(\frac{q}{q-1})} \underset{n\to\infty}{\longrightarrow} 0 \quad . \tag{10}$$

THEOREM 3 (on asymptotic normality). *Let assumptions* A1, A2, B, *and* C *be satisfied, and in addition assume that*

(1) $J(c)$ *is twice-differentiable,* $\nabla^2 J(c^*) > 0$ *and there exists a constant* $L \in (0, \infty)$ *such that for any* $c \in R^N$

$$\|\nabla^2 J(c) - \nabla^2 J(c^*)\| \le L \, \|c - c^*\| \quad ;$$

(2) $\Gamma_n = n^{-1}\Gamma$ *and the matrix* $[\frac{1}{2} I - \Gamma \nabla^2 J(c^*)]$ *is stable;*

(3) $\{\zeta_n\} (n = \ldots, -1, 0, 1, \ldots)$ *is a stationary (in a limited sense) ergodic sequence of random vectors such that*

$$\sum_{n=1}^{\infty} n^{-\alpha_0} \psi_n < \infty \quad , \tag{11}$$

where

$$\alpha_0 = \begin{cases} \min \{0 \, , \min_j \mathrm{Re}\, \lambda_j(B) \, , C \,\mathrm{sign}\, [\min_j \mathrm{Re}\, \lambda_j(B)]\} \, , \text{ if } \min_j \mathrm{Re}\, \lambda_j(B) \ne 0 \, , \\ 0 \text{ if } \min_j \mathrm{Re}\, \lambda_j(B) \overset{\Delta}{=} \mathrm{Re}\, \lambda_\alpha(B) = 0 \text{ and } \mathrm{Im}\, \lambda_\alpha(B) = 0 \, , \\ -C \text{ otherwise} \, ; \, C > 1/2 \, , \, B \overset{\Delta}{=} I - \nabla^2 J(c^*)\Gamma \, , \, \psi_n \overset{\Delta}{=} \psi_{k+n,k}^{(\frac{p}{p-1})} \, . \end{cases}$$

Then $\sqrt{n}\,(c_n - c^*) \sim N(0, V)$, *i.e., the distribution* $\sqrt{n}\,(c_n - c^*)$ *converges to a normal distribution, and the matrix* V *satisfies the equation*

$$[\frac{1}{2} I - \Gamma \nabla^2 J(c^*)]V + V[\frac{1}{2} I - \Gamma \nabla^2 J(c^*)]^{\mathrm{T}} = -\Gamma \Theta \Gamma^{\mathrm{T}} \quad ,$$

$$0 < \Theta = \Theta_0 + \sum_{i=1}^{\infty} (\Theta_i + \Theta_i^{\mathrm{T}}) \, , \; \Theta_i \overset{\Delta}{=} \Theta_{i+k,k} \, , \; k = \ldots, -1, 0, 1, \ldots \quad .$$

Note. The matrix $V = V(\Gamma)$ characterizes the rate of convergence of algorithm (3)–(4); the highest convergence rate is obtained with $\Gamma = \Gamma^* \overset{\Delta}{=} [\nabla^2 J(c^*)]^{-1}$, i.e., for any permissible $\Gamma = \Gamma^T > 0$,

$$V(\Gamma) \geq V(\Gamma^*) = [\nabla^2 J(c^*)]^{-1} \Theta [\nabla^2 J(c^*)]^{-1} \quad . \tag{12}$$

Now let us consider some specific examples of dependent noise.

(1) $\{\xi_n\}$ is *the martingale difference* ($E\{\xi_n / \mathbf{F}_{n-1}\} \overset{a.s.}{=} 0$). Then from Theorem 1, if $\sum_{n=1}^{\infty} \gamma_n^2 \sigma_n^2 < \infty$, then $c_n \underset{n\to\infty}{\overset{a.s.}{\longrightarrow}} c^*$; from Theorem 2 if $\gamma_n \sigma_n^2 \underset{n\to\infty}{\longrightarrow} \infty$, then $\|c_n - c^*\|_{L_2} \underset{n\to\infty}{\longrightarrow} 0$; from Theorem 3, if $\{\xi_n\}$ is a stationary ergodic sequence, then $\sqrt{n}\,(c_n - c^*) \sim N(0, V)$, $\Theta = \Theta_{1,1}$.

(2) $\{\xi_n\}$ is a sequence with *strong mixing*. Let $\{\xi_n\}$ be a sequence satisfying either the condition of uniform strong mixing, i.e.,

$$\varphi_n \overset{\Delta}{=} \sup_{k\geq 1} \sup_{\substack{A \in \sigma(\xi_{n+k}), P(A)>0 \\ B \in \mathbf{F}_k}} |P(B/A) - P(B)| \underset{n\to\infty}{\longrightarrow} 0$$

or the condition of strong mixing, i.e.,

$$\alpha_n \overset{\Delta}{=} \sup_{k\geq 1} \sup_{\substack{A \in \sigma(\xi_{n+k}) \\ B \in \mathbf{F}_k}} |P(B/A) - P(B)\,P(A)| \underset{n\to\infty}{\longrightarrow} 0 \quad .$$

Using the inequalities

$$\psi_{k,m}^{(q)} \leq 2\varphi_{k-m}^{1-(1/p)} \|\xi_k\|_p \quad ,$$

$$\psi_{k,m}^{(q)} \leq 2(2^{1/q} + 1)\alpha_{k-m}^{(1/q)-(1/p)} \|\xi_k\|_p \, , \; 1 \leq q \leq p < \infty \quad ,$$

Theorems 1–3 make it possible to assert that

(a) if for some $q \in [\frac{p}{p-1}, p]$ we have

$$r_n' \overset{\Delta}{=} \begin{Bmatrix} \text{either} & \sum_{m=1}^{\infty} \gamma_{m+n} \|\xi_{m+n}\|_{L_q} \varphi_m^{1-(1/q)} \\ \text{or} & \sum_{m=1}^{\infty} \gamma_{m+n} \|\xi_{m+n}\|_{L_q} \alpha_m^{(1/2)-(1/q)} \end{Bmatrix} \underset{n\to\infty}{\longrightarrow} 0$$

and $\sum_{n=1}^{\infty} (\gamma_n^2 \sigma_n^2 + \gamma_n \sigma_n r_n') < \infty$, then $c_n \underset{n\to\infty}{\overset{a.s.}{\longrightarrow}} c^*$;

(b) if $\gamma_n \sigma_n^2 \underset{n\to\infty}{\longrightarrow} 0$ and for some $q \in [\frac{p}{p-1}, p]$, $\tilde{q} \in [\frac{q}{q-1}, p]$ either

$$\|\xi_n\|_{L_q} \sum_{t=1}^{\infty} \gamma_{t+n} \|\xi_{t+n}\|_{L_{\tilde{q}}} \varphi_t^{1-(1/\tilde{q})} \underset{n\to\infty}{\longrightarrow} 0 \quad ,$$

or

$$\|\zeta_n\|_{L_q} \sum_{t=1}^{\infty} \gamma_{t+n} \|\zeta_{t+n}\|_{L_{\tilde q}} \alpha_t^{1-(1/q)-(1/\tilde q)} \underset{n\to\infty}{\longrightarrow} 0 \quad ,$$

then $\|c_n - c^*\|_{L_2} \underset{n\to\infty}{\longrightarrow} 0$;

(c) if $\{\zeta_n\}$ is a stationary ergodic sequence and either

$$\sum_{n=1}^{\infty} n^{-\alpha_0} \varphi_n^{1-(1/p)} < \infty$$

or

$$\sum_{n=1}^{\infty} n^{-\alpha_0} \alpha_n^{1-(2/p)} < \infty \quad ,$$

then $\sqrt{n}\,(c_n - c^*) \sim N(0, V)$.

(3) $\{\zeta_n\}$ is a stationary (in a broad sense) sequence *with restricted spectral density*, i.e.,

$$\Theta_n = \frac{1}{2\pi} \int_{-\pi}^{\pi} e^{t\lambda n} F(\lambda) d\lambda \, , \; \|F(\lambda)\| \le F_+ < \infty \quad .$$

Then by virtue of Theorem 2:

(4) $\{\zeta_n\}$ is the *ARMA model*, i.e., $\{\zeta_n\}$ is the sequence of inputs of the stable linear filter:

$$\sum_{t=0}^{M} B_t \, \zeta_{n+t} = \sum_{l=0}^{L} D_l \, \xi_{n+l} \, , \; \mathrm{E}\{\xi_n / \mathbf{F}_{n-1}\} \overset{\text{a.s.}}{=} 0 \, , \; \sup_n \|\xi_n\|_{L_2} < \infty \quad .$$

Then from Theorem 1, if $\sum_{n=1}^{\infty} \gamma_n^2 < \infty$, then $c_n \underset{n\to\infty}{\overset{\text{a.s.}}{\longrightarrow}} c^*$; from Theorem 2, $\|c_n - c^*\|_{L_2} \underset{n\to\infty}{\longrightarrow} 0$; from Theorem 3, if $\{\zeta_n\}$ is a stationary ergodic sequence, then

$$\sqrt{n}\,(c_n - c^*) \sim N(0, V) \quad .$$

$$0 < \Theta = \Big(\sum_{t=0}^{M} B_t\Big)^{-1} \Big(\sum_{t=0}^{L} D_t\Big) \, \mathrm{E}\,\{\xi_1 \xi_1^{\mathrm{T}}\} \Big(\sum_{t=0}^{L} D_t^{\mathrm{T}}\Big) \Big(\sum_{t=0}^{M} B_t^{\mathrm{T}}\Big)^{-1}$$

More details of the results given in this section may be found in [12].

We shall now consider how different the best rate (12) attained by the linear algorithm is from the maximum feasible rate. In the general case the problem of finding the maximum feasible rate is quite difficult. However, it can easily be found in the case of quadratic optimization problems, since for the function $J(c) =$

$\frac{1}{2}(c-c^*)^T\bar{A}(c-c^*)$ we have

$$\bar{A}=\bar{A}^T>0\ ,\ y_n=\bar{A}(c_{n-1}-c^*)+\zeta_n \quad ,$$

and the problem of finding the point c^* is reduced to that of estimating a shift parameter

$$z_n=m+\zeta_n\ ,\ z_n\overset{\Delta}{=}y_n-\bar{A}c_{n-1}\ ,\ m=-\bar{A}c^*$$

with respect to the Kramer–Rao inequality

$$\mathrm{E}\{(m_n-m)(m_n-m)^T\}\geq \mathrm{E}^{-1}\{\nabla \ln p_{z_1,\ldots,z_n}\nabla^T \ln p_{z_1,\ldots,z_n}\} \quad .$$

In the case of the ARMA model, the right-hand side is easily calculated and the following estimate is obtained:

$$\mathrm{E}\{(c_n-c^*)(c_n-c^*)^T\}\geq \frac{1}{n}[\nabla^2 J(c^*)]^{-1}(\sum_{t=0}^{M}B_t)^{-1}(\sum_{t=0}^{L}D_t)I_\Phi^{-1}\times$$

$$\times(\sum_{t=0}^{L}D_t)^T(\sum_{t=0}^{M}B_t^T)^{-1}[\nabla^2 J(c^*)]^{-1}+\vartheta(\frac{1}{n}) \quad , \tag{13}$$

and it is assumed that there exists a non-singular Fisher matrix

$$0<I_\Phi\overset{\Delta}{=}\mathrm{E}\{\nabla \ln p_\xi \nabla^T \ln p_\xi\}<\infty \quad .$$

Note that the inequality $\mathrm{E}\{\xi\xi^T\}\geq I_\Phi^{-1}$ holds and the equality occurs only in the case of a normal distribution ξ. This means that the rate of convergence (12) of linear algorithm (3)–(4) attains its maximum feasible rate (13) only for normal processes. In all other cases this rate cannot be achieved by a linear algorithm.

3. NONLINEAR RECURSIVE ALGORITHMS

Now let us consider procedure (3). Without any loss of generality, taking $\Gamma_n\overset{\Delta}{=}n^{-1}$, we can assume that

$$c_n=c_{n-1}-n^{-1}\varphi_n(y_n)\ ,\ c_0=\bar{c}\in R^N \quad . \tag{14}$$

In addition to A1, let the following conditions be satisfied:

(A2) $J(c)$ is differentiable, and for any x , z , $a\in R^N$ and $n=1,2,\ldots,$ we have

$$(x-z)^T[\varphi_n(\nabla J(x)+a)-\varphi_n(\nabla J(z)+a)]\geq k_{n-1}\|x-z\|^2 \quad ,$$

$$\|\varphi_n(\nabla J(x)+a)-\varphi_n(\nabla J(z)+a)\|\leq K_{n-1}\|x-z\|^2 \quad ,$$

where $k_{n-1} \overset{\text{a.s.}}{>} 0$ and $K_{n-1} \overset{\text{a.s.}}{\geq} 0$ are $\mathbf{F}_{n-1}$-measurable random variables such that for some $\varepsilon > 0$ we have (almost surely)

$$\lim_{m\to\infty} \inf_{n\geq m} (k_{n-1} - n^{-1}\varepsilon K_{n-1}) > 0 \quad ;$$

(B') $\mathrm{E}\{\varphi_n(\zeta_n)\} = 0$ and for some $p \geq 2$ we have $\|\varphi(\zeta_n)\|_{L_p} < \infty$ for all $n = 1,2,\ldots$.

We shall introduce the following notation:

$$\tilde{\psi}_{k,m}^{(q)} \overset{\Delta}{=} \|\mathrm{E}\{\varphi_k(\zeta_k)/\mathbf{F}_m\}\|_{L_q}, \; k = m+1, \ldots, \; q \in [\frac{p}{p-1}, p] \; ,$$

$$\tilde{\Theta}_{k,m} \overset{\Delta}{=} \mathrm{E}\{\varphi_k(\zeta_k)\varphi_m^{\mathrm{T}}(\zeta_m)\}, \; \tilde{\sigma}_n^2 \overset{\Delta}{=} \mathrm{Sp}\, \tilde{\Theta}_{n,n} \quad . \tag{15}$$

The following analogs of Theorems 1–3 then hold for nonlinear procedures (14).

THEOREM 4 (on almost sure convergence). *Let assumptions* A1, A2' *and* B' *be satisfied. If the series*

$$\sum_{t=1}^{n} t^{-1}\varphi_t(\zeta_t) \tag{16}$$

converges (component-wise) almost surely as $n \to \infty$, *then* $c_n \xrightarrow[n\to\infty]{\text{a.s.}} c^*$ *and it is sufficient that*

$$\tilde{r}_n \overset{\Delta}{=} \sum_{m=n+1}^{\infty} m^{-1}\tilde{\psi}_{k,m}^{(2)} \xrightarrow[n\to\infty]{} 0, \; \sum_{n=1}^{\infty} (n^{-2}\tilde{\sigma}_n^2 + n^{-1}\tilde{\sigma}_n \tilde{r}_n) < \infty \quad .$$

THEOREM 5 (on mean-square convergence). *Let assumptions* A1, A2' *and* B' *be satisfied. If*

$$n^{-2} \sum_{t=1}^{n} \sum_{k=1}^{n} tk\, \tilde{\Theta}_{t,k} \xrightarrow[n\to\infty]{} 0 \quad ,$$

then $\|c_n - c^*\|_{L_2} \xrightarrow[n\to\infty]{} 0$ *and it is sufficient that* $n^{-1}\tilde{\sigma}_n^2 \xrightarrow[n\to\infty]{} 0$, *and, for some* $q \in [\frac{p}{p-1}, p]$, *that*

$$\|\varphi_n(\zeta_n)\|_{L_q} \sum_{t=n+1}^{\infty} t^{-1}\tilde{\psi}_{t,n}^{(\frac{q}{q-1})} \xrightarrow[n\to\infty]{} 0 \quad .$$

THEOREM 6 (on asymptotic normality). *Let assumptions* A1, A2' *and* B' *be satisfied and in addition assume that*

(1) *$J(c)$ is twice-differentiable, $\nabla^2 J(c^*) > 0$ and in the neighborhood of the point c^*, $\nabla^2 J(c)$ satisfies the Lipschitz condition;*

(2) *there exist Borelian functions $\psi_n(x) \overset{\Delta}{=} \mathrm{E}\{\varphi_n(x+\zeta_n)/\mathbf{F}_{n-1}\}$, which are $\mathbf{F}_{n-1}$-measurable and such that*

$$\psi_n(0) \overset{\text{a.s.}}{=} 0 \, (n=1,2,\ldots), \ \sup_n \mathrm{E}\{\|\varphi_n(x+\zeta_n)\|^2\} \le \mathrm{Const}\,(1+\|x\|^2) \quad ,$$

$$\sup_n \mathrm{E}\{\|\psi_n(x)\|^2\} \le \mathrm{Const}\,\|x\|^2 \quad ,$$

$\psi_n(x)$ is differentiable at the point $x = 0 \, (n=1,2,\ldots)$, there exists (almost surely) a limit $\lim_{n\to\infty} n^{-1}\sum_{t=1}^{n}\psi_t'(0) \overset{\Delta}{=} \psi'(0)$, and the matrix $B \overset{\Delta}{=} [\frac{1}{2} I - \psi'(0)\, \nabla^2 J(c^*)]$ is stable;

(3) *$\{\varphi_n\}$, $\varphi_n \overset{\Delta}{=} \varphi_n(\zeta_n)$, is a stationary (in the restricted sense) ergodic sequence of random vectors such that*

$$\sum_{n=1}^{\infty} n^{-\alpha_0}\,\tilde{\psi}_n < \infty$$

(where α_0 is the same number as in Theorem 3 with $B = I - \nabla^2 J(c^*)\psi'(0)$),

$$\tilde{\psi}_k \overset{\Delta}{=} \tilde{\psi}_{k+n,k}^{(\frac{p}{p-1})} = \|\mathrm{E}\{\varphi_{n+k}(\zeta_{n+k})/\mathbf{F}_k\}\|_{L_{(\frac{p}{p-1})}} \ .$$

Then $\sqrt{n}\,(c_n - c^*) \sim N(0,V)$, where matrix V satisfies the equation

$$BV + VB^{\mathrm{T}} = -\tilde{\Theta}$$

$$0 < \tilde{\Theta} = \tilde{\Theta}_0 + \sum_{i=1}^{\infty}(\tilde{\Theta}_i + \tilde{\Theta}_i^{\mathrm{T}}) \, , \ \tilde{\Theta}_i \overset{\Delta}{=} \tilde{\Theta}_{i+k,k} \ . \tag{17}$$

Making use of the lower estimate for solving equations (17), [13], we obtain

$$V \ge [\nabla^2 J(c^*)]^{-1} R [\nabla^2 J(c^*)]^{-1} \quad , \tag{18}$$

where

$$R = [\psi'(0)]^{-1}\tilde{\Theta}[\psi'(0)]^{-1} \quad .$$

If we assume that the condition

$$\sup_n n^{-1}\sum_{t=1}^{\infty} \mathrm{E}\{\|\psi_t'(0)\|\} < \infty \quad ,$$

is satisfied, then from the Lebeg theorem on majorized convergence [13] we have

$$\psi'(0) = \lim_{n\to\infty} n^{-1} \sum_{t=1}^{n} \mathrm{E}\{\psi_t'(0)\} \quad ,$$

and thus $R = \lim_{n\to\infty} R_n$, where

$$R_n = n[\frac{1}{\sqrt{n}} \sum_{t=1}^{n} \mathrm{E}\{\psi_t'(0)\}]^{-1}\mathrm{E}\{(\frac{1}{\sqrt{n}} \sum_{t=1}^{n} \varphi_t(\zeta_t))(\frac{1}{\sqrt{n}} \sum_{t=1}^{n} \varphi_t^{\mathrm{T}}(\zeta_t))\}[\frac{1}{\sqrt{n}} \sum_{t=1}^{n} \mathrm{E}\{\psi_t'(0)\}^{\mathrm{T}}]^{-1} \quad .$$

According to the generalized Cauchy–Bunyakovski inequality [13],

$$R_n \geq nI_{\Phi}^{-1}(p(\zeta_1, \ldots, \zeta_n)) \quad , \tag{19}$$

where

$$I_{\Phi}(p(\zeta_1, \ldots, \zeta_n)) = \mathrm{E}\{\nabla \ln p(\zeta_1, \ldots, \zeta_n) \nabla^{\mathrm{T}} \ln p(\zeta_1, \ldots, \zeta_n)\}$$

and the equality in (18) is obtained with

$$\varphi_t(x) = \varphi_t^{*}(x) \overset{\Delta}{=} D\nabla_{\zeta_t} \ln p(\zeta_t \mid \zeta_1, \ldots, \zeta_{t-1})|_{\zeta_t = x} \quad ,$$

$$D = -[\nabla^2 J(c^{*})]^{-1} \lim_{n\to\infty} nI_{\Phi}^{-1}(p(\zeta_1, \ldots, \zeta_n)) \quad . \tag{20}$$

Thus, the optimal (with respect to its asymptotic convergence rate) algorithm (14) has the form

$$c_n = c_{n-1} - n^{-1}D \nabla_{\zeta_n} \ln p(\zeta_n \mid \zeta_1, \ldots, \zeta_{n-1})|_{\zeta_n = y_n} \quad . \tag{21}$$

Transformation (20) is rather difficult to realize if the sequence has some arbitrary dependence. However it can be done in the case of ARMA-type models [14], when procedure (21) takes the form

$$\sum_{t=0}^{L} D_t \tilde{y}_{n-t} = \sum_{l=0}^{M} B_l y_{n-l} \quad ,$$

$$c_n = c_{n-1} + n^{-1}[\nabla^2 J(c^{*})]^{-1}(\sum_{l=0}^{M} B_l)^{-1} (\sum_{t=0}^{L} D_t) I_{\Phi}^{-1}(p_{\xi})\nabla \ln p_{\xi}(\tilde{y}_n) \quad ,$$

$$0 < I_{\Phi}(p_{\xi}) = \mathrm{E}\{\nabla \ln p_{\xi} \nabla^{\mathrm{T}} \ln p_{\xi}\} < \infty \quad , \tag{22}$$

$$\tilde{y}_{-t} = 0(t = \overline{1,L}) \, , \; y_{-l} \overset{\Delta}{=} 0 \; (l = \overline{1,M}) \, , \; c_0 = \bar{c} \in R^N \quad .$$

Algorithm (22) consists of two parts: a preliminary procedure, observation "whitening", and a conventional part connected with recursive estimation of the extreme point. Thus, procedure (22) represents a realizable form of an asymptotically optimal recursive optimization, assuming dependent noise of the ARMA-model type.

REFERENCES

1. M. Driml and M. Nedoma. Stochastic approximations for continuous random processes. In *Transactions of the Second Prague Conference on Information Theory*, 1960, p. 145.
2. T.P. Krasulina. On stochastic approximation under random processes with continuous time. *Theory of Probability and its Applications*, 16(4)(1971)688–695.
3. O.Yu. Kultichitskii. Sufficient conditions for the convergence of stochastic approximation algorithms for random dependent processes with continuous time. *Kybernetika*, (6)(1979)14–126.
4. A.N. Borodin. Stochastic approximation procedure for observations satisfying the condition of weak dependence. *Theory of Probability and its Application*, 24(1)(1979)34–51.
5. L. Ljung. Analysis of recursive stochastic algorithms. *IEEE Transactions on Automatic Control*, AC-22(1977)551–575.
6. H.J. Kushner and D.S. Clark. *Stochastic Approximations for Constrained and Unconstrained Systems*. Springer-Verlag, Berlin, 1978.
7. A.P. Korostelev. *Stochastic Recursive Procedures*. Nauka, Moscow, 1984.
8. D.C. Farden. Stochastic approximation with correlated data. *IEEE Transactions on Information Theory*, IT-21 (1981)105–113.
9. A.S. Nemirovskii. On a stochastic approximation procedure under dependent noise. *Izvestia Akademii Nauk, USSR, Tekhnicheskaya Kybernetika*, (1)(1981)14–25.
10. V. Solo. Stochastic approximation with dependent noise. *Stochastic Processes and Their Application*, 13(1982)157–170.
11. Ya.Z. Tsypkin and A.S. Poznyak. Recursive algorithms for optimization under indeterminancy. *Tekhnicheskaya Kybernetika, VINITI, Moscow*, 16(1983)3–70.
12. A.S. Poznyak and D.O. Tchickin. Gradient procedures for stochastic approximation with dependent noise and their asymptotic behavior. *Systems Sciences*, 6(1984)51–79.

13. I.P. Devyaterikov and A.S. Poznyak. *Optimization of Mechanical Systems under Indeterminancy*. MFTI, Moscow, 1984.

14. A.S. Poznyak and Ya.Z. Tsypkin. Optimal algorithms for optimization under correlated noise. *Journal of Computational Mathematics and Mathematical Physics*, (6)(1984)806–822.

RANDOM SEARCH AS A METHOD FOR OPTIMIZATION AND ADAPTATION

L.A. Rastrigin
Politechnic Institute
Riga, USSR

1. PROBLEM STATEMENT

Consider the optimization problem:

$$\min_{x \in S} Q(x) \quad , \tag{1}$$

where a function $Q(x)$ is specified by algorithms (or by measurements) and $X = (x_1, \ldots, x_n)$ is a vector of parameters to be optimized. Let $X^* = (x_1^*, \ldots, x_n^*)$ be the optimal solution of the problem and S be the set of all admissible solutions which in the general case is specified by a system of inequalities and equalities:

$$S: \begin{cases} h_i(X) \geq 0 & (i = 1, \ldots, m) \\ g_j(X) = 0 & (j = 1, \ldots, k) \end{cases} \quad . \tag{2}$$

In the general case the functions $h_i(X)$ and $g_j(X)$ are also specified by algorithms.

2. RANDOM SEARCH

We shall now consider how to construct algorithms for solving problem (1). Random-search algorithms appear particularly promising.

Consider the procedures ψ that make up the algorithm $\vartheta(\vartheta = \psi\psi \cdots \psi \ldots)$:

$$X_0 \xrightarrow{\psi} X_1 \xrightarrow{\psi} \cdots \xrightarrow{\psi} X_N \xrightarrow{\psi} X_{N+1} \xrightarrow{\psi} \cdots \xrightarrow{\psi} X^* \quad ,$$

where the transformation $\rightarrow$ is performed at each step by the algorithm ψ. It is convenient to define this algorithm for the increment:

$$\Delta X = \psi_\zeta(X) \quad , \tag{3}$$

where ΔX is the change in the parameters X to be optimized in the search process and ψ_ζ is the random transformation operator.

Most conventional methods for constructing algorithms ψ are based on *a priori* information on the structure of the functional $Q(X)$ and constraints S. However this information is absent (or almost so) when $Q(X)$ and S are specified by algorithms, which makes it necessary to use heuristic methods.

3. CONSTRUCTION OF HEURISTIC SEARCH ALGORITHMS

Heuristic methods are based in the overwhelming majority of cases on human experience. How can such methods be used to construct algorithms ψ?

Let us start from ordinary human experience. It is first necessary to have a mechanism by which actions can be evaluated. Such a mechanism should be based on the notions "good" and "bad" and incorporate a scale built upon these notions. Next, one should have a set of alternative actions by which it is possible to change the state of the object under consideration. And, finally, there should be a mechanism for the heuristic choice of an action under various specified conditions.

We shall illustrate this using the heuristic trial-and-error method (or, more precisely, the heuristic random trial and error correction method). Here estimates for the minimization problem are of the form:

$$\text{"good"}: \quad (\Delta Q_N < 0) \wedge (X_N \in S) \tag{4}$$

$$\text{"bad"}: \quad (\Delta Q_N \geq 0) \vee (X_N \notin S) \quad ,$$

where ΔQ_N is the increment of the function to be minimized at the **N**-th step:

$$\Delta Q_N = Q(X_N) - Q(X_{N-1}) \quad . \tag{5}$$

There are two alternative actions:

$$\text{random trial}: \quad \zeta \tag{6}$$

$$\text{error correction}: \quad \tau \quad ,$$

where ζ represents a random change in the parameters X to be otimized and τ implies a return to the previous state.

Here the heuristic choice of an action is of the form: "if the present point is good then make a random trial; if it is bad return to the previous point". For the optimization problem this takes the form:

$$\psi_\zeta : \begin{cases} \zeta & \text{for } (\Delta Q_N < 0) \wedge (X_N \in S) \\ \tau(X_N) & \text{for } (\Delta Q_N \geq 0) \vee (X_N \notin S) \end{cases} \quad , \tag{7}$$

where $\tau(X_N)$ denotes a return to the point preceding X_N.

The final form of the algorithm is then as follows:

$$\Delta X_{N+1} = \begin{cases} a\,\Theta & \text{for } (\Delta Q_N < 0) \wedge (X_N \in S) \\ -\Delta X_N & \text{for } (\Delta Q_N \geq 0) \vee (X_N \notin S) \end{cases} \quad , \tag{8}$$

where Θ (the step direction) is a unit random vector uniformly distributed in the space of parameters $\{X\}$ to be optimized, and $\alpha > 0$ is the step size. Algorithm (8) can be described as an ***algorithm with nonlinear tactics*** [1], and is an efficient optimization algorithm.

4. ADAPTATION HEURISTICS

Consider the problem of adapting algorithm (8) to the specific situation occurring in the course of optimization. There are two possibilities: to modify the step size α or to modify the distribution of Θ.

Modification of the step size is based on the followingheuristic: "Failure is the result of over-stepping the target". This rather rough heuristic is true in the one-dimensional case; for $n > 1$ failure can also arise through an incorrect choice of step direction Θ. The adaptation algorithm constructed on this basis is of the form:

$$\alpha_{N+1} = \begin{cases} \delta_1 \alpha_N & \text{for} \quad \Delta Q_N < 0 \\ \delta_2 \alpha_N & \text{for} \quad \Delta Q_N \geq 0 \end{cases} , \tag{9}$$

where the coefficients δ_1 and δ_2 ($\delta_1 > -1$; $0 < \delta_2 < 1$) are related as follows [2]:

$$\delta_1^p \delta_2^{1-p} = 1 \quad . \tag{10}$$

Here p is the probability that the random step direction Θ will turn out to be successful ($\Delta Q_N < 0$).

Although this heuristic is quite rough it nevertheless ensures a geometrical rate of convergence [2] for algorithm (8).

The distribution $p(\Theta)$ of the random vector Θ may be modified using the heuristic "increase the probability of a successful step direction. If with a uniform distribution the probability of a successful step direction depends only on the current point (e.g., in the linear case this probability is equal to 1/2), then it is natural to increase this probability on the basis of information obtained from the previous history of the search process.

This heuristic may be implemented in different ways, e.g., by introducing a "history" vector W such that $E\Theta = W$. Here the direction and value of this vector reflect an estimate of the direction in which the target X^* lies and the degree of confidence in this estimate, respectively. For example,

$$W_N = kW_{N-1} - \alpha \Delta X_N \Delta Q_N \quad , \tag{11}$$

where $0 < k \leq 1$ is a memory coefficient and $\alpha > 0$ is a coefficient reflecting the degree to which new information is taken into account.

5. HEURISTICS BASED ON BIOLOGICAL SYSTEMS

One class of heuristics that can be used in the construction of optimization algorithms is generated by models describing the behaviour of biological systems. Biological models of living systems at all levels of organisation, from the neuronal level to the population level, provide ample heuristic opportunities for the construction of efficient random search algorithms.

5.1. Neuronal Heuristics

At the neuronal level the use of models of the learning process allows us to construct a matrix random search algorithm [3] in which not only the parameters to be optimized but also their transformation matrix are randomly changed. This simulates the change in the synaptic resistance of neurons during learning.

5.2. Behavioural Heuristics

These diverse heuristics are easily identified by an elementary analysis of their behaviour and yield efficient random search algorithms. A number of these are described below.

5.2.1. A "linear tactics" heuristic ("if an action is successful, repeat it; if not, select another action at random") yields a random search algorithm (or a random descent algorithm):

$$\Delta X_{N+1} = \begin{cases} \alpha \Theta & \text{for } (\Delta Q_N \geq 0) \vee (X_N \notin S) \\ \Delta X_N & \text{for } (\Delta Q_N < 0) \wedge (X_N \in S) \end{cases} \quad . \tag{12}$$

This operates well far from the optimum and is a stochastic generalization of steepest descent and alternate parameter change algorithms.

5.2.2. A heuristic describing the behaviour of a mouse in a T-shaped maze in terms of a Markovian model [4], underlies random-search "automata" algorithms [5,6]. Here each parameter is optimized by its own "automaton", which carries out a one-dimensional random search.

5.3. Evolutionary Heuristics

Evolutionary heuristics are associated with the dynamics of population phenomena, which can be described by evolutionary theories at different levels of aggregation. Darwin's heuristic "the survival of the fittest" is of fundamental importance. We will give the basic algorithms.

5.3.1. A population of points $\{X_i\}, X_i \in S$, evolves in the following way. New points are "born":

$$X_{ij} = X_i + \alpha\, \Theta_j \; (j = 1, \ldots, k) \quad , \qquad (13)$$

where Θ is a unit random vector which simulates random mutations, while the value of the parameter α simulates the mutagenesis intensity. Selection reduces to the "dying out" of points X_{ij} for which $X_{ij} \not\in S$ and $Q(X_{ij})$ takes its largest value. This algorithm can be used to solve complex multiextremum problems.

5.3.2. A population of "automata" $\{A_i\}$ consists of stochastic automata with linear tactics [7], each of which is characterized by three numbers:

$$A_i = \langle m_i , p_i , q_i \rangle (i = 1, \ldots, m) \;\; m > n \quad , \qquad (14)$$

where m_i is the amount of accumulated information, and p_i and q_i are the probabilities of translation to a better state or a worse state, respectively. These parameters simulate the "genes" of the automaton. The automaton's efficiency is estimated from its performance during the optimization of the j-th parameter:

$$\Delta x_j^{N+1} = A_i (\Delta Q_N)(j = 1, \ldots, n) \quad . \qquad (15)$$

the "birth" of a new automaton is simulated according to the laws of genetics: "genes" values p_i , q_i are inherited with equal probability from the parents, and automata with high efficiency (fitness) have a higher chance of reproducing them automata with low efficiency (which eventually "die out"). Such evolutionary strategies allow us to construct automata capable of solving a given optimization problem in the most efficient way.

5.3.3. A population of algorithms $\{\psi_i\}$ evolves in an analogous way. Here, for example, the step size α_i in (8) and the vector W_i in (11) represent genes. The increment ΔW_i in the optimization process (see (11) determines the way in which the algorithm ψ_i will adapt during its lifetime. The constraint imposed on ΔW simulates the degree of "conservatism" of ψ_i. It turns out that a good population of algorithms should include algorithms with both large and small values of $|\Delta W|$, simulating the division of the biological population into male and female individuals. Populations of algorithms with only a small range of $|\Delta W|$ values are significantly less efficient in solving optimization problems.

The parameter optimization algorithms considered above assumed that the set S is continuous, i.e. $S \in R^k$. In practice, however, it is necessary to optimize not only the parameters, but also the structure of the object under study.

6. STRUCTURAL RANDOM SEARCH

Let the factors X to be optimized in problem (1) also determine a structure, i.e.,

$$X = \langle V, C \rangle \quad , \tag{16}$$

where V is the structure of an object and $C = (c_1, \ldots, c_n)$ are its parameters.

Problem 81) should now be written as follows:

$$\min_{\substack{V \in S_v \\ C_v \in S_{cv}}} Q(V, C) \quad , \tag{17}$$

where S_v is a set of admissible structures and S_{cv} is a set of values of parameters C_v for structure V.

We shall consider, as before, the recursive algorithms Ψ for solving this roblem ($\Psi = \psi\psi \cdots \psi \cdots$):

$$\Delta V = \psi_{\xi} v(V, C_v) \; ; \; \Delta C_v = \psi_{\xi} c(V, C_v) \quad , \tag{18}$$

where ΔV is the change in the structure of the object, $\psi_{\xi} v$ is the random search operator for the structure, and $\psi_{\xi} c$ is the random search operator for parameters discussed in the previous paragraphs.

It is obvious that it is again necessary to use heuristic methods to construct such algorithms.

A block diagram of the structural search process is presented in Fig. 1. Two routes for adapting the structure V and parameters C are shown. The second route should operate significantly faster than the first since otherwise the efficience of the structure V could be estimated with error. Henceforth we will suppose that at each step in the search for the structure V the total optimization cycle over parameters C is carried out, i.e., the problem

$$Q(V, C) \to \min_{C \in S_{cv}}$$

is completely solved.

We will distinguish between the case in which the cardinality of the set of structures S_v is small, and the case in which it is large, since these cases require different approaches to adaptation:

$$|S_v| \text{ is } \begin{cases} \text{small (2–3), then use alternative adaptation} \\ \text{large, then use evolutionary adaptation} \end{cases}$$

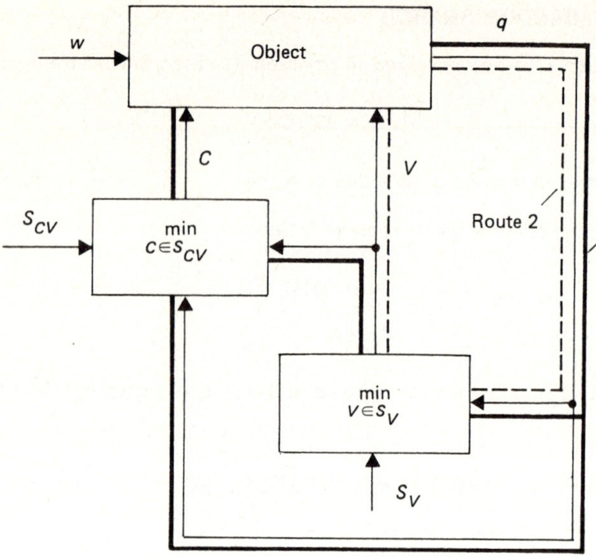

Figure 1.

7. ALTERNATIVE ADAPTATION

The "two-armed bandit" heuristic makes it possible to solve the problem of choosing between two alternative structures:

$$S_v = \{V_1, V_2\} \tag{19}$$

while changing the properties of both the medium and the object itself. It is convenient to represent the algorithm in the form of a graph (see Fig. 2). The vertices of the graph represent the alternative structures V_1 and V_2 while transitions are characterized by the success (+) or failure (-) of the previous step and by the probability of random transformation, which can be defined in various ways.

The probability p that structure V_1 is better than V_2 is

$$p = Bep\{V_1 > V_2 \mid \vec{V}^N, \vec{q}^N\} \quad , \tag{20}$$

where $\vec{V}^N$, $\vec{q}^N$ represent information on the history of the change in structure and the values of the efficiency characteristic. This probability ca be estimated as follows:

$$\hat{p}_N = \frac{1}{2}\left[1 - \Phi\left(\frac{\hat{Q}_1^N - \hat{Q}_2^N}{(\hat{\sigma}_1^2 + \hat{\sigma}_2^2)^{1/2}}\right)\right] \quad , \tag{21}$$

where $p(t)$ is a Lagrange function, $\hat{Q}_i^N$ is an estimate of the efficiency of i-th

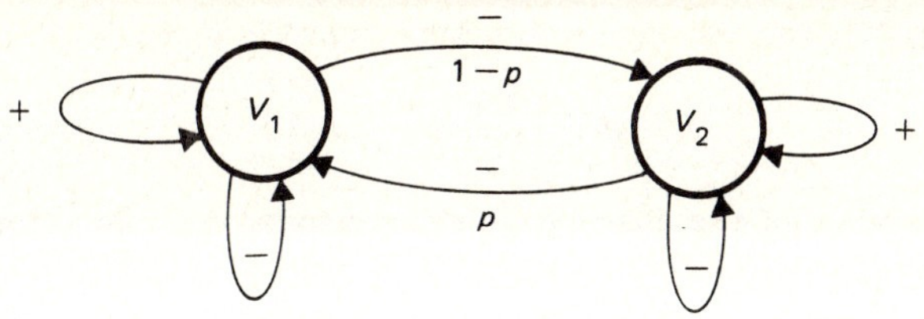

Figure 2.

alternative at the N-th step of the process, and $\hat{\sigma}_i^2$ is an estimate of its variance. These estimates are obtained by standard stastical methods. The resulting random search algorithm maintains the structure which optimizes the characteristic Q at a particular instant of time. When the best structure is changed, the algorithm chooses the other alternative. This algorithm is easily generalized to the multi-alternative case.

The pattern-recognition heuristic can also be used to solve the multi-alternative adaptation problem. To do this it is, however, necessary to describe the object for adaptation as follows:

$$Y = F^0(V) \tag{22}$$

where F^0 is an operator and Y represents information on the state of the object

$$Y = (y_1, \ldots, y_m) \quad . \tag{23}$$

This information also contains data on the efficiency characteristic:

$$q = q(Y) \quad . \tag{24}$$

Let the object have l alternative structures:

$$\{V_1\} \; (i =, \ldots, l) \quad . \tag{25}$$

Then the decision rule for choosing a structure is of the form:

$$j = \alpha(Y, V_i) \in \{1, \ldots, l\} \quad , \tag{26}$$

where j is the number of the alternative recommended by rule α for an object F^0 of structure V_i in state Y.

It is not difficult to see that this rule is a rule for solving the l-class pattern recognition problem and, therefore, the theory and methods of pattern recognition can be used to construct it. A set of sequences (l in number) provide the source information for constructing rule α:

$$Y^{Z_i}(V_i)\ , \ (Z_i = 1, \ldots, N_i \,; i = 1, \ldots, l) \tag{27}$$

The sequences are sets of observations (N_i in number) of the state of the object in different structures. Each element of these sequences is then matched with the corresponding best structure. The decision rule α can now be constructed in the standard way.

We shall now illustrate the performance of this algorithm for the adaptation of alternative (search) optimization algorithms. In this case $V_1, \ldots, V_l$ are alternative algorithms while Y is the information obtained in the search process (for example, Y could describe a trajectory in the space of parameters $\{X\}$ to be optimized – see Fig. 3). The process then switches from one algorithm to another (see Fig. 4).

8. EVOLUTIONARY ADAPTATION

An algorithm for the evolutionary adaptation of structure should be based on the evolution heuristic and can be reduced to simulation of its three components – heredity, variability and selection. We have the following graph of the structure to be adapted:

$$X = \Gamma = \langle A\ , B \rangle \quad , \tag{28}$$

where A is a set of vertices of the graph Γ and B is a set of its arcs. The quality characteristic is given by

$$\min_{\Gamma \in \Omega} Q(\Gamma) \tag{29}$$

and its optimal value Γ^* should be determined during the adaptation process. Here Ω is a set of admissible graphs which ensure the normal functioning of the object.

The graph adaptation algorithm reduces to the following:

1. Simulation of the descendants of the graphs in the preceding generation:

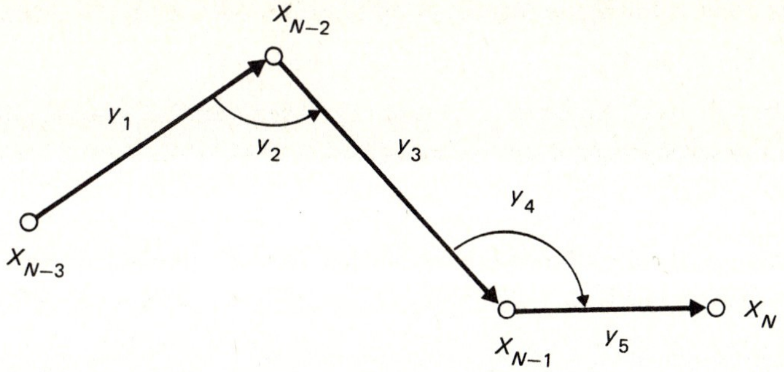

Figure 3:

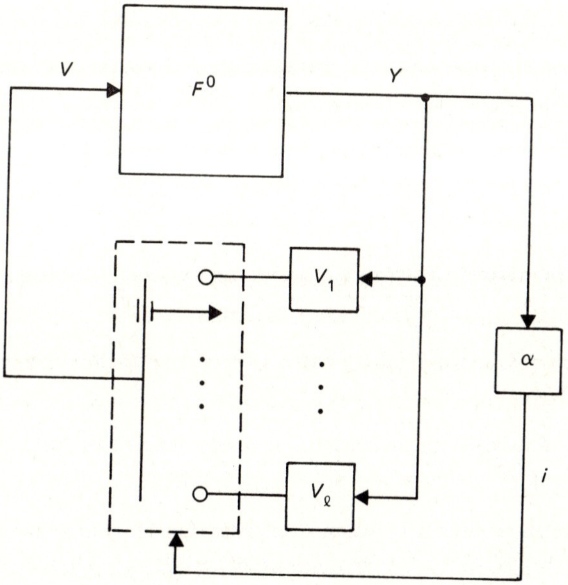

Figure 4:

$$\Gamma_{N+1}^{(i,j)} = \Gamma_N^{(i)} + \Delta\Gamma_{N+1}^{(j)} \quad , \tag{30}$$

where $\Delta\Gamma_{N+1}^{(j)}$ is the j-th random change in the i-th graph, simulating random mutation.

Such changes can involve the introduction of a new vertex or the elimination of an old vertex, arc switching, etc.

2. Selection (for the next evolutionary stage) of the graphs with the minimum values of the characteristic $Q(\Gamma)$ (the rest "die out").

The larger the number of "descendants" left in the population, the greater the global character of the adaptation algorithm. As the population and the number of evolutionary stages increase, the algorithm converges to the optimum solution Γ^*.

It should be noted that to evaluate the efficiency $Q(\Gamma)$ it is usually necessary to simulate the behaviour of the object possessing the structure specified by the graph Γ. The procedure described above is therefore often called *evolutionary simulation* [8].

REFERENCES

1. L.A. Rastrigin. *Adaptation of Complex Systems*. Zinatne, Riga, 1981 (in Russian).
2. L.A. Rastrigin, K.K. Ripa and G.S. Tarasenko. *Adaptation of the Random Search Method*. Zinatne, Riga, 1978 (in Russian).
3. S.N. Grinchenko and L.A. Rastrigin. Bionic random search algorithms. In *Voprosy Kibernetiki. Zadachi i Metodi Adaptivnogo Npravlenia*. Moscow, 1981, pp. 137-147 (in Russian).
4. R. Bush and F. Mosteller. *Stochastic Learning Models*. Fizmatgiz, Moscow, 1962 (in Russian).
5. L.A. Rastrigin and K.K. Ripa. *Automatic Theory of Random Search*. Zinatne, Riga, 1973 (in Russian).
6. U.I. Neymark. Automata optimization. Izvestia Vusov (in Russian): *Radiofizika*, 7(1972), 967-971.

V.I. Muhin, U.I. Neymark and E.I. Ronin. Optimization of aautomata with evolutionary adaptation (in Russian). *Prob. Sluchaini Poisk (Riga)*, 2(1973) 83-97.

8. L. Fogel, A. Owns and M. Walsh. *Artificial Intelligence and Evolutionary Modelling*. Miz, Moscow, 1969 (in Russian).
9. A.P. Bukatova. *Evolutionary Modelling and Its Applications*. Nauka, Moscow, 1979 (in Russian).
10. B.P. Korobkov and L.A. Rastrigin. Randomized methods for graph cutting (in Russian). *Izvestia AN SSSR, Tehnicheskaya Kibernetika*, 3(1982), 163-172; 4(1982), 120-126.
11. A.B. Glaz and L.A. Rastrigin. Solving rules with adaptive structure (in Russian). *Izvestia AN SSSR, Tehnicheskaya Kibernetika*, 6(1974), 192-201.

LINEAR-QUADRATIC PROGRAMMING PROBLEMS WITH STOCHASTIC PENALTIES: THE FINITE GENERATION ALGORITHM

R.T. Rockafellar[1] *and R.J.-B. Wets*[1]

Much of the work on computational methods for solving stochastic programming problems has been focused on the linear case, and with considerable justification. Linear programming techniques for large-scale deterministic problems are highly developed and offer hope for the even larger problems one obtains in certain formulations of stochastic problems. Quadratic programming techniques have not seemed ripe for such a venture, although the ultimate importance of quadratic stochastic programming has been clear enough.

There is another kind of approach, however, in which quadratic stochastic programming problems are no harder to solve than linear ones, and in some respects easier. In this approach, for which the theoretical groundwork has been laid in Rockafellar and Wets [1], the presence of quadratic terms is welcomed because of their stabilizing effect, and such terms are even introduced in iterative fashion. The underlying stochastic problem, whether linear or quadratic, is replaced by a sequence of deterministic quadratic programming problems whose relatively small dimension can be held in check. Among the novel features of the method is its ability to handle more kinds of random coefficients, for instance a random technology matrix.

In this paper we present a particular case of the problem and method in [1] which is especially easy to work with and capable nevertheless of covering many applications. This case falls in the category of stochastic programming with simple recourse. It was described briefly by us in [2], but with the theory in [1] now available, we are able to derive precise results about convergence and the nature of the stopping criterion that can be used. This is also the one case that has been implemented so far and for which numerical experience has gained. For a separate report on the implementation, see King [3].

For the purpose at hand, where duality plays a major role and the constructive use of quadratic terms must be facilitated, the format for stating the problem is crucial. The following deterministic model in linear-quadratic programming serves as the starting point:

[1] *This work was supported in part by a grant from the Air Force Office of Scientific Research at the University of California, Davis.*

$$(\mathrm{P_{det}})\qquad \text{maximize } f(x) = \sum_{j=1}^{n}[c_j x_j - \tfrac{1}{2} r_j x_j^2] - \sum_{k=1}^{\ell} \rho(v_k; p_k, q_k)$$

$$\text{subject to } 0 \le x_j \le s_j \text{ for } j = 1, \dots, n,$$

$$\sum_{j=1}^{n} a_{ij} x_j \le b_i \text{ for } i = 1, \dots m,$$

$$v_k = \sum_{j=1}^{n} t_{kj} x_j - h_k \text{ for } k = 1, \dots, \ell,$$

where ρ is a *penalty* function depending on two parameters p_k and q_k and having the form shown in Figure 1, namely

$$\rho(v_k; p_k, q_k) = \begin{cases} 0 & \text{for } v_k \le 0, \\ \frac{1}{2} v_k^2 p_k & \text{for } 0 \le v_k \le p_k q_k, \\ q_k v_k - \frac{1}{2} p_k q_k^2 & \text{for } v \ge p_k q_k. \end{cases} \tag{0.1}$$

This is convex in v_k, so the object function f in $(\mathrm{P_{det}})$ is concave; *it is assumed that* p_k, q_k, r_j *and* s_j *are nonnegative.* For $p_k = 0$, one takes

$$\rho(v_k; p_k, q_k) = \begin{cases} 0 & \text{for } v_k \le 0, \\ q_k v_k & \text{for } v_k \ge 0. \end{cases} \tag{0.2}$$

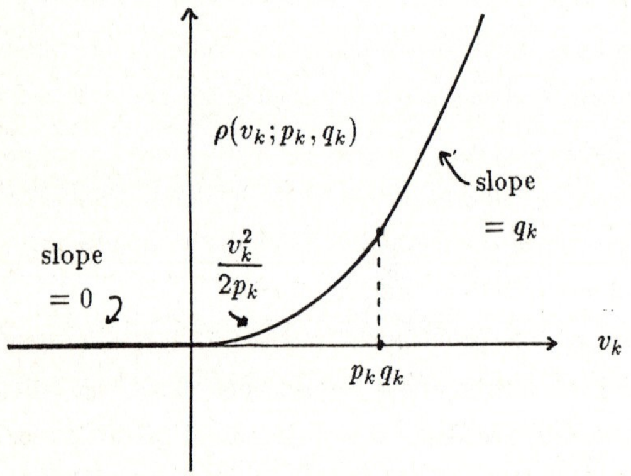

FIGURE 1

The penalty terms in $(\mathrm{P_{det}})$ represent a weakened incorporation of constraints

$$\sum_{j=1}^{n} t_{kj} x_j \le h_k \quad \text{for } k = 1, \dots, \ell \tag{0.3}$$

into the problem. They vanish as long as these constraints are satisfied, but charge a positive cost when they are violated. The cost grows linearly in the special case of (0.2), but otherwise it first passes smoothly through a quadratic phase.

The stochastic programming problem (P_{sto}) that we want to consider is obtained by allowing t_{kj}, h_k, p_k and q_k all to be random variables and replacing each penalty term by its expectation. (In any one application, of course, only a few of these variables might actually be random.) The interpretation is that the x_j's are decision variables whose values must be fixed here and now. The constraints $0 \le x_j \le s_j$ and

$$\sum_{j=1}^{n} a_{ij}x_j \le b_i \text{ for } i = 1, \ldots, m \tag{0.4}$$

are known at the time of this decision, but about the random variables in question there is only statistical information (their distributions). The constraints (0.3) therefore cannot be enforced in the selection of the x_j's without severe consequence. Instead of trying to guard against all possible violations by being extremely conservative, we imagine there is a way of coping with violations of the constraints (0.3), if they should occur. Some recourse action is considered to be possible after the values of the random variables have been realized, and this recourse has an associated cost which depends on the extent of violations. This cost is represented by the penalty terms $\rho(v_k; p_k, q_k)$, and its expectation is subtracted from the here-and-now expression in the x_j's that is being maximized.

Besides the direct applications of this model, we see it as potentially valuable in problems that until now have been formulated deterministically, but in which some of the data may be rather uncertain. By putting such problems in the form of (P_{sto}) it should be possible, even with every crude guesswork about penalty costs and probabilities, to gain some appreciation of how the choice of the x_j's should be modified to hedge against the uncertainties. Certainly this ought to be better than merely assigning specific values to the fuzzy data.

We mention again that although our basic problem is nominally quadratic (a formulation that sidesteps the "piecewise" nature of the penalty terms will be recorded later, in §3), we are also very much concerned with the linear case where $r_j = 0$ and $p_k = 0$. Our plan is first to display a method whose characteristics are most attractive in the strictly quadratic case where $r_j > 0$ and $p_k > 0$, and then apply it to problems lacking in strict quadraticity by means of the proximal point technique [4], [1].

1. OPTIMALITY CONDITIONS AND DUALITY

The approach we are taking depends very much on duality. A subproblem of a certain dual problem will explicitly be solved at every iteration. The Lagrange multipliers in this process will generate the optimizing sequence for the primal problem.

For the deterministic problem (P_{det}), the appropriate dual would be

$$(\mathrm{D_{det}})\qquad \text{minimize}\quad g(y,z)=\sum_{i=1}^{m} b_i y_i+\sum_{k=1}^{\ell}\left[h_k z_k+\tfrac{1}{2}p_k z_k^2\right] + \sum\nolimits_{j=1}^{n} \rho(w_j;r_j,s_j)$$

$$\text{subject to}\quad 0\le y_i \quad\text{for}\quad i=1,\dots,m,$$

$$0\le z_k\le q_k \quad\text{for}\quad k=1,\dots,\ell,$$

$$w_j=c_j-\sum_{i=1}^{m} y_i a_{ij}-\sum_{k=1}^{\ell} z_k t_{kj}\quad\text{for}\quad j=1,\dots,n.$$

Here ρ is the same function as before (cf. Fig. 1), except that the symbols for the variables have been switched:

$$\rho(w_j;r_j,s_j)=\begin{cases}0 & \text{for } w_j\le 0\\ \frac{1}{2}w_j^2/r_j & \text{for } 0\le w_j\le r_j s_j\\ s_j w_j-\frac{1}{2}r_j s_j^2 & \text{for } w_j\ge r_j w_j.\end{cases}\tag{1.1}$$

The terms $\rho(w_j;r_j,s_j)$ in $(\mathrm{D_{det}})$ are to be viewed as penalty representation replacements for constraints

$$\sum_{i=1}^{m} y_i a_{ij}+\sum_{k=1}^{\ell} z_k t_{kj}\ge c_j \quad\text{for}\quad j=1,\dots,n.\tag{1.2}$$

This form of duality is a special case of the scheme used in monotropic programming [5]. It results from the conjugacy between the convex functions

$$\begin{aligned}\varphi_k(v_k)&=\rho(v_k;p_k,q_k),\\ \psi_k(z_k)&=\begin{cases}\frac{1}{2}p_k v_k^2 & \text{if } 0\le z_k\le q_k,\\ \infty & \text{otherwise.}\end{cases}\end{aligned}\tag{1.3}$$

One can show that as long as the constraints $(\mathrm{P_{det}})$ are consistent, one has

$$\max\ (\mathrm{P_{det}}) \;=\; \min\ (\mathrm{D_{det}}).$$

In the stochastic case we are directly concerned with in this paper, the appropriately modified primal and dual problems are

$$(\mathrm{P_{sto}})\qquad \text{maximize}\quad f(x)=\sum_{j=1}^{n}\left[c_j x_j-\tfrac{1}{2}r_j x_j^2\right]-E\Big\{\sum_{k=1}^{\ell}\rho(\underset{\sim}{v}_k;\underset{\sim}{p}_k,\underset{\sim}{q}_k)\Big\}$$

$$\text{subject to}\quad 0\le x_j\le s_j\quad\text{for}\quad j=1,\dots,n,$$

$$\sum_{j=1}^{n} a_{ij}x_j\le b_i\quad\text{for}\quad i=1,\dots,m,$$

$$\underset{\sim}{v}_k=\sum_{j=1}^{n}\underset{\sim}{t}_{kj}x_j-\underset{\sim}{h}_k\quad\text{for}\quad k=1,\dots,\ell,$$

$$\text{(D}_{\text{sto}}) \quad \text{minimize } g(y,\underset{\sim}{z}) = \sum_{i=1}^{m} b_i y_i + E\{\sum_{k=1}^{\ell} [\underset{\sim}{h}_k \underset{\sim}{z}_k + \tfrac{1}{2}\underset{\sim}{p}_k \underset{\sim}{z}_k^2]\} + \sum_{j=1}^{n} \rho(w_j; r_j, s_j)$$

$$\text{subject to } 0 \le y_i \text{ for } i = 1,\dots,m,$$

$$0 \le \underset{\sim}{z}_k \le \underset{\sim}{q}_k \text{ for } k = 1,\dots,\ell,$$

$$w_j = c_j - \sum_{i=1}^{m} y_i a_{ij} - E\{\sum_{k=1}^{\ell} \underset{\sim}{z}_k \underset{\sim}{t}_{kj}\}.$$

The random variables in these problems have been indicated by $\sim$; the symbol E denotes mathematical expectation.

In order to avoid minor technical complications that have no real importance in our present task of setting up a computational framework for ($\mathrm{P_{sto}}$), we shall rely henceforth on two assumptions.

(A1) There is at least one vector x satisfying

$$0 \le x_j \le s_j \text{ for } j = 1,\dots,n, \text{ and } \sum\nolimits_{j=1}^{n} a_{ij} x_j \le b_i \text{ for } i = 1,\dots,m.$$

(A2) The given random variables $\underset{\sim}{t}_{pj}, \underset{\sim}{h}_k, \underset{\sim}{p}_k, \underset{\sim}{q}_k$ take on only finitely many values.

Only (A2) needs comment. We are assuming that whatever the "true" distribution of these variables might be, we are treating them here in terms of finitely many values to which probability weights have been assigned. Such a discrete distribution might be obtained by approximating a continuous distribution, or by sampling a continuous distribution, or empirically. For now, that need not matter; the question of the source of the discrete distribution and how it might be "improved" is quite separate. The important thing is that we impose no further conditions on the random variables. Aside from (A2), their distribution can be completely arbitrary. In particular a joint distribution is allowed; the variables do not have to be independent.

THEOREM 1. *Under assumptions* (A1) *and* (A2), *problems* ($\mathrm{P_{sto}}$) *and* ($\mathrm{D_{sto}}$) *both have optimal solutions, and*

$$\max(\mathrm{P_{sto}}) = \min(\mathrm{D_{sto}}).$$

Moreover in the strictly quadratic case where $r_j > 0$ and $\underset{\sim}{p}_k > 0$, the following conditions are necessary and sufficient in order that $\bar{x}$ be optimal for ($\mathrm{P_{sto}}$) *and* $(\bar{y}, \underset{\sim}{\bar{z}})$ *optimal for* ($\mathrm{D_{sto}}$):

$$\sum_{j=1}^{n} a_{ij}\bar{x}_j - b_i \le 0,\ \bar{y}_i \ge 0,\ \textit{and}\ (\sum_{j=1}^{n} a_{ij}\bar{x}_j - b_j)\bar{y}_i = 0, \tag{1.4}$$

$$\overline{x}_j = \rho'(\overline{w}_j; r_j, s_j) \;\; for \;\; \overline{w}_j = c_j - \sum_{i=1}^{m} \overline{y}_i a_{ij} - E\{\sum_{k=1}^{\ell} \overline{\underset{\sim}{z}}_k \underset{\sim}{t}_{kj}\}, \tag{1.5}$$

$$\overline{\underset{\sim}{z}}_k = \rho'(\overline{\underset{\sim}{v}}_k; \underset{\sim}{p}_k, \underset{\sim}{q}_k) \;\; for \;\; \overline{\underset{\sim}{v}}_k = \sum_{j=1}^{n} \underset{\sim}{t}_{kj} \overline{x}_j - \underset{\sim}{h}_k. \tag{1.6}$$

In these relations the derivatives ρ' refer to the first argument indicated, not the parameter arguments. Thus

$$\rho'(w_j; r_j, s_j) = \begin{cases} 0 & \text{if } w_j \leq 0 \\ w_j / r_j & \text{if } 0 \leq w_j \leq r_j s_j \\ s_j & \text{if } w_j \geq r_j s_j \end{cases} \tag{1.7}$$

and likewise, with just a change of notation,

$$\rho'(v_k; p_k, q_k) = \begin{cases} 0 & \text{if } v_k \leq 0 \\ v_k / p_k & \text{if } 0 \leq v_k \leq p_k q_k \\ q_k & \text{if } v_k \geq p_k q_k. \end{cases}$$

It is clear then that (1.5) entails $0 \leq \overline{x}_j \leq s_j$, and (1.6) entails $0 \leq \underset{\sim}{z}_j \leq \underset{\sim}{q}_j$. This is why these basic requirements do not appear explicitly in the theorem along with the feasibility and complementary slackness conditions (1.4).

Formula (1.5) serves as a means of obtaining the optimal solution to ($\mathrm{P_{sto}}$) from the optimal solution to ($\mathrm{D_{sto}}$), or an approximately optimal solution to ($\mathrm{P_{sto}}$) from an approximately optimal one for ($\mathrm{D_{sto}}$), the mapping being continuous. Formula (1.6), on the other hand, says that the component $\overline{\underset{\sim}{z}}$ of an optimal solution to ($\mathrm{D_{sto}}$) is a random variable expressible in terms of the known random variables $\underset{\sim}{t}_{kj}, \underset{\sim}{h}_k, \underset{\sim}{p}_k, \underset{\sim}{q}_k$, and the (nonrandom) optimal solution $\overline{x}$ to ($\mathrm{P_{sto}}$). More generally, by means of this formula as applied to various nonoptimal vectors x that arise in the solution process, it is possible economically to represent (and store in a computer) some of the elements $\underset{\sim}{z}$ that will be needed in the solution process.

PROOF OF THEOREM 1. The duality will be obtained from a minimax representation in terms of the sets

$$X = \{x = (x_1, \ldots, x_n) \mid 0 \leq x_j \leq s_j\}, \tag{1.9}$$

$$Y = \{y = (y_1, \ldots, y_m) \mid 0 \leq y_j\}, \tag{1.10}$$

$$Z = \{\underset{\sim}{z} = (\underset{\sim}{z}_z, \ldots, \underset{\sim}{z}_\ell) \mid 0 \leq \underset{\sim}{z}_k \leq \underset{\sim}{q}_k\}, \tag{1.11}$$

and the function L on $X \times Y \times Z$ defined by

$$\begin{aligned} L(x, y, \underset{\sim}{z}) = & \sum_{j=1}^{n} [c_j x_j - \tfrac{1}{2} r_j x_j^2] + \sum_{i=1}^{m} y_i [b_i - \sum_{j=1}^{n} a_{ij} x_j] \\ & + \sum_{k=1}^{\ell} E\{\underset{\sim}{z}_k [\underset{\sim}{h}_k - \sum_{j=1}^{n} \underset{\sim}{t}_{kj} x_j] + \tfrac{1}{2} \underset{\sim}{p}_k \underset{\sim}{z}_k^2\}. \end{aligned} \tag{1.12}$$

Here because of assumption (A2) we could think of each of the random variables as functions on a single finite probability space Ω, or equivalently as vectors indexed by $\omega \in \Omega$. Then in (1.1) we could write $0 \leq z_{\omega k} \leq q_{\omega k}$ for all ω and k, while in (1.12) we could write

$$E\{\underset{\sim}{z}_k[\underset{\sim}{h}_k - \sum_{j=1}^{n} \underset{\sim}{t}_{kj}x_j] + \tfrac{1}{2}\underset{\sim}{p}_k\underset{\sim}{z}_k^2\} = \sum_{\omega\in\Omega}(z_{\omega k}[\pi_\omega h_{\omega k} - \sum_{j=1}^{n}\pi_\omega t_{\omega kj}x_j] + \tfrac{1}{2}\pi_\omega p_{\omega k}z_{\omega k}^2),$$

where $\pi_\omega > 0$ is the probability weight assigned to the element ω of Ω. This makes it plain that Z, like X and Y, is simply a finite-dimensional convex polyhedron, although the dimension may be very large, and L is a quadratic function which is concave in x and convex in $(y, \underset{\sim}{z})$.

It is easily verified that

$$\inf_{(y,\underset{\sim}{z})\in Y\times Z} L(x,y,\underset{\sim}{z}) = \begin{cases} f(x) & \text{if } x \text{ is feasible in } (\mathrm{P_{sto}}), \\ -\infty & \text{otherwise}, \end{cases} \tag{1.14}$$

$$\sup_{x\in X} L(x,y,\underset{\sim}{z}) = \begin{cases} g(y,\underset{\sim}{z}) & \text{if } (y,\underset{\sim}{z}) \text{ is feasible in })(\mathrm{D_{sto}}), \\ \infty & \text{otherwise}, \end{cases} \tag{1.15}$$

where $f(x)$ and $g(y, \underset{\sim}{z})$ are the objective functions specified for ($\mathrm{P_{sto}}$) and ($\mathrm{D_{sto}}$). Thus ($\mathrm{P_{sto}}$) and ($\mathrm{D_{sto}}$) are the primal and dual problems associated with the minimax problem for L on $X \times (Y \times Z)$. Because L is quadratic concave-convex, and the sets X and $Y \times Z$ are convex polyhedra, we may conclude from generalized quadratic programming theory (see [1, Theorem 1]) that if the optimal value in either problem is finite, or if both problems have feasible solutions, then both problems have optimal solutions and max($\mathrm{P_{sto}}$) =min ($\mathrm{D_{sto}}$). This is indeed the case here, because ($\mathrm{D_{sto}}$) trivially has feasible solutions, and our assumption (A1) guarantees that ($\mathrm{P_{sto}}$) has feasible solutions.

The optimality conditions (1.4), (1.5), (1.6), are just a restatement of the requirement that $(\bar{x}, \bar{y}, \underset{\sim}{\bar{z}})$ be a saddlepoint of L on $X \times (Y \times Z)$. For instance, the part of the saddlepoint property that corresponds to maximization in $\underset{\sim}{z}$ decomposes into

$$\bar{z}_{\omega k} \in \operatorname*{argmin}_{0\leq z_{\omega k}\leq q_{\omega k}} \{z_{\omega k}[h_{\omega k} - \sum_{j=1}^{n} t_{\omega kj}\bar{x}_j] + \tfrac{1}{2}p_{\omega k}z_{\omega k}^2\}.$$

In terms of the conjugate convex functions in (1.3) and the notation

$$\bar{v}_{\omega k} = \sum_{j=1}^{n} t_{\omega kj}\bar{x}_{\omega k} - h_{\omega k},$$

this can be written as

$$\bar{z}_{\omega k} \in \operatorname*{argmin}_{z_{\omega k}\in R}\{\psi_{\omega k}(z_{\omega k}) - \bar{v}_{\omega k}z_{\omega k}\},$$

or $0 \in \partial\psi_{\omega k}(\bar{z}_{\omega k}) - \bar{v}_{\omega k}$, and then equivalently as $\bar{v}_{\omega k} \in \partial\psi_{\omega k}(\bar{z}_{\omega k})$ or $\bar{z}_{\omega k} \in \partial\varphi_{\omega k}(\bar{v}_{\omega k})$. The latter reduces to $\bar{z}_{\omega k} = \varphi'_{\omega k}(\bar{v}_{\omega k})$ and condition (1.6) when $\varphi_{\omega k}$ is differentiable, as is the case when $p_{\omega k} > 0$. The derivation of (1.5) from the saddle point property is similar. □

This formulation of stochastic programming duality differs somewhat from the one in our basic paper [1]. In order to facilitate the application of the results in [1] to the present context, an explanation of the connection is needed. In [1], problem (P_{sto}) is associated with a different minimax problem, namely for

$$L_0(x, \underset{\sim}{z}) = \sum_{j=1}^{n}[c_j x_j - \tfrac{1}{2}r_j x_j^2] + \sum_{k=1}^{\ell} E\{\underset{\sim}{z}_k[\underset{\sim}{h}_k - \sum_{j=1}^{n} \underset{\sim}{t}_{kj}x_j] + \tfrac{1}{2}\underset{\sim}{p}_k \underset{\sim}{z}_k^2\} \tag{1.16}$$

on $X_0 \times Z$, where Z is still the set in (1.11) but X_0 is the set of feasible solutions to (P_{sto}):

$$X_0 = \{x = (x_1, \ldots, x_n) \mid 0 \le x_j \le s_j, \sum_{j=1}^{n} a_{ij}x_j \le b_i\}. \tag{1.17}$$

This leads to the dual problem

$$\text{minimize } g_0(\underset{\sim}{z}) \text{ over all } \underset{\sim}{z} \in Z, \tag{D^0_{sto}}$$

where

$$g_0(\underset{\sim}{z}) = \min_{y \in Y} g(y, \underset{\sim}{z}). \tag{1.18}$$

Indeed, one has in parallel to (1.14), (1.15), that

$$\min_{\underset{\sim}{z} \in Z} L_0(x, \underset{\sim}{z}) = f(x) \text{ for all } x \in X_0, \tag{1.19}$$

and by quadratic programming duality (using (A1))

$$\begin{aligned} \max_{x \in X_0} L_0(x, \underset{\sim}{z}) &= \max_{x \in X} \inf_{y \in Y} L(x, y, \underset{\sim}{z}) \\ &= \min_{y \in Y} \max_{x \in X} L(x, y, \underset{\sim}{z}) = \min_{y \in Y} g(y, \underset{\sim}{z}) \text{ for all } \underset{\sim}{z} \in Z. \end{aligned} \tag{1.20}$$

(Actually in [1] one has minimization in the primal problem and maximization in the dual, but that calls for only a minor adjustment.) Obviously, then, the pairs $(\bar{y}, \bar{\underset{\sim}{z}})$ that solve (D_{sto}) are the ones such that $\bar{\underset{\sim}{z}}$ solves (D^0_{sto}) and $\bar{y}$ provides the corresponding minimum (1.18).

2. FINITE GENERATION ALGORITHM IN THE STRICTLY QUADRATIC CASE

The basic idea of our computational procedure is easy to describe. We limit attention for the time being to the strictly quadratic case where $r_j > 0$ and $\underset{\sim}{p}_k > 0$, because we will be able to show in section 4 that problems that are not strictly quadratic can be made so as part of an additional iterative process. This limitation also simplifies the exposition and helps us focus on the results we believe to be the most significant. It is not truly necessary, however. A more general version of what follows could likewise be deduced from the fundamental theory in [1].

In problem (D_{sto}) we minimize a certain convex function $g(y, \underset{\sim}{z})$ over $Y \times Z$, where Y and Z are the convex polyhedra in (1.10) and (1.11). As we have seen in the proof of Theorem 1, this corresponds to finding a saddlepoint $(\overline{x}, \overline{y}, \overline{\underset{\sim}{z}})$ of the function L in (1.12) relative to $X \times (Y \times Z)$, where X is the polyhedron in (1.9). Indeed, if $(\overline{y}, \overline{\underset{\sim}{z}})$ is optimal for (D_{sto}), then the $\overline{x}$ obtained from formula (1.5) gives us the saddlepoint. This $\overline{x}$ is the unique optimal solution to (P_{sto}).

The trouble is, however, that because of the potentially very high dimensionality of Z (whose elements $\underset{\sim}{z}$ have components $z_{\omega k}$ for $k = 1, \ldots, \ell$ and all $\omega \in \Omega$, with Ω possibly very large), we cannot hope to solve (D_{sto}) directly, even though it is reducible in principle to a quadratic programming problem. What we do instead is develop a method of descent which produces a minimizing sequence $\{(\overline{y}^\nu, \overline{\underset{\sim}{z}}^\nu)\}_{\nu=1}^\infty$ in (D_{sto}) and at the same time, by formula (1.5), a maximizing sequence $\{\overline{x}^\nu\}_{\nu=1}^\infty$ in (P_{sto}).

In this method we "generate Z finitely from within". Let Z be expressed as

$$Z = Z_1 \times \ldots \times Z_\ell \quad \text{with} \quad Z_k = \{\underset{\sim}{z}_k \mid 0 \leq \underset{\sim}{z}_k \leq \underset{\sim}{q}_k\}. \tag{2.1}$$

At iteration ν we take a finite subset $\widehat{Z}_k^\nu$ of Z_k, and instead of minimizing $g(y, \underset{\sim}{z})$ over $Y \times Z$ we minimize it over $Y \times Z^\nu$, where

$$Z^\nu = Z_1^\nu \times \ldots \times Z_\ell^\nu \quad \text{with} \quad Z_k^\nu = \text{co}\{0, \widehat{Z}_k^\nu\}. \tag{2.2}$$

By employing a parametric representation of the convex hull $\text{co}\{0, \widehat{Z}_k^\nu\}$ and keeping the number of elements in $\widehat{Z}_k^\nu$ small, which turns out always to be possible, we are able to express this subproblem as one of quadratic programming in a relatively small number of variables. This subproblem is deterministic in character; the coefficients are certain expectations in terms of the given random variables $\underset{\sim}{t}_{kj}, \underset{\sim}{h}_k, \underset{\sim}{p}_k$ and the chosen random variables in $\widehat{Z}_k^\nu$.

The details of the subproblem will be explained in due course (§3). First we state the algorithm more formally and establish its convergence properties.

FINITE GENERATION ALGORITHM (*version under the strict quadraticity assumption that $r_j > 0$ and $\underset{\sim}{p}_k > 0$.*)

Step 0 (Initialization). *Choose finite subsets $\widehat{Z}_k^1 \subset Z_k$ for $k = 1, \ldots, \ell$. Set $\nu = 1$.*

Step 1 ((Quadratic Programming Subproblem). *Calculate an optimal solution $(\overline{y}^\nu, \overline{\underset{\sim}{z}}^\nu)$ to the problem of minimizing $g(y, \underset{\sim}{z})$ over $Y \times Z^\nu$, where Z^ν is given by (2.2). Denote the minimum value by $\overline{\alpha}_\nu$. Define $\overline{x}^\nu$ from $(\overline{y}^\nu, \overline{\underset{\sim}{z}}^\nu)$ by formula (1.5).*

Step 2 (Generation of Test Data). *Define $\underset{\sim}{z}^\nu$ from $\overline{x}^\nu$ by formula (1.6). Set $\alpha_\nu = L_0(\overline{x}^\nu, z^\nu)$ in (1.16).*

Step 3 (Optimality Test). *Define* $\varepsilon_\nu = \overline{\alpha}_\nu - \alpha_\nu \geq 0$. *Then* $\overline{x}^\nu$ *is an* ε_ν*-optimal solution to* $(\mathrm{P_{sto}})$, $(\overline{y}^\nu, \overline{z}^\nu)$ *is an* ε_ν*-optimal solution to* $(\mathrm{D_{sto}})$, *and*

$$\overline{\alpha}_\nu \geq \max(\mathrm{P_{sto}}) = \min(\mathrm{D_{sto}}) \geq \alpha_\nu .$$

(Stop if this is good enough.)

Step 4 (Polytope Modification). *For each* $k = 1, \ldots, \ell$, *choose a finite set* $\widehat{Z}_k^{\nu+1} \subset Z_k$ *whose convex hull contains both* $\overline{z}^\nu$ *and* z^ν. *Replace* ν *by* $\nu+1$ *and return to Step 1.*

Note the very mild condition in Step 4 on the choice of $\widehat{Z}_k^{\nu+1}$. One could simply take

$$\widehat{Z}_k^{\nu+1} = \{\overline{z}_k^\nu, z_k^\nu\}, \tag{2.3}$$

or at the opposite extreme,

$$\widehat{Z}_k^{\nu+1} = \widehat{Z}_k^\nu \cup \{z_k^\nu\}. \tag{2.4}$$

Another possibility would be

$$\widehat{Z}_k^{\nu+1} = \widehat{Z}_k^1 \cup \{\overline{z}_k^\nu, z_k^\nu\} \tag{2.5}$$

in all iterations, with $\widehat{Z}_k^1$ selected initially to provide a certain richness of representation. Although the number of elements of $\widehat{Z}_k^\nu$ (which determines the dimensionality of the quadratic programming subproblem in Step 1) would continue to grow indefinitely under (2.4), it stays fixed under (2.3) or (2.5).

For the statement of our convergence result we introduce the vector norms

$$\|x\|_r = \Big[\sum_{j=1}^{n} r_j x_j^2\Big]^{1/2}, \tag{2.6}$$

$$\|z\|_p = \Big[\sum_{k=1}^{\ell} p_k z_k^2\Big]^{1/2}, \tag{2.7}$$

and matrix norm

$$\|T\|_{p,r} = \max\{z \cdot Tx \mid \|z\|_p \leq 1, \|x\|_r \leq 1\}. \tag{2.8}$$

THEOREM 2. *Under the strict quadraticity assumption that* $r_j > 0$ *and* $p_k > 0$, *the sequence* $\{\overline{x}^\nu\}_{\nu=1}^\infty$ *produced by the finite generation algorithm converges to the unique optimal solution* $\overline{x}$ *to* $(\mathrm{P_{sto}})$. *Moreover it does so at a linear rate, in the following sense.*

Let σ *be an upper bound to the range of the (finitely discrete) random variable* $\|T\|_{p,r}$ *in (2.8), where* T *is the matrix with entries* t_{kj}. *Let* $\tau \in [0,1)$ *be the factor defined by*

$$\tau = \begin{cases} \sigma^2 & \text{if } \sigma^2 \leq \frac{1}{2}, \\ 1 - (1/4\sigma^2) & \text{if } \sigma^2 \geq \frac{1}{2}. \end{cases} \tag{2.9}$$

Then in terms of the values

$$\bar{\alpha} = \max(\mathrm{P_{sto}}) = \min(\mathrm{D_{sto}}) \quad \textit{and} \quad \bar{\varepsilon}_\nu = \bar{\alpha} - \bar{\alpha}_\nu \leq \varepsilon_\nu \tag{2.10}$$

one has

$$\bar{\varepsilon}_{\nu+\mu} \leq \tau^\mu \varepsilon_\nu \quad \textit{for all} \quad \nu = 1, 2, \ldots, \quad \textit{and} \quad \mu = 1, 2, \ldots, \tag{2.11}$$

$$\|\bar{x} - \bar{x}^{\nu+\mu}\|_r \leq [2\tau^\mu \varepsilon_\nu]^{1/2} \quad \textit{for all} \quad \nu = 1, 2, \ldots, \quad \textit{and} \quad \mu = 1, 2, \ldots. \tag{2.12}$$

Observe well that in (2.11) and (2.12) the estimates are claimed for *all* ν and μ, *not* just when ν is sufficiently large. Most convergence results are not of such type, so this is rather surprising, especially in view of the fact that the factor $\tau \in [0, 1)$ can in principle, at least, be estimated in advance of computation, right from the given data. Moreover τ does not depend on any data in the problem other than $\underset{\sim}{t}_{kj}, \underset{\sim}{p}_k$ and r_j. In the special case of nonrandom t_{kj} and p_k (the only random variables in the problem being $\underset{\sim}{h}_k$ and $\underset{\sim}{q}_k$), one can simply take $\sigma = \|T\|_{p,r}$.

PROOF OF THEOREM 2. The procedure specified here is a special case of the algorithm presented in [1], as can be seen in the following way. In calculating a pair $(\bar{y}^\nu, \bar{\underset{\sim}{z}}^\nu)$ that minimizes $g(y, \underset{\sim}{z})$ over $Y \times Z^\nu$ in the subproblem in Step 1, we obtain a solution $\bar{\underset{\sim}{z}}^\nu$ to the different subproblem of [1], in which $g_0(\underset{\sim}{z})$ is minimized over Z^ν (with g_0 the function in (1.18)). The number $\bar{\alpha}_\nu$ is the optimal value in both subproblems, and $\bar{x}^\nu$ furnishes the saddle point $\bar{x}^\nu, \bar{y}^\nu, \bar{\underset{\sim}{z}}^\nu$ to L on $X \times (Y \times Z^\nu)$ in the present formulation, but also the saddlepoint $(\bar{x}^\nu, \bar{\underset{\sim}{z}}^\nu)$ to L_0 on $X \times Z^\nu$, as required by Step 1 of the algorithm as formulated in [1].

The elements $\underset{\sim}{z}^\nu$ and α_ν calculated in Step 2 satisfy

$$\underset{\sim}{z}^\nu = \underset{\underset{\sim}{z} \in Z}{\operatorname{argmin}}\, L_0(\bar{x}^\nu, \underset{\sim}{z}), \qquad \alpha_\nu = \min_{\underset{\sim}{z} \in Z} L_0(\bar{x}^\nu, \underset{\sim}{z}). \tag{2.13}$$

Thus these are the same as the elements calculated in the version of Step 2 in [1] (except for a notational switch between maximization and minimization). Of course they are given here by closed formulas, whereas in the far more general setting of [1] they might have to be calculated by solving a large collection of quadratic programming subproblems in the random components $z_{\omega k}$.

The updated polyhedron $Z^{\nu+1}$ does contain $\bar{\underset{\sim}{z}}^\nu$ and $\underset{\sim}{z}^\nu$ under the conditions in Step 4, as required by the conditions in the more general version of Step 4 in [1].

Thus all the conditions in Theorem 5 of [1] are fulfilled, and the stated convergence properties follow, provided that we reconcile the choice of σ given here with the corresponding one in [1]. The condition specified in [1, Theorem 5] is that

$$\sigma^2 \|z\|_{\underset{\sim}{p}}^2 \geq \|\underset{\sim}{T}^* z\|_{r^{-1}}^2 \tag{2.14}$$

for all realizations of the random vector $\underset{\sim}{p}$ and matrix $\underset{\sim}{T}$ and all possible choices of the vector z. Here we are using the notation $r^{-1} = (r_1^{-1}, \ldots, r_n^{-1})$. The norm $\|\cdot\|_{r^{-1}}$ is the dual of the norm $\|\cdot\|_r$ in (2.6), so

$$\|\underset{\sim}{T}^* \underset{\sim}{z}\|_{r^{-1}} = \max\{(\underset{\sim}{T}^* \underset{\sim}{z}) \cdot x \mid \|x\|_r \leq 1\}.$$

(T^* = transpose of T.) Therefore one has

$$\max\{\|\underset{\sim}{T}^* z\|_{r^{-1}} \mid \|z\|_{\underset{\sim}{p}} \leq 1\} = \|\underset{\sim}{T}\|_{\underset{\sim}{p},r}$$

as defined in (2.8). This shows that (2.14) is equivalent to

$$\sigma \geq \|\underset{\sim}{T}\|_{\underset{\sim}{p},r}$$

and the proof of Theorem 2 is thereby completed. □

3. SOLVING THE QUADRATIC PROGRAMMING SUBPROBLEM.

Returning now to the elucidation of the finite generation algorithm and how it may be implemented, we demonstrate that the subproblem in Step 2 can be represented easily as an ordinary quadratic programming problem of relatively low dimension and thereby solved using standard codes. Explicit notation for the elements of the finite sets $\widehat{Z}_k^\nu$ selected from Z_k is now needed. Let us suppose that

$$\widehat{Z}_k^\nu = \{\underset{\sim}{\widehat{z}}_{k\alpha}^\nu \mid \alpha = 1, \ldots, m_\nu\}. \tag{3.1}$$

This yields

$$Z_k^\nu = \mathrm{co}\{0, \widehat{Z}_k^\nu\} = \{\underset{\sim}{z}_k = \sum_{\alpha=1}^{m_\nu} \lambda_{k\alpha} \underset{\sim}{\widehat{z}}_{k\alpha} \mid \lambda_{k\alpha} \geq 0, \sum_{\alpha=1}^{m_\nu} \lambda_{k\alpha} \leq 1\}. \tag{3.2}$$

In Step 2 we want to minimize the objective $g(y, \underset{\sim}{z})$ in ($\mathrm{D_{sto}}$) not over all of $Y \times Z$ (the variables w_j standing for linear expressions in y and $\underset{\sim}{z}$), but only over $Y \times Z^\nu$. By virtue of (3.2) we can substitute for the elements $\underset{\sim}{z}$ of interest in this subproblem certain linear expressions in the parameters $\lambda_{k\mu}$. In this way we get the function

$$\begin{aligned} g^\nu(y, \lambda) = & \sum_{i=1}^{m} b_i y_i + \sum_{j=1}^{n} \rho(w_j; r_j s_j) \\ & + E\{\sum_{k=1}^{\ell} [\underset{\sim}{h}_k (\sum_{\alpha=1}^{m_\nu} \lambda_{k\alpha} \underset{\sim}{\widehat{z}}_{k\alpha}^\nu) + \tfrac{1}{2} \underset{\sim}{p}_k (\textstyle\sum_{\alpha=1}^{m_\nu} \lambda_{k\alpha} \underset{\sim}{\widehat{z}}_{k\alpha}^\nu)^2]\}, \end{aligned} \tag{3.3}$$

where

$$w_j = c_j - \sum_{i=1}^{m} y_i a_{ij} - E\{\sum_{k=1}^{\ell} (\sum_{\alpha=1}^{m_\nu} \lambda_{k\alpha} \underset{\sim}{\widehat{z}}_{k\alpha}^\nu) \underset{\sim}{t}_{kj}\}. \tag{3.4}$$

But these complicated expressions can greatly be reduced by carrying the expectation operation through the sums to get explicit coefficients for the parameters $\lambda_{k\mu}$. Specifically, let

$$\overline{h}^{\nu}_{k\alpha} = E\{\underset{\sim}{h}_k \hat{\underset{\sim}{z}}^{\nu}_{k\mu}\}, \tag{3.5}$$

$$\overline{p}^{\nu}_{k\alpha\beta} = E\{\underset{\sim}{p}_k \hat{\underset{\sim}{z}}^{\nu}_{k\alpha} \hat{\underset{\sim}{z}}^{\nu}_{k\beta}\}, \tag{3.6}$$

$$\overline{t}^{\nu}_{k\alpha j} = E\{\hat{\underset{\sim}{z}}^{\nu}_{k\alpha} \underset{\sim}{t}_{kj}\}. \tag{3.7}$$

Then

$$\begin{aligned} g^{\nu}(y,\lambda) = &\sum_{i=1}^{m} b_i y_i + \sum_{j=1}^{n} \rho(w_j; r_j, s_j) \\ &+ \sum_{k=1}^{\ell} \Big[\sum_{\alpha=1}^{m_\nu} \overline{h}^{\nu}_{k\alpha} \lambda_{k\alpha} - \tfrac{1}{2} \textstyle\sum_{\alpha,\beta=1}^{m_\nu,m_\nu} \overline{p}^{\nu}_{k\alpha\beta} \lambda_{k\alpha} \lambda_{k\beta}\Big], \end{aligned} \tag{3.8}$$

where

$$w_j = c_j - \sum_{i=1}^{m} y_i a_{ij} - \sum_{k=1}^{\ell} \sum_{\alpha=1}^{m_\nu} \lambda_{k\alpha} \overline{t}_{k\alpha j}. \tag{3.9}$$

Finally let us observe that the penalty expression $\rho(w_j; r_j, s_j)$ in these formulas, as given by (1.1), satisfies

$$\begin{aligned} \rho(w_j; r_j, s_j) = \text{minimum of } & s_j w_{1j} + \tfrac{1}{2} w_{2j}^2 / r_j \\ \text{subject to } & w_{1j} \geq 0, w_{1j} + w_{2j} \geq w_j. \end{aligned} \tag{3.10}$$

Moreover

$$\begin{aligned} \rho'(w_j; r_j, s_j) = \text{Lagrange multiplier } (\geq 0) \text{ for the constraint} \\ w_{1j} + w_{2j} \geq w_j \text{ in (3.10)}. \end{aligned} \tag{3.11}$$

With these facts in mind we pose the quadratic programming problem

$$(\mathrm{D}^{\nu}) \quad \begin{aligned} \text{minimize } & \sum_{i=1}^{m} b_i y_i + \sum_{j=1}^{n} \big[s_j w_{1j} + \tfrac{1}{2} w_{2j}^2 / r_j\big] \\ & + \sum_{k=1}^{\ell} \Big[\sum_{\alpha=1}^{m_\nu} \overline{h}^{\nu}_{k\alpha} \lambda_{k\alpha} - \tfrac{1}{2} \textstyle\sum_{\alpha,\beta=1}^{m_\nu,m_\nu} \overline{p}^{\nu}_{k\alpha\beta} \lambda_{k\alpha} \lambda_{k\beta}\Big] \\ \text{subject to } & y_i \geq 0, w_{1j} \geq 0, \lambda_{k\alpha} \geq 0, \\ & \sum_{\alpha=1}^{m_\nu} \lambda_{k\alpha} \leq 1 \text{ for } k = 1, \ldots, \ell, \end{aligned}$$

$$\sum_{i=1}^{m} y_i a_{ij} + \sum_{k=1}^{\ell} \sum_{\alpha=1}^{m_\nu} \lambda_{k\alpha} \overline{t}_{k\alpha j} + w_{1j} + w_{2j} \geq c_j \text{ for } j = 1, \ldots, n. \tag{3.12}$$

We then have the following implementation.

SUBALGORITHM (for Step 2). *Given the sets $\hat{Z}_k^\nu$ in the notation (3.1), calculate the coefficients (3.5), (3.6), (3.7), for the quadratic programming problem* (D^ν). *Solve* (D^ν) *by any method, getting from the optimal solution values $\overline{y}_i^\nu, \overline{w}_{1j}^\nu, \overline{w}_{2j}^\nu$ and $\overline{\lambda}_{k\alpha}^\nu$ the elements*

$$\underset{\sim}{z}_k = \sum_{\alpha=1}^{m_\nu} \overline{\lambda}_{k\alpha}^\nu \hat{\underset{\sim}{z}}_{k\alpha}^\nu .$$

The minimum value in (D^ν) *is the desired α_ν, and the Lagrange multiplier vector obtained for the constraints (3.12) in* (D^ν) *is the desired approximate solution $\overline{x}^\nu$ to* ($\mathrm{P}_{\mathrm{sto}}$).

Thus it is not actually necessary in Step 2 to invoke formula (1.5) to get $\overline{x}^\nu$. Instead, $\overline{x}^\nu$ can be obtained as a byproduct of the solution procedure used for the minimization.

4. APPLICATION TO PROBLEMS THAT ARE NOT STRICTLY QUADRATIC.

If in the given problem ($\mathrm{P}_{\mathrm{sto}}$) it is not true that $r_j > 0$ and $\underset{\sim}{p}_k > 0$ for all j and k, we use the proximal point technique [4] (as adapted to the Lagrangian $L_0(x, z)$ in (1.16)) to replace ($\mathrm{P}_{\mathrm{sto}}$) by a sequence of problems ($\mathrm{P}_{\mathrm{sto}}^\mu$), $\mu = 1, 2, \ldots$, that do have the desired character. To each problem ($\mathrm{P}_{\mathrm{sto}}^\mu$) we apply the finite generation algorithm as above, but with a certain stopping criterion in Step 3 that ensures finite termination. This is done in such a way that the overall doubly iterative procedure still converges at a linear rate.

To obtain the problems ($\mathrm{P}_{\mathrm{sto}}^\mu$), we introduce alongside the given values r_j and $\underset{\sim}{p}_k$ some other values $\overline{r}_j > 0, \overline{\underset{\sim}{p}}_k > 0$ and set

$$r_{*j} = r_j + \eta \overline{r}_j, \qquad \underset{\sim}{p}_{*k} = \underset{\sim}{p}_k + \eta \overline{\underset{\sim}{p}}_k, \tag{4.1}$$

where $\eta > 0$ is a parameter value that wil play a role in theory but can be held fixed for the purpose of computation. We also introduce elements

$$\overline{x}_*^\mu = (\overline{x}_{*1}^\mu, \ldots, \overline{x}_{*n}^\mu) \text{ and } \overline{\underset{\sim}{z}}_*^\mu = (\overline{\underset{\sim}{z}}_{*1}^\mu, \ldots, \overline{\underset{\sim}{z}}_{*\ell}^\mu),$$

which are to be thought of as estimates for the optimal solution values in ($\mathrm{P}_{\mathrm{sto}}$) and ($\mathrm{D}_{\mathrm{sto}}$). In terms of these we set

$$c_{*j}^\mu = c_j - \eta \overline{r}_j \overline{x}_{*j}^\mu \text{ and } \underset{\sim}{h}_{*k}^\mu = \underset{\sim}{h}_k - \eta \underset{\sim}{p}_k \overline{\underset{\sim}{z}}_{*k}^\mu . \tag{4.2}$$

Then

$$\begin{array}{l} (\mathrm{P}_{\mathrm{sto}}^\mu), (\mathrm{D}_{\mathrm{sto}}^\mu) \text{ are the problems obtained by replacing} \\ r_j, \underset{\sim}{p}_{*k}, c_j \text{ and } \underset{\sim}{h}_k \text{ in } (\mathrm{P}_{\mathrm{sto}}), (\mathrm{D}_{\mathrm{sto}}) \text{ by } r_{*j}, \underset{\sim}{p}_{*k}, c_{*j}^\mu \text{ and } \underset{\sim}{h}_{*k}^\mu . \end{array} \tag{4.3}$$

These modified problems are, of course, strictly quadratic: one has $r_{*j} > 0$ and $\underset{\sim}{p}_{*k} > 0$.

MASTER ALGORITHM.

Step 0 (Initialization). *Choose* $\bar{x}_*^1 \in X$ *and* $\bar{\underset{\sim}{z}}_*^1 \in Z$. *Set* $\mu = 1$.

Step 1 (Finite Generation Algorithm). *Apply the finite generation algorithm in the manner already described to the strictly quadratic problems* $(\mathrm{P}_{\mathrm{sto}}^{\mu})$ *and* $(\mathrm{D}_{\mathrm{sto}}^{\mu})$ *in (4.3). Terminate in Step 3 when the stopping criterion given below is satisfied.*

Step 2 (Update). *For the elements* $\bar{x}^{\nu}$ *and* $\bar{\underset{\sim}{z}}^{\nu}$ *with which Step 1 terminated, set* $\bar{x}_*^{\mu+1} = \bar{x}^{\nu}$ *and* $\bar{\underset{\sim}{z}}_*^{\mu+1} = \bar{\underset{\sim}{z}}^{\nu}$. *Replace* μ *by* $\mu + 1$ *and return to Step 1.*

The stopping criterion is as follows. In terms of the norm

$$\|(x, \underset{\sim}{z})\| = \left[\|x\|_r^2 + E\{\|\underset{\sim}{z}\|_{\underset{\sim}{p}}^2\}\right]^{1/2} \tag{4.4}$$

and a sequence of values θ_μ with

$$\theta_\mu > 0, \quad \sum_{\mu=1}^{\infty} \theta_\mu < \infty, \tag{4.5}$$

we define the function

$$\varepsilon_{*\mu}(x, \underset{\sim}{z}) = \theta_\mu^2 \min\{1, (\eta/2)\|(x, \underset{\sim}{z}) - (\bar{x}_*^{\mu}, \bar{\underset{\sim}{z}}_*^{\mu})\|^2\}. \tag{4.6}$$

We stop in Step 3 of the finite generation algorithm when the computed elements $\varepsilon_\nu, \bar{x}^{\nu}$ and $\bar{\underset{\sim}{z}}^{\nu}$ satisfy

$$\varepsilon_\nu \leq \varepsilon_{*\mu}(\bar{x}^{\nu}, \bar{\underset{\sim}{z}}^{\nu}). \tag{4.7}$$

This stopping criterion will eventually be satisfied, when ν is high enough; the only exception is the case where $\bar{x}_*^{\mu}$ happens already to be an optimal solution $\bar{x}$ to $(\mathrm{P}_{\mathrm{sto}})$ and $\bar{\underset{\sim}{z}}_*^{\mu}$ the $\bar{\underset{\sim}{z}}$-component of an optimal solution $(\bar{y}, \bar{\underset{\sim}{z}})$ to $(\mathrm{D}_{\mathrm{sto}})$. (See [1, §6] for details.)

THEOREM 3. *If the master algorithm is executed with the specified stopping criterion (4.7), then the sequences* $\{\bar{x}_*^{\mu}\}_{\mu=1}^{\infty}$ *and* $\{\bar{\underset{\sim}{z}}_*^{\mu}\}_{\mu=1}^{\infty}$ *converge to particular elements* $\bar{x}$ *and* $\bar{\underset{\sim}{z}}$, *where* $\bar{x}$ *is an optimal solution to* $(\mathrm{P}_{\mathrm{sto}})$ *and, for some* $\bar{y}$, *the pair* $(\bar{y}, \bar{\underset{\sim}{z}})$ *is an optimal solution to* $(\mathrm{D}_{\mathrm{sto}})$. *Moveover there is a number* $\beta(\eta) \in [0,1)$ *such that* $(\bar{x}_*^{\mu}, \bar{\underset{\sim}{z}}_*^{\mu})$ *converges to* $(\bar{x}, \bar{\underset{\sim}{z}})$ *at a linear rate with modulus* $\beta(\eta)$.

PROOF. This is an immediate specialization of Theorem 6 of [1] to the case at hand, the path of specialization having been established already in the proof of Theorems 1 and 2. □

The theory of proximal point technique in [4], as applied in the derivation of Theorem 3, shows actually that linear convergence is obtained at the rate

$$\beta(\eta) = \gamma\eta/[1 + (\gamma\eta)^2]^{1/2} \in [0, 1) \tag{4.8}$$

where $\gamma \geq 0$ is a number depending only on the data in the original problems (P_{sto}) and (D_{sto}), not on $\eta, \bar{r}_j$ or $\bar{\underset{\sim}{p}}_k$. In particular $\beta(\eta) \to 0$ as $\eta \to 0$. Thus an arbitrarily good rate of convergence can be obtained (in principle) for the outer algorithm (master algorithm) simply by choosing the parameter value η *small* enough.

At the same time, however, the choice of η affects the convergence rate in the inner algorithm (finite generation algorithm). That rate corresponds by (2.12) to a number $\tau(\eta)^{1/2} \in [0,1)$ defined by (2.9) in terms of an upper bound $\sigma(\eta)$ for $\| \underset{\sim}{T} \|_{\underset{\sim}{p}_*, r_*}$, where $\underset{\sim}{p}_*$ and r_* are vectors consisting of the parameters in (4.1). Thus $\sigma(\eta)^2$ is an upper bound for the expression

$$\frac{[z \cdot \underset{\sim}{T} x]^2}{[\|z\|^2_{\underset{\sim}{p}} + \eta \|z\|^2_{\bar{\underset{\sim}{p}}}]\,[\|x\|^2_r + \eta \|x\|^2_{\bar{r}}]}$$

over all possible choices of the vectors $x \in R^n$ and $z \in R^\ell$ and all possible values taken on by the random variables $\underset{\sim}{T}, \underset{\sim}{p}$ and $\bar{\underset{\sim}{p}}$. It follows that $\tau(\eta) \to 0$ as $\eta \to \infty$ but $\tau(\eta) \to 1$ as $\eta \to 0$. Thus an arbitrarily good rate of convergence can be obtained (in principle) for the inner algorithm by choosing η *large* enough, but too small a choice could do damage.

This trade-off between the outer and inner algorithms in the choice of η could be a source of difficulty in practice, although we have not had much trouble with the problems tried so far. (See King [3].)

References

1. R.T. Rockafellar and R.J.-B. Wets, "A Lagrangian finite generation technique for solving linear-quadratic problems in stochastic programming," Math. Programming Studies (1985), to appear.
2. R.T. Rockafellar and R.J.-B. Wets, "A dual solution procedure for quadratic stochastic programs with simple recourse," in: V. Pereyra and A. Reinoza, eds., *Numerical Methods* (Lecture Notes in Mathematics 1005), Springer Verlag, Berlin, 1983, 252-265.
3. A. King, "An implementation of the Lagrangian finite generation method," in: Y. Ermoliev and R.J.-B. Wets, eds., *Numerical Techniques for Stochastic Programming Problems*, Springer Verlag, to appear.
4. R.T. Rockafellar, "Monotone operators and the proximal point algorithm," SIAM Journal on Control and Optimization **14**(1976), 877-898.
5. R.T. Rockafellar, *Network Flows and Monotropic Optimization*, Wiley-Interscience, New York, 1984.

CONVERGENCE OF STOCHASTIC INFIMA: EQUI-SEMICONTINUITY

G. Salinetti
Università di Roma "La Sapienza"
Roma, Italy

1. Introduction and Problem Setting.

Many stochastic optimization problems focuse the attention on the map

$$\text{Inf}_{x\in X} f(x,\omega) \tag{1.1}$$

defined on a given probability space (Ω,A,μ) where:

- for every $\omega\in\Omega$, the map $x\mapsto f(x,\omega)$ is a lowersemicontinuous function on a given topological space (X,τ) valued in the extended reals $\bar{R}$, i.e. the epigraph of f, epi f = $\{(x,\alpha)\in X\times R:\ f(x,\omega)\leq\alpha\}$, is a closed subset of the product space $X\times R$,

- the closed valued multifunction $\omega\mapsto \text{epi}\, f(.,\omega)$ of Ω into $X\times R$ is measurable, i.e. for every closed subset F of $X\times R$, the pre-image

$$\{\omega\in\Omega:\ \text{epi}\, f(.,\omega)\cap F \neq \emptyset\} \tag{1.2}$$

belongs to the σ-algebra A.

The function $(x,\omega)\mapsto f(x,\omega)$ with the above properties will be always referred as *normal integrand*, the closed valued multifunction $\omega\mapsto \text{epi} f(.,\omega)$ as *epigraphical multifunction* associated to it . The map (1.1) on the probability space (Ω,A,μ) is the stochastic infimum of the normal integrand f or simply *stochastic infimum*.

The attention devoted to stochastic infima in their different aspects and, in particular to their probability distribution, finds its main motivations in stochastic optimization. In itself the probability distribution of a stochastic infimum is the objective in the so called "distribution problem" in stochastic programming.

But, more generally, the problem of the probability distribution of a stochastic infimum is still crucial in more sophisticated optimization problems. Many of them, for instance stochastic programs with recourse, certain classes of Markov decision problems, stochastic control problems in discrete time and others can be cast in the following abstract form:

$$\text{Find } x_1 \in X_1 \text{ that minimizes } \int u(Q(x_1,\omega))\mu(d\omega)$$

where u is a scaling, e.g. an utility function, and

$$Q(x_1,\omega) = \mathrm{Inf}_{x_2 \in X_2} g(x_1,x_2,\omega)$$

with $g: (X_1 \times X_2) \times \Omega \to \overline{R}$ is , as usual in the applications, a normal integrand.

If u(.) is linear, we may restrict our attention to comparing expectations of the random variables $\{Q(x_1,.),\ x_1 \in X_1\}$, but more generally is the whole distributions of the stochastic infima $Q(x_1,.)$, $x_1 \in X_1$ which are of interest; intimately connected with that it is also the probability distribution of the normal integrand $Q(.,.)$ and often of the associated stochastic infimum.

Unless we are in very particular cases, the only possible approach to solve this type of problems is via approximations.

This and related questions lead to study the following convergence question:

Given a sequence of normal integrands $\{f_n : X \times \Omega \to \overline{R},\ n=1,\ldots\}$ (1.3)
and an associated limit integrand f, *find minimal conditions that guarantee the convergence in distribution of the random variables* $Z_n(.) = \mathrm{Inf}_{x \in X} f_n(x,.)$ *to* $Z(.) = \mathrm{Inf}_{x \in X} f(x,.)$.

Answers to these convergence questions pass through the notion of convergence in distribution of integrands. The "traditional" approach which looks at the stochastic infimum (1.1) as functional of the stochastic process $\{f(x,.),\ x \in X\}$, in some sense, is not sufficient to deal with the type of convergence questions posed here. This is mainly due to the fact that the convergence of the infima is intrinsically tied to the convergence of the epigraphs (and then here to the whole trajectories of the processes) with their topological properties; actually the same notion of epi-convergence had as one of its main motivations just the need to substantiate the convergence of infima in approximation schemes for solving optimization problems.

It is thus inescapable to approach convergence in distribution of integrands basically relying on the topology of the epigraphs as done in [3].

On the other hand, more recently, [2], [7], the same notion of epi-convergence, or its symmetric version of hypo-convergence, revealed to be the appropriate notion to study convergence of probability measures.

Convergence of infima and convergence of probability measures, as key ingredients of convergence of stochastic infima, find then naturally in the epi-convergence an appropriate and fruitful tool to determine the minimal setting for convergence of stochastic infima.

From a more general point of view, the convergence theory for normal integrands can be regarded itself as an extension of the classical convergence of stochastic processes which, together with the alternative approach to weak convergence of probability measures, gives an extended setting to deal with convergence of functionals of stochastic processes.

2. Epi-convergence of semicontinuous functions and hypo-convergence of probability measures.

The basic idea here is to look at convergence of functions through the convergence of sets which inherit the topological properties of the functions: thus convergence of lowersemicontinuous functions as convergence of the corresponding (closed) epigraphs and convergence of uppersemicontinuous functions as convergence of the corresponding (closed) hypographs.

For that we basically rely on [3]. Here we just sketch the key elements of the topological construction, give the basic assumptions and introduce the notations.

In a topological space Y let $\mathcal{F}(Y)$ denote the space of all closed subsets of Y including the empty set. For the following it is sufficient to assume that Y is locally compact, Hausdorff and second countable. Let $\mathcal{T}$ be the topology of $\mathcal{F}(Y)$ generated by the subbase of open sets:

$$\{\mathcal{F}^K, K \in \mathcal{K}(Y)\} \quad \text{and} \quad \{\mathcal{F}_G, G \in \mathcal{G}(Y)\} \tag{2.1}$$

where $\mathcal{K}(Y)$ and $\mathcal{G}(Y)$ are the hyperspaces of compact and open subsets of Y respectively, and for any subset Q of Y,

$$\mathcal{F}^Q = \{F \in \mathcal{F}(Y): F\cap Q \neq \emptyset\} \quad \text{and} \quad \mathcal{F}_Q = \{F \in \mathcal{F}(Y): F\cap Q \neq \emptyset\}.$$

The topology $\mathcal{T}$ of $\mathcal{F}(Y)$ essentially inherits the properties of the topology of Y, in particular here $(\mathcal{F}(Y), \mathcal{T})$ is regular and compact; for Y separable $\mathcal{T}$ has a countable base [3, Proposition 3.2], [4].

Let now LSC(X) be the space of the lowersemicontinuous (l.sc.) functions on the topological space X with values in the extended reals and let E denote the space of the epigraphs. These are closed subsets of $Y=X\times R$.

The topology $\mathcal{T}$ on $\mathcal{F}(X\times R)$, restricted to E, gives a topology - called *epi-topology* and denoted *epi* - on E and correspondently on LSC(X). The topological space (LSC(X),epi) or (E,epi) is regular and compact if X is separated and locally compact [3,Corollary 4.3]. Moreover, if X is separable the topology epi has a countable base.

For a family $\{f; f_n, n=1,\ldots\}$ in LSC(X), *epi-convergence* of $\{f_n\}$ to f, denoted $f_n \overset{epi}{\to} f$, means that epi $f_n \to$ epi f in the topology $\mathcal{T}$ of E. A local characterization of epi-convergence can also be given. More rigorously $f_n \overset{epi}{\to} f$ if and only if the following conditions are satisfied:

for any $x \in X$, any subsequence $\{f_{n_k}, k=1,\ldots\}$ and any sequence $\{x_k, k=1,\ldots\}$ converging to x in X we have

$$f(x) \leq \liminf_k f_{n_k}(x_k); \tag{2.2}$$

for every $x \in X$ there exists a sequence $\{x_n, n=1,\ldots\}$ converging to x such that

$$\limsup_n f_n(x_n) \leq f(x). \tag{2.3}$$

This characterization immediately points out that epi-convergence is related to the pointwise convergence, here denoted $f_n \overset{p}{\to} f$, but it is neither implied nor does it imply pointwise convergence. However epi-convergence and pointwise convergence are equivalent only on equi-lowersemicontinuous subsets of LSC(X) [3,Theorems 2.18 and 4.6].

Recall that [3, Definition 2.17] a subset Q of LSC(X) is *equi-lowersemicontinuous* (equi-l.sc.) *at* $x \in X$ if for any $\varepsilon>0$ sufficiently small there corresponds a neighbourhood V of x such that for all $f \in Q$ we have

$$\operatorname{Inf}_{y\in V} f(y) \geq \min(\varepsilon^{-1}, f(x)-\varepsilon). \tag{2.4}$$

Q is said to be *equi-l.sc.* if (2.4) holds for all $x \in X$.

The local characterization of epi-convergence (2.2) and (2.3) together with (2.4) easily gives the following particular case of [3,Theorem 2.18], of direct use in the following:

2.5 THEOREM - *Let* $f_n \overset{epi}{\to} f$. *Then for any* $x \in X$ *such that* $f(x)>-\infty$, $f_n(x) \to f(x)$ *if and only if* $\{f;f_n,n=1,\ldots\}$ *is equi-l.sc. at* x.

The notion of epi-convergence and the engendered epi-topology on LSC(X) have their counterpart in the mirror setting of *uppersemicontinuous* (u.sc.) functions, i.e. functions with closed hypographs, where for $f:X \to \overline{R}$, hypo f = $\{(x,\alpha)\in X\times R:\ f(x)\geq\alpha\}$.

Hypo-convergence of uppersemicontinuous functions then means convergence of their hypographs and all the results , with the necessary adaptations, remain true: the notion of equi-l.sc. becames *equi-u.sc.* and it provides the minimal setting for equivalence between hypo-convergence and pointwise convergence for uppersemicontinuous functions. We do not restate these results here and, when necessary, we refer to results on epi-convergence just specifying that they have to be interpreted in the "hypo-version".

An interesting case of uppersemicontinuous functions to which all the above applies is given by probability measures on topological spaces when restricted to closed subsets: then hypo-convergence and the standard notion of weak convergence of probability measures are equivalent; equi-u.sc. and tightness are equivalent notions.

This approach to weak convergence of probability measures is presented in [7] for locally compact, Hausdorff and second countable spaces and to that we basically refer to obtain results for convergence of stochastic infima. A more general treatment dealing with the case when the space is metric and separable is forthcoming.

In the probability space $(Y, \mathcal{B}(Y), P)$ where Y is a locally compact, Hausdorff and second countable space, $\mathcal{B}(Y)$ the Borel field of Y and P a probability measure, we look at $\mathcal{B}(Y)$ as generated by $\mathcal{F}(Y)$, the closed subsets of Y. When $\mathcal{F}(Y)$ is equipped with the $\mathcal{T}$ topology, as described at the beginning of this section, the restriction of the probability measure P to $\mathcal{F}(Y)$ is uppersemicontinuous on the topological space $(\mathcal{F}(Y), \mathcal{T})$ [7, Proposition 3.1].

For a family $\{P; P_n, n=1,\dots\}$ of probability measures on $\mathcal{B}(Y)$, convergence can then be approached in terms of hypo-convergence of their restrictions, say $\{D; D_n, n=1,\dots\}$ on the topological space $(\mathcal{F}(Y), \mathcal{T})$.

Hypo-convergence of the restrictions, denoted $D_n \overset{hypo}{\to} D$ is shown to be equivalent to the weak convergence of the probability measures $\{P_n\}$ to P, denoted $P_n \overset{w}{\to} P$ [7, Theorem 4.1]. Results on epi-convergence in their "hypo-version" all apply. The equi-u.sc. of the sequence of the restrictions $\{D_n\}$ at $F \in \mathcal{F}(Y)$, since all the D_n's are finite, simply means that for any $\varepsilon>0$ there exists a neighbourhood V(F) in the topology $\mathcal{T}$ of $\mathcal{F}(Y)$ such that for all n

$$D_n(F') < D_n(F) + \varepsilon, \qquad \text{for all } F' \in V(F) \tag{2.6}$$

or equivalently

$$P_n(F') < P(F) + \varepsilon, \qquad \text{for all } F' \in V(F). \tag{2.7}$$

Abusing in the language, condition (2.7) will be referred as *equi-u.sc. of the probability measures* $\{P_n\}$ *at* $F \in \mathcal{F}(Y)$.

Theorem 2.5, restated in its "hypo-version", by equivalence of the hypo-convergence $D_n \overset{hypo}{\to} D$ of the restrictions with the weak convergence of the probability measures $P_n \overset{w}{\to} P$ and (2.7), becames:

2.8 THEOREM - *If* $P_n \overset{w}{\to} P$ *then for any* $F \in \mathcal{F}(Y)$ *we have* $P_n(F) \to P(F)$ *if and only if* $\{P; P_n, n=1,\ldots\}$ *is equi-u.sc. at* F.

3. Convergence in distribution of normal integrands.

On the probability space (Ω,A,μ), a normal integrand $f:X\times\Omega \to \overline{R}$, as defined in section 1, can be regarded as map on Ω taking values in LCS(X), the space of the lowersemicontinuous functions on X; its measurability (1.2) can be expressed in terms of the measurability of the map $\psi:\Omega \to LSC(X)$ with respect to σ-algebras A of Ω and the Borel field of LSC(X) generated by the open -or closed- subsets of the epi-topology of LSC(X).

Equivalently, and we follow this line here to avoid complicated notations, we can look at the normal integrand f,directly, through its epigraphical multifunction

$$\omega \to \text{epi}\, f(.,\omega) \tag{3.1}$$

regarded as map of Ω into the space $F(X\times R)$ of all closed subsets of $X\times R$. To avoid complicated notations in the following we put $F_o=F(X\times R)$ and denote T_o the topology T, as described in section 2, for F_o.

It is not difficult to show that the measurability (1.2) is equivalent to the measurability of the map $\text{epi}\, f: \Omega \to F_o$ with respect to the σ-algebras A of Ω and the Borel field $B(F_o)$ generated by the closed subsets of the topological space (F_o,T_o) [6].

Even if many results can be extended to a more general setting, we assume that X is a finite dimensional euclidean space: the setting is the same as in [6] where this approach to convergence of normal integrands has been introduced and to which we will refer for the following.

Thus the map $\text{epi}\, f(.,.):\Omega \to F_o$ induces a probability measure P on $B(F_o)$ and we refer to it as the *probability measure of the normal integrand* f.

Consider now a family $\{f; f_n, n=1,\ldots\}$ of normal integrands with induced probability measures $\{P; P_n, n=1,\ldots\}$. Recall from section 2 that (F_o,T_o)

is regular, compact and admits a countable base; then it is metrizable.

Thus the weak convergence of the probability measures $\{P;\ P_n,\ n=1,\dots\}$ induced by the normal integrands can be approached in the standard context of "weak convergence on metric spaces"; according with the known definition of weak convergence (see for example [1]), the sequence of probability measures $\{P_n\}$ is said to *weakly converge* to P if

$$P_n(B) \to P(B) \quad \text{for all } B \in \mathcal{B}(\mathcal{F}_o) \text{ with } P(\mathrm{bdy}B)=0 \tag{3.2}$$

where bdyB denotes the boundary of B in the topological space $(\mathcal{F}_o, \mathcal{T}_o)$.

3.3 REMARK - In view of the applications and aiming at a direct use of the probability measures of normal integrands, especially for what is concerning weak convergence, it is relevant to observe that P on $\mathcal{B}(\mathcal{F}_o)$ is uniquely determined by a much simpler function called *distribution function*, denoted T and defined on $\mathcal{K}(X\times R)$, the compact subsets of X×R, by

$$T(K) = \mu(\{\omega\in\Omega:\ \mathrm{epi}\, f(.,\omega)\cap K \neq \emptyset\})$$

[6]. The name "distribution function" finds its justification in the properties of T which can be regarded as extensions of the properties of the distribution functions of random vectors [6].

For a family of normal integrands $\{f;\ f_n,\ n=1,\dots\}$ with probability measures $\{P;\ P_n,\ n=1,\dots\}$ and distribution functions $\{T;\ T_n,\ n=1,\dots\}$ the weak convergence (3.2) can be shown to be equivalent to the convergence of the sequence $\{T_n\}$ to T "pointwise on the continuity set" of T [1,Sections 1 and 3]. Thus the weak convergence $P_n \overset{w}{\to} P$ will be also equivalently referred as *convergence in distribution* of $\{f_n\}$ to f.

Consider now the family of probability measures $\{P;\ P_n,\ n=1,\dots\}$ restricted to the space $F(\mathcal{F}_o)$, the closed subsets of $(\mathcal{F}_o,\mathcal{T}_o)$. The space $F(\mathcal{F}_o)$ equiped with the topology $\mathcal{T}$ is again regular, compact and second countable. In view of the results of section 2 the restrictions of $\{P;P_n,n=1,\dots\}$ to $F(\mathcal{F}_o)$ are uppersemicontinuous on $(F(\mathcal{F}_o),\mathcal{T})$ and their hypo-convergence is equivalent to the weak convergence $P_n \overset{w}{\to} P$: pointwise convergence on subsets of $F(\mathcal{F}_o)$, when needed, is guaranteed by equi-uppersemicontinuity (Theorem 2.8).

This is a crucial point in convergence of stochastic infima; in view of that next theorem gives a characterization of equi-u.sc. on particular elements of $\mathcal{F}(F_o)$.

According with section 2, for G and K respectively open and compact subsets of X×R, the sets F_G and F^K are open subsets of (F_o, T_o); thus their complements in F_o, respectively F^G and F_K are closed, i.e.

$$F^G \in \mathcal{F}(F_o) \text{ and } F_K \in \mathcal{F}(F_o). \tag{3.4}$$

By the standard characterizazation of weak convergence [1], we have:

3.5 PROPOSITION - *Let* $P_n \overset{W}{\to} P$. *Then:*

$$\limsup P_n(F^G) \leq P(F^G) \text{ for any G open in X×R} \tag{3.6}$$

$$\limsup P_n(F_K) \leq P(F_K) \text{ for any K compact in X×R.} \tag{3.7}$$

3.8 THEOREM - *Let* $P_n \overset{W}{\to} P$ *and let* G *be an open subset of* X×R *such that* $P(F^{clG}) = P(F^G)$. *Then* $\{P_n\}$ *is equi-u.sc. at* F^G *if and only if for any* $\varepsilon>0$ *there exists* $C \subset G$, C *open and relatively compact such that for all* n *we have:*

$$P_n(F^C) < P_n(F^G) + \varepsilon. \tag{3.9}$$

PROOF- a) the condition is necessary. Since G is open, for the topological properties of X×R there always exists an increasing sequence $\{G_i, i=1,\dots\}$ of open relatively compact sets such that $G = \cup G_i$, so that, as immediate to see, we have $F^{\cup G_i} = \cap F^{G_i} = F^G$. The sets F^{G_i}, as closed subsets of F_o decrease to $\cap F^{G_i} = F^G$, i.e. as elements of $\mathcal{F}(F_o)$ they converge to F^G.

If $\{P_n\}$ is equi-u.sc. at F^G, for any $\varepsilon>0$ there exists i' such that, for all n, we have

$$P_n(F^{G_{i'}}) < P_n(F^G) + \varepsilon.$$

Since $G_{i'}$ is open relatively compact and $G_{i'} \subset G$, (3.9) is satisfied and the necessary part is proved.

b) the condition is sufficient. Suppose now that (3.9) holds. By complementation we have for all n

$$P_n(F_G) \leq P_n(F_C) + \varepsilon$$

and taking lim sup in both sides

$$\limsup P_n(F_G) \leq \limsup P_n(F_C)+\varepsilon \leq \limsup P_n(F_{clC})+\varepsilon \leq$$

$$\leq P(F_{clC}) +\varepsilon \leq P(F_{clG}) +\varepsilon = P(F_G) + \varepsilon \tag{3.10}$$

where the third inequality follows from (3.7) and the last equality from the assumption $P(F_{clG}) = P(F_G)$.

The repeated use of the argument for every $\varepsilon>0$ shows that $P(F^G) \leq \liminf P_n(F^G)$ and by by (3.6) of Proposition 3.5 we actually have

$$P(F^G) = \lim P_n(F^G).$$

This means pointwise convergence at F^G; equi-u.sc. of $\{P_n\}$ at F^G follows then from Theorem 2.8 and the theorem is proved.

4. Convergence of stochastic infima.

We consider now stochastic infima of normal integrands and their convergence in distribution.

For a normal integrand $f:X\times\Omega \to \overline{R}$, the stochastic infimum

$$\omega \to Z(\omega) = \mathrm{Inf}_{x\in X} f(x,\omega) \tag{4.1}$$

is a (possibly extended) random variable on (Ω,A,μ). For, just observe that for any real z, setting $H(z)=\{(x,\alpha)\in X\times R: \alpha \leq z\}$ the subset of Ω defined by

$$\{\omega\in\Omega: Z(\omega)<z\} = \{\omega\in\Omega: \mathrm{epi}\, f(.,\omega)\cap H(z)\neq\emptyset\} = \{\omega\in\Omega : \mathrm{epi}\, f(.,\omega)\in F_{H(z)}\} \tag{4.2}$$

is measurable, i.e. belongs to A, because $\omega\to \mathrm{epi}\, f(.,\omega)$ is a closed valued measurable multifunction.

Let P be the probability measure induced by the integrand f. By (4.2) the distribution function Φ of $Z(.)$ can be directly expressed in terms of P as follows:

$$\Phi(z) = \mu(\{\omega\in\Omega: Z(\omega)<z\}) = P(F_{H(z)}) \tag{4.3}$$

Relation (4.3) clarifies, as natural to expect, that P completely determines Φ. It clarifies also that, when aiming at convergence of infima, epi-convergence in distribution of the normal integrands as introduced in section 3 is, in some sense, an inescapable condition.

For the family $\{f;f_n,n=1,\ldots\}$ of normal integrands with probability measures $\{P,P_n,n=1,\ldots\}$, let $\{Z;Z_n,n=1,\ldots\}$ be the corresponding family of stochastic infima and $\{\Phi;\Phi_n,n=1,\ldots\}$ their distribution functions.

The basic convergence question (1.3) can now be reformulated as follows:

$$\text{Given that } P_n \overset{w}{\to} P, \text{ find the minimal set of conditions for the convergence in distribution } Z_n \overset{d}{\to} Z, \text{ i.e. for } \Phi_n(z) \to \Phi(z) \quad \text{for every } z \in \operatorname{cont}\Phi \tag{4.4}$$

where contΦ denotes the set of points where Φ is continuous.

By definition of Φ and Φ_n as in (4.3), since $H(z)$ is open in $X\times R$, Proposition 3.5 immediately implies:

4.5 PROPOSITION - *If* $P_n \overset{w}{\to} P$ *then for every real* z *we have*

$$\Phi(z) \leq \liminf \Phi_n(z). \tag{4.5}$$

Still as consequence of weak convergence $P_n \overset{w}{\to} P$, whenever $F_{H(z)}$ is a P-continuity set, i.e. $P(\text{bdy}_{H(z)})=0$ we also have

$$\lim \Phi_n(z) = \lim P_n(F_{H(z)}) = P(F_{H(z)}) = \Phi(z).$$

Unfortunately even when $z\in\operatorname{cont}\Phi$ we do not have $P(\text{bdy}F_{H(z)})=0$; the converse is true. In this case however, that is when $z\in\operatorname{cont}\Phi$, we have

$$P(F_{\text{cl}H(z)}) = P(F_{H(z)}) \tag{4.6}$$

as it can be immediately derived from (4.3).

To solve the convergence question (4.4) in its full generality it is necessary to go back to the hypo-convergence of the probability measures and to its relations with pointwise convergence, basically to Theorem 2.8.

4.7 THEOREM - *Suppose that* $P_n \overset{w}{\to} P$. *Then* $Z_n \overset{d}{\to} Z$ *if and only if anyone of the following conditions is satisfied:*

(i) for every $z\in\operatorname{cont}\Phi$, $\{P_n\}$ *is equi-u.sc. at* $F^{H(z)}$;

(ii) for every $z\in\mathrm{cont}\,\Phi$ *and any* $\varepsilon>0$ *there exists* $C\subset H(z)$, C *open and relatively compact such that for all n we have*

$$P_n(F^C) < P_n(F^{H(z)}) + \varepsilon \tag{4.8}$$

PROOF - Condition (i) simply restates Theorem 2.8 with $B = F^{H(z)}$. As (ii) is concerned, for $z\in\mathrm{cont}\,\Phi$, relation (4.6) holds; then equi-u.sc. at $F^{H(z)}$ can be characterized as in Theorem 3.8 and (ii) is proved.

Just using complementation arguments, necessary and sufficient conditions more amenable to a direct use can be derived.

4.9 THEOREM - *If* $P_n \overset{w}{\to} P$ *then* $Z_n \overset{d}{\to} Z$ *if and only if for every* $z\in\mathrm{cont}\,\Phi$, *any* $\varepsilon>0$ *and any* $\delta>0$ *there exists* $K\subset \mathrm{cl}H(z)$, K *compact, such that for all* n *we have:*

$$P_n(F_{\mathrm{cl}H(z-\delta)}) < P_n(F_K) + \varepsilon \tag{4.10}$$

PROOF - Suppose $Z_n \overset{w}{\to} Z$. Then by theorem 4.7(ii) taking the complements in (4.8), observing that $\mathrm{cl}H(z-\delta)\subset H(z)$ and clC is compact, we have for all n

$$P_n(F_{\mathrm{cl}H(z-\delta)}) \le P_n(F_{H(z)}) < P_n(F_C)+\varepsilon < P_n(F_{\mathrm{cl}C})+\varepsilon.$$

This shows (4.10)with clC=K.

Suppose now that (4.10) holds. By Proposition 4.5 it is sufficient to show

$$\limsup \Phi_n(z) \le \Phi(z). \tag{4.11}$$

Let $z\in\mathrm{cont}\,\Phi$ and $\varepsilon>0$; let $\delta>0$ be such that $z+\delta\in\mathrm{cont}\,\Phi$ and $\Phi(z+\delta)<\Phi(z)+\varepsilon/2$. Then by (4.10) there exists $K\subset\mathrm{cl}H(z+\delta)$, K compact, such that for all n we have

$$P_n(F_{\mathrm{cl}H(z)}) < P_n(F_K) + \varepsilon/2;$$

taking the lim sup in both sides we have

$$\begin{aligned}\limsup P_n(F_{\mathrm{cl}H(z)}) &\le \limsup P_n(F_K)+\frac{\varepsilon}{2} \le P(F_K)+\frac{\varepsilon}{2} \\ &\le P(F_{\mathrm{cl}H(z+\delta)})+\frac{\varepsilon}{2} \le P(F_{H(z+\delta)})+\frac{\varepsilon}{2} \le P(F_{H(z)})+\varepsilon\end{aligned} \tag{4.12}$$

where the second inequality is due to Proposition 3.5 and the equality to (4.6). Since $P_n(F_{H(z)})\le P_n(F_{\mathrm{cl}H(z)})$, (4.12) implies (4.11) and that completes the proof.

Condition (4.10) rewritten in the more explicit form

$$\mu(\{\omega: \mathrm{epi}\, f_n(.,\omega)\cap\mathrm{cl}H(z-\delta)\neq\emptyset\}) \le \mu(\{\omega:\mathrm{epi}\, f_n(.,\omega)\cap K\neq\emptyset\})+\varepsilon \tag{4.13}$$

immediately reveals its connection with the existence of tight sequences of distribution optimal solutions in the sense that will be now specified.

4.14 REMARK - Observe that in Theorem 4.9 the compact set K can be replaced by the closed set $K_z=\{(x,\alpha): x\in C, \alpha\leq z\}$ where C is a compact subset of X and the proof of the theorem remains valid. Then the condition (4.10) or equivalently (4.13) can be replaced by

$$\mu\{\omega: \text{epi}\, f_n(.,\omega) \cap \text{clH}(z-\delta) \neq \emptyset\} \leq \mu\{\omega: \text{epi}\, f_n(.,\omega) \cap K_z \neq \emptyset\} +\varepsilon$$

It is known that if $f:X\times\Omega \to \overline{R}$ is a normal integrand then for any measurable map $x:\Omega \to X$, the map $\omega \to f(x(\omega),\omega):\Omega \to \overline{R}$ is measurable.

A map $y:\Omega \to X$ will be said *distribution optimal solution for the normal integrand* f if y(.) is measurable and the measurable map $\omega \to f(y(\omega),\omega)$ has the same probability distribution as $Z(\omega)=\text{Inf}_{x\in X} f(x,\omega)$.

4.15 COROLLARY - *Suppose that* $P_n \overset{w}{\to} P$. *Let* $\{x_n(.),n=1,\ldots\}$ *be a sequence of distribution optimal solutions. If* $\{x_n(.)\}$ *is tight then* $Z_n \overset{d}{\to} Z$.

PROOF. Tightness for $\{x_n(.)\}$ means tightness of the corresponding probability measures [1], i.e. for any $\varepsilon>0$ there exists a compact subset C of X such that

$$\mu\{\omega:x_n(\omega)\in C\} > 1-\varepsilon \qquad \text{for every } n \tag{4.16}$$

Let z contΦ, $\varepsilon>0$, $\delta>0$. Then (4.16) implies:

$$\mu\{\omega: \text{epi}\, f_n(.,\omega)\cap \text{clH}(z-\delta)\neq \emptyset\}\leq\mu\{\omega: Z_n(\omega)\leq z-\delta\}=\mu\{\omega:f_n(x_n(\omega),\omega)\leq z-\delta\}$$

$$\leq\mu\{\omega:f_n(x_n(\omega),\omega)\leq z-\delta, x_n(\omega)\in C\}+\varepsilon \leq \mu\{\omega:\text{epi}\, f_n(.,\omega)\cap K_z\neq \emptyset\} +\varepsilon.$$

The result follows then from Theorem 4.9 with Remark 4.14.

In general we cannot expect the existence of distribution optimal solutions even when $Z(\omega)>-\infty$ almost surely. A more general result can be given in terms of distribution δ-optimal solutions defined as follows.

For the normal integrand f, given $\delta>0$, the map $y:\Omega\to X$ is said to be *distribution δ-optimal solution for* f if it is measurable and for any real z we have

$$\mu\{\omega: Z(\omega)\leq z-\delta\} \leq \mu\{\omega: f(y(\omega),\omega)\leq z\}+\delta.$$

It can be proved that for any $\delta>0$, distributions δ-optimal solutions always for the family of normal integrands $\{f;f_n,n=1,...\}$ if $P_n \overset{w}{\to} P$, $Z_n \overset{d}{\to} Z$ and $Z>-\infty$ almost surely: the existence argument relies on the existence of measurable selectors (see for example[5]) for the multifunctions

$$\Gamma_n^{\delta}(\omega) = \begin{cases} \text{epi } f_n(\cdot,\omega)\cap \text{clH}(Z_n(\omega)+\delta) & \text{if } Z_n(\omega)>-\infty \\ X\times R & \text{otherwise} \end{cases} \quad (4.17)$$

and on the fact that since $Z_n(\cdot) \overset{d}{\to} Z(\cdot)$ and $Z>-\infty$ almost surely, for every $\delta>0$ there always exists a closed bounded interval $[z^-,z^+]$ such that $\mu\{\omega: Z_n(\omega)\in[z^-,z^+]\} > 1-\delta$. In (4.17) $\text{clH}(Z_n(\omega)+\delta)=\{(x,\alpha):\alpha< Z_n(\omega)+\delta\}$.

With the same arguments used in the proof of Corollary 4.15 we have:

4.18 COROLLARY - *If* $P_n \overset{w}{\to} P$ *and* $Z>-\infty$ *almost surely, then* $Z_n \overset{d}{\to} Z$ *if for every* $\delta>0$ *there exists a tight sequence of distribution* δ*-optimal solutions.*

Among the classes of problems which satisfy the above condtions it is noteworthy the case of inf-compact normal integrands, a restrictive situation which however already covers a large number of applications. This is related to [5,Sec.7]. Recall that a l.sc. function $g:X\to\bar{R}$ is *inf-compact* if and only if for every real z, epi $g\cap\text{clH}(z)$ is a compact subset of $X\times R$.

A sequence of normal integrands $\{f;f_n,n=1,...\}$ is said to be *equi-almost uniformly inf-compact* if, for every real z and any $\varepsilon>0$ there exists a compact subset K of $X\times R$ such that for all n we have

$$\mu\{\omega: \text{epi } f_n(\cdot,\omega)\cap\text{clH}(z)\subset K\} > 1-\varepsilon. \quad (4.19)$$

It is easy to show that (4.19) implies (4.10) of Theorem 4.9. Then we have:

4.20 COROLLARY - *Suppose that* $\{f;f_n,n=1,...\}$ *is a family of equi-almost uniformly inf-comact normal integrands. If* $P_n \overset{w}{\to} P$ *then* $Z_n \overset{d}{\to} Z$.

A special case of (4.19) is when for every real z and any $\varepsilon>0$ there exists a compact subsets C of X such that

$$\mu\{\omega: \text{dom } f_n(\cdot,\omega)\subset C\} > 1-\varepsilon \quad (4.20)$$

where $\text{dom } f_n(\cdot,\omega) = \{x\ X: f_n(x,\omega)<+\infty\}$ [6,Sec.7].

It is relevant to observe that condition (4.20) is satisfied by stochastic processes with l.sc. realizations on a compact set. These, under rather broad conditions, can be regarded as normal integrands. For them then, when converging in distribution in the sense of normal integrands, we immediately obtain convergence of the corresponding stochastic infima.

In this connection it has to be observed that convergence in distribution of stochastic processes regarded as normal integrands is closely connected but not equivalent to the convergence in distribution in the classical sense of the convergence of the finite dimensional distribution functions. These relations are also examined in [6]. This new setting however seems to be particularly appropriate to deal with convergence of functionals of stochastic processes: a more complete treatment is foreseen.

REFERENCES

1. P.Billingsley, *Convergence of Probability Measures*, J.Wiley,New York, 1968.

2. E.De Giorgi, On a definition of Γ-convergence of measures, in *Multifunctions and Integrands: Stochastic Analysis, Approximation and Optimization*, ed.G.Salinetti,Lecture Notes in Math. Vol.1091 Springer Verlag(1984),Berlin.

3. S.Dolecki, G.Salinetti and R.J-B Wets, Convergence of functions: Equi-semicontinuity, *Trans.Amer.Math.Soc.*,276(1983), 409-429.

4. Z. Frolik, Concerning topological convergence of sets, *Czechoslovak Math. J.*, 10(1960), 168-180.

5. R.T.Rockafellar, Integral functionals, normal integrands and measurable selections, in *Nonlinear Operators and Calculus of Variations*, eds.Gossez et al., Lecture Notes in Mathematics Vol.543,Springer Verlag (1976), Berlin.

6. G.Salinetti and R.J-B Wets, On the convergence in distribution of measurable multifunctions(random sets), normal integrands, stochastic processes and stochastic infima, *Mathematics of Operation Research* ,to appear, (CP-82-87,IIASA, Laxenburg,Austria).

7. G.Salinetti and R.J-B Wets, On the hypo-convergence of probability measures,in Proc. of *Optimization and Related Fields*, Erice 1984 (Tech.Rep.n.1,85,DSPSA).

GROWTH RATES AND OPTIMAL PATHS IN STOCHASTIC MODELS OF EXPANDING ECONOMIES

A.D. Slastinikov and E.L. Presman
Central Economical– Mathematical Institute (CEMI)
Moscow, USSR

1. INTRODUCTION

In this paper we investigate von Neumann–Gale (NG) stochastic models of expanding economies. These models are not only interesting theoretically, but also provide the basis (at least in the deterministic case) for numerous experimental calculations. Two of the most important notions in deterministic NG-models are those of the von Neumann growth rate (N-growth rate) and the von Neumann path (N-path), which characterize the asymptotic behavior of a broad class of optimal paths (more details can be found, e.g., in [1,2]). However, many difficulties arise when it is attempted to generalize these notions to the stochastic case. The present paper is therefore devoted to a study of two different ways of defining analogues of the N-growth rate and the N-path in stochastic NG-models.

2. PROBLEM DESCRIPTION

We shall first give the stochastic version of the NG-model. Let $(\Omega, \mathbf{F}, P)$ be a given probability space with a sequence of σ-fields $\mathbb{F} = \{\mathbf{F}_t\}_{t \geq 0}$. $\mathbf{F}_t$ is usually interpreted as the σ-field of events containing all the information available up to time t. The uncertainty in the model is represented by an exogeneous $\mathbb{F}$-adapted homogeneous Markov process $s = \{s_t\}_{t \geq 0}$ with a finite set of states S. [A process $\{x_t\}_{t \geq 0}$ is said to be $\mathbb{F}$-*adapted* if the x_t are $\mathbf{F}_t$-measurable ($t \geq 0$).] This process describes the influence of various random factors (such as the environment, production uncertainty, etc.) on the system. Let us assume that the process s cannot be decomposed (i.e., each state is accessible from every other state).

The state of the system is described by a vector $x \in R^n_+$; the transition to the next state is given by a function $f(s, x, u, \xi)$ which prescribes the next state of the system from the current state $x \in R^n_+$, random event $s \in S$, control $u \in U(s, x) \subset U$ and the next random event $\xi \in S$. Taking, for example, the stochastic version of the von

Neumann model with input matrices $A(s)$ and output coefficients $B(s,\xi)$, we can put $f(s,x,u,\xi) = B(s,\xi)u$, $U(s,x) = \{u \in R_+^m : A(s)u \le x\}$. [For vectors $x_i = (x_i^1, \ldots, x_i^n)$, $i = 1,2$, the inequality $x_1 \le x_2$ implies $x_1^k \le x_2^k$ for every k.] We shall introduce the point-to-set mapping $a(s,x,\xi) = f(s,x,U(s,x),\xi)$ to represent production (technological) correspondence in stochastic NG-models.

Given some $\mathbf{F}_0$-measurable x_0, a pair of $\mathbb{F}$-adapted processes $V = \{u_t\}_{t\ge0}$ and $X = \{x_t\}_{t\ge0}$ will be called (respectively) a *program* and a *path generated by* V, if $u_t \in U(s_t, x_t)$, $x_{t+1} = f(s_t, x_t, u_t, s_{t+1})$ $(t \ge 0)$. The program $V = \{u_t\}_{t\ge0}$ will be called a *stationary Markov process* if $u_t = u(s_t, x_t)$ *for all* $t \ge 0$ *and some measurable function* $u(s,x)$.

The requirement that the process in this definition be $\mathbb{F}$-adapted means, in particular, that we must choose non-anticipative programs (i.e., programs which do not depend on the future states of the system and process S).

3. STOCHASTIC ANALOGUES BASED ON "BALANCED GROWTH"

One way of constructing stochastic versions of the N-growth rate and the N-path is to generalize the notion of "balanced growth" to the stochastic case. Recall that the path $\{x_t\}_{t\ge0}$ in a deterministic NG-model is said to be balanced if $x_t = \alpha^t x_0$ $(t \ge 0)$, where $\alpha > 0$. Further, the largest possible value of α is called the N-growth rate and the corresponding balanced path is called the N-path. In the stochastic case, following Radner [3], we shall call the path $X = \{x_t\}_{t\ge0}$ balanced if $x_0 = \tilde{x}_0$, $x_t = \lambda_1 \cdot \cdots \cdot \lambda_t \cdot \tilde{x}_t$ $(t \ge 1)$, where $\lambda = \{\lambda_t\}_{t\ge0}$ is an $\mathbb{F}$-adapted stationary scalar process and $\tilde{X} = \{\tilde{x}_t\}_{t\ge0}$ is an $\mathbb{F}$-adapted stationary process on the unit simplex of R_+^n (we shall also set $X = (\lambda, \tilde{x})$). Thus a balanced path corresponds to system evolution with stationary proportions $\tilde{X}$ and stationary growth λ. The set of all balanced paths will be denoted by B. This set B needs to be quite "broad", so we assume the possibilty of arbitrary randomization. Formally, we suppose that the σ-fields $\mathbf{F}_t$ are sufficiently "rich", i.e., $\mathbf{F}_t$ contains the σ-field generated by the sequence η_0, s_0, η_1, $s_1, \ldots, \eta_t$, s_t, where $\{\eta_t\}_{t\ge0}$ are independent (both mutually and on the process S) random variables uniformly distributed on $[0,1]$.

Given some increasing continuous function $F: R_+ \to R$, we define the long-run growth rate of the path $X = (\lambda, \tilde{X}) \in B$ as follows:

$$v_F(X) = \lim_{T\to\infty} \frac{1}{T} \sum_{t=1}^{T} F(\lambda_t) \tag{1}$$

(this definition may be shown to be true by an ergodic theorem if $\mathrm{E}F(\lambda_1) < \infty$).

In order to derive our main results we shall make a number of assumptions (most of which will be expressed in terms of the mapping $a(s, x, \xi)$ introduced earlier). Define $S_*^2 = \{(s, \xi) \in S \times S: q(\xi \mid s) > 0\}$, where the $q(\xi \mid s)$ are the transition probabilities of process S. Then we assume the following:

(A1) $a(s, \lambda x, \xi) = \lambda a(s, x, \xi)$ for all $\lambda > 0$, $x \in R_+^n$, $(s, \xi) \in S_*^2$;

(A2) $f(s, x, u, \xi)$ is continuous in $u \in U$ for all $x \in R_+^n$, $(s, \xi) \in S_*^2$, and $U(s, x)$ is a metric compact set for all $x \in R_+^n$, $s \in S$;

(A3) $a(s, x, \xi)$ is an upper-semicontinuous mapping (in x) for all $(s, \xi) \in S_*^2$;

(A4) $a(s, 0, \xi) = \{0\}$ for all $(s, \xi) \in S_*^2$;

(A5) $U(s, x_1) \supset U(s, x_2)$ if $x_1 \geq x_2$ for all $s \in S$;

(A6) $a(s, x, \xi) \cap \text{int } R_+^n = \phi$ for all $x \neq 0$, $(s, \xi) \in S_*^2$.

Note that conditions (A1)–(A5) are natural stochastic analogues of the usual conditions for deterministic NG-models (see, e.g., [1]). In what follows we shall assume that the process s is stationary.

THEOREM 1. *If assumptions* (A1)–(A2) *are satisfied, then there exist a number* v_F^* *and a path* $X_F^* = (\lambda^*, \tilde{X}^*) \in B$ *such that*

$$v_F^* = v_F(X_F^*) = \sup_{(\lambda, \tilde{x}) \in B} EF(\lambda_1) = EF(\lambda_1^*)$$

and $v_F(x) \leq v_F^*$ (a.s.) *for any* $X \in B$.

Proof of this theorem is based on the ideas put forward in [3]. The essence of this result is that there exists a balanced path X_F^* with a long-run growth rate v_F (defined in (1)) such that, firstly, v_F^* is non-random and, secondly, v_F^* exceeds (a.s.) the long-run growth rates for any balanced path X. The number v_F^* and path X_F^* may be considered as analogues of the N-growth rate and N-path. Note that in the deterministic case $v_F^* = F(\alpha^*)$, where α^* is the N-growth rate in the deterministic NG-model. In the stochastic case v_F^* possesses the following properties:

THEOREM 2. *If assumptions* (A1)–(A6) *are satisfied, then for any path* $\{x_t\}_{t \geq 0}$ *we have*

$$\overline{\lim_{T \to \infty}} \left[\sum_{t=1}^{T} F\left(\frac{|x_t|}{|x_{t-1}|}\right) - Tv_F^*\right] < \infty \quad \text{(a.s.)} \quad .$$

[For $x = (x^1, \ldots, x^n) \in R_+^n$, $|x|$ should be interpreted as $\sum_{i=1}^{n} x^i$.]

Let us consider the following optimization problem:

$$E\left\{\sum_{t=1}^{T} F\left[\frac{|x_t|}{|x_{t-1}|}\right] + Q\left[s_T, \frac{x_T}{|x_T|}\right] \mid s_0 = s, x_0 = x\right\} \to \max , \tag{2}$$

where $Q(s, y)$ is a given function which is continuous in y and the maximum is taken over all paths $\{x_t\}_{t=0}^{T}$. The optimal value of the objective functional in (2) will be denoted by $V_T^{F,Q}(s, x)$.

THEOREM 3. *If assumptions* (A1)–(A6) *are satisfied, then*

$$\sup_{T \geq 1} \max_{s \in S, x \neq 0} |V_T^{F,Q}(s, x) - Tv_F^*| < \infty .$$

A similar result for the deterministic case was obtained in [2]. In the stochastic case the proofs of Theorems 2 and 3 are based on the method proposed in [4].

Thus, as in the deterministic case, v_F^* defines the asymptotic behavior of the optimal value of the objective functional in problem (2).

Now we shall consider the "logarithmic" case in which $F(y) = \log y$. Here it turns out that the optimal paths in (2) are close to the path $X_{\log}^*$ (this is the basis for regarding $X_{\log}^*$ as a stochastic analogue of the N-path).

Let $a^m(\xi_0, x_0; \xi_1, \ldots, \xi_m) = \{y_m \in R_+^n: \quad y_1 \in a(\xi_0, x_0, \xi_1), y_2 \in a(\xi_1, y_1, \xi_2), \ldots, y_m \in a(\xi_{m-1}, y_{m-1}, \xi_m)\}$ denote the set of states of the system which can be reached in m steps from the initial states (ξ_0, x_0) under the successive occurrence of random events $\xi_1, \ldots, \xi_m$. Let $q(\xi \mid s)$ be the transition probabilities of the process s. Then

(A6') For any $\xi_0 \in S$ and $x_0 \neq 0$ there exist an integer $m \geq 1$ and a sequence $\xi_1, \ldots, \xi_m \in S$ such that $\prod_{t=1}^{m} q(\xi_t \mid \xi_{t-1}) > 0$ and

$$a^m(\xi_0, x_0; \xi_1, \ldots, \xi_m) \cap \text{int } R_+^n \neq \phi .$$

The essence of this condition of stochastic primitiveness is that there is a positive probability of reaching a strictly positive state from any initial state (ξ_0, x_0) in a finite number of steps.

(A7) For any $(s, \xi) \in S_*^2$, $x_1, x_2 \in R_+^n$, $y_1 \in a(s, x_1, \xi)$, $y_2 \in a(s, x_2, \xi)$, there exists a $\tilde{y} \in a(s, x_1 + x_2, \xi)$ such that $\tilde{y} \geq y_1 + y_2$.

This quasiconcavity condition is slightly weaker than the corresponding concavity condition for deterministic NG-models.

(A8) For any $(s, \xi) \in S_*^2$, $x \neq 0$, y_1, $y_2 \in a(s, x, \xi)$, $y_1 \neq y_2$, there exists a $\tilde{y} \in a(s, x, \xi)$ such that $\tilde{y} \geq \frac{1}{2}(y_1 + y_2)$ and $\tilde{y} \neq \frac{1}{2}(y_1 + y_2)$.

(A condition for almost strict convexity of the set $a(s, x, \xi)$.)

THEOREM 4. *If assumptions* (A1)–(A4), (A6'), (A7) *and* (A8) *are satisfied, then there exists a distribution of initial states* x_0 *such that the path* $X_{\log}^*$ *is generated by a stationary Markov process.*

Let us now consider the following optimization problem:

$$E\{\log \Phi(s_T, x_T) \mid s_0 = s, x_0 = x\} \to \max \quad , \tag{3}$$

where a given function $\Phi(s, x)$ is superlinear (in x) and strictly positive on the unit simplex of R_+^n. This problem is a particular case of the general problem (2). We shall formulate two additional requirements: a "free disposal" condition

(A9) For any $x \in R_+^n$, $(s, \xi) \in S_*^2$, $y \in a(s, x, \xi)$, the relation $0 < y' < y$ implies $y' \in a(s, x, \xi)$;

and a condition for uniform strict convexity

(A10) For any $\varepsilon > 0$ there exists a $\rho = \rho(\varepsilon) > 0$ such that for all $(s, \xi) \in S_*^2$ and $|x_1| = |x_2| = 1$, $|x_1 - x_2| \geq \varepsilon$, the relations $y_2 \in a(s, x_1, \xi)$, $y_2 \in a(s, x_2, \xi)$ imply $y_1 + y_2 + w \in a(s, x_1 + x_2, \xi)$ for some $w \in R_+^n$, $|w| \geq \rho$.

THEOREM 5. *If assumptions* (A1)–(A4) *and* (A6)–(A10) *are satisfied and* $\{x_t^T\}_{t=0}^T$ *is the optimal path in problem* (3), *then for any* $\varepsilon > 0$, $s \in S$, $x \in \text{int } R_+^n$, *there exists an* $L = L(\varepsilon, \phi, s, x)$ *such that for all periods* $0 \leq t \leq T - L$ *we have*

$$E\{ \left| \frac{x_t^T}{|x_t^T|} - \frac{x_t^*}{|x_t^*|} \right| \mid s_0 = s, x_0^T = x_0^* = x\} \leq \varepsilon \quad ,$$

where $\{x_t^*\}_{t \geq 0}$ *is the balanced path* $X_{\log}^*$.

This result can be regarded as a stochastic analogue of the "turnpike theorems" for deterministic NG-models. A similar theorem has been established for non-Markov-type models in [5].

4. STOCHASTIC ANALOGUES – ANOTHER APPROACH

The approach described above (based on generalization of the "balanced growth" notion to the stochastic case) is not completely satisfactory. Indeed, if one is interested in the following optimization problem:

$$E\{\Phi(s_T, x_T) \mid s_0 = s, x_0 = x\} \to \max \quad , \tag{4}$$

where $\Phi(s, x)$ is a given function with a positive degree of homogeneity α (e.g., it is linear or superlinear), then, in contrast to the deterministic case, neither the asymptotic behavior of the optional value of the objective functional nor that of the optimal paths is connected with characteristics of the type v_F^* and X_F^* (see above).

Another way of defining stochastic analogues of the N-growth rate and N-path can be considered for such problems. This approach may be summarized as follows. Since the functional in (4) may be written in the multiplicative form

$$|x_0|^{\alpha}\, \mathrm{E}\, \{\Phi(s_T, \frac{x_T}{|x_T|}) \prod_{t=1}^{T} |x_t|^{\alpha}\, |x_{t-1}|^{-2}\, | s_0, x_0\} \quad ,$$

it seems natural enough to suppose that there exists a number λ such that the optimal value of the objective functional in problem (4) displays growth of order λ^T (as $T \to \infty$). This number will be called the ***model growth rate*** (stochastic N-growth rate). It turns out that for the model described in Section 1 such a number does indeed exist; it depends not on the given function $\Phi(s, x)$ but only on its degree of homogeneity and may be obtained as the solution of a certain one-step stationary problem.

So let us consider the model described above. In contrast to Section 2 we shall not assume that process s is stationary. Define a functional space $W_\alpha = \{\Phi(s, x) : S \times R_+^n \to R^+\}$ such that $\Phi(s, x)$ is concave, upper-semicontinuous and homogeneous with positive degree $\alpha > 0$ (in x) ($0 < \alpha \leq 1$) and an operatoar

$$\Gamma\, \varphi(s, x) = \sup_{u \in U(s,x)} \sum_S \Phi(\xi, f(s, x, u, \xi)) q(\xi \mid s) \quad .$$

It is easy to prove that under assumptions (A1)–(A4) and (A7), the operator Γ maps W_α into W_α. One of the main results is the following existence theorem:

THEOREM 6. *If assumptions* (A1)–(A4) and (A7) *are satisfied, then there exist a* $\lambda_\alpha > 0$ *and a* $G_\alpha \in W_\alpha$ *such that* $G_\alpha \equiv 0$ *and* $\Gamma G_\alpha = \lambda_\alpha G_\alpha$ ($0 < \alpha \leq 1$).

The proof of this theorem can be found in [6]; we shall only outline its main features here. We consider the space of functions $\{\Phi = \Phi_1 - \Phi_2,\ \Phi_1, \Phi_2 \in W_\alpha\}$ with norm $\|\Phi\| = \sum_S \int_{|x|=1} |\Phi(s, x)| dx$. It is then established that the operator Γ is continuous and the set $\{\Phi \in W_\alpha : \|\Phi\| = 1\}$ is compact with respect to $\|\cdot\|$-convergence. Finally, the fixed-point theorem is applied to the operator $L\Phi = (\Phi + \Gamma\Phi)/(1 + \|\Gamma\varphi\|)$.

In [6] it is proved that, under the additional assumption (A6'), the eigenvalue λ_α is unique and the corresponding eigenfunctions $G_\alpha(s, x)$ are strictly positive (if $x \neq 0$). Note that in the deterministic case $\lambda_\alpha = \lambda_*^\alpha$ for "nondegenerate" NG-models, where λ_* is the N-growth rate (see, e.g., [7]).

Now let $V_T^\varphi(s, x)$ be the optimal value of the objective function in problem (4).

THEOREM 7. *If assumptions* (A1)–(A4), (A6') *and* (A7) *are satisfied and* $0 < \inf_{|x|=1} \Phi(s, x) \leq \sup_{|x|=1} \Phi(s, x) < \infty$, *then*

$$0 < \inf_{T \geq 1} \inf_{s \in S, |x|=1} \lambda_\alpha^{-T} V_T^\varphi(s, x) \leq \sup_{T \geq 1} \sup_{s \in S, |x|=1} \lambda_\alpha^{-T} V_T^\varphi(s, x) < \infty \quad .$$

This result means that $V_T^\varphi(s, x)$ displays growth of order λ_α^T (for sufficiently large T). This property allows us to regard numbers λ_α as stochastic analogues of the N-growth rate. A similar property of the N-growth rate was established for the deterministic case in [2].

Now let $u_\alpha^*(s, x)$ be a measurable function (control) defined by the relation

$$\Gamma G_\alpha(s, x) = \sum_S G_\alpha(\xi, f(s, x, u_\alpha^*(s, x), \xi)) q(\xi \mid s) \quad ,$$

where G_α is an eigenfunction of the operator Γ. Let $X_\alpha^* = \{_\alpha x_t^*\}_{t \geq 0}$ be a path generated by the stationary Markov process $\{u_\alpha^*(s, x)\}$ and the initial distribution $_\alpha x_0^*$ on the unit simplex of R_+^n. It turns out that the path X_α^* possesses certain "turnpike" properties.

THEOREM 8. *Let assumptions* (A1)–(A4), (A6'), (A7) *and* (A10) *be satisfied. Then for any numbers* $\varepsilon, \delta, \eta > 0$ *there exists an integer* $L = L(\varepsilon, \delta, \eta)$ *such that for any path* $\{x_t\}_{t=0}^T$ *and all periods* $t = 0, 1, \ldots, T$ *(except perhaps L) the following relation holds:*

$$P(\{| \frac{x_t}{|x_t|} - \frac{_\alpha x_t^*}{|_\alpha x_t^*|} | \leq \varepsilon\} \cup B_t^\delta) \geq 1 - \eta \quad ,$$

where

$$B_t^\delta = \{\min (\frac{|x_t|}{|_\alpha x_t^*|}, \frac{|_\alpha x_t^*|}{|x_t|}) < \delta\} \quad .$$

In essence, this result states that for any $\eta > 0$ and almost all periods, the paths $\{x_t\}$ and X_α^* either have different orders of growth or are close to one another (on the unit simplex) with probability greater than $1 - \eta$. Note that for a certain class of NG-models, for example, when $a(s, x, \xi) = g(s, \xi) a(x)$ (where $g(s, \xi) \geq 0$, $a(x)$ is a point-to-set mapping of R_+^n into R_+^n) for the optimal path $\{x_t^T\}_{t=0}^T$ in (4) and $\delta > 0$, we have $P(B_t^\delta) = 1$ and, therefore, $\{x_t^T\}_{t=0}^T$ is close to the path X_α^* (on the simplex). This property of path X_α^* (which becomes the well-known "weak turnpike theorem" in the deterministic case – see, e.g., [1], allows us to regard the path X_α^* as a stochastic analogue of the N-path.

Thus for stochastic NG-models we have the set of N-growth rates $\{\lambda_\alpha : 0 < \alpha \le 1\}$. From simple examples it is easy to see that the numbers λ_α do not coincide with the numbers v_F^* introduced in Section 2. It would therefore be interesting to establish some relations between the N-growth rates obtained by different approaches, and also to study N-growth rates λ_α as function of parameter α. Preliminary results in this direction can be derived from the following statement.

THEOREM 9. *Let assumptions* (A1)–(A4), (A6') *and* (A7) *be satisfied. Then*

(1) *the function* $\log \lambda_\alpha$ *is convex in* α;

(2) $\lambda_\alpha^{1/\alpha}$ *is an increasing function*;

(3) $\frac{1}{\alpha} \log \lambda_\alpha \to v_{\log}^*$ (as $\alpha \to 0$), *where* $v_{\log}^*$ *is the "logarithmic"* N-*growth rate (i.e., the* N-*growth rate defined in Section 2 for the "logarithmic" case).*

Finally, it should be mentioned that although we make use of the finiteness of the set of states of Markov process s in the theorems formulated above, all of the results remain valid when process s is a sequence of independent, identically distributed random variables (with an arbitrary set of states) and $U(s, x)$ and $f(s, x, u, \xi)$ do not depend on parameter s (i.e., $U(s, x) = U(x)$, $f(s, x, u, \xi) = f(x, u, \xi)$).

5. CONCLUSION

Although deterministic NG-models have "global" indexes (the N-growth rate and N-path) which the asymptotic behavior of numerous classes of optimal paths, there are no such "global" indexes in the stochastic case. Thus, when studying the different types of extremal problems connected, for example, with objective functionals of the additive or multiplicative type, we need to define the notions of N-growth rate and N-path in different ways.

REFERENCES

1. V.L. Markov and A.M. Rubinov. *Mathematical Theory of Economic Dynamics and Equilibrium*. Nauka, Moscow, 1973 (in Russian).

2. I.V. Romanovsky. The asymptotic behavior of a discrete, deterministic, dynamic programming process with a continuous set of states (in Russian). *Optimalnoe Planirovanie*, 8(1967)171–193.

3. R. Radner. Balanced stochastic growth at the maximum rate. *Zeitschrift für Nationalökonomie*, Suppl. 1, (1971)39–52.

4. E.L. Presman and A.D. Slastnicov. Asymptotic behavior of the objective function in stochastic economic growth models (in Russian). In *Stochasticheskie Modeli i Upravlenie*, CEMI, Moscow, 1980, pp. 83–104.

5. I.V. Evstigneev. Homogeneous convex models in the theory of controlled random processes (in Russian). *Doklady Akademii Nauk SSSR*, 253(3)(1980)524–527.

6. E.L. Presman and A.D. Slastnikov. On an approach to growth rate notions in stochastic von Neumann–Gale models (in Russian). In *Modeli i Metody Stochasticheskoy Optimizaĉii*, CEMI, Moscow, 1983, pp. 123–152.

7. A.M. Rubinov. *Superlinear Multivalued Functions with Applications to Economic-Mathematical Problems*. Nauka, Leningrad, 1980 (in Russian).

EXTREMUM PROBLEMS DEPENDING ON A RANDOM PARAMETER

E. Tamm
Institute of Cybernetics
Tallin, USSR

1. INTRODUCTION

We shall consider the following nonlinear extremum problem in n-dimensional Euclidean space R^n:

$$\min_x \{f(x, \xi) \mid x \in \Gamma(\xi)\} \quad , \tag{1}$$

where ξ is an s-dimensional random parameter, f: $R^n \times R^s \to R^1$ and Γ is a multifunction from R^s to R^n.

The question of whether this problem has a measurable solution $x^*(\xi)$ then arises. In the literature there are a number of papers dealing with analogous questions for random equations in Banach space (see, e.g., [1–4]). Conditions for the existence of a measurable solution to (1) can be derived using methods similar to those adopted in these papers: it is sufficient to assume that $f(x, \xi)$ is continuous in both variables and that Γ is a measurable closed-valued multifunction.

Further, if $x^*(\xi)$ is proved to be measurable, i.e., it is an n-dimensional random vector, then the next step is to look for some information about its distribution. For every particular value of ξ, problem (1) is a deterministic extremum problem and for some values of ξ a solution may not necessarily exist. A lower bound for the probability that (1) has a solution is given in Section 2, where we also derive a Tchebycheff inequality-type estimate for this solution. In Section 3 these results are used to solve certain stochastic programming problems.

2. UNCONSTRAINED PROBLEM

Consider the minimization of a function depending on a random parameter ξ:

$$\min_x \{f(x, \xi) \mid x \in R^n\} \quad . \tag{2}$$

We shall address two main questions in connection with this problem:

(1) How can we estimate the probability that (2) has a solution $x^*(\xi)$?

(2) How can we estimate the distance $\|x^*(\xi) - x^*\|$, where x^* is a fixed point chosen by a certain rule?

It turns out that the most appropriate point for x^* is a local minimum of the function $Ef(x, \xi)$, i.e., $Ef(x^*, \xi) = \min_x \{Ef(x, \xi) \mid x \in R^n\}$. Here we assume the existence of x^*.

Suppose that $f(x, \xi)$ and the distribution of the random vector ξ satisfy the following conditions:

1. The function $f(x, \xi)$ is twice-differentiable in x, $f''_{xx}(x, \xi)$ satisfies the Lipschitz condition

$$\|f''_{xx}(x^1, \xi) - f''_{xx}(x^2, \xi)\| \le C(\xi)\, \|x^1 - x^2\|$$

and

$$u^T E f''_{xx}(x^*, \xi) u \ge m \|u\|^2 \quad \text{for all} \quad u \in R^n \quad .$$

2. The mathematical expectations $EC(\xi)$, $E\|f''_{xx}(x^*, \xi) - Ef''_{xx}(x^*, \xi)\|^2$ and variances $\sigma^2 C(\xi)$, $\sigma^2 f'_{x_i}(x^*, \xi)$, $i = 1, 2, \ldots, n$, are finite.

The probability that (2) has a solution and the distance between a local solution $x^*(\xi)$ and the point x^* are estimated in the following theorem:

THEOREM 1 [5]. *Let conditions 1 and 2 be satisfied. If there exist constants δ_1 and δ_2, $0 < \delta_1 < m$, $\delta_2 > 0$, such that the expression*

$$p(\delta_1, \delta_2) = 1 - \frac{E\|f''_{xx}(x^*, \xi) - Ef''_{xx}(x^*, \xi)\|^2}{\delta_1^2} -$$

$$- \frac{16[EC(\xi) + \delta_2]^2 \sum_{i=1}^{n} \sigma^2 f'_{x_i}(x^*, \xi)}{(m - \delta_1)^4} - \frac{\sigma^2 C(\xi)}{\delta_2^2}$$

is positive, then there exists a measurable set $M(\delta_1, \delta_2) \subset R^s$ such that

(1) *if $\xi \in M(\delta_1, \delta_2)$, then problem* (2) *has a local solution $x^*(\xi)$;*

(2) $P\{\xi \in M(\delta_1, \delta_2)\} \ge p(\delta_1, \delta_2)$;

(3) $P\{\|x^*(\xi) - x^*\| < \varepsilon$ *and* $\xi \in M(\delta_1, \delta_2)\} \ge p(\delta_1, \delta_2) - \dfrac{\sum_{i=1}^{n} \sigma^2 f'_{x_i}(x^*, \xi)}{\varepsilon^2 (m - \delta_1)^2}$ *for arbitrary $\varepsilon > 0$.*

Analogous results can also be obtained for the problem with equality constraints

[5]:

$$\min_x \{f(x,\xi) \mid g(x,\xi) = 0\}$$

and for the problem with inequality constraints [6]:

$$\min_x \{f(x,\xi) \mid h(x,\xi) \le 0\} \quad .$$

3. APPLICATIONS

Theorem 1 can be used in the minimization of mathematical expectation and probability functions.

Consider the problem

$$\min_x \{\mathrm{E}f(x,\xi) \mid x \in R^n\} \quad . \tag{3}$$

Formally this problem is a special case of an ordinary nonlinear unconstrained problem. However, existing methods for solving such problems are not very suitable for (3), because it is generally too troublesome to evaluate the values of the s-dimensional integral $\mathrm{E}f(x,\xi)$ and its derivatives. Furthermore, in many cases the distribution function of ξ is not known, and so it is actually impossible to calculate these values. Hence, to obtain some information about the solutions of problem (3), it is necessary to use observations of ξ in some way. If the number of observations available is practically infinite, then a procedure of the stochastic approximation type is usually recommended. If the number of observations is limited, another approach is more appropriate: instead of problem (3), solve the problem

$$\min_x \{\frac{1}{k}\sum_{i=1}^{k} f(x,\xi_i) \mid x \in R^n\} \quad , \tag{4}$$

where the ξ_i, $i = 1,2,\dots,k$, are independent observations of ξ. Problem (4) depends on the $k \times s$-dimensional random parameter $(\xi_1,\xi_2,\dots,\xi_k)$ and $\mathrm{E}[\frac{1}{k}\sum_{i=1}^{k} f(x,\xi_i)] = \mathrm{E}f(x,\xi)$. With the aid of Theorem 1 we can estimate the probability that (4) has a local solution $x_k^*(\xi_1,\xi_2,\dots,\xi_k)$ and the distance between $x_k^*(\xi_1,\xi_2,\dots,\xi_k)$ and x^*. Let the function $f(x,\xi)$ and the distribution of ξ satisfy conditions 1 and 2. Then

$$\|\frac{1}{k}\sum_{i=1}^{k} f(x^1,\xi_i) - \frac{1}{k}\sum_{i=1}^{k} f(x^2,\xi_i)\| \le \frac{1}{k}\sum_{i=1}^{k} C(\xi_i)\|x^1 - x^2\| \quad ,$$

$$\mathrm{E}[\frac{1}{k}\sum_{i=1}^{k} C(\xi_i)] = \mathrm{E}C(\xi) \, , \; \sigma^2[\frac{1}{k}\sum_{i=1}^{k} C(\xi_i)] = \frac{1}{k}\sigma^2 C(\xi) \quad ,$$

$$\mathrm{E}\|\frac{1}{k}\sum_{i=1}^{k} f''_{xx}(x^*,\xi_i) - \mathrm{E}f''_{xx}(x^*,\xi)\|^2 =$$

$$= \frac{1}{k^2}\mathrm{E}\|\sum_{i=1}^{k}[f''_{xx}(x^*,\xi_i) - \mathrm{E}f''_{xx}(x^*,\xi)]\|^2 \le$$

$$\le \frac{1}{k^2}\mathrm{E}\sum_{i=1}^{k}\|f''_{xx}(x^*,\xi_i) - \mathrm{E}f''_{xx}(x^*,\xi)\|^2 =$$

$$= \frac{1}{k}\mathrm{E}\|f''_{xx}(x^*,\xi) - \mathrm{E}f''_{xx}(x^*,\xi)\|^2$$

and analogously

$$\sigma^2[\frac{1}{k}\sum_{i=1}^{k} f'_{x_i}(x^*,\xi_i)] = \frac{1}{k}\sigma^2 f'_{x_i}(x^*,\xi)\,,\; i = 1,2,\ldots,n \quad .$$

As a corollary of Theorem 1, we can now state the following result:

THEOREM 2 [7]. *Let conditions 1 and 2 be satisfied. Then for arbitrary constants* δ_1 *and* δ_2, *such that* $0 < \delta_1 < m$, $\delta_2 > 0$, *and for sufficiently large k, there exists a measurable set* $M(k,\delta_1,\delta_2) \subset \underbrace{R^s \times \cdots \times R^s}_{k}$ *such that*

(1) *if* $(\xi_1,\xi_2,\ldots,\xi_k) \in M(k,\delta_1,\delta_2)$ *then problem* (4) *has a local solution* $x_k^*(\xi_1,\xi_2,\ldots,\xi_k)$;

(2) $P\{M(k,\delta_1,\delta_2)\} \ge p(k,\delta_1,\delta_2)$;

(3) $P\{\|x_k^*(\xi_1,\xi_2,\ldots,\xi_k) - x^*\| < \varepsilon$ *and* $(\xi_1,\xi_2,\ldots,\xi_k) \in M(k,\delta_1,\delta_2)\} \ge$

$$p(k,\delta_1,\delta_2) - \frac{\sum_{i=1}^{n}\sigma^2 f'_{x_i}(x^*,\xi)}{k\varepsilon^2(m-\delta_1)^2}$$ *for all* $\varepsilon > 0$, *where*

$$p(k,\delta_1,\delta_2) = 1 - \frac{1}{k}[\frac{\mathrm{E}\|f''_{xx}(x^*,\xi)-\mathrm{E}f''_{xx}(x^*,\xi)\|^2}{\delta_1^2} +$$

$$+ \frac{16[\mathrm{E}(\xi)+\delta_2]^2\sum_{i=1}^{n}\sigma^2 f'_{x_i}(x^*,\xi)}{(m-\delta_1)^4} + \frac{\sigma^2 C(\xi)}{\delta_2^2}$$

is positive.

In a similar way one can also find an approximate local solution for a probability function.

Consider the problem

$$\min_x \{P\{f(x,\xi) < 0\} \mid x \in R^n\} \quad . \tag{5}$$

For every fixed x the function $v(x,t) = P\{f(x,\xi) < t\}$ is the distribution function of the random variable $f(x,\xi)$. Therefore, to obtain an estimate of the function $v(x,0) = P\{f(x,\xi) < 0\}$ it is natural to use some method of estimating distribution functions. We shall use the Parzen estimate [8] $v_k(x,\xi_1,\ldots,\xi_k)$ of $v(x,0)$:

$$v_k(x,\xi_1,\ldots,\xi_k) = \frac{1}{kh}\sum_{i=1}^{k}\int_{-\infty}^{0} K\left(\frac{t-f(x,\xi_i)}{h}\right)dt \quad ,$$

where $h = h(k)$, $\lim_{k\to\infty} h(k) = 0$, $\lim_{k\to\infty} kh(k) = \infty$ and the differentiable function $K(\tau)$ satisfies the conditions

(a) $\int_{-\infty}^{\infty} K(\tau)d\tau = 1$,

(b) $\int_{-\infty}^{\infty} \tau K(\tau)d\tau = 0$, $\int_{-\infty}^{\infty} \tau^2 \,|K(\tau)|\, d\tau < \infty$.

Then the following problem is solved instead of problem (5):

$$\min_x \left\{\frac{1}{kh}\sum_{i=1}^{k}\int_{-\infty}^{0} K\left(\frac{t-f(x,\xi_i)}{h}\right)dt \;\middle|\; x \in R^n\right\} \quad . \tag{6}$$

Unfortunately, in the present case we only have

$$\lim_{k\to\infty} E\left[\frac{1}{kh}\sum_{i=1}^{k}\int_{-\infty}^{0} K\left(\frac{t-f(x,\xi_i)}{h}\right)dt\right] = v(x,0) \quad ,$$

and for this reason Theorem 1 is not directly applicable to problems (5) and (6). A relation between (5) and (6) is stated in [9] using the auxiliary problem

$$\min_k \left\{E\left[\frac{1}{kh}\sum_{i=1}^{k}\int_{-\infty}^{0} K\left(\frac{t-f(x,\xi_i)}{h}\right)dt\right] \;\middle|\; x \in R^n\right\} \quad . \tag{7}$$

REFERENCES

1. A.T. Bharucha-Reid. Fixed-point theorems in probabilistic analysis. *Bulletin of the American Mathematical Society*, 82(1976)641–657.

2. R. Kannan and H. Salehi. Measurability of the solutions of nonlinear equations. *Journal of Mathematical Analysis and Applications*, 53(1977)132–145.

3. A. Nowak. Random solution of equations. In *Transactions of the 8th Prague Conference on Information Theory*, 1978, pp. 77–82.

4. P.V. Kirilov. *Stochastic Equations*. Shtintsa, Kishinev, 1982 (in Russian).

5. E. Tamm. Inequalities for the solutions of nonlinear programming problems depending on a random parameter. *Mathematische Operationsforschung und Statistik, Series Optimization*, 11(1980)487–497.

6. E. Tamm. On the possibility of solving nonlinear programming problems with random parameters (in Russian). *Izvestiya Akademii Nauk Ehstonskoj SSSR, Seriya Fizika, Matematika*, 30(1981)220–225.

7. E. Tamm. Inequalities of the Chebycheff type for the solution of nonlinear stochastic E-models (in Russian). *Izvestiya Akademii Nauk Ehstonskoj SSSR, Seriya Fizika, Matematika*, 27(1978)448–450.

8. E. Parzen. On estimation of a probability density function and mode. *Annals of Mathematical Statistics*, 33(1962)1065–1076.

9. E. Tamm. On the minimization of a probability function (in Russian). *Izvestiya Akademii Nauk Ehstonskoj SSSR, Seriya Fizika, Matematika*, 28(1979)23–30.

ADAPTIVE CONTROL OF PARAMETERS IN GRADIENT ALGORITHMS FOR STOCHASTIC OPTIMIZATION

S.P. Urjas'ev
V. Glushkov Institute of Cybernetics
Kiev, USSR

1. INTRODUCTION

This paper is concerned with the development of iterative non-monotonic optimization algorithms for a broad class of stochastic programming problems. The following algorithms are considered:

(a) the quasi-gradient optimization algorithm involving projection onto an admissible domain;

(b) the Arrow—Hurwicz algorithm which searches for saddle points of convex-concave functions in the presence of noise;

(c) the generalized gradient algorithm which searches for Nash equilibria in non-cooperative many-person games.

Most of the problems under discussion are characterized by lack of complete information about objective and constraint functions (which are usually nonsmooth) and their derivatives. The central idea of the numerical methods considered here (which are called *stochastic quasi-gradient methods*) is to use random directions instead of precise values of gradients or their analogs. These random directions are statistical estimates of gradients (stochastic quasi-gradients). The resulting algorithms are derived from stochastic approximation algorithms. The first steps in this direction were taken by Robbins and Monro [1], and developed further by many other authors.

This approach was generalized by Ermoliev [2], who extended it to a broader class of optimization problems and introduced the notion of the stochastic quasi-gradient.

Adaptive procedures for controlling the parameters of the algorithms discussed in this paper and improving their practical characteristics are proposed. The term "adaptivity" as used here refers to the dependence of these parameters upon the process trajectory, as opposed to procedures in which the parameter values depend only on the number of iterations. The selection of step size controls and stopping criteria represent the main difficulties in the computer implementation of these methods, and are studied in [2–5]. This paper concentrates on the approach described in [6–9], which is further developed in [10].

The main features of the proposed approach are summarized briefly below.

Almost every iterative algorithm involves parameters that have to be controlled. Although criteria determining the controls that should be chosen are usually available, it is often very difficult to satisfy these criteria (to find the optimal control) in practice for computational reasons. However, it is possible to vary these criteria with respect to the parameters and, as a result, to compute their gradients or stochastic quasi-gradients. The resulting gradients (quasi-gradients) may be used to construct recursive procedures for parameter modification. Several gradient procedures are included in the algorithm, both in the basic space and with respect to the algorithm parameters, i.e., some adaptation of algorithm parameters occurs.

It is proved that algorithms with step-size rules of this type converge to the set of optimal points. Suggestions regarding the computer implementation of the algorithms are made.

2. A STOCHASTIC QUASI-GRADIENT ALGORITHM

Description. Assume that the problem is to minimize a convex (possibly nonsmooth) function

$$f(x) \longrightarrow \min_{x \in X} \quad ,$$

where X is a convex, closed, bounded subset of a separable Hilbert space H. In some fairly general classes of problems it is very difficult to compute exact values of the function and its gradients, but it is possible to find vectors which represent statistical estimates of these quantities. This occurs, for instance, in the minimization of functions of the form

$$f(x) = E_{\omega}\psi(x, \omega) = \int_{\omega \in \Omega} \psi(x, \omega) P(d\omega) \quad .$$

Recalling that in the most general case the generalized differential of the convex function $f(x)$ may be calculated using the formula

$$\partial f(x) = \int_{\omega \in \Omega} \partial_x \psi(x, \omega) P(d\omega) \quad ,$$

we may take $\partial_x \psi(x, \omega)$ as a set of vectors representing statistical estimates of gradients of the function $f(x)$. Denote the operatiion of projection onto the bounded convex set X by $\pi_X(\cdot)$, and let (Ω, A, P) be some probability space. Examine a sequence of random points generated by the formula

$$x^{s+1} = \pi_X(x^s - \rho_s \xi(s, x^s, \omega)) \quad , \tag{1}$$

where $x^0 \in H$ is an arbitrary point; $\rho_s \geq 0$, $s = 0,1, \ldots$ are step sizes; and $\xi(s, x, \omega)$, $s = 0,1, \ldots$ is a sequence of random functions defined on the probability space (Ω, A, P) and assuming values in a measurable space (H, B) (B is a Borel σ-algebra in H). It is required that the superpositions $\xi(s, \zeta(\omega), \omega)$ be random variables for any random value $\zeta : \Omega \to H$. The random functions $\xi(s, x, \omega)$ are jointly independent for any $s = 0,1, \ldots, x \in X$ and the relations

$$\mathrm{E}\xi(s, x, \omega) = \hat{f}_x(x) + b(s, x)$$

are satisfied, where E denotes mathematical expectation; vector $\hat{f}_x(x)$ belongs to the set of generalized gradients $\partial f(x)$ of the convex function $f(x)$; and $b(s, x)$, $s = 0,1, \ldots$ is a sequence of deterministic functions given on the space H. Thus, the functions $\xi(s, x, \omega)$, $s = 0,1, \ldots$ are statistical estimates of some generalized gradients of the function $f(x)$. The random functions $\xi(s, x, \omega)$ are called *stochastic quasi-gradients*. For simplicity we shall write

$$\xi(s, x^s, \omega) = \xi^s \quad , \quad b(s, x) = b^s \quad .$$

Construction of an Adaptive Step Size Control. To put algorithm (1) into practice, it is necessary to have some formulas for the computation of parameters ρ_s, $s = 0,1, \ldots$. For simplicity, we shall assume that $X = H$. Algorithm (1) allows a natural choice of the step size ρ_s using the condition for the function $\psi_s(\rho)$ to be a minimum with respect to ρ, where

$$\psi_s(\rho) = \mathrm{E}[f(x^s - \rho\xi^s) | x^s] \quad ,$$

and $\mathrm{E}[\cdot | \cdot]$ denotes conditional mathematical expectation. The values of the function $\psi_s(\rho)$ are usually very difficult to compute. Let us differentiate the function $f(x^s - \rho\xi^s)$ with respect to ρ at the point ρ_s:

$$\partial_\rho f(x^s - \rho_s \xi^s) = -\{<y, \xi^s> : y \in \partial f(x^{s+1}) = \partial f(x^s - \rho_s \xi^s)\} \quad ,$$

where $<\cdot,\cdot>$ denotes a scalar product. Since $\partial\psi_s(\rho_s) = \mathrm{E}[\partial_\rho f(x^s - \rho_s \xi^s) | x^s]$, we have $-E[<\xi^{s+1}, \xi^s> | x^s] \in \partial\psi_s(\rho_s)$.

The following gradient procedure:

$$\rho_{s+1} = \rho_s + \lambda_s <\xi^{s+1}, \xi^s> = \rho_s + \frac{\lambda_s}{\rho_s} <\xi^{s+1}, x^s - x^{s+1}> ,\ \lambda_s > 0 ,\ s = 0,1, \ldots$$

can be used to modify the step size ρ_s. In order to facilitate the proof of convergence, the above relation is rewritten in the form

$$\rho_{s+1} = \rho_s a_s^{<\xi^{s+1}, x^s - x^{s+1}> - \delta\rho_s} ,\ a_s > 1 ,\ \delta > 0 \quad , \tag{2}$$

where an additional element $-\delta\rho_s$ (which reduces the step size) is introduced into the exponent.

Convergence of the Algorithm It is possible to show [7,8] that under certain conditions the step sizes calculated using formula (2) satisfy the classical conditions for convergence of stochastic optimization algorithms [2]:

$$\rho_s \to 0 \text{ a.s.} ,\quad \sum_0^s \rho_s = \infty \text{ a.s.} ,\quad \sum_0^s \mathrm{E}\rho_s^2 < \infty \quad ,$$

and, moreover, that the condition $\rho_s / \rho_{s+1} \to 1$ holds a.s. The proof of convergence of method (1), (2) will therefore be reduced to a slight modification of the traditional proof [2] associated with the fact that the step size ρ_s depends not only upon the vectors $(x^0, \ldots, x^s, \xi^0, \ldots, \xi^{s-1})$ but also upon the vector ξ^s.

THEOREM 1. *Let $f(x)$ be a convex (possibly nonsmooth) function defined on a convex, closed, bounded subset X of a separable Hilbert space H. If the conditions*

$$\max_{x,y \in X} \|x - y\| = C_1 ;\ roma \sup_{s=0,1,\ldots,x \in X} \|\xi(s, x, \omega)\| < C_2 \text{ a.s.} \quad ;$$

$$\overline{\lim_{s \to \infty}} \sup_{x \in X} \|b(s, x)\| \le \bar{b} ;\ a_s = a ,\ s = 0,1, \ldots ;\ a > 1 \quad ;$$

$$\delta > C_2 \overline{\lim_{s \to \infty}} \sup_{x \in X} \inf_{h \in \partial f(x)} \|\xi(s, x, \omega) - h\| \text{ a.s.} \quad ,$$

are satisfied then for a sequence $\{x^s\}$ defined by relations (1), (2) *we have*

$$\overline{\lim_{s \to \infty}} f(\bar{x}^s) \le f(x^*) + \bar{b}C_1 \text{ a.s.} \quad ,$$

where

$$x^* \in X^* = \{x^* : f(x^*) = \min_{x \in X} f(x)\} ,\ \bar{x}^s = \sum_0^s \rho_l x^l / \sum_0^s \rho_l \quad .$$

If, in addition, $\sum_0^\infty |b^s| \rho_s < \infty$ *holds a.s., then*

$$\lim_{s \to \infty} f(x^s) = f(x^*) \text{ a.s.} \tag{3}$$

Note that if X is a compact set then the fact that $f(x^s) \to f(x^*)$ implies that the sequence $\{x^s\}$ converges to the set X^*, i.e., $\min_{y \in X} \|x^s - y\| \to 0$ almost surely as $s \to \infty$.

The convergence of the sequence $\{\bar{x}^s\}$ to the extreme set will be referred to as *Cesàro convergence*. This type of convergence was considered in [11,12].

Convergence Rate. The first estimates of the convergence rate of the algorithm obtained by modifying algorithm (1), (2) are given in [10]. It can be shown that if the function $f(x)$ is twice continuously differentiable, then, under the conditions specified in Theorem 1, the step size ρ_s in relation (2) satisfies the asymptotic relation

$$\rho_s = \frac{1}{\delta s} + o(\frac{1}{s}) \quad .$$

Assume that there exists a constant $B > 0$ such that $f(x) \geq f(x^*) + B\|x^* - x\|^2$. Then it is possible to demonstrate that

$$\mathrm{E}\|x^* - x^s\| = O(\frac{1}{s}) \quad .$$

Computer Implementation of the Algorithms. Theorem 1 is proved assuming that $a_s = \text{const.}$, $s = 0,1,\ldots.$ This assumption can result in very rapid changes in the step size ρ_s at each iteration. In practice it is desirable that the exponent in relation (2) should be divided by some value z_s representing the average of the values $|<\xi^{s+1}, x^s - x^{s+1}>|$ and to specify some bounds for the maximum step size variation. In numerical experiments, we used the following recursive relations to compute the step size:

$$T_s = <\xi^{s+1}, x^s - x^{s+1}> ; \; z_s = z_{s-1} + (|T_s| - z_{s-1})D \quad ;$$

$$\tilde{\rho}_{s+1} = \rho_s a^{t_s/z_s} \cdot \begin{cases} 1, & \text{if } T_s > 0 , \\ u, & \text{if } T_s \leq 0 ; \end{cases} \rho_{s+1}$$

$a = 2$, $u = 0.8$, $D = 0.2$, $z_{-1} = 0$.

The computer implementation of this algorithm and the results of numerical experiments are described at greater length in [7,8].

3. A STOCHASTIC ARROW–HURWICZ ALGORITHM

The problem of finding the saddle points of convex-concave functions often arises in mathematical economics. Constrained stochastic programming problems can also often be reduced to searching for the saddle point [2].

Let X, L be convex, closed, bounded sets in Hilbert spaces H_x, H_l, respectively, and $F(x, l): X \times L \to R$ be a continuous convex-concave (possibly nonsmooth) function.

To find the saddle point of function $F(x, l)$ we will make use of a generalization of the Arrow–Hurwicz algorithm [2, 12–14]:

$$x^{s+1} = \pi_X(x^s - \rho_s \xi^s), \quad l^{s+1} = \pi_L(l^s + \gamma_s \eta^s) \quad , \tag{4}$$

where ξ^s, η^s are stochastic quasi-gradients at the point (x^s, l^s) with respect to the first and second groups of components.

To control the step sizes, we may use relations similar to (2):

$$\begin{aligned} \rho_{s+1} &= \rho_s a_x^{\gamma_s <\xi^{s+1}, x^s - x^{s+1}> - \delta_s \rho_s \gamma_s} \\ \gamma_{s+1} &= \gamma_s a_l^{\rho_s <\eta^{s+1}, l^{s+1} - l^s> - \beta_s \rho_s \gamma_s} \end{aligned} \tag{5}$$

where $a_x > 1$, $a_l > 1$, $\delta_s > 0$, $\beta_s > 0$, $s = 0,1, \ldots$.

Since the sequence of points $\{(\bar{x}^s, \bar{l}^s)\}$ is a convex combination of points (x^s, l^s), $s = 0,1, \ldots$, it is clear that

$$(\bar{x}^s, \bar{l}^s) = (\sum_{j=1}^{s} \rho_j \gamma_j (x^j, l^j)) / \sum_{j=1}^{s} \rho_j \gamma_j$$

converges in functional to a set of saddle points. Let $\hat{F}_x(x^s, l^s)$ denote a generalized gradient of the function F with respect to x at the point (x^s, l^s).

We shall now formulate a theorem [9] for Cesàro convergence of algorithm (4), (5).

THEOREM 2. *Let X, L be convex, closed, bounded sets in separable Hilbert spaces H_x, H_l, respectively, and $F(x, l): X \times L \to R$ be a continuous convex-concave function. Assume that*

$$\|\xi^s\| < C_x \text{ a.s.}, \ \|\eta^s\| < C_l \text{ a.s.}, \ s = 0,1, \ldots; \ \lim_{s \to \infty} \delta_s > C_x^2; \ \lim_{s \to \infty} \beta_s > C_l^2 \quad ;$$

$$\overline{\lim_{s \to \infty}} \| E[\xi^s \mid x^s, l^s] - \hat{F}_x(x^s, l^s) \| \le b_x \text{ a.s.} ;$$

$$\overline{\lim_{s \to \infty}} \| E[\eta^s / x^s, l^s] - \hat{F}_l(x^s, l^s) \| \le b_l \text{ a.s.}$$

Define

$$\nu_s = \max \{ \bar{F}(\bar{x}^s) - \inf_{x \in X} \bar{F}(x), \ \sup_{l \in L} \underline{F}(l) - \underline{F}(\bar{l}^s) \} \quad ,$$

$$\bar{F}(x) = \sup_{l \in L} F(x, l), \ \underline{F}(l) = \inf_{x \in X} F(x, l) \quad .$$

Then

$$\overline{\lim_{s \to \infty}} \ \nu_s \le b_x K_x + b_l K_l \text{ a.s.} \quad ,$$

where K_x, K_l are the diameters of the sets X and L, respectively.

When implementing algorithm (4), (5) it is desirable to vary the coefficients a_x, a_l during the iterative process. Suggestions concerning the computer implementation of this algorithm are offered in [9], together with the results of some numerical experiments.

4. AN ALGORITHM WHICH SEARCHES FOR NASH EQUILIBRIA IN NONCOOPERATIVE MANY-PERSON GAMES

Many problems in mathematical economics can be reduced to the search for Nash equilibria. Here we use a generalized gradient algorithm to search for such equilibria. Applications of algorithms of this type for cases in which the objective function is smooth are discussed in [15,16]; algorithms for the nonsmooth case were developed in [17,18]. In what follows we shall look at a deterministic version of the algorithm; however, algorithms of this type can easily be extended to the stochastic case, in which only statistical estimates of objective functions and their gradients are available.

The computer implementation of these algorithms, like that of stochastic quasigradient algorithms, involves difficulties associated with the control of parameters. The suggested approach helps us to overcome these difficulties.

Statement of the Problem. Define an n-person game as the object $\gamma = (X, \{\psi_i\}_{i=1}^n)$, where

(a) X is a convex, closed, bounded set lying in the product of Hilbert spaces $H = H_1 \times \cdots \times H_n$;

(b) $\psi_i(x_1, \ldots, x_n) = \psi_i(x)$, $i = 1, \ldots, n$ are the players' payoff functions defined on X;

(c) a function $\Psi(x, y) \overset{\Delta}{=} \sum_{i=1}^{n} \psi_i(x_1, \ldots, x_{i-1}, y_i, x_{i+1}, \ldots, x_n)$ is jointly continuous in its variables on $X \times X$ and concave with respect to y on X for each $x \in X$;

(d) a function $\Phi(x, y) = \Psi(x, x) - \Psi(x, y)$ is concave with respect to x for each $y \in X$.

The point $x^* = (x_1^*, \ldots, x_n^*)$ is referred to as an *equilibrium point* of the n-person game if for $i = 1, \ldots, n$ we have

$$\psi_i(x^*) = \max_{y_i} \{\psi_i(x_1^*, \ldots, x_{i-1}^*, y_i, \ldots, x_n^*) \mid (x_1^*, \ldots, x_{i-1}^*, y_i, x_{i+1}^*, \ldots, x_n^*) \in X\} \quad .$$

The point $x^* \in X$ is defined as a *normalized equilibrium point* (n.e.p.) if $x^* \in Z(x^*)$, where $Z(x) = \operatorname{Arg}\max_{y \in X} \Psi(x, y)$. Let X^* denote the set of normalized equilibrium points. A normalized equilibrium point is always an equilibrium point, but the converse is not always true. The set X^* is not empty, convex, closed, and bounded under the above assumptions. To find an n.e.p., we use the following algorithm:

$$x^{s+1} = \pi_X(x^s + \rho_s g(x^s)) \, , \; s = 0,1 \, ,\ldots \quad , \tag{6}$$

where

$$g(x^s) \in \partial_y \Psi(x^s, y)|_{y=x^s} \quad .$$

With sufficiently natural conditions on the parameters of algorithm (6) it can be shown to converge in a Cesàro sense, i.e., to the set of normalized equilibrium points X^* of the sequence $\{\bar{x}^s\}$

$$\bar{x}^s = (\sum_0^s \rho_l x^l) / \sum_0^s \rho_l \quad .$$

THEOREM 3. *Let* γ *be a game satisfying conditions* (a)–(d), *and* $\Psi(x,y)$ *be a Lipschitz function with a constant* L *with respect to* y, *where* L *does not depend on* x. *If the step sizes* ρ_s *in* (6) *satisfy the conditions*

$$\rho_s \geq 0 \, , \; s = 0,1 \, ,\ldots; \;\; \rho_s \to 0 \text{ for } s \to \infty ; \;\; \sum_0^\infty \rho_s = \infty \quad , \tag{7}$$

then the value $\nu_s = -\Phi(\bar{x}^s, z(\bar{x}^s))$, where $z(\bar{x}^s) \in Z(\bar{x}^s)$, converges to zero with

$$\nu_s = \frac{D^2(X)}{2\sum_0^s \rho_i} + \frac{L^2 \sum_0^s \rho_i^2}{2\sum_0^s \rho_i} \quad ,$$

$D(X)$ being the diameter of the set X. The sequence $\{\bar{x}^s\}$ converges weakly to the set of normalized equilibrium points X^*, i.e., any weak limit point of the sequence $\{\bar{x}^s\}$ belongs to X^*.

If, for instance, we take $\rho_s = O(1/\sqrt{s})$, the algorithm convergence rate is estimated to be $\nu^s \leq O((\ln s)/\sqrt{s})$.

Adaptive Step Size Control. The step size ρ_s in algorithm (6) may be chosen using the condition for the function $\psi_s(\rho)$ to be a maximum with respect to ρ:

$$\psi_s(\rho) = \Psi(x^s, x^s + \rho g(x^s)) \quad .$$

Now calculate the generalized gradient of the function $\psi_s(\rho)$ with respect to ρ at the point ρ_s:

$$\partial\psi_s(\rho_s) = \{<q\,, g(x^s)>: q \in \partial_y \Psi(x^s, y)|_{y = x^{s+1}}\} \quad .$$

To modify the step size ρ_s, we resort to the following gradient procedure

$$\rho_{s+1} = \rho_s a^{<q^{s+1}, x^{s+1} - x^s> - \delta_s \rho_s}, \; a > 1, \; \delta_s > 0 \quad , \qquad (8)$$

$$q^{s+1} \in \partial_y \Psi(x^s, y)|_{y = x^{s+1}} \quad .$$

It can be shown that the step size ρ_s chosen according to the above procedure satisfies conditions (7) under certain assumptions.

THEOREM 4. *Assume that conditions* (a)–(d) *hold for game* γ *and that the function* $\Psi(x, y)$ *satisfies the Lipschitz condition on* $X \times X$ *with respect to* x *and* y *with constants* K *and* L , *respectively. If* $a > 1$, $\lim_{s \to \infty} \delta_s > KL$, *then the step size* ρ_s *in relation* (8) *satisfies conditions* (7); *consequently* $\nu_s \to 0$ *as* $s \to \infty$ *and the sequence* $\{\bar{x}^s\}$ *converges weakly to* X^*.

Let us briefly consider some points regarding the computer implementation of algorithm (6), (8). Firstly, the parameter a in relation (8) should be changed at each iteration, as for the stochastic quasi-gradient algorithm (1), (2).

Secondly, when values q^s , $g(x^s)$ are calculated in the absence of noise, then the following simplest recursive procedure is sufficient:

$$\rho_{s+1} = \rho_s \begin{cases} \beta_2 \, , & \text{if } < q^{s+1}, x^{s+1} - x^s > > 0 \, , \\ \beta_1 & \text{if } < q^{s+1}, x^{s+1} - x^s > \geq 0 \quad , \end{cases}$$

$0 < \beta_1 < 1 < \beta_2$, $\beta_1 \beta_2 < 1$ (for example, $\beta_1 = 0.5$, $\beta_2 = 1.5$).

In conclusion it should be noted that the proposed approach to the control of algorithm parameters may be successfully extended to other stochastic and non-stochastic algorithms.

REFERENCES

1. H. Robbins and S. Monro. A stochastic approximation method. *Annals of Mathematical Statistics*, 22(1951) 400–407.

2. Yu.M. Ermoliev. *Methods of Stochastic Programming*. Nauka, Moscow, 1976 (in Russian).

3. A.M. Gupal. *Stochastic Methods for Solving Nonsmooth Extremal Problems*. Naukova Dumka, Kiev, 1979 (in Russian).

4. H. Kesten. Accelerated stochastic approximation methods. *Annals of Mathematical Statistics*, 29(1958) 41–59.

5. G. Pflug. On the determination of the step size in stochastic quasigradient methods. Working Paper WP-83-025, International Institute for Applied Systems Analysis, Laxenburg, Austria, 1983.

6. S.P. Urjas'ev. A stepsize rule for direct methods of stochastic programming (in Russian). *Kibernetika*, 6 (1980) 96-98.

7. S.P. Urjas'ev. Step size rules for some nonmonotone methods of nonsmooth optimization (in Russian). Dissertation, V.M. Glushkov Institute of Cybernetics, Ukrainian Academy of Sciences, Kiev, 1982.

8. F. Mirzoakhmedov and S.P. Urjas'ev. Adaptive step size control for stochastic optimization algorithm (in Russian). *Zhurnal Vychisletelnoi Matematiki i Matematicheskoi Fiziki*, 6(1983) 1314–1325.

9. S.R. Urjas'ev. Arrow–Hurwicz algorithm with adaptively controlled step sizes (in Russian). *Operations Research and AMS*, 24(1984) 11–16.

10. A. Ruszczynski and W. Syski. A method of aggregate stochastic subgradients with on-line step size values for convex stochastic programming problems. In A. Prekopa and R. Wets (Eds.), *Stochastic Programming* (forthcoming).

11. R. Bruck. On weak convergence of an ergodic iteration for the solution of variational inequalities for monotone operators in Hilbert space. *Journal of Mathematical Analysis and Applications*, 61(1977) 159–161.

12. A.S. Nemirovski and D.B. Judin. *Complexity of Problems and Efficiency of Optimization Methods*. Nauka, Moscow, 1979 (in Russian).

13. B.T. Pojak. Methods of solving conditional extremum problems under random noise (in Russian). *Zhurnal Vychislitelnoi Matematiki i Matematicheskoi Fiziki*, 1(1979) 70–78.

14. E.A. Nurminski. *Numerical Methods for Solving Deterministic and Stochastic Minimax Problems*. Naukova Dumka, Kiev, 1979 (in Russian).

15. I.B. Rosen. Existence and uniqueness of an equilibrium point in concave n-person games. *Econometrica*, 33(1965) 520–534.

16. M.E. Primak. On one computational process which searches for equilibria (in Russian). *Kibernetika*, 1(1973) 91–96.

17. Yu.M. Ermoliev and A. Papin. An approach to the simulation of international oil trade. Working Paper WP-82-45, International Institute for Applied Systems Analysis, Laxenburg, Austria, 1982.

18. Yu.M. Ermoliev and S.P. Urjas'ev. On the search for Nash equilibria in many-person games (in Russian). *Kibernetika*, 3(1982) 85–88.

STOCHASTIC MODELS AND METHODS OF OPTIMAL PLANNING

A.I. Yastremski
Department of Cybernetics, Kiev State University, Kiev, USSR

Stochastic models provide a valuable means of representing and studying economic systems under conditions of incomplete information. The use of stochastic models makes it possible (1) to abandon the requirement that input data must be completely determined; (2) to establish a number of new results through the theoretical study of economic systems; (3) to improve economic plans (using applied stochastic models); and (4) to state and solve a number of new problems which, even in theory, cannot be formulated in a deterministic framework. These problems include optimal inventory control; the development of plans, and of measures ensuring the stability of such plans; the calculation of optimal expenditures after refinement of the required information; and theoretical and applied studies of the flexibility and adaptability of economic plans.

It is important to define a standard model which represents the characteristics of decision making under uncertainty in a sufficiently general form. The multistage linear stochastic optimal planning model

$$\mathrm{E}(c(\vartheta), x(\vartheta)) \longrightarrow \max, \quad A(\vartheta)x(\vartheta) \le (\vartheta) \pmod P$$

$$x(\vartheta) \ge 0 \pmod P, \; x = (x_1, \ldots, x_n), \quad x_j(\vartheta) \text{ is } M_j\text{–measurable } (j = \overline{1,n}) \qquad (1)$$

is just such a model. Here $A(\vartheta)$ is a random input–output matrix of production processes, ϑ is an event in the probability space (Θ, F, P), $b(\vartheta)$ is a random resource vector, $c(\vartheta)$ is a random vector composed of the economic efficiencies of production processes, $x_j(\vartheta)$ is the required production rate vector at the j-th stage of the decision process, M_j is a σ-subalgebra of the basic σ-algebra F, and describes the information available at the j-th stage of the decision process.

The two-stage model [1] is characterized by the conditions

$$\{1, \ldots, n\} = U_1 \cup U_2, \quad U_1 \cap U_2 = \phi, \quad M_j = \{\Theta, \phi\} \text{ if } j \in U_1, \quad M_j = F \text{ if } j \in U_2 .$$

The two-stage model is convenient in practice since it leads to a two-stage stochastic programming problem and can be solved by existing efficient numerical methods [2].

Existing methods for the qualitative analysis of stochastic models make it possible to study the theoretical properties of models, to generalize some theorems in mathematical economics, to test the stability of models, to determine the levels of resource deficiency, to examine the dependence of system efficiency upon the error level, and so on.

The analysis of linear stochastic models, like that of linear programming problems, is based on duality and optimality conditions. From the point of view of optimality conditions, problem (1) represents rather a special case. First, it is required that $x_j(\vartheta)$ be M_j-measurable, while the problem parameters are defined on the whole σ-algebra F. Second, the choice of the functional space over which the operator $b(\vartheta) - A(\vartheta)x(\vartheta)$ takes its values is of great significance. References [3–5] explore problems where the constraint operator takes values from the space L_∞. This assumption makes it possible to formulate a stochastic counterpart of the Slater condition which is used to prove the existence of stochastic Lagrange multipliers. The present paper explores problems where the constraint operator takes values from the space $L_p (1 < p < \infty)$. This makes it possible to study some classes of problems which are not restricted to the case of random variables constrained with probability 1. At the same time it is senseless to speak about the Slater condition in connection with such problems since, due to the properties of the L_p-norm, the cone of the L_p-random variables non-negative with probability 1 has no interior points. To formulate meaningful duality theorems and optimality conditions, it may at first sight seem that the assumption should be a little stronger, i.e., that solutions to the primal and dual problems must exist. However, using the fact that for a multistage linear stochastic program it is possible to write the dual problem in explicit form, we can obtain meaningful optimality conditions for such problems without strengthening the assumption. The dual problem of (1) is

$$\mathrm{E}(b(\vartheta), u(\vartheta)) \to \min \quad ,$$

$$\mathrm{E}([A^{\mathrm{T}}(\vartheta)u(\vartheta)]_j \mid M_j) \geq \mathrm{E}(c_j(\vartheta) \mid M_j) \; (\mathrm{mod}\, P) \; (j = \overline{1, n}) \tag{2}$$

$$u(\vartheta) \geq 0 \; (\mathrm{mod}\, P) \quad ,$$

where dual variables $u(\vartheta)$ are chosen from the space $L_q (1/q + 1/p = 1)$ and $E(\xi \mid M_j)$ is the conditional expectation with respect to the σ-subalgebra M_j. The formulation of problem (1) is based on the definition of dual problems applied to extremum problems in a Banach space [6], taking into account the specificity of problem (1) and the properties of conditional expectations.

Certain assumptions concerning the existence of bounded generalized solutions as discussed in [6]. Under these assumptions, a number of duality relationships and optimality conditions which admit meaningful economic interpretation can be established. Just as in the deterministic case, the optimal values of dual variables can be regarded as the optimal prices of the resources. Using the properties of quasi-differentiable functionals [7], it is possible to prove the marginal property for stochastic prices:

$$\mu(b(\vartheta)+\gamma e(\vartheta)) = \mu(b(\vartheta)) + \gamma \min_{u(\vartheta)\in U^*} \mathrm{E}(e(\vartheta), u(\vartheta)) + o(\gamma) \quad , \tag{3}$$

where $\mu(b(\vartheta))$ is the optimal value of the objective functional in (1). Equality (3) is an extension of the result established in [8] to the multistage problem in which the assumption that the probability measure is absolutely continuous is abandoned. Using (3), it is possible to prove a property similar to that of basic stability for a linear programming problem [2,9].

Reference [2] explores the following dynamic programming model:

$$\mathrm{E}(p(\vartheta), x(T)) \longrightarrow \max$$

$$x(\tau+1) = B(\vartheta,\tau)h(\tau,\vartheta) + b(\tau,\vartheta) \ (\mathrm{mod}\ P) ,$$

$$x(0) = x^0 , \ h(\tau,\vartheta) \geq 0 \ (\mathrm{mod}\ P) \quad ,$$

$$h(\tau, 0) \text{ is } M_\tau\text{-measurable} \quad .$$

The technological and economic growth rates $[\alpha(\tau), \beta(\tau)]$ and the rate of interest $[\rho_\tau(e(\vartheta))]$ are introduced in a natural fashion. They are defined by

$$\alpha(\tau) = \mathrm{E}(x^*(\tau+1), u^*(\tau,\vartheta))/\mathrm{E}(x^*(\tau), u^*(\tau,\vartheta)) \quad ,$$

$$\beta(\tau) = \mathrm{E}(x^*(\tau+1), u^*(\tau,\vartheta))/\mathrm{E}(x^*(\tau+1), u^*(\tau+1,\vartheta)) \quad ,$$

$$\rho_\tau(e(\vartheta)) = \lim \frac{\mathrm{E}(u_e(\tau+1,\vartheta), e(\tau+1))}{\mathrm{E}(u_e(\tau+1,\vartheta), \gamma e(\vartheta))} - 1 \quad ,$$

respectively, where $x^*(\tau)$ is the volume of production corresponding to the optimal solution, $u^*(\tau,\vartheta)$ are stochastic shadow prices, $\gamma e(\vartheta)$ is the production increment in direction $e(\vartheta)$ with rate γ during year τ, $e(\tau+1)$ is the production growth during year $\tau+1$ resulting from the increment $\gamma e(\vartheta)$ during year τ, and $u_e(\tau,\vartheta)$ are stochastic shadow prices similar to those appearing in (3) (marginal prices).

Relationships between α, β, and ρ are established which illustrate the nature of these values. Under assumptions which ensure that the duality relationships hold, it is possible to show that

$$\operatorname{sgn} \mathrm{E}(u^{*}(\tau+1,\vartheta)\,,\, b(\tau,\vartheta)) = \operatorname{sgn}\,(\alpha(\tau) - \beta(\tau)) \tag{4}$$

and

$$1 + \rho_\tau(e(\vartheta)) = \mathrm{E}(u_e(\tau+1,\vartheta)\,,\, e(\vartheta)) / \mathrm{E}(u_e(\tau+1,\vartheta)\,,\, e(\vartheta)) \quad . \tag{5}$$

Equality (4) shows that the price and production indices are related to each other through the "load" on the economy, while (5) reveals that the rate of interest is linked to the price index.

Using these general results, we can study specific cases of the general two-stage stochastic model, in particular the stochastic counterpart of the deterministic optimal planning model [10]

$$z \rightarrow \max\,,\; Bx \geq \alpha z\,,\; Ax \leq b\,,\; x \geq 0 \quad , \tag{6}$$

which is constructed under the assumption that the unrecoverable resources (matrix A) are deterministic, the recoverable resources (matrix B) are stochastic, and the plan x cannot be corrected. In this case (6) becomes the stochastic model

$$F(x) = \mathrm{E} \min_i \frac{1}{\alpha_i}[B(\vartheta)x]_i \rightarrow \max\,,\; Ax \leq b\,,\; x \geq 0 \quad . \tag{7}$$

Particular cases of model (7), including a stochastic model of cultivated area distribution and a stochastic input–output model, have been used in practice [11,12].

Stochastic methods are also useful for examining the interactive procedures by which decisions are made. One characteristic of optimal planning is that it is difficult to represent the outcomes of a plan by an explicitly specified function describing the priorities of the person responsible for creating the system. To compare plans, it is convenient to use the reflexive binary relation $x \quad y$, which should be interpreted as "plan x is no worse than plan y". Let D be a set of feasible plans. The problem of identifying a preferred plan can be stated in the following manner: find an $x^{*} \in D$ such that $x^{*} \quad x$ holds for all $x \in D$. The plan x^{*} is said to be the most highly preferred and the problem of finding this plan is conventionally written as

$$x \rightarrow \mathrm{pref}\,,\; x \in D \quad . \tag{8}$$

Under certain assumptions, such as the completeness, continuity, transitivity and convexity of relation $\quad$, and the compactness of D, a formal method for solving problem (8) has been proposed [2], which is based on the following assignment:

$$\xi^s = \begin{cases} \vartheta^s\,, & (x^s + \gamma_s \vartheta^s) \quad x^s \\ -\vartheta^s\,, & (x^s + \gamma_s \vartheta^s) \quad x^s \end{cases} \quad . \tag{9}$$

Here ϑ^s is an independent observation of the random vector ξ^s (which is uniformly distributed over the n-dimensional unit ball $\|x\| \leq 1$) obtained at iteration number s, and γ_s is a value which is calculated by a special rule. The formula

$$E(\xi^s \mid x^s) = a \frac{\nabla u(x^s)}{\|\nabla u(x^s)\|} , \tag{10}$$

is derived in [2], where $a > 0$, $u(x)$ satisfies the condition $u(x) \geq u(y) \iff x \quad y$. Formal computational methods based on (9) can be used to simulate real-life interactive procedures.

Note that (10) is proved in [2] under the assumption that $u(x)$ is concave and differentiable. It is not difficult to extend this proof to the case where $u(x)$ is quasiconvex and differentiable. Such generalizations of problem (8) and modifications of the methods evolved in [2] for these generalizations are described in [13].

REFERENCES

1. Yu.M. Ermoliev and A.I. Yastremski. *Stochastic Models and Methods in Economic Planning*. Nauka, Moscow, 1979 (in Russian).
2. Yu.M. Ermoliev. *Methods of Stochastic Programming*. Nauka, Moscow, 1976 (in Russian).
3. R.T. Rockafellar and R.J.-B. Wets. Non-anticipativity and L^1-martingales in stochastic optimization problems. *Mathematical Programming Study*, 6(1976) 170–187.
4. *Stochastic Models of Control and Economic Dynamics*. Nauka, Moscow, 1979 (in Russian).
5. Optimality conditions for the multi-step problem of random convex mapping control (in Russian). *Kybernetika*, (1)(1983) 83–87.
6. E.G. Golstein. *Duality Theory in Mathematical Programming*. Nauka, Moscow, 1971 (in Russian).
7. B.N. Pshenichnyi. *Necessary Conditions for an Extremum*. Nauka, Moscow, 1982 (in Russian).
8. B.M. Efimov. Optimal estimates under conditions of uncertainty (in Russian). *Ekonomiki i Matematicheski Metodi*, 6(1970) 21–24.

9. A.I. Yastremski. Marginal relations in the linear problem of multistage stochastic programming. *Cybernetics*, 16(1980)297–300.

10. L.V. Kantorovich. *Economic Calculations on the Best Use of Resources*. Academy of Sciences of the USSR, Moscow, 1959 (in Russian).

11. I.K. Fedorenko. A stochastic planning model of agricultural production and its application. *Cybernetics*, 16(1980)423–428.

12. A.I. Yastremski, V.I. Golovko and A.B. Demkovski. Research on the influence of information accuracy on the efficiency of multi-industrial systems, carried out using stochastic multi-industrial models (in Russian). *Operations Research and ACS*, 22(1983)12–20.

13. A.I. Yastremski and M.V. Mikhalevich. Stochastic search methods for finding the most preferred element and their interactive interpretation. *Cynbernetics*, 16(1980)893–898.

Section III
Problems with Incomplete Information

DIFFERENTIAL INCLUSIONS AND CONTROLLED SYSTEMS: PROPERTIES OF SOLUTIONS

A.V. Bogatyrjov
Institute of Control Sciences, Moscow, USSR

1. INTRODUCTION

Consider a differential inclusion of the form

$$\dot{x} \in R(x)\ ,\ x(t_0) \in X_0 \quad , \tag{1}$$

with a multivalued mapping $R\colon E^n \rightarrow 2^{E^n}$ on its right-hand side. Here 2^{E^n} denotes the space of non-empty subsets of E^n. These mathematical objects are attracting much attention today since they are useful in solving certain classes of problems. For example, they can be used in the investigation of controlled systems of the form

$$\dot{x} = f(x\ , u)\ ,\ u \in U\ ,\ x(t_0) \in X_0 \quad , \tag{2}$$

which by virtue of Filippov's Lemma [1], are reduced to the differential inclusion

$$\dot{x} \in f(x\ , U)\ ,\ x(t_0) \in X_0 \quad . \tag{3}$$

We arrive at a similar inclusion if u is some unknown functional parameter or noise rather than a control.

When investigating the properties of differential inclusions we naturally assume that the mapping R on the right-hand side of (1) satisfies certain requirements. In some cases, for instance [2–4], cnsideration is limited to a class of inclusions with convex $R(x)$ (or $R(t\ , x)$ for non-autonomous differential inclusions). One may treat the inclusions studied in these papers as generalizations of ordinary differential equations of the Caratheodory type. Not only do their solutions have all the features typical of the solutions of Caratheodory differential equations, but the inclusions themselves turn into the above differential equations if the mappings $R(t\ , x)$ are single-valued functions.

However, the sets $f(x\ , U)$ in differential inclusion (3) are not necessarily convex. In this paper we consider differential inclusions (1) with non-convex sets $R(x)$ which may conditionally be regarded as generalizations of ordinary differential equations whose right-hand sides are Lipschitzian functions. The reasons for this, which are treated in more detail below, are as follows. First, when a set $R(x)$ is a single point for all x the mapping R is a Lipschitzian function. Second, the family of solutions of such differential inclusions have properties which are stronger than those treated in [2–4]. A comparison of the properties of the solutions of an ordinary differential equation which has a Lipschitzian right-hand side with those of inclusion (1) is pointless because this differential equation has a unique solution.

2. DESCRIPTION OF THE CLASS OF DIFFERENTIAL INCLUSIONS

The form of the differential inclusion (3) suggests a transition from differential equations with Lipschitzian right-hand sides to differential inclusions (1).

A set of continuous functions $h_\alpha: E^n \to E^n$, $\alpha \in A$, is called an *exhaustive family of continuous selections* of multivalued mapping R if the following relations hold:

$$h_\alpha(x) \in R(x)\,,\ R(x) \subset \overline{\bigcup_\alpha h_\alpha(x)}\,,\ x \in E^n \quad . \tag{4}$$

Definition 1. A multivalued mapping $R: E^n \to 2^{E^n}$ with an exhaustive family of continuous selections satisfying the Lipschitz condition with a single constant K is said to be *K-dense*.

System (2) with a function f which satisfies the Lipschitz condition with constant K with respect to x may be reduced to a differential inclusion with a K-dense multivalued mapping on the right-hand side.

However, not for any differential inclusion (1) with a K-dense right-hand side it is possible to find a controlled system (2) such that their solutions coincide. For example, consider inclusion (1) with initial set $X_0 = [-1,1]$ whose right-hand side contains a mapping $R: E^1 \to 2^{E^1}$ of the following type:

$$R_1(x) = \begin{cases} \text{all rational points in the segment} & \\ [-1,1] \text{ except zero} & ,\ \text{if } x \neq 0 \\ [-1,1] & ,\ \text{if } x = 0 \end{cases} \quad .$$

Assume that the solutions of the differential inclusion

$$\dot{x} \in R_1(x)\,,\ x(t_0) \in [-1,1] \tag{5}$$

coincide with the solutions of some system (2). Since the set of solutions for inclusion (5) contains zero, this is also a solution of (2), which means that there exists a $u_0 \in U$ for which $f(x, u_0) = 0$. On the other hand (5) does not have any other constant solutions $x(t) \equiv c \neq 0$. Therefore $f(x, u_0) \neq 0$ if $x \neq 0$. Thus it is possible to find the point $x_0 \in [-1,1]$ at which the function $f(x, u_0)$ takes an irrational value. Let us denote the solution of the equation $\dot{x} = f(x, u_0)$ with initial condition $x(t_0) = x_0$ by $x(t)$. For any $t' > t_0$ the set

$$E = \{t \in [t_0, t']: f(x(t), u_0) \text{ is irrational number}\}$$

is of non-zero measure. Since $\dot{x}(t) \notin R(x(t))$ holds everywhere on E, $x(t)$ is not a solution of inclusion (5).

It can be observed that the condition of K-denseness is satisfied without requiring $R(x)$ to be convex, closed, bounded, measurable or to have any other property.

We shall now proceed directly to the results of the study. The following section is of an auxiliary nature.

3. CONNECTEDNESS OF A SET OF FIXED POINTS

Let X be a Banach space. As in the finite-dimensional case, we shall denote the space of nonempty subsets of X by 2^X. The point $x_0 \in X$ is called a *fixed point* of the multivalued mapping $F: X \to 2^X$ if $x_0 \in F(x_0)$.

Let us recall the definitions of metrical and linear connectedness. A set $A \subset X$ is said to be *metrically connected* if it is impossible to find two open sets B_1 and B_2 such that

$$B_1 \cap B_2 = \phi\,;\; B_i \cap A \neq \phi\,,\; i = 1,2\,;\; A \subset B_1 \cup B_2 \quad .$$

A set $A \subset X$ is said to be *linearly connected* if for all points there exists a continuous function q: $[0,1] \to A$ such that $q(0) = x_0$, $q(1) = x_1$.

Let us now define another type of connectedness.

Definition 2. A set $A \subset X$ is said to be *strongly linearly connected* if for all its points x_0, x_1 there is a linearly connected subset $G(x_0, x_1) \subset A$ containing these points such that

$$\operatorname{diam} G(x_0, x_1) \leq \omega(\|x_0 - x_1\|) \quad ,$$

where $\omega(t) \to 0$ as $t \to +0$.

In other words, neighboring points in a strongly linearly connected set may be connected by a continuous curve of small diameter.

Let L_I^1 denote a space of functions which are integrable over segment I with the usual norm

$$\|u\| = \int_I |u(\tau)|\,d\tau \quad .$$

A set $A \subset L_I^1$ is said to be *convex under switching* if for any functions $f_1, f_2 \in A$ and any measurable set $I_1 \subset I$ the inclusion $f \in A$ holds, where

$$f(t) = \chi_{I_1}(t) f_1(t) + (1 - \chi_{I_1}(t)) f_2(t) \quad .$$

Here χ_{I_1} is the characteristic function of set I_1.

THEOREM 1. *Let the multivalued mapping $F\colon L_I^1 \to 2^{L_I^1}$ have an exhaustive family of continuous selections (defined as in the finite-dimensional case), satisfying the single Lipschitzian constant $l_0 < 1/2$. In addition, let sets $F(u)$ be convex under switching for any $u \in L_I^1$. Then the fixed point set of the multivalued mapping F is strongly linearly connected.*

The proof relies upon the use of Theorem 1.2* and Lemma 1.1 from [5].

Remark 1. The assertion of Theorem 1 remains true if the norm in space L_I^1 is replaced by

$$\|u\|_{L_I^1} = \int_I \exp(-2LKt) \, |u(t)| \, dt \quad , \tag{6}$$

where K and L are positive constants.

Remark 2. The assertion of Theorem 1 remains true in any Banach space if the condition of convexity under switching is replaced by a convexity requirement and the demand that constant l_0 should satisfy $l_0 < 1$.

4. CONNECTEDNESS OF A FAMILY OF SOLUTIONS TO DIFFERENTIAL INCLUSIONS

Let C_I denote the space of continuous functions in segment I with the norm $\|x(\cdot)\|_{C_I} = \max_{t \in I} |x(t)|$, and A_I denote the space of functions which are absolutely continuous on I with the norm

$$\|x(\cdot)\|_{A_I} = |x(t_0)| + \int_I |\dot{x}(t)| dt \quad .$$

Theorem 1 may now be used to prove the following assertion:

THEOREM 2. *Assume that the right-hand side of the differential inclusion* (1) *is a K-dense multivalued mapping. Then for an arbitrary segment $I = [t_0, t_1]$ the family of solutions of this inclusion under the initial condition $x(t_0) = x_0$ is strongly linearly connected in the space A_I.*

The proof goes as follows. Consider the multivalued mapping $M_R\colon C_I \to 2^{L_I^1}$ which associates the set

$$M_R(x) = \{\nu \in L_I^1 : \nu(t) \in R(x(t)) \text{ almost everywhere at } I\}$$

with any continuous function $x(t)$. The non-emptines of the set $M_R(x)$ follows from the fact that the multivalued mapping $R(x(\cdot)): I \to 2^{E^n}$ has a continuous selection which may be taken to be $\nu(t)$. For an arbitrary function $u \in L_I^1$ the mapping $P: L_I^1 \to C_I$ may be obtained from the formula

$$P(u)(t) = x_0 + \int_{t_0}^{t} u(\tau)d\tau \quad .$$

Set

$$F(u) = M_R(P(u)) \quad . \tag{7}$$

It may easily be seen that the fixed points of the mapping $F: L_I^1 \to 2^{L_I^1}$ are the derivatives of the solutions of differential inclusion (1). Let us show that mapping F satisfies Theorem 1 (or, to be more accurate Remark 1). Let $x_0(t)$ be an arbitrary function which is continuous in the segment I. Consider a multivalued mapping $R(x_0(\cdot)): I \to 2^{E^n}$ and choose a function $f_0 \in L_I^1$ for which the relation $f_0(t) \in R(x_0(t))$ holds almost everywhere on I.

The fact that the multivalued mapping R possesses an exhaustive family of Lipschitzian selections allows us to construct a mapping $f^N: C_I \to L_I^1$ which has the following properties:

(1) $f^N(x)(t) \in R(x(t))$ almost everywhere on I for any continuous function $x(\cdot) \in C_I$;

(2) $|f^N(x)(t) - f^N(y)(t)| \le K\ |x(t) - y(t)|$ almost everywhere on I for any continuous functions $x(\cdot),\ y(\cdot) \in C_I$;

(3) $\|f^N(x_0) - f_0\|_{L_I^1} \to 0$ as $N \to \infty$.

Mappings f are actually constructed in the same way as in the proof of Theorem 2.1 in [5]. Let us define the mapping $g^N: L_I^1 \to L_I^1$ by the equality

$$g^N(u) = f^N(P(u)) \quad . \tag{8}$$

Consider norm (6) in space L_I^1. As shown in [6], property (2) guarantees that the mapping g^N satisfies the Lipschitz condition with the constant $1/2L < 1/2$. Furthermore, from property (1), g^N is a continuous selection of mapping (7), and property (3) yields

$$\|g^N(u_0) - f_0\|_{L_I^1} \to 0 \text{ as } N \to \infty \quad , \tag{9}$$

where $x_0 = P(u_0)$. Thus mapping (7) satisfies Remark 1, which proves the theorem.

Let us take an abitrary solution $x_0(t)$ of differential inclusion (1) under the initial condition $x_0(t_0) = x_0$, and let its derivative $\dot{x}_0(t)$ be $f_0(t)$, i.e., $f_0(t) = \dot{x}_0(t)$.

As in [5], it is possible to design mappings (8) whose fixed points $u^N = g^N(u^N)$ are the derivatives $\dot{x}_N$ of the solution of some differential equation

$$\dot{x} = a^N(t, x) \tag{10}$$

Here the functions $a^N(t, x)$ are measurable with respect to t and satisfy the Lipschitz condition with constant K with respect to x. In addition, the limit relation (9) and the fact that mappings (8) are contractive guarantee that $\|u^N(\cdot) - \dot{x}_0(\cdot)\|_{L_I^1} \to 0$ as $N \to \infty$. These points may be summarized in a separate lemma.

LEMMA 1 (on the approximating sequence). *Let the right-hand side of differential inclusion* (1) *be a K-dense multivalued mapping. Then for an arbitrary solution* $x(t)$ *of inclusion* (1) *under the initial condition* $x(t_0) = x_0$ *it is possible to find a sequence of functions*

$$a^N(\cdot,\cdot): I \times E^n \to E^n \ ; \ a^N(t, x) \in R(x) \quad ,$$

which is summable on I *for any* $x \in E^n$ *and satisfies with respect to* x *the Lipschitz condition with constant* K *such that for a sequence* x_N *of solutions* (10) *under the initial condition* $x_N(t_0) = x_0$ *the following relation holds:*

$$\|\dot{x}_N(\cdot) - \dot{x}(\cdot)\|_{L_I^1} \to 0 \text{ as } N \to \infty \quad . \tag{11}$$

With the help of Theorem 2 and Lemma 1 a more general result may be obtained.

THEOREM 3. *Assume that the right-hand side of the differential inclusion* (1) *is a K-dense multivalued mapping. Then if set* X_0 *is metrically (linearly, strongly linearly) connected, the family of solutions* $\sum(t_0, X_0)$ *of this inclusion with initial set* X_0 *is also metrically (linearly, strongly linearly) connected in spaces* C_I *and* A_I *on an arbitrary segment* $I = [t_0, t_1]$.

It should be noted that for ordinary differential equations satisfying the Peano condition only metrical connectedness of the family of solutions can be guaranteed. This is shown in [7].

5. THEOREMS ON BOUNDARY SOLUTIONS

The set

$$Z(t, t_0, X_0) = \{x(t): x(\cdot) \in \textstyle\sum(t_0, X_0)\}$$

is referred to as the *attainable set* of the differential inclusion (1) at time $t > t_0$. Any solution of inclusion (1) with K-dense right-hand side may be extended toward the entire half-line $t \geq t_0$. Therefore sets $\mathbf{Z}(t, t_0, X_0)$ are non-empty for any $t \geq t_0$.

The set of boundary points of the attainable set, i.e., those points in $\mathbf{Z}(t, t_0, X_0)$ which are not inner points, are denoted by $\partial\, \mathbf{Z}(t, t_0, X_0)$.

A solution $x(t)$ of the differential inclusion (1) is said to be a boundary solution for some segment (interval) of the real line if the relation $\dot{x}(t) \in \partial\, \mathbf{Z}(t, t_0, X_0)$ holds everywhere on this segment (interval).

Consider an arbitrary point $x^* \in \partial\, \mathbf{Z}(t^*, t_0, X_0)$. Then there exists a solution $\hat{x}(t)$ of the differential inclusion (1) for this point which satisfies the conditions $\hat{x}(t_0) \in X_0$ and $\hat{x}(t^*) = x^*$. The question of whether there exists a boundary solution among the $\hat{x}(t)$ has been studied by many authors. The most general result for a differential inclusion with convex-valued right-hand side is given in [4]. Making an additional assumption, it was proved in [8] that only boundary solutions can reach the boundary point.

As shown by the next theorem, a similar assertion holds for inclusion (1) with a K-dense right-hand side.

THEOREM 4. *Let the right-hand side of the differential inclusion* (1) *be a K-dense multivalued mapping. Then if the relation* $x(t') \in \partial\, \mathbf{Z}(t', t_0, X_0)$ *holds for some solution* $x(t_0)$ *of this inclusion under the initial condition* $x(t_0) \in X_0$ *at a time* $t' > t_0$, *the inclusion* $x(t) \in \partial\, \mathbf{Z}(t, t_0, X_0)$ *is true for any* $t \in [t_0, t']$.

The proof of this theorem with regard to Lemma 1 is identical to that of Theorem 3.1 in [5].

A result similar to Theorem 4 is given in [9], where the Lipschitz condition is somewhat weakened but it is required that sets $R(x)$ be compact.

We shall say that a solution $x(t)$ produces a transition from the initial set X_0 into the point x_1 during time T if the relations $x(t_0) \in X_0$, $x(t_0 + T) = x_1$ hold. The solution producing the transition from X_0 into x_1 within the minimal time T_0 is referred to as time-optimal, and T_0 is the optimal time.

THEOREM 5. *Assume that the right-hand side of* (1) *is a K-dense multivalued mapping and that the solution* $x(t)$ *is time-optimal for the transition from the initial set* X_0 *into the point* x_1. *Then, if the optimal time is* T_0, *the relation* $x(t) \in \partial\, \mathbf{Z}(t, t_0, X_0)$ *holds for all* $t_0 \leq t < t_0 + T_0$.

The proof of this theorem is identical to that of Theorem 3.2 in [5].

A somewhat weaker assertion for differential inclusions with convex-valued right-hand sides was presented in [10].

The above results hold for non-autonomous controlled systems $\dot{x} = f(t, x, u)$, $u \in U$, where the function f satisfies the Lipschitz condition with respect to x with the summable constant $k(t)$.

REFERENCES

1. A.F. Filippov. Some problems in the theory of optimal control (in Russian). *Vestnik Moskovskogo Universiteta, Seriya I, Matematika*, 2(1959)25–32.
2. A.F. Filippov. Differential equations with multivalued right-hand sides (in Russian). *Doklady Akademii Nauk USSR*, 151(1)(1963)65–68.
3. N. Kikuchi. On some fundamental theorems of contingent equations in connection with control problems. *RIMS Kyoto University*, 3(2)(1967)177–201.
4. J.L. Davy. Properties of the solution set of a generalized differential equation. *Bulletin of the Australian Mathematical Society*, 6(3)(1972) 379–398.
5. A.V. Bogatyrev. Fixed points and properties of solutions of differential inclusions (in Russian). *Izvestiya Akademii Nauk SSSR, Seriya Matematicheskaya*, 47(4)(1983)895–909.
6. C.J. Himmelberg and F.S. Van Vleck. Lipschitzian generalized differential equations. *Rendiconti del Seminario Matematico di Padua*, 48(1973)159–169.
7. M.F. Bokstein. Existence and uniqueness theorems for solutions of differential equations (in Russian). *Uchenye Zapiski, Moskva Universitet, Seriya Matematicheskaya*, 15(1939)3–72.
8. A. Plis. On the trajectories of orientor fields. *Bulletin de l'Academie Polonaise des Sciences*, 13(8)(1965)571–573.
9. K.A. Grasse. Some remarks on extremal solutions of multivalued differential equations. *Journal of Optimization Theory and Applications*, 40(2)(1983)221–235.
10. T. Wazewski. On an optimal control problem. In the *Proceedings of the Conference on Differential Equations and their Applications*, Prague, 1962, pp. 229–242.

GUARANTEED ESTIMATION OF REACHABLE SETS FOR CONTROLLED SYSTEMS

F.L. Chernousko
Institute of Mechanics Problems
Moscow, USSR

1. INTRODUCTION

Uncertainty is generally treated in one of two different ways: using a stochastic approach or a guaranteed approach. The stochastic approach is concerned with probabilities: each uncertain or unknown n-dimensional vector x is associated with some probability distribution $p(x)$. One of the most common of these is the Gaussian distribution

$$p(x) = c \exp (-D^{-1}(x - a) , (x - a)) \quad . \tag{1}$$

Here a is an n-dimensional vector of expected values, D is a symmetric, positive-definite $n \times n$ matrix, c is a scalar coefficient and $(\cdot , \cdot)$ is a scalar product of two vectors.

The guaranteed approach is concerned with the sets to which the unknown vectors belong. We denote this by $x \in M$, where M is a set in n-dimensional space. The guaranteed approach has some important advantages over the stochastic one. It leads to guaranteed results which hold for any individual set of circumstances, while the stochastic approach is to be preferred if many different possibilities have to be taken into account. Furthermore, the deterministic approach does not require any information on probability distributions, which is often not available in practical applications.

The deterministic approach has one main drawback: even if the initial sets to which the unknown vectors belong (the uncertainty sets) have simple shapes (e.g., parallelepipeds or spheres), basic operations with these sets, such as union and intersection, often lead to sets of complicated shape which require many parameters for their description. For instance, the intersection of several spheres is not a convenient set for further operations.

We therefore have to approximate these sets by means of sets which have some simple canonical shape described by a fixed number of parameters. Ellipsoids seem to be the most convenient approximating sets for a number of reasons:

(1) Ellipsoids can be fully described by only a small number of parameters: the vectors giving the foci and a symmetric, positive-definite matrix;

(2) Ellipsoids provide reasonably satisfactory approximations of arbitrary convex sets (we have the following result [1]: if Ω is a convex set in n-dimensional space with central symmetry, then there exists an ellipsoid E such that $E \subset \Omega$, $\sqrt{n}E \supset \Omega$);

(3) Ellipsoidal sets are similar to Gaussian distributions (1);

(4) The class of ellipsoids is invariant with respect to linear transformation.

A guaranteed approach to the treatment of uncertainties in dynamical systems has been used in, e.g. [2–6]. Ellipsoidal approximation is considered in [7–9], where some operations with ellipsoids are given, and a method of state estimation based on these operations is developed.

The present paper is devoted to a method for the two-sided approximation of attainable and uncertainty sets by means of ellipsoids which are optimal in terms of their volume. This method was originally suggested in [10–12], where optimal and suboptimal algebraic operations with ellipsoidal sets are given, and the differential equations of the approximating ellipsoids for dynamic systems are derived. These results are developed and summarized in [13–15]. Some properties of approximating ellipsoids are studied and some applications of the ellipsoid method are given in [15–23]. In this paper we present briefly the principal results obtained in [10–23].

2. REACHABLE SETS

We consider a controlled system described by differential equations, constraints and an initial condition:

$$\begin{gathered} \dot{x} = f(x, u, t), \; u(t) \in U(x(t), t) \\ x(s) \in M, \; t \geq s \quad . \end{gathered} \tag{2}$$

Here t represents time, x is an n-vector of state variables, u is an m-vector of controls, f is a given function, $U(x, t)$ is a given set of constraints in m-dimensional space, s is the initial time and M is a given initial set. The components of vector u in (2) may be either controls or disturbances. The state trajectories $x(t)$ of system (2) satisfy the following differential inclusion:

$$\begin{gathered} \dot{x} \in X(x, t), \; X(x, t) = f(x, U(x, t), t) \\ x(s) \in M, \; t \geq s \quad . \end{gathered} \tag{3}$$

We consider system (3) instead of (2). We are interested in the reachable or attainable sets for systems (2), (3). The reachable set $D(t, s, M)$ for system (2) (or (3)) is the set of all vectors $x(t)$ which are values of functions $x(\tau)$ satisfying (3) for $\tau \in [s, t]$. Reachable sets are essential in a number of problems in control and estimation theory.

Note the following important property of reachable sets:

$$D(t, s, M) = D(t, t_1, D(t_1, s, M)), \ t \geq t_1 \geq s \quad . \tag{4}$$

$\Omega^-(t)$ defined for $t \geq s$ will be called ***sub-reachable*** sets for system (2) (or (3)) if for all $t_1 \in [s, t]$ we have

$$\Omega^-(t) \subset D(t, t_1, \Omega^-(t_1)), \ t \geq t_1 \geq s \quad . \tag{5}$$

The reverse inclusion,

$$\Omega^+(t) \supset D(t, t_1, \Omega^+(t_1)), \ t \geq t_1 \geq s \quad , \tag{6}$$

defines ***super-reachable*** sets $\Omega^+(t)$. Properties (5), (6) are similar to the evolutionary property (4) of reachable sets.

Our aim is to obtain sub-reachable and super-reachable sets which provide simple two-sided bounds for reachable sets:

$$\Omega^-(t) \subset D(t, s, M) \subset \Omega^+(t) \quad . \tag{7}$$

3. ELLIPSOIDAL BOUNDS

Let $E(a, Q)$ denote an ellipsoid in n-dimensional space defined by the inequality

$$E(a, Q) = \{x : (Q^{-1}(x - a), (x - a)) \leq 1\} \quad . \tag{8}$$

Here a is an n-vector representing the foci of the ellipsoid and Q is a symmetric, positive-definite $n \times n$ matrix. Note that if $Q \to 0$ then the ellipsoid $E(a, Q)$ collapses into a point $x = a$.

Let the following two-sided ellipsoidal bounds hold for the sets X, M in (3):

$$E(A^-(t)x + f^-(t), G^-(t)) \subset X(x, t) \subset E(A^+(t)x + f^+(t), G^+(t)), \ t \geq s$$
$$\tag{9}$$
$$E(a_0^-, Q_0^-) \subset M \subset E(a_0^+, Q_0^+) \quad .$$

Here $A^{\pm}$, $G^{\pm}$ are given $n \times n$ matrices depending on t; $f^{\pm}$ are given n-vector functions of t; $a_0^{\pm}$ are given n-vectors; and $Q_0^{\pm}$ are given $n \times n$ matrices. The matrices $G^{\pm}$, $Q^{\pm}$ are symmetric and positive-definite. The estimates (9) imply that the reachable set of system (3) is bounded by the reachable sets of the following linear controlled systems with ellipsoidal constraints:

$$\dot{x} = A^{\pm}(t)x + f^{\pm}(t) + u \;,\; u \in E(0\;,\; G^{\pm}(t))$$
$$x(s) \in E(a_0^{\pm}\;,\; Q_0^{\pm})\;,\; t \geq s \quad . \tag{10}$$

We denote the reachable sets of systems (10) by $D^{\pm}(t\;,s\;,M)$. From (9) we then have

$$D^{-}(t\;,s\;,E(a_0^{-}\;,Q_0^{-})) \subset D(t\;,s\;,M) \subset D^{+}(t\;,s\;,E(a_0^{+}\;,Q_0^{+})) \quad . \tag{11}$$

The sets $D^{\pm}$ are not ellipsoids in the general case. We therefore introduce ellipsoidal approximations $E(a^{\pm}(t)\;,Q^{\pm}(t))$ which satisfy the following conditions:

(1) $a^{\pm}(s) = a_0^{\pm}\;,\; Q^{\pm}(s) = Q_0^{\pm}$; (12)

(2) $E(a^{-}(t),\; Q^{-}(t))$ are sub-reachable sets $\Omega^{-}(t)$ of system (10) for the minus index, and $E(a^{+}(t)\;,\; Q^{+}(t))$ are super-reachable sets $\Omega^{+}(t)$ of system (10) for the plus index;

(3) $\dot{v}^{-} \rightarrow \max\;,\; \dot{v}^{+} \rightarrow \min$ (13)

Conditions (13) mean that the volumes $v^{\pm}$ of ellipsoids $E(a^{\pm}(t),\; Q^{\pm}(t))$ change at a rate which is the highest (for v^{-}) or lowest (for v^{+}) possible for ellipsoids satisfying conditions 2. The main result can now be presented as the following theorem [10–14,16,17]:

THEOREM 1. *The ellipsoids defined by conditions* 1–3 *are unique, and their parameters* $a^{\pm}$, $Q^{\pm}$ *satisfy the following equations and initial conditions*:

$$\dot{a}^{\pm} = A^{\pm}a^{\pm} + f^{\pm}\;,\; a^{\pm}(s) = a_0^{\pm} \tag{14}$$

$$\dot{Q}^{-} = AQ^{-} + Q^{-}A^{\mathrm{T}} + 2G^{1/2}(G^{-1/2}Q^{-}G^{-1/2})^{1/2}G^{1/2} \tag{15}$$

$$\dot{Q}^{+} = AQ^{+} + Q^{+}A^{\mathrm{T}} + hQ^{+} + h^{-1}G$$
$$h = \{n^{-1}\,\mathrm{Tr}\,[(Q^{+})^{-1}G]\}^{1/2}\;,\; Q^{\pm}(s) = Q_0^{\pm} \quad . \tag{16}$$

Here we omit the dependence of A, f, G on t, and also the indices $^{-}$ and $^{+}$ after A, G in (15) and (16), respectively. A^{T} denotes the transpose of matrix A.

If the initial linear problems (14) can be solved for vectors $a^{\pm}(t)$ and the initial non-linear problems (15), (16) can be solved for matrices $Q^{\pm}(t)$, then we can obtain the desired estimates (7) in the form

$$E(a^{-}(t),Q^{-}(t)) \subset D(t\;,s\;,M) \subset E(Q^{+}(t),Q^{+}(t)) \quad . \tag{17}$$

If we have only internal (external) estimates (9) then we can obtain only internal (external) estimates (17). If system (2) is linear and similar to (10):

$$\dot{x} = A(t)x + f(t) + u \ , \ u \in E(0, G(t)) \ , \ x(s) \in E(a_0, Q_0) \tag{18}$$

then both systems (10) coincide with (18). In this case it is not necessary to put indices $\pm$ after A, G, f, a_0, Q_0 in (14)–(16), and $a^-(t) = a^+(t)$.

4. PROPERTIES OF ELLIPSOIDS

Equations (15), (16) can be simplified by means of the substitution

$$Q^{\pm}(t) = V(t)Z^{\pm}(t)V^{\mathrm{T}}(t) \quad , \tag{19}$$

where $V(t)$ is a non-degenerate $n \times n$ matrix.

THEOREM 2. *If $V(t)$ is the fundamental matrix of system* (18):

$$\dot{V} = A(t)V \ , \ V(s) = I \ , \ t \geq s \quad ,$$

where I is the $n \times n$ identify-matrix, then transformation (19) *reduces systems* (15), (16) *to the form*

$$\begin{aligned} \dot{Z} &= 2G_1^{1/2}(G_1^{1/2}Z^- G_1^{-1/2})^{1/2}G_1^{1/2} \\ \dot{Z}^+ &= h_1 Z_1^+ + h_1^{-1}G_1 \\ G_1 &= V^{-1}G(V^{-1})^{\mathrm{T}} \ , \ h_1 = \{n^{-1} \operatorname{Tr} [(Z^+)^{-1}G_1]\}^{1/2} \end{aligned} \tag{20}$$

corresponding to $A \equiv 0$. If $V = G^{1/2}$, transformation (19) reduces systems (15), (16) to the form

$$\begin{aligned} \dot{Z}^- &= A_1 Z^- + Z^- A_1^{\mathrm{T}} + 2(Z^-)^{1/2} \\ \dot{Z}^+ &= A_1 Z^+ + Z^+ A_1^{\mathrm{T}} + hZ^+ + h^{-1}I \\ A_1 &= G^{-1/2}[AG^{1/2} - d(G^{1/2})dt] \quad , \\ h &= [n^{-1} \operatorname{Tr} (Z^+)^{-1}]^{1/2} \end{aligned} \tag{21}$$

corresponding to $G = I$.

Theorem 2 makes it possible, without loss of generality, to take either $A = 0$ or $G = I$ in (15), (16), i.e., to consider simplified but equivalent systems (20) or (21) instead of (15), (16).

The following theorem gives the necessary and sufficient conditions under which the internal and external ellipsoids coincide:

THEOREM 3 [18]. *The internal and external ellipsoids* (17) *coincide* ($a^- \equiv a^+$, $Q^- \equiv Q^+$) *for all* $t \geq s$, *if and only if the systems* (10) *coincide so that the system* (2) *is reduced to* (18) *and*

$$AG + GA^{\mathrm{T}} - \dot{G} = \mu(t)G \ , \ t \geq s \ , \ Q_0 = \lambda_0^2 G(s) \ .$$

Here $\mu(t)$ is some arbitrary scalar function and λ_0 is a constant. Under these conditions the matrix Q for both ellipsoids is given by

$$Q(t) = \lambda^2(t)G(t) \ ,$$

$$\lambda(t) = \int_s^t \exp\left[\frac{1}{2}\int_{t_2}^t \mu(t_1)\mathrm{d}t_1\right]\mathrm{d}t_2 + \lambda_0 \exp\left[\frac{1}{2}\int_s^t \mu(t_1)\mathrm{d}t_1\right] \ .$$

Now we present some estimates of the volumes $v^{\pm}$ of the approximating ellipsoids (17).

THEOREM 4 [16]. *The following inequalities hold:*

$$\frac{v^+(t)}{v_0^+}\exp\left[-\int_s^t \mathrm{Tr}\, A^+(t_1)\mathrm{d}t_1\right] \leq \frac{v^{-(t)}}{v_0^-}$$

$$\cdot \exp\left[-\int_s^t \mathrm{Tr}\, A^-(t_1)\mathrm{d}t_1\right]\}^{\sqrt{n}} \ , \ v^-(t) \leq v^+(t) \ .$$

Here $v_0^{\pm}$ are the volumes of the initial ellipsoids $E(a_0^{\pm}, Q_0^{\pm})$ in (10). The following lower (for v^-) and upper (for v^+) bounds are given for the case $A^{\pm} = 0$ (which according to Theorem 2 implies no loss of generality):

$$v^-(t) \geq \{(v_0^-)^{1/n} + \int_s^t [v_G(t_1)]^{1/n}\mathrm{d}t_1\}^n$$

$$v^+(t) \leq v_0^+\{1 + [n^{-1}\,\mathrm{Tr}\,((Q_0^+)^{-1}G(s))]^{1/2} \cdot$$

$$\int_s^t \exp\left[\int_s^{t_1} (\mathrm{Tr}\,((G^{-1}(t_2)(\mathrm{d}G^{1/2}(t_2)/\mathrm{d}t_2)^2))^{1/2}\mathrm{d}t_2\right]\mathrm{d}t_1\}^n \ .$$

Here $v_G(t)$ is the volume of the ellipsoid $E(0, G(t))$. Theorem 4 allows us to evaluate the volumes $v^{\pm}$ without integrating systems (15), (16).

An important case arises if the initial point in (2), (3) is fixed: $x(s) = a_0$. In this case the initial ellipsoids $E(a_0^{\pm}, Q_0^{\pm})$ in (10) collapse into a point, and the initial conditions for both matrices $Q^{\pm}$ in (15), (16) are $Q_0 = 0$. From Theorem 2, we can consider this case using equations (21). The initial conditions for equations (21) are

$$Z^-(s) = Z^+(s) = 0 \ . \qquad (22)$$

Equations (15), (16), (21) have singularities if $Q^{\pm} \to 0$, $Z^{\pm} \to 0$. Theorem 5 establishes the asymptotic behavior of solutions in the vicinity of the initial point (22):

THEOREM 5 [18]. *Let matrix $A_1(t)$ from (21) have the following expansion in the neighborhood of the initial point:*

$$A_1(t) = A_{10} + A_{11}\vartheta + O(\vartheta^2)\,,\ \vartheta = t - s \geq 0 \quad .$$

Here A_{10}, A_{11} are constant matrices. Then the solutions of equations (21) under initial conditions (22) have the following expansions:

$$\begin{aligned} &Z^{\pm}(t) = \vartheta^2 I + \vartheta^3 B_0 + \vartheta^4 Z_4^{\pm} + O(\vartheta^5) \\ &Z_4^- = (7/12)B_0^2 + (2/3)B_1 \quad , \\ &Z_4^+ = (2/3)(B_0^2 + B_1) + (1/12)n^{-2}(\mathrm{Tr}\, B_0)^2 I - (1/6)n^{-1}(\mathrm{Tr}\, B_0)B_0 \\ &B_0 = (A_{10} + A_{10}^{\mathrm{T}})/2\,,\ B_1 = (A_{11} + A_{11}^{\mathrm{T}})/2 \quad . \end{aligned} \tag{23}$$

The expansions (23) are useful for starting the numerical integration of equations (15), (16) or (21) with initial conditions (22). The more general case in which the initial set M reduces to some r-dimensional set, $r < n$, is considered in [19,21].

The asymptotic behavior of ellipsoids as $t \to \infty$ has been studied in the case when the matrix A_1 in equations (21) is constant and diagonal:

$$A_1 = \mathrm{diag}\,\{\alpha_1, \ldots, \alpha_n\}\,,\ \alpha_1 \leq \alpha_2 \leq \cdots\ \alpha_n \quad . \tag{24}$$

We assume that the initial conditions for the matrices $Z^{\pm}$ are also diagonal so that the equations (21) have diagonal solutions

$$Z^{\pm}(t) = \mathrm{diag}\,\{[y_1^{\pm}(t)]^2, \ldots, [y_n^{\pm}(t)]^2\} \quad . \tag{25}$$

Here $y_i^{\pm} \geq 0$ are the semi-axes of the ellipsoids, $i = 1, \ldots, n$.

THEOREM 6 [18]. *All positive diagonal solutions (25) of equations (21), (24) for Z^- have the following limits as $t \to \infty$:*

$$\begin{aligned} &y_i^- \to +\infty\,,\ \text{if } \alpha_i \geq 0 \\ &y_i^- \to -\alpha_i^{-1}\,,\ \text{if } \alpha_i < 0 \quad . \end{aligned} \tag{26}$$

Positive diagonal solutions (25) of equations (21), (24) for Z^+ exhibit the same asymptotic behavior as $t \to \infty$

$$y_i^+ \to [-h^*(2\alpha_i + h^*)]^{-1/2} \text{ for } i = 1, \ldots, \nu$$
$$y_i^+ \to +\infty \text{ for } i = \nu + 1, \ldots, n \quad . \tag{27}$$

Here the integer ν and the number h^* are unique and defined by the conditions

$$\alpha_\nu(n+\nu) < \sum_{j=1}^{\nu} \alpha_j \le \alpha_{\nu+1}(n+\nu) \text{ if } \nu \ge 1 \quad ,$$
$$\nu = 0 \text{ if } \alpha_1 \ge 0 \, , \, h^* = -\frac{2}{n+\nu} \sum_{j=1}^{\nu} \alpha_j \ge 0 \quad . \tag{28}$$

It is interesting to compare these results with the asymptotic behavior of the reachable sets of the system

$$\dot{x}_1 = \alpha_i x_i + u_i \, , \, \sum_{i=1}^{n} u_i^2 \le 1 \, , \, i = 1, \ldots, n \quad . \tag{29}$$

Systems (29) corresponds to matrix (24), see (18) with $A = A_1$, $G = I$. For arbitrary initial conditions, the reachable set has a limit D_∞ as $t \to \infty$. The set D_∞ is convex, independent of the initial conditions and symmetric with respect to all axes x_i. The lengths of the semi-axes contained by D are equal to the semi-axes y_i^- of the limiting internal ellipsoid (26). The semi-axes of the limiting external ellipsoid (27) are greater than those of the limiting internal ellipsoid ($y_i^+ > y_i^-$), and in some cases we even have $y_i^+ \to \infty$ when y_i^- is finite. For instance, if $n = 2$ and $\alpha_1 < 3\alpha_2 < 0$, then from (26)–(28) we have $y_2^- = -\alpha_2^{-1}$, $y_2^+ = \infty$.

5. APPLICATIONS

We shall now briefly describe some possible applications of ellipsoidal estimates to various problems in dynamic systems.

1. If $u(t)$ in (2) is a control then the internal bound (17) can be used to evaluate control possibilities.

2. A procedure for obtaining admissible controls $u(t)$ that transform the linear system (18) from the initial state into some given terminal state $x_1 \in E(a^-(T), Q^-(T))$ which belongs to the internal ellipsoid is proposed in [24–26]. Both open-loop and closed-loop controls can be obtained.

3. If $u(t)$ in (2) is some bounded stochastic disturbance, then the external bound (17) can be used for guaranteed evaluation of possible perturbations caused by $u(t)$.

4. The estimates (17) can be used to obtain two-sided bounds for the cost functional in optimal control problems. We shall consider two such problems for system (18) with the initial condition $x(s) = a_0$. The boundary conditions at the termination time T and the cost functionals for these problems are

$$\begin{aligned} &(A)\ x(T)\ (\text{free}),\ J = F(x(T)) \to \min \\ &(B)\ x(T) \in N,\ T \to \min \end{aligned} \tag{30}$$

Here T is fixed for the cost minimization problem A and free for the time-optimal problem B. The scalar function $F(x)$ and the set N are given.

From (17) it is easy to derive the following bounds for the minimal values $J^{\mathfrak{x}}$ and $T^{\mathfrak{x}}$ of the functionals (30):

$$\begin{aligned} &(A)\ J^+ \le J^{\mathfrak{x}} = \min_{x \in D(T,s,a_0)} F(x) \le J^-,\ J^{\pm} = \min_{x \in E^-} F(x) \\ &(B)\ T^+ \le T^{\mathfrak{x}} \le T^-;\ E^{\pm} = E(a(T),\ Q^{\pm}(t)) \end{aligned} \tag{31}$$

Here $T^{\pm}$ are the times at which the ellipsoids $E^{\pm}$ first touch the set N. Similar estimates can be obtained for other classes of optimal control problems.

In order to calculate the bounds (31), it is necessary: (i) to obtain the approximating ellipsoids for $t \ge s$ and (ii) to calculate $J^{\pm}$, $T^{\pm}$ in (31). Here (i) does not depend on the cost functional; (ii) for problem A requires the solving of some standard nonlinear programming problems.

5. Estimates for differential games can be obtained with the help of Krasovky's extremal rule [2,3]. This rule holds for regular pursuit games; it says that if at time T the pursuer's reachable set $D_p(T)$ contains the evader's reachable set $D_e(T)$, then the pursuit can be terminated at $t = T$. Using estimates (17) we can obtain two-sided bounds for T: $T_1 \le T \le T_2$. Here T_1, T_2 are defined as the first times at which the inclusions $E_e^-(T_1) \subset E_p^+(T_1)$, $E_e^+(T_2) \subset E_p^-(T_2)$ hold.

6. GUARANTEED FILTERING

We assume now that the following observations of system (2) are available:

$$\begin{aligned} &y(t_i) = H(t_i)x(t_i) + v(t_i),\ s \le t_1 < t_2 \cdots \\ &v(t_i) \in E(0, L(t_i)),\ i = 1, 2, \ldots \end{aligned} \tag{32}$$

Here t_i are given observation times, the 1-dimensional vectors $y(t_i)$ are the results of the observations, $v(t_i)$ are the observation errors (bounded by ellipsoids), $H(t_i)$ are given $l \times n$ matrices, and $L(t_i)$ are given symmetric positive-definite $n \times n$ matrices. The vector $u(t)$ in (2) is an unknown disturbance. Let $P(t)$ be the set to which the state vector $x(t)$ of the system (2) belongs if the results of observations (32) with

$t_i < t$ are taken into account. The problem of guaranteed filtering is to estimate the set $P(t)$.

Using ellipsoidal estimates, we obtain the following bound for $P(t)$:

$$x(t) \in P(t) \subset E(a(t), Q(t)) \quad . \tag{33}$$

Here the vector $a(t)$ and the matrix $Q(t)$ are piecewise-continuous functions of time which satisfy initial conditions (14), (16), i.e.,

$$a(s) = a_0^+ , \; Q(s) = Q_0^+ \tag{34}$$

and differential equations (14), (16) for a^+, Q^+ within the intervals (t_i , t_{i+1}) between observations.

At the observation times t_i, $x(t)$ lies within the intersection of the ellipsoid $E(a(t_i -0), Q(t_i -0))$ and the ellipsoid E_i corresponding to the observation data (32). We define $E(a(t_i +0), Q(t_i +0))$ as an ellipsoid containing this intersection:

$$E(a(t_i +0) , Q(t_i +0)) \supset (E(a(t_i -0) , Q(t_i -0)) \cap E_i) \quad .$$

There are several ways of externally approximating the intersection of two ellipsoids by means of an ellipsoid [7–11,14,19]. Using one of them, we obtain the formulas

$$\begin{gathered} a(t_i +0) = \varphi_i (a(t_i -0) , Q(t_i -0) , y(t_i)) , \; i = 1,2 ,\ldots \\ Q(t_i +0) = \psi_i (a(t_i -0) , Q(t_i -0) , y(t_i)) \quad . \end{gathered} \tag{35}$$

The explicit form of functions φ_i, ψ_i is given in [11,19].

Thus, using differential equations (14), (16), initial conditions (34) and discontinuity conditions (35) at $t = t_i$, we can obtain the functions $a(t)$ and $Q(t)$ for all $t \geq s$. This procedure is a guaranteed method of filtering in the presence of external disturbances and observation errors for discrete-time models.

If the observations are continuous, we have

$$y(t) = H(t)x(t) + v(t) , \; v(t) \in E(0, L(t)) , \; t \geq s \tag{36}$$

instead of (32). The guaranteed method of filtering for this case was obtained in [22], and involves the integration of certain systems of ordinary differential equations

$$\dot{a} = \varphi(a , Q , t , y , \dot{y}) , \; \dot{Q} = \psi(a , Q , t , y , \dot{y})$$

with initial conditions (34). The explicit form of functions φ, ψ is given in [22]. Note that these functions depend on $\dot{y}$, and we require here that the results $y(t)$ of observations (36) be differentiable with respect to time.

7. CONCLUDING REMARKS

1. Numerical examples of ellipsoidal estimates of reachable sets are given in [11–14, 18–20].

2. Numerical examples demonstrating applications of the method of ellipsoids are given for optimal control problems in [14], for differential games in [23], and for guaranteed filtering in [19].

3. The method described above can also be applied to multistep discrete-time dynamic systems. In this case we deal with finite-difference equations instead of differential equations [11].

4. The method presented in this paper can be used to solve various problems in control and estimation theory. Proofs of all results and further details can be found in [10–23].

REFERENCES

1. F. John. Extremum problems with inequalities as subsidiary conditions. In *Studies and Essays*, presented to R. Courant on his 60th birthday. Interscience, New York, 1948, pp. 187–204.

2. N.N. Krasovsky. *Theory of Control of Motion*. Nauka, Moscow, 1968 (in Russian).

3. N.N. Krasovsky. *Game Problems of Meeting of Motions*. Nauka, Moscow, 1970 (in Russian).

4. F.C. Schweppe. *Uncertain Dynamic Systems*. Prentice-Hall, 1973.

5. A.B. Kurzhanski. *Control and Observation under Conditions of Uncertainty*. Nauka, Moscow, 1977 (in Russian).

6. F.L. Chernousko and A.A. Melikyan. *Game Problems of Control and Search*. Nauka, Moscow, 1978 (in Russian).

7. F.C. Schweppe. Recursive state estimation: unknown but bounded errors and system inputs. *IEEE Transactions on Automatic Control*, AC-13 (1968)22–28.

8. D.P. Bertsekas and I.B. Rhodes. Recursive state estimation for a set-membership description of uncertainty. *IEEE Transactions on Automatic Control*, AC-16(1971)117–128.

9. F.M. Schlaepfer and F.C. Schweppe. Continuous-time state estimation under disturbances bounded by convex sets. *IEEE Transactions on Automatic Control*, AC-17(1972)197–205.

10. F.L. Chernousko. Guaranteed estimates of undetermined quantities by means of ellipsoids. *Soviet Mathematics Doklady*, 21(1980)396–399.

11. F.L. Chernousko. Optimal guaranteed estimates of uncertainties by means of ellipsoids (in Russian). *Izvestiya Akademii Nauk USSR, Tekhnicheskaya Kibernetika*, 3(1980)3–11; 4(1980)3–11; 5(1980)5–11.

12. F.L. Chernousko. Ellipsoidal estimates of a controlled system's attainability domain. *Applied Mathematics and Mechanics*, 45(1)(1981)7–12.

13. F.L. Chernousko. Guaranteed ellipsoidal estimates of uncertainties in control problems. In *Proceedings of the 8th Triennial World Congress of IFAC*, Kyoto, 1981. Pergamon Press, Oxford, 1982. pp. 869–874.

14. F.L. Chernousko. Ellipsoidal bounds for sets of attainability and uncertainty in control problems. *Optimal Control Applications and Methods*, 382)(1982)187–202.

15. F.L. Chernousko. Guaranteed ellipsoidal estimates of state for dynamic systems. In *Random Vibrations and Reliability*, Proceedings of the IUTAM Symposium, Frankfurt/Oder, 1982. Akademie-Verlag, Berlin, 1983, pp. 145–152.

16. A.I. Ovseevich and F.L. Chernousko. Two-sided estimates on the attainability domains of controlled systems. *Applied Mathematics and Mechanics*, 46(5)(1982)590–595.

17. A.I. Ovseevich. Extremal properties of ellipsoids approximating reachable sets. *Problems of Control and Information Theory*, 12(1)(1983)43–54.

18. F.L. Chernousko. On equations of ellipsoids approximating reachable sets. *Problems of Control and Information Theory*, 12(2)(1983)97–110.

19. F.L. Chernousko, A.I. Ovseevich, B.R. Klepfish and V.L. Truschenkov. Ellipsoidal estimation of state for controlled dynamic systems. Preprint N 224, Institute for Problems of Mechanics, USSR Academy of Sciences, Moscow, 1983.

20. B.R. Klepfish. Numerical construction of ellipsoids approximating reachable sets (in Russian). *Izvestiya Akademii Nauk USSR, Tekhnicheskaya, Kibernetika*, (4)(1983)216–219.

21. B.R. Klepfish and A.I. Ovseevich. Asymptotic behavior of ellipsoids approximating reachable sets (in Russian). *Izvestiya Akademii Nauk USSR, Tekhnicheskaya Kibernetika*, (2)(1984)66–69 (in Russian).

22. A.I. Ovseevich, V.L. Truschenkov and F.L. Chernousko. Equations of continuous guaranteed state estimation for dynamic systems (in Russian). *Izvestiya Akademii Nauk USSR, Tekhnicheskaya Kibernetika*, (4)(1984)94–101.

23. B.R. Klepfish. Method of obtaining two-sided estimate of time of pursuit (in Russian). *Izvestiya Akademii Nauk USSR, Tekhnicheskaya Kibernetika*, (4)(1984)156–161.

24. V.A. Komarov. Estimates of a reachable set and construction of admissible controls for linear systems (in Russian). *Doklady Akademii Nauk USSR*, 268(1983)537–541.

25. V.A. Komarov. Estimates of a reachable set of linear non-autonomoous systems (in Russian). *Izvestiya Akademii Nauk USSR, Seriya Matematicheskaya*, 48(1984)865–879.

26. V.A. Komarov. Synthesis of bounded controls for linear non-autonomous systems (in Russian). *Automatics and Remote Control*, 10(1984)44–50.

METHODS OF GROUP PURSUIT

A.A. Chikrij
V.M. Glushkov Institute of Cybernetics, Kiev, USSR

1. INTRODUCTION

In this paper we review methods by which one evading object may be pursued by a group of controlled objects. These methods are based on various assumptions about the information available to the pursuers during the course of the game.

The quasilinear problem of group pursuit in the general case may be stated as follows. Consider a conflict-controlled system

$$\dot{z}_i = A_i z_i + \varphi_i(u_i, v), \; z_i \in \mathbb{R}^{n_i}, \; u_i \in U_i, \; v \in V, \; i = \overline{1,\nu} \quad , \tag{1}$$

where $\mathbb{R}^{n_i}$ is a finite-dimensional Euclidean space, A_i is a square matrix of order n_i, sets U_i and V are non-empty and compact, and functions $\varphi_i(u_i, v)$: $U_i \times V \rightarrow \mathbb{R}^{n_i}$ are continuous in the set of variables. The terminal set consists of sets $M_i^* = M_i^0 + M_i$, $i = \overline{1,\nu}$, where the M_i^0 are linear subspaces of $\mathbb{R}^{n_i}$ and the M_i are convex compact sets of orthogonal complements L_i of M_i^0 in the space $\mathbb{R}^{n_i}$. The game (1) terminates if $z_i(t) \in M_i^*$ for some i and $t > 0$.

It is easy to see that this formulation covers the pursuit of one evasive object (or a group of objects of which at least one must be caught) by a group of controlled objects.

2. POSITIONAL GROUP PURSUIT

Let π_i denote the operator of orthogonal projection of the space $\mathbb{R}^{n_i}$ onto the subspace L_i and consider the two subsets

$$\mathbf{A} = \{\alpha : \alpha = (\alpha_1, \ldots, \alpha_\nu), \; \alpha_i \in \mathbb{R}^1, \; \sum_{i=1}^{\nu} \alpha_i = 1, \; \alpha_i \geq 0\}$$

$$\Psi = \{p : p = (p_1, \ldots, p_\nu), \; p_i \in L_i, \; \|p_i\| = 1\} \quad .$$

For $p \in \Psi$, $\alpha \in \mathbf{A}$ we introduce the function

$$c(t,z,\alpha,p) = \sum_{i=1}^{\nu} \alpha_i(\vartheta_i^*(t)p_i, z_i) + \int_0^t \min_{v\in V} \max_{u_i\in U_i} \sum_{i=1}^{\nu} \alpha_i(\vartheta_i^*(\tau)p_i, \varphi_i(u_i, v))d\tau \quad ,$$

where $\vartheta_i^*(t)$ is a matrix conjugate to the fundamental matrix of the homogeneous system (1), $(\cdot,\cdot)$ represents a scalar vector product, and $z = (z_1, \ldots, z_\nu)$.

Let

$$\mu(t,z,\alpha,p) = c(t,z,\alpha,p) + \sum_{i=1}^{\nu} \alpha_i c_{M_i}(-p_i)$$

$$\lambda^*(t,z) = \min_{p\in\Psi} \max_{\alpha\in\mathbf{A}} \mu(t,z,p) \, , \; \lambda_*(t,z) = \max_{\alpha\in\mathbf{A}} \min_{p\in\Psi} \mu(t,z,\alpha,p) \quad ,$$

where $c_{M_i}(p_i)$ is a support function of the set M_i.

Take

$$T^*(z) = \inf\{t > 0: \lambda^*(t,z) = 0\} \, , \; T_*(z) = \inf\{t > 0: \lambda_*(t,z) = 0\} \quad .$$

In what follows we shall assume that the lower bounds in relation (2) are attained.

Condition 1. Mappings $\varphi_i(U_i, v)$, $i = \overline{1,\nu}$, $v \in V$, are convex-valued.

We shall write

$$K(M_i) = \{p_i: c_{M_i}(p_i) < +\infty\} \quad .$$

Conditon 2. Sets $K(M_i)$, $i = \overline{1,\nu}$, are non-empty and closed and functions $c_{M_i}(p_i)$ are continuous on $K(M_i)$.

We consider multivalued mappings

$$\Psi(t,z) = \{p \in \Psi: \lambda^*(t,z) = \max_{\alpha\in\mathbf{A}} \mu(t,z,\alpha,p)\} \quad ,$$

$$\mathbf{A}(t,z,p) = \{\alpha \in \mathbf{A}: \mu(t,z,\alpha,p) = \max_{\alpha\in\mathbf{A}} \mu(t,z,\alpha,p)\} \quad .$$

THEOREM 1. *Let Conditions 1 and 2 be satisfied. In addition, let the following conditions hold for any point z^1 such that $T^*(z^1) < T^*(z^0) < +\infty$:*

(a) *the set $\Psi(t,z)$ consists of a unique vector $p(t,z)$ for all t and for z from some neighborhood $\{T^*(z^1), z^1\}$;*

(b) *there exists an element α_0, $\alpha_0 \in \mathbf{A}(T^*(z^1), z^1, p(T^*(z^1), z^1))$,*
such that the function

$$\sum_{i=1}^{\nu} \alpha_{0i}(\vartheta_i^*(t)p_i(t,z_i), \varphi_i(u_i,v))$$

achieves its maximum in $u = (u_1, \ldots, u_\nu)$ on a unique vector $u(t,z,v)$ for all t

and for z from the neighborhood $\{T^*(z^1), z^1\}$ and $v \in V$.

Then, starting from the point z^0, *the game* (1) *can terminate in a time not greater than* $T^*(z^0)$.

Note. Uniqueness is necessary only for those components p_i and u_i which correspond to non-zero numbers α_{0i}, $i = \overline{1,\nu}$.

In the case when the pursuers and the evading object make discrete moves, the time $T^*(z)$ is the time at which the attainable domain of the evading object overlaps with the combination of the attainable domains of the pursuers [1–7]. Here the pursuers' controls depend only on the position. The pursuit procedure is an analog of Krasovski's rule of extremal aiming [1]. Numbers α_{0i} for each position represent the usefulness of the i-th pursuer in game (1). If one of the α_{0i} equals zero, then the corresponding player cannot affect the quality of the game; even without him the game will terminate not later than time $T^*(z)$.

The time $T_*(z)$ corresponds to the situation in which each of the players pursues the evading object individually and does not take into account the actions of other pursuers. The sufficient conditions for the game (1) to terminate in $T_*(z)$ are actually the conditions for regularity [1,7] of the ν games between each pursuer and the evasive object.

In positional group pursuit it is observed that, while each of the pursuers will eventually catch the evading object, this time decreases when they act together.

The positional pursuit scheme allows us to obtain positional analogs to Pontryagin's first direct method and method of alternating integral for the case of group pursuit [8]. The resulting statements are similar to Theorem 1; they are given in [4,6]. The above framework also encompasses the situation with a delay in information on states [6].

3. EFFICIENT METHODS OF GROUP PURSUIT

Methods in which information on the initial position z^0 and the control history of the evading object $\nu_t(\cdot) = \{\nu(s): s \in [0, t]\}$ is available to the pursuers at time t comprise a special class of methods of group pursuit called *efficient methods* [9–13].

Consider the multivalued mappings

$$W_i(t, v) = \pi_i \vartheta_i(t) \varphi_i(U_i, v), \; W_i(t) = \bigcap_{v \in V} W_i(t, v) \quad .$$

Condition 3. $W_i(t) \neq \phi$, $t \geq 0$, $i = \overline{1,\nu}$.

Let us choose fixed measurable selections $\gamma_i(t)$, $\gamma_i(t) \in W_i(t)$, $i = \overline{1,\nu}$, and let

$$\xi_i(t, z_i, \gamma_i(\cdot)) = \pi_i \vartheta_i(t) z_i + \int_0^t \gamma_i(t-\tau)\, d\tau \quad ,$$

$$\alpha_i(t, \tau, z_i, v, \gamma_i(\cdot)) = \begin{cases} \max\,[\alpha \geq 0: (W_i(t-\tau, v) - \gamma_i(t-\tau)) \cap \alpha(M_i - \xi_i(t, z_i, \gamma_i(\cdot))) \neq \phi], \\ \text{if } \xi_i(t, z_i, \gamma_i(\cdot)) \not\in M_i \\ 1/t \text{ , otherwise} \\ t \geq \tau \geq 0 \text{ , } v \in V \end{cases}$$

Consider the function

$$T(z, \gamma(\cdot)) = \inf\,\{t > 0: 1 - \min_{v(\cdot)} \max_i \int_0^t \alpha_i(t, \tau, z_i, v(\tau), \gamma_i(\cdot))d\tau = 0\}$$

$$\gamma(\cdot) = (\gamma_1(\cdot), \ldots, \gamma_\nu(\cdot)) \quad .$$

We shall assume that $t = +\infty$ if the equality in braces is not satisfied for any $t > 0$.

THEOREM 2. *Let Condition* 3 *be satisfied for some collection of measurable selections* $\gamma^0(\cdot)$ *and point* z^0*, the lower bound in expression* (3) *be attained, and* $T(z^0, \gamma^0(\cdot)) < +\infty$. *Then the game* (1) *starting from the specified initial position* z^0 *may terminate at the time* $T(z^0, \gamma^0(\cdot))$.

Pursuer controls which satisfy Theorem 2 are given by the Filippov–Kasten measurable choice theorem.

There are a number of modifications of the suggested scheme [6,9,10] which take into account special properties of game (1) or which try to direct the conflict-controlled system (1) to given points of set M_i.

As a result it is possible (in specific cases) to compute or estimate the time of group pursuit, to identify the pursuer controls, and to obtain explicit conditions or initial positions of the "encirclement" type which lead to a finite pursuit time [6,10,11]. The case in which the persistence of the pursuers and that of the evading object are different can be formalized by introducing an additional condition [9,10]. Pursuit problems with phase constraints on the state of the evading object which form a polyhedral set can be transformed into problems of group pursuit without constraints [6,9,11]. This includes a number of interesting classical examples of game situations [11].

One of the characteristics of group pursuit schemes is that they may be used to solve both problems in which it is required to search for moving objects with a discrete distribution of initial states and problems involving the pursuit of a group of evading objects by one controlled object.

4. PROGRAM GROUP PURSUIT AND COORDINATED PURSUIT

In a number of group pursuit problems it is possible to come to some conclusion about the result of game (1) and to formulate a pursuit algorithm given the initial position of the pursuers and the evading object without using current information [12]. We shall write

$$c_i(t, z_i, u_i(\cdot), v(\cdot), p_i) = (\vartheta_i^*(t)p_i, z_i) + \int_0^t (\vartheta_i^*(t-\tau)p_i, \varphi_i(u_i(\tau), v(\tau)))d\tau \quad ,$$

$$\lambda(t, z) = \max_{u_i(\cdot)} \min_{v(\cdot)} \max_{i} \min_{\|p_i\|=1} \{c_{M_i}(p_i) - c_i(t, z_i, u_i(\cdot), v(\cdot), p_i)\} \quad , \tag{4}$$

$$T(z) = \inf \{t > 0: \lambda(t, z) = 0\} \quad . \tag{5}$$

Assertion 1. Let the lower bound in relation (5) be attained for z^0 and $T(z^0) < +\infty$. Then game (1) starting from the state z^0 may terminate at or before time $T(z^0)$.

Note that the program method of group pursuit makes it possible to solve a number of problems in which the evading object has an advantage with respect to control resources over each pursuer and the methods proposed in Sections 1 and 2 do not address the issue of solvability of the group pursuit problem [10,12].

Let k, l be natural numbers such that $1 \le k \le l \le \nu$ and Condition 3 be satisfied for $i \in \overline{1,k}$. Let the following conditions hold for $i \in \overline{l+1,\nu}$:

Condition 4. Sets $M_i = \{m_i\}$ are singular for $i \in \overline{l+1,\nu}$.

Condition 5. Multivalued mappings $W_i(t, v)$, $t \ge 0$, $v \in V$, $i \in \overline{l+1,\nu}$, assume point values $w_i(t, v)$.

Let

$$\eta_i(t, z_i) = \pi_i \vartheta_i(t) z_i \quad ,$$

$$\beta_i(t, \tau, z_i, v) = \begin{cases} \max(\beta_i \ge 0: \beta_i(m_i - \eta_i(t, z_i)) = w_i(t, v)), & m_i \ne \eta_i(t, z_i) \\ 0, \ \beta_i(m_i - \eta_i(t, z_i)) \ne w_i(t, v) \ \forall \beta_i \ge 0, & m_i \ne \eta_i(t, z_i) \\ \|w_i(t-\tau, v)\| + 1/t, & m_i = \eta_i(t, z_i) \end{cases}$$

for $i \in \overline{l+1,\nu}$. We shall take

$$V(t) = \{v(\cdot): \max_{i \in \overline{1,k}} \int_0^t \alpha_i(t, \tau, z_i, v(\tau), \gamma_i(\cdot))d\tau < 1, \ \max_{i \in \overline{l+1,\nu}} \int_0^t \beta_i(t, \tau, z_i, \nu(\tau))d\tau < 1\}$$

and let $\tilde{\lambda}(t, z)$ denote the expression for $\lambda(t, z)$ which corresponds to a minimum in $v(\cdot)$ in (4) on $V(t)$ and a maximum on $i \in \overline{k+1,l}$. We thus obtain the time $\tilde{T}(z)$.

THEOREM 3. *Let Condition 3 hold for $i \in \overline{1,k}$, and Conditions 4 and 5 hold for $i \in \overline{l+1,\nu}$. Let the lower bound in an expression for $\tilde{T}(z^0)$ similar to* (5) *be attained* *at z^0 and $\tilde{T}(z^0) < +\infty$. Then starting from the initial position z^0 game (1) can terminate at or before time $\tilde{T}(z^0)$, with the pursuers using the following information:*

$$u_i(t) = u_i(z^0, v_t(\cdot)), \quad i \in \overline{1,k}$$

$$u_i(t) = u_i(z^0), \quad i \in \overline{k+1,l} \quad .$$

Here the pursuers $i \in \overline{l+1,\nu}$ play the role of phase constraints.

Each of the groups of pursuers with different dynamic possibilities is associated with particular controls, and despite the fact that the players have different information it is possible to combine their efforts for a successful group pursuit [12].

5. COMPARISON WITH PONTRYAGIN'S FIRST METHOD

Let $\nu = 1$. Since efficient methods for group pursuit are based on ideas close to those of L.S. Pontryagin, it is interesting to compare the methods with respect to the time of termination. This then allows us to clarify the role of information available to the pursuers about the control history of the evading object.

Consider the termination time given by the first direct method of Pontryagin [8]:

$$P(z) = \inf \{t \geq 0: \pi\vartheta(t)z \in M - \int_0^t W(t-\tau)d\tau\} \tag{6}$$

Here and elsewhere the index i is omitted since $\nu = 1$.

We now introduce the auxiliary scheme

$$\alpha(t, \tau, z, \gamma(\cdot)) = \begin{cases} \max\,[\alpha \geq 0: (W(t-\tau) - \gamma(t-\tau)) \cap \alpha(M - \xi(t, z, \gamma(\cdot)) \neq \phi], \\ \xi(t, z, \gamma(\cdot)) \bar{\in} M \\ 1/t, \quad \xi(t, z, \gamma(\cdot)) \in M \end{cases} \tag{7}$$

$$T_*(z, \gamma(\cdot)) = \inf \{t > 0: 1 - \int_0^t \alpha(t, \tau, z, \gamma(\cdot))d\tau = 0\} \tag{8}$$

and take $\Gamma = \{\gamma(\cdot): \gamma(t) \in W(t), t \geq 0, \gamma(t)$ is measurable.$\}$

Assertion 2. Let Conditon 3 be satisfied. Then for

$$\pi\vartheta(t)z \in M - \int_0^t W(t-\tau)d\tau, \; t \geq 0, \; z \in \mathbb{R}^n \quad ,$$

it is necessary and sufficient that there exist a measurable selector $\gamma(\cdot) \in \Gamma$ such that $\xi(t, z, \gamma(\cdot)) \in M$.

This suggests the inequality

$$\inf_{\gamma(\cdot)\in\Gamma} T(z, \gamma(\cdot)) \leq P(z)$$

for all $z \in \mathbb{R}^n$.

THEOREM 4. *Let Condition 3 be satisfied and the latest lower bound in expressions* (6), (8) *be attained. Then*

$$\min_{\gamma(\cdot)\in\Gamma} T_*(z, \gamma(\cdot)) = P(z) \quad \forall z \in \mathbb{R}^n \quad .$$

The scheme (7), (8) is called the functional form of Pontryagin's first methods.

Assertion 3. Let Condition 3 be satisfied, the mapping $\varphi(U, v)$, $v \in V$, be convex-valued and the set $M = \{m\}$ be singular. Then

$$P(z) = \min_{\gamma(\cdot)\in\Gamma} T(z, \gamma(\cdot)) \quad \forall z \in \mathbb{R}^n \quad .$$

Let Z be a non-empty set from $\mathbb{R}^n$. We shall write

$$\text{con } z = \{\lambda z: z \in Z, \lambda > 0\}$$

and let

$$S(t, v) = W(t, v) \overset{*}{-} \dot{W}(t), \ t \geq 0, \ v \in V$$

$$K(t, v) = \text{con } S(t, v)$$

$$K(t) = \bigcap_{v \in V} K(t, v) \quad ,$$

where $\overset{*}{-}$ is the operation of geometrical subtraction of sets [8]. We take

$$\Gamma(z) = \{\gamma(\cdot) \in \Gamma: \min_{\gamma(\cdot)\in\Gamma} T_*(z, \gamma(\cdot)) = T_*(z, \gamma(\cdot))\} \quad .$$

THEOREM 5. *Let Condition 3 be satisfied,* $M = \{m\}$ *and the lower bounds in expressions* (3), (6) *be attained. Then if in the initial state* z^0 *there exists a measurable selection* $\gamma^0(\cdot) \in \Gamma(z^0)$, *such that* $m - \xi(t, z^0, \gamma^0(\cdot)) \in K(t-\tau)$ *for all* $0 \leq \tau \leq t < P(z^0)$, *we have*

$$T(z^0, \gamma^0(\cdot)) < P(z^0) \quad .$$

We shall consider the function

$$\lambda(t, \tau, z, v, \gamma(\cdot)) = \max \{\lambda \geq 0: \lambda(m - \xi(t, z, \gamma(\cdot)) \in S(t-\tau, v)\} \quad ,$$

$$t \geq \tau \geq 0, \quad v \in V, \quad m \neq \xi(t, z, \gamma(\cdot)), \quad \gamma(\cdot) \in \Gamma \quad .$$

and the set

$$Q = \{z: \lambda(t, \tau, z, v, \gamma(\cdot)) + \alpha(t, \tau, z, \gamma(\cdot)) = \alpha(t, \tau, z, v, \gamma, (\cdot))\}$$

for all $0 \leq \tau \leq t < P(z)$, $v \in V$, $\gamma(\cdot) \in \Gamma$.

THEOREM 6. *Let Condition 3 be satisfied and $M = \{m\}$. If for a given point z^0, $z^0 \in Q$, we have $m - \xi(t, z^0, \gamma(\cdot)) \overline{\in} K(t-\tau)$ for all $0 \leq \tau \leq t < P(z^0)$, $\gamma(\cdot) \in \Gamma$, then*

$$\min_{\gamma(\cdot)\in\Gamma} T(z^0, \gamma(\cdot)) = P(z^0) \quad .$$

In the case of an arbitrary set M it is possible to formulate assertions analogous to Theorems 5 and 6.

Thus, under the conditions of Theorem 5 it is possible to reduce the termination time of Pontryagin's first method by using the control history of the evading object, under the conditions of Assertion 3 and Theorem 6, such information about the control history of the evading object does not affect the result of game (1) (at least in the framework of the schemes under consideration).

REFERENCES

1. N.N. Krasovski and A.I. Subbotin. *Positional Differential Games*. Nauka, Moscow 1974 (in Russian).
2. A.I. Subbotin and A.G. Chencov. *Optimization of Security in Control Problems*. Nauka, Moscow, 1981.
3. A.A. Chikrij. Quasi-linear differential many-person games. *Doklady Akademii Nauk SSSR*, (6)(1979)1306–1309.
4. A.A. Chikrij. A quasi-linear approach problem involving several persons. *Prikladraya Matematika i Mekhanika*, 3(1979)451–455.
5. S.I. Tarlinskij. On one linear differential approach game with several control objects. *Doklady Akademii Nauk SSSR*, 230(3)(1976)534–537.
6. A.A. Chikrij. Differential games with several pursuers. In *Mathematical Control Theory*, Banach Center Publications, 1983, pp. 81–107.
7. B.N. Pshenichnyj. Linear differential games. *Avtomatika i Telemekhanika*, (1)(1968)65–79.
8. L.S. Pontryagin. Linear differential games. *Mathematical Notes*, 112(1980)307–330.
9. A.A. Chikrij. Group pursuit under bounded coordinates of an evasive object. *Prikladnaya Matematika i Mekhanika*, 6(1982)906–913.

10. N.L. Grigorenko. *Differential Games of Pursuit by Several Objects*. Moscow State University, Moscow, 1983.

11. B.N. Pshenichnyj, A.A. Chikrij and I.S. Rappoport. Group pursuit in differential games. *Zhurnal Vysshej Tekhnicheskoj Shkoly, Leipzig*, (1)(1982)13–27.

12. A.A. Chikrij and M.V. Pticyk. Combination of efforts of pursuers with different dynamic possibilities. *Doklady Akademii Nauk Ukrainskoj SSR*, (1)(1984)73–76.

13. A.A. Chikrij and G.C. Chikriy. Group pursuit in different-differential games. *Differential Equations*, 20(1984)802–810.

AN AVERAGING PRINCIPLE FOR OPTIMAL CONTROL PROBLEMS WITH SINGULAR PERTURBATIONS

V.G. Gaitsgory
Leningrad Politechnic Institute

1. INTRODUCTION

There has been considerable interest in the possibility of extending the asymptotic averaging method developed in the theory of differential equations [1] to optimal control problems (see, e.g., [2–4]). This paper presents new results of this kind which allows us to apply the averaging method to the analysis of optimal control problems with singular perturbations, which have attracted the attention of many mathematicians throughout the world (see overviews [5,6]). The results also prove to be useful in the optimization of controlled Markov chains with weak interactions between subsets of states.

We shall first give a statement of the problem and then describe the results.

Consider a system with singular perturbations over an extended time scale:

$$\dot{z} = \varepsilon f_1(z, y, u), \; z(0) = z_0 ; \; \dot{y} = f_2(z, y, u), \; y(0) = y_0 ; \; t \in [0, T\varepsilon^{-1}] \quad , \qquad (1)$$

where ε is a small parameter, $f_1(\cdot)$, $f_2(\cdot)$ are vector functions with values in R^{n_1} and R^{n_2}, respectively, and u is an arbitrary measurable vector function with values in R^m which satisfies the inclusion $u(t) \in U$. The family of such functions will be called the set of feasible controls. Let $V(z, s, Q)$ denote the set of values of

$$s^{-1} \int_0^s f_1(z, y(z, t), u(t)) dt$$

calculated along the trajectories $y(z, t)$ of the system

$$\dot{y} = f_2(z, y, u), \; y(0) \in Q, \; z = \text{const} \qquad (2)$$

for all feasible controls.

THEOREM 1. *Assume that $f_1(z, y, u)$, $f_2(z, y, u)$ are continuous in some neighborhood of the product of compact sets $\hat{P}_1 \times \hat{P}_2 \times U$ and satisfy the Lipschitz* *conditions in z and y uniformly with respect to u. Let the trajectories of system (1) satisfy the inclusions*

$$y(t) \in P_2 \subset \operatorname{int} \hat{P}_2 , \; z(t) \in P_1 \subset \operatorname{int} \hat{P}_1 , \; t \in [0, T\varepsilon^{-1}] \quad ,$$

where the P_i are closed, and let the following conditions be satisfied:

(1) *For any $z \in \hat{P}_1$, any feasible control $u(t)$ and initial conditions $y(0) = y_0 \in P_2$, system (2) has a unique solution $y(z, t, u(\cdot), y_0)$ and this solution is completely contained in P_2.*

(2) *For any $z \in \hat{P}_1$, any feasible control $u(t)$ and initial conditions $y(0) = y_0' \in \hat{P}_2$, $y(0) = y_0'' \in \hat{P}_2$, system (2) also has a unique solution and*

$$\| y(z, t, u(\cdot), y_0') - y(z, t, u(\cdot), y_0'') \| \leq L \| y_0' - y_0'' \| \, \xi(t) , \; L = \text{const} \quad , \tag{3}$$

where $\xi(t)$ is an arbitrary function such that $\int_0^\infty \xi(t) dt < \infty$.

(3) *If $\tilde{y}(t)$ is the solution to the system*

$$\dot{\tilde{y}}(t) = f_2(\tilde{z}(t), \tilde{y}(t), \tilde{u}(t)) , \; \tilde{y}(0) = \tilde{y}_0 \in \hat{P}_2 \quad ,$$

where $\tilde{z}(t)$ is continuous and $\tilde{z}(t) \in \hat{P}_1$, $t \in [0, t_1]$, then the estimate

$$\max_{t \in [0, t_1]} \| \tilde{y}(t) - y(z, t, \tilde{u}(\cdot), \tilde{y}_0) \| \leq L \max_{t \in [0, t_1]} \| \tilde{z}(t) - z \|$$

holds for any $z \in \hat{P}_1$.

Then

(i) *For any $Q \subset P_2$ there exists a limit in the Hausdorff metric $\rho(\cdot, \cdot)$*

$$\lim_{s \to \infty} \bar{V}(z, s, Q) = \bar{V}(z) , \; \rho(\bar{V}(z, s, Q), \bar{V}(z)) \leq L s^{-1/2} , \; s \geq s_0 > 0 \quad , \tag{4}$$

where $\bar{V}(z, s, Q)$ is the closure of $V(z, s, Q)$ and $\bar{V}(z)$ is a convex compact subset of R^{n_1} which does not depend on Q.

(ii) *For any trajectory $(z^0(t), y^0(t))$ of system (1) one can construct a solution $\xi^0(t)$ to the differential inclusion*

$$\dot{\xi} \in \varepsilon \bar{V}(\xi) , \; \xi(0) = z_0 \tag{5}$$

such that

$$\| z^0(t) - \xi^0(t) \| \leq L \varepsilon^{1/3} , \; L = \text{const} , \; t \in [0, T\varepsilon^{-1}] \quad . \tag{6}$$

Conversely, each solution $\xi^0(t)$ to (5) corresponds to some trajectory $(z^0(t), y^0(t))$ of (1) so that the distance between $\xi^0(t)$ and $z^0(t)$ is estimated by (6).

(iii) *If the function* $\xi(t)$ *in* (3) *is monotonic, then the following limit exists in the Hausdorff metric:*

$$\lim_{s \to \infty} \bar{Y}(z, s, Q) = \bar{Y}(z), \; \rho(\bar{Y}(z, s, Q), \bar{Y}(z)) \le L\xi(s) \quad , \tag{7}$$

where $\bar{Y}(z, s, Q)$ *is the closure of the reachable set of system* (2).

Under the additional condition

$$\text{Conv}\,\{f_1(z, \bar{Y}(z), U), f_2(z, \bar{Y}(z), U)\} = \{f_1(z, \bar{Y}(z), U), f_2(z, Y(z), U)\} \tag{8}$$

the set $\bar{V}(z)$ may be written in the explicit form

$$V(z) = \{\xi \colon \xi = f_1(z, y, u), f_2(z, y, u) = 0, u \in U\} \tag{9}$$

and the differential inclusion (5) is equivalent to the system

$$\dot{\xi} = \varepsilon f_1(\xi, \psi(\xi, u), u), \; u \in U, \; \xi(0) = z_0 \quad , \tag{10}$$

where $y = \psi(\xi, u) \in \hat{P}_2$ is a root of the system

$$f_2(\xi, y, u) = 0, \; u \in U, \; \xi \in P_1 \quad .$$

Recall that the distance between two subsets Ω_1 and Ω_2 in the Hausdorff metric is defined as follows:

$$\rho(\Omega_1, \Omega_2) = \max\,\{\sup_{\xi_1 \in \Omega_1} \rho(\xi_1, \Omega_2), \; \sup_{\xi_2 \in \Omega_2} \rho(\xi_2, \Omega_1)\} \quad ,$$

where

$$\rho(\xi, \Omega) = \inf_{\xi' \in \Omega} \|\xi - \xi'\|$$

and $\|\cdot\|$ represents the Euclidean norm.

Remark 1. Note that system (10) is traditionally used in the reduction of optimal control problems with singular perturbations [5,6]. Condition (8) is similar to that guaranteeing the existence of solutions to optimal control problems [7], with "fast" variables y playing the role of controls in addition to $u(t)$.

2. CONDITIONS FOR EXISTENCE AND CONVEXITY OF THE LIMIT SET

Statement (i) of Theorem 1 concerns the existence and convexity of the limit set $\bar{V}(z)$. This statement is ergodic in nature and can be applied to different problems in control theory. We shall devote the rest of this paper to proving its validity under conditions weaker than those given in Theorem 1.

Note first of all that the components of vector z can be regarded as constant parameters when studying limit properties of type (4). Therefore we shall omit them, writing system (2) in the form:

$$\dot{y} = f_2(y, u), \; y(0) \in Q \quad . \tag{11}$$

In the same way we shall define the set $V(s, Q)$ as the set of values of

$$s^{-1} \int_0^s f_1(y, u) dt \quad . \tag{12}$$

THEOREM 2. *Let the vector functions $f_1(y, u)$, $f_2(y, u)$ be continuous in some neighborhood of the product $\hat{P}_2 \times U$, where $\hat{P}_2, U$ are sets in R^{n_2} and R^m, respectively. Suppose also that the following conditions are satisfied:*

(1) *For any feasible control $u(t)$ and initial conditions $y(0) = y_0 \in \hat{P}_2$ there exists a unique solution $y(t, u(\cdot), y_0)$ to the system* (11). *Moreover this solution satisfies the inequality*

$$\|f_1(y(t, u(\cdot), y_0), u(t))\| \leq L = \text{const}, \; \forall t \in [0, s], \; \forall s \geq 0 \quad . \tag{13}$$

(2) *Let y_0', y_0'' be two arbitrary vectors of initial conditions from $\hat{P}_2$ and let $u'(t)$ be an arbitrary feasible control. Then it is possible to choose a feasible control $u''(t)$ which satisfies the inequality*

$$\|s^{-1} \int_0^s f_1(y(t, u'(\cdot), y_0'), u'(t)) dt - s^{-1} \int_0^s f_1(y(t, u''(\cdot), y_0''), u''(t)) dt\| \leq$$

$$\leq L s^{-1}, \; L = \text{const}, \; s > 0 \quad . \tag{14}$$

(3) *There exists a subset P_2 of the set $\hat{P}_2$ such that, for all feasible controls, all trajectories of system* (11) *with initial conditions from P_2 remain in the set $\hat{P}_2$. Then for any $Q \subset P_2$ the following Hausdorff limit exists:*

$$\lim_{s \to \infty} \bar{V}(s, Q) = \bar{V} \quad , \tag{15}$$

where $\bar{V}(s, Q)$ is the closure of $V(s, Q)$ and $\bar{V}$ is a convex compact set in R^{n_1} which does not depend on Q. *Moreover, for $s \geq s_0 > 0$ we have*

$$\rho(\bar{V}(s, Q), \bar{V}) \leq L s^{-1}, \; L = \text{const} \quad . \tag{16}$$

Remark 2. It is easy to see that the conditions of Theorem 2 will be satisfied if the conditions of Theorem 1 are met. The relation (14) follows from (13) with $u''(t) =$

$u'(t)$, while (13) is conditioned by the compactness of $\hat{P}_2$. It is also easy to find an example in which the conditions of Theorem 1 are not satisfied but the conditions of Theorem 2 are satisfied. This is the case if, for example, $n_2 = 1$, $f_2(y, u) \equiv 1$ and $f_1(y, u)$ is a periodic function of y.

3. AUXILIARY STATEMENTS

We shall now consider a number of auxiliary statements. It follows from condition 1 of Theorem 1 that there exists a compact set $P \subset R^{n_1}$ such that

$$V(s, Q) \subset P \subset R^{n_1}, \ \forall s > 0, \ \forall Q \subset P_2 \quad . \tag{17}$$

On the basis of the same condition 1 it is easy to establish the following statement:

Proposition 1. For any $Q \subset \hat{P}_2$, $s > 0$, $s' > 0$ we have

$$\rho(V(s, Q), V(s', Q)) \leq \frac{L|s - s'|}{\max(s, s')}, \ L = \text{const} \quad . \tag{18}$$

The inclusion (7) is in fact equivalent to condition (13), given in terms of the set $V(s, Q)$. Condition 2 of Theorem 1 can also be presented in these terms.

Proposition 2. Condition (2) of Theorem 1 is satisfied if and only if

$$\rho(V(s, Q'), V(s, Q'')) \leq Ls^{-1} \tag{19}$$

for any $Q' \subset \hat{P}_2$, $Q'' \subset \hat{P}_2$ and $s > 0$. Condition 3 of Theorem 1 can also be reformulated:

$$Y(s, Q) \subset \hat{P}_2, \ \forall Q \subset P_2, \ \forall s > 0 \quad , \tag{20}$$

where $Y(s, Q)$ is the reachable set of system (11).

To describe the dynamics of $V(s, Q)$ we shall introduce the set $W(s, Q)$ defined as follows:

$$W(s, Q) = \{(\xi, \eta): \xi = s^{-1} \int_0^s f_1(y(t, u(\cdot), y_0), u(t))dt \quad , \tag{21}$$

$$\eta = y(s, u(\cdot), y_0), \ y_0 \in Q, \ u(t), \text{ is feasible}\} \quad .$$

It is obvious that $V(s, Q)$ and $Y(s, Q)$ are the projections of $W(s, Q)$ on R^{n_1} and R^{n_2}, respectively.

Proposition 3. For any $Q \subset P_2$, $s > 0$, $s' > 0$, $\lambda \in (0,1)$ we have

$$Y(s + s', Q) = Y(s, Y(s', Q)) \tag{22}$$

$$V(s, Q) = \bigcup_{(\xi,\eta)\in W(\lambda s, Q)} (\lambda\xi + (1-\lambda)V((1-\lambda)s, \eta)) \quad . \tag{23}$$

Relation (22) is obvious. Relation (23) may be proved using the fact that for any feasible control

$$s^{-1}\int_0^s f_1(y(t, u(\cdot), y_0), u(t))dt = \lambda(\lambda s)^{-1}\int_0^{\lambda s} f_1(y(t, u(\cdot), y_0), u(t))dt +$$

$$+ (1-\lambda)[(1-\lambda)s]^{-1}\int_{\lambda s}^{s} f_1(y(t, u(\cdot), y_0), u(t))dt$$

together with the fact that a control constructed of "pieces" of feasible controls is also feasible.

To conclude this section we shall list some of the properties of the Hausdorff metric. Consider the set of all closed subsets of the compact set P. When endowed with the Hausdorff metric this set becomes a compact metric space (Blascheke's theorem [8]).

Let V_i, $i = 1,2$, be subsets of P and $\bar{V}_i$, Conv V_i, $i = 1,2$ be their closures and convex hulls, respectively. Then

$$\rho(V_1, V_2) = \rho(\bar{V}_1, \bar{V}_2) \tag{24}$$

$$\rho(\text{Conv } V_1, \text{Conv } V_2) \leq \rho(V_1, V_2) \tag{25}$$

$$\rho(\text{Conv } V, k^{-1}\sum_1^k V) \leq Lk^{-1}, \; L = \text{const}, \; k = 1,2, \ldots \quad . \tag{26}$$

Relation (24) is obvious, (25) is easily verified [9] and the proof of (26) is based on Caratheodory's theorem on the representations of convex hulls [9].

4. PROOF OF THEOREM 2

The theorem is proved in several stages.

(a) Using induction, let us verify that for any natural number k, $s > 0$ and $Q \subset P_2$, the following relation holds:

$$\rho(V(ks, Q), k^{-1}\sum_1^k V(s, Q)) \leq Ls^{-1} \quad , \tag{27}$$

where L is the constant from condition (2) of the theorem. It is clear that the formula holds for $k = 1$. Assume that it also holds for $k - 1$:

$$\rho(V(k-1)s, Q), (k-1)^{-1}\sum_1^{k-1} V(s, Q)) \leq Ls^{-1} \quad . \tag{28}$$

From Proposition 2 we have

$$\rho(\frac{k-1}{k}V((k-1)s,\eta),\frac{k-1}{k}V((k-1)s,Q))\le\frac{k-1}{k}\frac{L}{(k-1)s}=\frac{L}{ks}\quad.$$

Moreover, (28) leads to

$$\rho(\frac{k-1}{k}V((k-1)s,\eta),k^{-1}\sum_{1}^{k-1}V(s,Q))\le\frac{L}{s}\frac{(k-1)}{k}+\frac{L}{ks}=\frac{L}{s}\quad.$$

Using relation (23) from Proposition 3 with $\lambda=k^{-1}$, $(1-\lambda)=(k-1)k^{-1}$ we have

$$V(ks,Q)=\bigcup_{(\zeta,\eta)\subset W(s,Q)}(k^{-1}\zeta+(k-1)k^{-1}V((k-1)s,\eta))\Longrightarrow$$

$$\rho(V(ks,Q),\bigcup_{(\zeta,\eta)\subset W(s,Q)}(k^{-1}\zeta+k^{-1}\sum_{1}^{k-1}V(s,Q))\le Ls^{-1}\quad.\tag{29}$$

Since

$$\bigcup_{(\zeta,\eta)\subset W(s,Q)}(k^{-1}\zeta+k^{-1}\sum_{1}^{k-1}V(s,Q))=k^{-1}V(s,Q)+k^{-1}\sum_{1}^{k-1}V(s,Q)=$$

$$=k^{-1}\sum_{1}^{k}V(s,Q)\quad,$$

relation (29) is equivalent to the statement we wanted to verify.

(b) Let us now show that the following limit exists:

$$\lim_{s\to\infty}\mathrm{Conv}\,\bar{V}(s,Q)=\bar{V}\quad.\tag{30}$$

Extend inequality (27) by taking (25) into account:

$$\rho\,(\mathrm{Conv}\,V(ks,Q),\mathrm{Conv}\,k^{-1}\sum_{1}^{k}V(s,Q))\le Ls^{-1}\quad.\tag{31}$$

It follows from well-known results in convex analysis [10] that Conv $k^{-1}\sum_{1}^{k}V(s,Q)=\mathrm{Conv}\,V(s,Q)$, $\overline{\mathrm{Conv}}\,V(s,Q)=\mathrm{Conv}\,\bar{V}(s,Q)$, where $\overline{\mathrm{Conv}}\,V(s,Q)$ is the closure of the convex hull of $V(s,Q)$. Thus (31) can be rewritten in the form

$$\rho\,(\mathrm{Conv}\,V(ks,Q)\,\mathrm{Conv}\,V(s,Q))\le Ls^{-1},\ \forall k=1,2,\dots\quad.\tag{32}$$

Let $k_2\ge k_1$ be arbitrary natural numbers. Then using (32) we obtain

$$\rho(\mathrm{Conv}\,\bar{V}(s,Q),\mathrm{Conv}\,\bar{V}(\frac{k_2}{k_1}s,Q))\le\rho(\mathrm{Conv}\,\bar{V}(s,Q),\mathrm{Conv}\,\bar{V}(k_2s,Q))+$$

$$+ \rho(\text{Conv } \bar{V}(k_2 s , Q) , \text{Conv } \bar{V}(\frac{k_2}{k_1} s , Q)) \leq \frac{L}{s} + \frac{L}{(\frac{k_2}{k_1})s} \leq \frac{2L}{s} \quad . \tag{33}$$

As noted above, the space of closed subsets of P endowed with the Hausdorff metric is compact. Therefore to prove the existence of the limit (30) it is sufficient to establish that for any $\nu > 0$ we can find an $s_\nu > 0$ such that for $s'' \geq s' \geq s_\nu$ the following inequality is satisfied:

$$\rho(\text{Conv } \bar{V}(s , Q) , \text{Conv } \bar{V}(s'' , Q)) \leq \nu \quad . \tag{34}$$

This is easily verified using (33) and Proposition 1.

(c) Let us now evaluate the rate of convergence in (30). We have

$$\rho(\text{Conv } \bar{V}(s , Q) , \bar{V}) \leq \rho(\text{Conv } \bar{V}(s , Q) , \text{Conv } \bar{V}(ks , Q)) +$$
$$+ \rho(\text{Conv } \bar{V}(ks , Q) , \bar{V}) \leq Ls^{-1} + \rho(\text{Conv } \bar{V}(ks , Q) , \bar{V}) \quad .$$

Taking the limit on the right-hand side as $k \to \infty$ we obtan

$$\rho(\text{Conv } \bar{V}(s , Q) , \bar{V}) \leq Ls^{-1} \quad . \tag{35}$$

(d) Using the triangle inequality and (26), (27) we can write

$$\rho(\bar{V}(ks , Q) , \text{Conv } \bar{V}(s , Q)) \leq L_1 s^{-1} + L_2 k^{-1} \quad . \tag{36}$$

Let us take the integer part of $s^{1/2}$ as k. From Proposition 1 we have

$$\rho(\bar{V}(s , Q) , \bar{V}(s^{1/2}[s^{1/2}] , Q)) \leq \frac{Ls^{1/2}(s^{1/2} - [s^{1/2}])}{s} \leq Ls^{-1/2} \quad . \tag{37}$$

Using (35), (36) and the triangle inequality we obtain (16). The convexity and compactness of $\bar{V}$ follow from (30).

REFERENCES

1. N.N. Bogoljubov and U.A. Mitropol'sky. *Asymptotic Methods in the Theory of Nonlinear Oscillations*. Fizmatgiz, Moscow, 1963 (in Russian).

2. N.N. Moiseev. *Asymptotic Methods in Nonlinear Mechanics*. Nauka, Moscow, 1969 (in Russian).

3. V.A. Plotnikov. *Asymptotic Methods in Optimal Control Problems*. Izdatel'stvo Odesskogo Universiteta, Odessa, 1976 (in Russian).

4. F.L. Chernousko, A.D. Akulenko and B.K. Sokolov. *Control of Oscillations*. Nauka, Moscow, 1980 (in Russian).

5. P.V. Kokotovic, R.E. O'Malley and P. Sanuti. Singular perturbations and order reductions in control theory. *Avtomatica*, 12(1976).

6. A.B. Vasileva and M.G. Dmitriev. Singular perturbations and some optimal control problems. In *Preprints of the VII World IFAC Congress*, Helsinki, Vol.2, Pergamon Press, Oxford, 1979.

7. A.F. Filippov. On some questions of optimal control theory (in Russian). *Vestnik Moskovskogo Universiteta*, (2)(1959).

8. D.H. Hadwiger. *Vorlesungen über Inhalt, Oberfläche und Isoperimetrie*. Springer-Verlag, 1957.

9. V.G. Gaitsgory. Asymptotic behavior of the solutions of an optimal control problem with an averaged performance index (in Russian). *Dep. Tadjik NIINTI*, 20(226)TA-D83.

10. R.T. Rockafellar. *Convex Analysis*. Princeton University Press, Princeton, New Jersey, 1970.

ON A CERTAIN CLASS OF INVERSE PROBLEMS IN CONTROL SYSTEM DYNAMICS

M.I. Gusev
Institute of Mathematics and Mechanics
Ural Center of USSR Academy of Sciences, Sverdlovsk

This report is concerned with the inverse problem of control theory: to identify an initial state and a disturbance in the input of a dynamic system using measurements of the output. Such problems arise, for example, in the solution of one of the two basic problems of dynamics: assuming the motion of a mechanical system to be given, find the forces that generate this motion [1]. On the other hand, this problem is also closely related to problems of control and observation under conditions of uncertainty [2,3].

Let the motion of a controlled plant on the interval $[t_0, t]$ be described by the differential equation

$$\dot{x} = f(t, x, u(t)), \; x(t_0) = x^0, \; x \in R^n, \; u \in R^r \quad , \tag{1}$$

where the initial state x^0 and the function $u(\cdot)$ representing the disturbance (control) are assumed to be unknown. The available *a priori* information on $w = (x^0, u(\cdot))$ is restricted to the inclusion $w \in W$, where W is a given set in the corresponding functional space.

Let the output of the plant be given by the equation

$$y(t) = h(t, x(t), u(t)), \; t \in [t_0, t_1] \quad , \tag{2}$$

where $h: [t_0, t_1] \times R^n \times R^r \to R^m$ is continuous.

Let us assume that $f(t, x, u)$ is continuous and the following standard conditions are satisfied: there exist $c_1 > 0$, $c_2 > 0$ such that

(a) $\|f(t, x, u)\| \le c_1(1 + \|x\| + \|u\|)$

(b) $\|f(t, x, u) - f(t, x', u)\| \le c_2\|x - x'\|(1 + \|u\|)$

for all $t \in [t_0, t_1]$, $x, x' \in R^n$, where $\|\cdot\|$ is the Euclidean norm.

We will call a function from the space $L_2([t_0, t_1], R)$ (sometimes also written L_2^r) an *admissible control* (disturbance). Let $u(\cdot) \in L_2^r$. Then the absolutely continuous function $x(t)$ is said to be a solution of system (1) if

$$x(t) = x^0 + \int_{t_0}^{t} f(\tau, x(\tau), u(\tau))d\tau, \ t_0 \le t \le t_1 \quad .$$

For every $x^0 \in R^n$, $u(\cdot) \in L_2^r$, there exists a unique solution $x(t, x^0, u(\cdot))$.

Let X^0 be a bounded set in R^n and U be a bounded set in L_2^r. Then the set of solutions $\{x(\cdot, x^0, u(\cdot)) : x^0 \in X^0, u(\cdot) \in U\}$ is uniformly bounded and equicontinuous.

The problem is to identify x^0 and $u(\cdot)$ (or $u(\cdot)$, or some of the coordinates of $u(\cdot)$ on a given basis) from the output $y(\cdot)$. We may rewrite the problem in the following way. Let X, Y, Z be Banach spaces and operators $A: X \to Y$, $F: X \to Z$, a point $y \in Y$ and a set $W \subset X$ be given. It is necessary to find $z = Fw$ under the conditions $Aw = y$, $w \in W$. Here the operator A transforms the pair $w = (x^0, u(\cdot))$ into $y = y(\cdot)$, see (1), (2). The operator F may be given, for example, by $Fw = w$, $Fw = u(\cdot)$, or $Fw = P_L u(\cdot)$ (where P_L is a projection operator on the subspace L).

We will take the Hilbert space $R^n \times L_2^r$ as X, the space L_2^m as Y, and the space $R_n \times L_2^r$, or L_2^r, or a finite-dimensional subspace as Z.

Examples of *a priori* restrictions on $(x^0, u(\cdot))$ could be the following:

$$W = \{(x^0, u(\cdot)) : x^0 \in X^0, \ u(t) \in U, \ t \in [t_0, t_1] \text{ (a.e.)}\} \quad ,$$

where $X^0 \subset R^n$, $U \subset R^r$ are given sets, or

$$W = \{(x^0, u(\cdot)) : x^{0'} M x^0 + \int_{t_0}^{t_1} u'(t) R(t) u(t) dt \le \mu^2\} \quad ,$$

where M, $R(t)$ are positive definite matrices.

Note that if $Fw = x(t_1)$ we obtain a state estimation problem and if $W = \{(x^0, u(\cdot)) : u(t) \equiv u = \text{const}, \ u \in U\}$ and $Fw = u(\cdot)$ we have a parameter identification problem.

Due to the non-invertibility of A the solution of the problem may not be unique. Following [3], we shall call the set

$$Z(y) = \{z : z = Fw, \ Aw = y, \ w \in W\}$$

the information domain associated with y.

A characteristic feature of the problem is the instability of $Z(y)$ with respect to y. In other words, if y is given with an error not exceeding $\delta > 0$ ($\|y - y_\delta\| \le \delta$), then $Z(y_\delta)$ may have an arbitrarily large deviation from $Z(y)$ regardless of how small δ is taken to be. Therefore, in order to solve the problem, it is necessary to combine methods from the theory of ill-posed problems [4,5] and the theory of observation under certainty [2,3].

Let X, Y, Z be Hilbert spaces ($<\cdot,\cdot>_X$ represents the corresponding scalar product) and $W \subset X$ be weakly compact. Let $w^* \in X$, $\mu \in R$ be given such that $W \subset \{w: <w-w^*, w-w^*>_X \leq \mu^2\}$ (for example, $\mu = \sup \{\|w-w^*\|: w \in W\}$). Let us consider the following extension $Z_{\varepsilon,\alpha}(y)$ of $Z(y)$, which is called the *regularization* of $Z(y)$:

$$Z_{\varepsilon,\alpha}(y) = \{z = Fw: Aw + \xi = y\ ,\ <w-w^*, w-w^*>_X + \frac{1}{\varepsilon} <\xi,\xi>_Y \leq \mu^2 + \alpha\ ,\ w \in W\}\ , \tag{3}$$

where $\varepsilon > 0$, $\alpha > 0$ are regularization parameters.

Proposition 1

(a) $Z(y) \subset Z_{\varepsilon,\alpha}(y_\delta)$ if $\delta^2/\varepsilon \leq \alpha$,

(b) if A, F are weakly closed and F is completely continuous, then $h(Z_{\varepsilon,\alpha}(y_\delta), Z(y)) \to 0$ for $\varepsilon \to 0$, $\alpha \to 0$, $\delta^2/\varepsilon \leq \alpha$, where h is the Hausdorff metric.

LEMMA 1. *Let equations* (1), (2) *be given as follows:*

$$\dot{x} = A(t,x) + B(t,x)u(t)\ ,\ x(t_0) = x^0 \tag{4}$$

$$y(t) = h_1(t, x(t)) + h_2(t, x(t))u(t)\ , \tag{5}$$

where $\|A(t,x)\| \leq c_1(1+\|x\|)$, $\|B(t,x)\| \leq c_2$, $A(t,x)$, $B(t,x)$ *are continuous in* (t,x) *and Lipschitz in* x, *and* h_1, h_2 *are continuous. Then* $A: R^n \times L_2^r \to L_2^m$ *is weakly closed.*

F is weakly closed and completely continuous, for example, if $Fw = x^0$, if $Fw = P_L u(\cdot)$ and L is finite-dimensional, or if $Fw = x(t_1, x^0, u(\cdot))$ and system (1) is linear in u.

Proposition 2. Let A be a weakly closed operator and F be a linear continuous operator. Then for every $z^* \in Z$ and for $\delta^2/\varepsilon \leq \alpha$ we have

$$\rho(z^* \mid Z_{\varepsilon,\alpha}(y_\delta)) \to \rho(z^* \mid Z(y))$$

as $\varepsilon \to 0$, $\alpha \to 0$, where $\rho(\cdot \mid Z)$ is the support function of Z.

Let A, F be linear continuous operators and $W = \{w: <w, w>_X \leq \mu^2\}$. Then we may consider the following regularization of $Z(y)$ instead of $Z_{\varepsilon,\alpha}(y)$:

$$\tilde{Z}_{\varepsilon,\alpha}(y) = \{z = Fw: Aw + \xi = y\ ,\ <w-w^*, w-w^*>_X + \frac{1}{\varepsilon} <\xi,\xi>_Y \leq \mu^2 + \alpha\}\ .$$

Propositions 1,2 also hold for $\tilde{Z}_{\varepsilon,\alpha}(y)$.

The domain $\tilde{Z}_{\varepsilon,\alpha}(y)$ may be described as follows:

THEOREM 1. *The domain $\tilde{Z}_{\varepsilon,\alpha}(y_\delta)$ is an ellipsoidal set with the support function*

$$\rho(z^* \mid \tilde{Z}_{\varepsilon,\alpha}(y_\delta)) = (\mu^2 + \alpha - \langle y_\delta - Aw_*, (AA^* + \varepsilon E)^{-1}(y_\delta - Aw^*)\rangle_Y)^{1/2} \times \qquad (6)$$

$$(\langle z^*, FTF^* z^*\rangle_Z)^{1/2} + \langle z^*, \hat{z}\rangle_Z$$

where

$$\hat{z} = F\hat{w}(\varepsilon, \delta, w^*), \; \hat{w}(\varepsilon, \delta, w^*) = A^*(AA^* + \varepsilon E)^{-1} y_\delta + Tw^*, \; T = \varepsilon(A^*A + \varepsilon E)^{-1}$$

and operator A^* is adjoint to A.

Proposition 3. The point $\hat{w}(\varepsilon, \delta, w^*)$ and the operators appearing in (6) may be found as the solutions of the following variational problems:

$$\hat{w}(\varepsilon, \delta, w^*) = Sw^* = \arg\min \{\|Aw - y_\delta\|^2 + \varepsilon \|w - w^*\|^2\} \qquad (7)$$

$$Tu = \arg\min \{\|Aw\|^2 + \varepsilon \|w - u\|^2\} \qquad (8)$$

$$(AA^* + \varepsilon E)^{-1} x = \frac{1}{\varepsilon}(x - A\bar{w}) \quad , \qquad (9)$$

where $\bar{w} = \arg\min \{\|Aw - x\|^2 + \varepsilon \|w\|^2\}$. Thus $\hat{w}$ is the solution of problem (7) from Tikhonov's regularization method [4,5]. The center of symmetry $\hat{z} = F\hat{w}(\varepsilon, \delta, w^*)$ of $\tilde{Z}_{\varepsilon,\alpha}(y_\delta)$ may be considered to be the best regularized minimax estimate [3] of z.

In particular, if $Fw = w$, we have that $\tilde{Z}_{\varepsilon,\alpha}(y)$ is an ellipsoid ($w^* = 0, \hat{w} = \hat{w}(\varepsilon, \delta, 0)$):

$$\tilde{Z}_{\varepsilon,\alpha}(y_\delta) = \{w \in X\colon \langle w - \hat{w}, \frac{1}{\varepsilon}(A^*A + \varepsilon E)(w - \hat{w})\rangle_X \le \mu^2 + \alpha - \frac{1}{\varepsilon}\langle y_\delta, y_\delta - A\hat{w}\rangle_Y\} \quad .$$

For systems (1), (2) which are linear in x, u, problems (7), (8) may be solved using standard methods from the theory of linear-quadratic optimal control.

If Z is infinite-dimensional, then $Z_{\varepsilon,\alpha}(y_\delta)$ does not converge to $Z(y)$ in the Hausdorff metric, but $\hat{w}(\varepsilon, \delta, w^*) \to \{w\colon Aw = y\}$ as $\varepsilon \to 0$, $\delta^2/\varepsilon \to 0$. The analogous convergence also holds for *a priori* restrictions.

In particular, let us consider the variational problem

$$\|Aw - y_\delta\|^2 + \varepsilon\|w - w^*\|^2 \to \min_w \, , \; w \in W \quad . \qquad (10)$$

THEOREM 2. *Let system* (1), (2) *be of the form* (3), (4) *and the assumptions of Lemma* 1 *be satisfied. Let W be weakly closed in $R^n \times L_2^T$. Then problem* (10) *has a* *solution* $\hat{w}(\varepsilon, \delta, w')$ *for* $\varepsilon > 0$, $w' \in X$, *and* $\hat{w} \to \{w: Aw = y, w \in W\}$ *as* $\varepsilon \to 0$, $\delta^2/\varepsilon \to 0$ *in the metric of space* $R^n \times L_2^r$.

IF F is uniformly continuous on every bounded set, then $\hat{z} = F\hat{w} \to Z(y)$ in the metric of Z.

Let system (1) be of the form (4) and all the coordinates of $x(t)$ be measurable, i.e., $y(t) = x(t)$. Let the measurement error be no greater than δ in the metric of L_∞^n:

$$\|x_\delta(\cdot) - x(\cdot)\|_{L_\infty^n} = \operatorname*{vrai\,max}_{t_0 \le t \le t_1} \|x_\delta(t) - x(t)\| \le \delta \quad .$$

Let us define $A_\delta(t) = A(t, x_\delta(t))$, $B_\delta(t) = B(t, x_\delta(t))$, and consider the optimal control problem

$$J = \int_{t_0}^{t_1} [(x_\delta - x)'Q(x_\delta - x) + \varepsilon u'Ru]dt + \varepsilon x^{0\prime}Mx^0 \to \min_{x^0, u(\cdot)} \tag{11}$$

for the linear system

$$\dot{x} = A_\delta(t) + B_\delta(t)u \, , \; x(t_0) = x^0 \quad , \tag{12}$$

where Q, R, M are positive definite matrices.

THEOREM 3. *Let* $\hat{w}(\varepsilon, \delta)$ *be the solution of problem* (11), (12) *under the constraint* $(x^0, u(\cdot)) \in W$, *where* W *is weakly closed. Then* $\hat{w}(\varepsilon, \delta) \to \{w: Aw = x(\cdot) \; w \in W\}$ *in the metric of* $R^n \times L_2^r$, *and* $A\hat{w}(\varepsilon, \delta) - x_\delta(\cdot) \to 0$ *in the metric of* L_∞^n *as* $\varepsilon \to 0$, $\delta \to 0$, $\delta^2/\varepsilon \to 0$.

Taking $W = R^n \times L_2^r$, the solution of (11), (12) may be obtained in explicit form as follows (where Q, R, M are assumed to be unit matrices):

$$\hat{w}(\varepsilon, \delta) = (x^0(\varepsilon, \delta), u_\delta^\varepsilon(\cdot))$$

$$u_\delta^\varepsilon(t) = \frac{1}{\varepsilon} B_\delta'(t)[h_\delta^\varepsilon(t) - K_\varepsilon(t)x_\delta^\varepsilon(t)] \, , \; x^0(\varepsilon, \delta) = (K_\varepsilon(t_0) + \varepsilon E)^{-1} h_\delta^\varepsilon(t_0)$$

$$\dot{K}_\varepsilon = \frac{1}{\varepsilon} K_\varepsilon B_\delta(t) B_\delta'(t) K_\varepsilon - E \, , \; K_\varepsilon(t_1) = 0$$

$$\dot{h}_\delta^\varepsilon = \frac{1}{\varepsilon} K_\varepsilon(t) B_\delta'(t) B_\delta(t) h_\delta^\varepsilon - x_\delta^\varepsilon(t) + K_\varepsilon(t) A_\delta(t) \, , \; h_\delta^\varepsilon(t_1) = 0$$

$$\dot{x}_\delta^\varepsilon = A_\delta(t) + \frac{1}{\varepsilon} B_\delta(t) B_\delta'(t)[h_\delta^\varepsilon(t) - K_\varepsilon(t) x_\delta^\varepsilon] \quad .$$

Remark 1. The case in which the right-hand side of (1), (2) depends on an *a priori* unknown constant vector of parameters c ($f = f(t, x, u, c)$, $h = h(t, x, u, c)$) may be reduced to the case under consideration by adding the equations $\dot{c} = 0$, $c(t_0) = c^0$

to system (1) and by choosing the triplet $(x^0, c^0, u(\cdot))$ as w. All of the propositions given in this paper also hold (with some modification) in this case. It should be noted that even if system (1), (2) is linear in x, u, the corresponding operator A is non-linear.

Remark 2. If Z is infinite-dimensional it is necessary to impose a stronger requirement concerning W in order for $Z_{\varepsilon,\alpha}(y_\delta)$ to converge to $Z(y)$ in the Hausdorff metric. We have convergence if, for example, $u(t)$ is a solution of the differential equation

$$\dot{u} = Du + Kv \ , \ u(t_0) = u^0, v(t) \in R^l \quad ,$$

with unknowns u^0 and $v(\cdot)$ restricted *a priori* by the inclusion $(u^0, v(\cdot)) \in V$, where V is weakly compact in $R^r \times L_2^l$.

Remark 3. If we have *a priori* restrictions on w we are faced with a variational problem (10) with constraints. Instead of solving this we can consider iterative process of the form $w_{k+1} = P_w S w_k$ (see [6]) representing a combination of the proximal point algorithm [7] and projection on W. In this case every step of the process reduces to the solution of problem (7) without constraints and calculation of the projection on W, which is easier than solving (10). For systems (1), (2) which are linear in x, u, the results given in [6] imply that the process w_k converges to the set $\{w: Aw = y \ , \ w \in W\}$ (under certain standard assumptions).

Remark 4. The problem of deducing the input of a dynamic system from measurements of the output has been considered in various papers. Most of these are concerned with the single-valued reconstruction of the input from precise measurements of the output (see, for example, [8–11]) i.e., with the question of the invertibility of A. The results of these investigations show that even if A^{-1} exists it is generally discontinuous.

We should note that there is a close relationship between the problem considered here and the problem of parameter identification (see [12,13]).

The problem of the *a priori* estimation [2,3] of input for linear systems is considered in [14], where questions of numerical solution are also discussed.

The stable reconstruction of input for systems with a completely observable state space vector is considered in [15].

Recursive procedures for estimating input for linear discrete-time systems are given in [16].

REFERENCES

1. N.E. Zukovski. *Theoretical Mechanics*. L. Gostehizdat, Moscow, 1952.
2. N.N. Krasovski. *Theory of Control of Motion*. Nauka, Moscow, 1968.
3. A.B. Kurzhanski. *Control and Observation under Conditions of Uncertainty*. Nauka, Moscow, 1977.
4. A.N. Tikhonov and W.J. Arsenin. *Methods of Solving Ill-Posed Problems*. Nauka, Moscow, 1975.
5. V.K. Ivanov, V.V. Vasin and V.P. Tanana. *Theory of Linear Ill-Posed Problems and its Applications*. Nauka, Moscow, 1978.
6. V.V. Vasin. *A Proximal Point Algorithm with Projection for Convex Programming Problems*. Ural Scientific Center, 1982.
7. R.T. Rockafellar. Monotone operators and the proximal point algorithm. *SIAM Journal on Control and Optimization*, 14(1976)877–898.
8. M.S. Nikolski. On perfectly observable systems. *Differential Equations*, 7(1971)631–638.
9. L.M. Silverman. Inversion of multivariable linear systems. *IEEE Transactions on Automatic Control*, AC-19(1969)270–276.
10. R.M. Hirshchorn. Invertibility of nonlinear control systems. *SIAM Journal on Control and Optimization*, 17(1979)289–297.
11. H. Nijmeijer. Invertibility of affine nonlinear control systems: a geometric approach. *System and Control Letters*, 2(3)(1982)163–168.
12. A.I. Isakov. On dual relations in problems of identification and observation. *Avtomatika i Telemekhanika*, (8)(1975).
13. A.B. Kurzhanski. On the estimation of the dynamics of control systems. Presented at the Eighth IFAC Congress, Kyoto, Japan, 1981.
14. N.E. Kirin. *Sequential Estimate Methods for the Optimization of Control Systems*. Leningrad State University, 1975.
15. A.V. Krjazimski and Ju.S. Osipov. On positional modelling of control in dynamic systems. *Izvestiya Akademii Nauk SSSR, Technicheskaya Kybernetika*, (2)(1983)51–60.
16. J.D. Glover. The linear estimation of completely unknown signals. *IEEE Transactions on Automatic Control*, AC-19(1969)766–767.

SIMULTANEOUS ESTIMATION OF STATES AND PARAMETERS IN CONTROL SYSTEMS WITH INCOMPLETE DATA

N.F. Kirichenko and A.S. Slabospitsky
Kiev State University, Kiev, USSR

1. INTRODUCTION

This paper is concerned with the following closely related optimization problems in dynamical systems: optimal stability, minimax state estimation, minimax parameters identification and pattern recognition. Some investigations of these questions are reported in [1–4].

2. OPTIMAL STABILITY [5,6]

Using Liapunov's definition of the stability of a trajectory and Chetaev's definition of the practical stability of a trajectory, we introduce the concept of *optimal practical stability* and present conditions under which systems are optimal in this sense.

Let some process be described by the set of equations

$$\frac{dx(t)}{dt} = F(x(t), u(t), f_2(t), t), \quad x(t_0) = f_1 \quad , \tag{1}$$

where x is the state vector, u is the control vector, f_2 is the vector of disturbances, and f_1 is the vector of disturbances to the initial data. Assume that $f = (f_1, f_2(\tau),\ \tau \in [t_0, t])$ takes values in the set $G_t(\lambda)$, which has an *a priori* known structure, and λ is the size characteristic of this set. We shall consider some closed set Γ in state space. Let $\lambda_t(u)$ be the maximum value of λ, which for a given control u satisfies the condition

$$x(t) \in \Gamma, \ \forall f \in G_t(\lambda_t(u)) \quad .$$

Then the optimal stability problem is formulated in the following way: find the optimal control $u(\tau)$, $\tau \in [t_0, t]$, with respect to the criterion

$$J(u) = \lambda_t(u) \to \max_u \quad . \tag{2}$$

THEOREM 1. *If system* (1) *and the corresponding sets* $G_t(\lambda)$ *and* Γ *are of the form*

$$\frac{dx(t)}{dt} = A(t,u)x(t),\ x(t_0) = f \tag{3}$$

$$G_t(\lambda) \triangleq \{x : x^* S x \leq \lambda^2\}$$

$$\Gamma \triangleq \{x : |l_s^* x| \leq 1,\ s = \overline{1,N}\}\ , \tag{4}$$

then the maximum value of criterion (2) *is given by the solution of the following optimal control problem:*

$$\frac{dP(t)}{dt} = A(t,u)P(t) + P(t)A^*(t,u),\ P(t_0) = S^{-1} \tag{5}$$

$$\tilde{J}(u) = \max_{s=\overline{1,N}} l_s^* P(t) l_s \to \min_u \ .$$

Here l_s *are given vectors,* S *is a positive-definite matrix, and* $(\cdot)^*$ *denotes transposition.*

COROLLARY 1. *If we consider the functional*

$$J_1(u) = \min_{t\in[t_0,t_1]} \lambda_t(u) \to \max_u \tag{7}$$

instead of (2), *then Theorem 1 holds with* (6) replaced by the criterion

$$\tilde{J}_1(u) = \max_{t\in[t_0,t_1]} \max_{s=\overline{1,N}} l_s^* P(t) l_s \ .$$

THEOREM 2. *Let* $A(t,u) = u(t)$, $N = 1$, *in system* (3). *Then, using the regularization functional*

$$\tilde{J}(u) = l_1^* P(t_1) l_1 + \mu \int_{t_0}^{t_1} \mathrm{tr}\,(A(t)P(t)A^*(t))dt \ ,$$

the solution of the optimal control problem for system (5) *is of the form*

$$A(t) = \frac{l_1 l_1^*}{\|l_1\|^2(t-t_1)-\mu},\ \mu > 0 \ .$$

THEOREM 3. *If in* (3) *we have*

$$A(t,u) = A_1(t) + B(t)C(t),\ C(t) = u(t),\ N = 1 \ ,$$

then for system (5) *the functional*

$$\tilde{J}(u) = l_1^* P(t_1) l_1 + \mu \int_{t_0}^{t_1} \mathrm{tr}\,(C(t)P(t)C^*(t))\mathrm{d}t\ ,\ \mu > 0$$

attains a minimum with

$$C(t) = \frac{1}{\mu} B^*(t)\psi(t)$$

$$\frac{\mathrm{d}\psi(t)}{\mathrm{d}t} = \frac{1}{\mu}\psi(t)B(t)B^*(t)\psi(t) - A^*(t)\psi(t) - \psi(t)A(t)\ ,\ \psi(t_1) = -l_1 l_1^* \quad .$$

THEOREM 4. *Let system* (1) *be of the form*

$$\frac{\mathrm{d}x(t)}{\mathrm{d}t} = A(t)(x(t) + f_3(t)) + f_2(t)\ ,\ \ A(t) = u(t)\ ,\ \ x(t_0) = f_1 \quad .$$

Take

$$G_t(\lambda) = \{f : f_1^* S_0 f_1 + \int_{t_0}^{t} (f_2^*(\tau)S_1(\tau)f_2(\tau) + f_3^*(\tau)S_2(\tau)f_3(\tau))\mathrm{d}\tau \le \lambda^2\} \quad , \tag{8}$$

and let the set Γ *be defined by formula* (4). *Then the functional* (7) *attains a maximum with*

$$A(t) = -P(t)S_2(t)\ ,\ \ t \ge t_0\ ,\ \forall l_s\ ,\ s = \overline{1,N}$$

$$\frac{\mathrm{d}P(t)}{\mathrm{d}t} = -P(t)S_2(t)P(t) + S_1^{-1}(t)\ ,\ P(t_0) = S_0^{-1} \quad .$$

THEOREM 5. *If system* (1) *is of the form*

$$\frac{\mathrm{d}x(t)}{\mathrm{d}t} = A(t)x(t) + B(t)(C(t)x(t) + f_3(t)) + f_2(t)\ ,\ \ B(t) = u(t)$$

and all of the other conditions of Theorem 4 are satisfied, then,

$$B(t) = -P(t)C^*(t)S_2(t)\ ,\ t \ge t_0\ ,\ \forall l_s\ ,\ s = \overline{1,N} \tag{9}$$

$$\frac{\mathrm{d}P(t)}{\mathrm{d}t} = A(t)P(t) + P(t)A^*(t) - P(t)C^*(t)S_2(t)C(t)P(t) + {}$$

$$+ S_1^{-1}(t)\ ,\ P(t_0) = S_0^{-1} \quad .$$

Proofs of these theorems may be obtained using the generalized Hölder inequality and Pontryagin's maximum principle. Numerical methods for the solution of the optimal stability problem for systems with incomplete data in more complicated cases are given in [7].

3. MINIMAX STATE ESTIMATION [5,6,8,9]

Consider the dynamic system

$$\frac{dx(t)}{dt} = A(\alpha, t)x(t) + f_2(t), \quad y(t) = C(t)x(t) + f_3(t) \quad , \tag{10}$$

where $f_1, f_2(\tau), f_3(\tau), \tau \in [t_0, t]$, belong to set G (see (8)), α is the parameter vector, and the state estimate $\hat{x}(t)$ is determined as a solution of the system

$$\frac{d\hat{x}(t)}{dt} = A(t)\hat{x}(t) - K(t)(y(t) - C(t)\hat{x}(t)), \quad \hat{x}(t_0) = 0 \quad . \tag{11}$$

THEOREM 6. *The optimal estimate* $\hat{x}(t)$ *from the class of solutions of system* (11) *with performance criterion*

$$J(K) = \sup_{G_t(\lambda)} |l^*(x(t) - \hat{x}(t))|^2 \to \min_K \tag{12}$$

is attained on $K(t) = B(t)$, *where* $B(t)$ *is defined by formula* (9).

THEOREM 7. *For system* (10) *the minimal value of functional* (12) *is*

$$J(B) = l^*P(t)l \times (\lambda^2 + \frac{1}{4}k^*(t)P(t)k(t) - h(t)) \quad , \tag{13}$$

where

$$\frac{dh(t)}{dt} = y^*(t)S_2(t)y(t) - \frac{1}{4}k^*(t)S_1^{-1}(t)k(t), \quad h(t_0) = 0$$

$$\frac{dk(t)}{dt} = -A^*(t)k(t) - 2C^*(t)S_2(t)y(t) - P^{-1}(t)S_1^{-1}(t)k(t), \quad k(t_0) = 0 \tag{14}$$

$$\frac{dP(t)}{dt} = A(t)P(t) + P(t)A^*(t) + S_1^{-1}(t) - P(t)C^*(t)S_2(t)C(t)P(t), \quad P(t_0) = S_0^{-1} \quad .$$

The optimal value of criterion (13) and equations (14) may be used to state a problem involving simultaneous estimation parameters and the states of the system, having already formulated the minimax likelihood principle: choose the parameters of the system α, including matrices S_0, S_1, S_2 from the available set of values such that the functional

$$J_1(\alpha) = \|P(t)\|_* (\lambda^2 + \frac{1}{4}k^*(t)P(t)k(t) - h(t))$$

attains a maximum in α, where $\|\cdot\|_*$ is a spectral norm.

The proof of Theorem 7 is based on Bellman's optimality principle and the following lemma:

LEMMA 1. (generalized Pifagor formula). *If vectors* x , φ *satisfy the relations*

$$Ax + \varphi = a \ , \quad \varphi' G \varphi + x' K x = \lambda^2$$

$$\hat{x} \ , \ \hat{\varphi} = \arg \min_{Ax + \varphi = a} (\varphi' G \varphi + x' K x) \ ,$$

then

$$(x - \hat{x})' (A' G A + K)(x - \hat{x}) = \lambda^2 - \hat{\varphi}' G \hat{\varphi} - \hat{x}' K \hat{x} \ .$$

4. OPTIMAL PARAMETER IDENTIFICATION [10–12]

We shall consider the optimal estimation of an unknown parameter vector α for a system described by the following set of equations:

$$\frac{dx(t)}{dt} = F(x, u, t)\alpha + \varphi(x, u, t) + C(x, u, t) f(t) \ , \ t \geq t_0 \ , \tag{15}$$

where $f(t)$ represents the disturbances.

Suppose that the pair $(\alpha \ ; f(\tau), \tau \in [t_0, t])$ can only take values in the available domain $G_t(\lambda)$. We want to find the minimax estimate $\hat{\alpha}(t)$, i.e.,

$$\hat{\alpha}(t) = \arg \inf_{\bar{\alpha} \in G_t^\alpha(\lambda)} \sup_{\alpha \in G_t^\alpha(\lambda)} \| \bar{\alpha} - \alpha \| \ .$$

Here $G_t^\alpha(\lambda)$ is the *a posteriori* information about the domain of available values for α up to time t, which are compatible with the next trajectory measurement and control actions. In what follows we shall suppose that strict identification of α in individual directions is not possible.

THEOREM 8. *The minimax* $\hat{\alpha}(t)$ *for system* (15) *with*

$$G_t(\lambda) \triangleq \{\alpha, f : (\alpha - \alpha_0)' S_0 (\alpha - \alpha_0) \int_{t_0}^{t} f'(\tau) S(\tau) f(\tau) d\tau \leq \lambda^2(t)\}$$

is obtained from

$$\frac{d\hat{\alpha}(t)}{dt} = R(t) F'(x, u, t) \tilde{S}(t) [\dot{x}(t) - F(x, u, t)\hat{\alpha}(t) - \tag{16}$$

$$\varphi(x, u, t)] \ , \ \hat{\alpha}(t_0) = \alpha_0$$

$$\frac{dR(t)}{dt} = -R(t) F'(x, u, t) \tilde{S}(t) F(x, u, t) R(t) \ , \ R(t_0) = S_0^{-1} \tag{17}$$

together with an estimate of the identification error in direction l ($\|l\| = 1$):

$$\sup_{\alpha \in G_t^\alpha(\lambda)} [l'\alpha - l'\hat{\alpha}(t)]^2 \leq l' R(t) l \ \times$$

$$\times [\lambda^2(t) - \alpha_0' S_0 \alpha_0 + \hat{\alpha}'(t) R^{-1}(t) \hat{\alpha}(t) -$$

$$- \int_{t_0}^{t} (\dot{x}(\tau) - \varphi(x, u, \tau))' \tilde{S}(\tau)(\dot{x}(\tau) - \varphi(x, u, \tau)) d\tau] \quad ,$$

where S_0, $S(t)$ are positive-definite matrices, the vector α_0 is known, $\lambda^2(t)$ is non-decreasing and such that $\lambda^2(t_0) > 0$, and we define

$$\tilde{S}(t) \triangleq (C(x, u, t) S^{-1}(t) C'(x, u, t))^+ \quad .$$

Here $(\cdot)^+$ represents pseudo-inversion.

In the case when only the domains of available values Ω for matrices S_0, S are given, the values of these matrices are selected by maximizing the following functional:

$$\|R(t)\| . [\lambda^2(t) - \alpha_0' S_0 \alpha_0 + \hat{\alpha}'(t) R^{-1}(t) \hat{\alpha}(t) -$$

$$- \int_{t_0}^{t} (\dot{x}(\tau) - \varphi(x, u, \tau))' \tilde{S}(\tau)(\dot{x}(\tau) - \varphi(x, u, \tau)) d\tau] \to \max_{S_0, S \in \Omega}$$

on (16), (17).

Let $C(x, u, t) = E$ (where E is the identity matrix) and $S(t) = S$ for all $t \geq t_0$. Then we have the following result:

THEOREM 9. *In order that* $\lim_{t \to \infty} \operatorname{diam} G_t^{\alpha}(\lambda) = 0$, *it is sufficient that*

$$\lambda^{-2}(t) \int_{t_0}^{t} \|F(x, u, \tau) l\|^2 d\tau \underset{t \to \infty}{\longrightarrow} \infty$$

uniformly in l (here $\|l\| = 1$ and $\operatorname{diam} G_t^{\alpha}(\lambda) \triangleq \sup_{\alpha_1, \alpha_2 \in G_t^{\alpha}(\lambda)} \|\alpha_1 - \alpha_2\|$).

COROLLAARY 2. *Let $\lambda^2(t)$ be bounded and*

$$\overline{\lim_{t \to \infty}} \, (\lambda^2(t) - (\alpha - \alpha_0)' S_0 (\alpha - \alpha_0) + \int_{t_0}^{t} f'(\tau) S(\tau) f(\tau) d\tau) > 0 \quad .$$

Then the condition

$$\int_{t_0}^{t} \|F(x, u, \tau) l\|^2 d\tau \underset{t \to \infty}{\longrightarrow} \infty \; \forall l \; (\|l\| = 1)$$

is necessary and sufficient to obtain

$$\lim_{t \to \infty} \operatorname{diam} G_t^{\alpha}(\lambda) = 0 \quad .$$

In practice $F(x, u, t)$ and $C(x, u, t)$ often have the following structure:

$$F(x, u, t) = \mathrm{diag}\,(g_1^*(x, u, t), \ldots, g_n^*(x, u, t)) \quad ,$$

$$C(x, u, t) = \mathrm{diag}\,(c_1^*(x, u, t), \ldots, c_n^*(x, u, t)) \quad .$$

The identification problem for system (15) is then split up into n estimation tasks for one-dimensional systems

$$\frac{dx_i(t)}{dt} = g_i^*(x, u, t)\alpha_i + \varphi_i(x, u, t) + c_i^*(x, u, t)f_i(t) \tag{18}$$

$$(\alpha_i - \alpha_{0i})^* S_{0i}(\alpha_i - \alpha_{0i}) + \int_{t_0}^{t} f_i^*(\tau)S_i(\tau)f_i(\tau)d\tau \le \lambda_i^2(t)\,,\quad i = \overline{1,n} \quad , \tag{19}$$

where

$$x^*(t) = (x_1(t), \ldots, x_n(t))\,,\quad \alpha^* = (\alpha_1^*, \ldots, \alpha_n^*)$$

$$\varphi^*(x, u, t) = (\varphi_1(x, u, t), \ldots, \varphi_n(x, u, t)) \quad ,$$

$\|c_i(x, u, t)\| > 0 \;\; \forall i$, matrices S_{0i}, $S_i(\tau)$, $t \in [t_0, t]$, are positive definite, and α_{0i} is known.

The following algorithm solves the identification problem for (18), (19):

$$\frac{d\hat{\alpha}_i(t)}{dt} = [\dot{x}_i(t) - g_i^*(x, u, t)\hat{\alpha}_i(t) - \varphi_i(x, u, t)]\gamma_i(t)k_i(t)\,,\quad \hat{\alpha}_i(t_0) = \alpha_{0i}$$

$$\frac{dR_i(t)}{dt} = -\gamma_i(t)k_i(t)k_i^*(t)\,,\quad R_i(t_0) = S_{0i}^{-1}\,,\quad i = \overline{\overline{1,n}}$$

where

$$k_i(t) \triangleq R_i(t)g_i(x, u, t)$$

$$\gamma_i(t) \triangleq (c_i^*(x, u, t)S_i^{-1}(t)c_i(x, u, t))^{-1} \quad .$$

The above result may be transferred to discrete systems

$$x(k+1) = F(x, u, k)\alpha + \varphi(x, u, k) + C(x, u, k)f(k)\,,\quad k = 1,2, \ldots$$

with the following available domain of values for (α, f):

$$G_N(\lambda) \triangleq \{\alpha, f : (\alpha - \alpha_0)^* S_0(\alpha - \alpha_0) + \sum_{k=1}^{N} f^*(k)S(k)f(k) \le \lambda^2(N)\} \quad .$$

The estimation algorithm and the estimate of the identification error in this case have the form

$$\hat{\alpha}(k+1) = D(k)\hat{\alpha}(k) + R(k+1)F'(x, u, k+1)\tilde{S}(k+1) \times$$

$$\times [x(k+2) - \varphi(x, u, k+1)], \quad \hat{\alpha}(0) = \alpha_0$$

$$R(k+1) = D(k)R(k), \quad R(0) = S_0^{-1}$$

$$\sup_{\alpha \in G_N^{\alpha}(\lambda)} [l'\alpha - l'\hat{\alpha}(N)]^2 \leq l'R(N)l \times \alpha \in G_N^{\alpha}(\lambda)\{\lambda^2(N) - \alpha_0'S_0\alpha_0 + \hat{\alpha}'(N)R^{-1}(N)\hat{\alpha}(N) -$$

$$- \sum_{k=1}^{N} [x(k+1) - \varphi(x, u, k)]'\tilde{S}(k)[x(k+1) - \varphi(x, u, k)]\} \quad .$$

Here

$$D(k) \triangleq E - R(k)W(k+1)[S(k+1) + W'(k+1)R(k)W(k+1)]^{-1}W'(k+1)$$

$$W'(k+1) \triangleq C'(x, u, k+1)\tilde{S}(k+1)F(x, u, k+1)$$

$$\tilde{S}(k) \triangleq (C(x, u, k)S^{-1}(k)C'(x, u, k))^{+} \quad .$$

For the linear dynamic system

$$x(k+1) = Ax(k) + Bu(k) + f(k), \quad k = 1,2, \ldots \tag{20}$$

with

$$G_N(\lambda) \triangleq \{A, f : \mathrm{tr}\,\{(A - A_0)'(A - A_0)\} + \sum_{k=1}^{N} f'(k)f(k) \leq \lambda^2(N)\} \quad , \tag{21}$$

the identification procedure for matrix A is simplified:

$$\hat{A}(N+1) = \{\hat{A}(N) + (x(N+2) - Bu(N+1))x'(N+1)R(N)\}P(N), \quad \hat{A}(0) = A_0 \tag{22}$$

$$R(N+1) = R(N)P(N), \quad R(0) = E \quad , \tag{23}$$

where

$$P(N) \triangleq E - Q(N)/(1 + \mathrm{tr}\,(Q(N)))$$

$$Q(N) \triangleq x(N+1)x'(N+1)R(N) \quad .$$

Consider system (20) with matrix $B \equiv 0$ and positive model errors and trajectory measurement. In this case the following results are obtained concerning the asymptotic behavior of the algorithm:

THEOREM 10. *If system* (20) *is asymptotically identifiable* (*i.e.*, $\lim_{N \to \infty} \|A - \hat{A}(N)\| = 0$), *then*

$$\lim_{N\to\infty} \frac{\sum_{k=1}^{N} \|f(k)\|^2}{\sum_{k=1}^{N} \|x(k)\|^2} = 0 \quad .$$

THEOREM 11. *For asymptotic identifiability of linear system* (20) *it is sufficient that*

$$\lim_{N\to\infty} \frac{\sum_{k=1}^{N} \|f(k)\|^2}{\min_{\|l\|=1} \sum_{k=1}^{N} (l^* x(k))^2} = 0 \quad .$$

Theorems 10 and 11 demonstrate a direct relation between the convergence of the estimation procedure and the model errors and observations of the phase state vector.

THEOREM 12. *Let linear system* (20) *be asymptotically stable and the disturbances* $f(k)$ *satisfy the conditions*

$$\lim_{N\to\infty} N^{-1} \sum_{k=1}^{N} f(k+j) f^*(k) = Q\delta(j) \quad \forall j$$

$$\lim_{N\to\infty} N^{-1} \sum_{k=1}^{N} f(k) =.0 \quad .$$

Then system (20) *is asymptotically identifiable. Here* $Q > 0$ *and* $\delta(j)$ *is the Kronecker delta.*

The convergence conditions of the procedure for the estimation of matrix B are given in the following theorem:

THEOREM 13. *If the control actions* $u(k)$ *and model errors* $f(k)$ *satisfy the conditions*

$$\lim_{N\to\infty} N^{-1} \sum_{k=1}^{N} u(k) u^*(k) = U (U > 0)$$

$$\lim_{N\to\infty} N^{-1} \sum_{k=1}^{N} \|f(k)\|^2 = 0 \quad ,$$

then the input matrix B *is asymptotically identifiable.*

5. PATTERN RECOGNITION

Solution of pattern recognition problems allows us to make effective use of optimal estimation methods.

Consider m classes of patterns with known sets of prototypes $x^{ij} \in \mathbb{R}^n$, $j = \overline{1,r_i}$, for category i. It is necessary to determine from the observations

$$y(k) = H(k)x + c(k)f(k), \; k = 1,2 \cdots$$

the class to which pattern x belongs. Here $H(k)$, $c(k)$ are given matrices, and $f(k)$ satisfies the condition

$$\sum_{k=1}^{N} f^*(k)S(k)f(k) \le \mu^2(N)$$

where $S(k) > 0$ and $\mu^2(N)$ is known.

If rank $(x^{11}, \ldots, x^{1r_1}, x^{21}, \ldots, x^{2r_2}, \ldots, x^{m1}, \ldots, x^{mr_m}) = p < n$, then reduction of feature space may be carried out and the prototypes transformed to $\alpha^{ij} \in \mathbb{R}^p$. The original prototypes may be presented in the form

$$x^{ij} = X\alpha^{ij}, \; i = \overline{1,m}, \; j = \overline{1,r_i} \quad ,$$

where rank $(X) = p$ and $X = (x^{i_1j_1}, x^{i_2j_2}, \ldots, x^{i_pj_p})$.

Finally, the following recursive procedure solves our problem:

$$\dot{I}(k+1) = \{i: A_i(k+1) \ne \phi, \; i \in \dot{I}(k)\}, \; \dot{I}(0) = \{1,2,\ldots,m\}$$

$$A_i(k+1) = \{j: \alpha^{ij} \in \bar{G}^{\alpha}_{k+1}, \; j \in A_i(k)\}, \; i \in \dot{I}(k), \; A_i(0) = \{1,2,\ldots,r_i\}, \; i = \overline{1,m}$$

$$\hat{\alpha}(k+1) = D(k)\hat{\alpha}(k) + R(k+1)X^*M^*(k+1)\tilde{S}(k+1)y(k+1), \; \hat{\alpha}(0) = 0$$

$$R(k+1) = D(k)R(k), \; R(0) = S_0^{-1} \quad ,$$

where $\dot{I}(k)$ is an index set of available classes after observation $y(i)$, $i = \overline{1,k}$, and

$$D(k) \triangleq E - R(k)W(k+1)[G(k+1) + W^*(k+1)R(k)W(k+1)]^{-1}W^*(k+1)$$

$$W^*(k+1) \triangleq c^*(k+1)\tilde{S}(k+1)H(k+1)X, \; \tilde{S}(k) \triangleq (c(k)S^{-1}(k)c^*(k))^+$$

$$\bar{G}^{\alpha}_{k+1} \triangleq \{\alpha: (\alpha - \hat{\alpha}(k+1))^*R^{-1}(k+1)(\alpha - \hat{\alpha}(k+1)) \le \lambda^2(k+1) +$$

$$+ \hat{\alpha}^*(k+1)R^{-1}(k+1)\hat{\alpha}(k+1) - \sum_{q=1}^{k+1} y^*(q)\tilde{S}(q)y(q)\}$$

$$\lambda^2(k) \triangleq \mu^2(k) + \max_{i,j} \{(\alpha^{ij})^* S_0 \alpha^{ij}\} \quad .$$

The problem is solved if at some time N the set $\dot{I}(N)$ contains a unique index of unknown class.

The structure of the algorithm will remain unchanged if for the measurement errors $f(k)$ we have

$$|f_i(k)| \le \Delta f_i \ , \ i = \overline{1,n} \ ; \ H(k) = c(k) = E \quad ,$$

where the Δf_i are given, and $f^*(k) = (f_1(k), \dots, f_n(k))$. Only the form of domain $\bar{G}^a_{k+1}$ is changed:

$$\bar{G}^a_{k+1} \triangleq \{a: d_i^-(k+1) \le l_i^* a \le d_i^+(k+1) \ , \ i = \overline{1,n}\} \quad ,$$

$$d_i^-(k+1) = \max \{d_i^-(k) \ , \ y_i(k+1) - \Delta f_i\} \ , \ d_i^-(0) = \min_{q,j} (l_i^* a^{qj})$$

$$d_i^+(k+1) = \min \{d_i^+(k) \ , \ y_i(k+1) + \Delta f_i\} \ , \ d_i^+(0) = \max_{q,j} (l_i^* a^{qj})$$

$$X^* = (l_1, l_2, \dots, l_n) \ , \ (\|l_i\| > 0 \ , \ y^*(k) = (y_1(k), \dots, y_n(k))) \quad .$$

Now consider the case of scalar observations

$$y(k) = \varphi^*(k) a + f(k) \ , \ k = 1,2, \dots$$

with a bounded measurement error $|f(k)| \le 1 \ \forall k$.

Suppose that the i-th class contains only one prototype a^i, $i = \overline{1,m}$.

THEOREM 14. *If there exists an N such that the relation*

$$(a^i - a^j)^* R^{-1}(N)(a^i - a^j) > 4\lambda^2(N) \quad ,$$

holds for all pairs $i \ , \ j \ (i \ne j)$, *then there exists a unique index* i_0 *such that*

$$(a^{i_0} - \hat{a}(N))^* R^{-1}(N)(a^{i_0} - \hat{a}(N)) \le \lambda^2(N) + \hat{a}^*(N) R^{-1}(N) \hat{a}(N) - \sum_{k=1}^{N} y^2(k) \quad ,$$

i.e., pattern a is recognized precisely. Here

$$\lambda^2(N) \triangleq \max_{i = \overline{1,m}} \|a^i\|^2 + N \quad .$$

Using Theorem 14, a condition for the ordering of the *a priori* observations is formulated in the following way:

$$\varphi(i_{k+1}) = \arg \max_{\varphi \in \Phi \setminus \Phi_k} \min_{i \ne j} (a^i - a^j)^* [R^{-1}(k) + \varphi\varphi^*](a^i - a^j) \ , \ k = 0,1, \dots ,$$

where

$$\Phi \triangleq \{\varphi: \varphi(1) \ , \ \varphi(2) \ , \dots \}$$

$$\Phi_k \triangleq \{\varphi: \varphi(i_1) \ , \ \varphi(i_2), \dots, \varphi(i_k)\} \ , \ \Phi_0 = \phi \quad .$$

REFERENCES

1. N.N. Krasovski. *Control Theory of Motion. Linear Systems.* Nauka, Moscow, 1968.

2. A.B. Kurzhanski. *Control and Observation under Conditions of Uncertainty.* Nauka, Moscow, 1977.

3. F.L. Chernousko. Guaranteed estimates of uncertain values using ellipsoids. *Doklady Akademii Nauk SSSR*, 251(1980)51–54.

4. B.N. Pshenichnyi and V.G. Pokotilo. On the accuracy of minimax estimates of observations. *Doklady Akademii Nauk Ukrainskoj SSR, Seriya A*, (3)(1982)62–64.

5. N.F. Kirichenko. *Introduction to the Stabilization Theory of Motion.* Vyshcha shkola, Kiev, 1978.

6. F.G. Garashchenko, N.F. Kirichenko and V.I. Lyashko. Practical stability, minimax estimation and adaptability of dynamic systems. Preprint 82-39, Institute of Cybernetics, Kiev, 1982.

7. F.G. Garashchenko. On a numerical approach to solution of stability problems within a finite time interval. *Doklady Akademii Nauk Ukrainskoj SSR, Seriya A*, (11)(1981)78–81.

8. B.N. Bublik, N.F. Kirichenko and A.G. Nakonechnyj. Minimax estimates and controllers of dynamic systems. Preprint 78-31, Institute of Cybernetics, Kiev, 1978.

9. N.F. Kirichenko. Minimax control and estimation of dynamic systems. *Avtomatika*, (1)(1982)32–39.

10. N.F. Kirichenko, A.G. Nakonechnyi and V.A. Navrodski. Minimax recursive estimates of the parameters of dynamic systems. *Doklady Akademii Nauk Ukrainskoj, SSR, Seriya A*, (11)(1978)1021–1025.

11. N.F. Kirichenko and A.S. Slabospitsky. On identifiability conditions for dynamic systems. *Doklady Akademii Nauk Ukrainskoj SSR, Seriya A*, (10)(1979)849–852.

12. A.S. Slabospitsky. Identifiability conditions of stable dynamic systems in the presence of uncertainty. *Doklady akademii Nauk Ukrainskoj SSR, Seriya A*, (2)(1982)72–74.

APPROXIMATE SOLUTIONS OF DIFFERENTIAL GAMES USING MIXED STRATEGIES

A.F. Kleimenov, V.S. Patsko and V.N. Ushakov
Institute of Mathematics and Mechanics
Sverdlovsk, USSR

This paper is concerned with the numerical solution of differential games using mixed strategies. Mixed strategies are defined [1] as functions which associate a probability measure with each position of the game. In a discrete control scheme these strategies could be realized by means of stochastic approximation procedures.

Let a conflict-controlled system be described on the interval $[t_*, \vartheta]$ by the equation

$$\frac{dx}{dt} = f(t, x, u, v), \; x \in R^n, \; u \in P, \; v \in Q \quad . \tag{1}$$

Here x is the phase vector, u is the control parameter of the first player, v is the control parameter of the second one, and P and Q are compact sets in R^p and R^q, respectively. The function f is continuous with respect to all its variables and Lipschitzian in x. We also assume that the function f satisfies a condition concerning the extension of solutions.

Let M be a compact set in R^n. The aim of the first player is to direct the system (1) into set M at time ϑ. The aim of the second player is to prevent this from happening. The mixed strategies of the first and second players are the functions which asociate probability measures $\mu(t, x)$ on P and $\nu(t, x)$ on Q with each position (t, x).

It is known [1] that for any initial position (t_0, x_0) there exists either a mixed strategy for the first player which solves the approach problem, or a mixed strategy for the second player which solves the evasion problem. Therefore it is important to construct the set W^0 of all initial positions from which the approach problem can be solved.

According to [1,2], the set W^0 is the maximal stable bridge and can be determined as the maximal closed set in the space of positions (t, x) which satisfies the conditions:

1. $W^0_\vartheta \subset M$;

2. $W^0_{\tau_*} \subset \pi(\tau_* ; \tau^* , W^0_{\tau^*})$ for all τ_*, τ^* such that $t_* \le \tau_* < \tau^* \le \vartheta$.

Here

$$W^0_\tau = \{x \in R^n : (\tau , x) \in W^0\} \quad ,$$

$$\pi(\tau_* ; \tau^* , W^0_{\tau^*}) = \bigcap_{l \in S} \mathbf{X}_l (\tau_* ; \tau^* , W^0_{\tau^*}) \quad ,$$

$$S = \{l \in R^n : \|l\| = 1\} \quad ,$$

$$X_l (\tau_* ; \tau^* , W^0_{\tau^*}) = \{x \in R^n : \mathbf{X}_l (\tau^* ; \tau_* , x) \cap W^0_{\tau^*} \ne \phi\} \quad .$$

$\mathbf{X}_l (\tau^* ; \tau_* , x_*)$ denotes the set of all points $y \in R^n$ for which there exists a solution $x(t)$ ($t \in [\tau_* , \tau^*]$, $x(\tau_*) = x_*$) of the differential inclusion

$$\dot{x} \in F_l(t , x) \; , \; F_l(t , x) = G_l(t , x) \cap O \tag{2}$$

$$G_l(t , x) = \{g \in R^n : l' g \le \max_{\mu \in \{\mu\}} \min_{\nu \in \{\nu\}} \int_P \int_Q l' f(t , x , u , v) \mu(du) \nu(dv)\}$$

such that $x(\tau^*) = y$. Here $\{\mu\}$ and $\{\nu\}$ are sets of probability measures on P and Q and O is a closed ball of sufficiently large radius.

Using the results of [2] we can establish that W^0 is the limit of the systems of sets $\{W_{t_i} : t_i \in \Gamma_m\}$, where Γ_m, $m = 1,2 ,\ldots,$ is a subdivision of the interval $[t_* , \vartheta]$ and its diameter $\Delta(\Gamma_m)$ approaches zero as $m \to \infty$.

For every subdivision $\Gamma_m = \{t_1 = t_* , t_2 , \ldots , t_{N(m)} = \vartheta\}$, the system $\{W_{t_i} : t_i \in \Gamma_m\}$ is defined by the recursive relations

$$W_{t_i} = \bar{\pi}(t_i ; t_{i+1} , W_{t_{i+1}}) \; , \; i = 1,2 , \ldots , N(m) - 1$$

$$W_{t_{N(m)}} = M_{\varepsilon_{N(m)}} \quad ,$$

where

$$\bar{\pi}(t_i ; t_{i+1} , W_{t_{i+1}}) = \bigcap_{l \in S} X_l (t_i ; t_{i+1} , W_{t_{i+1}}) \tag{3}$$

$$X_l (t_i ; t_{i+1} , W_{t_{i+1}}) = \{x \in R^n : X_l (t_{i+1} ; t_i , x) \cap W_{t_{i+1}} \ne \phi\}$$

$$X_l (t_{i+1} ; t_i , x) = x + (t_{i+1} - t_i) F_l (t_i , x) \quad .$$

$\varepsilon_{N(m)}$ denotes some non-negative number and M_α is a closed α-neighborhood of the set M.

These relations can be used as a basis for algorithms designed to compute the set W^0. We shall discuss a numerical procedure for system (1) with $f(t,x,u,v) = A(t)x + \varphi(t,u,v)$.

Let the dynamics of the system be described by the quasi-linear differential equation

$$\dot{x} = A(t)x + \varphi(t,u,v), \; x \in R^n, \; u \in P, \; v \in Q \quad . \tag{4}$$

Assume that the terminal set $\mathbf{M}$ depends on only k coordinates, i.e., $\mathbf{M} = \{x \in R^n : (x)_k \in M\}$, where $(x)_k$ is the vector of chosen coordinates. Let $\mathbf{M}$ be a closed, convex and bounded set.

Let $Z(\vartheta, t)$ denote the fundamental Cauchy matrix of solutions to (4), and $Z_k(\vartheta, t)$ be a submatrix consisting of k lines corresponding to the chosen coordintes of the vector x. Making the substitution

$$y(t) = Z_k(\vartheta, t)x(t)$$

we obtain the equivalent game of order k [1]:

$$\dot{y} = Z_k(\vartheta, t)\varphi(t,u,v), \; y \in R^k, \; u \in P, \; v \in Q \tag{5}$$

with terminal set M. The sections $\mathbf{W}_t^0$ and W_t^0 of the maximal stable bridges for games (4) and (5) are connected by the relation

$$\mathbf{W}_t^0 = \{x \in R^n : Z_k(\vartheta, t)x \in W_t^0\} \quad .$$

The convexity of the set M implies the convexity of the sections $\mathbf{W}_t^0$ and W_t^0.

Consider a subdivision Γ_m of the segment $[t_*, \vartheta]$. Define

$$D(t_{i+1}, u, v) = Z_k(\vartheta, t_{i+1})\varphi(t_{i+1}, u, v) \quad ,$$

and approximate the convex set M by a polytope M^*. Replace the compact sets P and Q by collections of finite points P^* and Q^*. Let μ^* and ν^* be probability measures on P^* and Q^*, respectively, and $\{l_s\}$ be the net of unit vectors in R^k.

It is known that in the case of a quasi-linear system (5) with a convex terminal set we can take an infinite ball instead of O in relations (2), (3). Applying relations (2), (3) to system (5), we have that the section W_{t_i} which approximates the section $W_{t_i}^0$ is the intersection of the half-spaces

$$\Lambda_i^s = \{y \in R^n : l_s' y \le \eta_i(l_s)\}, \; l_s \in \{l_s\} \quad .$$

Here

$$\eta_i(l_s) = \rho(l_s, W_{t_{i+1}}) + (t_{i+1} - t_i)\xi(l_s, t_i)$$

$$\xi(l_s, t_i) = \max_{\mu^* \in \{\mu^*\}} \min_{\nu^* \in \{\nu^*\}} \int_{P^*} \int_{Q^*} (-l_s')D(t_i, u, v)\mu^*(du)\nu^*(dv) \quad ,$$

and $\rho(l_s, W_{t_{i+1}})$ denotes the value of the support function of the set $W_{t_{i+1}}$ on the vector l_s.

The procedure outlined above has been formulated as a computer program for the $k = 2$. The graphs of the sections $W_{t_i}^0$ calculated for a concrete differential game are given below.

Let us consider the following system:

$$\frac{d^2 z}{dt^2} = L(v)u \quad ,$$

$$z = \begin{pmatrix} z_1 \\ z_2 \end{pmatrix}, \; L(v) = \begin{pmatrix} \cos v & -\sin v \\ \sin v & \cos v \end{pmatrix}, \; u = \begin{pmatrix} u_1 \\ u_2 \end{pmatrix} \quad . \tag{6}$$

Equation (6) describes the motion of a point of unit mass on the plane (z_1, z_2) under the action of a force $h(t)$. The force $h(t)$ has the same absolute value as the control vector $u(t)$ but forms an angle $v(t)$ with this vector. We assume that the control vector $u(t)$ at each time t may be selected from a given set P consisting of four vectors:

$$u^{(1)} = (1,0)\,,\; u^{(2)} = (0,1)\,,\; u^{(3)} = (-1,0)\,,\; u^{(4)} = (0,-1) \quad .$$

The angle between the force and the control at time t may take any value from the segment $Q = [-\beta, \beta]$, where $0 < \beta < \pi/2$.

Let the performance index be given by

$$\gamma = \|z(\vartheta)\| = (z_1^2(\vartheta) + z_2^2(\vartheta))^{1/2} \quad . \tag{7}$$

The aim of the first player (who governs the control u) is to move the point as close as possible to the origin at time ϑ.

Set $x_1 = z_1$, $x_2 = z_2$, $x_3 = \dot{z}_1$, $x_4 = \dot{z}_2$. Making the subdivisions $y_1 = x_1 + (\vartheta - t)x_3$, $y_2 = x_2 + (\vartheta - t)x_4$ leads to the system

$$\dot{y} = (\vartheta - t)L(v)u \tag{8}$$

with performance index

$$\gamma = \|y(\vartheta)\| = (y_1^2(\vartheta) + y_2^2(\vartheta))^{1/2} \quad . \tag{9}$$

In our example the order of the equivalent game (5) is equal to 2 and the dynamics of the game are described by equation (8). Take a level set of function (9) as the terminal set for this game. Note that it is a circle.

This differential game has been simulated on a computer, taking $\beta = \pi/6$, $\vartheta = 4.8$. Figure 1 shows sections W_t^0 of the set W^0 at times $t = 0$, 1.8, 3.3. Here the radius of the terminal set is 2.45.

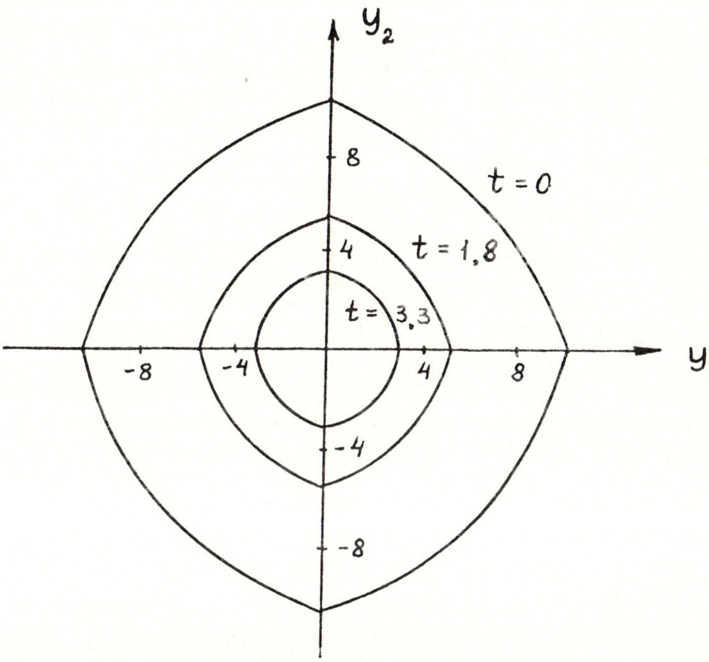

Figure 1. Sections of the set W^0 at times $t = 0$, $t = 1.8$, $t = 3.3$.

The optimal mixed strategies and corresponding motions were also calculated. We note again that in our game the first player tries to minimize the performance index (9), while the second tries to stop him. It is shown that our game satisfies the generalized regularity condition [3], which allows us to use the method of program construction [1].

Mixed strategies which lead to solutions have been constructed according to a scheme using a unified guide. The motion of the unified guide is such that it lies on the appropriate stable bridge. The corresponding control may be found from an extremum condition leading to the position of the guide.

The trajectory generated by optimal mixed strategies on the part of both players from the initial point $z_1(0) = 3.07$, $z_2 = 5.57$, $\dot{z}_1(0) = 0.8$, $\dot{z}_2(0) = 0.2$ on the plane (z_1, z_2) is illustrated in Figure 2. The calculated value of the performance index (7) is

2.47, i.e., approximately equal to the value of the game at the initial point.

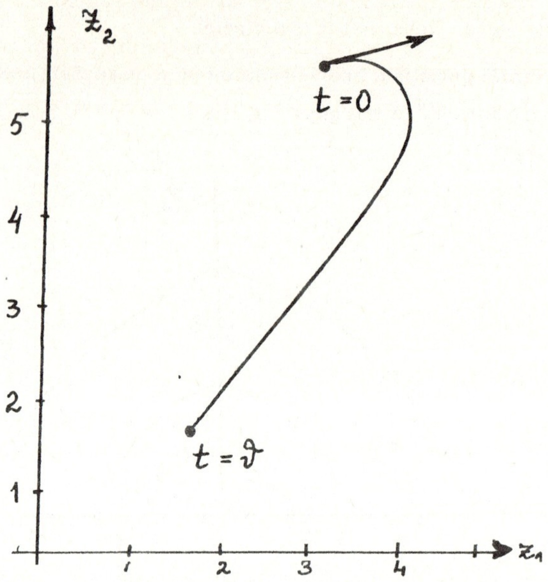

Figure 2. A trajectory generated by optimal mixed strategies for both players.

REFERENCES

1. N.N. Krasovski and A.I. Subbotin. *Positional Differential Games*. Nauka, Moscow, 1974 (in Russian).

2. N.N. Krasovski. On the problem of unification of differential games (in Russian). *Doklady Akademii Nauk SSSR*, 226(1976)1260–1263.

3. N.N. Krasovski. Extremal constructions for a differential game (in Russian). *Doklady Akademii Nauk SSSR*, 235(1977)1260–1262.

ON THE SOLUTION SETS FOR UNCERTAIN SYSTEMS WITH PHASE CONSTRAINTS

A.B. Kurzhanskii
Institute of Mathematics and Mechanics
Sverdlovsk, USSR

This report deals with multistage inclusions that describe a system with uncertainty in the model or in the inputs [1,2]. In particular this may be a difference scheme for a differential inclusion [3]. The solution to these inclusions is a multivalued function whose cross-section at a specific instant of time is the "admissible domain" for the inclusion.

The problem considered here is to specify a subset of solutions that consists of those "trajectories" which satisfy an additional phase constraint. These solutions are said to be "viable" with respect to the phase constraint [3]. The cross section of the set of all viable solutions is the attainability domain under the state constraint. The derivation of evolution equations for the latter domain is the objective of this paper.

The problem posed here is purely deterministic. However, the techniques applied to its solution involve some stochastic schemes. These schemes follow an analogy between some formulae of convex analysis [4,5] and those for calculating conditional mean values for specific types of stochastic systems [6,7] which was pointed out in [8,9].

A special application of the results of this paper could be the derivation of solving relations for nonlinear filtering under set-membership constraints on the "noise" and the description of the analogies between the theories of "guaranteed" and stochastic filtering.

1. DISCRETE-TIME UNCERTAIN SYSTEMS

Consider a multistage process described by an n-dimensional recurrent inclusion

$$x(k+1) \in F(k, x(k)), \quad k \geq k_0 \geq 0 \tag{1.1}$$

where $k \in \mathbb{N}$, $x(k) \in \mathbb{R}^n$, $F(k, x(k))$ is a given multivalued map from $N \times \mathbb{R}^n$ into $\text{comp}\,\mathbb{R}^n$ ($\mathbb{N}$ is the set of natural numbers, $\text{comp}\,\mathbb{R}^n$ is the set of all compact subsets of $\mathbb{R}^n$).

Suppose the initial state $x(k_0) = x^0$ of the system is confined to a preassigned set:

$$x^0 \in X^0 \tag{1.2}$$

Where X^0 is given in advance. A trajectory solution of system (1.1) that starts from point x^0 at instant k_0 will be denoted as $x(k|k_0,x^0)$. The set of all solutions for (1.1) that start from x^0 at instant k^0 will be denoted as $X(k|k_0,x^0)$ ($k \in \mathbb{N}$, $k \geq k^0$) with further notation

$$X(k|k^0,X^0) = \cup\{X(k|k_0,x^0)|x^0 \in X^0\}, \quad (k \in \mathbb{N}, k \geq k^0)$$

Let $Q(k)$ be a multivalued map from $\mathbb{N}$ into comp$\mathbb{R}^m$ and $G(k)$ be a single-valued map from $\mathbb{N}$ to the set of $m \times n$-matrices. The pair $G(k)$, $Q(k)$, introduces a state constraint

$$G(k)x(k) \in Q(k), \quad k \geq k_0+1 \tag{1.3}$$

on the solutions of system (1.1).

The subset of $\mathbb{R}^n$ that consists of all the points of $\mathbb{R}^n$ through which at stage $s \in [k_0,\tau] = \{k : k_0 \leq k \leq \tau\}$ there passes at least one of the trajectories $x(k|k_0,x^0)$, that satisfy constraint (1.3) for $k \in [k_0,\tau]$ will be denoted as $X(s|\tau,k_0,x^0)$.

The aim of this report is to study the sets $X(\tau|\tau,k_0,X^0) = X(\tau,k_0,X^0)$ and their evolution in "time" τ.

In other words, if a trajectory $x(k|k_0,x^0)$ of equation (1.1) that satisfies the constraint (1.3) for all $k \in [k_0,s]$ is named "viable until instant τ" ("relative to constraint (1.3)"), then our objective will be to describe the evolution of the set of all viable trajectories of (1.1). Here at each instant $k > k^0$ the constraint (1.3) may "cut off" a part of $X(k|k,k^0,x^0)$ reducing it thus to the set $X(k,k^0,x^0)$.

The sets $X(k,k^0,x^0)$ may also be interpreted as "attainability domains" for system (1.1) under the state space constraint (1.3). The objective is to describe evolution of these domains.

2. THE ATTAINABILITY DOMAINS

From the definition of sets $X(s|\tau,k^0,x^0)$ it follows that the following properties are true.

LEMMA 2.1. *Whatever are the instants* t,s,k, $(t \geq s \geq k \geq 0)$ *and the set* $\mathbb{F} \in$ comp$\mathbb{R}^n$, *the following relation is true*

$$X(t,k,\mathbb{F}) = X(t,s,X(s,k,\mathbb{F})). \tag{2.1}$$

Here $X(t,k,\mathbb{F}) = \cup\{X(t,k,x) | x \in \mathbb{F}\}$.

LEMMA 2.2. *Whatever are the instants* s,t,τ,k,l $(t \geq s \geq l;\ \tau \geq l \geq k;\ t \geq \tau)$ *and the set* $\mathbb{F} \in \text{comp}\,\mathbb{R}^n$ *the following relations are true*

$$X(s | t,k,\mathbb{F}) = X(s | t,l,X(l | \tau,k,\mathbb{F}) . \tag{2.2}$$

Relation (2.2) shows that sets $X(k,\tau,X)$ satisfy a semigroup property which allows to define a generalized dynamic system in the space $2^{\mathbb{R}^n}$ of all subsets of $\mathbb{R}^n$.

In general the sets $X(s | t,k,\mathbb{F})$ need not be either convex or connected. However, it is obvious that the following is true

LEMMA 2.3. *Assume that the map* F *is linear in* x*:*

$$F^{(k)}x = A^{(k)}x + P$$

where $P \in \text{conv}\,\mathbb{R}^n$. *Then for any set* $\mathbb{F} \in \text{conv}\,\mathbb{R}^n$ *each of the sets* $X(s | t,k,\mathbb{F}) \in \text{conv}\,\mathbb{R}^n$ $(t \geq s \geq k \geq 0)$.

Here $\text{conv}\,\mathbb{R}^n$ stands for the set of all convex compact subsets of $\mathbb{R}^n$.

3. THE ONE-STAGE PROBLEM

Consider the system

$$z \in F(x),\ \ Gz \in Q,\ \ x \in X,$$

where $z \in \mathbb{R}^n$, $X \in \text{comp}\,\mathbb{R}^n$, $Q \in \text{conv}\,\mathbb{R}^m$, $F(\kappa)$ is a multivalued map from $\mathbb{R}^n$ into $\text{conv}\,\mathbb{R}^n$, G is a single-valued map from $\mathbb{R}^n$ into $\mathbb{R}^m$.

It is obvious that the sets $F(X) = \{\cup F(x) | x \in X\}$ need not be convex.

Let Z, Z^* respectively denote the sets of al solutions for the following systems:

(a) $z \in F(X)$, $Gz \in Q$,

(b) $z^* \in \text{co}F(X)$, $Gz^* \in Q$,

where $\text{co}F$ stands for the closed convex hull of $F(X)$.

The following statement is true

LEMMA 3.1. *The sets* Z, $\text{co}\,Z$, Z^* *satisfy the following inclusions*

$$Z \subseteq \text{co}\,Z \subseteq Z^* \tag{3.1}$$

Let $\rho(l | Z) = \sup\{l'z | z \in Z\}$ denote the support function [4] of set Z. Also denote

$$\Phi(l,p,q) = (l - G'p,q) + \rho(-p | Q) + (p,y)$$

Then the function $\Phi(l,p,q)$ may be used to describe the sets $\text{co}\,Z, Z^*$.

LEMMA 3.2. *The following relations are true*

$$\rho(l|Z) = \rho(l|\mathrm{co}\, Z) = \sup_q \inf_p \Phi(l,p,q) \qquad q \in \boldsymbol{F}(X),\ p \in \mathbb{R}^m \tag{3.2}$$

$$\rho(l|Z^*) = \inf_p \sup_q \Phi(l,p,q) \qquad q \in \boldsymbol{F}(X),\ p \in \mathbb{R}^m \tag{3.3}$$

It is not difficult to give an example of a nonlinear map $F(x)$ for which Z is nonconvex and the functions $\rho(l|\mathrm{co}\, Z)$, $\rho(l|Z^*)$ do not coincide, so that the inclusions $Z \subset \mathrm{co}\, Z$, $\mathrm{co}\, Z \subset Z^*$ are strict. For a linear-convex map $F(x) = Ax + P$ $(P \in \mathrm{conv}\, \mathbb{R}^n)$ their is no distinction between Z, $\mathrm{co}\, Z$, and Z^*:

LEMMA 3.3 *Suppose $F(x) = Ax + P$ where $P \in \mathrm{conv}\, \mathbb{R}^n$, A is a linear map from $\mathbb{R}^n$ in to $\mathbb{R}^n$. Then $Z = \mathrm{co}\, Z = Z^*$.*

4. THE ONE-STAGE PROBLEM – AN ALTERNATIVE APPROACH

The description of Z, $\mathrm{co}\, Z$, Z^* may be given in an alternative form which, however, allows to present these sets as the intersections of some varieties of multivalued maps.

Indeed, whatever are the vectors l, p $(l \neq 0)$ it is possible to present $p = Ml$ where M belongs to the space $\mathbb{M}^{m \times n}$ of real matrices of dimension $m \times n$. Then, obviously,

$$\rho(l|Z) = \sup_q \inf_M \Phi(l,Ml,q) = \rho(l|\mathrm{co}\, Z) \qquad q \in \boldsymbol{F}(X),\ M \in \mathbb{M}^{m \times n},$$

$$\rho(l|Z^*) = \inf_M \sup_q \Phi(l,Ml,q) \qquad q \in \boldsymbol{F}(X),\ M \in \mathbb{M}^{m \times n} \tag{4.1}$$

or

$$\rho(l|Z^*) = \inf \{ \Phi(l,Ml) \mid M \in \mathbb{M}^{m \times n} \}, \tag{4.2}$$

where

$$\Phi(l,Ml) = \{ \Phi(l,Ml,q) \mid q \in \mathrm{co}\, F(x) \} =$$

$$= \rho((E - G'M)l,\ \mathrm{co}\, F(X)) + \rho(-Ml \mid Q) + (Ml, y).$$

From (4.1) it follows that

$$Z \subseteq \bigcup_{q \in \boldsymbol{F}(X)} \bigcap_M R(M,q) \subseteq \bigcap_M \bigcup_{q \in \boldsymbol{F}(X)} R(M,q), \tag{4.3}$$

where

$$R(M,q) = (E_n - MG)q + -MQ.$$

Similarly (4.2) yields

$$Z^* \subseteq \bigcap_M \bigcup_{q \in \mathrm{co}\boldsymbol{F}(X)} \{(E_n - MG)q - MQ\}. \tag{4.4}$$

Moreover a stronger assertion holds.

THEOREM 4.1. *The following relations are true*

$$Z = \bigcup_{q \in \boldsymbol{F}(X)} \bigcap_M R(M,q) \tag{4.5}$$

$$Z^* = \bigcap_M R(M, \mathrm{co}\boldsymbol{F}(X)) \tag{4.6}$$

where $M \in \mathbb{M}^{m \times n}$.

Obviously for $\boldsymbol{F}(x) = AX + \boldsymbol{P}, (X, P \in \mathrm{co}\,\mathbb{R}^n)$ we have $F(X) = \mathrm{co}\boldsymbol{F}(X)$ and $Z = Z^* = \mathrm{co}Z$.

This *first* scheme of relations may serve to be a basis for constructing recurrent filtering procedures. Another recurrent procedure could be devised from the following *second* scheme. Consider the system

$$z \in \boldsymbol{F}(x) \tag{4.7}$$

$$Gx \in Q, \tag{4.8}$$

for which we are to determine the set Z of all vectors z consistent with inclusions (4.7), (4.8). Namely, we are to determine the restriction $\boldsymbol{F}_Q(x)$ of $F(x)$ to set Q. Here we have

$$F_Q(x) = \begin{cases} \boldsymbol{F}(x) \text{if} x \in Q \\ \phi \quad \text{if} x \in Q \end{cases}$$

where $Y = \{x : Gx \in Q\}$.

LEMMA 4.1 *Assume* $\boldsymbol{F}(x) \in \mathrm{comp}\,\mathbb{R}^n$ *for any* x *and* $Q \in \mathrm{comp}\,\mathbb{R}^m$. *Then*

$$\boldsymbol{F}_Q(x) = \bigcap_L (\boldsymbol{F}(x) - LGx + LQ)$$

over all $n \times m$ *matrices* L, *(*$L \in \mathbb{M}^{n \times m}$*).*

Suppose $x \in Y = \{x : Gx \in Q\}$, $\{0\}_m \in \mathbb{R}^m$. Then $\{0\}_m \in Q - Gx$ and for any $(m \times n)$ matrix L we have $\{0\}_n \in L(Q - Gx)$. However, if $x \overline{\in} Y$, then $\{0_m\} \overline{\in} (Q - Gx)$, since $\cap L(Q - Gx) \subset L_m(-Gx + Q)$. Here

$$L_m = \begin{pmatrix} E_m \\ \{0\}_{m,n} \end{pmatrix}$$

is an $(n \times m)$-matrix and $\{0\}_{m,n}$ is a matrix of dimension $(n-m)\times m$. Then it follows that for $x \in Y$

$$F(x)=F(x)+\bigcap_L L(Q-Gx)\subseteq\bigcap_L(F(x)+L(Q-Gx))\subseteq F(x)$$

On the other hand, suppose $x \overline{\in} Y$. Let us demonstrate that in this case

$$\bigcap_L\{F(x)+L(Q-Gx)\}=\phi.$$

Denote $A=F(x)$, $B=Q-Gx$. For any $\lambda>0$ we then have

$$\bigcap_L(A+LB)\subseteq(A+\lambda L_m B)\cap(A-\lambda L_m B)$$

Since $\{0\}_m \overline{\in} B$ we have $\{0\}_n \overline{\in} L_m B$. Therefore there exists a vector $l \in \mathbb{R}^n$, $l \neq 0$ and a number $\gamma>0$ such that

$$(l,x)\geq\gamma>0 \quad \text{for any} \quad x\in L_m B,$$

Denote

$$L=\{x:(l,x)\geq\gamma\}.$$

Then $L \supseteq L_m B$ and

$$(A+\lambda L_m B)\cap(A-\lambda L_m B)\subseteq(A+\lambda L)\cap(A-\lambda L)$$

Set A being bounded there exists a $\lambda>0$ such that

$$(A+\lambda L)\cap(A-\lambda L)=\phi.$$

Hence

$$\bigcap_L(A+LB)=\phi$$

and the Lemma is proved.

5. STATISTICAL UNCERTAINTY. THE ELEMENTARY PROBLEM

Consider the system

$$z=q+\xi, \quad Gz=v+\eta, \tag{5.1}$$

where

$$q\in F(x),\ v\in Q,\ x\in X$$

and ξ,η are independent gaussian random vectors with zero means ($E\xi=0, E\eta=0$) and with variances $E\xi\xi'=R$, $E\eta\eta'=N$, where $R>0$, $N>0$ ($R\in\mathbb{M}_n, N\in\mathbb{M}_m$).

Assuming at first that the pair $h=\{q,v\}$ is fixed, let us find the conditional mean $E(z\,|\,y=0, h=h^*)$ under the condition that after one realization of the values ξ,η the relations

$$z=q+\xi,\quad y=-Gz+v+\eta=0$$

are satisfied. After a standard calculation we have

$$\bar{z}_{y,h}=E(z\,|\,y=0,\ h=h^*)=q+PG'N^{-1}(-Gq-Gv)+v\ ,$$

where $P^{-1}=R^{-1}+G'N^{-1}G$.

After applying a well-known matrix transformation [6]

$$P=R-RG'Q^{-1}GR,\ Q=N+GRG'\ ,$$

we have

$$E(z\,|\,y=0,\ h=h^*)=(E-RG'Q^{-1}G)q-RG'Q^{-1}v\ . \tag{5.1}$$

The matrix of conditional variances is

$$E((z-\bar{z}_{y,h})'(z-\bar{z}_{y,h}))=P_y\ .$$

It does not depend upon h and is determined only by q,v and the element $\Lambda=RG'K^{-1}G$. Therefore it makes sense to consider the sets

$$W(\Lambda,q)=\cup\{z_{y,h}\,|\,v\in Q\}$$

$$W(\Lambda)=\cup\{z_{y,h}\,|\,q\in F(X),\ v\in Q\}$$

and

$$W_*(\Lambda)=\cup\{z_{y,h}\,|\,q\in \mathrm{co}F(X),\ v\in Q\}$$

of conditional mean values. Each of the elements of these sets has one and the same variance P_y. The sets $W_*(\Lambda)$ and $W(\Lambda,q)$ are obviously convex while $W(\Lambda)$ may not be convex.

LEMMA 5.1 *The following inclusions are true* ($Z\subset Z^*$)

$$Z\subset W(\Lambda),\ Z^*\subset W^*(\Lambda),\ W(\Lambda)\subset W^*(\Lambda)\ . \tag{5.2}$$

It can be sen that $W(\Lambda,q)$ has exactly the same structure as $R(M,q)$ of (4.3) (with only Λ substituted by M). Hence for the same reason as before we have

$$Z \subseteq \bigcup_{q \in \mathbf{F}(X)} \bigcap_{D \in \mathbf{D}} W(\Lambda,q) = \bigcap_{D \in \mathbf{D}} W(\Lambda) \tag{5.3}$$

$$Z^* \subseteq \bigcup_{q \in \mathrm{co}\mathbf{F}(X)} \bigcap_{D \in \mathbf{D}} W(\Lambda,q) = \bigcap_{D \in \mathbf{D}} W_*(\Lambda) \tag{5.4}$$

where the intersections are taken over the class $\mathbf{D}$ of all possible pairs $D = \{R,N\}$ of nonnegative matrices R,N of respective dimensions. However, a property similar to that of Lemma 4.1 happens to be true. Namely if by $\mathbf{D}(\alpha,\beta)$ we denote the class of pairs $\{R,N\}$ where $R = \alpha E_n$, $N = \beta E_m$, $\alpha > 0$, $\beta > 0$, then the element X will depend only upon two parameters α,β.

THEOREM 5.1 *Suppose matrix G is of full rank m. Then the following equalities are true*

$$Z = \cap\{W(\Lambda) \mid D \in \mathbf{D}(1,\beta),\ \beta > 0\} \subseteq \mathrm{co}Z \tag{5.5}$$

$$\subseteq \cap\{W_*(\Lambda) \mid D \in \mathbf{D}(1,\beta),\ \beta > 0\} = Z_* .$$

Here it suffices to take the intersections only over a one-parametric variety $D \in \mathbf{D}(1,\beta)$.

There are some specific differences between this scheme and the one of §4. These could be traced more explicitly when we pass to the calculation of support functions $\rho(l \mid Z)$, $\rho(l \mid Z^*)$ for Z, Z^*.

LEMMA 5.2 *The following inequality is true*

$$\rho(l \mid Z^*) = f^{**}(l) \le f(l) = \inf\{\Phi(l, Fl) \mid D \in \mathbf{D}(1,\beta),\ \beta > 0\} \tag{5.6}$$

where $f^{**}(l)$ is the second conjugate to $f(l)$ in the sense of Fenchel [4].

Moreover if we substitute $\mathbf{D}(1,\beta)$ in (5.6) for a broader class $\mathbf{D}$ then an exact equality will be attained, i.e.

$$\rho(l \mid Z^*) = f^{**}(l) = \inf\{\Phi(l, Fl) \mid D \in \mathbf{D}\} \tag{5.7}$$

More precisely, we come to

THEOREM 5.2 *Suppose matrix G is of full rank m. Then equality (5.7) will be true together with the following relation*

$$\rho(l \mid Z) = \rho(l \mid \mathrm{co}Z) = \sup\inf\{\Phi(l, Fl, q) \mid d \in \mathbf{D} \mid q \in \mathbf{F}(X)\} \tag{5.8}$$

Problems (5.7), (5.8) are "stochastically dual" to (3.3, (3.2).

The results of the above may now be applied to our basic problem for multistage systems.

6. SOLUTION TO THE BASIC PROBLEM

Returning to system (1.1)–(1.3) we will seek for the sequence of sets $X[s] = \boldsymbol{X}(s, k_0, X^0)$ together with two other sequences of sets. These are

$$X^*[s] = \boldsymbol{X}^*(s, k_0, X^0)$$

the solution set of the system

$$X_{k+1} \in \mathrm{co}\boldsymbol{F}(k, X^*[k]), \ \boldsymbol{X}^*[k_0] = X^0 \tag{6.1}$$

$$G(k+1)X_{k+1} \in Q(k+1), \ k \geq k_0 \tag{6.2}$$

and $\boldsymbol{X}_*[s] = X_*(s, k_0, X^0)$ which is obtained due to the following relations:

$$\boldsymbol{X}_*[s] = \mathrm{co}\, Z[s]$$

where $Z[\tau]$ is the solution set for the system

$$Z(k+1) \in \boldsymbol{F}(k, \mathrm{co}\, Z[k]), \ Z[k^0] = X^0,$$

$$G(k+1)Z(k+1) \in Q(k+1), \ k \geq k_0.$$

The sets $\boldsymbol{X}_*[\tau]$, $X^*[\tau]$ are obviously convex. They satisfy the inclusions

$$X[\tau] \subseteq \boldsymbol{X}_*[\tau] \subseteq X^*[\tau]$$

where each of the sets $\boldsymbol{X}[\tau]$, $X_*[\tau]$, $\boldsymbol{X}^*[\tau]$ lies within

$$Y(\tau) = \{x : G(\tau)x \in Q(\tau)\}, \quad \tau \geq k_0 + 1,$$

The set $X^*[\tau]$ may therefore be obtained for example by either solving a sequence of problems (6.1), (6.2) (for every $k \in [k_0, s-1]$ with $\boldsymbol{X}^*[k_0] = X^0$) (the first scheme of §4) or by finding all the solutions $\bar{x}[k] = \bar{x}(k, k_0, x^0)$ of the equation

$$x(k+1) \in (\mathrm{co}\, F)_{Y(k)}(k, x(k)), \ x(k_0) \in X^0, \tag{6.3}$$

that could be prolongated until the instant $\tau + 1$ and finding the relation of this set to $X[\tau]$, $X_*[\tau]$, and $X^*[\tau]$.

Following the first scheme of §4 we may therefore consider the recurrent system

$$z(k+1) = (I_n - M(k+1)G(k+1))F^0(k, S(k)) + M(k+1)Q(k+1) \tag{6.4}$$

$$S(k) = \{\cap Z(k) | M(k)\}, \ k > k_0, \ S(k_0) = X^0, \tag{6.5}$$

where $M(k+1) \in \mathbb{R}^{m \times n}$.

From Theorem 4.1 we may now deduce the result

THEOREM 6.1 *The solving relations for $X[s]$, $X_*[s]$, $X^*[s]$ are as follows*

$$X[s] = S(s) \quad \text{for} \quad F^0(k, S(k)) = F(k, S(k)) \tag{6.6}$$

$$X^*[s] = S(s) \quad \text{for} \quad F^0(k, S(k)) = \text{co}F(k, S(k)) \tag{6.7}$$

$$X_*[s] = \text{co}S(s) \quad \text{for} \quad F^0(k, S(k) = F(k, \text{co}S(k)) . \tag{6.8}$$

It is obvious that $X[\tau]$ is the exact solution while $X_*[\tau]$, $X^*[\tau]$ are convex majorants for $X[\tau]$. Clearly by interchanging and combining relations (6.7), (6.8) from stage to stage it is possible to construct a variety of other convex majorants for $X[\tau]$. However for the linear case they all coincide with $X[\tau]$.

LEMMA 6.1 *Assume $F^0(k, S(k)) = A(k)S(k) + P(k)$ with $P(k)$, X^0 being closed and compact. Then $X[k] = X^*[k] = X_*[k]$ for any $k \geq k_0$.*

Consider the system

$$Z(k+1) = (I_n - M(k+1)G(k+1))F^0(k, Z(k)) - M(k+1)Q(k+1), Z(k_0) = X^0, \tag{6.9}$$

denoting its solution as

$$Z(k; M_k(\cdot)) \quad \text{for} \quad F^0(k, Z) = F(k, Z)$$

$$Z_*(k, M_k(\cdot)) \quad \text{for} \quad F^0(k, Z) = F(k, \text{co}\, Z)$$

$$Z^*(k, M_k(\cdot)) \quad \text{for} \quad F^0(k, Z) = \text{co}F(k, Z)$$

Then the previous suggestions yield the following conclusion

THEOREM 6.2 *Whatever is the sequence $M_s(\cdot)$, the following solving inclusions are true*

$$X[s] \subseteq Z(s, M_s(\cdot)) \tag{6.10}$$

$$X_*[s] \subseteq Z_*(s, M_s(\cdot))$$

$$X^*[s] \subseteq Z^*(s, M_s(\cdot)), \quad s > k_0,$$

with $Z(s, M_s(\cdot)) \subseteq Z_*(s, M_s(\cdot)) \subseteq Z^*(s, M_s(\cdot))$.

Hence we also have

$$X[s] = \cap \{ Z(s, M_s(\cdot)) | M_s(\cdot)) \} \tag{6.11}$$

$$X_*[s] = \cap \{ Z_*(s, M_s(\cdot)) | M_s(\cdot)) \} \tag{6.12}$$

$$X^*[s] = \cap \{ Z^*(s, M_s(\cdot)) | M_s(\cdot)) \} \tag{6.13}$$

over all $M_s(s)$.

However a question arises which is whether (6.11)–(6.13) could turn into exact equalities.

LEMMA 6.2 *Assume the system (1.1), to be linear:* $\mathbf{F}(k,x) = A(k)x + P(k)$ *with sets* $P(k)$, $Q(k)$ *convex and compact. Then the inclusions (6.11)–(6.13) turn into the equality*

$$X[s] = \mathbf{X}^*[s] = \cap\{Z_s(\cdot,M_s(\cdot))\} = \cap\{Z_s^*(\cdot,M_s(\cdot))\} \tag{6.14}$$

Hence in this case the intersections over $M(k)$ could be taken either in each stage as in Theorem 6.1 (see (6.6), (6.7)) or at the final stage as in (6.14).

Let us now follow the second scheme of §4, considering the equation

$$x(k+1) \in \tilde{F}_{Y(k)}(k,x(k)), \quad x(k_0) \in X^0, \tag{6.15}$$

and denoting the set of its solutions that starts at $x^0 x(k_0) \in X^0$ as $X^0(k,k_0,x^0)$ with

$$\cup\{x^0(k,k_0,x^0) | x^0 \in X^0\} = X^0(k,k_0,X^0) = X^0[k].$$

According to Lemma 4.1 we substitute (6.15) by the equation

$$x(k+1) \in \bigcap_L (\tilde{F}^*(k,x(k)) - LG(k)x(k) + LY(k)) \qquad x(k_0) \in X^0,$$

and the calculation of $X^0[k]$ should thence follow the procedure

$$\tilde{X}[k+1] = \bigcup_{x \in \tilde{X}(k)} \bigcap_L (\tilde{F}(k,x) - LG(k)x + LQ(k)), \qquad X(k_0) = X^0. \tag{6.16}$$

Denote the whole solution "tube" for $k_0 \le k \le s$ as $\tilde{X}^s_{k_0}[\cdot]$. Then the following assertion will be true.

THEOREM 6.3 *Assume* $\tilde{X}^s_{k_0}[k]$ *to be the cross-section of the tube* $\tilde{X}^s_{k_0}[\cdot]$ *at instant* k. *Then*

$$X[s] = \tilde{X}^{s+1}_{k_0}[s] \text{ if } \tilde{F}(k,x) = \mathbf{F}(k,x)$$

$$X^* = \tilde{X}^{s+1}_{k_0}[s] \text{ if } \tilde{F}(k,x) = \text{co}F(k,x)$$

Here $\tilde{X}^s_{k_0}[s] \supseteq \tilde{X}^{s+1}_{k_0}[s]$ and the set $\tilde{X}^s_{k_0}[s]$ may not lie totally within $Y(s)$.

The solution of equation (6.16) is equivalent to finding all the solutions for the inclusion

$$x(k+1) \in \bigcap_L (\tilde{F}(k,x) - LG(k)x + LQ(k)) \qquad x(k_0) \in X^0 \tag{6.17}$$

Equation (6.17) may be substituted by a system of "simpler" inclusions

$$x(k+1) \in \tilde{F}(k,x) - L(k)G(k)x + L(k)Q(k) \qquad x(k_0) \in X^0 \tag{6.18}$$

for each of which the solution set for $k_0 \le k \le s$ will be denoted as

$$\tilde{X}^s_{k_0}(\cdot, k_0, X^0, L(\cdot)) = \tilde{X}^s_{k_0}[\cdot, L(\cdot)]$$

THEOREM 6.4 *The set $X^s_{k_0}[\cdot, L(\cdot)]$ of viable solutions to the inclusion*

$$x_{k+1} \in \tilde{F}(k, x(k)) \qquad x(k_0) \in X^0$$

$$G(k)x(k) \in Q(k), \qquad k_0 \le k \le s$$

is the restriction of set

$$X^{s+1}_{k_0}[\cdot] = \bigcap_L \tilde{X}^{s+1}_{k_0}[\cdot, L]$$

defined for stages $[k_0, \ldots, s+1]$ to the stages $[k_0, \ldots, s]$. The intersection is taken here over all constant matrices L.

However a question arises, whether this scheme allows also to calculate $X^s_{k_0}[s]$. Obviously

$$X^s_{k_0} \subseteq \bigcap_{L[\cdot]} \tilde{X}^{s+1}_{k_0}[s, L[\cdot]] \tag{6.19}$$

over all sequences $L[\cdot] = \{L(k_0), L(k_0+1), \ldots, L(s+1)\}$.

THEOREM 6.5 *Assume $\tilde{\boldsymbol{F}}(k,x)$ to be linear-convex: $\tilde{\boldsymbol{F}}(k,x) = A(k)x + \boldsymbol{P}(k)$, with $P(k)$, $Q(k)$ convex and compact. Then (6.19) turns to be an equality.*

7. SOLUTION TO THE BASIC PROBLEM. "STOCHASTIC" APPROXIMATIONS

The calculation of $X[s]$, $X_*[s]$, $X^*[s]$ may be also performed on the basis of the results of §5. Namely system (6.6), (6.7) should now be substituted by the following

$$Z(k+1) = (I_n - F(k+1)G(k+1))F^0(k, H(k)) - F(k+1)Q(k+1) \tag{7.1}$$

$$H(k+1) = \{\cap Z(k+1) | D(k+1) \in \boldsymbol{D}(1,\beta)\} \tag{7.2}$$

$$F(k+1) = R(k)G'(k+1)K^{-1}(k+1), \; F(k_0) = X^0 \tag{7.3}$$

$$K(k+1) = N(k+1) + G(k+1)R(k)G'(k+1)$$

$$D(k+1) = \{R(k), N(k+1)\}$$

THEOREM 7.1 *Assume that in theorem 6.1 $S(k)$ is substituted by $H(k)$ and $M(k)$ by $F(k)$. Then the result of this theorem remains true.*

If set $Q(k)$ of (1.3) is of specific type

$$Q(k) = y(k) - \tilde{Q}(k)$$

where $y(k)$ and $\tilde{Q}(k)$ are given, then (1.3) is transformed into

$$y(k) \in G(k)x(k) + \tilde{Q}(k) \tag{7.4}$$

which could be treated as an equation of observations for the uncertain system (7.1). Sets $X[s]$, $X_*[s]$, $X^*[s]$ therefore give us the guaranteed estimates of the unknown state of system (1.1) on the basis of an observation of vector $y(k)$, $k \in [k_0, s]$ due to equation (7.4). The result of Theorem 7.1 then means that the solution of this problem may be obtained via equations (7.1)–(7.3), according to formulae (6.8)–(6.10) with $M(k)$, $S(k)$ substituted respectively by $F(k)$, $H(k)$. The deterministic problem of nonlinear "guaranteed" filtering is hence approximated by relations obtained through a "stochastic filtering" approximation scheme.

REFERENCES

1. Krasovskii, N.N. Control under Incomplete Information and Differential Games, Proc. Intern. Congress of Mathematicians, Helsinki, 1978.

2. Kurzhanskii, A.B. *Control and Observation under Conditions of Uncertainty*, Nauka, Moscow, 1978.

3. Aubin, J.-P., and A. Cellina. *Differential Inclusions*. Springer-Verlag, Heidelberg, 1984.

4. Rockafellar, R.T. *Convex Analysis*. Princeton University Press, 1970.

5. Ekeland, I. and R. Teman. Analyse Convexe et problèmes Variationelles. Dunod, Paris, 1974.

6. Albert, A. Regression and the Moore–Penrose Pseudo-Inverse. *Mathematics in Science & Engineering Series*, 94(1972)(ISBN 0-12-048450-1), Academy Press.

7. Davis, M. *Linear Estimation and Control Theory*. London Chapman-Hall, 1977.

8. Kurzhanskii, A.B. Evolution Equations in Estimation Problems for Systems with Uncertainty. IIASA Working Paper WP-82-49, 1982.

9. Koscheev, A.S. and A.B. Kurzhanski. Adaptive Estimation of the Evolution of Multisgate Uncertain Systems. Ivestia Akad. Nauk. SSSR Teh. Kibernetika ("Engineering Cybernetics"), No. 2, 1983.

EXISTENCE OF A VALUE FOR A GENERAL ZERO-SUM MIXT GAME

J.P. LEPELTIER - UNIVERSITE DU MAINE - FACULTE DES SCIENCES - LE MANS - FRANCE

We suppose the evolution of the system described by a stochastic differential equation

(*) $dx_t = f\ (t, x, u)\ dt + \sigma\ (t, x)\ dB_t\ , x(o) = x_o$

where B is a m-dimensional brownian motion defined on a probability space $(\Omega, \mathcal{a}, \mu)$.

A first player J_1 chooses a stopping time S, while a second player chooses a continuous strategy u. There is an associated cost which acts until S, constituated of a continuous cost, and a terminal cost. J_1 (resp. J_2) looks for maximize (resp. minimize) this cost J (S, u). Under smooth assumptions we prove that this game is closed (or has a value) i. e. :

$$\inf_u \sup_S J\ (S, u) = \sup_S \inf_u J\ (S, u).$$

We take a model "in law", more precisely :

1.THE GAME MODEL

Let $\mathcal{C} = \{x : R_+ \to R^m$ continuous$\}$. If x denotes a member of $\mathcal{C}$, x_t denotes the value of x at t. Finally write $\underline{\underline{F}}_t$ for the σ-field generated by $(x_s, x \in \mathcal{C}, s \leq t)$.

The brownian motion (B_t) is separable and defined on a probability space $(\Omega, \mathcal{A}, \mu)$

Under Lipschitz and regularity conditions on σ the equation :

$dx_t = \sigma\ (t, x)\ dB_t,\ x\ (o) = x_o$

has a unique solution x_t and it induces a probability P_o on $(\mathcal{C}, \underline{\underline{F}}_\infty)$ by the formula :

$P_o\ (A) = \mu\ \{\omega : x\ (\omega) \in A)$

Now if $f : R_+ \times \mathcal{C} \times \mathcal{V} \to R^m$ is measurable, such that for all u, f(., ., u) is $\underline{\underline{F}}_t$ - adapted and such that $|f\ (t, x, u)| \leq K\ (1 + ||x||_t)$, if we define

$\mathcal{U} = \{u : R_+ \times \mathcal{C} \to \mathcal{V}$ compact metric space, predictable$\}$, by the Cameron-Martin (or Girsanov) formula we can define for all u in $\mathcal{U}$,

P_u on $(\mathcal{C}, \underline{\underline{F}}_\infty)$ by :

$$\left.\frac{dP_u}{dP_o}\right|_{\underline{\underline{F}}_t} = \exp\ (\zeta_t\ (u))\ \forall t$$

with :

$$\zeta_t\ (u) = \int_o^t f\ (s,\ u_s)\ a_s^{-1}\ dx_s - \frac{1}{2}\int_o^t f\ (s,\ u_s).a_s^{-1}\ f\ (s,\ u_s)\ ds$$

$(a = \sigma\ \sigma^*)$, and under P_u the process coordinate on $\mathcal{C}$ is solution of the differential stochastic equation (*).

Finally let $\mathcal{T} = \{T,\ \underline{\underline{F}}_t$ - stopping times$\}$. The payoff corresponding to the strategy T (for J_1) and u (for J_2) is :

$$J\ (T,\ u) = E_u\ \left(\int_o^T e^{-\alpha s}\ c\ (s,\ u_s)\ ds + Y_T\right)$$

with c positive bounded and Y $\underline{\underline{F}}_t$ - adapted, right continuous bounded.

Define for any $\underline{\underline{F}}_t$ stopping time T, u in $\mathcal{U}$

$$\overline{X}\ (u,\ T) = \underset{v \in \mathcal{D}(u,\ T)}{\text{P-ess inf}}\ \underset{S \geq T}{\text{P-ess sup}}\ E_v\ \left(\int_o^S e^{-\alpha s}\ c\ (s,\ v_s)ds + Y_S/\underline{\underline{F}}_T\right)$$

$$(\mathcal{D}(u,\ T) = \{v \in \mathcal{U} |\ v \underline{=} u \text{ on} [\![o,\ T [\![\})$$

We notice that :

$$\overline{X}(u,\ T) = \int_o^T e^{-\alpha s}\ c(s,\ u_s)\ ds + \underset{v \in \mathcal{U}}{\text{P-ess inf}}\ \underset{S \geq T}{\text{P-ess sup}}\ E_v\ \left(\int_T^S e^{-\alpha s}\ c(s,\ v_s)ds + Y_S/\underline{\underline{F}}_T\right)$$

$$= \int_o^T e^{-\alpha s}\ c\ (s,\ u_s)\ ds + \widehat{W}\ (T)$$

The family $(\widehat{W}\ (T),\ T \in \mathcal{T})$ is called <u>upper-value of the game</u>

We wish first aggregate $\widehat{W}\ (T)$, i. e. prove that there exists an optional process $\widehat{W}$ such that $\widehat{W}\ (T) = \widehat{W}_T$ a. e. $\forall$ T.

2. <u>AGGREGATION OF $\widehat{W}$</u>

For this we need the fundamental result of Dellacheric-Lenglart. From their terminology we call $\mathcal{T}$-system any family $(X\ (T),\ T \in \mathcal{T})$ of random functions such that :

i) X (T) = X (T') a. e. or (T = T') for any T, T'

ii) X (T) is $\underline{\underline{F}}_T$-measurable for any T

Theorem 1 [1]. Any $\mathcal{C}$-system X upper right semi-continuous i.e.

$$X(T) \geq \limsup_n X(T_n) \text{ a. e. if } T_n \searrow T$$

can be aggregated by an upper right semi-continuous process.

We prove easily that $\hat{W}(T)$ is upper right semi-continuous, using the facts that an infimum of upper-right semi-continuous functions is upper-right semi-continuous, and that the P-ess inf is always attained by a countable infimum.

Then with the help of

Lemma 2. For all $u \in \mathcal{U}$, for all stopping times T_1, T_2, $T_1 \leq T_2$ we have :

$$E_u(\bar{X}^u_{T_2}/\underline{F}_{T_1}) = \underset{v \in \mathcal{D}(u,T_2)}{\text{P-ess inf}} \ \underset{S \geq T_2}{\text{P-ess sup}} \ E_v\left(\int_0^S e^{-\alpha s} c(s, v_s)\, ds + Y_S/\underline{F}_{T_1}\right)$$

where :

$$\bar{X}^u = \int_0^{\cdot} e^{-\alpha s} c(s, u_s)\, ds + \hat{W}$$

result based on the properties of increasing or decreasing filtration which allow to inverse ess inf or ess sup with conditional expectation, we can prove by a technical proof the :

Theorem 3. $\bar{X}^u$ is lower right semi-continuous in expectation i. e. if $T_n \searrow T$ $E_u(\bar{X}^u_T) \leq \liminf_n E_u(\bar{X}^u_{T_n})$, then lower right semi-continuous.

Proof of the lemma.

We notice easily that for all v in $\mathcal{D}(u, T_2)$, the family $\left(E_v\left(\int_0^S e^{-\alpha s} c(s, v_s)\, ds + Y_S/\underline{F}_{T_2}\right), S \geq T_2\right)$ is a supremum lattice. Therefore we have for all v in $\mathcal{D}(u, T_2)$:

$$(1)\ E_u\left(\underset{S \geq T_2}{\text{P-ess sup}}\ E_v\left(\int_0^S e^{-\alpha s} c(s, v_s)\, ds + Y_S/\underline{F}_{T_2}\right)/\underline{F}_{T_1}\right)$$

$$= \underset{S \geq T_2}{\text{P-ess sup}}\ E_u\left(E_v\left(\int_0^S e^{-\alpha s} c(s, v_s)\, ds + Y_S/\underline{F}_{T_2}\right)/\underline{F}_{T_1}\right)$$

$$= \underset{S \geq T_2}{\text{P-ess sup}}\ E_v\left(\int_0^S e^{-\alpha s} c(s, v_s)\, ds + Y_S/\underline{F}_{T_1}\right),$$

since P^u and P^v are the same on $\underline{F}_{T_2}$.

On the other hand, the family

$$\left(\underset{S \geq T_2}{\text{P-ess sup}}\ E_v\left(\int_0^S e^{-\alpha s} c(s,v_s) ds + Y_S/\underline{F}_{T_2}\right),\ v \in \mathcal{D}(u, T_2)\right)$$

is also and infimum lattice. From this fact we can write :

$$(2)\quad E_u(\bar{X}^u_{T_2}/\underline{\underline{F}}_{T_1}) = \underset{v \in \mathcal{D}(u,T_2)}{\text{P-ess inf}}\, E_u\Big(\underset{S \geq T_2}{\text{P-ess sup}}\, E_v\Big(\int_o^S e^{-\alpha s} c(s, v_s)\, ds + Y_S/\underline{\underline{F}}_{T_2}\Big)/\underline{\underline{F}}_{T_1}\Big)$$

Now using (1) and (2) the proof of the lemma is established.

<u>Proof of the theorem</u>

From the lemma when $T_1 = o$, $T_2 = T$ we get :

$$E_u(\bar{X}^u_T) = \inf_{v \in \mathcal{D}(u,\, T)} \sup_{S \geq T} E_v\Big(\int_o^S e^{-\alpha s} c(s, v_s)\, ds + Y_S\Big)$$

$$(3)\qquad \leq \inf_{v \in \mathcal{D}(u,\, T_n)} \sup_{S \geq T} E_v\Big(\int_o^S e^{-\alpha s} c(s, v_s)\, ds + Y_S\Big)$$

(when (T_n) is decreasing to T), since $\mathcal{D}(u, T_n) \subset \mathcal{D}(u, T)$.

Let v be in $\mathcal{D}(u, T_n)$, then :

$$E_v\Big(\int_o^S e^{-\alpha s} c(s, v_s)\, ds + Y_S\Big) = E_v\Big(\mathbb{1}_{(S<T_n)}\Big(\int_o^S e^{-\alpha s} c(s, v_s)\, ds + Y_S\Big) + \mathbb{1}_{(S \geq T_n)}\Big(\int_o^S e^{-\alpha s} c(s, v_s)\, ds + Y_S\Big)\Big)$$

and since v and u are the same until T_n :

$$E_v\Big(\int_o^S e^{-\alpha s} c(s, v_s)\, ds + Y_S\Big) = E_u\Big(\mathbb{1}_{(S<T_n)}\Big(\int_o^S e^{-\alpha s} c(s, u_s)\, ds + Y_S - \int_o^{T_n} e^{-\alpha s} c(s, u_s)\, ds - Y_{T_n}\Big)\Big) + E_v\Big(\int_o^{S \vee T_n} e^{-\alpha s} c(s, v_s)\, ds + Y_{S \vee T_n}\Big)$$

Then :

$$\sup_{S \geq T} E_v\Big(\int_o^S e^{-\alpha s} c(s, v_s)\, ds + Y_S\Big) \leq \sup_{S \geq T} E_u\Big(\mathbb{1}_{(S<T_n)}\Big(\int_o^S e^{-\alpha s} c(s, u_s)ds + Y_S - \int_o^{T_n} e^{-\alpha s} c(s, u_s)\, ds - Y_{T_n}\Big)\Big) + \sup_{S \geq T_n} E_v(C^v_S + Y_S)$$

Taking the infimum on v of $\mathcal{D}(u, T_n)$, and using at left the inequality (3) and at right the lemma 2, we finally obtain :

$$E_u(\bar{X}^u_T) \leq \sup_{S \geq T} E_u\left(\mathbb{1}_{(S<T_n)}\left(\int_0^{S\wedge T_n} e^{-\alpha s} c(s, u_s)\, ds + Y_{S\wedge T_n} - \int_0^{T_n} e^{-\alpha s} c(s, u_s)\, ds - Y_{T_n}\right)\right) + E_u(\bar{X}^u_{T_n})$$

For any $\varepsilon > 0$, we can choose $S_n \geq T$ such that :

$$E_u(\bar{X}^u_T) \leq E_u\left(\mathbb{1}_{(S_n<T_n)}\left(\int_0^{S_n\wedge T_n} e^{-\alpha s} c(s, u_s)\, ds + Y_{S_n\wedge T_n} - \int_0^{T_n} e^{-\alpha s} c(s, u_s)\, ds - Y_{T_n}\right)\right) + \varepsilon + E_u(\bar{X}^u_{T_n})$$

Since $T_n \searrow T$, $S_n \wedge T_n \to T$, with Lebesgue's theorem we get :

$$E_u(\bar{X}^u_T) \leq \liminf_n E_u(\bar{X}^u_{T_n}) + \varepsilon \quad \forall\, \varepsilon > 0 \text{ and the result.}$$

Finally with the theorem 1, and with the fact that any optional process lower right semi-continuous in expectation is lower right semi-continuous ([2] for example) we get the main result of this part, i.e. there exists a right continuous process $\hat{W}$ such that

$\hat{W}_T = \hat{W}(T)$ a. e. for all stopping time T

This process is now used to construct stopping times which realize the ε-value.

3. EXISTENCE OF THE VALUE

For all $\varepsilon > 0$, all stopping time T, let

$$D^\varepsilon_T = \inf(s \geq T,\ \hat{W}_s \leq Y_s + \varepsilon)$$

We have the :

Proposition 4

We have for all u of $\mathcal{U}$, for all stopping time T :

$$\bar{X}^u_T \leq E_u(\bar{X}^u_{D^\varepsilon_T} / \underline{\underline{F}}_T)$$

(i. e. $\bar{X}^u_T$ is like a supermartingale (w. r. to P_u) between T and D^ε_T)

Proof

For all stopping time $U \leq D^\varepsilon_T$, $v \in \mathcal{D}(u, U)$, if Z^v is the P^v

Snell's envelope of $\int_0^{\cdot} e^{-\alpha s} c(s, v_s)\, ds + Y$, we have :

$$\bar{X}^u_U = \underset{v \in \mathcal{D}(u,\, U)}{\text{P-ess inf}}\ Z^v_U \leq \underset{v \in \mathcal{D}(u,\, D^\varepsilon_T)}{\text{P-ess inf}}\ Z^v_U$$

since $\mathcal{D}(u, U) \subset \mathcal{D}(u, D^\varepsilon_T)$. Then :

$$\bar{X}^u_U \leq Z^v_U \text{ a. e.} \qquad \forall\, v \in \mathcal{D}(u, D^\varepsilon_T),\ \forall\, U \leq D^\varepsilon_T$$

Let $T \leq t < D^\varepsilon_T$, by the definition of D^ε_T we have :

$$\bar{X}^u_t > \int_0^t e^{-\alpha s} c(s, u_s)\, ds + Y_t + \varepsilon = \int_0^t e^{-\alpha s} c(s, v_s)\, ds + Y_t + \varepsilon \quad \forall\, v \in \mathcal{D}(u, D^\varepsilon_T)$$

Finally we get

$$Z^v_t > \int_0^t e^{-\alpha s} c(s, v_s)\, ds + Y_t + \varepsilon \quad \forall\, v \in \mathcal{D}(u, D^\varepsilon_T)$$

Now if we define :

$$D^{\varepsilon, v}_T = \inf\left(t \geq T,\ Z^v_t \leq Y_t + \int_0^t c(s, v_s)\, e^{-\alpha s}\, ds + \varepsilon\right)$$

we get finally :

$$D^{\varepsilon, v}_T \geq D^\varepsilon_T \qquad \forall\, v \in \mathcal{D}(u, D^\varepsilon_T)$$

Then using results on optimal stopping, for example N. El Karoui [2] we obtain for all $v \in \mathcal{D}(u, D^\varepsilon_T)$ (since $Z^v_{t \wedge D^{\varepsilon, v}_T}$ has the P_v-martingale property between T and $D^{\varepsilon, v}_T$)

$$Z^v_T = \underset{S \geq D^\varepsilon_T}{\text{P-ess sup}}\ E_v\left(\int_0^S c(s, v_s)\, e^{-\alpha s}\, ds + Y_S / \underline{\underline{F}}_T\right)$$

and then using Lemma 2 :

$$\bar{X}^u_T \leq \underset{v \in \mathcal{D}(u,\, D^\varepsilon_T)}{\text{P-ess inf}}\ Z^v_T = \underset{v \in \mathcal{D}(u,\, D^\varepsilon_T)}{\text{P-ess inf}}\ \underset{S \geq D^\varepsilon_T}{\text{P-ess sup}}\ E_v\left(\int_0^S e^{-\alpha s} c(s, v_s)\, ds + Y_S / \underline{\underline{F}}_T\right)$$

$$(4) \qquad = E_u\left(\bar{X}^u_{D^\varepsilon_T} / \underline{\underline{F}}_T\right)$$

From this inequality we have easily the main result in :

Theorem 8 The mixt game has a value

Proof Since $\hat{W}$ and Y are right continuous we have :

(5) $\hat{W}_{D_T^\varepsilon} \leq Y_{D_T^\varepsilon} + \varepsilon$

With (4) and (5) we have easily :

$$\bar{X}_T^u \leq E_u \left(\int_0^{D_T^\varepsilon} e^{-\alpha s} c(s, u_s)\, ds + Y_{D_T^\varepsilon} / \underline{\underline{F}}_T \right) + \varepsilon \qquad \forall u \in \mathcal{U}$$

Then for $T = 0$, $D_T^\varepsilon \equiv D^0$

$$\bar{X}_0 = \inf_u \sup_T E_u \left(\int_0^T e^{-\alpha s} c(s, u_s)\, ds + Y_T \right)$$

$$\leq E_u \left(\int_0^{D^0} e^{-\alpha s} c(s, u_s)\, ds + Y_{D^0} \right) + \varepsilon \qquad \forall u \in \mathcal{U}$$

Then

$$\bar{X}_0 \leq \inf_{u \in \mathcal{U}} E_u \left(\int_0^{D^0} e^{-\alpha s} c(s, u_s)\, ds + Y_{D^0} \right) + \varepsilon$$

and finally

$$\bar{X}_0 \leq \sup_T \inf_u E_u \left(\int_0^T e^{-\alpha s} c(s, u_s)\, ds + Y_T \right)$$

Since the converse inequality is always true we have the final result.

REMARKS

1. This kind of technic has been already used by M.A. MAINGUENEAU and myself [3] to study the Dynkin game without the "Mokobodski's assumption (aggregation and supermartingale behaviour of the upper value of the game).

2. We can conjecture that in the markovian case, the conditional value $\hat{W}$ is markovian, and finally the D^ε markovian.

REFERENCES

[1] C. DELLACHERIE :

E. LENGLART

Sur des problèmes de régularisation, de recollement et d'interpolation en théorie générale des Processus. Séminaire de Probabilités de l'Université de Strasbourg - Lecture Notes - Springer Verlag (1982)

[2] N. EL KAROUI :

Cours sur le Contrôle Stochastique - Ecole d'Eté de Probabilités de St Flour IX - Lecture Notes in Maths 876 - Springer Verlag (1979)

[3] J.P. LEPELTIER - M.A. MAINGUENEAU :

Le jeu de Dynkin en théorie générale sans hypothèse de Mokobodski Stochastics (1984) - Volume 13 - p. 25-44.

POSITIONAL MODELING OF STOCHASTIC CONTROL IN DYNAMICAL SYSTEMS

Yu.S. Osipov and A.V. Krjazhimskii
Institute of Mathematic and Mechanics
Sverdlovsk, USSR

1. INTRODUCTION

This paper deals with the construction of physically realizable regularizing operators for one class of inverse dynamical problems. These are problems of the following type: given a measurement of the trajectory of a control system, find the unknown control function acting in the system. The conditions under which such problems can be solved, and some solution algorithms, have been found for various classes of control systems (see, for instance, [1–4]). These problems are connected with non-parametric estimation problems [5], and also arise in control theory [4,6].

In many cases the solution operators for inverse problems prove to be physically realizable. This means that the value (at an arbitrary time instant) of the control function calculated by the operator does not depend on future measurements. This property is important from the practical point of view, because it is then possible to organize the calculation of the control function in real time. In this paper we shall consider only physically realizable operators.

Inverse dynamical problems turn out to be ill-posed (a small distance between the trajectories does not imply a small distance between the corresponding controls). Hence if the measurements are not precise a regularizing operator is needed [7,8].

Thus we have the problem of constructing a physically realizable regularizing operator for an inverse dynamical problem. Such operators have been constructed for several classes of finite-dimensional control systems with deterministic controls (using the approach of [9,10]) in [11]. In the present paper we shall discuss the case of stochastic controls. The suggested method is based on some principles of control from the theory of positional differential games [12,13].

2. STATEMENT OF THE PROBLEM

Let us consider the control system governed by the differential equation

$$\dot{x} = f(t, x, u) \quad .$$

Here x is an n-dimensional state vector, the time t varies within a given interval $T = [t_0, \vartheta]$, and the k-dimensional control vector u takes values within a given compact set Q. The function f is continuous and satisfies the local Lipschitz condition with respect to the second variable, i.e.,

$$|f(t, x_1, u) - f(t, x_2, u)| \leq B_c \cdot |x_1 - x_2|$$

$$(t \in T, u \in Q, x_1, x_2 \in R^n, |x_1|, |x_2| \leq c, c > 0, B_c = \text{const}) \quad .$$

Function f also satisfies the growth condition

$$|f(t, x, u)| \leq K_0 \cdot (1 + |x|), K_0 = \text{const} \quad .$$

Here and elsewhere the norm of a finite-dimensional vector is taken to be Euclidean.

Let a probability space $(\Omega, \mathbf{A}, P)$ be fixed; hereafter all random variables are defined on this probability space, and all random processes are defined on the time interval T. An n-dimensional random variable x_0 such that $\mathrm{E}|x_0|^2 < \infty$ is assumed to be fixed; x_0 corresponds to the distribution of the state of the system at the initial time instant t_0. Let also a family $(\mathbf{A}_t)$, $t \in T$, of σ-subalgebras of the σ-algebra $\mathbf{A}$ such that $\mathbf{A}_{t_1} \subset \mathbf{A}_{t_2}$ for $t_1 \leq t_2$ be fixed; x_0 is assumed to be $\mathbf{A}_{t_0}$-measurable.

We shall define a control as an arbitrary measurable random process $u = u(t)$ with values in Q compatible with the family $(\mathbf{A}_t)$ ($u(t)$ is $\mathbf{A}_t$-measurable for each $t \in T$). A motion generated by the control u is defined as an n-dimensional measurable random process $x = x(t)$ such that with probability 1 for all $t \in T$

$$x(t) = x_0 + \int_{[t_0 t]} f(\tau, x(\tau), u(\tau)) \lambda(d\tau) \quad .$$

The integral should be interpreted in the usual sense (see [14, pp. 241,242]), i.e., all realizations of the process are integrated. Here and elswhere λ is the Lebesque measure on T. The above assumptions concerning function f allow us to show easily that for each control u a motion x exists and is unique (in the sense that the realizations of two motions coincide with probability 1).

Let x_* be a fixed motion. Denote by U_* the set of all controls generating x_*. The problem considered below is to find a physically realizable operator which gives a good approximation of one of the controls from U_* using perturbed (not precise) measurements of the motion x_*. However, U_* may be very large. Using an approach from the theory of ill-posed problems, we will introduce a selection principle by which only those controls from U_* which minimize a certain functional will be approximated.

Let us be more formal. Introduce the space $L^2 = L^2(T \times \Omega, \lambda \times P, R^k)$ which is assumed to be separable, and a functional J on L^2 of the form

$$J(u) = \mathrm{E} \int_T \gamma(t, u(t))\lambda(dt)$$

where $\gamma(t, u)$: $T \times R^k \to R^1$ is continuous, convex in the second variable and satisfies the condition

$$\gamma(t, u) \leq K_1(1 + \|u\|^2), \; K_1 = \text{const} \quad .$$

Let

$$J_* = \inf \{J(u): u \in U_*\} \quad ,$$

$$U_{**} = \{u \in U_*: J(u) = J_*\} \quad .$$

Introduce a class Θ of n-dimensional random processes $\xi = \xi(t)$ compatible with the family $(\mathbf{A}_t)$; we shall call the elements of Θ *measurements*. We say that a measurement ξ is h-precise ($h > 0$) if for each $t \in T$ we have

$$(\mathrm{E}\, |\xi(t) - x_*(t)|^2)^{1/2} \leq h \quad .$$

An operator D mapping the set of all measurements into the set of all controls is said to be *physically realizable* if for each $\xi_1, \xi_2 \in \Theta$, $t \in T$, such that $\xi_1(\tau) = \xi_2(\tau)$ for all $\tau \in [t_0, t]$, we have $D\xi_1(\tau) = D\xi_2(\tau)$ for λ-almost all $\tau \in [t_0, t]$. In terms of the general theory of stochastic systems (see [15]), D determines a control system (in which the measurements play the role of controls).

A family (D_h), $h > 0$, of physically realizable operators is said to be *regularizing* if for each family (ξ_h), $h > 0$, of measurements, where ξ_h is h-precise, we have

$$J(D_h \xi_h) \to J_* \text{ as } h \to 0 \quad .$$

If in addition $U_{**} \neq \phi$ and

$$\mathrm{dist}_{L^2}(D_h \xi_h, U_{**}) \to 0 \text{ as } h \to 0 \quad ,$$

then (D_h) is said to be *strongly regularizing*. The problem is to find a regularizing (or strongly regularizing) family of physically realizable operators.

3. SOLUTION APPROACH

Under certain conditions a solution of the problem can be obtained by modeling a parallel motion as in the theory of positional differential games.

Introduce an auxiliary control system

$$\dot{z} = g(t, z, \xi, v) \quad ,$$

which we shall call the *model*; here $z \in R^n$, g is continuous, $\xi \in R^n$, $v \in Q$. A *strategy* (for the model) is defined as an arbitrary continuous function $S = S(t, z, \xi)$: $T \times R^n \times R^n \longrightarrow Q$. An *approximating strategy* is defined as a pair $D = (S, G)$, where S is a strategy, and $G = \{\tau_0, \ldots, \tau_m\}$, $t_0 = \tau_0, \ldots,$

$$\tau_m = \vartheta \, , \; \tau_{i+1} - \tau_i = \Delta \quad ,$$

is a uniform partition of the interval T. For each measurement ξ we define the motion $z = z(t)$ of the model generated by the approximating strategy $D = (S, G)$ using Euler splines:

$$z(t_0) = \xi(t_0) \quad ,$$

$$z(t) = z(\tau_i) + g(\tau_i, z(\tau_i), \xi(\tau_i), v_i)(t - \tau_i)$$

$$(\tau_i < t \leq \tau_{i+1}) \quad ,$$

$$v_i = S(\tau_i, z(\tau_i), \xi(\tau_i)) \quad .$$

The control $v(t) = v_i$, $\tau_i < t \leq \tau_{i+1}$, is called the *realization* of the approximating strategy D by the measurement ξ. We will consider each approximating strategy D as the operator which associates with every measurement ξ the corresponding realization of D. It is clear that D is physically realizable.

In this section we shall assume the following conditions:

(A1) $f(t, x, u) = f_1(t, x) + f_2(t, x)u$

and Q is convex.

It follows from the theory of optimization in Hilbert spaces that condition (A1) implies that U_{**} is non-empty and, if J is strictly uniformly convex, contains a single element (class of λ-equivalent elements).

Introduce the model

$$\dot{z} = f(t, \xi, v) \quad .$$

The solution of the problem (under conditon (A1)) may be constructed using the following theorem:

THEOREM 1. *Let the approximating strategy* $D_h = (S_h, G_h)$ *be given by the conditions:*

(a) $S_h(t, z, \xi)$ *is a minimum of the function*

$$\delta_h(u) = (z - \xi)' f_2(t, \xi)u + \alpha(h)\gamma(t, u) \quad ;$$

(b) $G = \{\tau_{0,h}, \ldots, \tau_{m,h}\}$, $\Delta_h = \tau_{i+1,h} - \tau_{i,h} \leq ch$;

(c) $\alpha(h)$, $h / \alpha(h) \to 0 + (h \to 0)$.

Then

(1) *family* (D_h) *is regularizing;*

(2) *if* J *is strictly uniformly convex, then* (D_h) *is strongly regularizing.*

The proof of the theorem has two main steps. First it is shown that, for the motions z_h of the model generated by D_h and ξ_h, the functional

$$\Lambda_h(t) = \mathrm{E}[\max_{\tau \leq t} |z_h(\tau) - x_*(\tau)|^2 +$$

$$+ \alpha(h) \int_{[t_0 t]} (\gamma(\tau, D_h \xi_h(\tau)) - \gamma(\tau, u_*(\tau)))\lambda(d\tau) \quad ,$$

where u_* is an arbitrary element of U_{**}, satisfies the condition

$$\Lambda_h(t) \leq \varepsilon(h), \ \varepsilon(h) / \alpha(h) \to 0 \text{ as } h \to 0 \quad .$$

Here we have Krasovskii's idea of extemal construction from the theory of positional differential games. The second step is to prove the theorem using the above condition. This may be done with the aid of some modifications of Tikhonov's method from the theory of ill-posed problems [7].

4. NONLINEAR CONTROL SYSTEMS

Now consider a system which is nonlinear in control, i.e., for which condition (A1) is not satisfied. Let $J \equiv 0$. Assume that $|x_0| \leq B$ (B = const) with probability 1. We shall first give a brief description of the problem. Consider the auxiliary control system

$$\dot{y} = \omega, \ |\omega| \leq W, \ y \in R^n \quad ,$$

where W is a constant such that $|f(t, x, u)| \leq W$ for all $t \in T$, $u \in Q$, $x \in N$; N is a compact set containing the values of all deterministic motions (on T) of the system with initial states (at time t_0) in the B-neighborhood of zero. Then x_* is the motion of the auxiliary system generated by the control $\omega_*(t) = f(t, x_*(t), u_*(t))$, where $u_* \in U_*$. This control can be found approximately by means of the strongly regularizing family (D_h) using Theorem 1 with $\gamma(t, \omega) = |\omega|^2$. Then for small h and h-precise ξ, the control $D_h \xi$ is close to ω. in L^2. We shall define the value $\bar{v}_{h,\xi}$ of the operator F_h for the measurement ξ by the following condition. For each $t \in T$, $\omega \in \Omega$, $\bar{v}_{h,\xi}(t, \omega)$ is a minimum of the function

$$\beta_{h,\xi}(u) = |v_h(t, \omega) - f(t, z_{h,\xi}(t, \omega), u)| \quad ,$$

where $z_{h,\xi}$ is the motion of the model for the auxiliary system generated by the approximating strategy D_h and the measurement ξ.

THEOREM 2. *Let*

(a) *Ω be a compact metric space;*

(b) **A** *be the expansion of the Borel σ-algebra of Ω;*

(c) *the measure P be non-atomic.*

Then the family (F_h) is strongly regularizing.

The proof of the theorem is based on the properties of a generalized control [20].

5. ADDITIONAL CONDITIONS FOR APPROXIMATING CONTROLS

Let us consider some special additional conditions for approximating controls. Suppose that $x.$ is a Markov process (we consider Markov processes in the narrow sense. Note that each feedback (Markov control) $u(t, x)$: $T \times R^n \to Q$ which is continuous in the second variable generates the Markov motion

$$x(t) = x_0 + \int_{[t_0 t]} f(\tau, x(\tau), u(\tau, x(\tau)))\lambda(d\tau)$$

$$(t \in T)(\text{mod } P) \quad .$$

Suppose that each measurement ξ has the form $\xi = x. + \zeta$, where ζ, $\zeta(t_0) = 0$, belongs to a given class Γ of n-dimensional Markov processes which are independent on $x.$. Then the combined motion measurement process $\varphi. = (x., \xi)$, $\xi \in \Theta$, is a Markov process. For a given measurement ξ we consider the class of all controls v such that the combined motion measurement proces $\varphi = (x, \xi)$, where x is the motion generated by v, is a Markov process. Each control v with this property is said to be *compatible* with the measurement ξ. Let **M** denote the class of all physically realizable operators D such that for each measurement ξ the control $D\xi$ is compatible with ξ. Condiser the problem of constructing a regularizing (strongly regularizing) family within the class **M**.

Under various assumptions certain operators from **M** may be represented by strategies. We shall introduce the model $\dot{z} = f(t, z, v)$, which is a precise copy of the system. The motion generated by the strategy S and the measurement ξ we define as an

n-dimensional measurable random process x such that with probabiliy 1 for all $t \in T$

$$x(t) = x_0 + \int_{[t_0,t]} f(\tau, x(\tau), S(\tau, x(\tau), \xi(\tau)))\lambda(d\tau) \quad .$$

The control $v(t) = S(t, x(t), \xi(t))$ is said to be a *realization* of the strategy S by the measurement ξ. If the motion x is unique for each ξ, then we consider S to be the physically realizable operator which associates with each mesurement ξ the corresponding realization of the strategy S. We shall introduce the following conditions:

(B1) Each measurement ξ has continuous realizations with probability 1.

(B2) Ω is a separable metric space, **A** contains the σ-algebra of Ω, and the measure P is regular.

(B3) For each $y \in R^n$ and measurable function $\xi: T \to R^n$, the Cauchy problem

$$\dot{x} = f(t, x, S(t, x, \xi(t))), \; x(t_0) = y$$

has a unique solution on the interval T.

THEOREM 3. *Let*

(a) *the strategy S satisfy condition* (B3);

(b) *one of the conditions* (B1), (B2) *be satisfied. Then S is an operator from* **M**.

The solution of the problem is given by the following theorem:

THEOREM 4. *Let*

(a) *condition* (A1) *and one of the conditions* (B1), (B2) *be satisfied;*

(b) *the strategies S_h, $h > 0$, be given by the following condition: $S_h(t, z, \xi)$ is a minimum of the function*

$$\kappa_h(u) = (z - \xi)' f_2(t, z)u + \alpha(h)\gamma(t, u) \quad ;$$

(c) *the strategies S_h, $h > 0$, satisfy condition* (B3).

Then

(1) *the family (S_h) of operators from* **M** *is regularizing;*

(2) *if J is strictly uniformly convex, then (S_h) is strongly regularizing.*

The proof follows the same general lines as the proof of Theorem 1.

REFERENCES

1. R.W. Brockett and M.D. Mesarovic. The reproducibility of multivariable control systems. *Journal of Mathematical Analysis and Applications*, 11(1965)548–563.

2. L.M. Silverman. Inversion of multivariable linear systems. *IEEE Transactions on Automatic Control*, AC14(1969)270–276.

3. V.T. Bozuchov and P.M. Kolesnikov. Identification of the output of systems with distributed parameters (in Russian). *Izvestiya Akademii Nauk SSSR, Tekhnicheskaya Kibernetika*, 3(1982)168–174.

4. B.N. Petrov, P.D. Krutko and E.P. Popov. Construction of control algorithms as an inverse dynamical problem (in Russian). *Doklady Akademii Nauk SSSR*, 244(5)(1979)1078–1081.

5. I.A. Ibragimov and R.Z. Chasminskiy. On the non-parametric estimation of linear functionals in the presence of Gaussian white noise (in Russian). *Teoriya Veroyatnosti i ee Primeneniya*, 24(1)(1984)19–32.

6. P.D. Krutko and E.P. Popov. Analytical construction of optimal regulators and inverse dynamical problems in controlled systems (in Russian). *Izvestiya Akademii Nauk SSSR, Tekhnicheskaya Kibernetika*, 3(1982)182–193.

7. A.N. Tikhonov and V.Ya. Arsenin. *Methods for the Solution of Ill-Posed Problems*. Nauka, Moscow, 1979 (in Russian).

8. V.K. Ivanov, V.V. Vasin and V.P. Tanana. *Theory of Ill-Posed Problems and Its Applications*. Nauka, Moscow, 1978 (in Russian).

9. Yu.S. Osipov and A.V. Krjazhimskiy. On the dynamical solution of operator equations (in Russian). *Doklady Akademii Nauk SSSR*, 269(3)(1983)552–556.

10. A.V. Krjazhimskiy, V.J. Maximov and Yu.S. Osipov. On positional simulation in dynamical systems (in Russian). *Prikladnaya Matematika i Mekhanika*, 47(6)(1981)883–889.

11. A.V. Krjazhinskiy and Yu.S. Osipov. Simulation of control in dynamical systems (in Russian). *Izvestiya Akademii Nauk SSSR, Tekhnicheskaya Kibernetika*, 2(1982)51–60.

12. N.N. Krasovskiy. *Game Problems of Encounter of Motions*. Nauka, Moscow, 1970 (in Russian).

13. N.N. Krasovskiy and A.J. Subotin. *A Differential Game with a Positional Strategy*. Nauka, Moscow, 1979 (in Russian).

14. I.I. Gikhman and A.V. Skorohod. *Introduction to the Theory of Stochastic Processes*. Nauka, Moscow, 1965 (in Russian).

15. R. Rishel. Necessary and sufficient dynamic programming conditions for continuous-time stochastic optimal control. *SIAM Journal on Control*, 8(4)(1970)559–571.

16. F.P. Vasiliev. *Methods of Solving Extremal Problems*. Nauka, Moscow, 1981 (in Russian).

USE OF THE H-CONVEX SET METHOD IN DIFFERENTIAL GAMES

V.V. Ostapenko
V.M. Glushkov Institute of Cybernetics, Kiev, USSR

The foundations of the theory of differential games and methods for solving the associated problems are quite well-established. However, in the general case, the solution of such problems is difficult or requires a large investment of computer memory and time, while the characteristics of differential games often call for players to act rapidly, with only small computers to support them. The development of fairly efficient methods for solving certain classes of such problems is therefore of great importance.

This work is an investigation of the same type as [1–3] and develops methods proposed in [4]. We give a method of solving approach and evasion problems for a sufficiently broad class of linear games with a fixed termination time. The method owes its name to the fact that the notion of H-convexity is used to describe the sets of initial positions favourable to one or other player. Recall the following definition [5]:

Definition 1. Let X be a Euclidean finite-dimensional space and $H \subset \{x^* \in X: \|x^*\| = 1\}$. The term *H-convex half-space* will be used to describe a half-space of the form $\{x \in X: \langle x, x^* \rangle \leq c\}$, where $x^* \in H$, c is a number. A set is referred to as *H-convex* if it can be represented as the intersection of a number of H-convex half-spaces.

Assume that $C(\tau)$, $\tau \in [0, t]$, is an integrable family of linear operators acting from X into X. H represents the set of all unit vectors for which the following conditions are satisfied:

(a) $C^*(\tau)x^* = \lambda(\tau \mid x^*)x^*$ for any $\tau \in [0, t]$;

(b) if x^* is fixed the function $\lambda(\cdot \mid x^*)$ has the same sign for all $\tau \in [0, t]$.

LEMMA 1. *Let M be an H-convex set and for some closed convex set W let*

$$\int_0^t C(\tau)d\tau W \subset M \quad .$$

Then

$$\int_0^t C(\tau)W\mathrm{d}\tau \subset M \quad .$$

LEMMA 2. *Let M be an H-convex set. Then*

$$\int_0^t C(\tau)M\mathrm{d}\tau = \int_0^t C(\tau)\mathrm{d}\tau M \quad .$$

The above lemmas are used to prove the basic results of this paper.

Assume that Z, L are Euclidean spaces, $\dim L \leq \dim Z$, $\varphi: L \to Z$ are linear mappings, and $\pi: Z \to L$. The dynamics of the game are described by the equation

$$\dot{z} = Az + \varphi B(u, v) \quad ,$$

where $z \in Z$, $u \in U$, $v \in V$, and U and V are compact sets. The termination set and the set of phase constraints are specified in the form

$$M_L = \{z \in Z: \pi z \in M\}, \; N_L = \{z \in Z: \pi z \in N\} \quad ,$$

where $M \subset N$ are given closed sets in the space L.

The goal of player P (the pursuer) is to ensure that the inclusions $z(t) \in M_L$, $z(\tau) \in N_L$, are satisfied for all $\tau \in [0, t]$, where t is the fixed time of termination of the game. The goal of player E (the evader) is to try to prevent these inclusions being satisfied. Set

$$P^{\bullet}_{N,t}(M) = \bigcap_{v \in V} \bigcup_{u \in U} \{z \in N_L: \pi e^{At} z + \int_0^t \pi e^{A(t-\tau)} \varphi \mathrm{d}\pi B(u, v) \in M\} \quad .$$

Let H_t denote the set of all $x^* \in L$ such that

(a) $(\pi e^{A\tau}\varphi)^* x^* = \lambda(\tau \mid x^*) x^*$ for all $\tau \in [0, t]$;

(b) if x^* is fixed the function $\lambda(\cdot \mid x^*)$ has the same sign on the whole interval.

THEOREM 1. *Let M be an H-convex set. Then, if* $z \in P^{\bullet}_{N,t}(M)$, *there exists a function* $u_*: V \to U$ *such that*

(a) $u_*(v(\tau))$ *is an admissible control for player* P *when* $v(\tau)$ *is an admissible control for player* E;

(b) *the inclusion* $\pi z(t) \in M$ *holds for the trajectory* $z(\tau)$ *starting in* z *and corresponding to controls* $u_*(v(\tau))$ *and* $v(\tau)$;

(c) *if N is an H-convex set and* $Az \in \varphi L$, *then* $\pi z(\tau) \in N$ *for all* $\tau \in [0, t]$.

THEOREM 2. *Let* $B(U, v)$ *be an H-convex set for all* $v \in V$. *If* $z \,\overline{\in}\, P^{\bullet}_{N,t}(M)$, *then either* $\pi z \,\overline{\in}\, N$ *or there exists* $v_* \in V$ *such that* $\pi z(t) \,\overline{\in}\, M$ *for the trajectory* $z(\tau)$

starting in z *and corresponding to some arbitrary admissible control* $u(\tau)$ *for player* P *and to control* $v(\tau) \equiv v_*$ *for player* E.

Note that the value of $u_*(v)$ for each $v \in V$ is estimated as the solution of the inclusions

$$\pi e^{At} z + \pi \int_0^t e^{A(t-\tau)} \varphi B(u_*(v), v) \in M \quad ,$$

$$u_*(v) \in U \quad ,$$

and the value of $v_* \in V$ is estimated from the condition

$$\pi e^{At} z + \int_0^t \pi e^{A(t-\tau)} \varphi d\tau B(U, v_*) \subset L \setminus M \quad .$$

If $M, U, V, B = u + v$ are polyhedra, then the values of $u_*(v)$ and v_* can be estimated by solving a system of linear inequalities.

The strategies described in Theorems 1 and 2 are special cases of ε-strategies [3], and therefore we have the following result:

COROLLARY 1. *Assume that for any* $v \in V$, *the sets* $B(U, v)$, M, *and* N *are* H*-convex and either* $N = L$ *or* $AZ \subset \varphi L$. *Then*

$$\tilde{P}_{N_{L,t}}(M_L) = P^*_{N,t}(M) \quad ,$$

where $\tilde{P}_{N_{L,t}}(M_L)$ *is the set of all initial positions from which player* P *can terminate the game in his own favor by playing* ε*-strategies* [3].

We shall now look at the case $Z = L$, taking π and φ as identity operators. Let H_A denote the set of eigenvectors of the matrix A^*. Then for each t we can take $H_A = H_t$. If $A = \mathrm{diag}\,\{\alpha_1, \ldots, \alpha_n\}$, then a set of the form $\{z = (z_1, \ldots, z_n): a_i \le z_i \le b_i\}$ is H_A-convex, where a_i can assume the value $-\infty$ and b_i can assume the value $+\infty$.

If the operators A, φ, π are selected properly, Theorems 1 and 2 can be applied to games whose dynamics are described by equations $\ddot{x} = D\dot{x} + B(u, v)$, $\ddot{x} = Dx + B(u, v)$, and so on.

Theorem 1 can be partly extended to games in a Banach space. Consider the evolutionary equation

$$\frac{d}{dt} x = Ax + u + v \, , \; x \in X \, , \; u \in U \, , \; v \in V$$

where X is a reflexive separable Banach space, U and V are closed bounded sets in X, U is convex, and A is a linear operator with a domain of definition which is dense in X.

The termination set M and the set of phase constraints are convex subsets in X. The strongly measurable functions $u(\tau)$ and $v(\tau)$ are admissible controls for players P and E.

Suppose that there exists a $\beta > 0$ such that for any $m = 1,2,\ldots$ and sufficiently large n we have $\|(E - n^{-1}A)^{-m}\| \le (1 - n^{-1}\beta)^{-1}$, where E: $x \to x$ is the identity mapping. Then from the Hille–Yosida theorem there exists a semi-group of linear operators $G(t)$ such that the solution of the corresponding evolutionary equation with initial condition $x(0) = x_0$ can be represented in the form

$$x(t) = G(t)x_0 + \int_0^t G(t-\tau)[u(\tau) + v(\tau)]d\tau \quad .$$

Assume that H_A is the set of unit eigenvectors of the operator A^*, and

$$P^*_{N,t}(M) = \bigcap_{v \in V} \bigcup_{u \in U} \{x \in N: G(t)x + \int_0^t G(t-\tau)d\tau[u+v] \in M\} \quad .$$

THEOREM 3. *Let M and N be H_A-convex sets. Then, if $x \in P_{N,t}(M)$, there exists a mapping $u_*: V \to U$ such that*

(a) *for any control $v(\tau)$ which is admissible for player* E, *the control $u_*(v(\tau))$ is admissible for player* P;

(b) *the inclusions $x(t) \in M$ and $x(\tau) \in N$ hold for all $\tau \in [0,t]$, where*

$$x(s) = G(s)x + \int_0^s G(s-\tau)[u_*(v(\tau)) + v(\tau)]d\tau \quad .$$

Consider the following example. Let Ω be an open domain in R^n and Γ be its boundary. For any function $x(y)$, $y = (y_1, \ldots, y_n) \in \Omega$, such that

$$x \in L_2(\Omega), \quad \frac{\partial}{\partial y_i} x \in L_2(\Omega)$$

we will define an operator A, putting

$$Ax = \sum_{i,j=1}^{n} \frac{\partial}{\partial y_i} \frac{\partial}{\partial y_j} x \quad ,$$

for $y \subset \Omega$ and

$$\alpha x + \beta \frac{\partial}{\partial \nu} x = 0 \, , \; \alpha, \beta \ge 0$$

for $y \in \Gamma$. The operator A is self-conjugate and in the space $L_2(\Omega)$ generates an orthonormal basis w_j consisting of eigenvectors

$$Aw_j = \lambda_j w_j \ , \ \lambda_j < \lambda_{j+1} \ , \ \lambda_j \to \infty \ , \ j = 1,2,\ldots \quad .$$

In this case

$$G(t)x = \sum_{j=1}^{\infty} e^{-\lambda_j t} <x, w_j> w_j \quad ,$$

where

$$<x, w_j> = \int_{\Omega} x(y) w_j(y) dy \quad .$$

Sets of the form

$$M = \{x \in L_2(\Omega) : a_j \leq <x, w_j> \leq b_j \ , \ j = 1,2, \ldots, \} \quad ,$$

are H_A-convex, where a_j and b_j can assume values $\pm\infty$. Let us consider the construction of the mapping $u_{\cdot}(v)$. For each v the value of $u_{\cdot}(v)$ can be obtained by solving the following system of inequalities:

$$a_j \leq e^{-\lambda_j t} x_0^j + \int_0^t e^{-\lambda_j (t-\tau)} d\tau [u_j(v) + v_j] \leq b_j \ , \ j = 1,2, \ldots \quad ,$$

where $x_0^j = <x(0), w_j>$, $u_j = <u, w_j>$, $v_j = <v, w_j>$.

Consider the finite system of inequalities $j = 1, \ldots, m$. There exists a mapping $u_{\cdot}^m(v)$ which satisfies this system for any v, and if $v(\tau)$ is admissible the control $u_{\cdot}^m(v(\tau))$ is also admissible. Furthermore, the sequence $u_{\cdot}^m(v)$ converges weakly to $u_{\cdot}(v)$ for any v.

Theorems 1 and 2 may be extended in part to games with a non-fixed termination time. Returning to the notation adopted in the earlier part of this paper, put

$$T^{\cdot}_{N,t}(M) = \bigcap_{v \in V} \bigcup_{\substack{u \in U \\ 0 \leq \vartheta \leq t}} \{z \in N_L : \pi e^{A\vartheta} z + \int_0^t \pi e^{A(\vartheta - \tau)} \varphi d\tau B(u, v) \in M\} \quad .$$

Here M, N and $B(U, v)$ are assumed to be convex sets for all $v \in V$; we have $\pi e^{A\tau}\varphi = \lambda(\tau) E_L$, where $E_L : L \to L$ is an identity mapping; and the function $\lambda(\tau)$ is assumed to have the same sign for all $\tau \in [0, t]$.

THEOREM 4. *Let $z \in T^{\cdot}_{N,t}(M)$ and $Az \in \varphi L$. Then there exists a mapping $u_{\cdot} : V \to U$ such that*

(a) *$u_{\cdot}(v(\tau))$ is an admissible control for player* P *if $v(\tau)$ is an admissible control of player* E;

(b) *the inclusion* $\pi z(t_*) \in M$ *holds for the trajectory* $z(\tau)$ *starting from* z *and corresponding to* $u_*(v(\tau))$ *and* $v(\tau)$, *where* $t_* \leq t$ *and* $\pi z(\tau) \in N$ *for all* $\tau \in [0, t_*]$.

THEOREM 5. *Let* $z \bar{\in} T^{\bullet}_{N,t}(M)$. *Then either* $\pi z \bar{\in} N$, *or there exists a* $v_* \in V$ *such that for the trajectory* $z(\tau)$ *starting from* z *and corresponding to the arbitrary control* $u(\tau)$ *and to* $v(\tau) \equiv v_*$, *we have* $\pi z(\tau) \bar{\in} M$ *for all* $\tau \in [0, t]$.

COROLLARY 2. *Let* $AZ \subset \varphi L$. *Then*

$$\tilde{T}_{N_L,t}(M_L) = T^{\bullet}_{N,t}(M) \quad .$$

Here $\tilde{T}_{N_L,t}(M_L)$ is the set of initial positions from which player P can terminate his game at or before time t by playing an ε-strategy.

It should be noted that the result stated as Corollary 2 was obtained in [6] for $N = Z$, with π and φ as identity operators.

The method described above provides a basis for the solution of practical problems such as the development of mathematical methods for controlling water transport in irrigation system channels [7]. A supervisor controlling an irrigation system from his control room acts as one player; the water consumers are regarded as his opponents. Water requests are usually submitted in good time, for instance, one day ahead. For various reasons, however, these requests are constantly being modified and the supervisor finds himself operating within the framework of the theory of differential games, with the opponent's action being unknown in advance. To make the most use of an irrigation system it suffices to maintain certain levels of water in the channels. Thus, we have to solve a confinement problem, which is a special case of the above problem with phase constraints.

REFERENCES

1. L.S. Potryagin. Linear differential games of pursuit. *Mathematical Collection, New Series*, 112(3)(1980)307–330.

2. N.N. Krasovkski and A.I. Subbotin. *Positional Differential Games*. Nauka, Moscow, 1974

3. B.N. Pshenichnyi and M.I. Sagajdak. On differential games with fixed time. *Kibernetika*, (6)(1970)54–63.

4. V.V. Ostapenko. Linear differential games where key operators admit simple control. *Doklady Akademii Nauk, USSSR*, 261(1981)808–810.

5. V.G. Boltjanski and P.S. Soltan. *Combinatorial Geometry of Various Classes of Convex Sets*. Shtiinca, Kishinev, 1978.

6. P.B. Gusjatnikov and E.S. Polovinkin. A linear differential game with a simple matrix. *Differencial'nye Uravnenija*, 16(1980)1360–1369.

7. V.E. Danil'chenko, P.I. Kovalenko, V.V. Ostapenko and A.P. Jakovleva. A game approach to control of water flow in irrigation system channels. *Avtomatika*, (3)(1983)53–63.

A LINEAR DIFFERENTIAL PURSUIT GAME

L.S. Pontryagin
Steklov Institute of Mathematics, USSR Academy of Sciences, ul. Vavilova 42, 117333 Moscow, USSR

The differential game described by the equation

$$\dot{z} = Cz - u + v \tag{1}$$

was studied in [2], where complete proofs of the results given in [1] may be found. Here z is the phase vector of the game in n-dimensional vector space R, C is a linear mapping of the space R into itself, and u and v are controls, i.e., vector functions of time t which are not known in advance. Vectors u and v satisfy the inclusions

$$u \in P \quad , \quad v \in Q \quad , \tag{2}$$

where P and Q are convex compact subsets of the space R and have arbitrary dimension. The game is considered finished when the point z enters a given closed convex set M from R. Control u is called the *pursuer control* and v the *evader control*.

In pursuit problems the control v is a function of time t, $v = v(t)$, and is not known in advance; the problem is to choose the control u as a function of t in such a way as to finish the game as quickly as possible. This is done at time t using information on $z(s)$ and $v(s)$ for $s \leq t$.

The most natural way to solve this problem is to try to choose the control $u(t)$ at any time t in such a way that the distance from the point $z(t)$ to the set M decreases as rapidly as possible. However, this turns out to be impossible. We have to use another method to estimate the rate of approach of the point $z(t)$ to the set M. We shall construct a convex set $W(\tau)$, $\tau \geq 0$, $W(0) = M$, and define the minimal value $\tau = T(z)$ for which a point $e^{\tau C} z$ belongs to the set $W(\tau)$. It is evident that the point $w = e^{\tau C} z$ lies on the boundary of the set $W(\tau)$ and depends on z. Let $\psi(w)$ be a unit exterior normal to the surface $\partial W(\tau)$ at the point w. The resulting function $T(z)$ is an estimation function for the time of approach of the point z to the set M.

If the value of $T(z)$ decreases during the game and finally becomes equal to zero then the game comes to an end. It can be proved that the rate of decrease of the function $T(z)$ during the game is not less than the rate of increase of the time t. Thus a game beginning at the point z_0 will finish at a time not greater than the value $T(z_0)$. It is important that an incorrect choice of evader control $v(t)$ gives an advantage to the pursuer, i.e., will accelerate the end of the game.

An important deficiency of [2] is that we use knowledge of the function $v(s)$ for $t \le s \le t + \varepsilon$, where $\varepsilon > 0$ is any given arbitrary small value, to find control $u(t)$. This is called *discrimination* of the evader control.

This deficiency is overcome in [2] under some natural assumptions on the smoothness of certain sets.

Since we use stronger assumptions here, the present paper is not simply a generalization of [2] but eliminates the discrimination of the control $v(t)$ and allows us to define optimal control $u(t)$ more constructively.

Let us recall the construction of convex set $W(\tau)$ given in [2]. First of all we introduce some natural operations over convex sets from the space R.

1. If X and Y are convex sets from the space R, and α and β are real numbers, then we define the convex set

$$Z = \alpha X + \beta Y \tag{3}$$

of all vectors $z = \alpha x + \beta y$, where $x \in X$, $y \in Y$. Hence we can define the Riemann integral

$$\int_{s_0}^{s_1} X(s) ds \tag{4}$$

Here it is assumed that the convex set-valued mapping $X(s)$ is continuous in real parameter s, $s_0 \le s \le s_1$. In (3) we consider only non-negative α, β.

2. Define the geometrical difference

$$Z^* = X \quad Y \tag{5}$$

of two convex sets X and Y from the space R. The set Z^* consists of all vectors $z^* \in R$ such that $Y + z^* \subset X$. Note that the sets (3–5) are convex and are also compact if X and Y are compact.

3. Define the set $W(\tau)$ in the form of an alternating integral

$$W(\tau) = \int_{M,0}^{T} (P(\tau)d\tau \quad Q(\tau)d\tau) \quad , \tag{6}$$

where $P(\tau) = e^{\tau C}P$, $Q(\tau) = e^{\tau C}Q$. To evaluate this we define an alternating sum of convex sets $(A, X_1, \dots, X_n, Y_1, \dots, Y_n)$. We set

$$A_0 = A \ , \ A_i = (A_{i-1} + X_i) \ Y_i \ , \quad i = 1, \dots, n \quad . \tag{7}$$

Let $(r_0, r_1, \dots, r_n)$ be a partition of the interval

$$0 = r_0 < r_1 < \cdots < r_n = r \quad . \tag{8}$$

We set (see (4))

$$X_i = \int_{r_{i-1}}^{r_i} P(\tau) d\tau \ , \quad Y_i = \int_{r_{i-1}}^{r_i} Q(\tau) d\tau \quad . \tag{9}$$

We consider the alternating sum A_n (see (7)) for set $A = M$, with X_i, Y_i given by formula (9), as an approximate value of alternating integral (6). It can be proved that alternating sum (9) has a limit if the maximal length of intervals from partition (8).tends to zero. This limit is the value of the alternating integral (6).

In [2] it is proved that if a function $v(s)$ is known on the interval $t \leq s \leq t + \varepsilon$ then we can choose the control $u(t)$ on the same interval in such a way that the inequality

$$T(z(t+\varepsilon) < T(z(t)) - \varepsilon$$

holds. For this we choose the control $u(t)$ in such a way that the difference

$$P(z(t+\varepsilon)) - T(z(t))$$

has its largest absolute value. Hence we solve some nontrivial variational problem with discrimination of evader control on every time interval of length ε.

In the simple case considered in [2] (see §6, p.325), the set M is a linear vector subspace. Consider an orthogonal complement L of dimension ν to the subspace M in the space R. Let π be the orthogonal projection of the space R onto the subspace L, and consider the sets

$$P(\tau) = \pi e^{\tau C} P \ , \quad Q(\tau) = \pi e^{\tau C} Q \quad . \tag{10}$$

Suppose that the set

$$S(\tau) = P(\tau) \ \ Q(\tau) \tag{11}$$

has dimension ν for $0 < \tau < T$. We distinguish between two separate cases:

1. $P(\tau) = Q(\tau) + S(\tau)$ (the exhaustive case)

2. $P(\tau) \neq Q(\tau) + S(\tau)$.

Consider the convex set

$$\bar{W}(\tau) = \int_0^\tau S(t)dt \quad . \tag{12}$$

We define the estimating function $\bar{T}(z)$ as the minimal value of τ for which the inclusion

$$\pi e^{\tau C} z \in \bar{W}(\tau) \tag{13}$$

holds.

In the present paper we give a way of constructing the pursuit control $u(t)$ without discrimination of the evader control $v(t)$ under certain differentiability conditions. In particular, we suppose that the $\bar{W}(\tau)$ are convex sets with smooth boundaries and that the boundaries of the sets $P(\tau)$ and $Q(\tau)$ do not contain linear segments.

Consider the support function $c(\bar{W}(\tau) - \pi e^{\tau C} z, \psi)$ of convex set $\bar{W}(\tau) - \pi e^{\tau C} z$, where ψ is a unit vector. This support function is greater than or equal to zero for any ψ if

$$\pi e^{\tau C} z \in \bar{W}(\tau) \quad , \tag{14}$$

and has negative values for some ψ if inclusion (14) does not hold. We denote the minimum of this function by

$$-\bar{F}(z, \tau) = \min_{\psi} c(\bar{W}(\tau) - \pi e^{\tau C} z, \psi) \quad . \tag{15}$$

When point $\pi e^{\tau C} z$ reaches the set $\bar{W}(\tau)$ the function $\bar{F}(z, \tau)$ changes sign from positive to negative. The value of $\bar{T}(z)$ is the smallest positive root of the equation

$$\bar{F}(z, \tau) = 0 \quad . \tag{16}$$

The derivative

$$\bar{G}(z, \tau) = \frac{\partial \bar{F}}{\partial \tau}(z, \tau) \tag{17}$$

is nonpositive when the point $\pi e^{\tau C} z$ reaches the set $\bar{W}(\tau)$. If the inequality $\bar{G}(z, \tau) \neq 0$ holds at this time then $\bar{T}(z)$ is a smooth function of z in a neighborhood of this point. If $\bar{G}(z, \tau) = 0$ then function $\bar{T}(z)$ may be discontinuous.

If u and v are known functions then z is a function of parameter t and $\tau = \bar{T}(z)$ is also a function of t. This means that relation (16) is an identity with respect to t.

Differentiating the identity (16) in t we get the relation

$$\dot{\tau}\bar{G}(z,\tau) + \dot{z}\frac{\partial \bar{F}}{\partial z} = 0 \quad .$$

Hence for $\bar{G} \neq 0$ we have

$$\dot{\tau} = \frac{\left(\frac{\partial \bar{F}}{\partial z}, \dot{z}\right)}{-\bar{G}(z,\tau)} \quad . \tag{18}$$

Let $\bar{\psi}(t)$ be the unit vector which minimizes the support function (15) and $s(\psi,\tau)$ be the point on the boundary of the convex set $S(\tau)$ which maximizes the scalar product

$$(s,\psi) \ , \quad s \in S(\tau) \quad .$$

Then function $\bar{G}$ has the form

$$\bar{G}(z,\tau) = (\pi e^{\tau C} Cz - s(\psi,\tau),\psi) \quad , \tag{19}$$

and formula (18) becomes

$$\dot{\tau} = \frac{(\pi e^{\tau C}(Cz - u + \gamma),\psi)}{-(\pi e^{\tau C} Cz - s(\psi,\tau),\psi)} \quad . \tag{20}$$

It is clear from formula (20) that we can choose the control u in such a way that $\dot{\tau} \leq -1$. Take the value of u which minimizes $\dot{\tau}$. The corresponding value of $\dot{\tau}$ is less than or equal to -1. It is evident that $u(t)$ maximizes the scalar product $(\pi e^{\tau C} u,\psi)$. This value of $u = u_{\text{opt}}$ is said to be optimal and is the value of the control chosen during the pursuit process if $\bar{G} \neq 0$.

If we choose control $u(t)$ according to this rule and function $\bar{G}$ tends to zero then the value of $\dot{\tau}$ is defined by the same relation (18). Here we have to consider two different cases. The control v_{opt} is said to be optimal if it maximizes the scalar product $(\pi e^{\tau C} v,\psi)$. Consider the exhaustive case. If the control v is optimal on some time interval and $\bar{G} = 0$ at the initial time t_0, then $\dot{\tau} = -1$ and $\bar{G} = 0$ for all t from this interval. If $v \neq v_{\text{opt}}$ and $\bar{G} = 0$ then the point $z(t)$ leaves the surface $\bar{G}(z,\tau) = 0$ in a small neighborhood of t_0. Moreover, the function τ displays the following behavior:

$$\tau_0 - \tau = A(t - t_0)^{1/k+1} + O((t - t_0)^{1/k+1}) \quad , \tag{21}$$

where k is the multiplicity of the root τ_0 of equation (16). Two cases can arise if the point $z(t)$ arrives at the surface $\bar{C}(z,\tau) = 0$: τ changes continuously or displays a jump. In the first case the behavior of τ has the following form:

$$\tau - \tau_0 = A(t_0 - t)^{1/k+1} + O((t_0 - t)^{1/k+1}) \quad . \tag{22}$$

In the non-exhaustive case the behavior of the trajectory may be considered in a similar way with some small differences.

Hence for an optimal choice of $u(t)$ the solution $z(t)$ of the differential game always satisfies the following condition:

$$\frac{d}{dt} T(z(t)) \leq -1 \quad .$$

In the case of the alternating integral we set $P(\tau) = e^{\tau C} P$, $Q(\tau) = e^{\tau C} Q$. Let $L(P(\tau))$ be the affine support of convex set $P(\tau)$. If the vector $\psi(w)$ is not orthogonal to the space $L(P(\tau))$ then we choose the control $u(t)$ which maximizes the function

$$(e^{\tau C} u, \psi(w)) \ , \quad u \in P \quad . \tag{23}$$

This relation defines a unique control u which is the best pursuit strategy. If the vector $\psi(w)$ is orthogonal to the space $L(P(\tau))$ at the time t_0 then rule (23) does not give us the opportunity to choose control $u(t)$ and it must be selected in some other way.

In the general case consider the support function

$$c(W(\tau) - e^{\tau C} z, \psi) \ , \quad |\psi| = 1 \quad . \tag{24}$$

It is clear that this support function is greater than or equal to zero if $e^{\tau C} z \in W(t)$ and has a negative value if this inclusion does not hold. Define

$$-F(z, \tau) = \min_{\psi} c(W(\tau) - e^{\tau C} z, \psi) \quad . \tag{25}$$

Hence the value $T(z)$ is the smallest positive root of the equation

$$F(z, \tau) = 0 \tag{26}$$

with respect to τ. Set $G(z, \tau) = \partial F / \partial z$.

We choose the optimal control $u(t)$ in the following way. Since τ is a root of equation (26) we differentiate it in t and obtain the relation

$$\dot{\tau} = -\frac{\left[\frac{\partial F}{\partial z}, \dot{z}\right]}{G(z, \tau)} \ , \tag{27}$$

which is similar to (18). We choose the control $u(t)$ in such a way that the value of $\dot{\tau}$ given by relation (27) is minimal. This approach is similar to the choice of optimal control $u(t) = u_{opt}(t)$ given previously.

It can be proved that $\dot{\tau} \leq 1$ if we use this rule. Hence the estimating function $T(z(t))$ decreases more quickly than t increases.

The control $v(t)$ which maximizes $\dot{\tau}$ (see (27)) for any given $u(t)$ is called the optimal evader control and is denoted by $v_{opt}(t)$. This optimal control $v_{opt}(t)$ does not depend on the choice of control u.

Relation (27) is meaningful only if $G \neq 0$. It can be proved that

$$G = (e^{\tau C} Cz, \psi(w)) - (e^{\tau C}(u_{opt} - v_{opt}, \psi(w))) \quad . \tag{28}$$

If $G \neq 0$ then formula (27) has the form

$$\dot{\tau} = \frac{(e^{\tau C}(Cz - u_{opt} + v), \psi)}{-(e^{\tau C}(Cz - u_{opt} + v_{opt}), \psi)} \quad . \tag{29}$$

Hence $\dot{\tau} \leq -1$ and $\dot{\tau} = -1$ if $v = v_{opt}$.

It can be proved that $\dot{\tau} = -1$ if $G = 0$ and $v = v_{opt}$. This fact does not follow from (29). If $v = v_{opt}$ on some time interval and $G = 0$ at the initial time t_0 then $G = 0$, $\dot{\tau} = -1$ and ψ = const. all over this interval.

If $v \neq v_{opt}$ and $G = 0$ then point $z(t)$ leaves the surface $G = 0$ in a small neighborhood of t_0. Moreover, the behavior of function τ is described by formula (21).

When vector $\psi(w)$ becomes orthogonal to the subspace $L(P(\tau))$ the control u displays a jump. We would therefore have to choose it in a different way were it not for the fact that it can be proved that this orthogonality disappears and we can take the rule for choosing the optimal control u given earlier.

The relation

$$\frac{d}{dt} T(z(t)) \leq -1$$

holds for all of the methods of choosing the pursuit control $u(t)$ mentioned here, i.e., the rate of decrease of function $T(z(t))$ is not less than the rate of increase of t.

REFERENCES

1. L.S. Pontryagin. *Doklady Akademii Nauk SSSR* 175 (1967) 764–766.

2. L.S. Pontryagin. *Mathematichesky Sbornik* 112 (1980) 307–330.

METHODS OF CONSTRUCTING GUARANTEED ESTIMATES OF PARAMETERS OF LINEAR SYSTEMS AND THEIR STATISTICAL PROPERTIES

B.N. Pshenichnyi and V.G. Pokotilo
V. Glushkov Institute of Cybernetics, Kiev, USSR

This paper is concerned with the properties of guaranteed estimates of unknown parameters of linear systems. In the theory of guaranteed or minimax estimation [1,2] as opposed to mathematical statistics, the nature of the perturbations is assumed to be uncertain and we consider either the problem of finding estimates that minimize the estimation error under the worst (from the viewpoint of an observer) possible perturbations from some *a priori* known set, or the problem of finding the whole set of parameters compatible with the observed signal.

When there is no sufficiently complete description of the random perturbations the guaranteed approach can also be used for stochastic systems. In this case the assumption of random noise implies that the guaranteed estimates have additional properties. In this paper (see also [3–6]) we obtain sufficient conditions for the convergence of these estimates to the actual values of the unknown parameters. We consider the case when we have geometrical constraints, implying that the perturbations are bounded at each instant of time. In this case to develop an exact description of the information sets mentioned above is a very cumbersome nonlinear programming problem. A method of approximating guaranteed estimates which only requires the solution of a linear programming problem is suggested, and examples are given.

1. DEFINITION OF GUARANTEED ESTIMATES

Assume that the signal

$$y(t) = \vartheta(t)z + w(t)\ , \ t \in [0,T] \tag{1}$$

is measured.

Here $\vartheta(t)$ is a known, deterministic, $(m \times n)$-matrix which is continuous in $[0,T]$, $z \in R^n$ is an unknown vector of parameters, and the $w(\cdot)$ are indefinite perturbations which satisfy the inclusion

$$w(t) \in W(t) \subset W \subset R^m\ , \ \forall t \in [0,T] \quad , \tag{2}$$

where W, $W(t)$ are convex compact sets, and the multivalued mapping $W(\cdot)$ is continuous in the Hausdorff metric, $0 \in W(t)$, $t \geq 0$.

Let $C_m[0,T]$ be a space of continuous m-functions in $[0,T]$ (the elements of this space will be denoted by $f_T = \{f(t), t \in [0,T]\}$), and $V_m[0,T]$ be the space of the m-dimensional functions with bounded variation on $[0,T]$.

If $f_T \in C_m[0,T]$ and $\psi \in V_m[0,T]$ then

$$<\varphi, f_T> = \int_0^T f'(t) d\varphi(t) \quad ,$$

where f' is the transpose of f.

The expressions

$$D = \{w_T \in C_m[0,T]: w(t) \in W(t), t \in [0,T]\} \tag{3}$$

$$E(z) = \{y_T \in C_m[0,T]: y(t) = \vartheta(t)z + w(t): w_T \in D\} \tag{4}$$

define the sets of admissible perturbations and outputs of system (1).

Assume also that the set

$$\Phi(\psi) = \{\varphi \in V_m[0,T]: \int_0^T \vartheta'(t) d\varphi(t) = \psi\} \tag{5}$$

is non-empty.

Problem I (*a priori* estimation). Let $\psi \in R^n$. Find $\varphi_0 \in V_m[0,T]$ such that the equality

$$\sup_{\substack{y_T \in E(z) \\ z \in R^n}} |\psi' z - <\varphi_0, y_T>| = \inf_{\varphi \in V_m[0,T]} \sup_{\substack{y_T \in E(z) \\ z \in R^n}} |\psi' z - <\varphi, y_T>|$$

is satisfied.

Problem II (*a posteriori* estimation). Let the signal $y_T^* \in E(z)$ be observed. Define the set

$$Z(T; y^*(\cdot)) = Z(y_T^*) = \{z^* \in R^n: y_T^* \in E(z^*)\} \quad .$$

The set $Z(T; y^*(\cdot))$ wll be called the information set compatible with the observed signal (see [2]).

Using duality theory we can show that the following relations hold (see [2]):

$$\kappa(T) = \rho[\varphi_0] = \int_0^T \rho(d\varphi_0(t) \mid W(t)) =$$

$$\inf \{ \int_0^T \rho(d\varphi(t) \mid W(t)) : \varphi \in \Phi(\psi) \} \tag{6}$$

$$\rho(\psi \mid Z(T; y^*(\cdot))) = \inf \{ \int_0^T \rho(-d\varphi(t) \mid W(t)) + \langle \varphi, y_T^* \rangle : \varphi \in \Phi(\psi) \} \quad .$$

Here $\varphi_0 \in V_m[0,T]$ is the solution of problem I; $\rho(\psi \mid X) = \sup \{ \psi' x : x \in X \}$ is the support function of the set X, and the integral

$$\int_0^T \rho(d\varphi(t) \mid W(t)) = \rho[\varphi]$$

coincides with the support function of the set D (see [2, p. 100]).

2. CONSISTENCY OF GUARANTEED ESTIMATES

In this section we address the following question: What are the properties of guaranteed estimates in situations where the perturbations can be simulated by random processes? We consider a probability space $\{\Omega, \sum, P\}$ and make additional assumptions about system (1) and perturbations (2), (3). We assume that the perturbations are outcomes of the random process $\{w(t)\ ,\ t \geq 0\}$ and satisfy the specified constraints with probability one. By U_a^b, $0 \leq a \leq b \leq \infty$, we denote all of the σ-algebras generated by the process $\{w(t)\ ,\ t \geq 0\}$.

Condition 1. The equality

$$\sup_{t \geq 0} q(\vartheta(t)z \mid W(t)) = \sup_{t \geq T} q(\vartheta(t)z \mid W(t))$$

for any $z \in R^n$ holds. Here $q(x \mid V) = \inf \{ \alpha \geq 0 : x\alpha^{-1} \in V \}$ is a gauge function (Minkovsky functional) of the set V.

Condition 2. The random process $\{w(t)\ ,\ t \geq 0\}$ has zero mean and a covariance matrix $Q(t\ ,t') = \mathrm{E}[w(t)w'(t')]$ such that $|\ \mathrm{sp}\ Q(t\ ,t')| \downarrow 0$ for $|t - t'| \to \infty$ and t, $t' \geq 0$, and

$$\int_1^\infty \frac{1}{\tau^3} \int_0^\tau \int_0^\tau |\mathrm{sp}\ Q(t\ ,t')|\ dt dt' d\tau < \infty \quad .$$

Condition 3. For any $z \in R^n$, $z \neq 0$, we have

$$\overline{\lim_{t \to \infty}} \|\vartheta(t) z\| > 0 \quad .$$

Conditon 4. The random process $\{w(t), t \geq 0\}$ is completely regular so that

$$\alpha(\tau) = \sup_{\substack{E_1 \in U_0^t, E_2 \in U_{t+\tau}^\infty \\ t \geq 0}} |P(E_1 \cap E_2) - P(E_1)P(E_2)| \longrightarrow 0$$

$$\tau \longrightarrow \infty \quad .$$

Condition 5. For any $\varepsilon > 0$, $\eta \in R^m$, $\|\eta\| = 1$, $t \geq 0$, we have

$$P\{\rho(-\eta \mid W(t)) + \eta' w(t) < \varepsilon\} \geq J(\varepsilon) > 0 \quad .$$

THEOREM 1. *Let Conditions 1 and 2 be satisfied. Then for any $\varepsilon > 0$ there exists a $\varphi^\varepsilon \in V_m[0, T]$ such that*

$$\int_0^T \rho(d\varphi^\varepsilon(t) \mid W(t)) < \kappa(T) + \varepsilon \quad ,$$

and the estimate

$$z_0^\varepsilon(\psi, T) = \int_0^T y'(t) d\varphi^\varepsilon(t)$$

is strongly consistent.

Proof. Let $\kappa(T_*, T)$ denote the value of the lower bound in (6) under the additional assumption that $\varphi(t) \equiv 0$ for $t \in [0, T_*)$. This means that the information corresponding to $t \geq T_*$ is used in constructing the estimate. From the duality theorems which characterize the solution of the problem of moments (6) (see [1]) and Condition 1, it follows that

$$\lim_{T \to \infty} \kappa(T_*, T) = \kappa_0$$

does not depend on T_*.

We fix ε and select the sequence $\{T_i ; i = 0,1,2, \ldots\}$ in such a way that

$$T_0 = 0 \; ; \; \kappa(T_{i-1} + 1, T_i) < \kappa_0 + \varepsilon \quad .$$

It is known that the solution of problem (6) can be chosen in the form of n pulses such that

$$d\varphi_* / dt = \sum_{i=1}^n \varphi_{*i} \delta(t - t_i) \quad ,$$

where $\delta(\cdot)$ is a δ-function. Denote the moments and values of these pulses for $T = T_{i-1}+1$, $T = T_i$ by t_{ij} and φ_{ij}, respectively, where $j = 1,2,\ldots,n$ and define $\varphi^\varepsilon \in V_m[0,T]$ as follows:

$$d\varphi^\varepsilon(t)/dt = \frac{1}{N(T)} \sum_{i\geq 1}^{N(T)} \sum_{j\geq 1}^{n} \varphi_{ij}\delta(t-t_{ij})$$

$$N(T) = \max\{i : T_i \leq T\}$$

It follows from the choice of the sequence $\{T_i\}$ that

$$\rho[\varphi^\varepsilon] = \frac{1}{N(T)} \sum_{i=1}^{N(T)} \sum_{j=1}^{n} \|\varphi_{ij}\| = \frac{1}{N(T)} \sum_{i=1}^{N(T)} \kappa(T_{i-1},T_i) < \kappa_0 + \varepsilon \leq \kappa(T) + \varepsilon \quad .$$

We show that the estimate $z_0^\varepsilon(\psi,T) = \langle\varphi^\varepsilon, y(\cdot)\rangle$ is strongly consistent. Actually we have

$$z_\cdot^\varepsilon(\psi,T) = \frac{1}{N(T)} \sum_{i=1}^{N(T)} \sum_{j=1}^{n} \varphi_{ij} y(t_{ij}) = \psi' z + \frac{1}{N(T)} \sum_{i=1}^{N(T)} v_i \quad ,$$

where

$$v_i = \sum_{j\geq 1}^{n} \varphi_{ij} w(t_{ij}) \quad .$$

Under the conditions of the theorem we have $E[v_i] \equiv 0$, $i = 1,2,\ldots,$ and $E[v_i v_j] \leq n\ (\kappa_0+\varepsilon)^2\, \mathrm{Sp}\, Q(i,j)$ with

$$\sum_{k=1}^{\infty} \frac{1}{k^3} \sum_{i=1}^{k} \sum_{j=1}^{k} |\mathrm{Sp}\, Q(i,j)| < \infty \quad .$$

The assertion of the theorem then follows from the theorem of stability of almost certainly bounded second-order processes [7, p. 510].

THEOREM 2. *Let Conditions 2–5 be satisfied. Then*

$$Z(T; y^*(\cdot)) \to \{z\} \text{ as } T \to \infty$$

with probability one. (Convergence takes place in the Hausdorff metric).

This theorem follows from Theorem 3 below.

3. APPROXIMATION OF INFORMATION SETS

It is difficult to obtain an exact solution to the problem of *a posteriori* guaranteed estimation with geometrical constraints on perturbations in even the simplest cases. There is therefore a natural interest in the approximation of information sets (see, e.g., [8,9]). It is shown below that we can approximate the support functions of informatin sets by solving a linear programming problem while retaining the consistency of the estimates.

Consider a function ψ, $\|\psi\| = 1$, time instants $t_i \in [0, T]$, $i = 1,2, \dots, N$, and let vectors $\varphi_i \in R^m$, $i = 1,2, \dots, N$, be fixed. Then it is easy to see that the function

$$\gamma(\psi, T) = \inf \{ \sum_{i=1}^{N} \alpha_i [\rho(-\varphi_i \mid W(t_i)) + \varphi_i' y^*(t_i)]: \sum_{i=1}^{N} \alpha_i \varphi_i \vartheta(t_i) = \psi: \alpha_i \geq 0 \}$$

majorizes the support function of the information set $Z(T, y^*(\cdot))$.

THEOREM 3. *Let the assumptions of Theorem 2 be satisfied. Then for any* ψ, $\|\psi\| = 1$, *it is possible to find a partition* $\{t_i\}$ *of the segment* $[0, T]$ *and vectors* $\varphi_i \in R^m$ *such that*

$$\lim_{T \to \infty} \gamma(\psi, T) = \psi' z$$

with probability one.

We shall give two auxiliary statements without proofs.

LEMMA 1. *Let* $\{w(t), t \geq 0\}$ *be a completely regular random process,* U_a^b, $0 \leq a \leq b \leq \infty$ *be a system of* σ*-algebras related to it,* $t_i \to \infty$, $t_{i+1} > t_i$, $i = 1,2, \dots,$ *and* $A_i \in U_{t_{i-1}}^{t_i}$ *such that*

$$P(A_i) < 1 - \varepsilon, \ \varepsilon > 0, \ i = 1,2, \dots .$$

Then

$$P(\bigcap_{i \geq 1}^{N} A_i) \to 0 \text{ as } N \to \infty .$$

LEMMA 2. *Let Condition 3 be satisfied. Then there exists a* $M > 0$ *such that for any sequence* S_i, $i = 1,2, \dots,$ *it is possible to find a set of time instants*

$$J_s = \{t_{ij}: j = 1,2, \dots, n; i = 1,2, \dots \}$$

such that the following conditons hold:

(1) $t_{ij} \to \infty$ *for* $\tau \to \infty$, $j = 1,2,\ldots,n$;

(2) $|t_{ij} - t_{ik}| \geq S_i$, $i = 1,2,\ldots$; $j \neq k$;

(3) *the set of equations*

$$\sum_{j=1}^{n} \varphi'_{ij} \vartheta(t_{ij}) = \psi'$$

has a solution φ_{ij} *bounded uniformly on* i, $\|\varphi_{ij}\| \leq M$.

Proof of Theorem 3. Let $S_i \to \infty$ as $i \to \infty$ and J_s be a set of time instants defined by Lemma 2.

Then

$$0 \leq \gamma(\psi_j T) - \psi' z \leq \min_i \{\sum_{j=1}^{n} [\rho(-\varphi_{ij} \mid W(t_{ij})) + \varphi_{ij} w(t_{ij})]: 1 \leq i \leq N(T)\} \quad ,$$

$$N(T) = \max \{N: t_{ij} \leq T \ , \ j = 1, \ldots, n \ , \ i = 1,2, \ldots, N\} \quad .$$

If

$$A_i(\varepsilon) = \{\sum_{j=1}^{n} \rho(-\varphi_{ij} \mid W(t_{ij})) + \varphi_{ij} w(t_{ij}) \geq \varepsilon\} \quad ,$$

then for sufficiently large i

$$\mathrm{P}(A_i(\varepsilon)) \leq 1 - \frac{1}{2}\pi(\varepsilon(\eta M)^{-1})^{\eta} \quad .$$

The assertion of the theorem then follows from Lemma 1.

4. EXAMPLES

Example 1. Let $\eta = 1$, $\vartheta(t) \equiv 1$ and $w(t) \in [-1,1]$. The relations

$$z_0(T) = \frac{1}{T} \int_0^T y(t)\mathrm{d}t$$

$$\rho(+1 \mid Z(T; y^*(\cdot))) = \min \{y^*(t) + 1 \ ; \ t \in [0, T]\}$$

define *a priori* and *a posteriori* estimates which are consistent under Conditions 1 and 2, respectively. Note that the *a priori* estimate is defined ambiguously in this case.

Example 2. Let $n = 2$, $\vartheta(t) = (1, t)$, $\psi' = (1, 0)$, $w(t) \in [-1,1]$; and $z' = (z_1, z_2)$.

We define sequences of times $\{t_i^{(j)}; j = 1,2; \ i = 1,2, \ldots\}$ such that $t_i^{(2)} > t_i^{(1)}$, $i = 1,2, \ldots$;

$$\{t_i^{(2)}; i = 1,2, \ldots\} \cap \{t_i^{(1)}; i = 1,2, \ldots\} = \phi$$

$$(t_i^{(1)}+t_i^{(2)})(t_i^{(2)}-t_i^{(1)})^{-1}<1+\varepsilon\ ;\ i=1,2,\dots\ .$$

The equalities

$$z_0^{\varepsilon}(T)=<\varphi^{\varepsilon}\ ,\ y_T>$$

$$\frac{d\varphi^{\varepsilon}(t)}{dt}=\frac{1}{N(T)}\sum_{i=1}^{N(T)}\frac{t_i^{(2)}\delta(t-t_i^{(1)})-t_i^{(1)}\delta(t-t_i^{(2)})}{t_i^{(2)}-t_i^{(1)}}$$

define an *a priori* estimate whose maximum error does not exceed the value $1+\varepsilon$. The optimal *a priori* estimate is defined in this case by the relations $z_0(T)=y(0)$, $\kappa(T)=1$.

Now consider the approximation of information sets compatible with the signal measured at times $t_i^{(j)}$; $j=1,2$; $i=1,2,\dots$. We obtain

$$\psi' Z(T;y^*(\cdot))-z_1\subset[-\gamma(-1,T)\ ,\ \gamma(l\ ,T)]\quad ,$$

where

$$\gamma(\sigma;T)=\min_i\ \{\frac{t_i^{(1)}+t_i^{(2)}}{t_i^{(2)}-t_i^{(1)}}+\sigma\frac{t_i^{(2)}y\,(t_i^{(1)})-t_i^{(1)}y\,(t_i^{(2)})}{t_i^{(2)}-t_i^{(1)}}\}\ ;\ \sigma=\pm1\quad .$$

Under the conditions of Theorem 2 $\gamma(+1,T)+\gamma(-1,T)\rightarrow0$ as $T\rightarrow\infty$ only if $(t_i^{(1)}+t_i^{(2)})(t_i^{(2)}-t_i^{(1)})^{-1}\le M<\infty$, $i=1,2,\dots$.

5. CONCLUSION

In order to construct guaranteed estimates it is only necessary to have the measurement of the signal and to be aware of the *a priori* constraints on the perturbations. The guaranteed estimates may prove to be too rough if there is any information about the distribution of the random perturbations. However, it is this very "roughness" that demonstrates the advantage of having such information available. This makes it possible to class the guaranteed estimation method with the so-called robust methods of statistics.

The consistency of the guaranteed estimates is connected with the conditions for weak dependence of the random processes simulating the perturbations (randomness of the noise) and with the stability properties at infinity of the ideal signal $y_0(t)=\vartheta(t)z$. These conditions are completely natural and are satisfied for a broad class of real systems.

Note also that the *a posteriori* estimates constructed in this way are monotonic (the information set is not extended by increasing the observation interval) and, therefore, consideration of the support functions of information sets gives a convenient rule for terminating the observation process.

REFERENCES

1. N.N. Krasovski. *Theory of Movement Control*. Nauka, Moscow 1968.
2. A.B. Kurzhanski. Control and Observation under Conditions of Uncertainty. Nauka, Moscow, 1977.
3. B.N. Pshenichny and V.G. Pokotilo. On the linear object observation problem. *Prikladnaja Matematika i Mekhanika*, 46(2)(1982)212–217.
4. V.G. Pokotilo. On the asymptotic properties of minimax estimates under stochastic perturbations. *Doklady Akademii Nauk SSSR*, 264(1982)1084–1086.
5. V.G. Pokotilo. On the asymptotic properties of *a priori* minimax estimates. *Prikladnaja Matematika i Mekhanika*, 46(1982)900–905.
6. B.N. Pshenichnyi and V.G. Pokotilo. Minimax approach to estimation of linear regression parameters. *Izvestija Akademii Nauk SSSR, Teknicheskaja Kibernetika*, (2)(1982)94–108.
7. M. Loeve. *Probability Theory*. Izd. In. Lit., Moscow, 1969.

STOCHASTIC AND DETERMINISTIC CONTROL: DIFFERENTIAL INEQUALITIES

N.N. Subbotina, A.I. Subbotin and V.E. Tret'jakov
Institute of Mathematics and Mechanics
Sverdlosk, USSR

1. INTRODUCTION

Two types of controlled processes are considered in this paper. The first includes deterministic processes described by ordinary differential equations, while the second comprises diffusion-type controlled processes described by Ito's stochastic equation. Problems of feed-back optimal control are considered and the properties of a value function are investigated for both types of controlled process. This function assigns to an initial position the guaranteed result which can be attained by choosing the optimal feedback strategy. The following fact is well-known in control theory and the theory of differential games: If the value function is sufficiently smooth in some region, then it satisfies a partial differential equation which is commonly called the *Bellman equation.*

For controlled processes of the diffusion type, if the noise acting on a system is non-degenerate, the corresponding Bellman equation is a non-degenerate parabolic equation with a unique solution for a given boundary condition. Thus, in this case the Bellman equation uniquely determines the value function.

However, in the case of a deterministic controlled system or a diffusion-controlled process with degenerate noise, attempts to use the Bellman equation to determine the value function run into considerable difficulties. One problem is that the Bellman equation for deterministic controlled systems is a first-order partial differential equation of Hamilton–Jacobi type. In general, this equation has no classical solution. Therefore, notions of generalized solution have to be introduced and existence theorems have been proved by various authors. In some cases the uniqueness of the generalized solutions may be proved. We shall return to this problem below.

In the present paper we suggest replacing the Bellman equation by two differential inequalities. These inequalities, together with a boundary condition, form necessary and sufficient conditions which the value function must satisfy; these conditions determine the behavior of the value function both at the points where this function is differentiable, and on singular sets. In regions where the value function has derivatives of the required order (of first order in the deterministic case and of second order in the stochastic case), these inequalities are equivalent to the Bellman equation. Therefore, the proposed inequalities can be viewed as a generalization of the Bellman equation.

2. PROBLEM FORMULATION

Let us turn to the formulation of the problems under consideration. First we shall consider a deterministic controlled system whose motion is described by the ordinary differential equation

$$\frac{dx}{dt} = f(t, x, u, v) \quad t \in T = [0, \vartheta] \quad , \tag{1}$$

where $\vartheta > 0$ is a fixed instant of time, $x \in R^n$ is an n-dimensional phase vector, u is a control parameter, and v is a disturbance (or the control of a second player). Suppose that $u \in P \subset R^p$, $v \in Q \subset R^q$, where P and Q are compact sets. The function f: $T \times R^n \times P \times Q \to R^n$ is taken to be continuous and to satisfy the Lipschitz condition with respect to the variable x. We shall assume that for all $(t, x, s) \in T \times R^n \times R^n$ the following minimax condition holds:

$$\min_{u \in P} \max_{v \in Q} \langle s, f(t, x, u, v)\rangle = \max_{v \in Q} \min_{u \in P} \langle s, f(t, x, u, v)\rangle \quad . \tag{2}$$

Here $\langle \cdot, \cdot \rangle$ denotes the inner product. The payoff functional $\gamma_*(x(\cdot))$ is defined by the equality: $\gamma_*(x(\cdot)) = \gamma(x(\vartheta))$, where the payoff function γ: $R^n \to R$ satisfies the Lipschitz condition. We shall adopt the concept of a differential game presented in [1,2]. As we proceed the notion of the value function of a positional differential game will assume considerable importance. This function assigns to the starting position of the game (t_0, x_0) the result $\rho^0(t_0, x_0)$ guaranteed to the first and second players if they choose the optimal feedback strategies.

Note that other concepts of differential games and other definitions of the value of a game are possible (see, for example, constructions connected with majorant and minorant games [3] and the constructions in [4,5]). It is important to note that the value of a game does not depend on the formalization of the differential game. Thus, the properties of the value function given below are valid in the framework of the above-mentioned concepts. With the value function readily available, the optimal strategy can be determined in a relatively simple way. This explains the importance of studying the value function.

3. INVESTIGATION OF THE VALUE FUNCTION

Investigation of the value function is usually connected with the Bellman–Isaacs equation, which in the problem under consideration has the following form [6]:

$$\frac{\partial \rho(t,x)}{\partial t} + \min_{u \in P} \max_{v \in Q} \langle \frac{\partial \rho(t,x)}{\partial x}, f(t,x,u,v) \rangle = 0 \tag{3}$$

$$\rho(\vartheta, x) = \gamma(x) \quad . \tag{4}$$

Here $\partial \rho / \partial x = (\partial \rho / \partial x_1, \ldots, \partial \rho / \partial x_n)$ is a column vector of partial derivatives of the function ρ with respect to x_i.

In the theory of differential games equation (3) holds at every point (t, x) where the value function is differentiable. Note that in the case under consideration the function ρ satisfies the Lipschitz conditon and, therefore, according to Rademacher's theorem, ρ is non-differentiable on the set of zero measure. Hence, the value function satisfies equation (3) almost everywhere and obeys the boundary condition (4) for all $x \in R^n$. However, these necessary conditions are not sufficient, since the number of functions satisfying equation (3) almost everywhere and obeying the boundary condition for all $x \in R^n$ may be infinite.

Let us turn to the formulation of the necessary and sufficient conditions which should be satisfied by the value function. Let Lip denote all the functions ρ: $T \times R^n \to R$ satisfying the Lipschitz condition, and take $\rho \in$ Lip, $t \in [0, \vartheta)$, $x \in R$, $h \in R^n$. We define the lower and upper derivatives of the function ρ at the point (t, x) in the direction $(1, h)$ by the relations

$$\partial_- \rho(t,x) | (h) = \liminf_{\delta \downarrow 0} \, [\rho(t+\delta, x+\delta h) - \rho(t,x)]\delta^{-1} \tag{5}$$

$$\partial_+ \rho(t,x) | (h) = \limsup_{\delta \downarrow 0} \, [\rho(t+\delta, x+\delta h) - \rho(t,x)]\delta^{-1} \quad .$$

Note that functions $h \mapsto \partial_- \rho(t,x)|(h)$ and $h \mapsto \partial_+ \rho(t,x)|(h)$ satisfy the Lipschitz condition.

We then have the following result [7,8]:

THEOREM 1. *For a function ρ: $T \times R^n \mapsto R$ to be the value function of the differential game* (1) (2), *it is necessary and sufficient that the following conditions be fulfilled:*

$$\rho \in \text{Lip} \, , \quad \rho(\vartheta, x) = \gamma(x) \ \text{ at } \ x \in R^n \tag{6}$$

$$\max_{v \in Q} \min_{h \in F_1(t,x,v)} \partial_- \rho(t,x)|(h) \le 0 \le \max_{u \in P} \min_{h \in F_2(t,x,u)} \partial_+ \rho(t,x)|(h) \tag{7}$$

$$\text{at } \ (t,x) \in [0, \vartheta) \times R^n \quad ,$$

where

$$F_1(t,x,v) = \text{co}\,\{f(t,x,u,v): u \in P\}, \quad F_2(t,x,u) = \text{co}\,\{f(t,x,u,v): v \in Q\} \tag{8}$$

Note that the inequalities (7) express the conditions of u-stability and v-stability of the function ρ in the infinitesimal form. A formal definition of the stability properties is given, for example, in [1,7]; these conditions express the property of non-deterioration of position. It is easy to show that at every point (t,x) where the function ρ is differentiable (i.e., almost everywhere), the inequalities (7) are equivalent to the Bellman equation (3).

Note also that the equalities (7) assume a form more convenient for verification when the function ρ can be represented in the form

$$\rho(t,x) = \min_{k \in K} \max_{l \in L} \varphi_{k,l}(t,x) \quad , \tag{9}$$

where K and L are finite sets, and functions $\varphi_{k,l}(\cdot): T \times R^n \mapsto R$ are continuously differentiable. Functions of the form (9) are known to be directionally differentiable and the formulae for the directional derivatives are also known [9]. Substituting these formulae into (7) leads to relatively simple inequalities.

Thus, Theorem 1 states that the value function is the generalized solution of the Bellman equation (3) in that it satisfies the inequalities (7). Note that different generalized solutions of partial differential equation of the Hamilton–Jacobi type (equation (3) is also of this type) are given, for example, in [10–13], where the existence and uniqueness of such solutions are investigated. In particular, the notion of a "viscosity solution" is proposed in [13,14]. Here we note only that it is eásy to prove the following assertion: any function satisfying conditions (6), (7) of Theorem 1 satisfies the definition of a "viscosity solution" for equation (3). The converse is also true.

4. DIFFUSION CONTROLLED PROCESSES

Now let us consider a diffusion controlled process described by Ito's stochastic differential equation [15,16]

$$\mathrm{d}\xi_t = f(t,\xi_t,u_t,v_t)\mathrm{d}t + \sigma \mathrm{d}W_t \quad . \tag{10}$$

Here σ is a constant $(n \times m)$-dimensiional matrix and W_t is an m-dimensional Wiener process defined on probability space $(\Omega, \mathbf{F}, \mu)$. The function f satisfies the conditions given above in connection with the deterministic controlled system. We will take the solution of the stochastic equation (10) in its strong sense [15,16].

We consider a stochastic differential game for system (10) and introduce the notion of the value function of a game. Differential games for diffusion systems have been studied by many authors; this idea was first put forward in [3,17]. The results presented in the present paper are obtained within the framework of the approach suggested in [18,19].

We shall consider a stochastic differential game for the classes of feedback strategies $U(t,x)$ and $V(t,x)$, i.e., Borel measurable functions $U: T \times R^n \mapsto P$, $V: T \times R^n \mapsto Q$. Let $(t_0, x_0) \in T \times R^n$ be the starting position, $\Delta = \{t_0 = \tau_0 < \tau_1 < \cdots < \tau_{k+1} = \vartheta\}$ be a partition of the segment $[t_0, \vartheta]$, the function $v_{(\cdot)}: [0, \vartheta] \times \Omega \mapsto Q$ be a non-anticipatory process, and $U: T \times R^n \mapsto P$ be a feedback strategy. We shall denote by $\xi_t(t_0, x_0, U, v_{(\cdot)}, \Delta)$ the random process ξ_t, $t_0 \le t \le \vartheta$, described by the stochastic equation

$$\xi_t = \xi_{\tau_i} + \int_{\tau_i}^{t} f(s, \xi_s, u_{\tau_i}, v_s) ds + \int_{\tau_i}^{t} \sigma dW_s \tag{11}$$

$$\xi_{\tau_0} = x_0 \text{ (a.e.)}, \quad u_{\tau_i} = U(\tau_i, \xi_{\tau_i}), \quad t \in [\tau_i, \tau_{i+1}), \quad i = 0, 1, \ldots, k \quad .$$

Let $\operatorname{diam} \Delta = \max_{0 \le i \le k} (\tau_{i+1} - \tau_i)$. The guaranteed result for strategy U at position (t_0, x_0) is defined by the following relation:

$$\Gamma_1(t_0, x_0, U) = \lim_{\operatorname{diam} \Delta \downarrow 0} \sup_{v_{(\cdot)}} \sup E_{t_0 x_0} \{\gamma(\xi_\vartheta(t_0, x_0, U, v_{(\cdot)}, \Delta))\} \quad , \tag{12}$$

where $E\{\cdot\}$ represents the mean value and $\gamma: R^n \mapsto R$ is a given Lipschitz function.

The optimal guaranteed result is defined by the relation

$$\rho_1(t_0, x_0) = \inf_U \Gamma_1(t_0, x_0, U) \quad . \tag{13}$$

The existence theorem for the value function of a stochastic differential game for the controlled diffusion process under consideration is given in [18]. This is expressed by the equality

$$\rho_1(t_0, x_0) = \rho_2(t_0, x_0) = \rho_0(t_0, x_0), \ (t_0, x_0) \in T \times R^n \quad , \tag{14}$$

where

$$\rho_2(t_0, x_0) = \sup_V \Gamma_2(t_0, x_0, V) \tag{15}$$

$$\Gamma_2(t_0, x_0, V) = \lim_{\operatorname{diam} \Delta \downarrow 0} \inf_{u_{(\cdot)}} \inf E_{t_0 x_0} \{\gamma(\xi_\vartheta(t_0, x_0, u_{(\cdot)}, V, \Delta))\} \quad . \tag{16}$$

Here $V: T \times R^n \mapsto Q$ is the feedback strategy and $u_{(\cdot)}: [t_0, \vartheta] \times \Omega \mapsto P$ is a non-anticipatory process.

The function ρ_0: $T \times R^n \mapsto R$ defined by relation (14) is called the ***value function*** of the stochastic differential game. Note that here, just as in the deterministic case, the function ρ_0 (14) coincides with the value function of the game defined according to another well-known framework for differential games [3]. Thus, the properties of the value function discussed below do not depend on the framework in which it has been constructed.

5. DIFFUSION PROCESSES WITH NON-DEGENERATE NOISE

Diffusion processes with non-degenerate noise have been studied particularly thoroughly in control theory and the theory of differential games. In this case σ is a square $(n \times n)$-dimensional matrix such that

$$a = (a_{ij}) = \frac{1}{2}\sigma\,\sigma^{\mathrm{T}}\,,\ i\ ,\ j\ = 1\,,\ldots,\ n \tag{17}$$

is positive definite. Here σ^{T} denotes the transpose of σ. In the non-degenerate case the value function is known to satisfy the Bellman equation

$$\frac{\partial \rho(t,x)}{\partial t} + \min_{u \in P}\max_{v \in Q} <\frac{\partial \rho(t,x)}{\partial x}\,,\, f(t\,,x\,,u\,,v)> + \sum_{i,j=1}^{n} a_{ij}\,\frac{\partial^2 \rho(t,x)}{\partial x_i\,\partial x_j} = 0 \tag{18}$$

and the boundary condition

$$\rho(\vartheta\,,x) = \gamma(x) \quad . \tag{19}$$

In the non-degenerate case the equations (18), (19) have a unique solution, and thus equation (18) completely determines the value function.

Thus, summing up, we can say that in two extreme cases, i.e., in the deterministic case (when there is no noise) and in the case when the noise is non-degenerate, we can formulate the necessary and sufficient conditions which the value function must satisfy. In the deterministic case these conditions contain the inequalities (7). In the case of non-degenerate noise, the main requirement is to satisfy equation (18).

6. THE INTERMEDIATE CASE

Let us consider the intermediate case in which σ is an arbitrary $(n \times m)$-dimensional matrix. We can write the Bellman equation for an arbitrary $(n \times m)$-dimensional matrix σ, but this may turn out to be degenerate, in which case it cannot be used for unique determination of the value function. This situation is similar to that arising with a deterministic system. To avoid this difficulty we shall take the same action as in the deterministic case and replace the Bellman equation by a pair of differential inequalities which express the stability properties of the value function in infinitesimal form.

Before proceeding to the formulation of new results we should first explain the meaning of the conditions of u-stability and v-stability. The original definitions of these notions for controlled stochastic processes are presented in [18]. Note that these definitions permit a number of different formulations, of which we give one below.

Let

$$A = \{\alpha : s \mapsto \alpha_s : T \mapsto \mathrm{rpm}(P) \,|\, \alpha \text{ is measurable}\} \quad , \tag{20}$$

where rpm(P) is a set of regular probability measures on P with a weak norm generating a topology which is equivalent to the weak-* topology of the space conjugate to the space $C(P)$ of continuous scalar funcions on P [20]. The elements of the set A are called *generalized program controls*. For any collection $(t_*, x_*, \alpha, v) \in T \times R^n \times A \times Q$ the solution $\xi_t = \xi_t(t_*, x_*, \alpha, v)$ of the equation

$$\xi_t = x_* + \int_{t_*}^{t} ds \int_P f(s, \xi_s, u, v)\alpha_s(du) + \int_{t_*}^{t} \sigma dW_s \;, \; t \in [t_*, \vartheta] \tag{21}$$

exists and is unique.

Definition 1. A function $\rho \in$ Lip satisfies the condition of u-stability if the inequality

$$E_{t_*,x_*}\{\min_{\alpha \in A} \rho(t, \xi_t(t_*, x_*, \alpha, v))\} \le \rho(t_*, x_*) \tag{22}$$

holds for all $(t_*, x_*, v) \in T \times R^n \times Q$, $t \in [t_*, \vartheta]$.

(Note that this achieves a minimum over α on the left-hand side of inequality (21)).

We can define the v-stability property of function $\rho \in$ Lip in a similar way. To do this it is necessary to introduce an analogous set of generalized controls with values in rpm (Q) and to replace the minimum in (22) by a maximum and the symbol $\le$ by the sign $\ge$.

Then the following assertion is valid [18]:

THEOREM 2. *For a function $\rho(t, x)$ to be the value function of a stochastic differential game, it is necessary and sufficient that this function be u-stable and v-stable and satisfies the boundary condition.*

We shall now try to write the stability conditions in infinitesimal form, i.e., to express the conditions in the form of differential inequalities.

7. FORMULATION OF RESULTS

Let $\rho \in \text{Lip}$, $(t, x) \in [0, \vartheta] \times R^n$, the set H be compact in R^n, $\eta = (\eta_1, \ldots, \eta_m)$ be an m-dimensional Gaussian random variable, $E\{\eta_i\} = 0$, $E\{\eta_i \eta_j\} = \delta_{ij}$, $i, j = 1, \ldots, n$, where δ_{ij} is the Kronecker delta.

We introduce the following quantities:

$$\tilde{\partial}_- \rho(t, x) | (H) = \liminf_{\delta \downarrow 0} [E\{\min_{h \in H} \rho(t + \delta, x + h \cdot \delta + \sqrt{\delta} \sigma \eta)\} - \rho(t, x)] \delta^{-1}$$

(23)

$$\tilde{\partial}_+ \rho(t, x) | (H) = \limsup_{\delta \downarrow 0} [E\{\max \rho(t + \delta, x + h \cdot \delta + \sqrt{\delta} \sigma \eta)\} - \rho(t, x)] \delta^{-1}$$

Note that in the case when the set H contains a unique element h, the quantities defined above can be taken as the lower and upper stochastic derivatives of the function ρ in the direction $(1, h)$.

If in some neighborhood of the point (t, x) the function ρ has a derivative with respect to t and first and second derivatives with respect to x_i, then the following equalities hold:

$$\tilde{\partial}_- \rho(t, x) | (H) = \frac{\partial \rho(t, x)}{\partial t} + \min_{h \in H} \left< \frac{\partial \rho(t, x)}{\partial x}, h \right> + \sum_{i,j=1}^{n} a_{ij} \frac{\partial^2 \rho(t, x)}{\partial x_i \partial x_j}$$

(24)

$$\tilde{\partial}_+ \rho(t, x) | (H) = \frac{\partial \rho(t, x)}{\partial t} + \max_{h \in H} \left< \frac{\partial \rho(t, x)}{\partial x}, h \right> + \sum_{i,j=1}^{n} a_{ij} \frac{\partial^2 \rho(t, x)}{\partial x_i \partial x_j} .$$

We shall now formulate our main result.

THEOREM 3. *For a function* ρ to be the value function of the stochastic differential game under consideration, it is necessary and sufficient that the conditions

$$\rho \in \text{Lip}, \quad \rho(\vartheta, x) = \gamma(x), \quad x \in R^n \tag{25}$$

$$\max_{v \in Q} \tilde{\partial}_- \rho(t, x) | (F_1(t, x, v)) \le 0 \le \min_{u \in P} \tilde{\partial}_+ \rho(t, x) | (F_2(t, x, u))$$

(26)

$$t \in [0, \vartheta), \quad x \in R^n$$

be satisfied, where

$$F_1(t, x, v) = \text{co}\{f(t, x, u, v): u \in P\}, \quad F_2(t, x, u) = \text{co}\{f(t, x, u, v): v \in Q\} .$$

Some comments on Theorem 3 are in order. Two inequalities (26) appear in the conditions of the theorem. The left-hand inequality expresses the condition for u-stability in infinitesimal form while the right-hand inequality expresses the property of

v-stability. Note that inequalities (26) can be treated as a generalization of the Bellman equation (18). Indeed, if the required derivatives of the function ρ exist in some neighborhood of the point (t, x), then using equalities (24) and (2) we obtain that equation (18) holds at the point (t, x). Thus, in the case of non-degenerate noise, and if the value function ρ_0 is twice-differentiable, (26), (24), and (2) may be used to show that equation (18) holds for ρ_0 at all points $(t, x) \in [0, \vartheta) \times R^n$. If we take another extreme case, i.e., the deterministic system (1), then conditions (25), (26) of Theorem 3 formally become conditions (6), (7) of Theorem 1.

Theorems 1–3 give necessary and sufficient conditions for the value function and therefore can be utilized to demonstrate the coincidence of functions ρ with value function ρ_0 constructed in different ways. These theorems are formulated for differential games with a fixed termination time ϑ. Analogous results can be obtained for other types of differential games.

REFERENCES

1. N.N. Krasovski and A.I. Subbotin. *Positional Differential Games*. Nauka, Moscow, 1974 (in Russian).

2. N.N. Krasovski. Differential games. Approximation and formal models (in Russian). *Matematicheskie Sbornik*, 107(7)(1978).

3. W.H. Fleming. The convergence problem for differential games. *Journal of Mathematical Analysis and Applications*, 3(1961).

4. L.S. Pontryagin. Linear differential games. II. *Soviet Mathematics Doklady*, 8(1967)910–912.

5. B.N. Pshenichnyi. The structure of differential games. *Soviet Mathematics Doklady*, 10(1969)70–72.

6. R. Isaacs. *Differential Games*. J. Wiley, New York, 1965.

7. A.I. Subbotin. Generalization of the main equation of differential game theory. *Journal of Optimization Theory and Applications*, 43(1984).

8. A.I. Subbotin and N.N. Subbotina. Properties of the potential of a differential game (in Russian). *Prikladnaya Mathematika i Mekhanika*, 46(1982).

9. V.F. Demyanov. *Minimax: Directional Differentiation*. Leningrad University Press, Leningrad, 1974.

10. S.N. Kruzkov. Generalization solutions of nonlinear first-order equations with many independent variables. I (in Russian). *Matematicheskie Sbornik*, 70(3)(1966).

11. N.V. Krylov. *Controlled Diffusion Processes*. Springer-Verlag, New York, 1980.

12. M.M. Hrustalev. Necessary and sufficient conditions of optimality in the form of the Bellman equation. *Soviet Mathematics Doklady*, 19(1978).

13. M.G. Crandall and P.L. Lions. Viscosity solutions of Hamilton–Jacobi equations. *Transactions of the American Mathematical Society*, 277(1983).

14. E.N. Barron, L.C. Evans and R. Jensen. Viscosity solutions of Isaacs' equations and differential games with Lipschitz conditions. *Journal of Differential Equations*, 53(1984).

15. R.S. Liptzer and A.N. Shiryaev. *Statistics of Random Processes*. Nauka, Moscow, 1974 (in Russian).

16. I.I. Gikman and A.V. Skorohod. *Stochastic Differential Equations*. Springer-Verlag, Berlin, 1972.

17. P. Bellman. *Dynamic Programming*. Princeton University Press, Princeton, New Jersey, 1957.

18. N.N. Krasovski and V.E. Tret'jakov. A saddle point of a stochastic differential game (in Russian). *Doklady Akademii Nauk SSSR*, 254(1980).

19. V.E. Tret'jakov. A program synthesis in a stochastic differential game (in Russian). *Doklady Akademii Nauk SSSR*, 270(1983).

20. J. Warga. *Optimal Control of Differential and Functional Equations*. Academic Press, 1972.

THE SEARCH FOR SINGULAR EXTREMALS

M.I. Zelikin
Moscow State University
Moscow, USSR

1. INTRODUCTION

The use of techniques developed for exterior differential systems in the calculus of variations seems quite promising. If the Legendre condition is non-degenerate this approach leads to differential forms on the jet manifold [1]. In this paper we consider the case when the Legendre condition is identically degenerate; this enables us to obtain much more effective results using differential forms defined on the original manifold.

2. PROBLEM FORMULATION

Let Ω be a smooth n-dimensional manifold and ω be a differential k-form on Ω. The tangent space to Ω at a point x is $T_x\Omega$, and the tangent bundle is $T\Omega$. Let $G_k(T\Omega)$ be the Grassmann bundle, its fibres $G_k(T_x\Omega)$ being the Grassmann manifolds of (oriented) k-planes in the tangent space $T_x\Omega$. Let us suppose that we have an open subset $K_x \subset G_k(T_x\Omega)$ in each fibre of $G_k(T\Omega)$ such that the multivalued function $x \mapsto K_x$ is lower semicontinuous [2]. Let A be a smooth $(k-1)$-dimensional compact submanifold of Ω.

We shall consider the set of piecewise-smooth k-dimensional manifolds W with a boundary which is diffeomorphic to A, and σ^1-mappings $f: W \longrightarrow \Omega$ [3] such that the restriction $f\,|_{\partial W}$ is a diffeomorphism of ∂W on A. The local coordinates on W are $t = (t_1, \ldots, t_k)$, while those on Ω are $x = (x_1, \ldots, x_n)$. The mapping f is defined by $x = x(t)$ in the coordinates t, x. Let W_i, $i = 1, \ldots, N$ denote the cells of W of maximum dimension defined by a piecewise-smooth structure of W and its mapping f; in addition let $V = \bigcup_{i=1}^{N} W_i$. A pair (W, f) is said to be an *admissible pair* if $x(t) \in C^2(V)$ and

$$\operatorname{Im} \dot{x}(t) \in K_{x(t)} \quad \forall t \in V \tag{1}$$

Here Im $\dot{x}(t)$ is an oriented tangent plane to $f(W)$ at the point $x(t)$ (the orientation being induced by the choice of coordinate system).

Problem 1. Minimize $F = \int_{f(W)} \omega$ for all admissible pairs (W, f).

We shall say that the C^1-minimum of F is attained on the pair $(\hat{W}, \hat{f})$ if there exists a $\sigma > 0$ such that for all admissible pairs $(\hat{W}, f)$ with the property

$$|x(t) - \hat{x}(t)| < \sigma ; \|\mathrm{Im}\ \dot{x}(t) - \mathrm{Im}\ \dot{\hat{x}}(t)\|_G < \sigma \quad \forall t \in V \cap \hat{V} \quad , \tag{2}$$

the inequality $F(\hat{W}, f) \geq F(\hat{W}, \hat{f})$ is satisfied. Here $\|\cdot\|_G$ is a standard metric on the Grassmann manifold. If the second inequality in (2) is omitted we shall refer to the C-minimum of F.

LEMMA 1. *The Euler equation for F is equivalent to the following condition: the value of the form* $\mathrm{d}\omega$ *on* $(k+1)$*-vectors which contain the plane* Im $\dot{x}(t)$ *is equal to zero.*

Proof. Let $I = \{i_1, \ldots, i_k\}$ be a subset of the set $\{1, \ldots, n\}$. We shall take $I^{(i)}$ to be the same as I, except that the element $i \in I$ is omitted, and $I_j^{(i)}$ to be the same as I, except that the element $i \in I$ is replaced by j. Let $J = \{1, \ldots, k\}$; $D\{x_I\}/D\{t_J\}$ be a jacobian corresponding to variables x_I and t_J; and $\mathrm{d}x_I = \mathrm{d}x_{i_1} \wedge \cdots \mathrm{d}x_{i_k}$. We then have $\omega = \sum_I P_I\, \mathrm{d}x_I$,

$$\mathrm{d}\omega = \sum_I \sum_{j \in I} \frac{\partial P_I}{\partial x_j} \mathrm{d}x_j \wedge \mathrm{d}x_I$$

$$\mathrm{d}\omega\{\frac{\partial x}{\partial t_1}, \ldots, \frac{\partial x}{\partial t_k}, \xi\} = \sum_I \sum_{j \in I} \frac{\partial P_I}{\partial x_j} \begin{vmatrix} \xi_j & \frac{\partial x_j}{\partial t_1} & \cdots & \frac{\partial x_j}{\partial t_k} \\ \xi_{i_1} & \frac{\partial x_{i_1}}{\partial t_1} & \cdots & \frac{\partial x_{i_1}}{\partial t_k} \\ \xi_{i_k} & \frac{\partial x_{i_k}}{\partial t_1} & \cdots & \frac{\partial x_{i_k}}{\partial t_k} \end{vmatrix} .$$

The Laplace expansion of determinants in the first column gives us

$$\mathrm{d}\omega\{\frac{\partial x}{\partial t_1}, \ldots, \frac{\partial x}{\partial t_k}, \xi\} = \sum_{i=1}^{n} \xi_i [-\sum_{\substack{I \in i \\ j \bar{\in} I}} \frac{\partial P_I}{\partial x_j} \frac{D\{x_{I_j^{(i)}}\}}{D\{t_J\}} + \sum_{\substack{i \bar{\in} I \\ j \in I}} \frac{\partial P_I}{\partial x_j} \frac{D\{x_I\}}{D\{t_J\}}] \quad . \tag{3}$$

The Euler equation for F is

$$f_{x_i} = \sum_I \frac{\partial P_I}{\partial x_i} \frac{D\{x_I\}}{D\{t_J\}} \quad ,$$

$$f_{\frac{\partial x_i}{\partial t_\alpha}} = \sum_{i \in I} P_I \frac{D\{x_{I^{(i)}}\}}{D\{t_{J^{(\alpha)}}\}} (-1)^{\gamma+\alpha} \quad .$$

Here γ corresponds to the position of index i in I. We then have

$$\frac{\partial}{\partial t_\alpha}[f_{\frac{\partial x_i}{\partial t_\alpha}}] = \sum_{j=1}^{n}\sum_{i\in I}\frac{\partial P_I}{\partial x_j}\frac{\partial x_j}{\partial t_\alpha}\frac{D\{x_{I(i)}\}}{D\{t_J(\alpha)\}}(-1)^{\gamma+\alpha} .$$

The last formula does not contain the second derivatives of $x(t)$ due to the identity

$$\det \left\| \begin{matrix} \frac{\partial}{\partial t_1} & \frac{\partial}{\partial t_2} & \cdots & \frac{\partial}{\partial t_k} \\ \frac{\partial x_1}{\partial t_1} & \frac{\partial x_1}{\partial t_2} & \cdots & \frac{\partial x_1}{\partial t_k} \\ & & \cdots & \\ \frac{\partial \dot{x}_{k-1}}{\partial t_{1_{i/y}}} & \frac{\partial \dot{x}_{k-1}}{\partial t_2} & & \frac{\partial x_{k-1}}{\partial t_k} \end{matrix} \right\| \equiv 0 ,$$

which is easily verified. Finally, we have

$$-\sum_{\alpha=1}^{k}\frac{\partial}{\partial t_\alpha}[f_{\frac{\partial x_i}{\partial t_\alpha}}] + f_{x_i} =$$

$$= \sum_{\alpha=1}^{k}\sum_{j=1}^{n}\sum_{i\in I}\frac{\partial P_I}{\partial x_j}\frac{\partial x_j}{\partial t_\alpha}\frac{D\{x_{I(i)}\}}{D\{t_{J(\alpha)}\}}(-1)^{\gamma+\alpha+1} + \sum_{I}\frac{\partial P_I}{\partial x_i}\frac{D\{x_I\}}{D\{t_J\}} = 0 .$$

Evaluating the first term with $j \in I$, $j \neq i$, gives zero, corresponding to the Laplace expansion of determinants with two equal columns. The result obtained taking $j = i$ in the first term, i.e.,

$$-\sum_{i\in I}\frac{\partial P_I}{\partial x_i}\frac{D\{x_I\}}{D\{t_J\}} ,$$

cancels with the corresponding elements of the second term. Thus we have

$$\sum_{\alpha=1}^{k}\sum_{j\bar{\in} I}\sum_{i\in I}\frac{\partial P_I}{\partial x_j}\frac{\partial x_j}{\partial t_\alpha}\frac{D\{x_{I(i)}\}}{D\{t_J(\alpha)\}}(-1)^{\gamma+\alpha+1} + \sum_{i\in I}\frac{\partial P_I}{\partial x_i}\frac{D\{x_I\}}{D\{t_J\}} = 0 . \tag{4}$$

The first term in (4) is the expansion of the determinant

$$-\frac{D\{x_{I_j(i)}\}}{D\{t_J\}}$$

in the j-th row and hence the left-hand side of (4) coincides with the coefficient of ξ_i in (3). This proves Lemma 1.

3. NECESSARY CONDITION FOR OPTIMALITY

THEOREM 1. *For (W, f) to be a C^1-minimum of problem 1 it is necessay that*

$$d\omega = 0 \text{ at all points of } f(W) \quad . \tag{5}$$

Proof. Let us take a point of smoothness of the manifold $f(W)$ and choose a local coordinate system **A** with this point as the origin ($t = 0$, $x(0) = 0$) and such that the inverse image of $f(W)$ in the chart **A** is a k-plane. To simplify the notation we will sometimes ignore the distinction between the image and the inverse image of chart **A**. Let $X = \operatorname{Im} \dot{x}(0)$ and $\xi_1, \ldots, \xi_k$ be a basis of X. Let H be a subspace of $T_0\Omega$ such that $H \oplus X = T_0\Omega$ and $h_1, \ldots, h_{n-k}$ is a basis of H. The following result is obvious:

LEMMA 2. *Consider a k-plane A in $T_0\Omega$, a $(k-1)$-plane B such that $(A \supset B)$, and a vector φ, such that $\varphi \in A$, $\varphi \in B$. Let $\varphi_n \in T_0\Omega$, $\varphi_n \to \varphi$ as $n \to \infty$, and the k-plane A_n be such that $A \supset B$, $\varphi_n \in A_n$. (The orientation of A is induced by the orientation of (φ, B); the orientation of A_n is induced by that of (φ_n, B)). Then $\|A_n - A\|_G \to 0$.*

We shall prove condition (5) by induction. From Lemma 1, condition (5) is valid for any polyvector containing X. Let us suppose that it also holds for any polyvector which has at least an $(s+1)$-dimensional intersection with X, i.e., we assume that

$$d\omega\{\xi_1, \ldots, \xi_{\sigma+1}, \eta_1, \ldots, \eta_{k-\sigma}\} = 0 \text{ for } \sigma \geq s \text{ if } \xi_i \in X \quad . \tag{6}$$

We shall prove that $d\omega\{\xi_1, \ldots, \xi_s, \eta_1, \ldots, \eta_{k-s+1}\} = 0$ if $\xi_i \in X$. Note that we can restrict ourselves to taking $s \geq 2k+1-n$, since the dimension of the intersection of the k-plane X with any $(k+1)$-plane in n-dimensional space $T_0\Omega$ is equal to at least $(2k+1-n)$. Hence $n-k \geq k-s+1$, and we can select the subset $\{h_1, \ldots, h_{k-s+1}\}$ from the set $\{h_1, \ldots, h_{n-k}\}$. With the help of this subset we shall build $(k-s+1)$ simplexes in the inverse image of chart **A**. Let us take the k-dimensional simplex D_0 in X with vertices $\{0, \vartheta\xi_1, \ldots, \vartheta\xi_k\}$, where ϑ is a scalar parameter. Let Δ_0 be a $(k+1)$-dimensional simplex with base D_0 and vertex

$$\eta_1 = \frac{\vartheta}{k+1} \sum_{i=1}^{k} \xi_i + \alpha\vartheta h_1 \quad .$$

The vector η_1 is chosen in such a way that its projection onto X along H falls in the centre of D_0. Hence, by Lemma 2, all the side faces of Δ_0 belong to K_0 for all sufficiently small $\alpha > 0$. Let D_1 be a side face of Δ_0 with vertices $\{0, \vartheta\xi_1, \ldots, \vartheta\xi_{k-1}, \eta_1\}$. Now take a simplex Δ_1 with base D_1 and vertex

$$\eta_2 = \frac{1}{k+1} \left[\sum_{i=1}^{k-1} \vartheta\xi_i + \eta_1 \right] + \alpha\vartheta h_2 \quad .$$

Again, the projection of η_2 falls in the centre of D_1. By decreasing the value of α (if necessary) we can secure the inclusion of all the side faces of Δ_1 in K_0. Proceeding in this way we obtain a chain $Z = (\Delta_0 + \Delta_1 + \ldots + \Delta_{k-s})$. Since the number of steps is finite, we can find an $\alpha > 0$ which secures the inclusion of $\partial Z \setminus D_0$ (the side surface of Z) in K_0. Let us fix such an α and let $\vartheta \to 0$. Then the chain Z is transformed homothetically and contract to 0. In view of the lower-semicontinuity of the mapping $x \mapsto K_x$, we conclude that the side surface of Z is admissible for all sufficiently small ϑ. We take $f(W)$ with D_0 replaced by $\partial Z \setminus D_0$ as a variation of the pair (W, f). For sufficiently small $\alpha > 0$ this variation is contained in any C^1-neighborhood of the pair (W, f). Hence

$$\Delta F = \int_{\partial Z \setminus D_0} \omega - \int_{D_0} \omega \geq 0 \quad . \tag{7}$$

We have

$$\Delta F = \int_{\partial Z} \omega = \sum_{i=0}^{k-s} \int_{\Delta_i} d\omega = \sum_{i=0}^{k-s} d\omega\{\eta_1, \ldots, \eta_{i+1}, \vartheta\xi_1, \ldots, \vartheta\xi_{k-i}\} + o(\vartheta^{k+1}) \quad .$$

By hypothesis (6), only one term in this sum is not equal to zero:

$$\Delta F = d\omega\{\eta_1, \ldots, \eta_{k-s+1}, \vartheta\xi_1, \ldots, \vartheta\xi_s\} + o(\vartheta^{k+1}) =$$

$$= \alpha^{k-s}\vartheta^{k+1} d\omega\{h_1, \ldots, h_{k-s+1}, \xi_1, \ldots, \xi_s\} + o(\vartheta^{k+1}) \geq 0 \quad .$$

On dividing by ϑ^{k+1} and letting $\vartheta \to 0$, we obtain

$$d\omega\{h_1, \ldots, h_{k-s+1}, \xi_1, \ldots, \xi_s\} \geq 0 \quad . \tag{8}$$

The same construction with the vectors $h_1, \ldots, h_{k-s+1}$ in a different order implies that (8) is valid for any ordering of $h_1, \ldots, h_{k-s+1}$. Hence

$$d\omega\{h_1, \ldots, h_{k-s+1}, \xi_1, \ldots, \xi_s\} = 0 \quad .$$

Thus we have that $d\omega = 0$ at any point of smoothness of $f(W)$. Since such points are dense and the coefficients of $d\omega$ are continuous, relation (5) holds at all points of $f(W)$. This result is invariant with respect to the choice of coordinate system. This proves Theorem 1.

4. SUFFICIENT CONDITIONS FOR OPTIMALITY

Definition 1. The differential k-form ω is said to be a *monom* if it can be represented as an exterior product of independent, totally integrable 1-forms

$$\omega = \psi_1 \wedge \cdots \wedge \psi_k \quad . \tag{9}$$

Definition 2. The annihilator of ω (denoted by Ann ω) is the Serre subbundle of $T\Omega$ for all vector fields v. Substitution in ω gives zero $(k-1)$-forms:

$$\text{Ann}\,\omega = \{v \in C^\infty(\Omega, T\Omega) \mid v \rfloor \omega = 0\} \quad .$$

Here $v \rfloor \omega$ is the contraction of the vector field v by the differential form ω. The fibre of the bundle Ann ω at the point x will be denoted by $\text{Ann}_x\,\omega$. Using the results of Cartan [4], it is easy to see that the necessary and sufficient condition for ω to be a monom is $\dim \text{Ann}_x\,\omega \equiv n-k$ and the distribution of subspaces $\text{Ann}_x\,\omega$ is integrable. If ω is a monom then there exists a local coordinate system on Ω such that

$$\omega = a(y)\mathrm{d}y_1 \wedge \cdots \wedge \mathrm{d}y_k \quad . \tag{10}$$

The function $s(y) = 1/a(y)$ will be called the *integrating factor* of ω. In view of the integrability of the distribution $\text{Ann}_x\,\omega$, there exists a foliation $\mathbf{L}$ whose leaves $\mathbf{L}_x$ are $(n-k)$-dimensional manifolds which are tangent to $\text{Ann}_x\,\omega$ at every point x. These leaves $\mathbf{L}_x$ will be called *integral surfaces* of ω.

Definition 3. We shall say that the function $\varphi(x, t)$ (where t is a parameter) attains its local maximum at the point $x = x(t)$ uniformly with respect to $t \in E$ if there exists a $\sigma > 0$ such that $\varphi(x, t) \le \varphi(x(t), t)$ for any $t \in E$ and any x satisfying $|x - x(t)| < \sigma$.

Definition 4. We shall say that ω is *positively-definite* on K, i.e., $\omega|_K > 0$, if $\omega\{\xi\} > 0$ for all $\xi \in K$.

THEOREM 2. *Let $(\hat{W}, \hat{f})$ be an admissible pair, the form ω be a monom, and ω be positively definite on K. Then if $s|_{\mathbf{L}_{\hat{x}(t)}}$ attains its local maximum at the point $\hat{x}(t)$ uniformly with respect to $t \in W$, we have that $(\hat{W}, \hat{f})$ is a C-minimum for problem* 1.

Proof. Let us consider a system of local charts $\mathbf{A}_i$ covering $\hat{f}(\hat{W})$ such that in any chart of this system ω has the form (10). The compactness of $\hat{f}(\hat{W})$ enables us to choose a finite number of such charts. Consider any admissible manifold $f(\hat{W})$ that lies in a sufficiently small neighborhood U_σ of $\hat{f}(\hat{W})$. It follows from the condition of admissibility (1) and the condition $\omega|_K > 0$ that the projection of a k-plane which is tangent to $f(\hat{W})$ onto the plane defined in $\mathbf{A}_i$ by the basis $\partial/\partial y_1, \ldots, \partial/\partial y_k$ is non-degenerate. Thus $f(\hat{W})$ in $\mathbf{A}_i$ can be parameterized using $y_1, \ldots, y_k$, i.e., its equation in $\mathbf{A}_i$ is $y_i = y_i(y_1, \ldots, y_k)$, $i = k+1, \ldots, n$. Using the condition $\omega|_K > 0$, it is easy to show that U_σ is fibred into the integral surfaces of ω, and the functional in question can be represented in the form

$$\int \omega = \sum_i \int_{G_i} a_i(y) dy_1 \wedge \cdots \wedge dy_k \quad ,$$

where the domains G_i are the same for all admissible pairs $(\hat{W}, \hat{f})$ with $f(\hat{W}) \subset U_\sigma$. For any such a pair we have

$$a_i(y_1, \ldots, y_k, \hat{y}_{k+1}(y_1, \ldots, y_k), \ldots, \hat{y}_n(y_1, \ldots, y_k)) \le$$
$$\le a_i(y_1, \ldots, y_k, y_{k+1}, \ldots, y_n) \quad .$$

Hence a C-minimum is attained at $(\hat{W}, \hat{f})$. This proves the theorem.

Remark. Let ω be a *monom* and $\omega|_K > 0$. Then from Theorem 1, $d\omega = 0$ implies that the vector grad $a(y)$ is orthogonal to the plane $\mathrm{Ann}_y\ \omega$; this is the first-order necessary condition for $s|_{L_y}$ to attain its maximum at the point y. The condition $d\omega = 0$ gives $(n-k)$ finite relations $\partial a / \partial y_j = 0$, $j = k+1, \ldots, n$, which for a general position define the smooth k-dimensional submanifold of Ω which is the only candidate for the role of optimal solution.

5. DEGENERATE MULTIPLE INTEGRAL MINIMIZATION PROBLEMS

Let us apply the above results to the problem of minimizing a multiple integral.

Problem 2. $\Phi = \int_G f(t, x, \dot{x})\, dt \to \inf$; $x|_{\partial G} = \varphi$, where $t \in G \subset \mathbb{R}^k$, $x \in \mathbb{R}^n$, $\dot{x} \in \mathrm{Lin}(\mathbb{R}^k, \mathbb{R}^n)$, $dt = dt_1 \wedge \cdots \wedge dt_k$, and φ is a given function on ∂G. It is well known that the second-order necessary condition for $\hat{x}(t)$ to be a weak minimum in problem 2 is the Hadamard–Legendre condition [5]:

$$\sum_{i,j=1}^{n} \sum_{\alpha,\beta=1}^{n} \frac{\partial^2 f(t, \hat{x}(t), \dot{\hat{x}}(t))}{\partial\left(\frac{\partial x_i}{\partial t_\alpha}\right) \partial\left(\frac{\partial x_j}{\partial t_\beta}\right)} \xi_i \xi_j \lambda_\alpha \lambda_\beta \ge 0 \quad . \tag{11}$$

The situation in which the condition (11) is degenerate has not been studied. In this paper we consider the totally degenerate case when the biquadratic form (11) is identically zero. The above theory is concerned exactly with this case. Indeed:

Definition 5. The extremal $\hat{x}(t)$ of the functional Φ is said to be *degenerate* on the set $B \subset G$ if for any $t \in B$ the biquadratic form (11) is identically zero for all $\xi \in \mathbb{R}^n$, $\lambda \in \mathbb{R}^k$.

Let $\delta^2\Phi$ be the second variation of the functional Φ.

Proposition 1. The extremal $\hat{x}(t)$ of the functional Φ is degenerate on B iff $\delta^2\Phi$ can be written as an integral of a differential k-form on the graph of the mapping $y(t)$ in $B \times \mathbb{R}^n$.

Here y is a variation of x, which is an argument of $\delta^2\Phi$.

Proof. Let $dt^{(\alpha)}$ be an exterior product of dt_i, omitting the factor dt_α: $dt^{(\alpha)} = dt_1 \wedge \cdots \wedge dt_{\alpha-1} \wedge dt_{\alpha+1} \wedge \cdots \wedge dt_k$. Let $dt^{(\alpha,\beta)}$ be the same product omitting dt_α and dt_β. The integrand in $\delta^2\Phi$ has the form

$$\sum_{i,j=1}^{n} \sum_{\alpha,\beta=1}^{k} [a_{ij\alpha\beta}(t) \frac{\partial y_i}{\partial t_\alpha} \frac{\partial y_j}{\partial t_\beta} + b_{ij\alpha}(t) \frac{\partial y_i}{\partial t_\alpha} y_j + c_{ij}(t) y_i y_j] dt \quad , \tag{12}$$

where, since the second-order mixed derivatives are independent of the order of differntiation,

$$a_{ij\alpha\beta} = a_{ji\beta\alpha} \quad . \tag{13}$$

The third term in (12) is a differential form; the second also turns out to be a differential form if we recall that

$$\frac{\partial y_i}{\partial t_\alpha} dt_1 \wedge \cdots \wedge dt_k = (-1)^{\alpha-1} dy_i \wedge = dt^{(\alpha)} \quad .$$

The first term can be written as a differential form iff $a_{ij\alpha\beta}$ is skew-symmetric about i , j:

$$a_{ij\alpha\beta} = - a_{ji\alpha\beta} \quad . \tag{14}$$

Indeed, if an integrand of $\delta^2\Phi$ is a differential form, then the summation of the first term in (12) can be written in the form

$$A_{ij\alpha\beta} dy_i \wedge dy_j \wedge dt^{\alpha+\beta} = (-1)^{\alpha+\beta} A_{ij\alpha\beta} [\frac{\partial y_i}{\partial t_\alpha} \frac{\partial y_j}{\partial t_\beta} - \frac{\partial y_i}{\partial t_\beta} \frac{\partial y_j}{\partial t_\alpha}] dt \quad .$$

To conclude the proof of Proposition 1 it remains only to note that the condition

$$\sum_{i,j=1}^{n} \sum_{\alpha,\beta=1}^{k} a_{ij\alpha\beta} \xi_i \xi_j \lambda_\alpha \lambda_\beta \equiv 0 \text{ for } \xi \in \mathbb{R}^n \ , \ \lambda \in \mathbb{R}^k \tag{15}$$

is equivalent to relation (14).

Thus, the exploration of degenerate extremal points leads to the problem of minimizing integrals of differential forms. A simple consequence of Theorem 1 is:

THEOREM 3. *Let $\hat{x}(t)$ be the extremal surface of the functional Φ which is degenerate on an open set $B \subset G$. Then for $\hat{x}(t)$ to be optimal in problem 2 it is necessary that $d\omega = 0$ for all $(t, \hat{x}(t))$, $t \in B$.*

Here the differential k-form ω is the integrand of the functional $\delta^2\Phi$, and is defined using Proposition 1 and the assumptions of Theorem 3.

REFERENCES

1. P.A. Griffiths. *Exterior Differential Systems and the Calculus of Variations*, Birkhäuser, Stuttgart, 1983.

2. C. Kuratowski. *Topology I* (2nd edn.), Warsaw, 1948.

3. J.R. Mankres. *Elementary Differential Topology*. Lectures given at Massachusetts Institute of Technology, Autumn, 1961. Revised edn. published as *Annals of Mathematical Studies*, 54, Princeton University Press, Princeton, New Jersey, 1966.

4. E. Cartan. *Les systèmes différentielles extérieurs et leurs applications géométriques*. Hermann, Paris, 1945.

5. R. Klötzler. *Mehrdimensionale Variationsrechnung*. Berlin, 1971.

ON THE SMOOTHNESS OF THE BELLMAN FUNCTION IN OPTIMAL CONTROL PROBLEMS WITH INCOMPLETE DATA

L.F. Zelikina
Central Economic-Mathematical Institute (CEMI)
Moscow, USSR

1. INTRODUCTION

Studies of the feedback control problem in optimal control and differential games as well as in sequential control with incomplete data rely heavily on the smoothness of the integral functionals defined on solutions of ordinary differential equations with discontinuous right-hand sides:

$$\dot{x} = F(x)\,,\; x(0) = x_0\,;\; x \in \Omega \subset \mathbb{R}^n \quad . \tag{1}$$

Here Ω is an open connected set and $F\colon \Omega \to \mathbb{R}^n$ is a measurable locally bounded mapping, i.e., for every compact set $D \subset \Omega$ there exists a constant $K > 0$ such that $|F(x)| \leq K$ (a.e. in D).

An absolutely continuous vector function $x(t) = x_{x_0}(t)$, $t \in [0, t_1)$, is said to be a solution of system (1) if $x(0) = x_0$ and

$$\frac{dx}{dt} \in K_{x(t)}\{F\} \quad \text{a.e. in } [0, t_1) \quad , \tag{2}$$

where

$$K_x\{F\} = \bigcap_{N, mesN=0} \; \bigcap_{x \in U} c_0\{F(U \setminus N)\} \quad .$$

Here the intersection is taken over all N of measure zero and U is an arbitrary neighborhood of x. (This definition is due to Filippov [1].) It is easy to show that if $F(x)$ is continuous, then $K_x\{F\} = F(x)$ and $x(t)$ is a standard solution of (1).

The smoothness of integral functionals defined on solutions of systems of type (1) has previously been considered only in the case of optimal feedback controls. Most results are concerned with time-optimal problems, provided the assumptions of regularity are satisfied for the optimal feedback control. Thus Boltjanski [2] has proved the smoothness of the Bellman function on cells of maximum dimension, i.e., at points where the optimal control is continuous.

The smoothness of the Bellman function on switching hypersurfaces has been proved by Satimov [3] and Trynkin [4]. In [5], Trynkin gives conditions which guarantee the smoothness of the Bellman function on universal and semi-universal manifolds of co-dimension 1. For arbitrary integral functionals, the Bellman function is known to be smooth at points where the optimal control is continuous (see [6]). The question of the smoothness of the Bellman function at points of discontinuity of the optimal control was considered by Pressman and Sonin [7] in the framework of the theory of sequential control problems with incomplete data. It was conjectured that for problems with Poisson jumps the Bellman function is smooth if the payoff function is smooth.

In this paper we prove the differentiability of an integral functional at points of generalized controls on the discontinuity manifold of the optimal control (of any co-dimension). Note that no assumptions regarding the optimality or regularity of the feedback control are made.

2. FUNCTIONALS OF THE TIME TRANSITION TYPE

Let $B \subset \mathbb{R}^n$ be a manifold (terminal manifold) and S be an open connected subset of $\mathbb{R}^n$ such that for every $x_0 \in S$ there exists a $t > 0$ satisfying $x_{x_0}(t) \in B$. Denote by $\vartheta_{x_0} = \vartheta_{x_0}(x(\cdot))$ the time at which the trajectory $x_{x_0}(t)$ first encounters B. Let $F(x)$ be discontinuous on a smooth manifold M and let the trajectories of the differential inclusion (2) be able to move along M, i.e., for any $x \in M$ we have $T_x M \cap K_x \neq \phi$, where $T_x M$ is the tangent plane to M at x. Let aff K_x be the affine span of the set K_x and Γ_x be the subspace passing through the point x parallel to aff K_x. Let $\tau_{x_0}(x(\cdot))$ be the time at which the trajectory $x_{x_0}(t)$ first encounters the manifold M; and let $\Theta(x) = \vartheta(x)|_M$. Assume that for all solutions of system (1) the values of $\vartheta_{x_0}(x(\cdot))$ and $\tau_{x_0}(x(\cdot))$ are the same. (This is the case if, for instance, the solution of (1) is assumed to have a unique right-hand side.) In this case the functionals $\vartheta_{x_0}(x(\cdot))$ and $\tau_{x_0}(x(\cdot))$ turn out to be single-valued functions of x_0, which will be denoted by $\vartheta(x_0)$ and $\tau(x_0)$.

THEOREM 1. *Let the following conditions be satisfied*:

1. $\Gamma_x \oplus T_x M = \mathbb{R}^n$;
2. $\tau(x) \in \mathrm{Lip}\,(M)$;
3. $\Theta(x) \in C^1(M)$.

Then the function $\vartheta(x)$ is differentiable at each point $x \in M$ and

$$\operatorname{grad} \vartheta(x) = y + \operatorname{grad} \Theta(x) \quad , \tag{3}$$

where vector y is such that

$$\begin{aligned} &<y\,,\tau> = 0 \quad \text{for any} \quad \tau \in T_x M \\ &<y + \text{grad}\,\Theta\,,\gamma> = 0 \quad \text{for any} \quad \gamma \in \Gamma x \quad . \end{aligned} \tag{4}$$

The following lemmas are used in the proof of Theorem 1:

LEMMA 1. *Let $\Gamma_x \oplus T_x M = \mathbb{R}^n$. Then system (4) has a unique solution.*

LEMMA 2. *Let $F(x)$ be continuous at a point x_0 and the function $\vartheta(x)$ be differentiable at this point. Then*

$$\frac{\partial \vartheta}{\partial x}(x_0)\, F(x_0) = -1 \quad .$$

Set $R(x) = T_x M \cap K_x$. Condition 1 clearly implies that $R(x)$ is a single-valued vector field.

LEMMA 3. *Let $\Gamma_x \oplus T_x M = \mathbb{R}^n$ for $x \in M$. Then $R(x)$ is continuous on M.*

The proofs of Lemmas 1 and 2 are trivial; Lemma 3 is a simple consequence of the fact that an upper-semicontinuous multivalued mapping with a single-valued image is continuous.

LEMMA 4. *Let $x_0 \in M$, $\Gamma_{x_0} \oplus T_{x_0} M = \mathbb{R}^n$ and $\Theta(x) \in C^1(M)$. Then*

$$<\frac{\partial \Theta}{\partial x}(x_0) + y_0\, \kappa> = -1 \tag{6}$$

for any $\kappa \in K_{x_0}$.

Proof. Let us apply Lemma 2 to functions $R(x)$, $\Theta(x)$ defined on the manifold M. We have

$$<\frac{\partial \Theta}{\partial x}(x_0)\,,\, R(x_0)> = -1$$

In view of the equality $<y_0\,,\, R(x_0)> = 0$, the last equation can be rewritten in the form

$$<\frac{\partial \Theta}{\partial x}(x_0) + y_0\,,\, R(x_0)> = -1 \quad . \tag{7}$$

Now, since any vector $\kappa \in K_{x_0}$ can be represented in the form $\kappa = R(x_0) + \gamma$, where $\gamma \in \Gamma_{x_0}$, Lemma 4 follows from (7) and (4).

Proof of Theorem 1. Let $y(x_0)$ denote the point at which the trajectory $x_{x_0}(t)$ first encounters M. This implies that $\vartheta(x) = \tau(x) + \Theta(y(x))$. For $x_0 \in M$ we have $\tau(x_0) = 0$, $y(x_0) = x_0$ and

$$\Delta\vartheta = \vartheta(x_0+\Delta x)) - \vartheta(x_0) = \tau(x_0+\Delta x) + \Theta(y(x_0+\Delta x)) - \Theta(x_0) \quad . \tag{8}$$

It follows from condition 3 that

$$\Theta(y(x_0+\Delta x)) - \Theta(x_0) = \frac{\partial\Theta}{\partial x}(x_0)\Delta y + o(\Delta y) \quad .$$

Condition 2 implies that $o(\Delta y) = o(\Delta x)$. Since $y_0 \perp T_{x_0}M$ we have $<y_0, \Delta y> = o(\Delta y)$ and the right-hand side of (8) can be rewritten in the form

$$\Delta\vartheta = \tau(x_0+\Delta x) + <\frac{\partial\Theta}{\partial x}(x_0) + y_0, \Delta y> + o(\Delta x) \quad . \tag{9}$$

We shall now calculate Δy:

$$\Delta y = y(x_0+\Delta x) - y(x_0) = x_0 + \Delta x + \int_0^{\tau(x_0+\Delta x)} F(x_{x_0+\Delta x}(t))dt - x_0 \quad . \tag{10}$$

On making the substitution $t = \tau(x_0+\Delta x)_S$ in the last integral, we obtain

$$\Delta y = \Delta x + \tau(x_0+\Delta x)\int_0^1 F_s ds \quad , \tag{11}$$

where F_s is the velocity vector at the point $x_{x_0+\Delta x}(\tau(x_0+\Delta x)s)$. Substituting the expression (11) for Δy in (9) leads to

$$\Delta\vartheta = <\frac{\partial\Theta}{\partial x}(x_0) + y_0, \Delta x> + \tag{12}$$
$$+ \tau(x_0+\Delta x)\{1 + <\frac{\partial\Theta}{\partial x}(x_0) + y_0, \int_0^1 F_s ds>\} + o(\Delta x) \quad .$$

Let U^σ be a closed, convex, σ-neighborhood of the set K_{x_0}. The upper-semicontinuity of the mapping $x \mapsto K_x$ means that for any $\sigma > 0$ there exists a $\delta > 0$ such that $K_x \subset U^\sigma$ for any x satisfying $|x - x_0| < \delta$. The local boundedness of $F(x)$ and condition 2 imply that there exists a $\delta_1 > 0$ such that for any Δx satisfying $|\Delta x| < \delta_1$ and any $t \in [0, \tau(x_0+\Delta x)]$, we have $|x_0 - x_{x_0+\Delta x}(t)| < \delta$. Note that $x_{x_0+\Delta x}$ is the Filippov solution of system (1) and hence $F(x_{x_0+\Delta x}(t)) \in K_{x_{x_0+\Delta x}(t)}$ almost everywhere in $[0, \tau(x_0+\Delta x)]$. Thus $F(x_{x_0+\Delta x}(t)) \in U^\sigma$ (a.e.). We shall now consider the vector $A_{\Delta x} = \int_0^1 F_s ds$ which appears on the right-hand side of (12). We have

$$A_{\Delta x} = \int_0^1 F_s\, ds = \lim \sum_i \Delta s_i F_{s_i} \quad .$$

Since $\sum \Delta s_i = 1$ and $\Delta s_i \geq 0$, the integral sum is the convex linear combination of the vectors F_{s_i}. The points S_i can be chosen in such a way that the inclusion $F(x_{x_0+\Delta x}(t)) \in U^\sigma$ is satisfied. The convexity of U^σ implies that the integral sum $\sum_i \Delta s_i F_{s_i}$ belongs to U^σ and hence it follows from the fact that U^σ is closed that the limit of these sums also belongs to U^σ, i.e., $A_{\Delta x} \in U^\sigma$ for all Δx such that $|\Delta x| < \delta_1$. Thus from Lemma 4 we have

$$\{1 + \langle \frac{\partial \Theta}{\partial x}(x_0) + y_0 \, , A_{\Delta x} \rangle\} = o(1)$$

and from condition 2

$$\tau(x_0 + \Delta x)\{1 + \langle \frac{\partial \Theta}{\partial x}(x_0) + y_0 \, , A_{\Delta x} \rangle\} = o(\Delta x) \quad .$$

Finally, we have

$$\Delta \vartheta = \langle \frac{\partial \Theta}{\partial x}(x_0) + y_0 \, , \Delta x \rangle + o(\Delta x) \quad .$$

This proves the theorem.

The following result is a direct consequence of Theorem 1:

THEOREM 2. *Consider the time-optimal problem*

$$\dot{x} = f(x, u)$$

$$u \in U \subset \mathbf{R}^k \, , \; x(0) = x_0 \, , \; x(T) \in B \quad ;$$

$$T \rightarrow \inf \quad .$$

Assume that in the region Ω we have a synthesis of extremal paths containing a smooth universal manifold M and such that all the conditions of Theorem 1 are satisfied. Then the Bellman function is differentiable at all points in M, and its gradient is of the form (3)–(4).

Let us suppose that the discontinuity manifold M for some time-optimal problem is isolated, i.e., there exists a neighborhood of M which contains no points of discontinuity apart from those contained in M. Then the Bellman function is continuously differentiable if some additional assumptions are made. The exact formulation of this theorem is given in [8].

3. THE CASE OF ARBITRARY INTEGRAL FUNCTIONALS

Let us consider the integral functional

$$\omega = \int_0^{\vartheta_{x_0}(x(\cdot))} f(x_{x_0}(t))dt \quad , \tag{13}$$

where $f: \Omega \to \mathbb{R}^1$ is a measurable, locally bounded mapping. As before, $\vartheta_{x_0}(x(\cdot))$ is the time at which the trajectory $x_{x_0}(t)$ first encounters B. Additional difficulties arise here when $f(x)$ is discontinuous on the same manifold M as $F(x)$. Indeed, the restriction of $f(x)$ to the discontinuity manifold could lead to a "bad" (in particular, unmeasurable) function. In the case of generalized controls this difficulty is associated with an open set of initial values of x_0. To define the functional in such cases let us consider the solution $(x(t), y(t))$ of the extended system of differential equations:

$$\dot{x} = F(x)\, ,\ x(0) = x_0$$
$$\dot{y} = f(x)\, ,\ y(0) = 0 \quad . \tag{14}$$

Here $(\dot{x}, \dot{y}) \in K_{(x,y)}\{F, f\}$. Define

$$\omega(x_{x_0}(\cdot)) = y(\vartheta_{x_0}(x(\cdot)) \quad .$$

Assume that for all solutions of system (14) the value of $y(\vartheta_{x_0}(x(\cdot)))$ will be the same. (This is the case, if, for instance, the solution is assumed to have a unique right-hand side.) In this case the functional $\omega(x_{x_0}(\cdot))$ turns out to be a single-valued function of x_0, which will be denoted by $\omega(x_0)$. Let $F(x)$ be discontinuous on the smooth manifold M and let there exist a neighborhood of M such that all trajectories of (2) from this neighborhood reach M and move along M.

Our aim is to study the smoothness property of ω on M. Since $K_{(x,y)}\{F, f\}$ depends only on x, we shall denote it by $\mathbf{K}_x$. Let Γ_x be the subspace passing through the point x parallel to aff $\mathbf{K}_x$ and let $\tau(x)$ be the time at which the manifold $\mathbf{M} = M \times \mathbb{R}^1$ is first encountered, starting from the point x and moving along the trajectories of system (14). Let $T_x\mathbf{M}$ be the tangent plane to $\mathbf{M}$ at the point x and $\Omega(x)$ be the restriction of the function $\omega(x)$ to M: $\Omega(x) = \omega(x)|_M$.

THEOREM 3. *Suppose that the following conditions are satisfied:*

1. $\Gamma_x \oplus T_x\mathbf{M} = \mathbb{R}^{n+1}$;

2. $\tau(x) \in \mathrm{Lip}\,(M)$;

3. $\Omega(x) \in C^1(M)$.

Then the function $\omega(x)$ *is differentiable at each point* $x \in M$ *and* $\mathrm{grad}\,\omega = \nu + \mathrm{grad}\,\Omega$, *where the vector* ν *satisfies the system*

$$<\xi, \nu> = 0 \quad \text{for any } \xi \in T_x M$$

$$\gamma_0 + <\nu + \mathrm{grad}\,\Omega, \gamma> = 0 \quad \text{for any } (\gamma, \gamma_0) \in \Gamma_x \quad .$$

The above theorem is due to Zelikina and Zelikin.

4. ANOTHER PROBLEM

Let us now consider the following problem:

$$\dot{\eta} = (\mu - \lambda)\alpha^1 + (\lambda - \mu)\alpha^2 \quad , \tag{15}$$

$$\dot{z} = z[p_1(\eta)\alpha^1 + p_2(\eta)\alpha^2]\,;\; \alpha^1 + \alpha^2 = 1\,;\; \alpha^1, \alpha^2 \geq 0$$

$$F_{k+1}(t, \eta) = \int_t^u z(s)[p_1(\eta)F_k(s, \Gamma^1\eta)\alpha^1 + p_2(\eta)F_k(s, \Gamma^2\eta)\alpha^2]ds \rightarrow \min \quad , \tag{16}$$

$$F_0 \equiv 0 \quad ,$$

where $\mu > \lambda$,

$$p_1(\eta) = p_2(-\eta) = \frac{\lambda e^\eta + \mu}{1 + e^\eta}$$

$$\Gamma^1\eta = \eta - \ln\frac{\mu}{\lambda}\,,\; \Gamma^2\eta = \eta + \ln\frac{\mu}{\lambda}$$

Γ^1 is a jump of the first type and Γ^2 is a jump of the second type. In addition $p_j\alpha^j$ is the probability density for jumps of the type $j-s$ under control $\alpha = (\alpha^1, \alpha^2)$ and $z(s)$ is the unconditional probability that there are no jumps before some time s.

This problem is similar to the "two-armed bandit" problem (see, e.g., [7]), the main difference lying in the fact that in the classical version we have to maximize the functional (16). In this case we find the strategies which maximize the probability of the event: "The number of jumps on the time interval $[t, u)$ is not greater than K". This leads to a new phenomenon: from the second step ($K > 1$) we obtain a non-differentiable optimization problem.

THEOREM 4. *The optimal feedback control in problem* (15)–(16) *for any* $k > 0$ is

$$\alpha^1(s, \eta) = \begin{cases} 1, \text{ if } \eta > 0 \\ 0, \text{ if } \eta < 0 \\ 1 \text{ or } 0, \text{ if } \eta = 0 \end{cases} .$$

An explicit formula for the Bellman function is obtained for any $K > 0$. For instance, if $K = 1$ we have

$$F_1(t, \eta) = 1 - \frac{[e^{-\mu(u-t)} + e^{|\eta|} e^{-\lambda(u-t)}]}{1 + e^{|\eta|}} .$$

It is easy to see that $F_1(t, \eta)$ is non-differentiable at points on the line $\eta = 0$ despite the fact that the payoff function (16) is smooth, and hence the conjecture suggested in [7] is not correct. The non-differentiability of the Bellman function in our case is due to the violation of condition 2 of Theorem 3.

REFERENCES

1. A.F. Filippov. Differential equations with discontinuous right-hand sides. *Matematicheskie Sbornik*, 51(1960).
2. V.G. Boltjanski. *Mathematical Methods of Optimal Control*. Nauka, Moscow, 1969 (2nd revised edition).
3. N.Yu. Satimov. On the smoothness of the Bellman function for linear optimal control problems. Differencial'nye Uravneniya, 12(1973)2176–2179.
4. Yu.v. Trynkin. Sufficient conditions for optimality in time-optimal problems. Ph.D. Thesis, Moscow University, 1983 (in Russian).
5. Yu.V. Trynkin. On the smoothness of the Bellman function (in Russian). *Vestnik Moskovskogo Universiteta, Seriya I. Matematika, Mekhanika*, (1)(1981).
6. W.H. Fleming and R.W. Rishel. *Deterministic and Stochastic Optimal Control*. Springer-Verlag, 1975.
7. E.L. Presman and J.M. Sonin. *Sequential Control with Incomplete Data*. Nauka, Moscow, 1982 (in Russian).
8. L.F. Zelikina. On the question of regular feedback control. *Doklady Akademii Nauk SSSR*, 267(3)(1982).

Lecture Notes in Control and Information Sciences

Edited by M. Thoma

Vol. 43: Stochastic Differential Systems
Proceedings of the 2nd Bad Honnef Conference
of the SFB 72 of the DFG at the University of Bonn
June 28 – July 2, 1982
Edited by M. Kohlmann and N. Christopeit
XII, 377 pages. 1982.

Vol. 44: Analysis and Optimization of Systems
Proceedings of the Fifth International
Conference on Analysis and Optimization of Systems
Versailles. December 14–17, 1982
Edited by A. Bensoussan and J. L. Lions
XV, 987 pages, 1982

Vol. 45: M. Arató
Linear Stochastic Systems
with Constant Coefficients
A Statistical Approach
IX, 309 pages. 1982

Vol. 46: Time-Scale Modeling of Dynamic Networks
with Applications to Power Systems
Edited by J. H. Chow
X, 218 pages. 1982

Vol. 47: P. A. Ioannou, P. V. Kokotovic
Adaptive Systems with Reduced Models
V, 162 pages. 1983

Vol. 48: Yaakov Yavin
Feedback Strategies for Partially
Observable Stochastic Systems
VI, 233 pages, 1983

Vol. 49: Theory and Application of Random Fields
Proceedings of the IFIP-WG 7/1
Working Conference
held under the joint auspices of the
Indian Statistical Institute
Bangalore, India, January 1982
Edited by G. Kallianpur
VI. 290 pages. 1983

Vol. 50: M. Papageorgiou
Applications of Automatic Control Concepts
to Traffic Flow Modeling and Control
IX, 186 pages. 1983

Vol. 51: Z. Nahorski, H.F. Ravn, R.V.V. Vidal
Optimization of Discrete Time Systems
The Upper Boundary Approach
V, 137 pages 1983

Vol. 52: A. L. Dontchev
Perturbations, Approximations and Sensitivity Analysis
of Optimal Control Systems
IV, 158 pages. 1983

Vol. 53: Liu Chen Hui
General Decoupling Theory of Multivariable
Process Control Systems
XI, 474 pages. 1983

Vol. 54: Control Theory for Distributed
Parameter Systems and Applications
Edited by F. Kappel, K. Kunisch,
W. Schappacher
VII, 245 pages. 1983.

Vol. 55: Ganti Prasada Rao
Piecewise Constant Orthogonal Functions
and Their Application to Systems and Control
VII, 254 pages. 1983.

Vol. 56: Dines Chandra Saha, Ganti Prasada Rao
Identification of Continuous
Dynamical Systems
The Poisson Moment Functional
(PMF) Approach
IX, 158 pages. 1983.

Vol. 57: T. Söderström, P. G. Stoica
Instrumental Variable Methods
for System Identification
VII, 243 pages. 1983.

Vol. 58: Mathematical Theory of
Networks and Systems
Proceedings of the MTNS-83 International
Symposium
Beer Sheva, Israel, June 20–24, 1983
Edited by P. A. Fuhrmann
X, 906 pages. 1984

Vol. 59: System Modelling and Optimization
Proceedings of the 11th IFIP Conference
Copenhagen, Denmark, July 25-29, 1983
Edited by P. Thoft-Christensen
IX, 892 pages. 1984

Vol. 60: Modelling and Performance
Evaluation Methodology
Proceedings of the International Seminar
Paris, France, January 24–26, 1983
Edited by F. Bacelli and G. Fayolle
VII, 655 pages. 1984

Vol. 61: Filtering and Control of Random
Processes
Proceedings of the E.N.S.T.-C.N.E.T. Colloquium
Paris, France, February 23–24, 1983
Edited by H. Korezlioglu, G. Mazziotto, and
J. Szpirglas
V, 325 pages. 1984

Lecture Notes in Control and Information Sciences

Edited by M. Thoma and A. Wyner

Vol. 62: Analysis and Optimization
of Systems
Proceedings of the Sixth International
Conference on Analysis and Optimization
of Systems
Nice, June 19–22, 1984
Edited by A. Bensoussan, J. L. Lions
XIX, 591 pages. 1984.

Vol. 63: Analysis and Optimization
of Systems
Proceedings of the Sixth International
Conference on Analysis and Optimization
of Systems
Nice, June 19–22, 1984
Edited by A. Bensoussan, J. L. Lions
XIX, 700 pages. 1984.

Vol. 64: Arunabha Bagchi
Stackelberg Differential Games
in Economic Models
VIII, 203 pages, 1984

Vol. 65: Yaakov Yavin
Numerical Studies
in Nonlinear Filtering
VIII, 273 pages, 1985.

Vol. 66: Systems and Optimization
Proceedings of the Twente Workshop
Enschede, The Netherlands, April 16–18, 1984
Edited by A. Bagchi, H. Th. Jongen
X, 206 pages, 1985.

Vol. 67: Real Time Control of Large Scale Systems
Proceedings of the First European Workshop
University of Patras, Greece, Juli 9–12, 1984
Edited by G. Schmidt, M. Singh, A. Titli,
S. Tzafestas
XI, 650 pages, 1985.

Vol. 68: T. Kaczorek
Two-Dimensional Linear Systems
IX, 397 pages, 1985.

Vol. 69: Stochastic Differential Systems –
Filtering and Control
Proceedings of the IFIP-WG 7/1 Working Conference
Marseille-Luminy, France, March 12-17, 1984
Edited by M. Metivier, E. Pardoux
X, 310 pages, 1985.

Vol. 70: Uncertainty and Control
Proceedings of a DFVLR International Colloquium
Bonn, Germany, March, 1985
Edited by J. Ackermann
IV, 236 pages, 1985.

Vol. 71: N. Baba
New Topics in Learning Automata
Theory and Applications
VII, 231 pages, 1985.

Vol. 72: A. Isidori
Nonlinear Control Systems:
An Introduction
VI, 297 pages, 1985.

Vol. 73: J. Zarzycki
Nonlinear Prediction
Ladder-Filters for Higher-Order
Stochastic Sequences
V, 132 pages, 1985.

Vol. 74: K. Ichikawa
Control System Design based on
Exact Model Matching Techniques
VII, 129 pages, 1985.

Vol. 75: Distributed Parameter
Systems
Proceedings of the 2nd International
Conference, Vorau, Austria 1984
Edited by F. Kappel, K. Kunisch,
W. Schappacher
VIII, 460 pages, 1985.

Vol. 76: Stochastic Programming
Edited by F. Archetti, G. Di Pillo,
M. Lucertini
V, 285 pages, 1986.

Vol. 77: Detection of
Abrupt Changes in Signals
and Dynamical Systems
Edited by M. Basseville,
A. Benveniste
X, 373 pages, 1986.

Vol. 78: Stochastic
Differential Systems
Proceedings of the 3rd Bad Honnef
Conference, June 3–7, 1985
Edited by N. Christopeit, K. Helmes,
M. Kohlmann
V, 372 pages, 1986.

Vol. 79: Signal
Processing for Control
Edited by K. Godfrey, P. Jones
XVIII, 413 pages, 1986.

Vol. 80: Artificial Intelligence
and Man-Machine Systems
Edited by H. Winter
IV, 211 pages, 1986.